OILFIELD PROCESSING OF PETROLEUM
VOLUME ONE: NATURAL GAS

OILFIELD PROCESSING OF PETROLEUM VOLUME ONE: NATURAL GAS

Francis S. Manning, Ph.D., P.E.
Professor of Chemical Engineering
The University of Tulsa
Tulsa, Oklahoma

and

Richard E. Thompson, Ph.D., P.E.
Professor and Chairman of Chemical Engineering
The University of Tulsa
Tulsa, Oklahoma

With
Contributions
by

William P. Manning, Ph.D., P.E.
Coastal Chemical Company
Author of Chapter 7 "Gas Sweetening"
Coauthor of Chapter 8 "Dehydration Using Glycol"

and

Paul Buthod, M.S., P.E.
Professor Emeritus of Chemical Engineering and Petroleum Refining
The University of Tulsa
Coauthor of Appendix 4 "OPSIM"

Copyright© 1991 by
PennWell Corporation
1421 South Sheridan Road
Tulsa, Oklahoma 74112-6600 USA

800.752.9764
+1.918.831.9421
sales@pennwell.com
www.pcnnwcllbooks.com
www.pennwell.com

Marketing Manager: Julie Simmons
National Account Executive: Barbara McGee

Library of Congress Cataloging-in-Publication Data

Manning, Francis S.
   Oilfield processing of petroleum / Francis S. Manning and Richard E. Thompson:
  with contributions by William P. Manning and Paul Buthod
     p. cm.
   Includes bibliographical references and index.
   Contents: v. 1. Natural gas.
   ISBN 10: 0-87814-343-2 (v. 1)
   ISBN 13: 978-0-87814-343-6
   1. Petroleum—Refining. I. Thompson. Richard E., Ph.D.
II. Title. III. Title: Oil Field processing of petroleum.
TP690.M288   1991   90–22260
665.5'3—dc20   CIP

IN MEMORIAM

Francis Fyles Manning, D.D.S.
(1891–1972)

Eileen Manning, née Robinson
(1895–1987)

DEDICATED TO

Helen Eileen Manning,

Frank C. Manning

and

Marilyn Thompson

Contents

Preface

Oilfield Processing of Natural Gas is the first book in a three-volume series on the various surface unit operations commonly used in production facilities. Crude oil and oilfield waters will be treated separately in Volumes 2 and 3, respectively.

This book will hopefully serve three needs. First, in the form of typed notes, the current material has been used as a text for a senior-level, petroleum engineering design course on surface production and processing. And, the authors do appreciate the numerous suggestions from the Tulsa University seniors who used these notes.

Second, this book material has been used in short courses for engineers and foremen working in field handling of natural gas. It is hoped that this book will help engineers in other disciplines learn petroleum production concepts.

Thirdly, this book should serve as a refresher and handbook for all engineers interested in natural gas processing.

The mathematical background required to use this book has been purposely kept to a minimum so as to make it easily readable and immediately useful. Where advantageous, current computer simulation has been identified but computer expertise is not required.

The authors express their gratitude and thanks to the University of Tulsa for providing the opportunity and environment to write this book. The University of Tulsa enjoys many long-standing close relationships with the petroleum industry. In fact, so many petroleum-industry engineers helped so much that it is impossible to document every kindness.

Nevertheless the authors are most pleased to thank the following friends and companies for providing up-to-date information, and reviewing drafts:

Amoco Production Company	Boyd George
	Bob MacCallum
	Terry McCarthy
	Russ McGalliard
Coastal Chemical Company	Don Ballard
	Bill Manning
	Ray Veldman
Conoco Inc.	Garvin Fryar
	Jim Hunt
	Robin McGlynn
	Gene Morrison
	Joe Provine
	Bill Schula
	Duane Wilson
EG&G Chandler Engineering	Patrick Gibson
	Bob Kahmann
	Michael Morrison
Flow Con	Mike Hein
Dow Chemical Company	Tom Bacon
Dresser-Rand	Martin Stern
Gas Processors Suppliers Association	Carl Sutton
	Mark Sutton
Hughes Anderson	Ed Flaxbart
S. H. Landes Inc	Spencer Landes
Maloney-Crawford	Bob Hodgson
	Hal Wood
NATCO	Bill Ball
	B. E. Harrell
Occidental Oil & Gas	Ron Brunner
	Wayne Fling
	Jan Herbert
	Dan Kemp
Oklahoma Natural Gas	R. M. Nicholson
The Pro Quip Corp	Ron Key
	Don Love
Radco Inc	Reed Melton
	Shannon Melton
T. H. Russell Company	Tom Russell
Solar Turbines	Lawrence J. Serra
Phillips Petroleum Company	Dr. Charles Cook
	Michael Olbrich
	Don Parrish
Sensidine	Ron Roberson
Zeochem	Bob Trent

While all these friends were exceedingly helpful some contributions demand individual recognition. Dr. Boyd George and Dr. Russ McGalliard individually reviewed, critiqued, and proofread Chapters 1 through 9. Dr. Bill Manning wrote Chapter 7, and coauthored Chapter 8. Dr. Jim Hunt critiqued Chapter 10, while Robin McGlynn's detailed comments constituted a de facto revision of Chapter 12. Dr. Mike Hein reviewed Chapter 13 in great detail and Joe Provine critiqued Chapter 14. Professor Paul Buthod codeveloped the OPSIM simulation package presented in Appendix 4.

The authors have collectively and individually taught numerous short courses world-wide for Amoco Production, OGCI, Rike Service Inc., and Texaco. This experience proved invaluable.

The authors would like to thank all their T.U. colleagues, in particular Dr. E. T. Guerrero, for encouraging them to teach a senior-level petroleum engineering course on surface production. Virginia Wood of T.U. Petroleum Abstracts Service and Jim Murray in the Sidney Born Technical Library cheerfully found every reference requested.

The authors acknowledge the help, and infinite patience provided by Sue Rhodes Sesso, Jay Kilburn, and Don Karecki of PennWell Books.

The authors thank their wives, Ardis Manning and Marilyn Thompson for their patience, understanding, and encouragement. All too often the authors spent the weekend at the office.

Finally the authors find that mere words are woefully inadequate to express their appreciation and gratitude to Nelda Whipple. Nelda typed and retyped with astonishing speed, accuracy, and cheerfulness numerous drafts, revisions, and changes in format. Without her contribution this book would have remained an impossible dream.

Chapter 1

Introduction and Scope

Oil and gas production involve a number of surface unit operations between the wellhead and the point of custody transfer or transport from the production facilities. Collectively these operations are called field handling or oilfield processing. Accordingly, *oilfield processing* is defined as the processing of oil and/or gas for safe and economical storage and/or transport by pipeline, tanker, or truck. Oil-field processing also includes *water treatment*, whether produced waters for disposal and/or reinjection, or additional injection waters used for formation flooding or reservoir-pressure maintenance.

The present Volume 1 describes *oilfield processing of natural gas*. Process descriptions, design methods, operating procedures, and troubleshooting are covered in detail. Crude oil and oilfield waters are treated separately in Volumes 2 and 3, respectively. The wellhead is the starting point, and the ultimate destination is either a sales or reinjection pipeline for the natural gas, and a pipeline or storage tank for the natural-gas liquids (NGL). While NGL production and N_2 rejection are not usually major objectives, they are also reviewed.

In this volume "conditioning," "processing," and "handling" are used synonymously to refer to all oil-field operations. "Cleaning," "treating," dehydration," and "HC dew-point control" describe specific operations as follows:

Cleaning:	Removing liquids and solids such as sand, pipeline dirt, reservoir fines, corrosion products and inhibitors, liquid or free water, salt, drilling mud
Treating:	Sweetening or removal of acid gases (H_2S and/or CO_2)
Dehydration:	Drying or removing water vapor or controlling H_2O dew point
HC dew-point control	Recovery of ethane and heavier hydrocarbons as condensate.

Oil-field processing generally consists of two distinct categories of operations:

1. Separation of the gas-oil-brine wellstream into its individual phases,
2. Removal of impurities from the separated phases to meet sales/transportation/reinjection specifications and/or environmental regulations.

Phase separation is a very important part of field handling that is discussed in the volume on crude-oil treating, and it is not repeated here. *Instrumentation, controls, pressure relief*, and *flaring* are also discussed in Volume 2. *Corrosion* is treated with oil-field waters in Volume 3. *Safety* and *environmental concerns* are addressed throughout all three volumes.

Obviously the selection and operation of field handling equipment depends very strongly on the *volume* and *characteristics* of the stream produced at the wellhead. Accordingly, this volume starts by discussing:

- characterization of natural gas and its products
- phase and thermodynamic behavior of natural gas
- water-natural gas phase behavior.

Then the *individual unit operations* commonly used in field handling of natural gas are described. These include:

- basic field-processing schemes
- prevention of hydrate formation
- sweetening (removing H_2S, CO_2)
- dehydration using TEG, solid desiccants, LTX, and $CaCl_2$
- condensate recovery and HC dew-point control
- compression
- flow measurement
- heating and cooling
- pipeline transport of natural gas.

Next, design or sizing calculations are presented for two wellstreams—a large-volume offshore and a limited-volume onshore. These *case histories* are used to illustrate how field processing objectives, equipment, and operation can vary.

Finally, the following *auxiliary* and *fundamental topics* are covered in separate appendices:

- glossary of terms and list of abbreviations used
- material balances
- energy balances
- OPSIM computer simulation
- conversion of units
- physical properties of fluids.

Review Questions

1. What is "oilfield processing?"
2. What are the goals (requirements) of field processing? What are the overriding factors?
3. Which of the following operations are involved in field processing:
   - separation of natural gas, crude oil, and brine
   - crude oil dehydration and desalting
   - gas sweetening and dehydration
   - dew-point control
   - oil pipelining
   - artificial lift
   - refining of crude oil
   - condensate recovery.
4. Define conditioning, processing, handling, cleaning, treating, dehydration, and dew-point control.
5. What is the difference between "dehydration" and "water dew-point control?"
6. Name the two distinct categories of oil-field processing operations.

Chapter 2

Characterization of Natural Gas and Its Products

As indicated in Chapter 1, the *conditioning* of natural gas for transportation and sale involves two process objectives:

1. *Separation* of the natural gas from free liquids (crude oil, brine) and entrained solids (sand),
2. *Removal of impurities* from the natural gas and any condensate formed to meet sales/reinjection specifications while observing all environmental regulations.

Sales specifications can be described most readily in terms of the composition and properties of the produced hydrocarbons. Also, the selection, design, and operation of the processes required to separate gas from liquid and to remove impurities depend on the wellstream properties. Accordingly, natural gas and its products are discussed using the following topics: spectrum of produced hydrocarbon fluids, natural-gas constituents, natural-gas compositions, heating values, analysis of natural gases, sampling, and product specifications.

SPECTRUM OF PRODUCED HYDROCARBON FLUIDS

The desirable constituents of crude oil and natural gas are hydrocarbons. These compounds range from methane (CH_4) at the low-molecular-weight end all the way up to paraffin hydrocarbons with 33 carbon atoms and polynuclear aromatic hydrocarbons with 20 or more carbon atoms (Hatch and Matar, 1977). *Natural gas* is principally methane. *Crude oil* is principally liquid hydrocarbons having four or more carbon atoms.

There is a tendency to regard crude oil as a liquid and natural gas as a gas and to consider production of the two phases as separate operations. However, in the reservoir, crude oil almost always contains dissolved methane and other light hydrocarbons that are released as gas when the pressure on the oil is reduced. As the gas evolves, the

remaining crude-oil liquid volume decreases; this phenomenon is known as *shrinkage*. The gas so produced is called *associated* or *separator gas* or *casinghead gas*. Shrinkage is expressed in terms of barrels of stock-tank oil per barrel of reservoir fluid. Crude-oil shrinkage is the reciprocal of *oil formation volume factor* (FVF).

Similarly, natural gas produced from a gas reservoir may contain small amounts of heavier hydrocarbons that are separated as a liquid called *condensate*. Natural gas containing condensate is said to be *wet*. Conversely, if no condensate forms when the gas is produced to the surface, the gas is called *dry*.

A spectrum of well fluids is actually produced, as noted by McCain (1973) and summarized in Table 2–1. The type of fluid produced depends on the phase diagram of the reservoir fluid and the reservoir temperature and pressure, as will be discussed in Chapter 3, Phase Behavior of Natural Gas.

Figure 2–1 depicts a typical gas-oil separation sequence (including incidental water and sand removal). Table 2–1 lists the five common types of wellstream fluids and summarizes typical yields and liquid properties. When crude oil is separated from its associated gas during production, the total gas evolved while reducing the oil to atmospheric pressure divided by the volume of the remaining crude oil is called the *gas-oil ratio* or *GOR*. The GOR is expressed as the total standard cubic feet of gas evolved per 60°F barrel of stock-tank or atmospheric-pressure oil (scf/bsto), in English engineering units and standard cubic meters of gas per cubic meter (or metric ton) of 15°C oil in SI or metric units. Standard conditions for natural gas are 60°F and 1 atm (English engineering units) and 15°C and 1 atm (SI or metric).

The total GOR depends on the number of stages used in the separation sequence, as well as the operating pressure of each stage. For three or more stages, the GOR approaches a limiting value. Optimization of the separation sequence usually involves either maximizing crude-oil yield or

Table 2–1 Petroleum Fluid Spectrum (after McCain, 1990)

	Typical Initial Gas-Oil Ratio		Stock-Tank Liquid	
Fluid Type	Bsto/Brf	scf/bsto	°API	Color
Low-shrinkage Crude Oil (Low-GOR) Black oil, Ordinary oil	>0.5	<2,000	<45	Very dark, often black
High-shrinkage Crude Oil (High-GOR) Volatile oil	<0.5	2,000–3,300	>40	Colored, usually brown
Retrograde Gas, Gas condensate	>0.35	3,300–50,000	50–60	Light or water white
Wet Gas	—	>50,000	>50	Water white
Dry Gas	—	—	—	—

minimizing recompression horsepower. This optimization is discussed fully in Volume II.

For wet natural gas, the liquid content is given in barrels of condensate per million standard cubic feet of gas (*bbl/MMscf*) or in U.S. gallons of condensate per thousand standard cubic feet (*GPM*).

The various types of produced hydrocarbons have been described using *total GOR* (McCain, 1973—Table 2–1) or composition (Gould and McDonald, 1979—Table 2–2). Of course the gas yield depends on the relative amounts of the various hydrocarbons present, and perhaps Penick (1983) summarizes these relationships best in Figure 2–2 (originally drawn by W. H. Speaker, Jr.).

Allen (1952) emphasizes that classifications like Table 2–1 are an oversimplification in the sense that GOR does not always reflect the condition of the fluid in the reservoir. For example, in the GOR range of 3000–7000 scf/bbl the reservoir fluid may be a *liquid or a dense fluid* (highly-

Table 2–2 Well-Fluid Type (Gould and McDonald, 1979)

	Mole Percent		
Fluid Type	Methane (C1)	Intermediate (C2–C6)	Heptanes+ (C7+)
Black Oil	30	35	35
Volatile Oil	55	30	15
Gas Condensate	70	22	8
Dry Gas	90	9	1

compressed gas) but still be a *black oil* on the surface. If the fluid is a dense fluid at reservoir temperature and pressure, then any liquid produced is usually defined as *condensate*. If the reservoir fluid is liquid, then the produced liquid is called a *volatile oil*. The difference in these two cases is explained in Chapter 3.

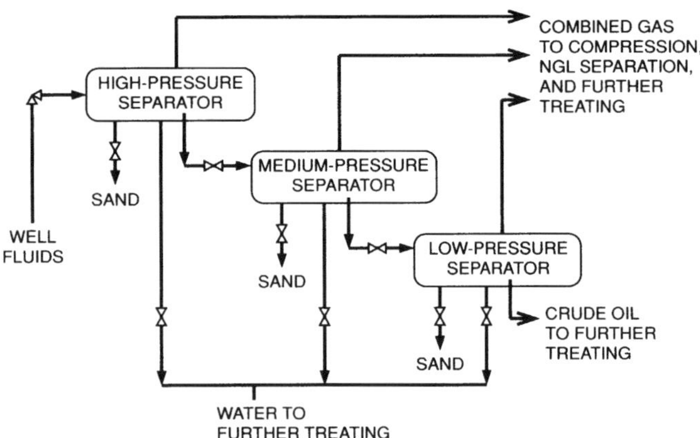

Figure 2–1. Gas-oil separator train.

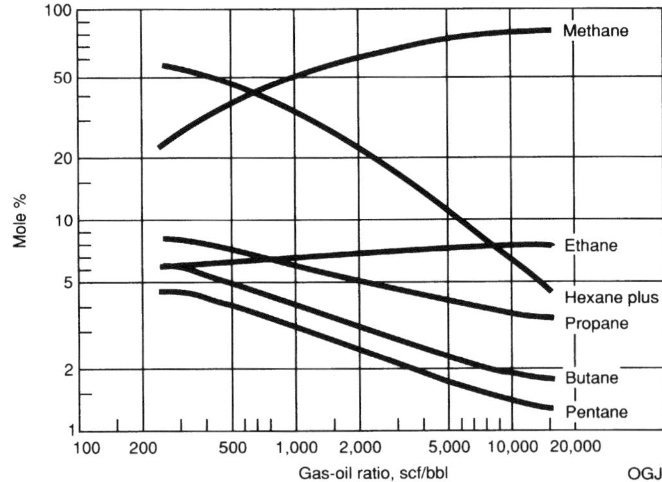

Figure 2–2. Typical reservoir composition (Penick, 1983).

Oil and gas can be produced simultaneously from a *gas-cap* reservoir; separator gas consists of gas originating in the reservoir *plus* evolved solution gas.

NATURAL-GAS CONSTITUENTS

As produced in the field, natural gas may contain the substances shown in Table 2–3. The primary constituents are usually the *paraffin hydrocarbons* (alkanes), methane through pentane, including both isobutane and isopentane. There is apparently very little or no neopentane. In some cases cyclopentane is present in very small amounts, as is indicated below. The amount of hexanes and heavier hydrocarbons ($C6+$) in natural gas is small also. This $C6+$ fraction can contain trace components such as *naphthenes* (methylcyclopentane, cyclohexane, etc.) and *aromatics* (benzene, toluene).

The main constituent of natural gas is *methane*, desirable as a primary fuel. Sales gas also contains smaller amounts of the heavier hydrocarbons listed in Table 2–3. Often a portion of the heavier hydrocarbons can be recovered profitably in a field-gas processing plant as one or more

Table 2–3 Constituents of Wellhead Natural Gas

Class	Component	Formula	Shorthand
Hydrocarbons	Methane	CH_4	C1
	Ethane	C_2H_6	C2
	Propane	C_3H_8	C3
	i-Butane	iC_4H_{10}	iC4
	n-Butane	nC_4H_{10}	nC4
	i-Pentane	iC_5H_{12}	iC5
	n-Pentane	nC_5H_{12}	nC5
	Cyclopentane	C_5H_{10}	
	Hexanes and heavier		C6+
Inert Gases	Nitrogen	N_2	N2
	Helium	He	
	Argon	A	
	Hydrogen	H_2	H2
	Oxygen	O_2	O2
Acid Gases	Hydrogen Sulfide	H_2S	H2S
	Carbon Dioxide	CO_2	CO2
Sulfur Compounds	Mercaptans	R-SH	
	Sulfides	R-S-R'	
	Disulfides	R-S-S-R'	
Water Vapor		H_2O	
Liquid Slugs	Free water or brine		
	Corrosion inhibitors		
	Methanol	CH_3OH	
Solids	Millscale and rust		
	Iron sulfide	FeS	
	Reservoir fines		

liquid products. These liquefiable components (or *condensate*) may be recovered as a single liquid stream that is transported to a separate plant for fractionation into salable products. Alternatively, in very large field units, fractionation is performed in the field. Common natural gas liquid (NGL) products are summarized below.

NGL Product	Use
Ethane	Petrochemical feedstock
Propane	Industrial and domestic fuel, petrochemical feedstock
Liquified Petroleum Gas (LPG, C3/C4 mix)	Industrial and domestic fuel, petrochemical feedstock
n-Butane	Gasoline additive for vapor-pressure control, petrochemical feedstock
i-Butane	Refinery feedstock for the alkylation process
Natural Gasoline (C5 and heavier)	Refinery feedstock for the reforming process

Throughout the world, natural gas itself is liquefied to allow transport by ocean-going vessels. The production of liquefied natural gas (LNG) is beyond the present scope.

Mass-spectographic analyses of U.S. natural gases (Miller and Hertweck, 1981) reveal that many contain finite amounts of cyclopentane. Most natural-gas analyses used in design and operations are chromatographic analyses that lump any cyclopentane present into the C6 fraction.

The permanent gases occurring in natural gas include *nitrogen, helium, argon, hydrogen, and oxygen*. Most natural gases contains some nitrogen, and a few have 30 mole percent or more. Nitrogen lowers the heat of combustion of the gas. Since natural gas is usually sold on the basis of energy content with a fixed minimum heating value, the nitrogen content is limited to fairly low amounts in commercially salable gas. The removal of nitrogen requires expensive cryogenic processing, so too high a nitrogen content may render a gas unsalable.

Mass-spectrographic analyses reveal that many natural gases contain a few hundredths of a percent of helium (Miller and Hertweck, 1981). Helium has no deleterious effect other than lack of heating value. Separated helium is very valuable and, in the U.S., successful lawsuits have awarded royalty payments for helium that could have been recovered economically from sales gases. The Miller and Hertweck (1981) analyses also report occasional small amounts of oxygen (not normally regarded as being present in natural gas), as well as argon and hydrogen. Chromatographic analyses probably lump all the inert gases as nitrogen.

Hydrogen sulfide and *carbon dioxide* are found in many

natural gases and may occur in very high percentages. In fact, essentially pure carbon dioxide is produced, dehydrated, and pipelined for CO_2 floods in the enhanced recovery of crude oil. Hydrogen sulfide and carbon dioxide are referred to as *acid gases* because they dissociate upon solution in water to form acidic solutions. Hydrogen sulfide is very toxic and corrosive, while carbon dioxide is corrosive.

A natural gas containing no hydrogen sulfide is said to be *sweet*. Conversely, a *sour* gas contains hydrogen sulfide. Strictly speaking, "sweet" and "sour" refer to both acid gases (CO_2 and H_2S) but are usually applied to hydrogen sulfide alone.

Sulfur compounds, other than H_2S, are present in minute amounts and can affect field processing. These compounds tend to concentrate in the condensate and sometimes require treating (or sweetening) of the liquid products.

Removal of hydrogen sulfide to very low content (4 ppmv or 0.25 gr/100 scf) is required in the field. Carbon dioxide can be tolerated to much higher levels, say 1–2%, as long as the heating value of the sales gas is satisfactory.

There are many so-called "treating" processes for *sweetening* natural gas. These processes are either batch, reactant-discarded processes for removing low amounts of hydrogen sulfide or continuous solvent-regenerated processes for large amounts. Batch processes are used when the consumable-chemical cost is not prohibitive. The principal continuous, solvent-regenerated treating processes use water solutions of chemical solvents, typically alkanolamines. Other sweetening processes such as physical solvents, mixed physical-chemical solvents, or membranes may be more economical in some cases, e.g., gas streams with high concentrations of acid gases (especially CO_2) and/or mercaptans.

Because these processes also remove appreciable amounts of other sulfur compounds and/or carbon dioxide, many gas streams need no further processing to meet total-sulfur and acid-gas specifications.

Hydrogen sulfide is an *extremely toxic* substance, as documented in Table 2–4. Fortunately, the familiar sulfurous smell can be detected at concentrations less than 1 ppmv. However, extended exposure at higher levels deadens the sense of smell, so that odor alone cannot be a reliable detector. Suitable detection methods include lead-acetate tape, absorptive photometric analysis, titration with hydrobromic acid and with iodine, and gas-detector sniffer tubes (Tiemstra, 1988). Merrill (1987) reviews continuous hydrogen sulfide monitoring systems.

Water or *brine* is undoubtedly present in many gas reservoirs but usually is not entrained to the surface. If free liquid water or brine is produced, a wellhead knockout vessel, or separator, is needed to prevent the water entering

Table 2–4 Toxicity of Hydrogen Sulfide

Concentration, (ppmv)	Human Reaction	Source
0.1 to 0.01	Threshold odor level	
0.9	Detection limit (OSHA ID 141)	OSHA
1	Characteristic "rotten-egg" odor clearly detected	
10	Permissible exposure level (PEL) employee's time weighted average (TWA) exposure in any 8-hour work shift of a 40-hour work week	OSHA
15	Short-term exposure limit (STEL) employee's 15-minute TWA exposure	OSHA
100–150	Loss of smell in 2 to 15 minutes	NSC
150–200	Loss of sense of smell in 2–15 minutes; hemorrhage and death in 8–48 hours	NSC
500–600	Coughing, collapse, unconsciousness within 2 minutes	NSC
600–1500	Collapse, unconsciousness, death within 2 minutes	NSC

NSC   National Safety Council (1982), "Industrial Safety Data Sheet No. 284, Hydrogen Sulfide," Chicago, IL 60611.

OSHA   Occupational Safety and Health Administration (1990), "General Industry Standards and Interpretations," Vol. 1 OSHA 2077, Change 53, Washington, DC (March 19).

the gathering lines. At high pressure and low temperature, natural gas and free liquid water form *solid hydrates* capable of plugging flow lines. Produced water also may contain methanol and/or corrosion inhibitors that have been injected in the well string.

Produced gas is regarded as being saturated with water vapor at the wellhead conditions of temperature and pressure, even if no liquid water is produced. Associated gas is regarded as being saturated with water vapor at the outlet from the gas-oil separator in which it is produced. Water condensed downstream of the wellhead or gas-oil separator will be essentially pure (fresh) water rather than saline brine.

Water vapor is a contaminant that must be removed by proper processing of the natural gas stream. Gas transmission lines often specify water contents of *7 lb/MMscf* (or a water dew point of 32°F). Triethylene glycol (TEG) is almost always used for such applications. Cryogenic plants require "bone-dry" gas (water dewpoints as low as −150°F). Solid dessicant dehydration is typically used for dewpoints below −25°F.

Formation solids are not produced with most natural gas. Nevertheless, *solids* are sometimes separated from the gas in process plant inlet separators. Usually these solids are mainly mill scale and rust from the pipe wall, along with iron sulfide (for sour gases). The solids interfere with treating and dehydration processes and should be removed in a scrubber, filter, or possibly a filter-coalescer separator.

Very deep, high-temperature, high-pressure sour gas, such as in southeast USA, may contain *solid sulfur* (Grizzaffi and Thompson, 1970). Sulfur is inert but must be properly separated to prevent downstream processing problems.

Mercury has been detected in natural gas streams (Seaton, 1978; Chiu, 1978), in concentrations from approximately 1 ppbw to 230 ppmw. Mercury usually causes no processing problems but has caused corrosion of aluminum tubes in heat exchangers (Seaton, 1978). Removal using sulfur-impregnated activated carbon, sulfur beds, or molecular sieves has been suggested. Bingham (1990) discusses the operating implications of mercury in natural gas, the corrosion mechanism of mercury and aluminum, mercury removal methods, and field detection and collection techniques in detail.

Recently (Anon., 1988; Slaton, 1988–89), *arsenic compounds* were reported present in natural gas produced from the Permian Abo sand in Chaves County, New Mexico, USA. The discovery was made indirectly by the detection of white powder on regulators in the Southern California Gas Co. system. The arsenic will be removed using a vertical copper-zinc adsorbent bed (Slaton, 1988–89). Arsines in gas from West Virginia and offshore Louisiana have also been reported. Arsenic compounds may be present in other gases as well, since the content appears to be very low.

Methanol (to prevent hydrate formation) and corrosion inhibitors are sometimes added downhole and so may be present in natural gas at the wellhead. Any compressed gas will contain some entrained lubricating oils. Field processing may also introduce additional contaminants such as glycols and amines. Mixtures of these liquids with the previously listed solid contaminants are called sludges, however several more graphic and less formal terms are used in the oil patch.

NATURAL-GAS COMPOSITIONS

The gas analyses shown in Table 2–5 span the composition ranges normally encountered. These analyses are typical of the data furnished to the designer of surface-processing equipment. As previously mentioned, the heavier hydrocarbons in these gases are regarded as recoverable liquids. The amount of potentially recoverable liquid is expressed as gallons liquid at 60°F, *if totally condensed*, per 1000 standard cubic feet of the gas (so-called *GPM*, not to be confused with gallons per minute). A gas is termed *lean* or *rich* as follows (Ewan et al., 1975):

Lean	< 2.5 GPM
Moderately-Rich	2.5–5 GPM
Very Rich	> 5 GPM

The above classification is based on ethane and heavier hydrocarbons (C2+) because ethane is sometimes regarded as a desirable feed for petrochemical processes and can be recovered as a liquid in expander-type gas processing plants. If ethane is not regarded as a valuable liquid component, the GPM can be based on propane and heavier hydrocarbons (C3+).

Table 2–5   Typical Gas Analyses (Mol Percent)

Component	Natural Gas	Natural Gas	Condensate Fluid	Separator Gas
N2	0.51	4.85	—	—
CO2	0.67	0.24	0.47	—
C1	91.94	83.74	82.13	59.04
C2	3.11	5.68	6.37	10.42
C3	1.26	3.47	4.09	15.12
i-C4	0.37	0.30	0.50	2.39
n-C4	0.34	1.01	1.85	7.33
i-C5	0.18	0.18	0.55	2.00
n-C5	0.11	0.19	0.67	1.72
C6	0.16	0.09	1.03	1.18
C7+	1.35	0.25	2.34	0.80
	100.00	100.00	100.00	100.00
C7+ Mol.Wt.	172	115	114	—
C7+ Sp.Gr.	0.803	0.744	0.765	—
C2+ GPM	2.5	3.2	5.5	12.3[a]

Source: Kaasa (1978) Kaasa (1978) Allen (1952) Unknown
a. Used properties of n-C8 for C7+ to calculate GPM C2+

Example 2–1. Confirm the GPM given in Table 2–5 for the first natural gas stream.

Solution: First find the scf/gal for the C7+ component.

$$\frac{\text{scf C7+}}{\text{gal liq C7+}} = (\text{SG C7+}) \frac{\text{lb H}_2\text{O}}{\text{gal}} \frac{\text{lb mol C7+}}{\text{lb C7+}} \frac{\text{scf}}{\text{lb mol C7+}}$$

$$= (0.803)(8.334)(1/172)(379.5) = 14.77$$

The table that follows details the calculations (which are most conveniently done with a spread-sheet program).

Component	Mol% = Vol%	scf/Mscf	scf/gal[a]	gal/Mscf[b]
N2	0.51	5.1	—	—
CO2	0.67	6.7	—	—
C1	91.94	919.4	—	—
C2	3.11	31.1	37.476	0.830
C3	1.26	12.6	36.375	0.346
iC4	0.37	3.7	30.639	0.122
nC4	0.34	3.4	31.790	0.107
iC5	0.18	1.8	27.393	0.066
nC5	0.11	1.1	27.674	0.040
C6	0.16	1.6	24.371	0.066
C7+	1.35	13.5	14.77	0.914
Sum	100.00	1000.0		2.491

Round off to 2.5 gal/1000 scf = 2.5 GPM   ←

Note: GPM C3+ = 2.491 − 0.830 = 1.6

a. From Appendix 6, Table A6–1.
b. Column 3 divided by column 4.

The weak link in gas analysis is often the composition of the C6+ or C7+ portion of the gas. For many purposes the small amount of the C6+ material renders its characterization unimportant. One important exception is when a full wellstream gas is to be transported in a pipeline over a long distance, such as from an offshore platform. Condensation of liquids in the line will cause a large pressure drop that must be anticipated if adequate platform compression is to be furnished and the proper pipe diameter selected. Very accurate characterization of the C6+ is needed for this application—predicting HC dew points (Wilson *et al.*, 1978). Accurate characterization of the C7+ fraction also is needed for prediction of reservoir behavior and for calculating heating values (Gill, 1987).

HEATING VALUES

While, in theory, heating value is simply the heat evolved on combustion, in practice the situation is complicated by the use of many bases: higher (gross) or lower (net); wet or dry; and real or ideal gas. Accordingly, these matters are reviewed first and then measurement of heating values is discussed.

The *heating value* is defined as the heat evolved when fuel is completely burned at some specified or standard conditions usually 60°F and 1 atm (14.696 psia) in the U.S. or the metric standard of 15°C (59°F) and 1 atm (14.69595 psia). Thermodynamically, the heating value is the enthalpy change on combustion when both the products and reactants are at the standard conditions.

Higher (Gross) or Lower (Net) Heating Value

When fuels containing hydrogen are burned the condition (liquid or vapor) of the H_2O produced by combustion must be specified. For example for methane burning in air

$$CH_4(g) + 2\ O_2(g) + 7.52\ N_2(g) =$$
$$CO_2(g) + 2\ H_2O(g/l) + 7.52 N_2(g)$$

Higher Heating Value (HHV) or Gross Heating Value (GHV) assumes that all H_2O produced by combustion and/or entering with the fuel and air leaves as condensed liquid.

Lower Heating Value (LHV) or Net Heating Value (NHV) assumes that all H_2O leaves as vapor or steam.

The GPSA (1987) Engineering Data Book gives heating values in the following units (p. 23–4).

NHV Btu/scf (ideal gas at 60°F, 14.696 psia)
       Btu/lbm liquid at 60°F
GHV Btu/scf (ideal gas at 60°F, 14.696 psia)
       Btu/lbm liquid at 60°F
       Btu/gal liquid at 60°F

Wet or Dry Basis

Unless specifically stated otherwise, heating values are given in terms of *scf* of *dry* natural *gas*. If the fuel gas is wet, then the water content must be known in order to specify the wet-basis heating value. If the fuel gas is saturated with water at 60°F (519.67°R) and 1 standard atm (14.696 psia), then

$$HV_{wet} = HV_{dry}\ (P_{std} - VP_{H_2O})/P_{std} + (50.4)\ VP_{H_2O}/P_{std}$$

where HV = gross heating value (Btu/scf ideal gas)
       P_{std} = base pressure for measuring scf
       VP_{H_2O} = vapor pressure of water at T_{std}.
       $50.4 = (\Delta H_{vap,H_2O})(P_{std})(MW_{H_2O})/(R)(T_{std})$
       MW_{H_2O} = molecular weight of water

At 60°F, $VP_{H_2O} = 0.25636$ psia (GPSA, 1986–Std 2172)

Ideal or Real Gas Basis

When the heating value is stated on a standard volume basis—scf or standard m^3—then the gas may be considered to be *ideal* ($Z = Pv/RT = 1$) or *real* ($Z \neq 1$). Accordingly

$$HV_{ideal} = (HV_{real})/(Z \text{ at std conditions})$$

Detailed procedures for these calculations are given in GPA (1986) Standard 2172–86 and ASTM (1981) Method D 3588.

Determination of Heating Values

In practice, the heating value may be either measured directly in a flow calorimeter, or the gas composition may value of natural gas by the chromatograph and by the calorimeter show excellent agreement—nearly always within 1 Btu/scf. A good chromatographic analysis should have an uncertainty of plus or minus 2 Btu/scf at the 95% confidence level.

Example 2–2. Calculate the gross and net heating values for the first gas in Table 2–5. Use a dry basis and ideal gas behavior.

Solution: As in Example 2–1, a tabular solution is used. Again, the calculations are most conveniently done by use of a spread-sheet program.

Component	Mol%	scf[a]	HHV Btu/scf[b]	Btu[c]	LHV Btu/scf[b]	Btu[c]
N2	0.51	0.0051	0.0	0.0	0.0	0.0
CO2	0.67	0.0067	0.0	0.0	0.0	0.0
C1	91.94	0.9194	1010.0	928.6	909.4	836.1
C2	3.11	0.0311	1769.6	55.0	1618.7	50.3
C3	1.26	0.0126	2516.1	31.7	2314.9	29.2
iC4	0.37	0.0037	3251.9	12.0	3000.4	11.1
nC4	0.34	0.0034	3262.3	11.1	3010.8	10.2
iC5	0.18	0.0018	4000.9	7.2	3699.0	6.7
nC5	0.11	0.0011	4008.9	4.4	3706.9	4.1
C6	0.16	0.0016	4755.9	7.6	4403.8	7.0
C7+	1.35	0.0135	9161.1[d]	123.7	8512.9[d]	114.9
Sum	100.00	1.0000		1181.3ⁱ		1069.6

a. Mol% divided by 100
b. Appendix 6, Table A6–1
c. Scf times heating value
d. Use dodecane heating value (closest molecular weight); data from API Technical Data Book—Petroleum Refining, Vol. 1, p. 1–95 (1985). Two comments can be made. First, if the gross heating value specification is of the order of 1000 Btu/scf, then recovery of NGL from this gas is acceptable. Also, if the content of C7+ is high, as in this gas, then the heating value of this fraction contributes much to the total heating value; an extended analysis of the fraction is justified.

be determined and the heating value calculated from the measured composition. In the U.S., the Federal Energy Regulatory Commission (FERC) generally accepts any reasonable method for determining Btu content upon which the parties to a gas sales contract agree. Because the composition measurement by gas chromatograph is quicker and easier than calorimetric determination, the calculated value is usually used. The recent GPA RP 181 (1986) summarizes accepted procedures for determining heating values, gives examples and suggestions for applying the methods, and amplifies the comments given above on calculation of heating values for custody transfer.

Detailed comparisons for determining the higher heating

Continuous Recording Calorimeter

Prior to 1975, the calorific values of gaseous fuels (with heating values between 475 and 3300 Btu/scf) were measured using water-flow calorimeters (ASTM D 900–55). Since 1975, the calorific values of natural gases with HHV from 900–1200 Btu/scf have been measured using a *continuous recording calorimeter* (ASTM D 1826–83). In this method all the heat released by combustion of the natural gas is imparted to a stream of air. The heating value is obtained by measuring the temperature rise of the air. The test-gas and heat-absorbing air streams are controlled in fixed volumetric proportion by metering devices similar to wet-test meters which are geared together and driven by a common electric motor. The flue gases (combustion products plus excess air) are kept separate from the heat-absorbing air and are cooled to a few degrees above the initial temperature of the fuel gas and air.

The calorimeter is standardized using standard methane of a known heating value (500–900 Btu/scf). The National Institute for Standards and Technology (formerly National Bureau of Standards, NBS) followed the reproducibility of three such calorimeters for four years. The calorimeters were standardized weekly and, one week after standardization, 95% of the errors were less than 0.3% with a few errors as large as 0.5%. Kersey (1983, 1989) summarizes proper installation, maintenance, and testing procedures.

ANALYSIS OF NATURAL GASES

Natural gases may be analyzed using mass spectroscopy and chromatography. While mass spectroscopy is more accurate, the equipment is far more expensive. Therefore gas (or gas-liquid) chromatography is used.

Gas Analysis by Mass Spectrometer

In ASTM (1983) Method D 2650, the molecular components are ionized and dissociated by electron bombardment. The resulting positive ions of different masses are accelerated in an electric field, separated magnetically, and the abundance of each mass present recorded. The observed data are processed by digital computer to arrive at the gas analysis.

Gas Analysis by Chromatography

Standard methods for analyzing natural gases by gas chromatography are described by both ASTM Method D 1945–81 and by GPA Std 2261–90. As shown in Figure 2–3, the basic gas chromatography system consists of a sample preparation system, sample valve, column(s), detector, and signal recorder. In the simplest terms, the sample is forced through the chromatographic column using a suitable carrier gas. Separated components are measured as "peaks" as they leave the column (Fig. 2–4). The column separates by distributing the gas components between two phases; one phase is the large surface area of an adsorbent in a stationary bed of particles and the other is the flowing carrier gas. In gas-solid chromatography (GSC), the stationary phase is a solid packing such as silica gel, alumina, or molecular sieves. In gas-liquid chromatography (GLC),

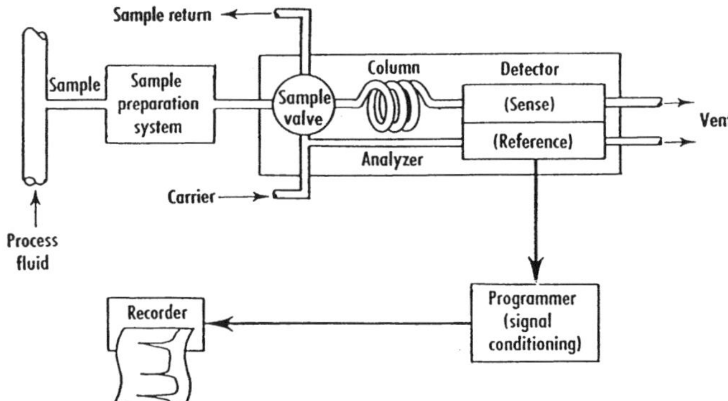

Figure 2–3. Basic chromatography system (Leisey *et al.*, 1977).

the stationary phase is a liquid, spread as a thin film over the inert particles.

Because different components have different affinities for the stationary phase, the carrier gas sweeps the gas components out of the column at different times. The thermal conductivity of the gas passing a detector is measured, and the signal is amplified electronically. When a carrier-gas/component binary passes the detector, the thermal conductivity changes from the base value of the carrier gas. The resulting signal is recorded as a "peak" (Figs. 2–4 and 2–5). The area underneath the peak is determined by electronic integration of the thermal-conductivity signal. The peak areas are converted to mole percent by use of response factors obtained from the analysis of a standard reference mixture of known composition. Heavy-end components (C6+, C7+) are detected as the total area beneath the peaks in the reverse-flow trace; their composition is calculated by peak area ratio from the total C5 mole percent.

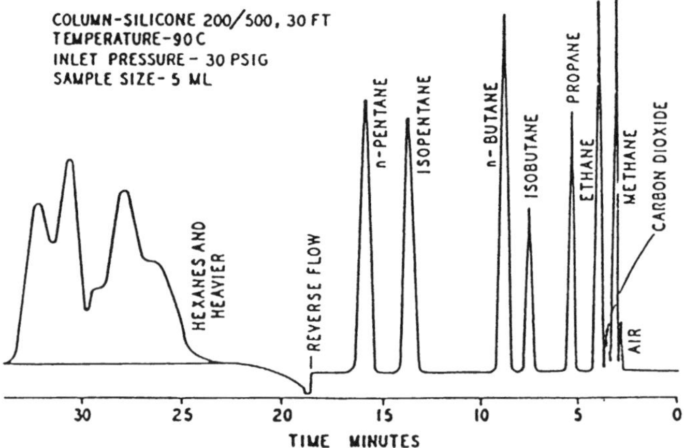

Figure 2–4. Chromatogram of natural gas (Silicone 200/500) (GPA, 1986, Std 2261–86).

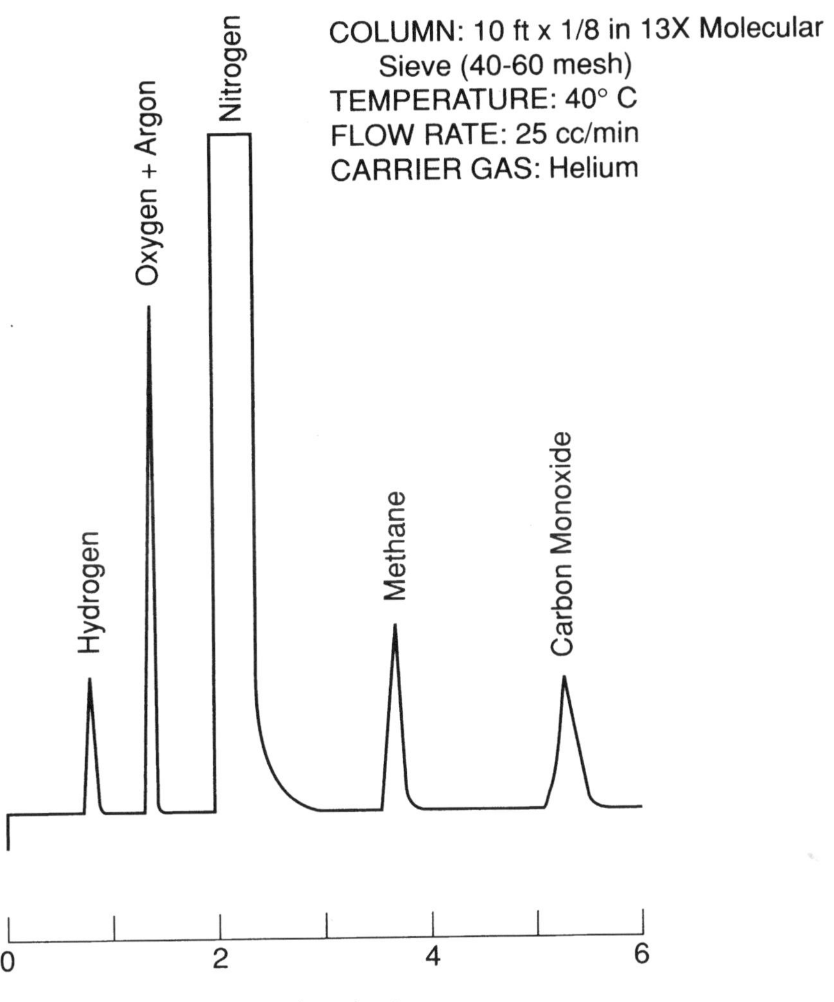

Figure 2–5.   Separation of noncondensable gases on molecular sieves (GPA Std 2261–90).

In practice, an adsorption column of Linde mole sieves is used to separate oxygen, nitrogen, and methane (Fig. 2–5). A silicone partition column is recommended for ethane, heavier hydrocarbons, and CO_2 (Fig. 2–4).

Table 2–6 presents repeatability and reproducibility statements for the GPA 2261 method. Testing included 6 samples analyzed by 21 laboratories and was completed in 1987. Repeatability is the precision expected using the same equipment and analyst. Reproducibility is the precision obtained using different laboratories, equipment, and analysts.

The grouping of the C6+ materials in natural gas as a single component is unsatisfactory for some purposes, such as the calculation of dew points. GPA Standard 2261–90 does not furnish any information concerning the nature of the C6+ fraction, but simply suggests a set of physical constants for that fraction as follows:

Physical Constants for Hexanes and Heavier	
Molecular weight	92.00
Summation factor, b, (GPA Std 2172–86)	0.2830
Compressibility factor, Z, (60 F, 1 atm)	0.9200
Specific gravity (air = 1.0)	3.1765
Scf gas/gal liq	23.2742
Btu/scf (gross, dry)	5065.83

A recent GPA standard (1986, GPA Std 2286–86) describes a tentative method for the extended analysis of natural gas using temperature-programmed chromatography. This method provides a much better description of the C6+ fraction, essentially by component up to C8 and by carbon-number group to C14. A dual hydrogen-flame ionization detector is used in addition to the thermal conductivity detector. Also, three chromatographic columns are used.

Table 2–6 Precision Criteria for Chromatographic Analyses of Natural Gas (GPA Std 2261–90)

Component	Mol % Range	Percent Relative	
		Repeat-ability	Reproduc-ibility
Nitrogen	1.0 – 7.7	2	7
Carbon Dioxide	0.14– 7.9	3	12
Methane	71.6 –86.4	.2	.7
Ethane	4.9 – 9.7	1	2
Propane	2.3 – 4.3	1	2
Isobutane	0.26– 1.0	2	4
n-Butane	0.6 – 1.9	2	4
Isopentane	.12– .45	3	6
n-Pentane	.14– .42	3	6
C6+	.10– .35	10	30

SAMPLING

While "any fool can grab a sample," collection of truly representative samples challenges experts. A moment's thought will confirm that any subsequent design, operating, or investment decision can be no better than the prior sampling and analyses.

In spite of its obvious importance, natural-gas sampling is seldom done well. Natural gas, flowing with accompanying condensate, cannot be properly sampled by withdrawal of a portion through a sample tap. Reflection on the complex nature of two-phase flow—with its varying flow patterns and liquid-holdup phenomenon, along with velocity gradients in the fluid—should convince one of the impossibility.

In the ideal case, a separator should be set up at the wellhead to collect condensate and meter each phase. Subsequent analysis of the separator gas and liquid allows recombination to obtain the well-stream analysis. This is the reason well-test separators are even installed offshore where space and weight are exceedingly expensive.

Poor sampling is generally caused by:

1. Ignorance of its economic and technical importance.
2. Inadequate training of personnel involved.
3. Ignorance and/or unwillingness to follow standard procedures.

While sampling in the field *do not*:

1. Use dirty cylinders that contain previous samples.
2. Sample when P and T are not stable.
3. Sample when P separator is different from P sample.
4. Dilute sample with air.
5. Fail to identify sample completely.

Sampling hydrocarbon fluids can be *hazardous*. Everybody involved in sampling must be familiar with and follow safe practices while handling *flammable fluids under pressure*. In the U.S., all sample containers must meet the specifications and label identification required by the Hazardous Material Regulations of the Department of Transportation.

DO NOT FILL LIQUID SAMPLE CONTAINERS COMPLETELY FULL. A 150-ml, stainless-steel container was filled full with gasoline and closed at 0 psig and 61°F. When heated to 105°F, the sample pressure was 3100 psig!

Gas Sampling

Hefley *et al.* (1985) present a GPA Work Group Report on natural gas sampling. This sampling program served as the basis for the 1986 revision of GPA Standard 2166–68 and examined all eight GPA 2166 sampling methods in great detail. GPA Standard 2166–86 indicates that good samples can be obtained by all eight approved methods provided extreme care is taken. However, the sampling project report (Hefley *et al.*, 1985) ranked the accuracy of the eight methods in the following order of decreasing accuracy:

1. Water Displacement
2. Continuous Purge
3. Purge and Fill
4. Glycol Displacement
5. Reduced Pressure
6. Floating Piston Cylinder
7. Helium filled to 5 psig
8. Evacuated Container

In practice *purge and fill* is the most popular method while the *floating piston cylinder* is finding increased use because composite samples can be obtained (Hefley, 1989).

Several problems can arise during sampling of natural gas:

1. Condensation of hydrocarbons due to temperature and pressure changes during sampling.
2. Entrainment of liquid droplets and mist.
3. Sample constituents can react with vessel container.
4. Some sample components may dissolve in displacement media.

Subsequent handling of the sample is, therefore, very important to assure that the relatively small amounts of condensable components remain in the gas phase and are removed to the chromatograph or mass spectrometer.

In a natural gas, hydrogen sulfide in small amounts may be overlooked entirely, to the later chagrin of the designer and operator. Hydrogen sulfide reacts rapidly with carbon steel and may disappear from the sample altogether. A stainless-steel sample cylinder should be used if the presence of H_2S is suspected. Even austenitic stainless-steel will absorb a small amount of H_2S. GPA Standard 2261 (1986) recommends that the gas be analyzed at its source for hydrogen sulfide content less than 3% by GPA Standard method 2377 (1986). Teflon-lined cylinders have also been used successfully.

The following precautions are recommended by Toole (1981):

1. Use a prefilter and/or "knockout" trap on the sample system just downstream from the source valve. This is mandatory on "wet" gas samples.
2. Use heated sample lines to prevent condensation from low ambient temperatures during sampling.
3. Keep sample lines as short as possible. This should be done for all sample grabbing, wet or dry.
4. Clean, dry, and evacuate sample cylinders before taking to the field. This prevents possible liquid carryover from a previous sample.
5. Purge sample cylinders carefully.
6. Empty cylinders through a non-metallic pigtail on the outlet for the expansion of pressured gas to the atmosphere.
7. For samples requiring calorimetric heating value determination and chromatographic analysis, use stainless steel or carbon steel, DOT3A and DOT3AA, cylinders, 300 cu.in. volume. Smaller cylinders may be used for GC analysis only; however, a calorimeter Btu determination provides good confirmation of the chromatograph analysis. Calculated Btu from the analysis should not be over ± 3 Btu from the measured value.
8. Keep sample cylinders upright while filling, with the purge line at the bottom. Do not lay the cylinders on the ground.
9. Use a line probe that extends at least a third of the way into the line for sampling. This prevents the entry of condensate and other contaminants into the sampling line.

Sampling by liquid displacement has been used in the past and is still followed by some organizations. It is not recommended for gases or liquids containing acid-gases such as H_2S and CO_2. These components dissolve in the liquid, usually water, and are lost. Therefore the resulting analysis is not representative.

Liquid Sampling

A GPA Work Group (McCann *et al.*, 1988) conducted a cooperative natural gas liquids sampling project for future revision of GPA (1983) Standard 2174–83. The following four sampling methods were judged acceptable:

1. Floating Piston Cylinder (GPA, 1983—Std 2174–83)
2. Water Displacement (total removal—80% hydrocarbons/20% displaced outage)
3. Water Displacement (partial removal—70% hydrocarbon/20% displaced outage/10% water remaining in cylinder)
4. Ethylene Glycol Displacement (total removal—80% hydrocarbon/20% displaced outage)

The following precautions are recommended for liquid sampling while using the floating piston cylinder (Toole, 1981).

1. Pressure on backside of piston should be higher than line pressure at beginning of operation.
2. Slowly bleed pressure down to enter sample, and maintain pressures on each side of the piston at nearly equal levels with just enough difference to move the pistons.
3. Do not bleed the pressure off the backside of the piston after sample has been taken, because this will flash some of the liquid into gas. Then you will need to resample.
4. Do not fill cylinder 100% full with sample. Leave at least 25% as a pressure buffer to take care of ambient temperature fluctuations.

The floating piston cylinder also can be used for "wet" gas sampling with excellent results in getting representative samples, according to Toole (1981).

1. Pressure up back of cylinder with line gas to full line pressure.
2. Connect cylinder to source (use of a line probe is mandatory) and open valves to cylinder.
3. Fill cylinder as you would in taking a liquid sample. Note: It is extremely important to avoid any pressure drops. The pressure differences should not exceed 2 psig maximum. This will prevent flashing of heavy components which may be in aerosol form. Sample should be slowly entered into cylinder.

Continuous Sampling

Welker (1985) recommends the following procedures and precautions for continuous or composite sampling of natural gas:

1. Sample point should "see" center one-third of the pipeline in an area of good velocity with minimum turbulence.
2. Sample probe, equipped with full-open ball or gate valve, must be kept away from pipeline fittings and orifice

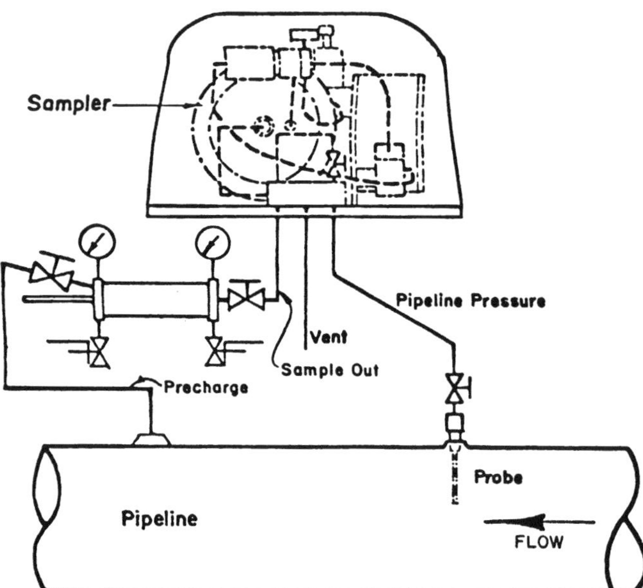

Figure 2–6. Sampler, sampler hook-up and manifold (Welker, 1985).

plates. The probe may be bevel or flat cut at end, but must be kept clear of free liquids and aerosols. Bevel may face upstream or downstream.

3. Probe construction: stationary or permanent, manually insertable or automatic insertion type probes may be used. The probe should be stainless steel so that it will not react with the sampled gas.

4. Sampler hook-up and manifold: The sampler should be mounted above the sample point as shown in Figure 2–6. Other precautions are:
   a. Line from probe to sampler should slope down to let any free liquids drain back into pipeline.
   b. Never sample a dead-end line.
   c. Any leaks in the line from the sampler to the cylinder will ''lose'' light ends preferentially.
   d. Any filters, drips, or regulators between the probe and sampler will invalidate the sample.

5. Sampler: should take a composite sample in the same way as an operator would take a spot sample. If the pipeline flow rate varies, the sampler should be actuated proportional to the flow. The sampler should be able to pump the sample into the cylinder and purge itself before pumping each new ''bite.''

Sample Containers

Proper cleaning and inspection of sample cylinders cannot be overemphasized. Several methods are available (Welker, 1985):

1. Volatile solvent and air dry
2. Steam clean and air dry
3. Evacuation
4. Evacuation and fill with 5 psig helium

Correct procedures must be followed rigorously in every detail.

PRODUCT SPECIFICATIONS

The objective of field processing is to provide transportable and/or salable fluids. There are two main products: natural gas and condensate (raw mix) or natural-gas liquids (NGL).

Natural Gas

Table 2–7 lists typical *natural-gas pipeline specifications*. These specifications are fixed by negotiation between seller and buyer and vary from case to case. Not all sales gases will have all the specifications shown in the table. Nevertheless, the ranges shown for each specification are typical.

Test methods used for determining the various compositions in Table 2–7 are given by the Gas Processors Association (GPA) and the American Society for Testing Materials (ASTM). The usual methods are included in Table 2–7.

The most important specifications are the first three. Note that both water and hydrogen sulfide must be removed to very low concentrations. Heating value is more complex—specifications usually range from 950 to 1200 Btu/scf. Gross heating values for a few paraffin hydrocarbons are:

Methane	1010 Btu/scf
Ethane	1770 Btu/scf
Propane	2516 Btu/scf
n-Butane	3262 Btu/scf

The most abundant component, methane, has a relatively low heating value. Therefore, by itself, methane cannot always fulfill the minimum heating value requirement when inerts (nitrogen and/or carbon dioxide) are present. However, enough heavier hydrocarbons are usually present to provide the required heating value, even with condensate recovery.

When there is a maximum heating value or dew-point specification, some of the heavier hydrocarbon constituents may have to be removed as condensate. Refrigeration to –30°F reduces the heating value—how much, of course, depends on the gas composition and pressure. Further reduction requires cryogenic processing.

Table 2–7 Natural Gas Pipeline Specifications

Characteristic	Specification (Goar and Arrington, 1978)	Test Method (Hefley, 1989)
Water content	4–7 lb/MMscf max.	ASTM (1986), D 1142
Hydrogen sulfide content	1/4 grain/100 scf max.	GPA (1968), Std 2265 GPA (1986), Std 2377
Gross heating value	950 Btu/scf min.	GPA (1986), Std 2172
Hydrocarbon dew point	15°F @ 800 psig max.	ASTM (1986), D 1142
Mercaptan content	0.2 grain/100 scf max.	GPA (1968), Std 2265
Total sulfur content	1–5 grain/100 scf max.	ASTM (1980), D 1072
Carbon dioxide content	1–3 mole percent max.	GPA (1990), Std 2261
Oxygen content	0–0.4 mole percent max.	GPA (1990), Std 2261
Sand, dust, gums, and free liquid	Commercially free.	
Delivery temperature, °F	120°F max.	
Delivery pressure, psia	700 psig min.	

The remaining specifications are met by suitable processing of the type mentioned previously. These processes are the subjects of later chapters.

Natural-Gas Liquids

Hydrocarbon condensate recovered from natural gas may be shipped without further processing or stabilized to produce a safely-transportable liquid. In the case of raw condensate, there are no particular specifications for the product other than the process requirements. Stabilized liquid, on the other hand, will generally have a vapor-pressure specification, since the product will be injected into a pipeline or transport pressure vessel which has definite pressure limitations.

Natural-gas liquid products are prepared by fractionation of the raw make into appropriate products, either at the field processing site or, perhaps more commonly, at a large central facility. In any case, product specifications are not so typical as those of sales gas, but depend heavily on the particular contract. The Gas Processors Suppliers Association (1987) presents specifications for commercial propane, LPG, and natural gasoline. These specifications represent a "broad industry consensus for minimum quality products" and do not cover all the possible NGL products.

Liquid product specifications generally include composition, vapor pressure, water content, and sulfur content.

Composition is usually measured by gas chromatograph (GPA Std 2177–84). If the composition of the C6+ fraction is important, extended analysis can be used (GPA Std 2286–86). Mass spectrographic analysis is more comprehensive, but also more expensive.

Vapor pressure is an easily measured indication of composition of the particular liquid product. Vapor pressure is important also because of U.S. Department of Transportation (DOT) regulations governing transport of vessels and cylinders. The Reid vapor pressure (RVP) method (ASTM Method D 323–82) is used for natural gasoline, and a modification of the method (ASTM Method D–1267–84) is used for LPG.

Water content is important primarily for propane, because propane dissolves more water than the other NGL. Low water concentrations are necessary to avoid the formation of hydrate and consequent storage and delivery problems. Tests include the GPA Cobalt Bromide Test (GPA Standard 2140–88) and the valve freeze method (ASTM Method D 2713–86).

Sulfur compounds in liquid fuels contribute to corrosion in storage and distribution systems. Total sulfur is measured by the method of ASTM Method D–2784–80. Potential corrosion due to hydrogen sulfide in propane and LPG is indicated by the copper-strip test (ASTM Method D–1838–84); this sensitive test is, unfortunately, not as easy to interpret as it superficially appears (GPSA, 1987, p. 2–2).

Review Questions

1. What are the two main purposes of natural gas conditioning?
2. Name the principal types of petroleum fluids produced from subsurface reservoirs.
3. Define: natural gas, crude oil, condensate, associated gas, separator gas, oil formation factor, shrinkage, GOR, GLR, volatile oil, wet gas, dry gas, NGL, LNG, sweet gas, sour gas, lean gas, rich gas.
4. Define heating value.
   - Distinguish between gross and net heating values.
   - Distinguish between dry and wet scf in reporting GHV.
   - Distinguish between real and ideal gas in reporting GHV.
5. How are heating values obtained?
   How accurate can a heating value measurement be?
6. Compare the two methods of obtaining heating values. Include accuracy, speed, and ease of measurement.
7. Identify the constituents of natural gas as produced in the field.
8. Discuss the constituents of wellhead natural gas. Include the following considerations: usual concentration, value, and potential problems.
9. What is the most toxic component in natural gas?
10. What are the two principal means of natural gas analysis? Compare these two methods (include accuracy, cost, and ease of measurement).
11. Define C6+ and C7+.
    How are "heavy ends" usually characterized?
    When is accurate characterization of heavy ends required?
12. Is an associated gas richer or leaner than a gas-well gas?
13. The proper sampling of hydrocarbon fluids is a simple task. True/False?
14. What factors cause poor sampling?
    Identify some common sampling errors.
15. Name four approved methods for sampling natural gas.
    List nine precautions recommended for sampling natural gas.
16. Name four approved methods for sampling NGL.
    List seven precautions recommended for sampling NGL.
17. What is continuous sampling?
    List some precautions recommended for continuous sampling.
18. How would you clean a sample cylinder?
19. Identify the common specifications for a sales gas.
    Discuss these specifications.
20. Discuss the important specifications for NGL products.
21. What safety precautions should be observed while sampling and transporting natural gas and NGL products?

Problems

1. Confirm the GPM shown in Table 2–5 for one of the gases (other than the first, which is done in Example 2–1).
2. Calculate the gross and net heating values for one of the gases in Table 2–5 (other than the first, which is done in Example 2–2).
3. At 60°F and 14.696 psia:

   For methane: GHV = 1010.0 Btu/scf ideal gas
   $Z = 0.998$
   For water: ΔH_{vap} = 1059.6 Btu/lb
   Calculate: NHV in Btu/scf ideal gas.
   Compare answer with:
   NHV = 909.4 Btu/scf (GPSA 1987, p. 23–4)
   GHV in Btu/scf wet gas (assume saturated)
   GHV in Btu/scf real gas.

4. When the demand for natural gas is unusually high (e.g., in winter) propane (C3) can be mixed with air and sold as a substitute for natural gas. This is called "propane shaving." Calculate the composition of the C3–air mixture to make it have a GHV similar to natural gas— say 1000 Btu/scf. The lower and higher flammability limits for C3 in air are 2.0 and 9.5 vol%, respectively. Is this C3–air mixture safe?
5. A reservoir fluid has a GOR of 2900 scf/bbl. How would you classify this fluid (oil, gas, condensate, or other)?

Nomenclature

ASTM = American Society for Testing and Materials
Brf = barrels of reservoir fluid
Bsto = barrels of stock-tank oil
DOT = U.S. Department of Transportation
FERC = Federal Energy Regulatory Commission
FVF = formation volume factor
GHV = gross or higher heating value
GLC = gas-liquid chromatography
GOR = gas oil ratio
GPM = gal liquefiable NGL per 1000 scf gas
GSC = gas-solid chromatography
HHV = higher or gross heating value
HV   = heating value
LHV = lower or net heating value
LNG = liquefied natural gas
MW   = molecular weight
NBS = National Bureau of Standards

NGL = natural gas liquids
NHV = net or lower heating value
NIST = National Institute for Standards and Technology
P = absolute pressure
ppbw = parts per billion by weight
ppmw = parts per million by weight
R = gas constant on a mol basis
RVP = Reid vapor pressure
T = absolute temperature
VP = vapor pressure
v = specific volume (ft^3/lb mol)
Z = compressibility factor = $P v / R T$
ΔH_{vap} = change in enthalpy upon vaporization (latent heat)

Subscripts

ideal = denotes ideal gas
H_2O = denotes water property
real = denotes real gas
std = denotes property at standard or reference conditions

References

Allen, J. C. (1952), "Classification of Oil and Gas Wells," *Oil & Gas J.*, Vol. 51, No. 12, pp. 355–365 (July 28).

Anonymous (1988), "Pilots Due to Scrub Arsenic from Abo Gas," *Oil & Gas J.*, Vol. 86, No. 32, p. 24 (Aug. 8).

ASTM (1980), Method ANSI/ASTM D 1072, "Total Sulfur in Fuel Gases," American Society for Testing and Materials, 1916 Race Street, Philadelphia, PA 19103.

ASTM (1980), Method ANSI/ASTM D 2784, "Sulfur in Liquefied Petroleum Gases (oxy-Hydrogen Burner or Lamp)," American Society for Testing and Materials, 1916 Race Street, Philadelphia, PA 19103.

ASTM (1981), Method ANSI/ASTM D1945, "Analysis of Natural Gas by Chromatography," American Society for Testing and Materials, 1916 Race Street, Philadelphia, PA 19103.

ASTM (1981), Method ANSI/ASTM D 3588, "Calculating Calorific Value and Specific Gravity of Gaseous Fuels," American Society for Testing and Materials, 1916 Race Street, Philadelphia, PA 19103.

ASTM (1982), Method ANSI/ASTM D 323, "Vapor Pressure of Petroleum Products (Reid Method)," American Society for Testing and Materials, 1916 Race Street, Philadelphia, PA 19103.

ASTM (1983), Method ANSI/ASTM D 1826, "Calorific Value of Gases in Natural Gas Range by Continuous Recording Calorimeter," American Society for Testing and Materials, 1916 Race Street, Philadelphia, PA 19103.

ASTM (1983), Method ANSI/ASTM D 2650, "Chemical Composition of Gases by Mass Spectrometry," American Society for Testing and Materials, 1916 Race Street, Philadelphia, PA 19103 (discontinued 1981).

ASTM (1984), Method ANSI/ASTM D 1267, "Vapor Pressure of Liquefied Petroleum (LP) Gases (LP-Gas Method)," American Society for Testing and Materials, 1916 Race Street, Philadelphia, PA 19103

ASTM (1984), Method ANSI/ASTM D 1838, "Copper Strip Corrosion by Liquefied Petroleum (LP) Gases," American Society for Testing and Materials, 1916 Race Street, Philadelphia, PA 19103

ASTM (1986), Method ANSI/ASTM D1142, "Water Vapor Content of Gaseous Fuels by Measurement of Dew Point Temperature," American Society for Testing and Materials, 1916 Race Street, Philadelphia, PA 19103.

ASTM (1986), Method ANSI/ASTM D 2713, "Dryness of Propane (Valve Freeze Method)," American Society for Testing and Materials, 1916 Race Street, Philadelphia, PA 19103.

ASTM (1987), Method ANSI/ASTM D 1265, "Sampling Liquefied Petroleum (LP) Gases," American Society for Testing and Materials, 1916 Race Street, Philadelphia, PA 19103.

Bingham, Mark D. (1990), "Field Detection and Implications of Mercury in Natural Gas," SPE Production Engineering, Vol. 5, No. 2, pp. 120–124 (May).

Chiu, Chen-hwa (1978), "Evaluate Separation for LNG Plants," *Hydro. Proc.*, Vol. 57, No. 9, pp. 266–272 (Sept).

Ewan, D. N., J. B. Lawrence, C. L. Rambo, and R. R. Tonne (1975), "Why Cryogenic Processing (Investigating the Feasibility of a Cryogenic Turbo-Expander Plant)," *Proc. 54th Ann. Conv. GPA*, Houston, pp. 130–135 (March 10–12).

GPA (1968), Std 2265–68, "Standard for Determination of Hydrogen Sulfide and Mercaptan Sulfur in Natural Gas," Gas Processors Association, 6526 East 60th St., Tulsa, OK 74145.

GPA (1983), Std 2174–83, "Method for Obtaining Liquid Hydrocarbon Samples Using a Floating Piston Cylinder," Gas Processors Association, 6526 East 60th St, Tulsa OK 74145

GPA (1984), Std 2177–84, "Analysis of Demethanized Hydrocarbon Liquid Mixtures Containing Nitrogen and Carbon Dioxide by Gas Chromatography," Gas Processors Association, 6526 East 60th St., Tulsa, OK 74145.

GPA (1986), Std 2166–86, "Obtaining Natural Gas Samples for Analysis by Gas Chromatography," Gas Processors Association, 6526 East 60th St., Tulsa, OK 74145.

GPA (1986), Standard 2172–86, "Calculation of Gross Heating Value, Relative Density and Compressibility of Natural Gas Mixtures from Compositional Analysis," Gas Processors Association, 6526 East 60th St., Tulsa, OK 74145.

GPA (1986), Standard 2286–86, "Tentative Method of Extended Analysis for Natural Gas and Similar Gaseous Mixtures by Temperature-Programmed Gas Chromatography," Gas Processors Association, 6526 East 60th St., Tulsa, OK 74145.

GPA (1986), Standard 2377–86, "Test for Hydrogen Sulfide and Carbon Dioxide in Natural Gas Using Length of Stain Tubes," Gas Processors Association, 6526 East 60th St., Tulsa, OK 74145.

GPA (1986), Tentative Reference Bulletin 181–86, "Tentative Reference Bulletin, Heating Value as a Basis for Custody Transfer of Natural Gas," Gas Processors Association, 6526 East 60th St., Tulsa, OK 74145.

GPA (1988), Standard 2140–88, "Liquefied Petroleum Gas Specifications and Test Methods," Gas Processors Association, 6526 East 60th St., Tulsa, OK 74145.

GPA (1990), Std 2261–90, "Analysis for Natural Gas and Similar Gaseous Mixtures by Gas Chromatography," Gas Processors Association, 6526 East 60th St., Tulsa, OK 74145.

GPSA (1987), Engineering Data Book, 10th Edition, Gas Processors Suppliers Association, 6526 East 60th St., Tulsa, OK 74145.

Gill, Eric (1987), "The History and Economics of Extended Hydrocarbon Analysis," *Proc. 66th GPA Annual Convention*, pp. 218–224, Denver, CO (March 16–18).

Goar, B. G., and T. O. Arrington (1978), "Processing Sour Gas—1: Guidelines for Handling Sour Gas," *Oil & Gas J.*, Vol. 76, No. 26, pp. 160–164 (June 26).

Gould, Thomas L., and Robert E. McDonald (1979), "Practical Approaches to Two-Phase Flow Problems in Producing Operations," *Proc. 26th Southwestern Petroleum Short Course*, pp. 161–172, Lubbock, TX (April 19–20).

Grizzaffi, L. P., and B. M. Thompson (1970), "Completing Gas Wells That Produce High H$_2$S," *World Oil*, Vol. 170, No. 7, pp. 70–74, (June).

Hatch, Lewis F., and Sami Matar (1977) "From Hydrocarbons to Petrochemicals, Part 2," *Hydro. Proc.*, Vol. 56, No. 6, pp. 189–194 (June).

Hefley, Carl (1989), Oxy NGL Inc., Tulsa, OK, Private Communication (June).

Hefley, Carl, W. J. Hines, and Frank W. Sattler (1985), "Work Group Report on Natural Gas Sampling for Revision of GPA Standard 2166–88," *Proceedings 64th Annual GPA Convention*, Houston, TX (March 18–20).

Hines, W. J. (1984), "The Column in Gas Chromatography," Presented at 11th GPA School of Gas Chromatography, University of Tulsa, Tulsa, OK (Aug. 6–10).

Kaasa, Oyvind (1978), "Optimum Conditions for Oil Recovery in a Three-Stage Process," M.S. Thesis, The University of Tulsa, Tulsa, OK.

Kersey, A. F. (1983), "Operation and Maintenance of the Cutler-Hammer Calorimeter," *Proc. Appalachian Gas Measurement. Short Course*, Coraopolis, PA, pp. 181–184 (Aug. 16–18).

Kersey, A. F. (1989), "Operation and Installation of a Recording Calorimeter," *Proc. of 64th International School of Hydrocarbon Measurement*, p. 93, University of Oklahoma, Norman, OK (May 16–18)

Kilmer, J. W. (1982), "Custody Transfer of Natural Gas by Heat Content," *Oil & Gas J.*, pp. 63–65 (Sept. 13).

Leisey, Frank, Gary Potter, and Ray McCoy (1977), "Measuring gas; Chromatography bests calorimetry," *Oil & Gas J.*, Vol. 75, No. 29, pp. 66–75 (July 18).

McCain, William D., Jr. (1990), *The Properties of Petroleum Fluids*, *2nd Ed.*, PennWell Publ. Co., Tulsa, OK, 1973.

McCann, Patrick M., Mark A. Scripsick, Chris Wilkins, and Carl G. Hefley (1988), "Work Group Report Natural Gas Liquids Sampling Project (for Revision of GPA Standard 2174)," *Proc. 67th Annual GPA Convention*. Dallas, TX (March 14–16).

Merrill, Kent E. (1987), "Continuous H$_2$S Monitoring Systems," *Proc. 34th Annual Southwestern Petroleum Short Course*, pp. 249–257, Texas Tech University, Lubbock, TX (April 22–23).

Miller, Richard D. and Floyd R. Hertweck, Jr. (1981), "Analyses of Natural Gases," *Bureau of Mines Information Circular 8890*, U.S. Department of the Interior.

Penick, Dudley P. (1983), "Factors to Use in Dealing with North Sea Gas," *Oil & Gas J.*, Vol. 81, No. 25, pp. 114–120 (June 20).

Seaton, Earl (1978), "Indonesia's Arun LNG Plant Nears Completion," *Oil & Gas J.*, Vol. 76, No. 11, pp. 60–64 (March 13).

Slaton, Mike (1988–89), "Ironing out arsenic-tainted wrinkle," *Southwest World Oil*, Vol. 36, No. 11, pp. 14–17 (Dec.–Jan.).

Tiemstra, E. (1988), "Comprehensive Monitoring Controls Hydrogen Sulfide in Gas Transmission," *Oil & Gas J.*, Vol. 86, No. 1, pp. 45–50 (Jan. 4).

Toole, John W. (1981), "Field Sampling of Natural Gas," presented at GPA School of Chromatography, Tulsa, OK (Aug. 3–7).

Welker, Thomas F. (1985), "Composite Sampling of Natural Gas," *Proceedings of 60th International School of Hydrocarbon Measurement*, pp. 257–264, University of Oklahoma, Norman, OK (April 16–18).

Wilson, Arild, R. N. Maddox, and J. H. Erbar (1978), "C6+ Fractions Affect Phase Behavior," *Oil & Gas J.*, Vol. 76, No. 34, pp. 76–81 (Aug. 21).

Chapter 3

Phase Behavior of Natural Gas

INTRODUCTION

Obviously, field processing of natural gas depends strongly on the behavior, characteristics, or properties of the well stream involved. It is important to know and predict the amount, composition, and density of any phases present in any process situation. Accordingly, hydrocarbon *phase behavior* is discussed, first for pure components and then for mixtures. The more commonly used *phase diagrams* are presented. Phase-equilibrium calculations are reviewed next, including both hand and computer methods. Phase equilibrium calculations require component K-values and densities. Two sources of these important data are reviewed: charts and equations of state.

Phase behavior in the presence of water is important also, since if water condenses a second liquid phase will be present. Water/hydrocarbon phase behavior is discussed in Chapter 4. At low temperatures, carbon-dioxide/hydrocarbon phase behavior is important, and this topic is summarized in Chapter 4 also.

PURE-SUBSTANCE PHASE DIAGRAMS

A simple or one-component system is defined as one that undergoes no changes in composition; in practice, this means either a *pure substance* or a homogeneous mixture. Frequently, air can be considered a pure substance even though it is a mixture of nitrogen, oxygen, carbon dioxide, and water vapor, plus minor amounts of other gases. In the compression or heat exchange of air, the air can be treated as homogeneous. Of course air must be considered a complex mixture when separated into oxygen and nitrogen, or when partially condensed into liquid and vapor phases having different compositions, or when one component (oxygen) reacts during combustion.

Everyday experience tells us the number of *independent thermodynamic variables* that can be assigned arbitrary values before the *state* or condition of a pure substance is known. For example, if we choose pure H_2O as the system and specify the temperature to be 20°C (68°F) and the pressure to be 1 atm (101.3 kPa, 14.696 psia), we know that the H_2O is liquid. In addition, all other properties (v, h, s, ρ, μ, etc.) are completely fixed also. Two variables have determined the state of the system.

We also know that, at one atmosphere, water boils at 212°F (100°C). Thermodynamics explains this experimental fact as follows: If the pressure is fixed at 1 atm, then the temperature must be 212°F for equilibrium to exist between liquid water and water vapor (steam). Again we have made two specifications: First, there are two phases (liquid and vapor) in equilibrium and, second, the pressure is 1 atm. Again two statements fix all the properties of the two phases (v_f, v_g, h_f, h_g, s_f, s_g, etc.). In fact, everything is known about both the liquid and the vapor phases except how much of each phase is present.

The above facts can be described from another viewpoint. We specify the liquid and vapor phases to be an equilibrium two-phase system. Then only one additional thermodynamic property or variable can be chosen before the system is fixed.

The number of choices required to specify or fix a system completely is called the *degrees of freedom* or *variance* (f). Gibbs' phase rule (Van Wylen and Sonntag, p. 534) is a convenient way of remembering the number of degrees of freedom:

$$f + \phi = C + 2 \qquad (3\text{–}1)$$

where f = number of degrees of freedom (the number of independent, intensive thermodynamic variables)

ϕ = number of equilibrium phases

C = number of components in system (by definition 1 for a pure substance).

Using the phase rule for a pure substance we find:

when $\phi = 1$, there are 2 degrees of freedom

    $\phi = 2$, there is 1 degree of freedom

    $\phi = 3$, there are no degrees of freedom.

In addition, for a pure substance it is impossible to have four different phases coexisting in equilibrium.

The phase behavior of pure substances can be described using diagrams. One example is the pressure-temperature (P-T) diagram (Fig. 3–1) which shows vapor, liquid, and solid phases. In many cases, pure substances exhibit several different solid phases; e.g., there are at least six different phases for ice (Van Wylen and Sonntag, p. 42). Note Figure 3–1 also summarizes our previous ''phase-rule'' conclusions. Single-phase loci ($\phi =1$) are areas (f = 2). Two-phase loci ($\phi = 2$) are lines (f = 1). The three-phase locus, or triple point, ($\phi = 3$) is a point (f = 0).

Because a pure substance has, at most, two degrees of freedom, we can use any pair of independent thermodynamic properties for the axes of our phase diagram. Some of the more popular diagrams are the pressure-volume (P-v), temperature-entropy (T-s), pressure-enthalpy (P-h, or log [P-h]), and Mollier or enthalpy-entropy (h-s) diagrams. Figure 3–2 shows examples of these diagrams.

The liquid phase above (higher P) and left (lower T) of the vapor-pressure curve, such as point A in Figure 3–1, is called ''sub-cooled'' or ''compressed'' liquid. Vapor below (lower P and/or higher T) the vapor-pressure curve, e.g., B, is ''superheated'' vapor. The vapor pressure curve ends at the critical point, C. Above this critical temperature it is no longer possible for a change in phase to occur as the pressure is increased.

Consider a superheated vapor at state B in Figure 3–1.

If the vapor is compressed at constant temperature (isothermally), as in a variable-volume cell in a constant-temperature bath, the path followed is the vertical dotted line shown. At first, the vapor is simply compressed until the vapor pressure curve is reached. Then further attempts at compression result in condensation. Note the *pressure* remains *constant* until *all* of the *vapor is condensed to liquid*. Only then does the pressure begin to rise again, and then very rapidly.

Let the process described above start at B, but now occur in three steps (B to D, D to E, and E to A) as shown by the dashed lines in Figure 3–1. Now a continuous path is followed with no visible phase change because we chose a temperature higher than the critical for the compression step. At such a temperature, we call the fluid a ''gas'' rather than a ''vapor,'' the distinction being that a vapor is condensible and a gas is not.

To the right of and above the critical point (higher T and/or P) on the P-T diagram, the fluid is called a ''supercritical fluid,'' or more simply, a ''dense fluid.''

Vapor-pressure curves (VP-T diagrams) can be plotted in various ways. One popular method is to plot log vapor pressure versus reciprocal absolute temperature because the resulting plot is very nearly a straight line. The abscissa is often plotted backwards and labeled as temperature rather than reciprocal absolute temperature, for convenience in use. The resulting temperature scale is highly nonlinear as shown in Figure 3–3. Another popular method is to plot log vapor pressure versus $1/(T^{\circ}F + 382)$, which is

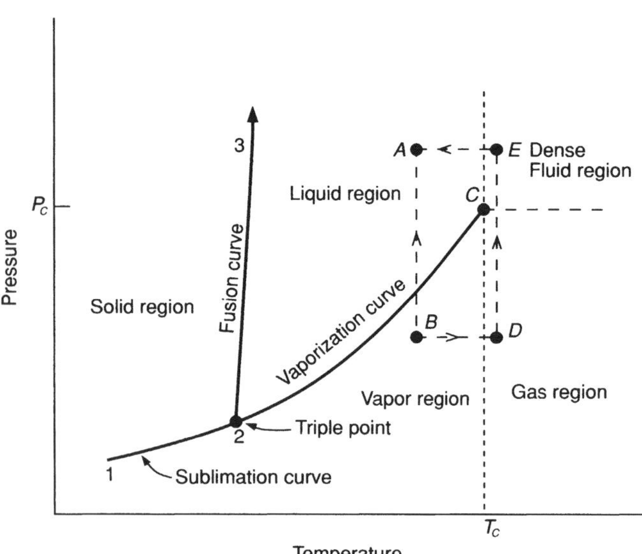

Figure 3–1.   Pressure-temperature diagram.

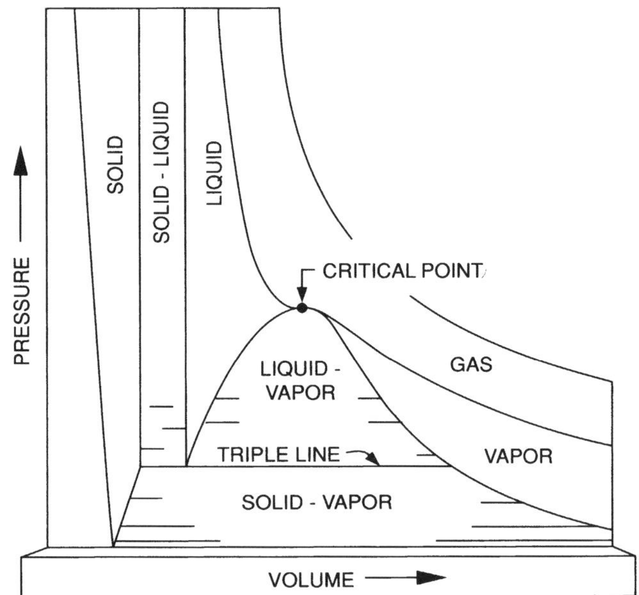

Figure 3–2a.   P-V diagram for pure substances that contract on freezing (Lee and Sears, 1963, p. 40).

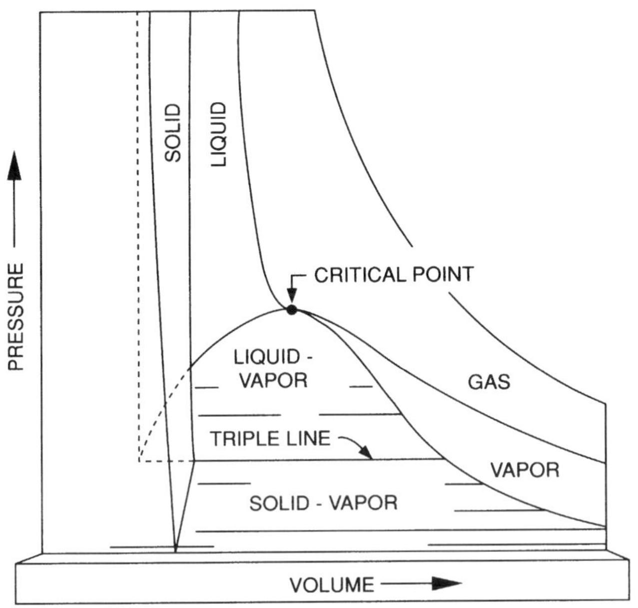

Figure 3–2b. P-V diagram for pure substances that expand on freezing (Lee and Sears, 1963, p. 40).

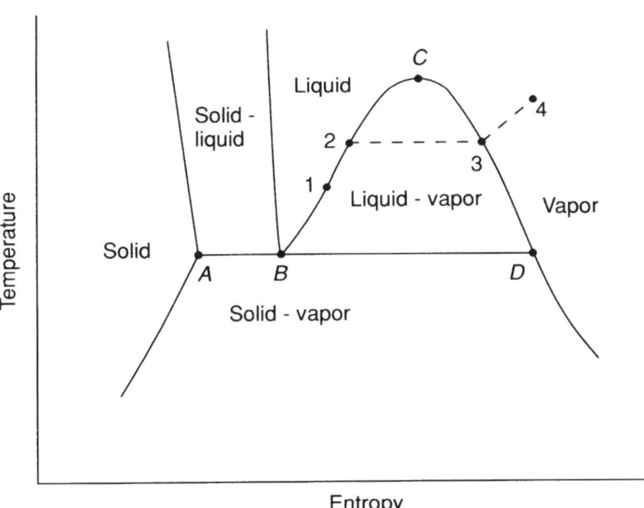

Figure 3–2c. Temperature-entropy (T-s) diagram for a single-component system (Smith and van Ness, 1975, p. 213).

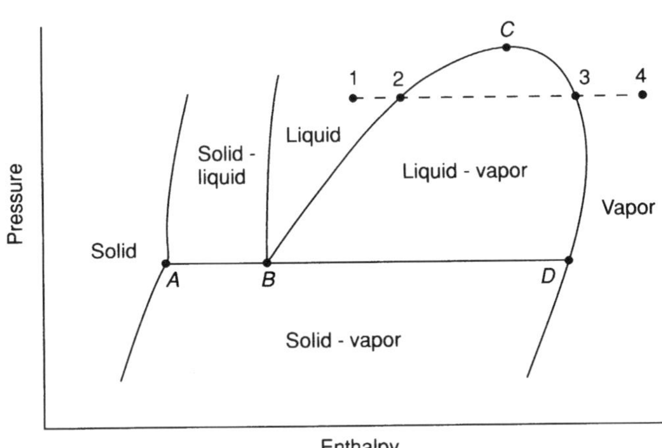

Figure 3–2d. Pressure-enthalpy (1n P-h) diagram for a single-component system (Smith and van Ness, 1973, p. 214).

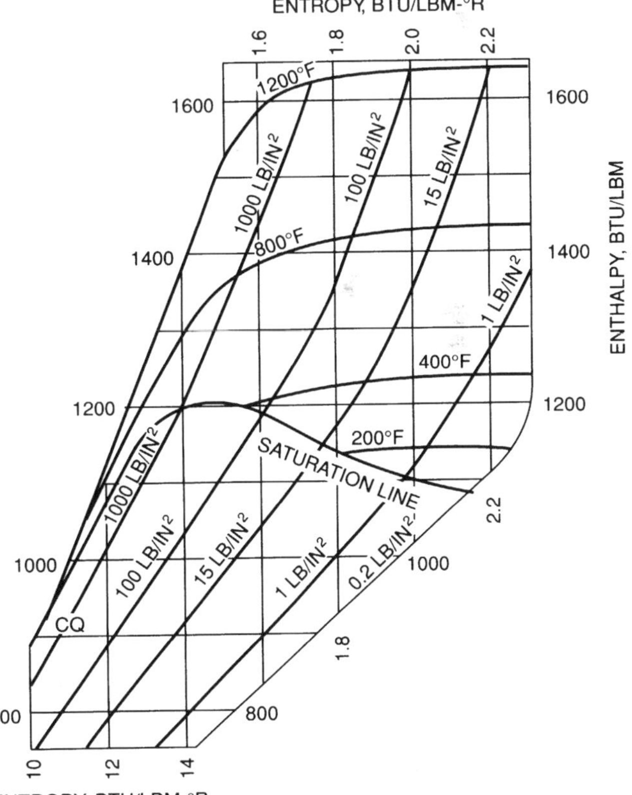

Figure 3–2e. The Mollier (h-s) diagram for water (Lee and Sears, 1963, p. 259).

Figure 3–2. Thermodynamic diagrams for pure substances.

Fig. 3-3
High-Temperature Vapor Pressures for Light Hydrocarbons
GPSA Engineering Data Book, p. 23-41

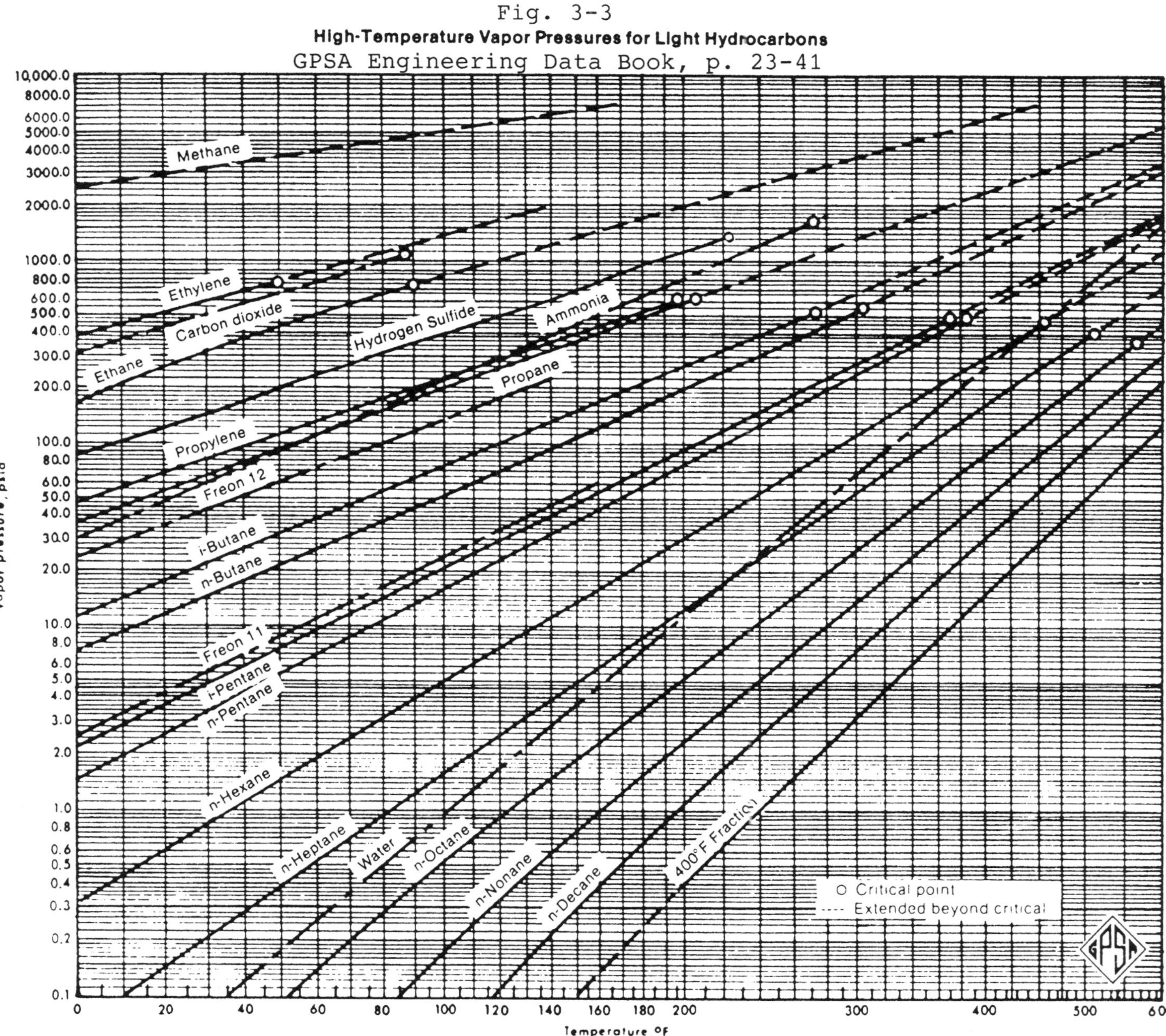

Figure 3–3. High-temperature vapor pressures for light hydrocarbons (GPSA, 1987, p. 23–41).

equivalent to the so-called Cox chart (Hougen *et al.*, 1962, p. 85).

TWO-COMPONENT SYSTEMS

Addition of a second pure component complicates the P-T diagram, although the P-v plot remains much the same. Figure 3–4 shows that the one-component, vapor-pressure *line* "becomes" an *envelope*. For a two-component, two-phase mixture the phase rule yields f = 2. Thus both P and T must be specified to define the state of the system; the two-phase locus is an area rather than a line, as Figure 3–4 also shows.

The location of this phase envelope depends on the vapor-pressure curves of the pure components and the composition of the mixture. Vapor pressure curves for the two pure components are shown also. The more volatile component, A, has its vapor-pressure line to the left (lower T) of the mixture phase envelope. The vapor-pressure curve for less-volatile pure B lies to the right (higher T) of the phase envelope. Notice that the critical pressure of the mixture, located at C, is much higher than the critical pressure of either pure component.

The *bubble-point line* is the locus at which vaporization begins. If the overall binary-mixture composition is given (one variable, since the second composition is found by difference), then f = 1 and the locus is a line, as shown. The *dew-point line* is the locus at which condensation begins for the given binary mixture. Again, this locus is a line.

The dew-point and bubble-point lines meet each other at the critical point, C. Any particular P-T diagram corresponds to a fixed composition for the overall mixture.

In addition to the bubble- and dew-point lines, Figure 3–4 shows a constant percent-vaporization line for 90 percent vaporized. This line converges to the critical point, as do all constant-percentage vaporization lines.

In Figure 3–4, A is a point in the subcooled liquid region. Suppose we heat this liquid at constant pressure. Point B is the bubble point of the mixture. As vaporization continues at constant pressure, the temperature rises. When the last drop of liquid vaporizes, the dew-point curve is reached (point D). In going at constant pressure from the bubble point to the dew point, the mixture undergoes a rise in temperature; it has a *boiling range, not a boiling point.* Further heating produces a superheated vapor, as at E.

Another new type of behavior arises with two-component mixtures. Line FG in Figure 3–4 shows a constant-temperature, or isothermal, compression at a temperature less than the critical. The dew point is first reached, then progressive condensation occurs until the mixture is all liquid at the bubble point (point G). This behavior is regarded as "normal."

Line HJ shows the new possibility. Start with vapor at a temperature higher than the critical (point H). If we compress isothermally, we reach the dew point and condensation results, as expected. Further increase in pressure at first results in more condensation, but a point is reached at which condensate begins to *decrease* in amount. With continued pressure increase, the dew point is again reached at point J. This phenomenon is known as *retrograde vaporization.*

Line KL depicts a second type of retrograde vaporization. In this case a decrease in temperature at constant pressure from point K at first yields liquid formation in increasing amount, but at some point the liquid begins to evaporate and the dew point is once again reached at point L. This is a bit confusing, because the fluid at point L is perhaps more liquid-like than vapor-like. Nevertheless, it is the fluid remaining above the liquid phase at dew-point L. One might term it a *dense phase.*

The highest pressure at which liquid and vapor can coexist in a mixture can be greater than the critical pressure. This pressure, shown in Figure 3–4, is referred to as the *criccondenbar.* Similarly, the highest temperature at which liquid and vapor can coexist for the mixture is called the *criccondentherm.*

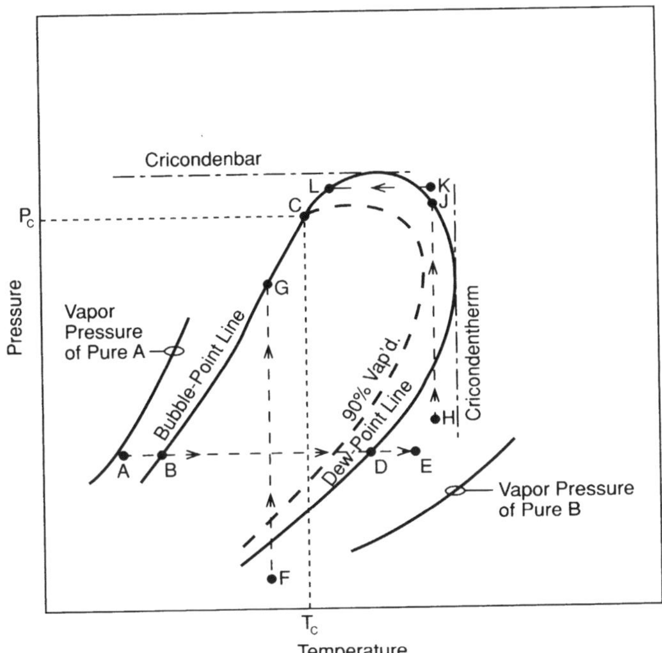

Figure 3–4. Phase envelope for a mixture of two components.

MULTICOMPONENT SYSTEMS

The basic phase behavior of multicomponent systems is the same as for binary mixtures. This is because compositions for all but one component, or C–1 variables, must be specified to define the mixture. Only two degrees of freedom remain—exactly the same as for a binary mixture. The P-T diagram for a multicomponent mixture of given composition has a shape very similar to that of a binary mixture. Phase envelopes for most naturally-occurring hydrocarbon mixtures are very broad because the components have a wide boiling range. Phase diagrams of typical full wellstreams are shown as examples.

Full Wellstreams

Figure 3–5 shows the general shape of the phase diagram for a *reservoir fluid*. The type of reservoir and fluid produced from it depend upon the location of the reservoir temperature and pressure relative to the phase diagram.

Line A-A′ shows the situation for a low-GOR crude oil, one in which the amount of gas evolved from the oil is quite low. Point A represents the reservoir pressure and temperature. As the oil flows out of the reservoir, into the wellbore, and up the producing string to the wellhead, both the temperature and the pressure decrease.

The fluid temperature decreases for two reasons. First, the temperature of the surrounding rock decreases toward the surface, so that heat transfer occurs between the hot well fluid and the cooler rock. Second, vaporization may take place also, and the required energy is supplied by the fluid itself, causing a drop in temperature.

Similarly, the pressure decreases for two reasons. As fluid travels up the string, the depth decreases and the hydrostatic head thus decreases. In addition, there is frictional pressure drop in the tubing.

The phenomena just described account for the general

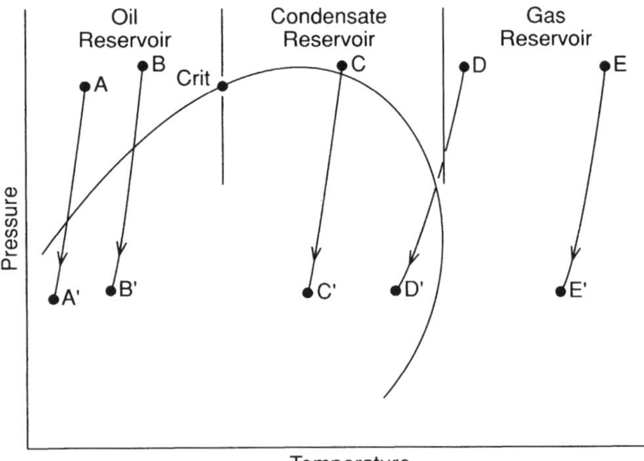

Figure 3–5. Typical phase diagram of a reservoir fluid (after McCain, 1990, and Sarssam, 1988).

shape of curve A-A′. The net effect is to bring the fluid below its bubble curve and into the two-phase region. A few crude oils do not reach the bubble-point curve and evolve no associated gas; such oils are referred to as "dead" oils.

Curve B-B′ corresponds to the situation of a high-GOR crude oil. More gas is evolved than for curve A-A′. If the reservoir fluid T and P lie in the region to the left (lower temperature) of the critical point, the produced fluid is termed a *crude oil*.

Reservoir pressure tends to decrease as the oil field is produced. Figure 3–5 shows that if point A falls below the bubble-point line, vaporization will occur in the reservoir itself. Valuable hydrocarbons may remain in the formation since the gas and liquid will not flow equally to the producing well. This explains why water injection is practiced in some fields almost from the beginning of production, to help maintain reservoir pressure.

Line C-C′ shows the behavior of a retrograde condensate fluid as it is produced, something between a natural gas and an oil. In this case, the relative amounts of both liquid and gas may be very high. If the reservoir temperature lies between the critical temperature and the cricondentherm, the reservoir is usually termed a *condensate reservoir* and the produced fluid is often called a *retrograde condensate gas*.

An important conclusion can be drawn from Figure 3–5. If the pressure in a *condensate reservoir* falls below the dew-point line during production, condensation will take place in the reservoir itself. Valuable heavier liquid components will likely remain in the reservoir and not be produced. Pressure maintenance by gas reinjection is sometimes practiced in such a reservoir. Reinjection and other producing strategies for gas condensate fields will be discussed in Chapter 5.

Suppose line B-B′ lies to the left and very close to the critical point and line C-C lies to the right and very close to the critical point. Now these two lines are very close to each other. Clearly, mere observation of the well-head separator GOR is not suffient evidence to decide the type of reservoir. Furthermore, the arbitrary nature of calling the reservoir a crude oil or a condensate reservoir becomes obvious.

Curve D-D′ depicts the situation for a "wet" natural gas. Reservoir conditions are in the gas or dense fluid region to the right of the cricondentherm. When produced, the fluid yields hydrocarbon condensate due to the temperature and pressure decrease. Point D′ is within the two-phase region, below the dew-point line. The arbitrary distinction between a condensate reservoir and a wet gas reservoir become clear.

Curve E-E′ represents the reservoir and producing

conditions for a ''dry'' natural gas. No hydrocarbon condensate is formed in the surface separator.

Note well: While Figure 3–5 illustrates and clarifies the classification of well fluids, the figure is, however, misleading in the sense that points A′, B′, C′, D′, and E′ actually represent approximately the same wellhead temperature, usually in the range of 100–140°F. Points A, B, C, D, and E will not be the same for the reason that the reservoir temperature and pressure are dependent on the depth of the producing formation. Nevertheless, these production lines will more or less coincide; it is the phase diagram of the reservoir fluid that will change its position relative to the line.

Another important point with respect to Figure 3–5 is that the shape of the phase diagram and the relative location of the critical point is highly dependent on the reservoir fluid composition. A dry natural gas has a fairly low range of components. Its phase diagram will be of relatively narrow width; the critical point will lie well down on the left side of the envelope. Conversely, a crude oil has a very wide range of components. An oil envelope will be very wide and the critical point near the top or somewhat to the right side of the envelope. Figures 3–6 and 3–7 show these features.

Source of Phase Diagrams

Phase diagrams are measured in the laboratory or predicted from equilibrium calculations as shown below. Very briefly, experimental measurements involve determining the locus of the dew and bubble points in a variable-volume cell. Determination may be visual, as in a windowed cell, or based on break points in the pressure-volume data. Jacoby and Yarborough (1967) and others describe apparatus and techniques.

Experimental *measurements* are extremely time-consuming and expensive. Such data are used to develop

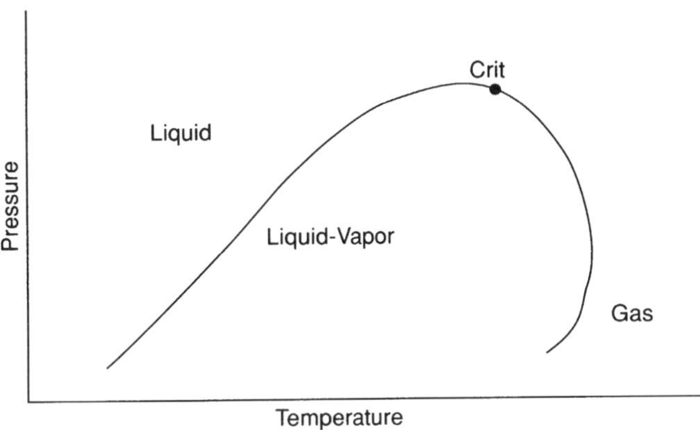

Figure 3–7. Typical phase diagram of a low-GOR crude oil.

calculational methods for predicting phase behavior. Invariably, computer methods use equations of state and require very complex calculations. Hand calculation methods must be much simpler to be practical, and even then can be time-consuming.

Computer methods are generally quite satisfactory for phase calculations. However, the calculations are only as good as the analyses of the streams being considered. In general, sampling and analysis require great care, as discussed in Chapter 2.

A second source of error in computer-generated phase diagrams lies in the characterization of the heavy ends. In natural gases, the quantity of heavy ends may not be large, but the characterization of them influences the location of the dew-point line. Phase diagrams developed from computer calculations for two simulated natural gases and one simulated separator gas are shown in Figure 3–8. The two natural gases, A and B, differ only in the identity of the small fraction of C7+ material as nC7 or nC8. Yet the dew-point locus is shifted several degrees.

Accurate location of the dew-point locus is important in at least two cases. First is the simulation of phase behavior in a reservoir, in which the calculations must represent the actual reservoir fluid as well as possible. Second is the pipeline flow in which small amounts of hydrocarbon condense. Liquid formation greatly increases the pressure drop in the pipeline and hence gas-compression requirements. An undersized compressor can limit the production from a field.

The bubble-point locus in Figure 3–8 is close to the methane vapor-pressure curve for both simulated natural gases, as it is for all gases that contain more than about 80 mole percent of methane. A separator or associated gas, which contains a much higher amount of heavier components, has a broader phase diagram, as is shown in Figure 3–8.

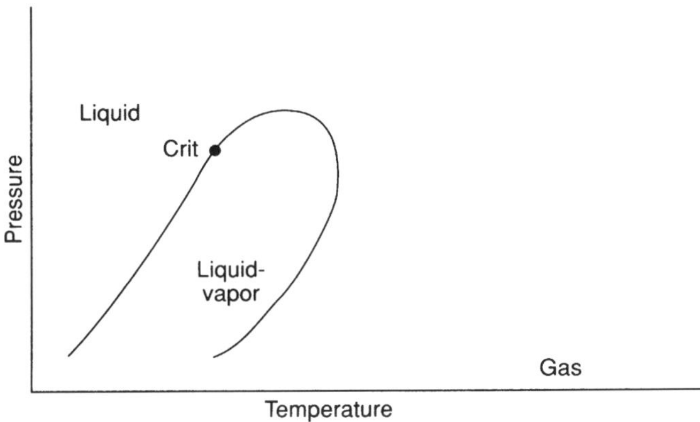

Figure 3–6. Typical phase diagram of a dry natural gas.

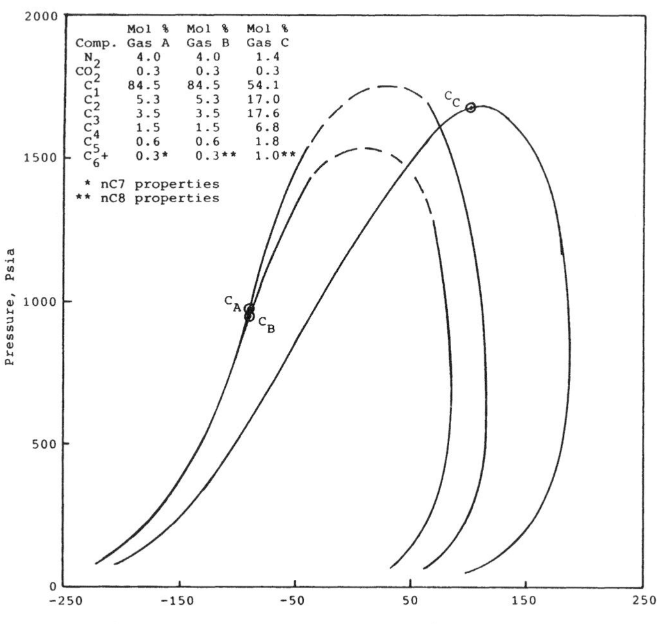

Figure 3–8. Phase diagrams for simulated natural gases.

Hand phase-equilibrium calculations can also be used but are not as accurate as computer calculations, as might be expected. However, hand calculations are satisfactory for many purposes. These methods are discussed below.

Phase Behavior in Separators

Figure 3–9 shows computer-generated phase diagrams for a separator feed (F) and the liquid (F_L) and vapor (F_V)

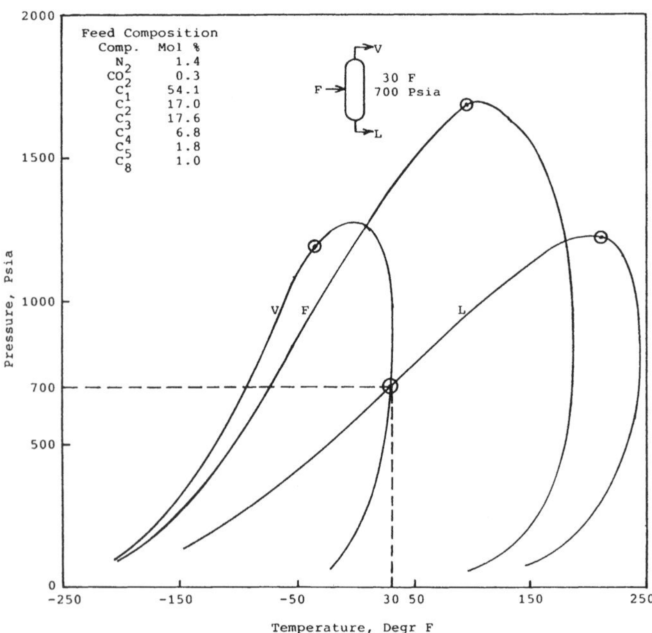

Figure 3–9. Phase diagrams for separator streams.

products. The separator products are assumed to leave in phase equilibrium with each other. The separator pressure and temperature are located at point A, where the products are saturated liquid and saturated vapor, respectively.

PHASE-EQUILIBRIUM CALCULATIONS

There are three basic phase-equilibrium calculations:

1. Bubble-point
2. Dew-point
3. Equilibrium-Flash

Types (1) and (2) determine phase envelopes and the temperature or pressure at which a given mixture will begin to vaporize or condense. Type (3) estimates the percentage vaporized and the equilibrium-phase compositions for mixtures that are partly vaporized or condensed.

The nomenclature for phase-equilibrium calculations is shown below.

Flash-Vaporization Nomenclature

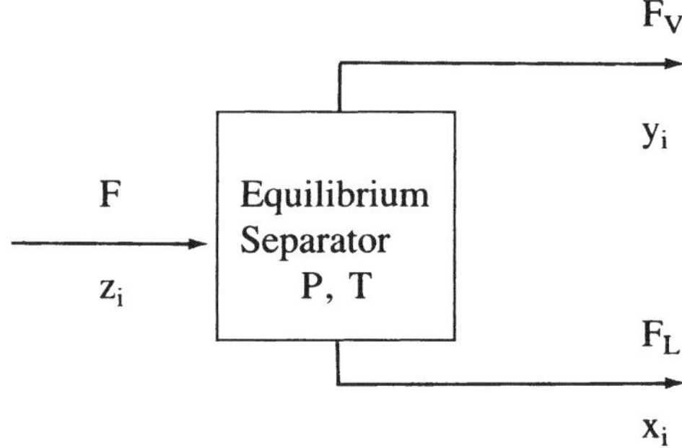

F = feed rate, mol/hr
F_L = liquid rate, mol/hr
F_V = vapor rate, mol/hr
z_i = mole fraction of component i in the feed
x_i = mole fraction of component i in the liquid
y_i = mole fraction of component i in the vapor
P = pressure of the flash vaporization
T = temperature of the flash vaporization

Phase-equilibrium calculations use a quantity, K_i, called the *vapor-liquid equilibrium ratio*, or more simply, the *K-value*.

$$K_i = y_i \,/\, x_i \qquad (3-2)$$

K_i's are functions of component identity, temperature, pressure, and composition. The vapor-liquid equilibrium ratio or K-value may be obtained from so-called K charts or calculated from equations of state using computer programs. The GPSA (1987) Section 25, Engineering Data Book presents three sets or types of K charts:

(i) K-values for specific binary systems: C1–C2, C1–C3, C1–nC4, C1–nC5, Cl-nC6, C1–nC7, N_2–C1, N_2–C2, and C1–CO_2. These charts are used for temperatures below –100°F.

(ii) K-values for various components based on different convergence pressures of hydrocarbon systems. These charts are described and used in the examples that follow.

(iii) K-values of high-boiling oil fractions. Poettman and Maylund (1949) developed K charts for heavier HC which are characterized by their normal boiling point (NBP) and characterization factor. Figure 2–5 shows a typical chart in which P is plotted against K at various temperatures. The GPSA (1987) presents charts for NBPs of 300, 400, 500, 600, 700, and 800°F.

The OPSIM computer program (Appendix 4) illustrates the calculation of K-values using the SRK equation of state.

Vapor-liquid equilibria will be discussed in more detail in Volume 2.

For simplicity, the subscript i is now omitted from x, y, z, and K. Each component present is understood to have its own value for each of these variables, and the summation sign refers to the summation for all components present.

Bubble-Point Temperature

Consider a liquid mixture of composition z at pressure P. Mole fraction z is used because the given composition is

Thus, our task is to find T such that when the K's are obtained for that T and the given P, the sum of Kz's is indeed equal to one, within an acceptable tolerance. The procedure is trial-and-error.

Example 3–1. A mixture has the composition below at 400 psia. What is the bubble-point temperature?

Component	z
C1	0.75
C3	0.24
nC6	0.01

Solution: Look up the component K's in the GPSA Data Book. Use a convergence pressure of 800 psia (justified later). Three trials are shown below. The first temperature tried, –100°F, gives a sum greater than 1.0. Therefore the second temperature chosen is lower, –120°F. Note that the K-values for nC6 are so low that they are simply taken to be zero. The sum at –120°F is still slightly less than 1.0. The final trial is for –126°F. Within the ability to read the K-values, the sum is sufficiently close to one. At 400 psia, the feed bubble-point is –126°F.

Comp.	z	$K^{-100°F}_{400\ psia}$	Kz	$K^{-120°F}_{400\ psia}$	Kz	$K^{-126°F}_{400\ psia}$	Kz
C1	0.75	2.0	1.500	1.45	1.087	1.33	0.998
C3	0.24	0.030	0.007	0.019	0.005	0.016	0.004
nC6	0.01		0.000		0.000		0.000
		~0		~0		~0	
Sum			1.507		1.092		1.002

regarded as the "feed." Actually, z plays the part of x, because the feed is liquid. The question is, at what temperature will the liquid feed begin to vaporize at the given pressure P? Calculations are based on the concept that the first tiny bubble of vapor formed is in equilibrium with the feed at essentially unchanged composition, z. The composition of the vapor bubble is, by the previous definition,

$$y = K z \qquad (3\text{–}3)$$

for each component. The sum of the mole fractions of any mixture must be identically one.

$$\Sigma\, y = \Sigma\, K z = 1.0 \qquad (3\text{–}4)$$

Convergence Pressure

The above K-values are based on a convergence pressure of 800 psia. This choice must be justified. First, the concept of convergence pressure is reviewed.

Convergence pressure of a multicomponent mixture is somewhat analogous to critical pressure of a binary mixture. However, in the GPSA Data Book method, it is used as a correlating parameter for estimating the effect of composition on K-values. There are 7 different sets of K-charts for 14 hydrocarbons (C1 through C10) and N_2, and H_2S in Section 25 of the Data Book, each set for a different convergence pressure. Figure 25–2 GPSA (1987) lists the components and convergence pressures, which vary from 800 to 10000 psia.

To find the convergence pressure, the multicomponent mixture is treated as a pseudobinary. By convention, the light component of the pseudobinary is the lightest hydrocarbon component present to at least 0.1 mol percent in the liquid phase. The heavy component is represented by the weight-average critical temperature and pressure of the remaining heavier (less volatile) components.

The technique for finding convergence pressures, which is described in GPSA (1987) on p. 25–4., is now illustrated using the above bubble-point calculation. A convergence pressure of 800 psia is used. Refer to steps 1–8, p. 25–4, GPSA (1987).

Example 3–2. Calculate the convergence pressure for Example 3–1.

Step 1.   Estimate the liquid-phase composition. Here, the liquid composition is the given feed composition.

Step 2.   Select the light component as the lightest hydrocarbon present with $x > 0.001$. Here methane is the light component ($x = 0.75$, which is $>>0.001$).

Step 3.   Calculate the weight-average critical temperature and pressure of the remaining heavier components.

Comp.	x	MW	g	w	Tc	Pc
C3	0.24	44.1	10.58	0.925	206.1	616.0
nC6	0.01	86.2	0.86	0.075	453.6	436.9
			$\Sigma g = 11.44$	1.000		

$$g = x\ MW$$

$$w = g\ /\Sigma g$$

$$T_{cm} = \Sigma\ w_i T_{ci} = 225°F \tag{3–5}$$

$$P_{cm} = \Sigma\ w_i P_{ci} = 603\ psia \tag{3–6}$$

Step 4.   Locate T_{cm}, P_{cm} on Figure 3–10 (GPSA, Fig. 25–11). Now, sketch in the locus of the pseudobinary critical envelope, as shown. This is done by comparison with the critical locus curves for actual binary mixtures shown in Figure 3–10.

Step 5.   Read the convergence pressure from the envelope at the temperature of the calculation. For the bubble-point T of $-126°F$, go vertically to the critical envelope just sketched. Read to the left to estimate the convergence pressure. For the present example, the envelope lies to the

right of the operating temperature. In such a case, the convergence pressure is chosen as the critical pressure of the light component, methane in this case. The convergence pressure is about 670 psia, but the lowest chart convergence pressure is 800 psia, thus verifying the choice of 800 psia. Steps 6, 7, and 8 need not be repeated, since the calculated and assumed convergence pressures are the same. If the assumed and calculated convergence pressures were not the same, these steps would be done, as follows.

Step 6.   Look up the K's for P = 400 psia, T = $-126°F$, P_k = 800 psia, as done in the example.

Step 7.   Carry out the bubble-point calculation.

Step 8.   Repeat steps 2–7 as necessary for convergence.

Bubble-Point Pressure

If temperature is given instead of pressure, the bubble-point pressure is calculated similarly.

Example 3–3. The mixture of Example 3–1 is at $-100°F$. What is the bubble-point pressure?

Solution: Trial-and-error is required, as before. A pressure is guessed and the K-values (at the given T and assumed P) read from the charts. The sum of the Kz's is determined and compared to one. The convergence pressure can be determined directly, since the temperature is known, not guessed. The convergence pressure is 800 psia, as in Example 3–1. Only the final trial is shown.

Comp.	z	$K^{-100°F}_{620\ psia}$	Kz	
C1	0.75	1.31	0.983	
C3	0.24	0.062	0.015	
nC6	0.01	~0	0.000	
Sum			0.998	Accept

Dew-Point Calculations

For dew-point calculations, the vapor feed composition is given. We wish to know at what temperature will this gas, or vapor, begin to condense at the given pressure. When the first tiny drop of liquid is condensed, the feed composition is essentially unchanged and the liquid is in

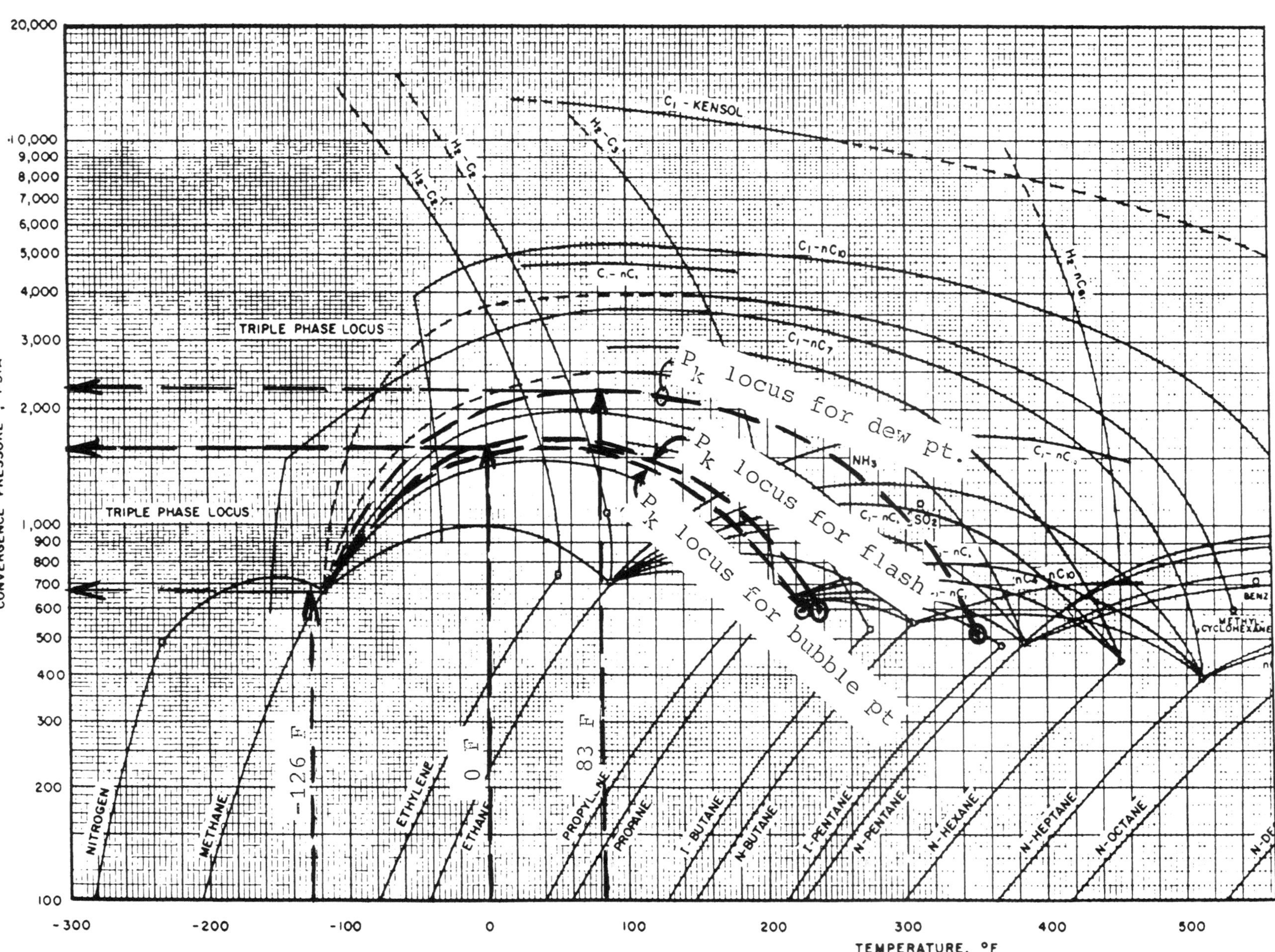

Figure 3–10. Determination of convergence pressure for example calculations (GPSA, 1987, p. 25–11).

29

equilibrium with the vapor. The liquid mol fraction of each component is:

$$x = z / K \qquad (3\text{--}7)$$

As before, the sum of the x's must be one.

$$\Sigma\, x = \Sigma\, z / K = 1.0 \qquad (3\text{--}8)$$

Trial-and-error is required; a temperature is guessed and the K's found from the charts. The summation of x's is computed and compared to one. The temperature is adjusted as necessary to satisfy the dew-point requirement.

Example 3–4. The mixture previously given in Example 3–1 is at 400 psia. What is its dew-point temperature? Assume a convergence pressure of 2000 psia.

Solution: Only the final trial at 83°F is shown below.

Comp.	z	$K^{83°F}_{400\ psia}$	x = z/K	Normalized x
C1	0.75	6.6	0.114	0.113
C3	0.24	0.49	0.490	0.488
nC6	0.01	0.025	0.400	0.399
Sum			1.004	1.000

Now check the assumed convergence pressure. The light component is methane. The weight-average critical temperature and pressure of the remaining liquid-phase C3 and nC6 are 358°F and 506 psia, respectively. Figure 3–10 indicates a convergence pressure of about 2200 psia, which, for the present conditions, represents a close-enough check of the assumed value.

A dew-point pressure can be calculated similarly, if the temperature is specified instead of the pressure.

Example 3–5. The mixture given in Example 3–4 is at 60°F. What is the lower dew-point pressure? The convergence pressure is 2000 psia, as in Example 3–4.

Solution: Only the last trial is shown.

Comp.	z	$K^{60°F}_{205\ psia}$	z/K	
C1	0.75	11.7	0.064	
C3	0.24	0.60	0.400	
nC6	0.01	0.0188	0.532	
Sum			0.996	Accept

Equilibrium Flash Vaporization (EFV) Calculations

Flash calculations combine the total stream material balance, the component material balance, and the equilibrium relation.

$$\text{Total stream balance} \quad F = F_L + F_V \qquad (3\text{--}9)$$

$$\text{Component balance} \quad Fz = F_L x + F_V y \qquad (3\text{--}10)$$

$$\text{Equilibrium relation} \quad y = Kx \qquad (3\text{--}11)$$

Eliminate y between the last two equations and solve for x

$$x = \frac{Fz}{F_L + F_V K} \qquad (3\text{--}12)$$

Let F be 1 mole. Then

$$x = \frac{z}{F_L + F_V K} \qquad (3\text{--}13)$$

where

$$F_V = 1.0 - F_L \qquad (3\text{--}14)$$

Finally,

$$\Sigma\, x = 1.0 \qquad (3\text{--}15)$$

The usual calculation is the so-called isothermal flash, which does not mean the feed is flashed isothermally from its initial condition, but that the temperature and pressure of the flash are specified. In this case, one can look up the K's. The z's are known. The unknown is F_L. We guess F_L, compute F_V by the next-to-last equation above, then calculate the x's for each component. The sum of the x's should be one within a specified tolerance; if it is not, F_L is adjusted and the calculation repeated.

In flash calculations, be sure the feed is indeed flashed at the specified P and T. Check to be sure that the mixture is above its bubble-point and below its dew-point temperature. Use the following relations.

$$\Sigma\, Kz > 1.0 \qquad \text{guarantees vapor is present} \quad (3\text{--}16)$$
$$\Sigma\, z/K > 1.0 \qquad \text{guarantees liquid is present} \quad (3\text{--}17)$$

If both sums are greater than one, proceed with the flash; otherwise only a single phase is present.

Example 3–6. The mixture of Example 3–5 is at 400 psia, 0°F. Find the percent vaporized and the compositions of the vapor and liquid phases. Assume a convergence pressure of 1500 psia and check as usual.

Solution: Look up the K-values at 0°F and 400 psia and obtain the check sums.

Comp.	z	$K_{400\ psia}^{0°F}$	z K	z/K
C1	0.75	4.4	3.300	0.170
C3	0.24	0.173	0.042	1.387
nC6	0.01	0.0048	0.000	2.083
Sum			3.342	3.640

Both Σ Kz and Σ z/K are greater than one; therefore, the feed is a flashed vapor-liquid mixture. Flash calculations are performed.

Comp.	z	$K_{400\ psia}^{0°F}$	$F_L = 0.15$ $F_V = 0.85$ $x = \dfrac{z}{F_L + KF_V}$	$F_L = 0.18$ $F_V = 0.82$ $x = \dfrac{z}{F_L + KF_V}$	$F_L = 0.179$ $F_V = 0.821$ $x = \dfrac{z}{F_L + KF_V}$
C2	0.75	4.4	0.193	0.198	0.1978
C3	0.24	0.173	0.808	0.746	0.7476
nC6	0.01	0.0048	0.065	0.054	0.0547
Sum			1.066	0.998	1.0001

Note the check sum must be very close to 1.0000 in order that the solution be precise. The x's are normalized by dividing each by 1.0001. Then y = Kx is calculated. If the sum of the y's is not quite 1.0, the y's also must be normalized.

	60°F	
Comp.	x	y = Kx
C1	0.1978	0.8704
C3	0.7475	0.1293
nC6	0.0547	0.0003
Sum	1.0000	1.0000

The weight-average critical temperature and pressure of the pseudo heavy-liquid component are 237°F and 593 psia, respectively. Figure 3–10 confirms the convergence pressure of 1500 psia very closely.

K-Value Behavior

The K charts in the GPSA Data Book are plotted as log K versus log pressure, for lines of constant temperature.

Ideal gas behavior and ideal liquid solutions may be assumed for paraffin hydrocarbon mixtures at low pressure. Now Raoult's Law gives the K-values.

$$K = VP / P \qquad (3–18)$$

where VP = vapor pressure of the component

P = total pressure

Taking logarithms, we obtain

$$\log K = \log VP - \log P \qquad (3–19)$$

At constant temperature, the vapor pressure is constant so that the latter equation has the form of a straight line, if plotted as log K versus log P. The intercept is log VP and the slope is -1. GPSA (1987) K-charts show this behavior at low pressures; the isotherms indeed asymptotically approach a straight line with slope -1 at low pressure. At higher pressures, the K-values deviate from this ideal behavior and approach a value of 1.0 at the convergence pressure. This behavior is typical of a true binary mixture, as is shown in Figure 3–11, for example.

Applicability of the Equilibrium Model

The liquid and vapor products from a separator are assumed to be in phase equilibrium. Similar assumptions are made about vapor and liquid streams in other process applications. For example, products from reboilers are assumed to be in equilibrium; vapors and liquids from distillation tower reflux accumulators are also assumed to be in equilibrium.

How good are these equilibrium assumptions? In general, the vapor and liquid from a separator approach equilibrium with each other. Entrainment of liquid droplets may cause small deviations. As long as sufficient contact time is provided and entrainment is not excessive, separator products are essentially in equilibrium (Plane, 1966). This model is used routinely in process calculations and yields reasonable results. The liquid leaving a reboiler is slightly superheated compared to the vapor, but the model is still applicable as long as the flow pattern matches that of the model, i.e., thorough mixing followed by efficient

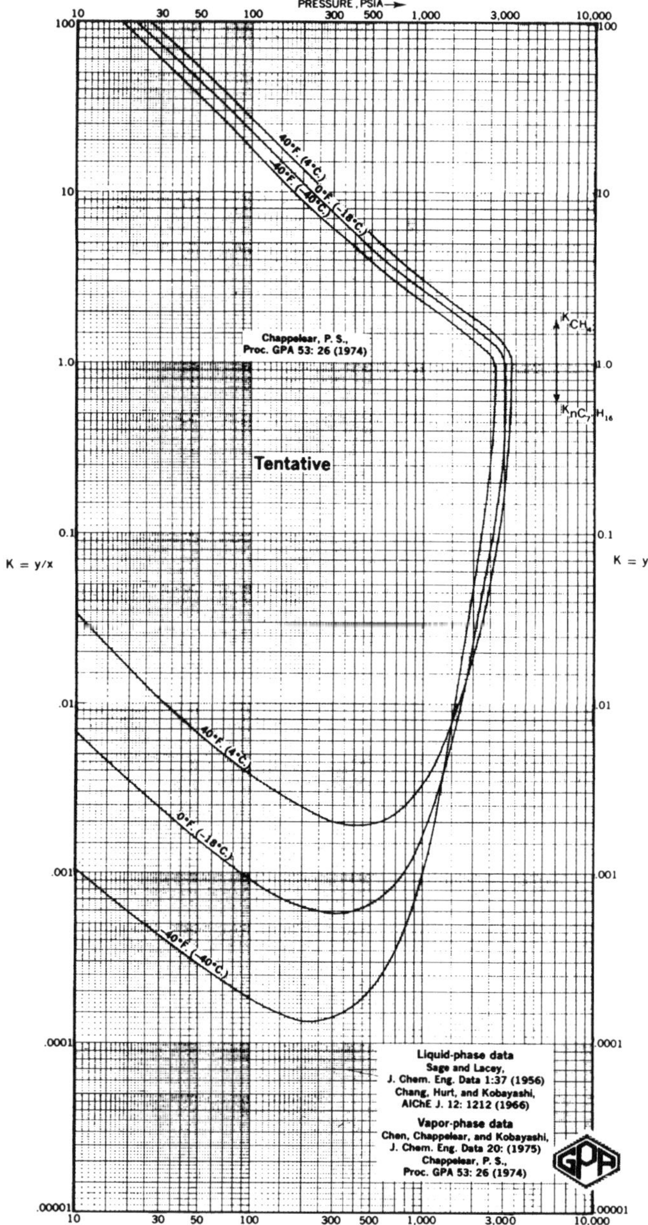

Figure 3–11. K-value behavior for a binary mixture (GPSA Eng. Data Book).

separation of the vapor and liquid. The same comments apply to reflux accumulators and similar devices.

K-VALUES FROM EQUATIONS OF STATE

The previous examples show how tedious hand vapor-liquid equilibrium calculations are. Further, K-values of the type given in the GPSA Data Book inadequately reflect the effect of composition. Convergence pressure is a good concept and a powerful correlating tool. Use of the liquid-phase composition in determining P_k is reasonable, because the molecules in the liquid are much closer to each other than in the vapor, and the interactive forces in the liquid should dominate phase behavior. Nevertheless, hand methods are limited in accuracy.

More sophisticated, and more accurate, calculations can be performed with digital computers. Computer phase-equilibrium calculations use equations of state, which reflect the behavior of natural-gas constituents very well. These equations are now discussed.

As previously stated, Gibbs' Phase Rule confirms that the state or condition of a one-phase, pure substance is known when only two independent properties are specified. An equation of state is merely a mathematical way of expressing a third property in terms of any two independent ones. The usual choice of variables is P, v, and T, and the equation is generally explicit in P.

$$P = P(T,v) \tag{3–20}$$

Gibbs' Phase Rule tells us that P, v, and T are related, but, unfortunately, it does **not** provide the mathematical equation.

Developing an equation of state involves a compromise between the simplicity of the equation (and hence its ease of use) and the accuracy with which it represents actual data. Hundreds of such equations have been proposed. A few of the more common equations of state are now presented, and then the application to mixtures (varying compositions) is summarized.

The ideal gas law is the simplest equation of state.

$$P = n R T / V \tag{3–21}$$

where P is absolute pressure (psia, atm, kPa)

V = volume (ft^3, cm^3, m^3)
n = moles (lbmol, gmol, kgmol)
T = absolute temperature (°R, K)
R = universal ideal-gas-law constant
    = 10.732 (psia)(ft^3)/(lbmol)(°R)
    = 82.057 (atm)(cm^3)/(gmol)(K)
    = 8.3145 (kPa)(m^3)/(kgmol)(K)
    = 8.3145 kJ/(kgmol)(K)
    = 1.9859 Btu/(lbmol)(°R)

The ideal gas law is very simple and R has the same value for all gases. Its disadvantage is that it represents real data poorly—e.g., it cannot predict liquid formation. The ideal gas law treats molecules as having zero volume and no interactive forces of attraction, poor assumptions at high pressure and/or low temperature.

Many mechanical engineering texts define the gas constant R on a unit mass or weight basis. Then,

$$P = m \, \Re \, T \, / \, V \qquad (3\text{--}22)$$

where m = mass of gas (lb, g, kg)
$\Re$ = R/MW
MW = molecular weight of gas.

From now on, the specific volume, $v = V/n = V/(m/MW)$, is used. The ideal gas law becomes

$$P = R \, T \, / \, v \qquad (3\text{--}23)$$

The van der Waals equation of state (Van Wylen and Sonntag, p. 392) attempts to allow for the size of the molecules by means of a factor b and for the attraction between the molecules by an additional term, a/v^2.

$$P = \frac{RT}{v - b} - \frac{a}{v^2} \qquad (3\text{--}24)$$

With two constants, a and b, in addition to R, the van der Waals equation is more accurate than the ideal gas law. If $\partial P/\partial v$ and $\partial^2 P/\partial v^2$ are assumed to be zero at the critical point (point of inflection in the Pv isotherm at the critical temperature), then:

$$a = 27 \, R^2 \, T_c^2 \, / \, 64 \, P_c \qquad (3\text{--}25)$$

and

$$b = R \, T_c \, / \, 8 \, P_c \qquad (3\text{--}26)$$

These formulae permit calculation of the van der Waals constants from the critical properties.

First presented in 1873, the van der Waals equation uses a theoretical approach to improve on the ideal gas law, and it can predict liquid formation! Unfortunately, the equation fits real data quite poorly, no matter how the constants a and b are determined. In the years that followed, many attempts were made to present improved equations of state. Improved accuracy always required additional complexity. Redlich and Kwong (1949) proposed two simple empirical modifications of van der Waals' equation that greatly improved its accuracy. The first change involved introduction of temperature dependency to the attractive force term. The second change was to replace the v^2 term by $v(v + b)$.

$$P = \frac{RT}{v - b} - \frac{a \, T^{-0.5}}{v(v + b)} \qquad (3\text{--}27)$$

Application of the critical isotherm conditions yielded

$$a = 0.4278 \, R^2 T_c^{2.5}/P_c \text{ and } b = 0.0867 \, RT_c/P_c \qquad (3\text{--}28)$$

The Redlich-Kwong equation represents real-gas data well at reduced temperatures above 1.0, but is not very accurate for vapors (reduced temperatures less than 1.0). Redlich and Kwong extended the use of their equation to mixtures and to computing fugacity and the real-gas deviation of enthalpy.

Soave (1972) modified the Redlich-Kwong equation by replacing the $a/T^{0.5}$ with a factor, a(T), that made the equation fit the vapor pressure curve in the subcritical temperature range. This improvement allowed the calculation of vapor-liquid K-values of good accuracy, heretofore not done by a single equation of state. Soave also introduced the use of interaction coefficients to provide better fits for gas mixtures containing polar substances such as carbon dioxide and hydrogen sulfide. The equation does not give very accurate liquid densities.

Peng and Robinson (1976) suggested a similar modification to the Redlich-Kwong equation.

$$P = \frac{RT}{v - b} - \frac{a(T)}{v(v + b) + b(v - b)} \qquad (3\text{--}29)$$

They claimed better liquid densities than Soave, but the densities are nevertheless relatively inaccurate. Proprietary interaction coefficients permit the equation to fit carbon dioxide and hydrogen sulfide K-values quite well.

Usdin and McAuliffe (1976) also modified the Soave treatment of the Redlich-Kwong equation by the introduction of a third parameter, d.

$$P = \frac{RT}{v - b} - \frac{a(T)}{v(v + d)} \qquad (3\text{--}30)$$

They further showed that all three parameters (a, b, and d) could be related to a single parameter, z_c^*, termed the critical compressibility parameter. They also claim to have fit liquid densities better for light-hydrocarbon mixtures. Buthod $et \, al.$ (1978) extended the use of the equation to heavy hydrocarbons, such as crude-oil boiling-point components. A general purpose computer program, suitable for use on microcomputers, based on the Usdin-McAuliffe modification of the Soave-Redlich-Kwong equation of state, has been developed (Thompson, 1986) and is used in this text (see Appendix 4).

Benedict, Webb, and Rubin (1940) developed an eight–constant equation of state that has been used rather widely. It is interesting to note that this very complex equation was implemented prior to the introduction of either the digital computer or the hand-held calculator.

$$\begin{aligned} P = \rho RT &+ (B_o RT - A_o - C_o/T^2) \, \rho^2 \\ &+ (bRT - a) \, \rho^3 + \alpha \, a \, \rho^6 \\ &+ c \, \rho^3 \, (1 + \gamma \, \rho^2) \, [\exp(-\gamma \, \rho^2)]/T^2 \qquad (3\text{--}31) \end{aligned}$$

where B_o, A_o, C_o, b, a, α, c, and γ are BWR constants and ρ is density (1/v)

Starling and Han (1972) proposed an eleven-constant, generalized BWR equation in which the constants are functions of v_c, T_c, and the acentric factor.

$$P = \rho RT + (B_o RT - A_o - C_o/T^2 + D_o/T^3 - E_o/T^4)\,\rho^2$$
$$+ (bRT - a - d/T)\,\rho^3 + \alpha(a + d/T)\,\rho^6$$
$$+ c\,\rho^3\,(1 + \gamma\,\rho^2\,)[\exp(-\gamma\,\rho^2)]/T^2 \quad (3\text{--}32)$$

where D_o, d, and E_o are the additional constants

The BWRS equation gives good thermodynamic properties, including liquid densities. However, the equation is cumbersome to use and difficult to solve since it is not a simple cubic equation like the Redlich-Kwong-based equations. Further, the coefficients, as well as interaction coefficients, must be available for all system components.

Mixtures

There is no problem in applying the ideal gas law to mixtures because R has the same value for all gases. For other equations of state the constants for the gas mixture are calculated from the pure-component values using complex mixture rules. For example the recommended mixing rules for the Peng-Robinson equation are:

$$b_m = \Sigma\, x_i\, b_i \quad (3\text{--}33)$$

$$a_m = \Sigma_i\, \Sigma_j\, x_i\, x_j\, a_{ij} \quad (3\text{--}34)$$

$$a_{ij} = (1 - \delta_{ij})\,(a_i\, a_j)^{.5} \quad (3\text{--}35)$$

where δ_{ij} = fitted binary interaction parameter

Summary

In general, equations of state are not appropriate for hand calculations due to their complexity. However, digital computers permit the practical use of these equations for calculating thermodynamic properties and vapor-liquid K-values. The natural gas and petroleum industries use the SRK and PR equations extensively. Gas- and liquid-phase chromatography provide the experimental data required to check the computer-calculated predictions. Agreement is generally excellent. Improvements in digital computers and chromatographic techniques have permitted the development of more accurate predictions, including three-phase equilibria with hydrocarbons, water, and carbon dioxide. Four-phase equilibria—gas, HC liquid, aqueous liquid (water and hydrate inhibitor), and solid hydrate—can be simulated also. This development continues.

COMPRESSIBILITY FACTOR, Z

As previously stated, complex equations of state are not suitable for hand calculations, which are still needed for quick estimates or in the absence of a computer or programmable calculator. A very useful approach is to retain the simple form of the ideal gas law and add a dimensionless correction factor, called the *Z-factor* or *compressibility factor*.

$$P\,v = Z\,R\,T \quad (3\text{--}36)$$

This equation, sometimes called the *real gas equation, defines Z*, because Z assumes the value required to maintain the equality. Fortunately, it is an experimental fact that all fluids exhibit approximately the same value of Z at the same reduced temperature and pressure. Reduced properties are defined as follows:

$$\text{Reduced temperature} = T_r = T/T_c \quad (3\text{--}37)$$

$$\text{Reduced pressure} = P_r = P/P_c \quad (3\text{--}38)$$

where T = absolute temperature (K or °R)
p = absolute pressure (not gauge)
T_c = critical temperature (K or °R)
P_c = critical pressure (not gauge).

Figure 3–12 (Lee and Sears, 1963) illustrates this equality of Z for different gases at the same reduced temperatures and pressures. This fact is called the Law or Theorem of Corresponding States and was originally proposed by van der Waals in 1873. The exact dependence of Z on P_r and T_r is not amenable to simple mathematical representation; therefore, we resort to graphs. The first Z-charts were prepared by Cope, Lewis, and Weber (1931); Brown, Souders, and Smith (1932); and Dodge (1932).

Dodge (1944, p. 160) estimated the difference between compressibility factors based directly on experimental data and those read from a chart. The average deviation was 2

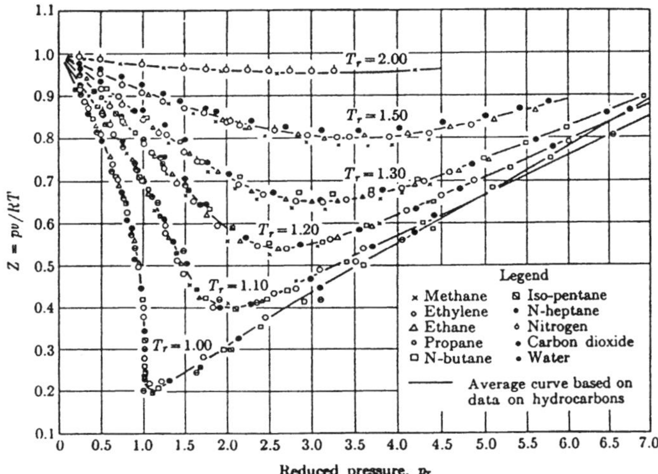

Figure 3–12. Generalized compressibility chart (Lee and Sears, 1963, p. 65).

percent for 263 individual cases; the worst case was ammonia (polar), where the average deviation was 7 percent.

The accuracy of the compressibility factor approach can be improved by adding a third factor to characterize the molecule of the gas being considered. Pitzer's correlation (1955) uses the acentric factor, ω, and is probably the most widely used.

$$Z = Z\ (P_r,\ T_r,\ \text{and}\ \omega) \qquad (3\text{--}39)$$

$$\text{where}\ \omega = -\log P_r - 1.0 \qquad \text{at}\ T_r = 0.70 \qquad (3\text{--}40)$$

Improvements in digital computers have made the equation of state approach more attractive than these three-factor Z-charts.

Compressibility factors can be used for liquids as well as gases.

Z-Charts

Since the early 1930s, many compressibility charts have been prepared. For natural gases, the GPSA (1987) recommends Figure 3–13 to 3–21. Figures 3–13, 3–14, and 3–15 are typical Z-charts in that Z is plotted against reduced pressure for various reduced temperatures. Figures 3–16 through 3–21 are more convenient because the Z-factors are plotted directly against pressure in psia for various temperatures in °F. Linear interpolation may be used between Figures 3–16 through 3–21 to obtain Z's for gases with specific gravities between 0.55, 0.60, 0.65, 0.70, 0.80, and 0.90.

Because natural gases are mixtures, pseudo-reduced temperatures and pseudo-reduced pressures are used in place of reduced pressures and temperatures. The word "pseudo" reminds us that the critical properties of mixtures depend very strongly on the composition of the mixture. Pseudocritical properties of conventional, relatively sweet natural gases can be estimated from Figure 3–22.

The GPSA (1987) Engineering Data Book provides the following guidelines:

1. In general, Figures 3–13 through 3–21 predict Z-factors within 2% for natural gases with less than 5% nitrogen, 2% carbon dioxide, and 2% hydrogen sulfide.
2. Errors of 10% can occur when the molecular weight exceeds 20 and the Z-factor is below 0.6.
3. Figures 3–13 through 3–21 are derived from several different sets of data that often differ by 1 or 2%.
4. Because real gases deviate less from the ideal gas law at low pressures, Figure 3–15 is usually reliable to 0.1%.
5. Figure 3–13 can be used for natural gases with as much as 50% nitrogen only if molar-average pseudocriticals are used.

The Wichert and Aziz (1972) method can be used to compute compressibilities for sour natural gases. This method uses the standard Z-charts, but adjusts the pseudocritical temperature and pressure as follows:

$$T_{cs} = T_c - \epsilon \qquad (3\text{--}41)$$

$$p_{cs} = p_c T_{cs}/[T_c + x_{H_2S}\ (1 - x_{H_2S})\ \epsilon] \qquad (3\text{--}42)$$

where T_{cs} = adjusted pseudocritical temperature (°R)
  p_{cs} = adjusted pseudocritical pressure (psia)
  ϵ = adjustment read from Figure 3–23
  x_{H_2S} = mol fraction of H_2S in sour gas

Example 3–7. Calculate the mass of 1 MMscf of a 0.650 gravity natural gas.

Solution: Assume that scf refers to the usual process calculational definition, one cubic foot of ideal gas at 14.696 psia, 60°F. Then no gas law calculations are required. From Appendix 2, the volume of a lbmol of gas is 379.5 scf. The molecular weight of the gas is (see App. 2)

$$MW = SG \times 28.964 = 0.650 \times 28.964$$
$$= 18.8\ \text{lb/lbmol}$$

Thus,
$$m = 1 \times 10^6\ \text{scf} \times 1\ \text{lbmol}/379.5$$
$$\text{scf} \times 18.8\ \text{lb/ lbmol}$$
$$= 4.96 \times 10^4\ \text{lb} \qquad \longleftarrow$$

Example 3–8. Estimate the density of the gas of Example 3–7 at 1500 psig, 50°F at an altitude of 1000 ft.

Solution: First convert the pressure to absolute. At 1000 ft the barometric pressure is about 14.2 psia. Round off to 14. The absolute pressure of the gas is 1500 + 14 = 1514 psia.

$$T = 50 + 459.67 = 510\ °R$$

The behavior is real, not ideal, at 1514 psia. Density is given by the formula shown below.

$$\rho = P\ MW\ /\ (Z\ R\ T)$$

Figure 3–18 gives, for this case, Z = 0.698.

$$\rho = (1514 \times 18.8)\ /\ (0.698 \times 10.732 \times 510)$$
$$= 7.45\ \text{lb/ft}^3 \qquad \longleftarrow$$

An alternative solution is obtained by using Figures 3–13 and 3–22.

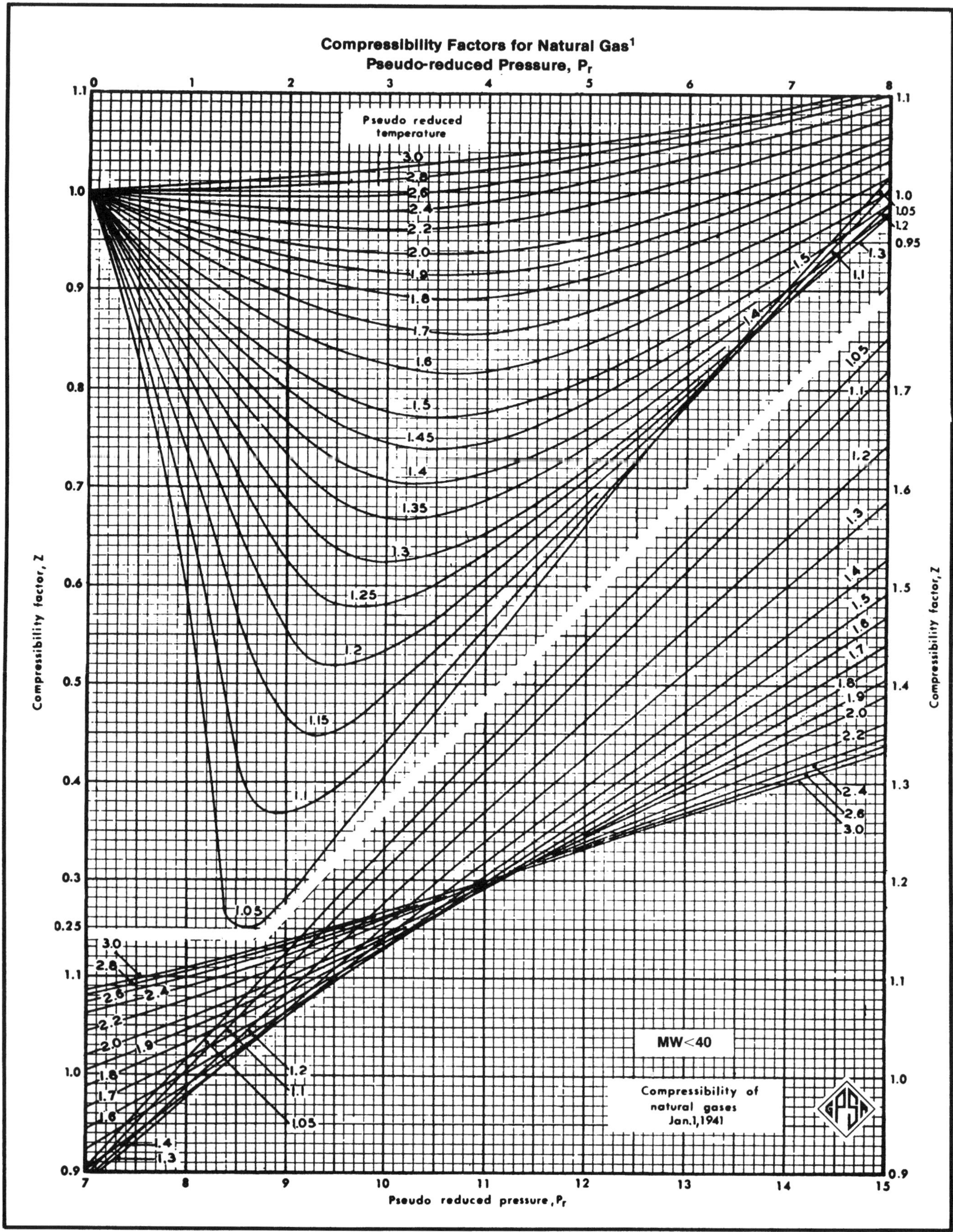

Figure 3–13. Compressibility factors for natural gas, high-reduced pressure range (GPSA, 1987).

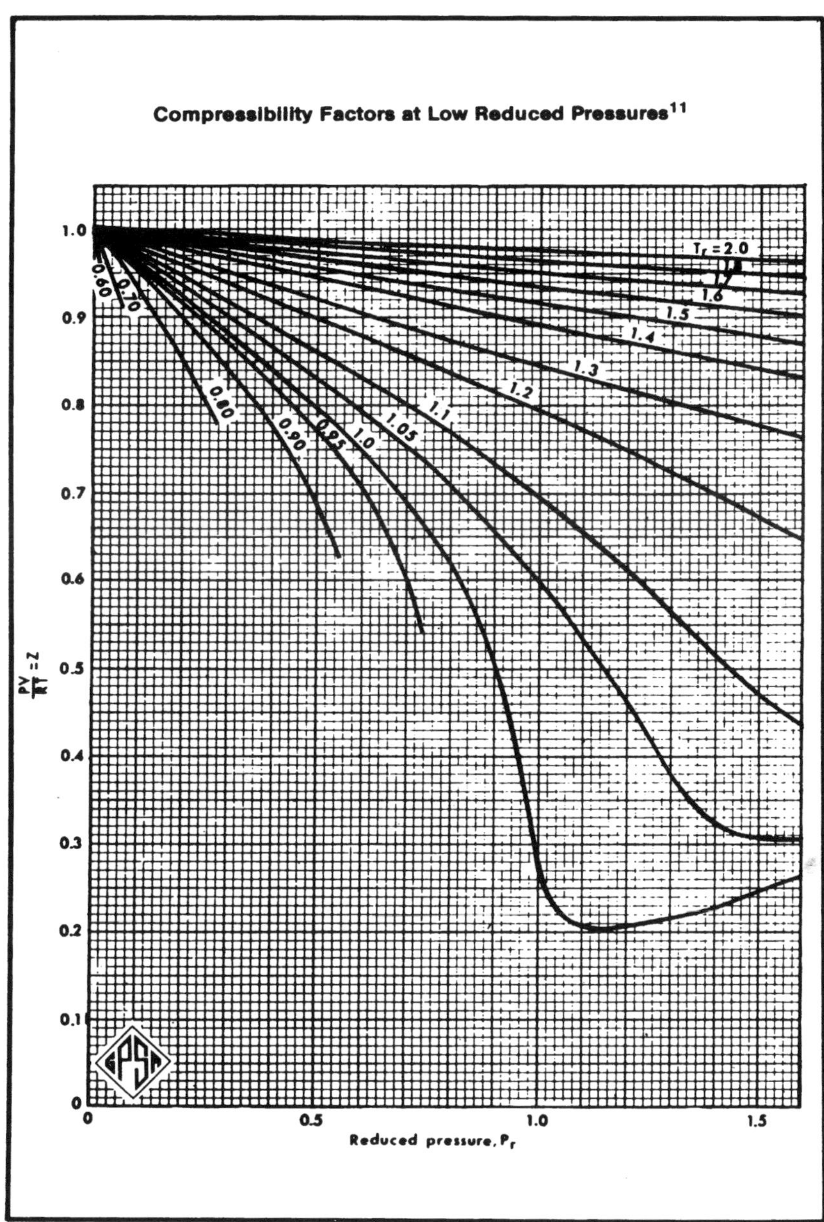

Figure 3–14. Compressibility factors for natural gas, intermediate-reduced pressure range (GPSA, 1987).

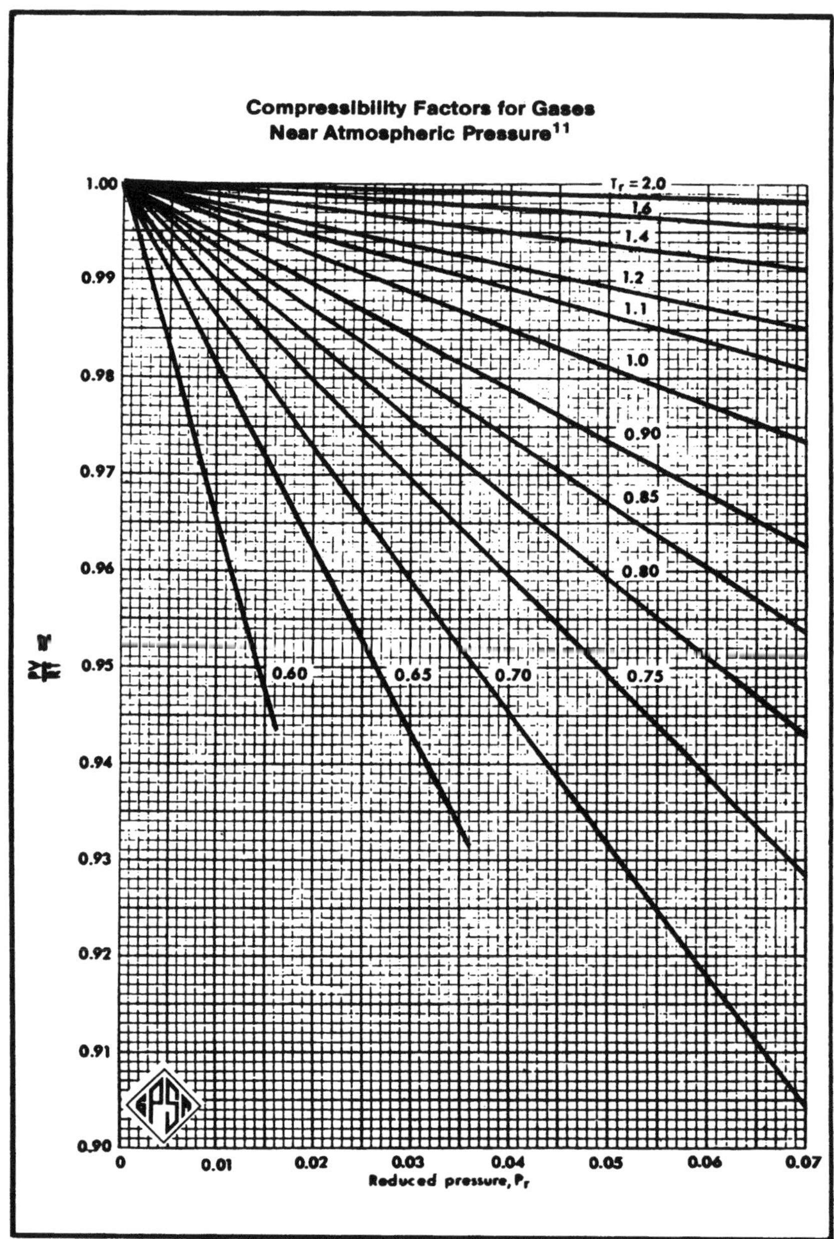

Figure 3–15. Compressibility factors for natural gas, low-reduced pressure range (GPSA, 1987).

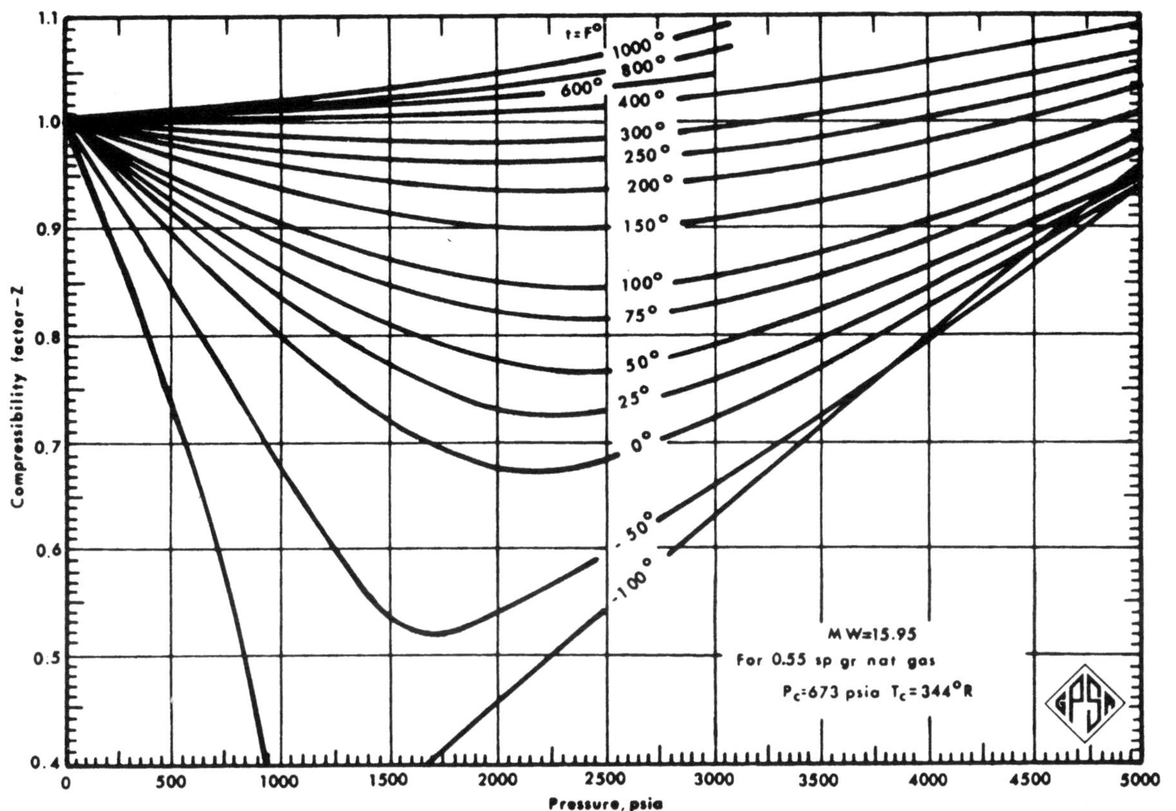

Figure 3–16. Compressibility of low-molecular-weight natural gases (GPSA, 1987).

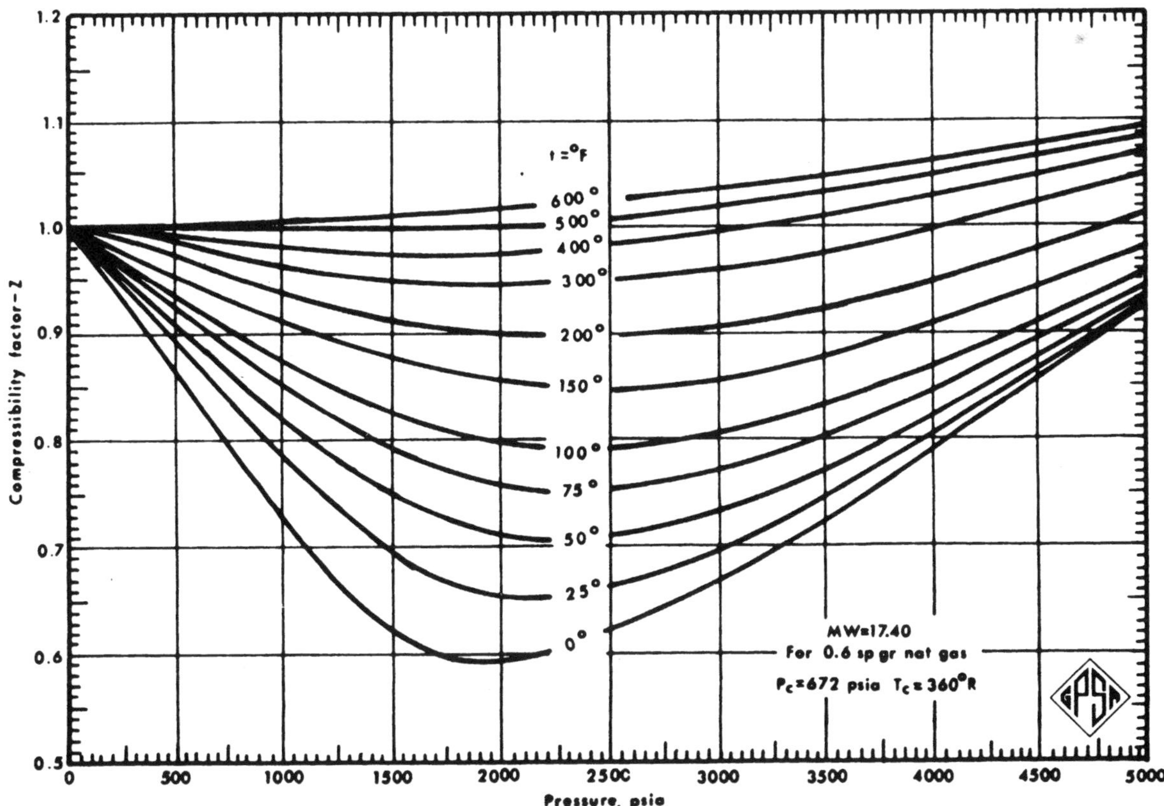

Figure 3–17. Compressibility of low-molecular-weight natural gases (GPSA, 1987).

39

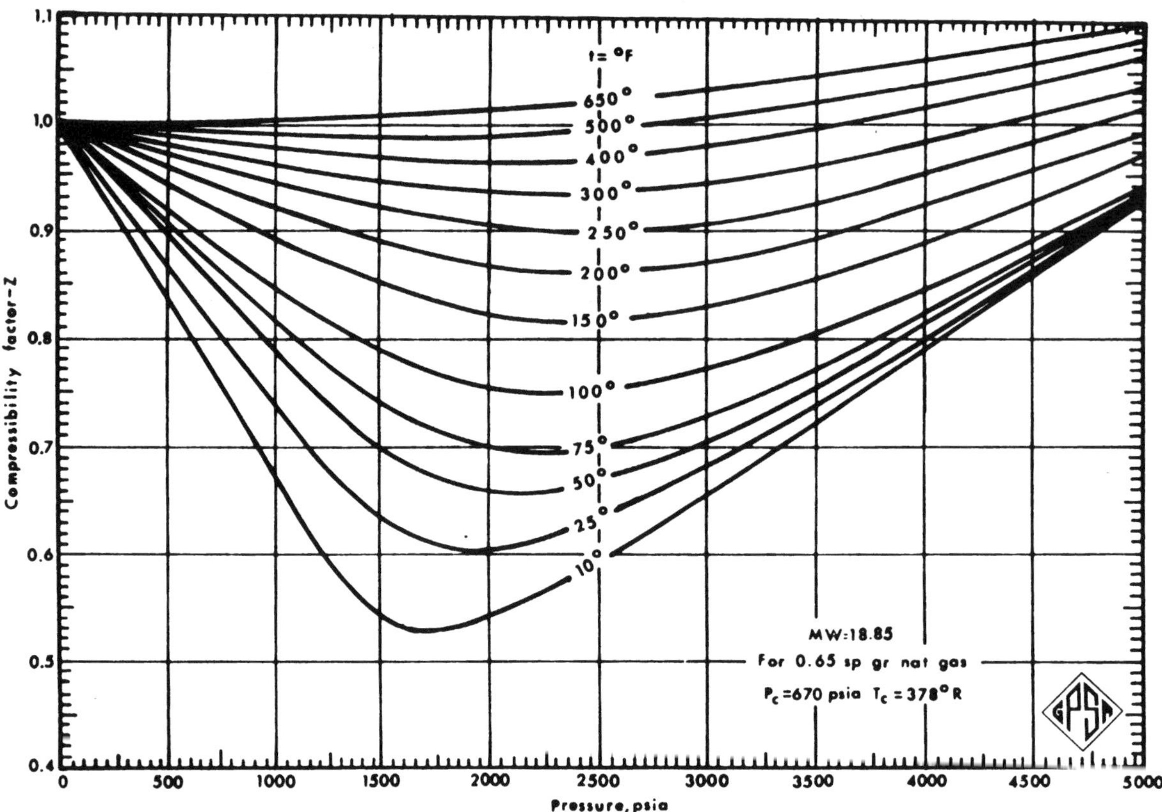

Figure 3–18. Compressibility of low-molecular-weight natural gases (GPSA, 1987).

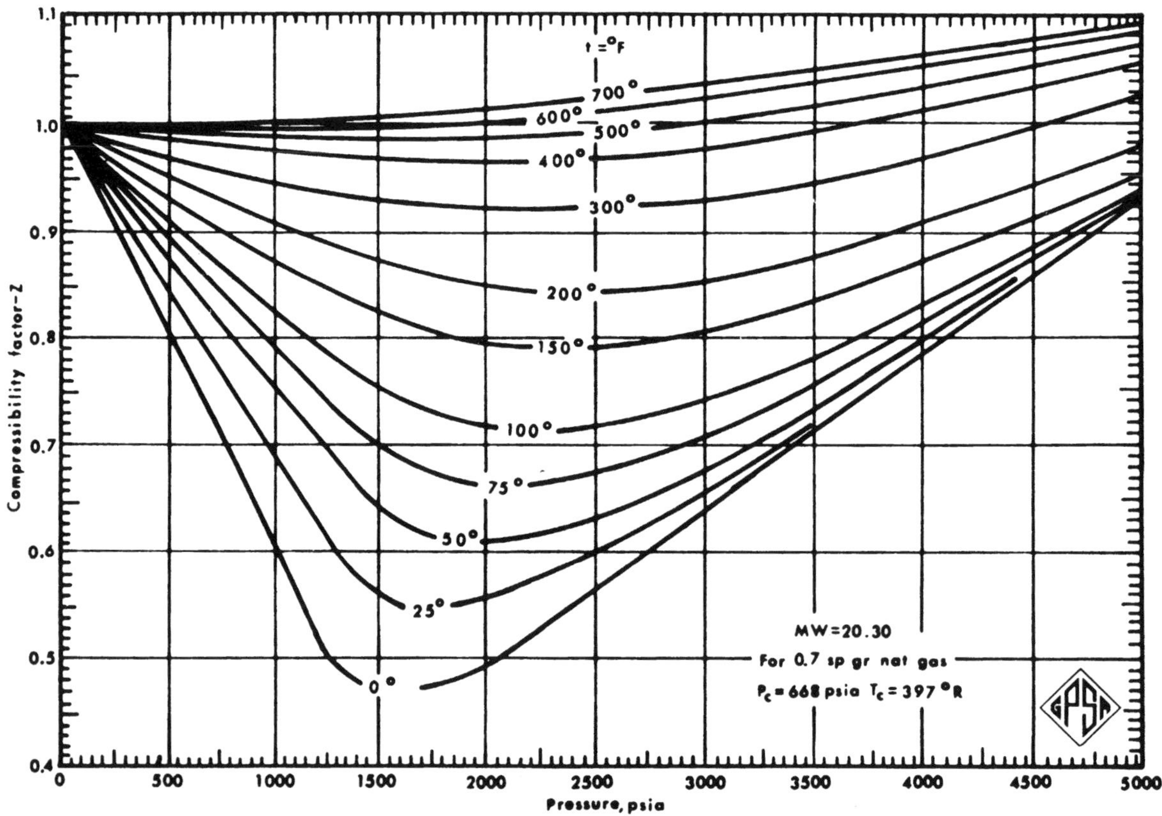

Figure 3–19. Compressibility of low-molecular-weight natural gases (GPSA, 1987).

40

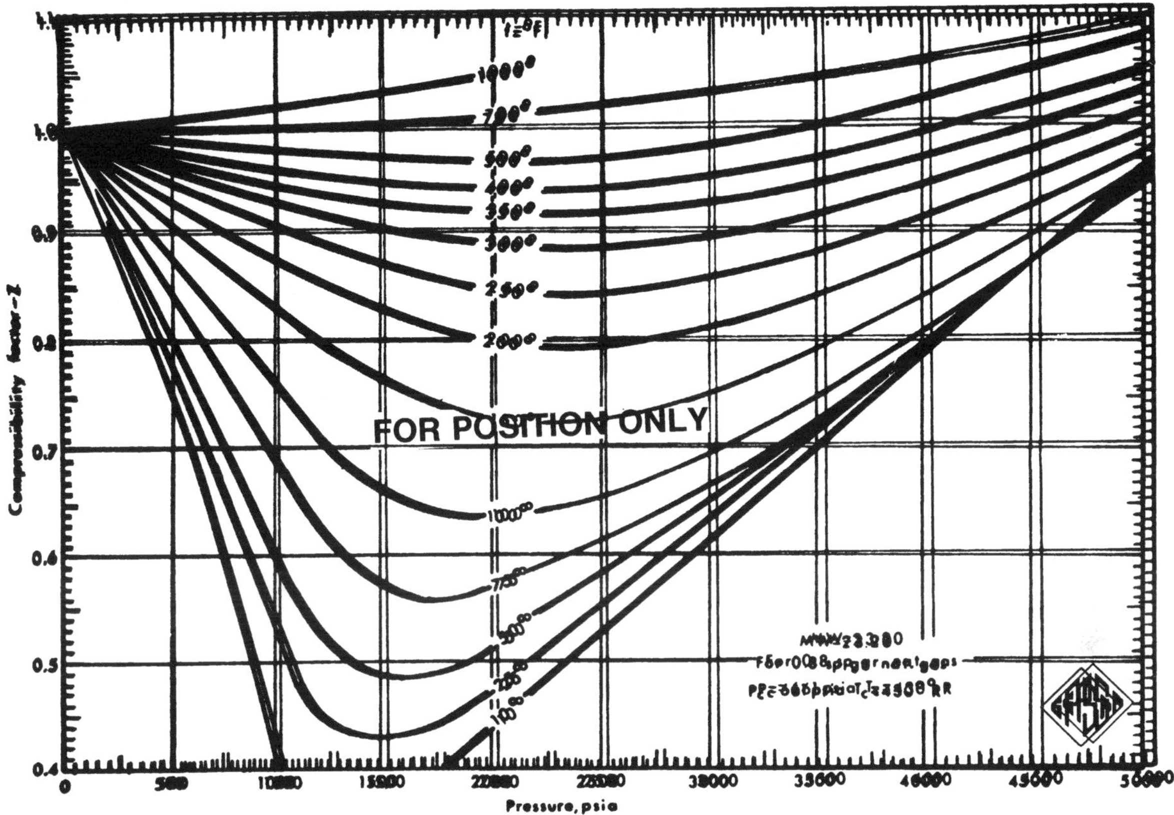

Figure 3–20. Compressibility of low-molecular-weight natural gases (GPSA, 1987).

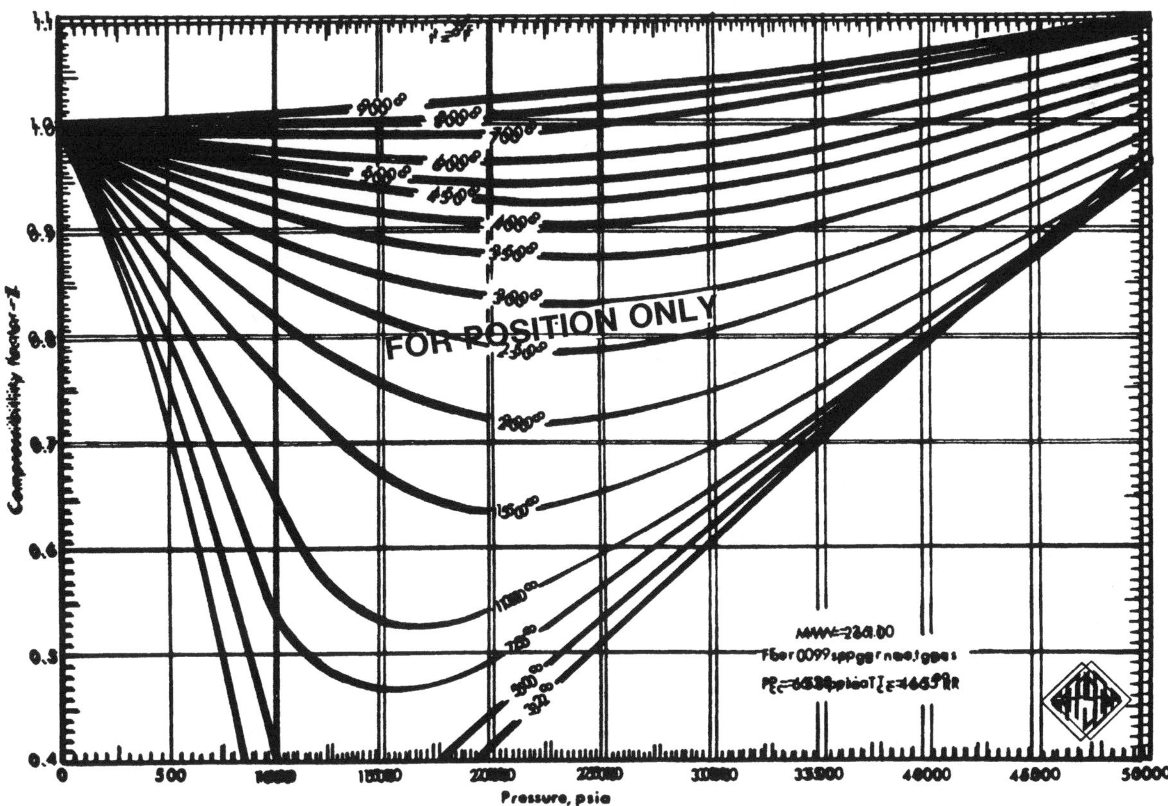

Figure 3–21. Compressibility of low-molecular-weight natural gases (GPSA, 1987).

Figure 3–22 gives $T_{pc} = 373°R$ and $P_{pc} = 670$ psia from the miscellaneous gases line.

The reduced conditions are:

$$T_r = T / T_{pc} = 510 / 373 = 1.37$$
$$P_r = P / P_{pc} = 1514 / 670 = 2.26$$

From Figure 3–13, $Z = 0.720$

$$\rho = (1514 \times 18.8) / (0.720 \times 10.732 \times 510)$$
$$= 7.22 \text{ lb/ft}^3 \qquad \longleftarrow$$

This result is in agreement with the first result within the accuracy of the correlations.

Example 3–9. The gas of Example 3–8 is determined to have 10 mol% H_2S. Reestimate the density.

Solution: Use Figure 3–23. $\epsilon = 17$ for 10% H_2S, 0% CO_2.

$$T_{cs} = 373 - 17 = 356°R$$
$$P_{cs} = (670 \times 356) / [373 + (0.1)$$
$$(1 - 0.1)(17)]$$
$$= 637 \text{ psia}$$
$$T_r = 510 / 356 = 1.43$$
$$P_r = 1514 / 637 = 2.38$$

From Figure 3–13, $Z = 0.758$

$$\rho = (1514 \times 18.8) /$$
$$(0.758 \times 10.732 \times 510)$$
$$= 6.86 \text{ lb/ft}^3 \qquad \longleftarrow$$

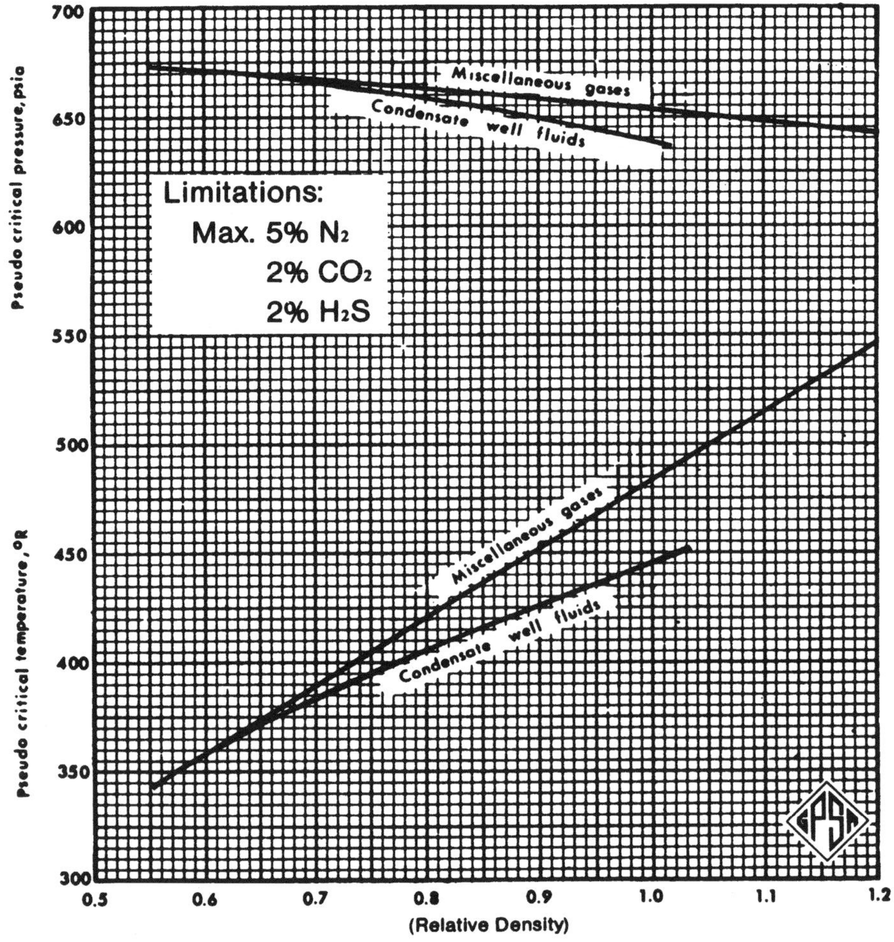

Figure 3–22. Pseudocritical properties of natural gases (GPSA, 1987).

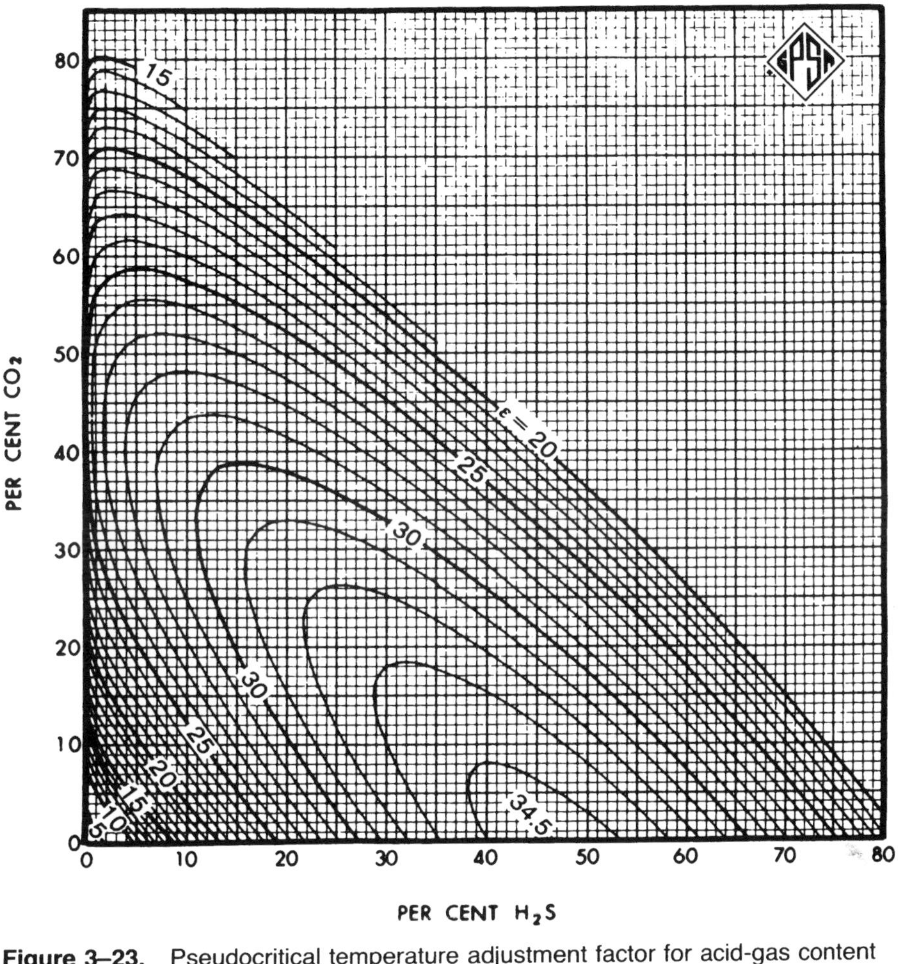

Figure 3–23. Pseudocritical temperature adjustment factor for acid-gas content (GPSA, 1987).

Review Questions

1. Define triple point, critical point, vapor pressure curve, degrees of freedom, saturated liquid, and saturated vapor.
2. Sketch a vapor-pressure curve for a pure material on P-T coordinates, with temperature along the abscissa or x-axis. Show the following: locus of saturated liquid-vapor mixtures, critical point, subcooled-liquid region, superheated-vapor region, gas region.
3. Define bubble point, dew point, pseudo-critical point, cricondentherm, cricondenbar, and retrograde behavior. Can a natural gas have two dew-point pressures at the same temperature? If yes, explain.
4. Sketch a phase envelope for a two-component mixture on P-T coordinates, with temperature along the abscissa or x-axis. Show the following features: liquid region, vapor region, critical point, bubble-point locus, dew-point locus, gas region, dense-phase region.

5. What is retrograde vaporization or condensation? Explain by sketching the phenomena on a mixture P-T diagram. Can these phenomena occur for a pure material?
6. State in words two important differences between the phase diagrams of crude oil and natural gas.
7. How do the bubble-point and dew-point loci vary with the composition of the natural gas? How does the accuracy of the C7+ analysis affect the estimation of dew points and bubble points?
8. Point out a difficulty with respect to phase behavior that can occur during the production life of:
   a. a crude-oil reservoir
   b. a condensate gas reservoir.
9. Sketch a line on a P-T diagram that represents the path of a well fluid as it flows from reservoir conditions, through the well bore, and to a surface separator.
   Now, sketch on this same diagram the phase envelope if the well fluid was:

   a. a low-GOR crude oil

   b. a retrograde condensate gas

   c. a dry natural gas.

10. Sketch on a P-T diagram a point representing the T and P conditions in a vapor-liquid phase separator.
Now superimpose on this graph the phase envelopes of the separator feed, the vapor product, and the liquid product.

11. Describe in words the concepts used in calculating:

   a. the bubble point of a mixture

   b. the dew point of a mixture.

12. What is convergence pressure and what is it used for?

13. What is an equation of state?
Name four such equations of state.
Compare these equations of state. Include simplicity, ease of use, and accuracy.

14. Vapor-liquid equilibrium calculations by the equation-of-state method are routinely done by hand calculation. True/False?

15. Equations of state generally provide very good estimations of liquid density. True/False? Of vapor density. True/False?

16. Gases approach ideal behavior as the compressibility factor Z approaches 0. True/False?

17. Gases approach ideal behavior for low temperature and pressure. True/False?

18. When can the ideal gas law be used?

19. The presence of acid-gas components has little effect on gas density (other factors remaining equal). True/False?

Problems

1. Pure methane is to be liquefied at atmospheric pressure (assume sea level). What will be the temperature of the methane in the atmospheric storage tank?

2. Pure methane is to be liquefied at atmospheric temperature (80°F). To what pressure must the methane be compressed?

3. A tank contains pure ethane at a temperature of 100°F. Can the ethane be liquefied (at any pressure)?

4. Pure propane is to be pipelined as a liquid. The maximum expected summertime temperature in the pipeline is 80°F. What is the minimum pressure that can be allowed in the pipe line?

5. A liquid mixture of 30 mol % ethane and 70 mol % propane is to be pipelined. What is the minimum pipeline pressure? Assume that the pipeline temperature is always below 100°F.

6. A pipeline carries 100 MMscfd of a 0.6 SG gas at 100°F and 800 psia. What is the minimum diameter of the pipeline if the gas velocity is not to exceed 70 ft/s.

7. A commercial grade propane containing 8 mol percent C2, 90 mol percent C3, and 2 mol percent i-C4 is being considered for use as a refrigerant. Does this commercial grade propane meet the GPA's maximum vapor pressure specification of 208 psig at 100°F?

8. Consider the following mixture:

   C2 20.0 wt percent

   C3 30.0 wt percent

   nC4 50.0 wt percent

   a. Compute the composition in mol percent and in vol percent (see app. 2).

   b. Estimate the dew-point pressure at 200°F. K-values may be found in GPSA (1987).

   c. Estimate the bubble-point temperature at 400 psia.

   d. What phases are present at 200°F and 400 psia? How much of each phase is present?

   e. What phases are present at 200°F and 600 psia? How much of each phase is present?

   f. Keep the temperature fixed at 200°F. Compute $\Sigma z_i/K_i$ and $\Sigma K_i z_i$ for P = 300, 400, 500, 600, 700, and 800 psia. Use a convergence pressure of 800 psia.

   g. Plot $\Sigma K_i z_i$ and $\Sigma z_i/K_i$ versus P. Cover the range 300 to 800 psia.

   h. Using the plot developed in part (g) verify the answers to parts (b), (d), and (e). Read the bubble-point and dew-point pressures at 200°F.

   i. Determine the pressure range for both the liquid and vapor phases to be present.

   j. Is there any pressure at which both $\Sigma K_i z_i < 1$ and $\Sigma z_i/K_i < 1$?

   k. Why are $\Sigma K_i z_i = 1$ and $\Sigma z_i/K_i = 1$ at 800 psia?

   l. Compare the above results with those obtained by a computer simulation package such as OPSIM.

   m. Do the K-value and computer methods agree? Why or why not?

9. a. Calculate the weight (or mass) of 1 MMscf of a 0.65 gravity natural gas.
Assume P standard = 14.696 psia and T standard = 60°F.

   b. Estimate the density (lb/ft^3) of this natural gas, SG = 0.65, at 1500 psig and 50°F.

10. Compare the Z-charts presented in this chapter (Figs. 3–13 through 3–21). One quantitative, but time-consuming, approach would be to estimate the compressibility factor, Z, f or the following conditions:

Gas gravity   SG = 0.6, 0.7, 0.8

Reduced temperature T_r = 0.7, 1.0, 2.0

Reduced pressure P_r = 0.5, 1.0, 2.0.

A computer simulation package, e.g., OPSIM, provides a much faster (and more accurate) method of estimating how Z varies with pressure, temperature, and gravity. Of course, such an equation-of-state approach does not compare Z-charts.

11. Estimate how the compressibility of a natural gas varies with the concentration of acid gases. Note that two approaches are available: Z-charts and simulation packages. A complete answer is very time-consuming because the effects of gas pressure and temperature should be included.

12. How many GPSA, AGA, TX, or OK scf are equivalent to 1 lb mol? Appendix 2 provides the temperatures and pressures for these standard cubic feet:
    - GPSA scf (60°F, 14.696 psia)
    - AGA scf (60°F, 14.73 psia)
    - TX scf (60°F, 14.65 psia)
    - OK scf (60°F, 14.65 psia)

13. Is the TX scf "bigger" than the LA (or Cajun) scf?
    - TX scf (60°F, 14.65 psia)
    - LA scf (60°F, 15.025 psia)

Nomenclature

A_o, B_o, C_o, a, b, c, α, γ	= original 8 BWR constants
a, b	= constants in van der Waals, SRK, PR equations of state
C	= number of components in system
D_o, d, E_o	= 9th, 10th and 11th BWRS constants, used in Starling-Han equation of state
F	= feed rate (mol/time)
F_L	= liquid rate (mol/time)
F_V	= vapor rate (mol/time)
f	= degrees of freedom or variance
h	= specific enthalpy (energy/mass)
K	= vapor-liquid equilibrium ratio (y/x)
m	= mass
MW	= molecular weight
NBP	= normal boiling point
n	= number of moles
P	= absolute pressure
P_k	= convergence pressure
R	= gas constant on mol basis
$\mathfrak{R}$	= gas constant on unit mass basis
s	= specific entropy (energy/mass-degree)
SG	= specific gravity of gas (MW / 28.9625)
T	= absolute temperature
V	= total volume of gas
VP	= vapor pressure
v	= specific volume (volume/mol)
w	= mass fraction
x	= mol fraction in liquid phase
y	= mol fraction in vapor phase
Z	= compressibility factor = P v/R T
z	= overall mol fraction in feed
$z_c{}^*$	= critical compressibility parameter
δ_{ij}	= fitted binary interaction parameter
ϵ	= adjustment to pseudo-critical temperature
μ	= viscosity
ρ	= density (mass/volume)
Σ	= indicates summation over all components present
ϕ	= number of phases
ω	= acentric factor

Subscripts

c	= denotes critical property
cs	= denotes adjusted critical property
f	= saturated liquid property in steam tables
g	= saturated vapor property in steam tables
i	= denotes component i
ij	= denotes a binary interaction parameter
m	= denotes property for a gas mixture
pc	= pseudo-critical property
r	= denotes reduced property
std	= denotes property at standard or reference conditions

References

Benedict, M., G. B. Webb, and L. C. Rubin (1940), "An Empirical Equation for Thermodynamic Properties of Light Hydrocarbons and Their Mixtures," *J. Chem. Phys.*, Vol. 8, pp. 334–345.

Brown, G. G., M. Souders, and R. L. Smith (1932), "Pressure-Volume-Temperature Relations of Paraffin Hydrocarbons," *Ind. Eng. Chem.*, Vol. 24, pp. 513–515.

Buthod, P., N. Perez, and R. Thompson (1978), "Program, Constants Developed for SRK Equation," *Oil Gas J.*, Vol. 76, No. 48, pp. 60–64 (Nov. 27).

Cope, J. Q., W. K. Lewis, and H. C. Weber (1931), "Generalized Properties of Higher Hydrocarbon Vapors," *Ind. Eng. Chem.*, Vol. 23, pp. 887–892.

Dodge, B. F. (1932), "Physico-Chemical Factors in High-Pressure Design," *Ind. Eng. Chem.*, Vol. 24, pp. 1353–1363.

Dodge, B. F. (1944), "Chemical Engineering Thermodynamics," McGraw-Hill Book Co., NY.

GPSA (1987), "Engineering Data Book," 10th Edition, Gas Processors Suppliers Association, 6526 East 60th St., Tulsa, OK 74145.

Hougen, O. A., K. M. Watson, and R. R. Ragatz (1962), "Chemical Process Principles, Part 1: Material and Energy Balances," John Wiley and Sons, Inc., NY, 2nd ed.

Jacoby, R. H., and L. Yarborough (1967), "PVT Measurements on Petroleum Reservoir Fluids and Their Uses," *Ind. Eng. Chem.*, Vol. 59, pp. 48–62 (Oct.).

Lee, John F., and F. W. Sears (1963), "Thermodynamics," Addison-Wesley Pub. Co., Reading, MA.

McCain, W. D., Jr. (1990), "The Properties of Petroleum Fluids," 2nd Ed., Pennwell Publ. Co., Tulsa, OK.

Peng, D-Y., and D. B. Robinson (1976), "A New Two-Constant Equation of State," *Ind. Eng. Chem. Fund.*, Vol. 15, pp. 59–64.

Pitzer, K. S., *et al.* (1955), "The Volumetric and Thermodynamic Properties of Fluids (II) Compressibility Factor, Vapor Pressure and Entropy of Vaporization," *J. Amer. Chem. Soc.*, Vol. 77, pp. 3433–3440.

Plane, A. S. (1966), "Comparison between Calculations by Computer and Measurement in the Field of Oil and Gas Quantities and Qualities Produced in Multi-stage Separation of Reservoir Crude," *J. Inst. Pet.*, Vol. 52, No. 506, pp. 31–37 (Feb.).

Poettman, F. H., and B. J. Maylund (1949), "Equilibrium Constants for High-Boiling Hydrocarbon Fractions of Varying Characterization Factors," *Petroleum Refiner,* Vol. 28, No. 7, pp. 101–102 (July).

Redlich, O., and J. N. S. Kwong (1949), "On the Thermodynamics of Solutions. V. An Equation of State. Fugacities of Gaseous Solutions," *Chem. Rev.*, Vol. 44, pp. 223–244.

Sarssam, S. (1988), "Condensate Fields—Processing Options," IBC Tech. Serv. Ltd., *Proc. Conf. on Development of Condensate Fields*, London, 21 pp.

Soave, G. (1972), "Equilibrium Constants from a Modified Redlich-Kwong Equation of State," *Chem. Eng. Sci.*, Vol. 27, pp. 1197–1203.

Starling, K. E., and M. S. Han (1972), "Thermodynamic Data Refined for LPG. Part 14. Mixtures," *Hydro. Proc.,* Vol. 51, No. 5, pp. 129–132 (May); "Part 15. Industrial Applications," *loc. cit.*, No. 6, pp. 107–115 (June).

Thompson, R. E. (1985), "Ch. E. Process Simulation on the Microcomputer," *Proc. ASEE Midwest Section Meeting*, Columbia, MO (March 21).

Usdin, E., and J. C. McAuliffe (1976), "A One Parameter Family of Equations of State," *Chem. Eng. Sci.*, Vol. 31, pp. 1077–1084.

Van Wylen, G. J., and R. E. Sonntag (1986), "Fundamentals of Classical Thermodynamics," John Wiley & Sons Inc., NY, 3rd ed.

Wichert, E., and K. Aziz (1972), "Calculate Z's for Sour Gases," *Hydro. Proc.*, Vol. 51, No. 5, pp. 119–122 (May).

Chapter 4

Water-Hydrocarbon Phase Behavior

INTRODUCTION

In the *liquid phase, water* and *hydrocarbons* are virtually *insoluble*, and this is well documented (GPSA, 1987, p. 20–3; API, 1983). Of course liquid water can and does form emulsions with crude oil and other liquid hydrocarbons as is discussed in detail in Volume 2. Accordingly attention is focused here on the gas-phase behavior of water and hydrocarbons. In addition the conditions for CO_2 frost formation are also reviewed because of its importance in cryogenic NGL recovery.

As produced at the wellhead, natural gases are nearly always saturated with water. When water-saturated natural gas flows in a pipeline the following problems can occur:

1. Liquid water can collect in pipelines and so increase the pressure drop and/or cause slug flow.
2. Free water also can freeze into ice and/or form solid hydrates and so reduce the gas flow or even plug the line completely.
3. Acid gases—H_2S and/or CO_2—dissolve in free water and can cause severe corrosion.

Water is called the *universal impurity* or *contaminant*. Water removal and/or inhibition of hydrate formation is therefore a basic part of gas gathering. Design of oilfield gathering lines and dehydration and hydrate-inhibition facilities requires two key phase-behavior predictions:

1. The water content of saturated natural gases.
2. The hydrate formation temperature and pressures.

Accordingly, attention is focused on these two aspects of water/ natural-gas behavior.

The water content of saturated natural gases is discussed by first describing and comparing the common measurement techniques and then reviewing alternative prediction methods. Finally, example calculations of the water content of both sweet and sour gases are presented.

Hydrates are reviewed by first describing their chemical nature and the thermodynamic conditions necessary for their formation. Then prediction techniques are summarized and example calculations of hydrate formation temperature and pressures are presented. Next, the overall phase behavior of water-hydrocarbon systems is outlined very concisely. Finally, the phase behavior of CO_2 and natural gas is summarized.

MEASUREMENT OF WATER CONTENT OF GASES

Sampling for accurate water dew point or content measurements require the following precautions (Chandler Engineering, 1986):

1. The gas sample must be representative.
2. The sampling line cannot contain free water.
3. While flowing in the sample line, the gas must be kept above its dew-point temperature.
4. If a glycol filter is installed, the sample gas must flow for five minutes—to saturate the filter with water—before the water content is measured.

The water content of natural gases can be measured by six different techniques: dew point, electrolysis, capacitance, conductivity, titration, and IR absorption. In addition, dew-point tubes can provide approximate estimates. IR absorption is not used very often, and so the other methods are now summarized.

Bureau of Mines Dew-Point Tester (ASTM D 1142—63)

As shown in Fig. 4–1, the *Bureau of Mines dew-point tester* consists of a high-pressure, stainless steel or nickel-plated brass chamber. Gas entering through the inlet valve, A, is directed by the deflector, B, onto the chilled, highly-polished stainless-steel mirror, C, and then leaves the

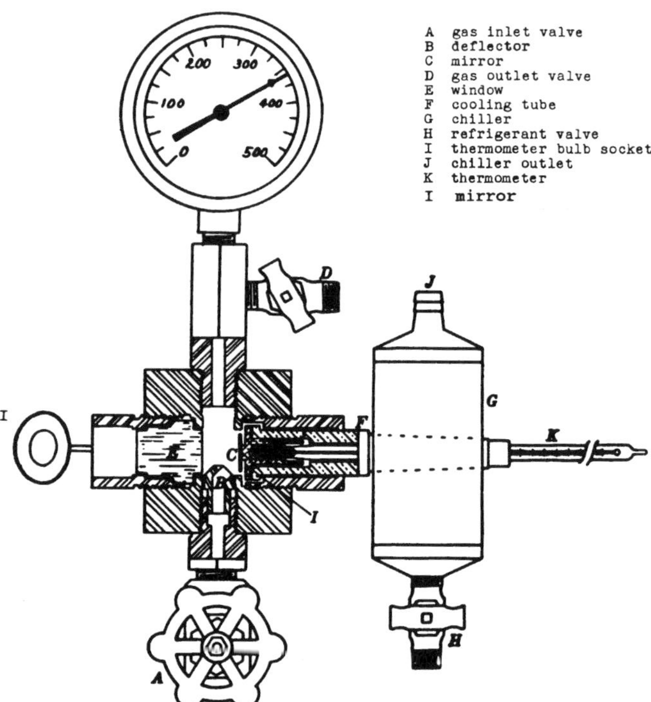

Figure 4–1. Bureau of Mines dew-point tester (Deaton and Frost, 1938; ASTM, 1963).

chamber through outlet valve, D. The mirror, C, is cooled by cooling tube, F, which is attached to the chiller, G. Refrigerant enters the chiller at valve, H, and leaves at J. Any dew on the mirror, C, is observed through the transparent lucite window, E. The temperature at which dew is observed is read using the calibrated thermometer, K. Mirror I permits simultaneous viewing of the mirror C and reading of the thermometer, K.

Dew-point measurement involves cooling the mirror, C, with a suitable refrigerant (C3 or Freon 12 down to –20°F, liquid CO_2 down to –90°F, dry ice/acetone down to −100°F, or liquid N_2 down to −200°F); flowing the sample gas over the polished mirror; and reading the temperature and pressure at which dew first appears and disappears on the mirror surface.

The following precautions will improve accuracy (Chandler Engineering, 1986):

1. Use an illuminated magnifier and/or an LED temperature readout if the lighting makes it difficult to observe condensation.
2. Purge the tester to remove all air.
3. Do not cool the mirror faster than 2°F/min when within 5°F of the dew point.
4. While observing the mirror and thermometer, record the temperature at which dew first forms.

5. Let the mirror warm up and observe the temperature at which the dew disappears.
6. Repeat steps 4 and 5 until the two temperatures agree within 2°F.
7. Take the average of the two temperatures as the dew point.

A water dew-point measurement can be confirmed by measuring the dew point of the same gas sample at several pressures. The observed dew points should change so the resulting water contents remain constant. This implies that uncertainties in the gas-sample pressure and/or any overlooked pressure drops will cause corresponding errors in the reported water content.

Liquid hydrocarbons, alcohols, or glycols also can condense on the mirror before the water dew point is reached. The following characteristics distinguish water dew points (Gibson, 1980; Chandler Engineering, 1986):

1. Water dew forms a distinct, opaque, grey circular spot in the center of the mirror (coldest spot). Water should not "wet" the mirror and should resist being blown off the mirror by increasing the gas flow. Ice crystals form an irregular white pattern against the previously-formed, grey water condensate. Barium sulfate and "water-cut" paste can confirm water dew points also.
2. In contrast, liquid-hydrocarbon condensates wet the mirror, expand in rainbow-like rings to cover all the mirror, and can be "blown off" or "streak" the mirror by a sudden increase in the sample gas flowrate.
3. Alcohol dew points appear as white spots with indistinct edges. Advanced alcohol spots are larger, increasingly white, and eventually form liquid drops that do not freeze.
4. Glycol dew points are darker, cover the entire mirror, and do not evaporate.

With the exception of the thermometer and pressure guage, the Bureau of Mines tester requires no calibration. The method is relatively inexpensive and easy to operate. However, this type of measurement can be time-consuming and cannot be recorded automatically. Accuracy can be very good but varies with operator skill and dedication.

Electrolysis Method

The electrolysis method involves adsorbing and electrolyzing the water vapor in the sample gas. The heart of this instrument is an electrolytic cell consisting of two 5–mil wires spirally wound throughout the inner wall of an insulating tube. A thin film of phosphorus pentoxide (P_2O_5) is applied between these two wires which are spaced 5 mil apart. As shown in Figures 4–2 and 4–3 (Mayeaux,

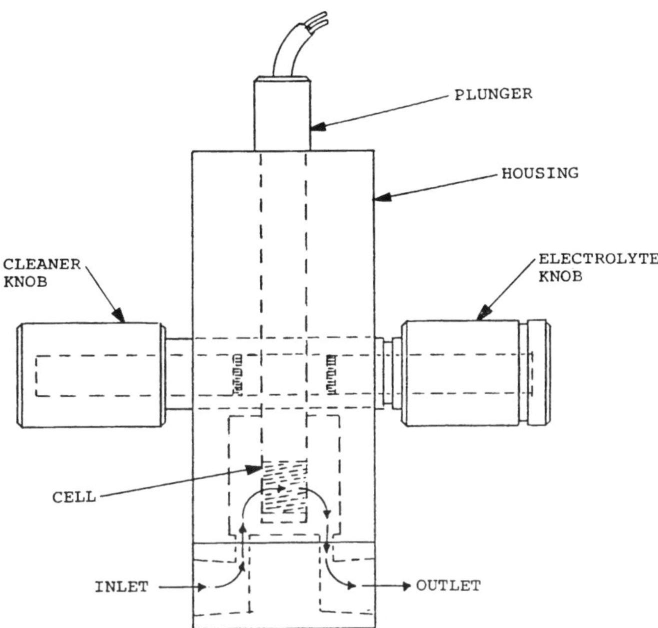

Figure 4–2. Electrolytic moisture analyzer cell assembly (Mayeaux, 1987; Ranarex, 1988), U.S. Patent No. 4,842,709.

1987, and Ranarex, 1988), the sample gas first flows into the cell, then passes sensing windows covered with a semipermeable membrane, and finally exits. Water vapor, in direct proportion to the sample-gas concentration, is adsorbed by the membrane, diffuses into the P_2O_5 film, and is electrolyzed quantitatively. The resulting current is, therefore, directly proportional to the water-vapor content of the sample gas.

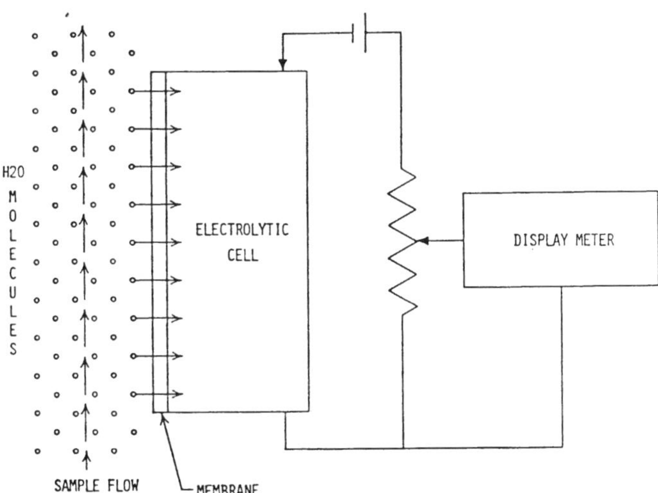

Figure 4–3. Electrolytic moisture analyzer principle of cell operation (Mayeaux, 1987; Ranarex, 1988), U.S. Patent No. 4,842,709.

Only the ammeter and flowmeter require calibration, and this makes the electrolytic method one of the most accurate and fundamental available. Ammeter calibration requires "calibration gas"—natural gas with a known water content. Moisture calibrators (e.g., Ranarex, 1988) provide a continuous supply of calibration gas by saturating the gas with water at 32°F. Water contents from 9.5 to 170 lb/MMscf can be obtained by varying the calibration gas pressure.

Note the following potential sources of errors:

1. Contamination of the cell or coating of the P_2O_5 strip by oil, condensate, glycol, compressor oil, etc.—anything that changes the adsorption of the water vapor (by far the most frequent source of failure). Such a contaminated cell will exhibit a low reading (0.25 lb H_2O/MMscf) that will not change when the sample gas flowrate is varied (Barnes, 1988).
2. Washout of the cell by excess water, alcohol, oil, methane, amine, etc., produces an essentially zero meter reading.
3. A dead short produces an offscale meter reading.

Models without a semipermeable membrane to protect the P_2O_5 membrane are far more susceptible to contamination and washout. Two additional warnings are worthwhile:

1. The electrolytic cell does not operate well below 32°F and should be temperature controlled if necessary.
2. Phosphoric acid can cause severe harm to skin and eyes. Extreme caution should be exercised when the electrolytic cell is cleaned and recoated.

Aluminum Oxide Humidity Sensor

The moisture sensor consists of a thin, porous layer or film of aluminum oxide (Al_2O_3) sandwiched between two electrodes. Panametrics (Scelzo, 1989) uses an aluminum base and a very thin gold coating as the electrodes, while EG&G Chandler Engineering (1985) employs electrodes coated with a proprietary alloy. Panametrics fabricates the sensor from a thin strip of pure aluminum which is anodized in sulfuric acid to form a surface layer of porous aluminum oxide. Then a layer of gold, thin enough to be porous, is evaporated over the aluminum oxide (Scelzo, 1989). This *sandwich sensor* is essentially a *capacitor*, with the Al_2O_3 the *dielectric*. When an AC voltage is applied, the resulting impedance varies with the amount of water adsorbed in the aluminum oxide film. In turn, the quantity of adsorbed water depends on the partial pressure of water vapor in the sample gas flowing around the sensor. A suitable electronic circuit converts the measured impedance to the desired units of water vapor content. The sensor or probe is built so that the water vapor equilibrates rapidly.

This capacitance method is used to measure water dew points ranging from -150 to $70°F$ with a response time of less than five seconds for a 63% step change in moisture content. As with the electrolysis method, contamination by pipescale, carbon, salt, and conductive liquids (glycol, methanol) can impede measurement. The sensor is not harmed by liquid slugs of condensate, methanol, glycol, and water. Proper cleaning restores the sensor (Scelzo).

Titration Method

The water content is determined by titration using a water-specific reagent—usually *Karl Fischer* reagent. As shown in Figure 4–4 (UGC Industries, 1983), the sample gas enters the reaction cell, A; bubbles through a small (0.5 mL) known quantity of liquid reagent; and exits via a reagent trap, a pressure-reducing regulator, and finally a flowmeter. A pair of platinum electrodes, D, sense the end-point of the titration—when the entering water vapor has exhausted the batch of liquid reagent. Then a fresh batch of reagent is injected into the reaction cell by pump, P, which is activated by solenoid valves B and C.

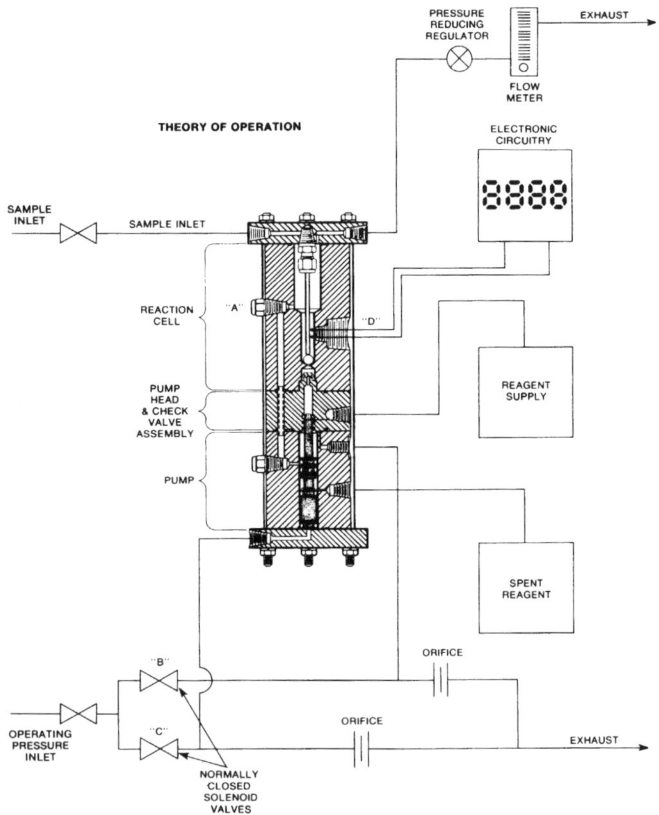

Figure 4–4. Moisture titrator (UGC Industries, 1983).

Electronic circuitry measures the time between end points and the sample-gas flowrate and pressure. Then it computes and then displays the water content. The entire cycle takes about two seconds.

Karl Fischer reagent is inert to hydrocarbons, carbon dioxide, glycols, amines, and most sulfur compounds, e.g., odorizing mercaptans. Hydrogen sulfide will cause the moisture titrator to read high by 0.7 lb H_2O/MMscf per grain H_2S/100 scf.

Conductivity Cell

The Hygromat measuring cell (Welker, 1989) consists of two stainless steel plates separated and electrically insulated from each other by a ceramic layer. The ceramic layer has eight holes which are partially filled with a *hygroscopic salt-glycerol solution*. Water is absorbed reversibly by the hygroscopic solution until equilibrium is reached with the surrounding natural gas. In turn the conductivity of the salt-glycerol solution increases as water is absorbed and decreases when water is desorbed. The sample gas is passed at 1/10 scfh through the meter. See Welker (1989) for further details.

Dew-Point Tubes

The dew-point tube uses a sampling pump the same as is used in the H_2S length-of-stain test, the equipment for which is shown in Figure 4–5a (ASTM, 1988, D 4810). A water detector tube is placed in the pump and 100 mL of pipeline gas is pulled through the tube. The detector tube (Fig. 4–5b) is filled with *magnesium perchlorate* contained on fine-grain silica gel (Sensidyne, 1984). Water vapor is adsorbed by magnesium perchlorate to produce an alkaline reaction that changes the color of Hammett's indicator (*crystal violet*). The water content in lb/MMscf is read directly from the length of the stain in the detector tube. Overall accuracy is ±25%. Alcohols, glycols, and amines cause high readings. The tube range is 6–80 lb H_2O/MMscf.

Comparison of Methods

The *Bureau of Mines dew-point* method is respected as the one defined by an *ASTM standard*. Equipment cost is low, but the method is labor intensive in that a single measurement requires approximately 15 minutes. Accuracy varies with operator skill. Uncertainties of ±1°F are attainable for dew points above 32°F but increase to ±4°F for dew points between -80 to $-100°F$.

The popularity of the electrolytic moisture analyzers is increasing significantly. Their accuracy compares favorably with the dew-point method. Advantages include light

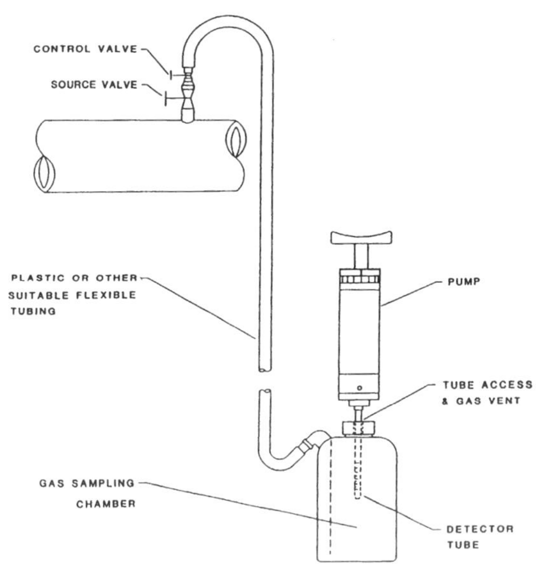

Figure 4–5a. Sampling equipment (ASTM, 1988).

PIPE LINE DEW POINT TUBE		No.6LP

1.	Performance			
	Calibration Scale	6-40 LB/MMCF (n=2)	Color Change	Yellow - Greenish Brown
	Measuring Range	6-40 LB/MMCF (n=2) 40-80 LB/MMCF (n=1)	Sampling Time	1 minute/pump stroke
	Detecting Limit	2 LB/MMCF (n=2)	Shelf Life	3 years

Figure 4–5b. Pipeline dew-point tube (Sensidyne, 1984).

weight, portability, continuous readings, fast response times, and ready interfacing with alarms and other process monitors. Improved methods of cleaning and recoating the P_2O_5 film reduce the most frequent disadvantages of contamination and washout.

The aluminum oxide sensor is relatively recent and exhibits accuracies and advantages similar to the electrolytic analyzer. It is especially suitable for very dry gases. However, response time is slow and removal of contaminants more difficult.

The titrator equipment is relatively expensive but not readily portable. The main advantages are accuracy (3% of reading) and immunity to contaminants, such as glycols and alcohols. The chief disadvantage is the hazardous nature of the Karl Fischer reagent, which creates a disposal problem.

The conductivity method also is relatively recent. Its advantages include long-term stability—typically better than 4°F dew point over six months. The 63% response time to sudden changes in gas humidity varies from 5 to 30 minutes depending on gas flow rate and pressure. One disadvantage is that conductivity varies with temperature,

and so the meter must be kept at constant temperature—usually 30°C (86°F).

Detector tubes provide an inexpensive approximate estimate. They can be used by nontechnical personnel with minimum training.

WATER CONTENT OF NATURAL GASES

Methods for predicting the saturated water vapor content of natural gases may be classified as follows:

1. *Charts for sweet gas*, for which the water content depends on pressure and temperature only.
2. *Charts for acid gas*, for which the water content depends on pressure, temperature, and acid-gas content.
3. *Equation-of-state methods*, with rigorous multicomponent equilibrium calculations performed on the computer.

The above three approaches are first reviewed and then compared.

Sweet Natural Gas Charts and Tables

Figure 4–6 presents the McKetta-Wehe (1958) correlation for the saturated water vapor content of sweet, lean, natural gases. Similar charts were prepared by Hammerschmidt (1939), McCartney, Boyd and Reid (1950), and Campbell (1976). Maddox and Erbar (1982) report that Figure 4–6 is reliable (about 5% average error) for gases containing up to 10% CO_2 and/or H_2S at pressures below 500 psia. This simple method becomes increasingly inaccurate with increasing pressure, temperature, and acid-gas content. While the McKetta-Wehe correlation includes corrections for gas gravity and salinity of the produced brine, these "corrections" are often ignored as "safety factors."

The Institute of Gas Technology (Bukacek, 1955) has extensive tables showing the water content of natural gas over wide ranges of temperature and pressure. These tables are used to legally define water contents of natural gases in specification sheets.

Acid-Gas Charts

The two most common correlations are those due to Robinson, Wichert, Moore, and Heidemann (1978) and Campbell (1976).

Robinson, Wichert, Moore, Heidemann Charts. Robinson *et al.* convert the carbon dioxide concentration into an equivalent H_2S concentration by multiplying the actual CO_2 composition by 0.75. Then they

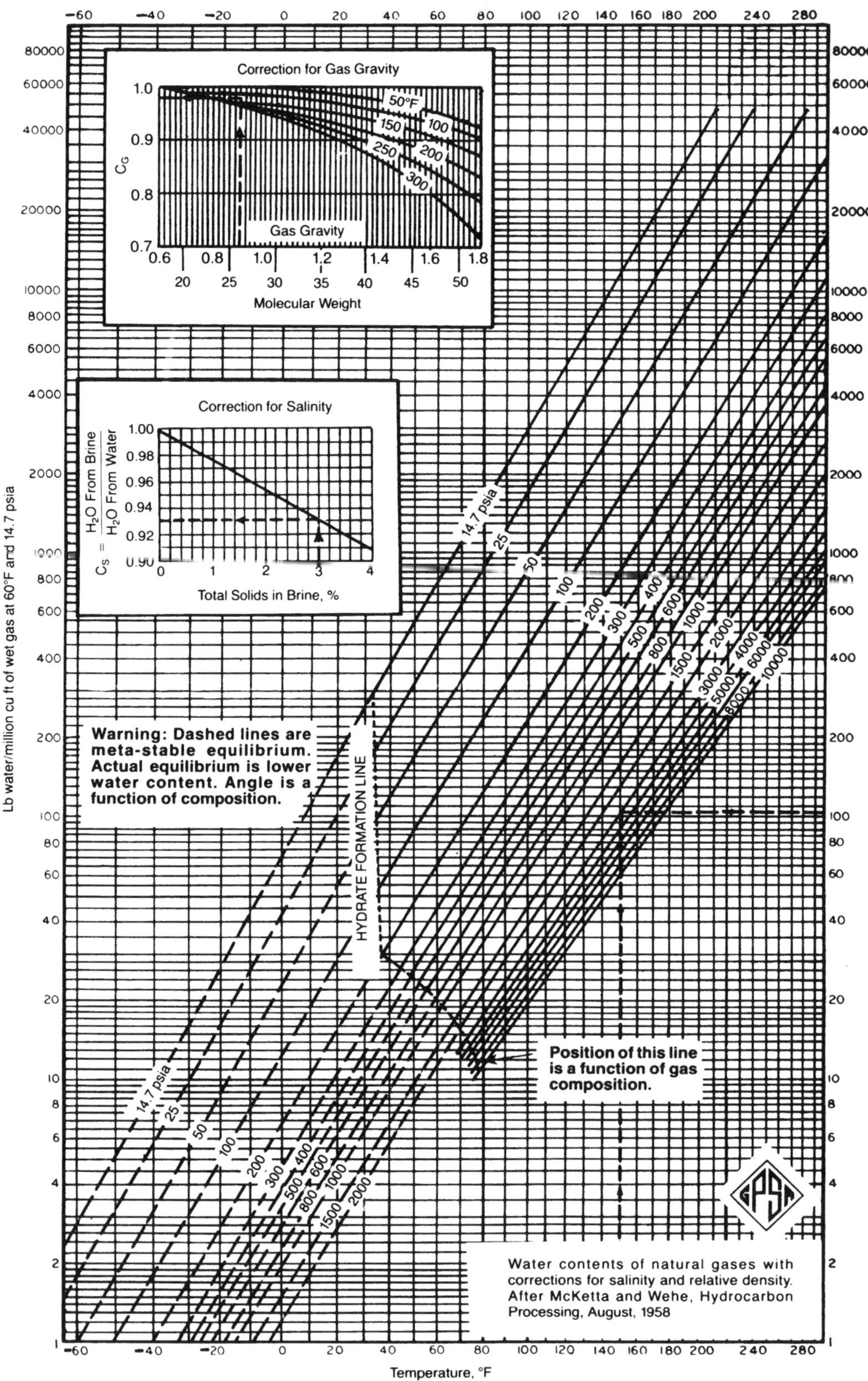

Figure 4–6.   Dew point of natural gas (McKetta and Wehe, 1958; GPSA, 1987, p. 20–4).

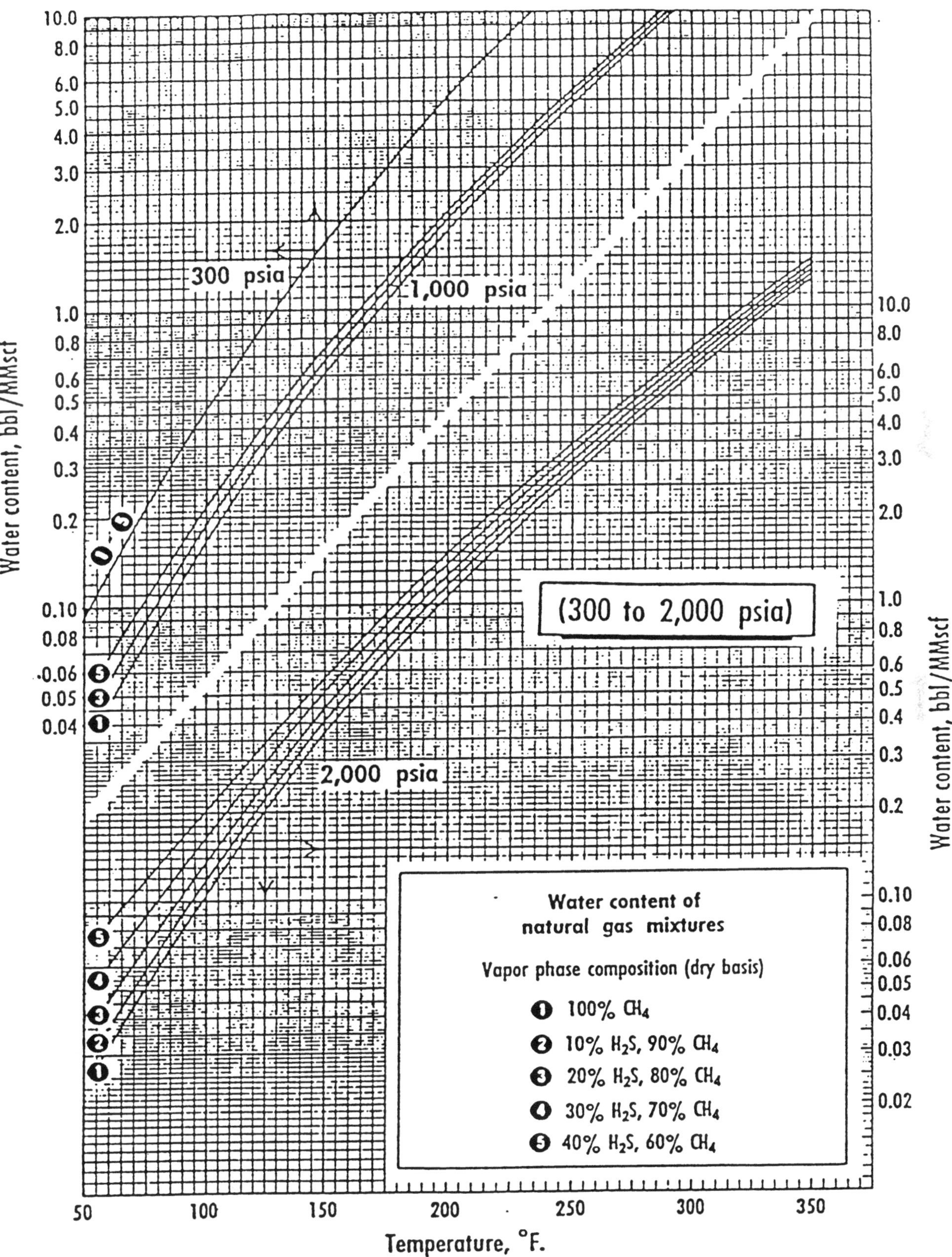

Figure 4–7a. Water content of sour gases (Robinson *et al.*, 1978).

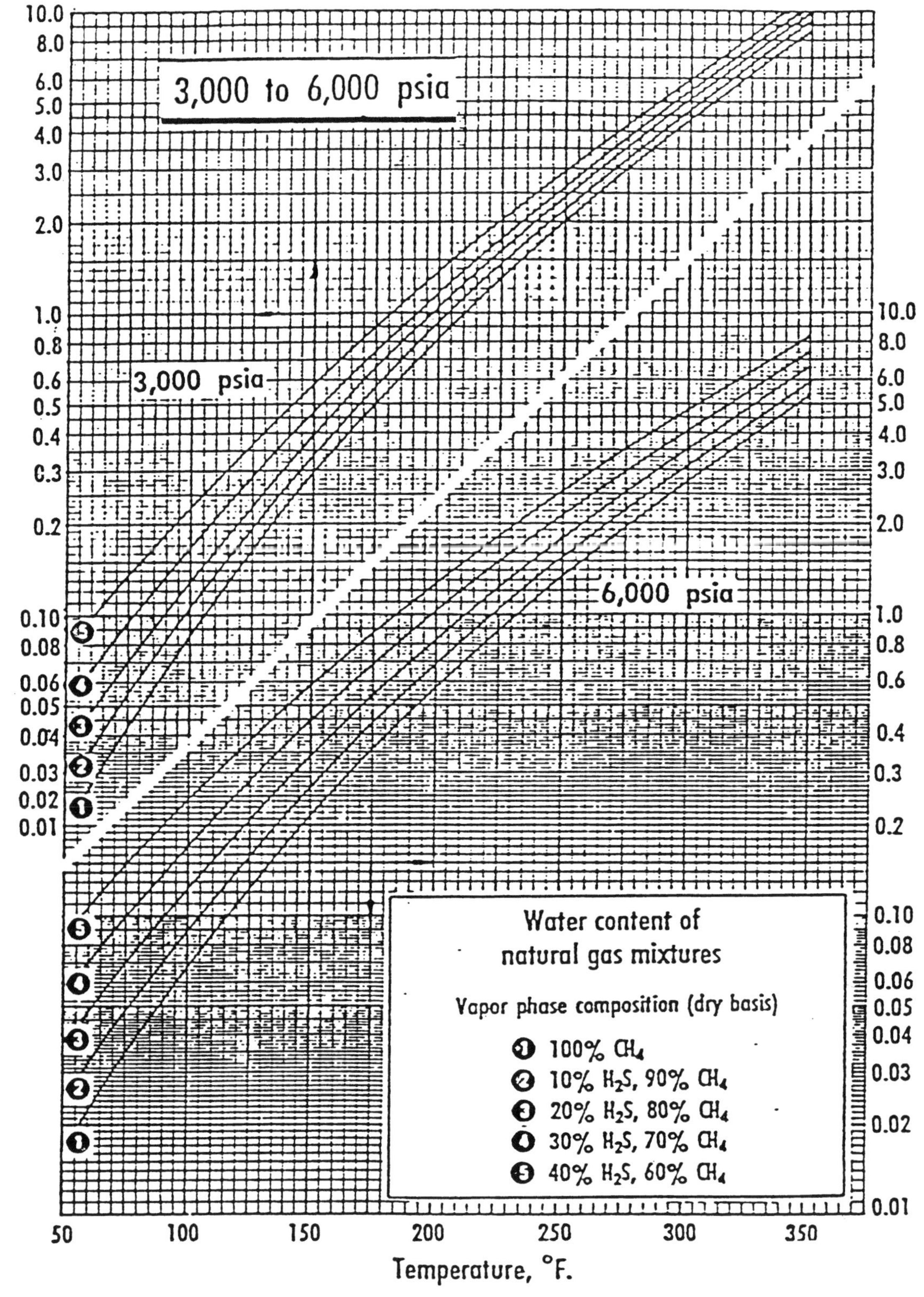

Figure 4–7b.   Water content of sour gases (Robinson *et al.*, 1978).

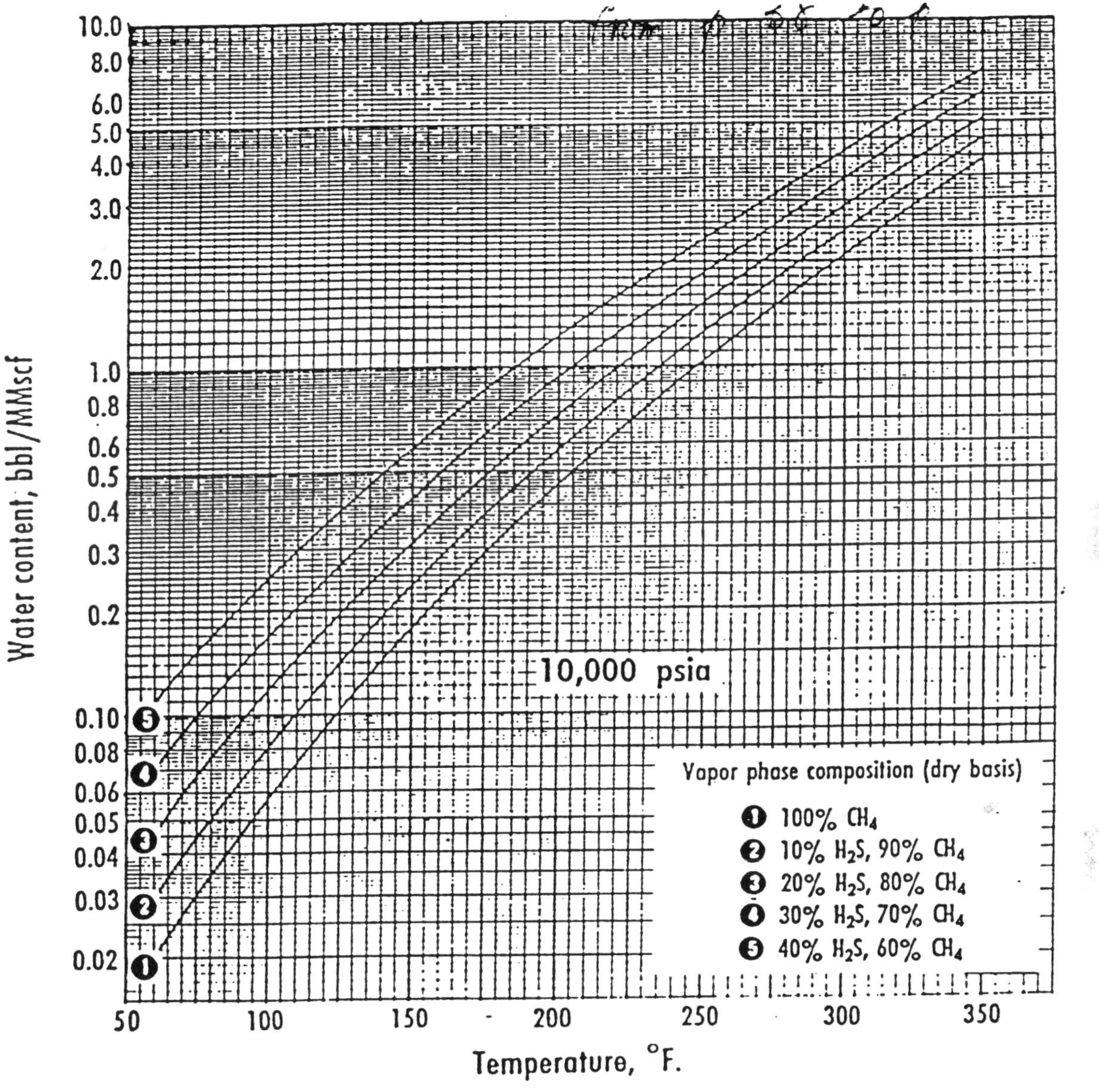

Figure 4–7c. Water content of sour gases (Robinson *et al.*, 1978).

present charts of water content (bbl/MMscf) as ordinate and temperature (°F) as abscissa for six pressures (300, 1000, 2000, 3000, 6000, and 10,000 psia) and five natural gas mixtures containing 0, 10, 20, 30, and 40% H_2S (Fig. 4–7a,b,c). Pressure interpolation should be logarithmic. This is an easy approach and covers a wide range of pressures (300–10,000 psia) and temperatures (50–375°F). The Robinson *et al.* charts are based on computer calculations using the SRK equation of state, and so are less empirical than the Campbell method.

Campbell Charts. The Campbell method (1976), uses the following empirical combination rule:

$$W = y_{hc}\, W_{hc} + y_{CO_2}\, W_{CO_2} + y_{H_2S}\, W_{H_2S} \qquad (4\text{--}1)$$

where W = water content of sour gas (lb H_2O/MMscf)
$\quad W_{hc}$ = water content of sweet gas (lb H_2O/MMscf) (can use McKetta-Wehe, Fig. 4–6)
$\quad y_{hc}$ = mol fraction of all components except H_2S and CO_2

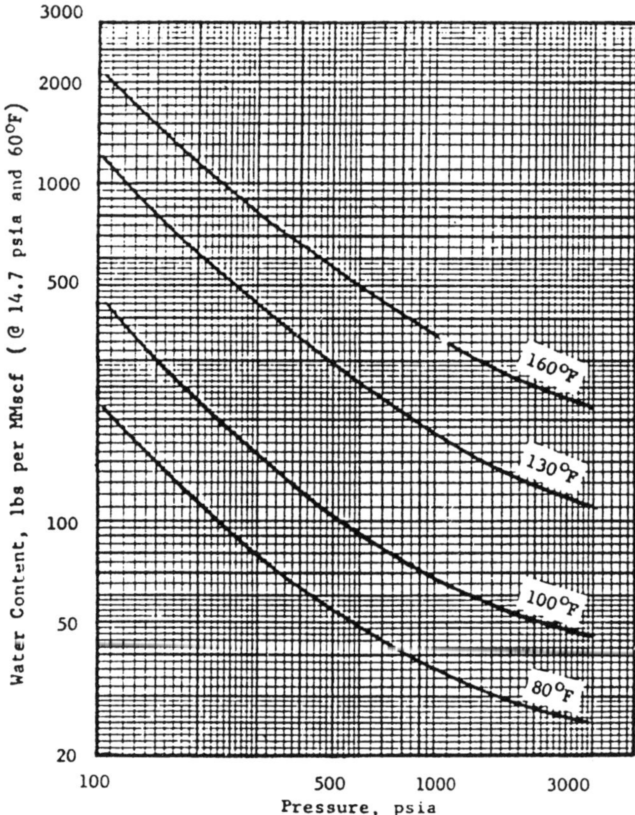

Figure 4–8. Effective water content of saturated CO_2 in natural gas mixtures (Campbell, 1976).

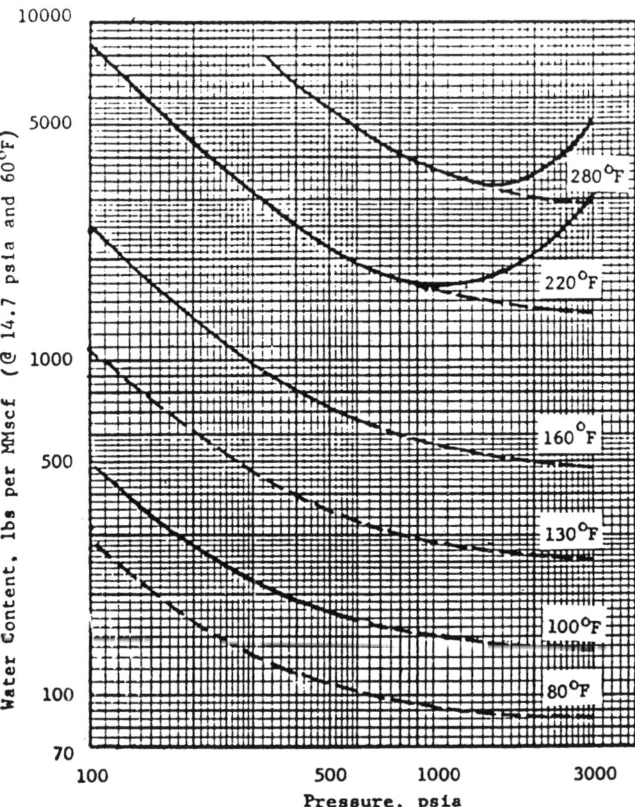

Figure 4–9. Effective water content of saturated H_2S in natural gas mixtures (Campbell, 1976).

W_{CO_2} = effective water content of CO_2 from Figure 4–8 (lb H_2O/MMscf)

y_{CO_2} = mol fraction of CO_2 in natural gas

W_{H_2S} = effective water content of H_2S from Figure 4–9 (lb H_2O/MMscf)

y_{H_2S} = mol fraction of H_2S in natural gas.

Equation-of-State Methods. The previous hand-calculation methods are unable to predict the distribution of hydrocarbons between the hydrocarbon liquid, aqueous liquid, and gas phases—or the water content of the hydrocarbon-liquid phase.

The water content of hydrocarbon-rich or inorganic-rich gas in equilibrium with a water-rich liquid can be made with the de Santis-Breedveld-Prausnitz, Nakamura-Breedveld-Prausnitz, or Peng-Robinson equations of state (Adler, Ozkardesh, and Spencer, 1980). This approach assumes that the water-rich liquid phase is 99 mol percent water. When acid gases, methanol, or glycols are present, the water-rich liquid can contain much less than 99 mol percent water.

A rigorous three-phase (gas or vapor; water-rich liquid; hydrocarbon-rich liquid) computer-based flash calculation can predict the distribution of all the components in all three phases. The Soave-Redlich-Kwong (SRK)-GPA*SIM, Peng-Robinson (PR)-Equi-Phase, and Parameters From Group Contributions (PFGC-MES)-Aqua*Sim methods have been used for such rigorous three-phase calculations (Maddox and Erbar, 1982). Acid gases, C6+ fractions, olefins, hydrogen, aromatics, and cycloparaffins are included. The water content in the hydrocarbon vapor and hydrocarbon-rich liquid phases and the hydrocarbon content of the water-rich liquid phase are predicted. The method was tested over the range 14.7–5000 psia and 70–600°F (Maddox and Erbar, 1982). The GPA*SIM, Equi-Phase, and Aqua*SIM computer programs are described by GPA (1987, p. 1–21), Nolte *et al.* (1985), and Wagner *et al.* (1985), respectively.

Maddox and Erbar (1982) and Nolte, Robinson, and Ng (1985) presented detailed comparisons between experimental and predicted water contents. In general, the agreement is excellent—usually within 5%.

Comparison

Methods that include the gas composition are better than those that do not, however simple and easy-to-use the latter may be. The equation-of-state approach is inherently superior since it incorporates partial miscibility (or distribution) of all components in all three phases. However, below 300 psia the sweet gas charts can be used. Above 300 psia the presence of H_2S and/or CO_2 can increase the saturated water content of a natural gas. The influence of the acid gases increases with increasing pressure, as is shown clearly in Figure 4–7 (Robinson *et al.*,1978).

Example 4–1. Estimate the water content of a sour natural gas (65% HC, 20% H_2S, 15% CO_2) at 100°F and 1500 psia. Use:
1. McKetta-Wehe plot (Fig. 4–6)
2. Robinson *et al.* charts (Fig. 4–7)
3. Campbell correlation (Figs. 4–6, 4–8, 4–9).

Check the results using the Peng-Robinson method by means of the HYSIM program (Hyprotech, 1988).

Solution: McKetta-Wehe Method (Sweet Gas)

$$W = 46 \text{ lb } H_2O/MMscf \text{ (Fig. 4–6)} \leftarrow$$

A composition is required to calculate the sweet-gas water by the HYSIM program (Hyprotech, 1988). A composition of 75% C1, 20% C2, and 5% C3 yields a specific gravity of 0.70. For this composition at 100°F and 1500 psia, the water content by HYSIM is:

$$W = 44.2 \text{ lb } H_2O/MMscf \leftarrow$$

This is a reasonable check.

Robinson *et al.* Method (Sour Gas)

Equivalent H_2S concentration
$$= y_{H_2S} + (0.75) \, y_{CO_2}$$
$$= 0.20 + (0.75)(0.15)$$
$$= .312$$

Use mixture 4 in Figure 4–7a, at 100°F:

$$W = 0.168 \text{ bbl } H_2O/MMscf \text{ at } 2000 \text{ psia}$$
$$W = 0.205 \text{ bbl } H_2O/MMscf \text{ at } 1000 \text{ psia}$$

Use logarithmic interpolation:

$$\frac{\ln W_{1500} - \ln W_{1000}}{\ln W_{2000} - \ln W_{1000}}$$

$$= \frac{\ln 1500 - \ln 1000}{\ln 2000 - \ln 1000}$$

$$\frac{\ln W_{1500} - \ln (.205)}{\ln (.168) - \ln (.205)}$$

$$= \frac{7.3132 - 6.9078}{7.6009 - 6.9078}$$

$$\frac{\ln W_{1500} - (-1.5847)}{(-1.7838) - (-1.5847)} = 0.585$$

$$\ln W_{1500} = -1.5847 + (.585)(-.1991)$$
$$= -1.7012$$
$$W_{1500} = .182 \text{ bbl } H_2O/MMscf$$
$$= 64 \text{ lb } H_2O/MMscf \leftarrow$$

Campbell Method (Sour Gas)

Given $y_{hc} = .65$; $y_{CO_2} = .15$; $y_{H_2S} = .20$

At 100°F and 1500 psia

$$W_{hc} = 46 \text{ lb } H_2O/MMscf \text{ (Fig. 4–6)}$$
$$W_{CO_2} = 56 \text{ lb } H_2O/MMscf \text{ (Fig. 4–8)}$$
$$W_{H_2S} = 142 \text{ lb } H_2O/MMscf \text{ (Fig. 4–9)}$$
But from eq. 4–1, W
$$= y_{hc} W_{hc} + y_{CO_2} W_{CO_2} + y_{H_2S} W_{H_2S}$$
$$= (.65)(46) + (.15)(56) + (.20)(142)$$
$$= 66.7 \text{ lb } H_2O/MMscf. \leftarrow$$
HYSIM (Hyprotech, 1988) gives 69.2 lbH_2O/MMscf for a 65% C1, 20% H_2S, and 15% CO_2 gas. Changing the sweet-gas components to 49% C1, 13% C2, and 3% C3 while maintaining 20% H_2S and 15% CO_2 increases the HYSIM water content to 74.3 lb H_2O/MMscf.

GAS HYDRATES

Natural gas molecules smaller than n-butane can react with liquid or free water to form crystalline, snow-like solid solutions called *hydrates*. Hydrates have specific gravities ranging from 0.96 to 0.98 and therefore float on water and sink in liquid hydrocarbons. They are 90 weight percent water; the other 10 weight percent is composed of one or more of the following compounds: methane, ethane, propane, iso-butane, n-butane, nitrogen, carbon dioxide, and hydrogen sulfide. Alone, n-butane does not form hydrates, but can contribute in a mixture (GPSA, 1987, p. 20–8). In solid hydrates the water or "host" molecules are linked together by hydrogen bonding into cage-like structures, called *clathrates*, that are stabilized by the inclusion of the natural-gas "guest" molecules. Smaller natural-gas molecules (C1, C2, H_2S, CO_2) form more-stable, body-centered cubic structures, while small quantities of larger molecules (C3, iC4) usually produce less-stable diamond lattices.

Hydrate formation causes many operating problems, such as partial or complete blocking of gas-gathering flowlines, fouling and plugging of heat exchangers, erosion of expanders, etc.

The thermodynamic conditions promoting hydrate formation are (GPSA, 1987, p. 20–6):

1. Presence of "free" or liquid water.
2. Low temperature.
3. High pressure.

Hydrate formation is accelerated by agitation (such as high velocities or other turbulence), pressure pulsations, "seed" hydrate crystals, and a suitable site for crystal formation, such as pipe elbows, orifice plates, thermowells, scale, and solid corrosion products.

Prediction of Hydrate Formation Conditions

The temperature and pressure at which hydrates form may be estimated by Katz's gas-gravity method, Katz's equilibrium constant method, Baillie and Wichert's chart for sour gases, and by equation-of-state approaches that require computer programs. These four methods are now summarized and compared. Campbell's empirical method is available elsewhere (Campbell, 1976).

Gas-Gravity Method

In Katz's (1945) gas-gravity method the temperature and/or pressure at which hydrates form is read directly from one graph (Fig. 4–10). The natural gas is characterized by a single parameter—the specific gravity (or average molecular weight). While this method is very simple, description of a natural gas by a single parameter is approximate. Loh, Maddox, and Erbar (1983) compared Katz's (1945) method with predictions using the Soave-Redlich-Kwong (SRK) equation of state. Excellent agreement was found for methane and natural gases with SG = 0.7 or less. Significant differences were observed for natural gases with SG between 0.9 and 1.0.

Equilibrium-Constant Method

In the equilibrium-constant or K-value method, hydrate formation conditions are computed in a manner directly analogous to the standard vapor-liquid dew-point calculations. Vapor-hydrate K-values were developed by Wilson, Carson, and Katz (1941) and Carson and Katz (1942) for the following hydrate-forming gases: CH_4, C_2H_6, C_3H_8, H_2S, and CO_2. Robinson and Ng (1975) presented

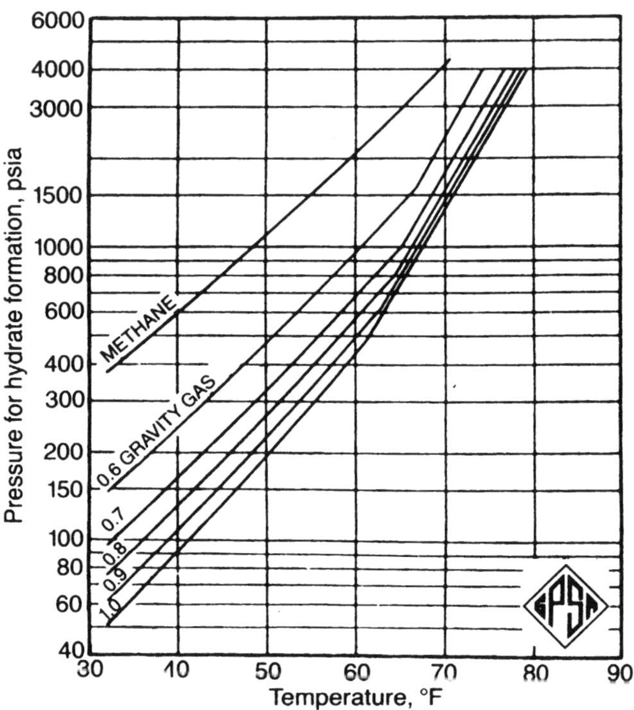

CAUTION: Figures 20-13 through 20-18 should only be used for first approximations of hydrate formation conditions. For more accurate determination of hydrate conditions make calculations with K_{v-s}.

Figure 4–10. Pressure-temperature curves for predicting hydrate formation (Katz, 1945; GPSA, 1987, p. 20–8).

a vapor-hydrate K-chart for i-C_4H_{10}, while Poettmann (1984) prepared a chart for n-C_4H_{10}. These vapor-hydrate K-values may be read from Figure 4–11 through 4–17.

Baillie and Wichert Method

Baillie and Wichert (1987) developed a graph (Fig. 4–18) using a large number of hydrate formation conditions generated by the HYSIM process simulator program (Hyprotech, 1988). Baillie and Wichert compared their graph (Fig. 4–18) with HYSIM predictions over the following ranges: total acid gases 1–70%; H_2S content 1–50%; H_2S/CO_2 ratios from 1:3 to 10:1. The graph agreed with the computer within ± 2°F for 75% of the comparisons and within ± 3°F for 90% provided the propane correction was made.

Figure 4–18 also can be used for sweet, conventional natural gases with C3 contents up to 10%. Baillie and Wichert compared their graph with the charts prepared by Loh, *et al.* (1983) and found agreement within ± 3°F.

Equation-of-State Methods

These methods are based on the fundamental equations of phase equilibrium—namely that when hydrates form the

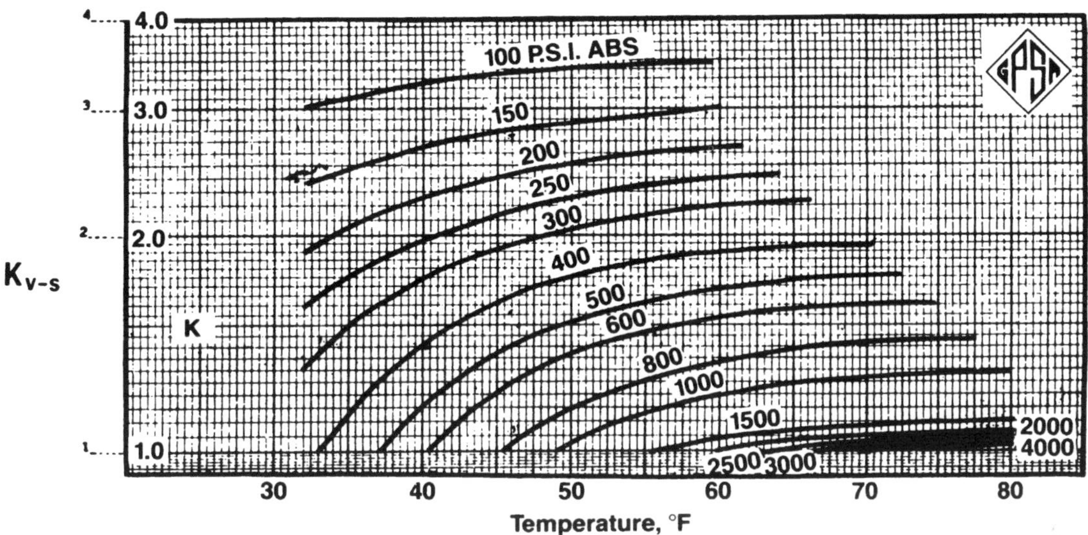

Figure 4–11.   Vapor-solid hydrate equilibrium constants for methane (Carson and Katz, 1942; GPSA, 1987, p. 20–11).

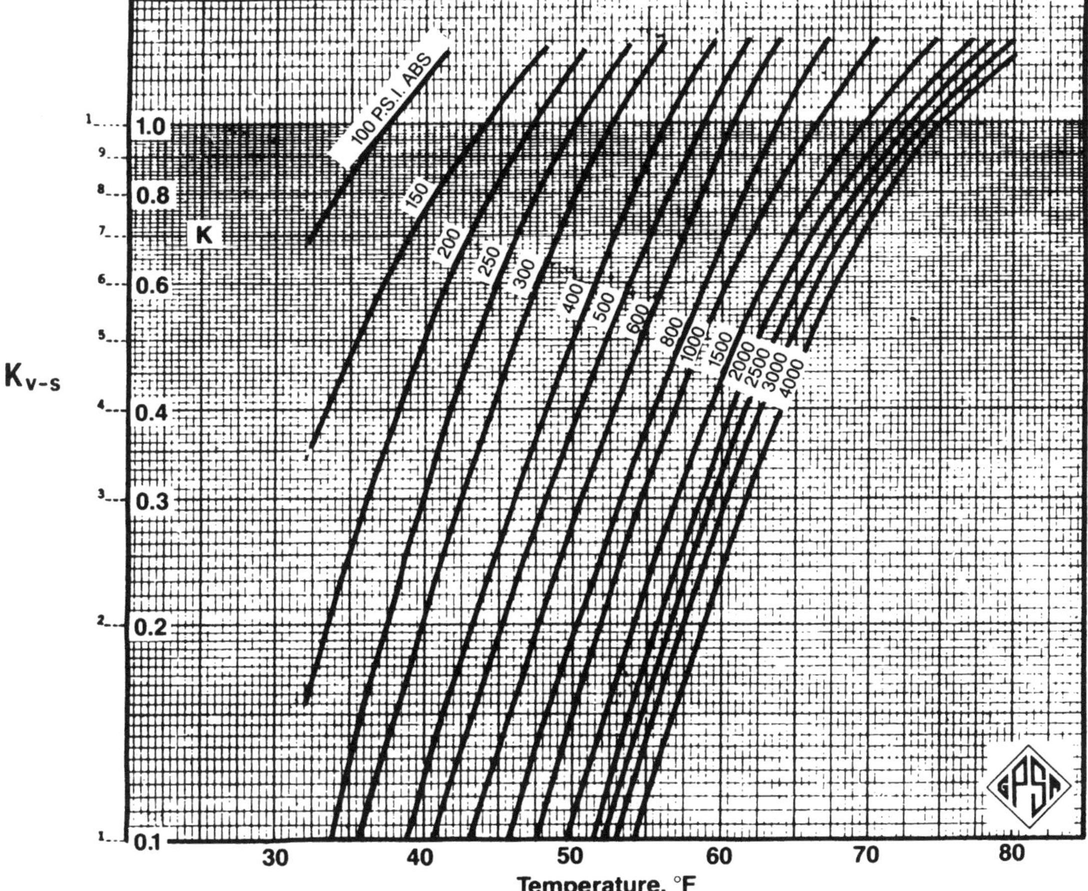

Figure 4–12.   Vapor-solid hydrate equilibrium constants for ethane (Carson and Katz, 1942; GPSA, 1987, p. 20–11).

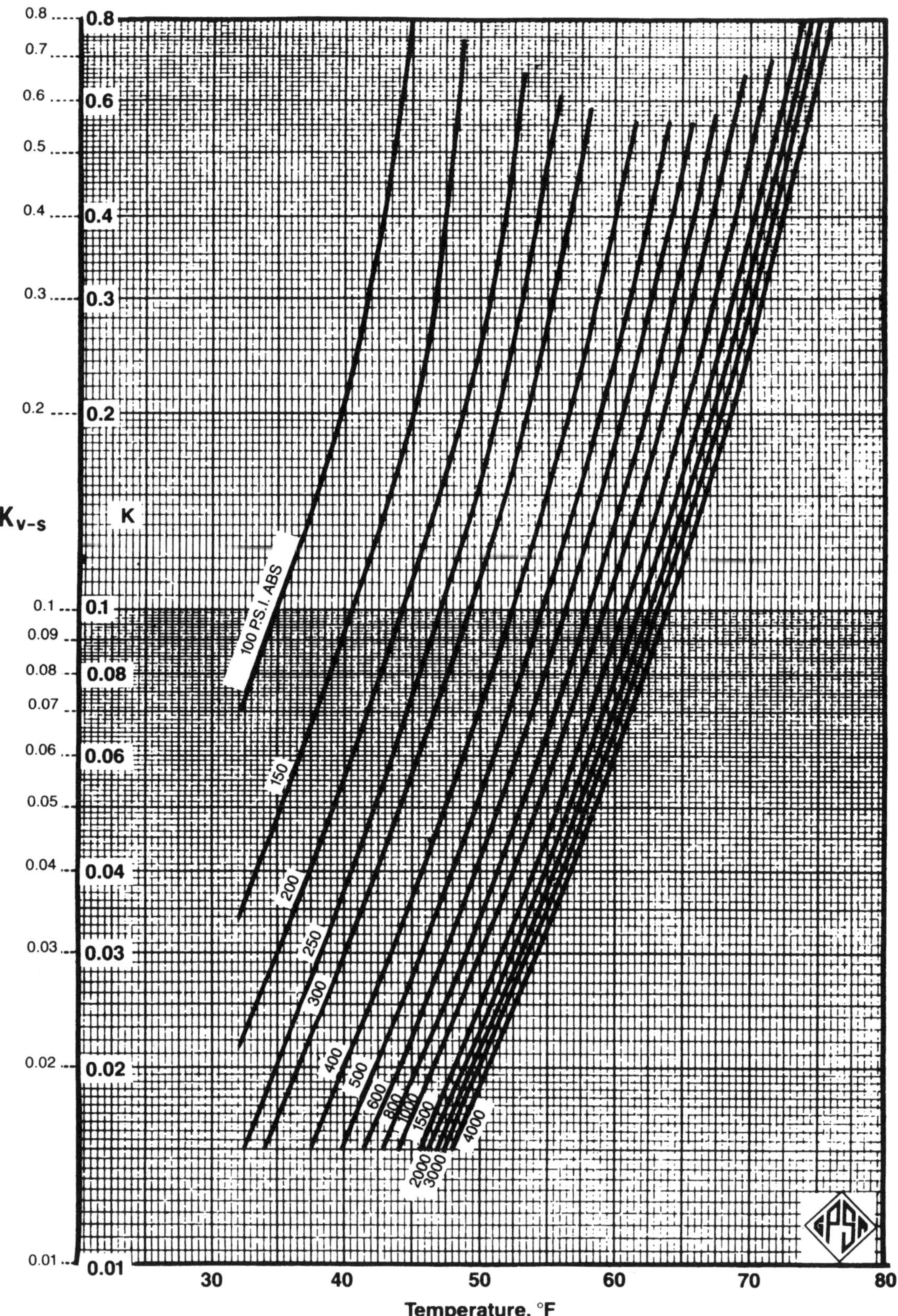

Figure 4–13. Vapor-solid hydrate equilibrium constants for propane (Carson and Katz, 1942; GPSA, 1987, p. 20–12).

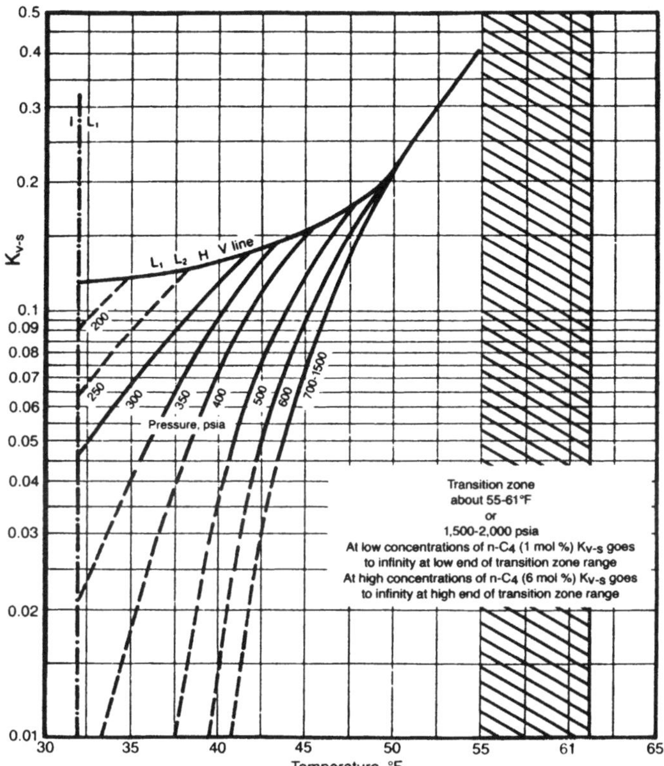

Figure 4–14. Vapor-solid hydrate equilibrium constants for n-Butane (Poettmann, 1984; GPSA, 1987, p. 20–14).

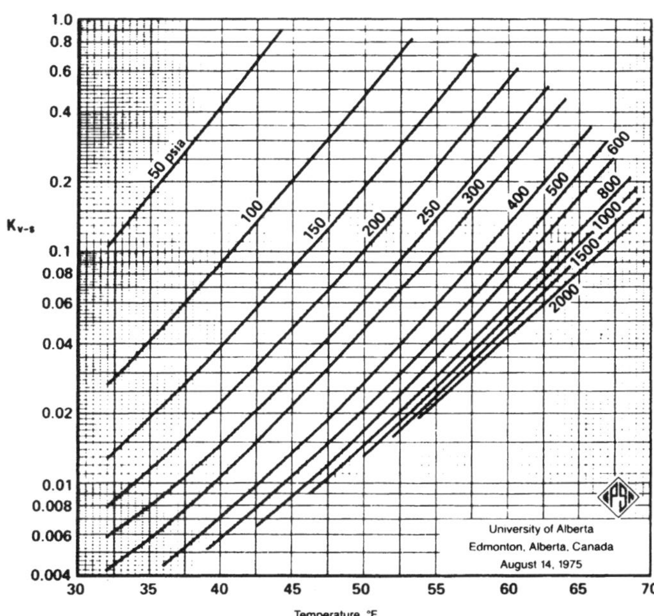

Figure 4–15. Vapor-solid hydrate equilibrium constants for isobutane (Robinson and Ng, 1975; GPSA, 1987, p. 20–13).

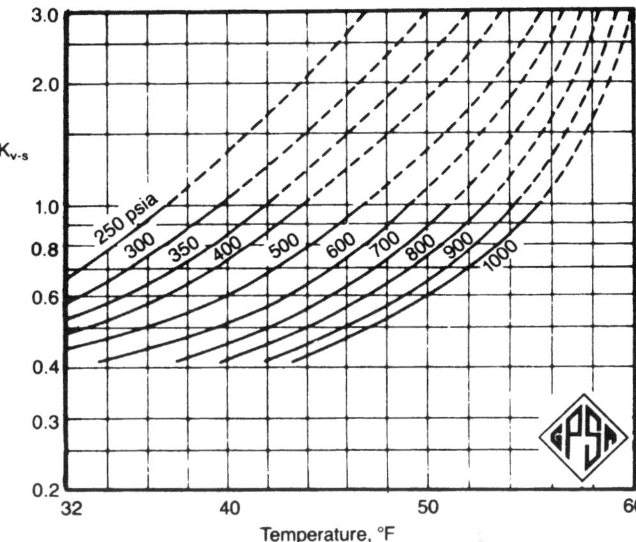

Figure 4–16. Vapor-solid equilibrium constants for carbon dioxide (Unruh and Katz, 1949).

temperature, pressure, and chemical potential of H_2O have the same values in the natural-gas, liquid-water, and solid-hydrate phases. The chemical potential of water in the solid hydrate phase is computed using the Platteeuw and van der Waals method as developed by Saito and Kobayashi and Nagata and Kobayashi, and extended to multicomponent gases (Parrish and Prausnitz, 1972; Robinson *et al.*, 1987).

The Colorado School of Mines (CSM) hydrate program uses the Soave-Redlich-Kwong equation of state to calculate the chemical potential of water in the gas phase (Sloan, 1985). Poettmann *et al.* (1989) developed equilibrium-

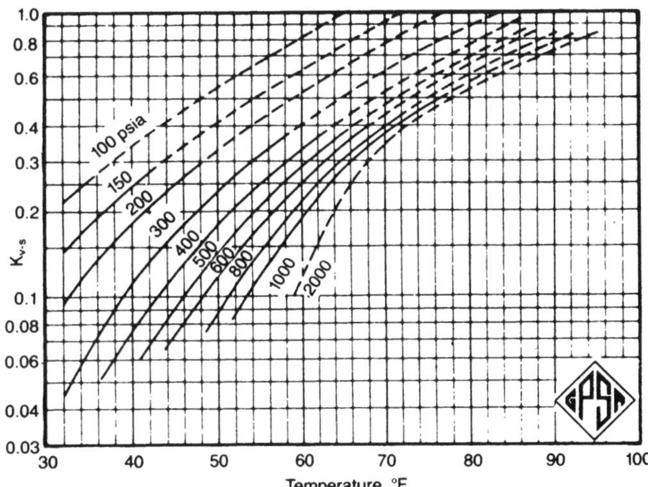

Figure 4–17. Vapor-solid equilibrium constants for hydrogen sulfide (Noaker and Katz, 1954; GPSA, 1987).

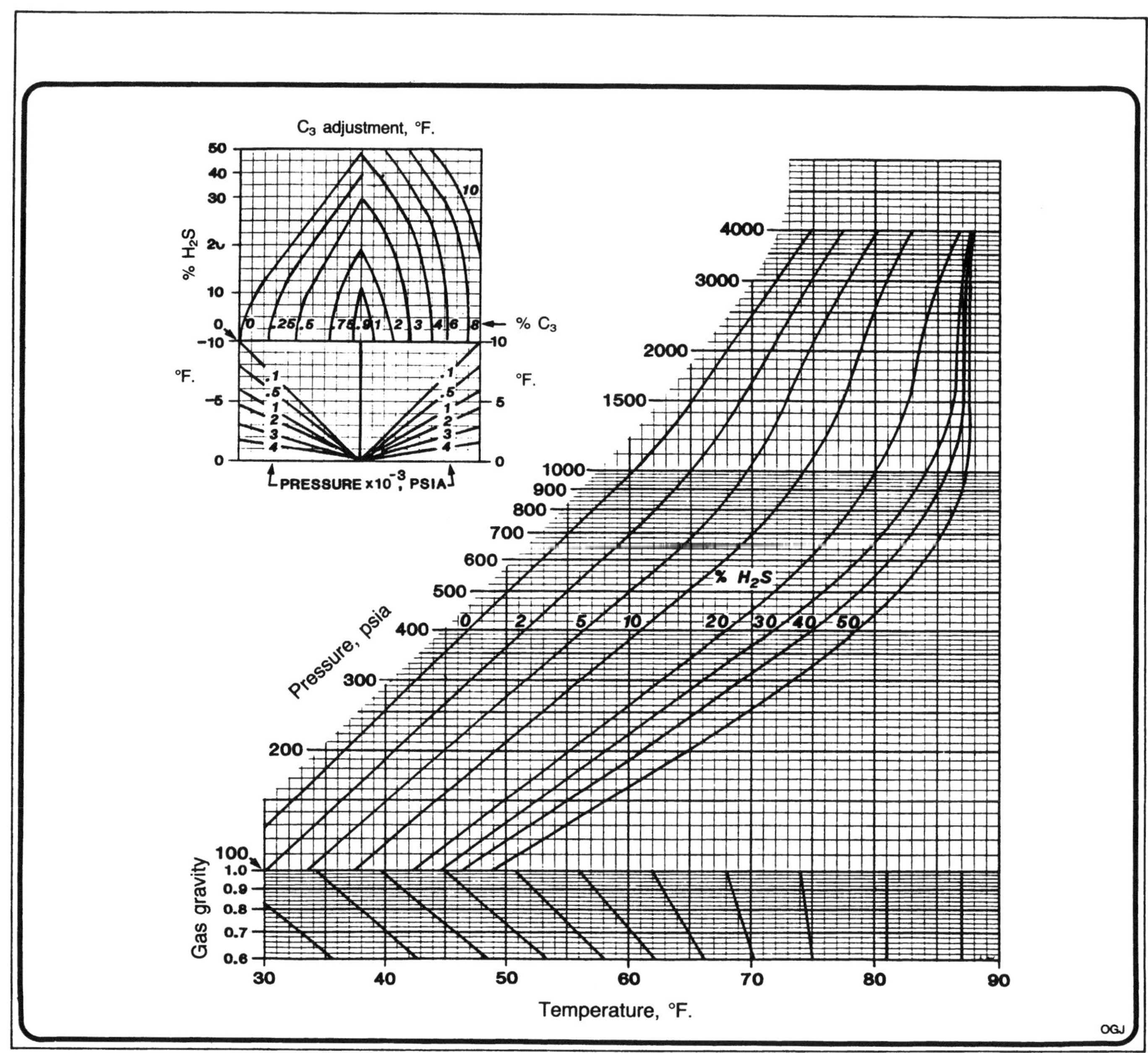

Figure 4–18.   Hydrate chart for sour natural gas (Baillie and Wichert, 1987).

vaporization (K_{V-S}) charts using the CSM hydrate program. These charts, which are an alternative to the Katz *et al.* charts, cover a wide range of pressure, temperature, and composition.

Ng and Robinson (1976) modified the Parrish and Prausnitz algorithm and their hydrate program Equi-Phase (Nolte *et al.*, 1985), which uses the Peng-Robinson equation of state, and it is available from the Gas Processors Suppliers Association (GPSA). The Aqua*SIM computer program (Wagner *et al.*, 1985) uses the PFGC equation of state.

Comparison

Sloan (1984) assesses the gas-gravity, equilibrium-constant, and equation-of-state methods as follows:

1. The gas-gravity method should be considered a first-order estimate.
2. The Katz *et al.* equilibrium-constant method represents a considerable improvement over the gas-gravity method.
3. The equation-of-state (or computer) methods are superior both in accuracy and ease of extrapolation.

Maddox and Erbar (1982, p. 14–34) report that both the Peng-Robinson and Parrish and Prausnitz (1972) computer methods appear to predict hydrate formation temperatures within 2°F.

Wilson *et al.* (1985) estimated incipient hydrate formation temperatures and pressures by four methods:

1. GPA equilibrium-constant or K-value method
2. Aqua*Sim (Wagner *et al.*, 1985)
3. Sloan (1985)
4. Robinson *et al.* (1987)

Figures 4–19 and 4–20 show their results for a lean gas and a rich gas, respectively. All four methods agree within 3°C (5°F).

Example 4–2. Estimate the hydrate formation temperature at 1000 psia for the following natural gas:

	Mol %	
N2	10.1	
C1	77.7	
C2	6.1	
C3	3.5	
iC4	0.7	
nC4	1.1	
C5+	0.8	(assume C6)
	100.0	

Use the gas-gravity, the equlibrium-constant, and the Baillie-Wickert methods. Check with the Peng-Robinson equation of state in the HYSIM simulator (Hyprotech, 1988).

Gas-Gravity Method

Compute the SG of the gas.

Component	z	MW	z.MW
N2	0.101	28	2.83
C1	0.777	16	12.43
C2	0.061	30	1.83
C3	0.035	44	1.54
iC4	0.007	58	0.41
nC4	0.011	58	0.64
C5+	0.008	86	0.69
			20.38

$$SG = 20.38/28.9625 = 0.70$$

Read hydrate formation temperature from Figure 4–10.

Hydrate formation temperature
$$= 65°F. \longleftarrow$$

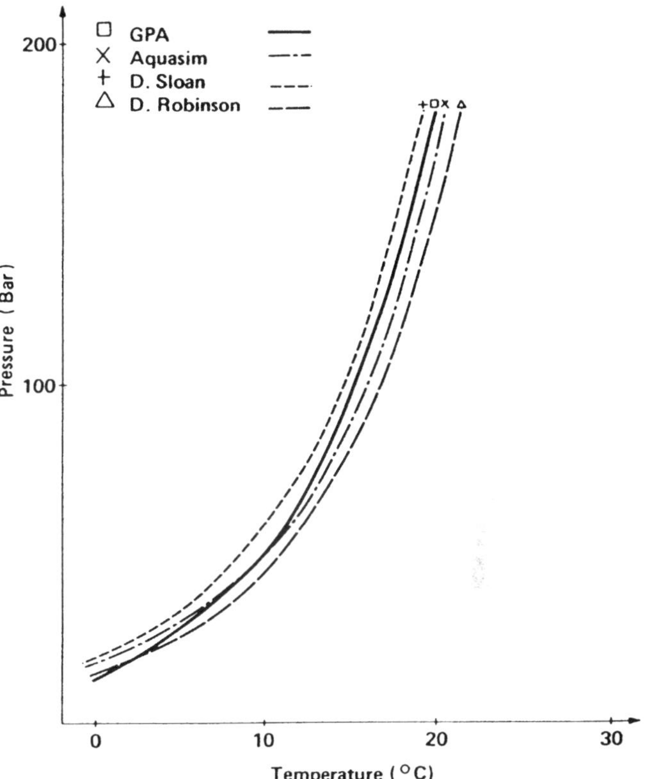

Figure 4–19. Comparison of hydrate-prediction methods (Wilson *et al.*, 1985).

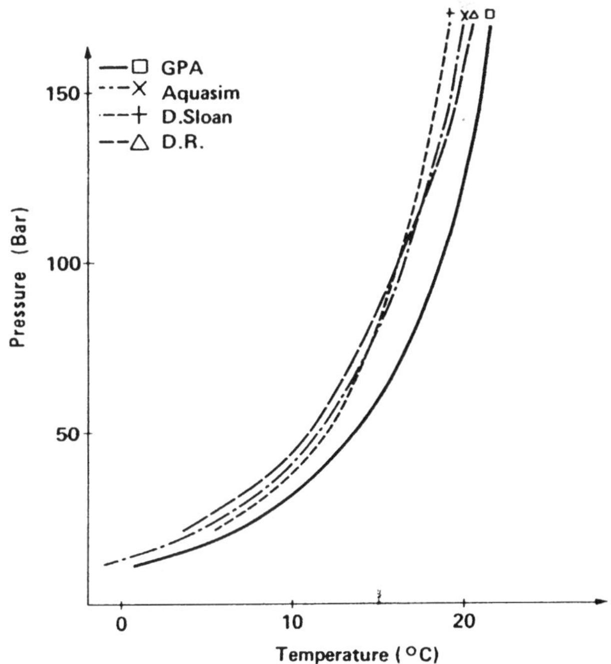

Figure 4–20. Comparison of hydrate-prediction methods (Wilson *et al.*, 1985).

Katz *et al.* Equilibrium-Constant Method

The method is as follows:

Step 1. Estimate a hydrate formation temperature.

Step 2. Look up $K_{V\text{-}S}$ from Figure 4–11 through 4–15.

Step 3. Compute $\Sigma\, z_i/K_{V\text{-}Si}$.
Repeat steps 1–3 until $\Sigma\, z_i/K_{V\text{-}Si} = 1$.
The calculations are summarized below.
Use result of first calculation for first trial.

Comp.	z	$K_{v-s}^{65°F}$	z/K_{v-s}	$K_{v-s}^{63°F}$	z/K_{v-s}
N2	0.101	∞	0.0	∞	0.0
C1	0.777	1.24	0.627	1.22	0.636
C2	0.061	0.90	0.068	0.77	0.079
C3	0.035	0.26	0.135	0.18	0.194
iC4	0.007	0.108	0.065	0.08	0.087
nC4	0.011	∞	0.0	∞	0.0
C5+	0.008	∞	0.0	∞	0.0
Sum			0.895		0.996
			Low		Accept

Hydrate formation temperature
$$= 63°F. \leftarrow$$

Note: $K_{V-S} = \infty$ for non-hydrate formers and for n-butane above 55°F.

Baillie-Wichert Method

Step 1. Compute gas gravity. As in the gas-gravity method, SG = 0.70

Step 2. Enter Figure 4–18 at 1000 psia; move horizontally to 0% H_2S; drop vertically to the horizontal SG = 0.7 line; follow sloping lines to the horizontal bottom temperature scale; and read 62°F.

Step 3. Add correction: interpolate for 3.5% C3; enter at 0% H_2S; move horizontally to 3.5% C3; drop vertically to 1×10^{-3} line and read horizontally a correction of +3 °F.

Step 4. Add correction.

Hydrate formation temp
$$= 62 + 3 + 65°F \leftarrow$$

Peng-Robinson Equation of State

The HYSIM program (Hyprotech, 1988) gives a hydrate formation temperature of 64.1°F $\leftarrow$

OVERALL PHASE BEHAVIOR

Maddox and Erbar (1982) summarized water-hydrocarbon phase behavior in phase diagrams similar to those sketched in Figure 4–21. These diagrams depict water-dew-point and hydrate curves superimposed on the water-free hydrocarbon gas phase diagram. This technique only approximates the true phase behavior, because the presence of water changes the hydrocarbon phase diagram, and *vice versa*. However, Maddox and Erbar state that rigorous three-phase calculations indicate that the hydrocarbon dew-point line and the water dew-point line are not affected greatly.

Figure 4–21a represents the usual behavior of a water-hydrocarbon mixture. There are five regions in which the various phases may occur, as shown. To the right of the

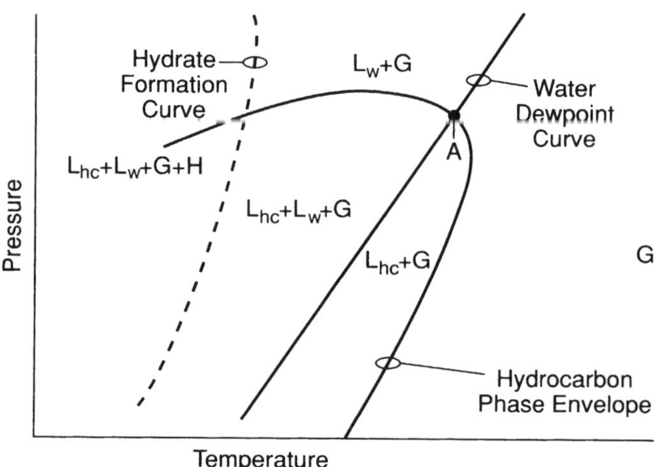

Figure 4–21a. Phase behavior of water-natural gas mixtures: usual case (after Maddox and Erbar, 1983).

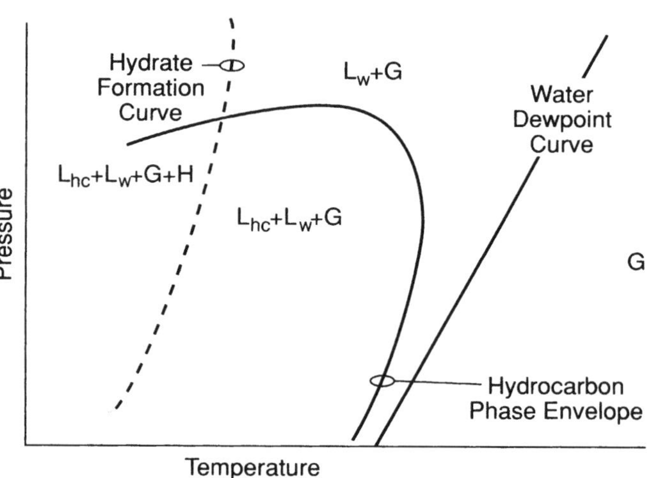

Figure 4–21b. Phase behavior of water-natural gas mixtures: high water content (after Maddox and Erbar, 1982).

dew-point curves, only the gas phase (G) is present. At pressures greater than that at point A, water-rich liquid condenses first (L_w + G) when the gas is cooled at constant pressure. The water dew point is said to control. At pressures below that at A, hydrocarbon-rich liquid condenses first (L_{hc} + G) when the gas is cooled. The hydrocarbon dew point controls. Further cooling yields water-rich liquid when the water dew-point curve is reached (L_{hc} + L_w + G). Additional cooling ultimately produces hydrate formation (L_{hc} + L_w + G + H).

Figure 4–21b shows the phase diagram for a mixture containing a relatively-large amount of water in the gas. In this case the water dew-point line controls at all pressures. The L_{hc} + G zone is not present.

The main usefulness of Figure 4–21 is in indicating the basic behavior of water and natural gas mixtures. If amounts and compositions of phases are desired, three-phase equilibrium calculations must be done on a computer. Maddox and Erbar (1982) note that the diagrams of Figure 4–21 apply only if significant amounts of acid gases are not present in the natural gas.

CARBON-DIOXIDE FROST POINT

Solid carbon dioxide can form when natural gases containing significant amounts of carbon dioxide are cooled below −70°F. Solid carbon dioxide can plug and/or damage cryogenic NGL recovery equipment, especially the expander and the demethanizer. Therefore, prevention of solid carbon dioxide formation is of great interest. Obviously, determination of the temperature and pressure conditions at which solid CO_2 forms, the *CO_2 frost point*, is important. Accordingly, the phase behavior of hydrocarbon/carbon-dioxide mixtures is now reviewed.

Because natural gas is usually predominantly methane, the CO_2-CH_4 (C1) binary is considered first. Figure 4–22 is a pressure-temperature (P-T) plot that shows the triple points for pure C1 (Point A) and for pure CO_2 (Point B). Also shown are the vapor pressure curves (two-phase, vapor-liquid equilibrium locus)—line AC_1 for pure C1 and line BC_2 for pure CO_2 . Note that the vapor-pressure lines start at the respective triple points and end at the critical points.

Dashed line C_1C_2 represents the locus of critical points for binary mixtures of C1 and CO_2. Finally, dotted line AB represents the L-V-S locus, where solid (pure CO_2), liquid (C1–CO_2 mixture), and vapor (C1–CO_2 binary) are in equilibrium.

In the figures that follow the same letter always represents the same point; *e.g.*, point A is always the C1 triple point.

Figure 4–23 shows bubble-point/dew-point phase envelopes (L-V equilibrium) for various compositions of

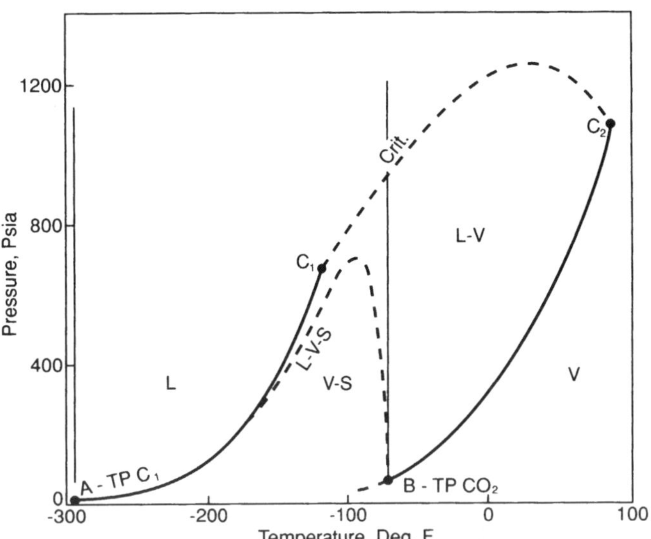

Figure 4–22.   P-T diagram for CO_2-C1 binary.

C1–CO_2 binaries. As expected, when the mol percent C1 in the C1–CO_2 binary increases, the L-V phase envelope moves closer to the C1 vapor-pressure curve. Note that the 29.5 mol percent C1 bubble-point and dew-point lines ''stop'' at the L-V-S locus (points J and I, respectively) and merge at the critical (point H).

Figure 4–24 shows the locus of the L-V-S equilibrium in great detail. Remember that the solid phase is essentially pure CO_2 while the L and V phases are binary mixtures. The hump-shaped L-V-S curve shows variation of pressure and temperature for the L-V-S equilibrium as the binary composition changes from 100% C1 (point A) to 100% CO_2 (point B).

The Gibbs Phase rule (see Chapter 3 states

$$f = C + 2 - \phi$$

Along the L-V-S locus there are three phases and two components, therefore f = 1. Fixing the composition of one phase (L or V) eliminates the one degree of freedom, and so specifies the state (P and T) completely.

Below the L-V-S locus pure solid CO_2 and gaseous C1–CO_2 exist. Figure 4–24 also shows how the solubility (mol percent) of CO_2 in the vapor phase varies with P and T; dashed line DE represents equilibrium between solid CO_2 and vapor containing 10 mol percent CO_2. Point D represents the P and T on the L-V-S locus where the vapor phase is 10 mol percent CO_2.

The area above the L-V-S locus represents solid CO_2 and liquid C1–CO_2 in equilibrium. Again, Figure 4–24 shows the CO_2 solubility in the liquid. Line FG is the equilibrium line for 90% C1–10% CO_2 liquid and solid CO_2.

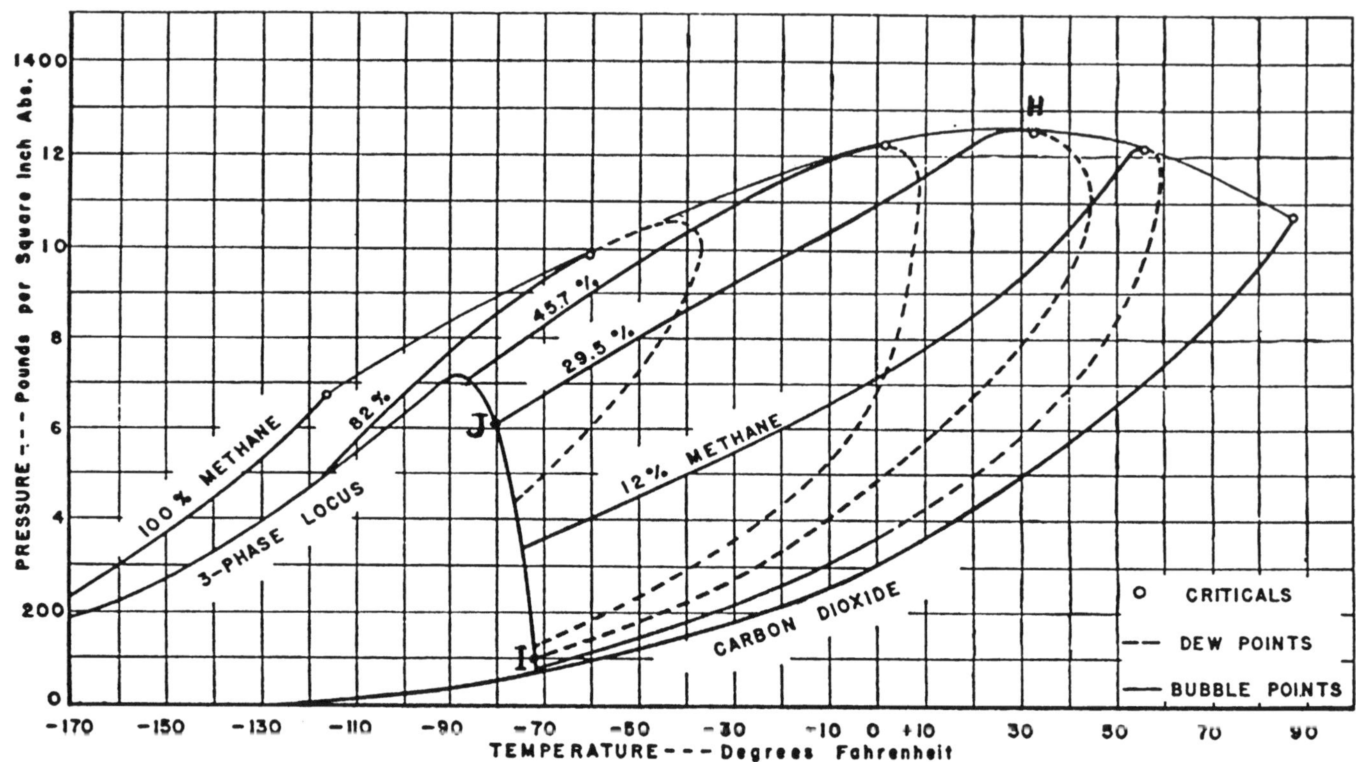

Figure 4–23. Pressure-temperature relations for carbon dioxide-methane system (Donnelly and Katz, 1954).

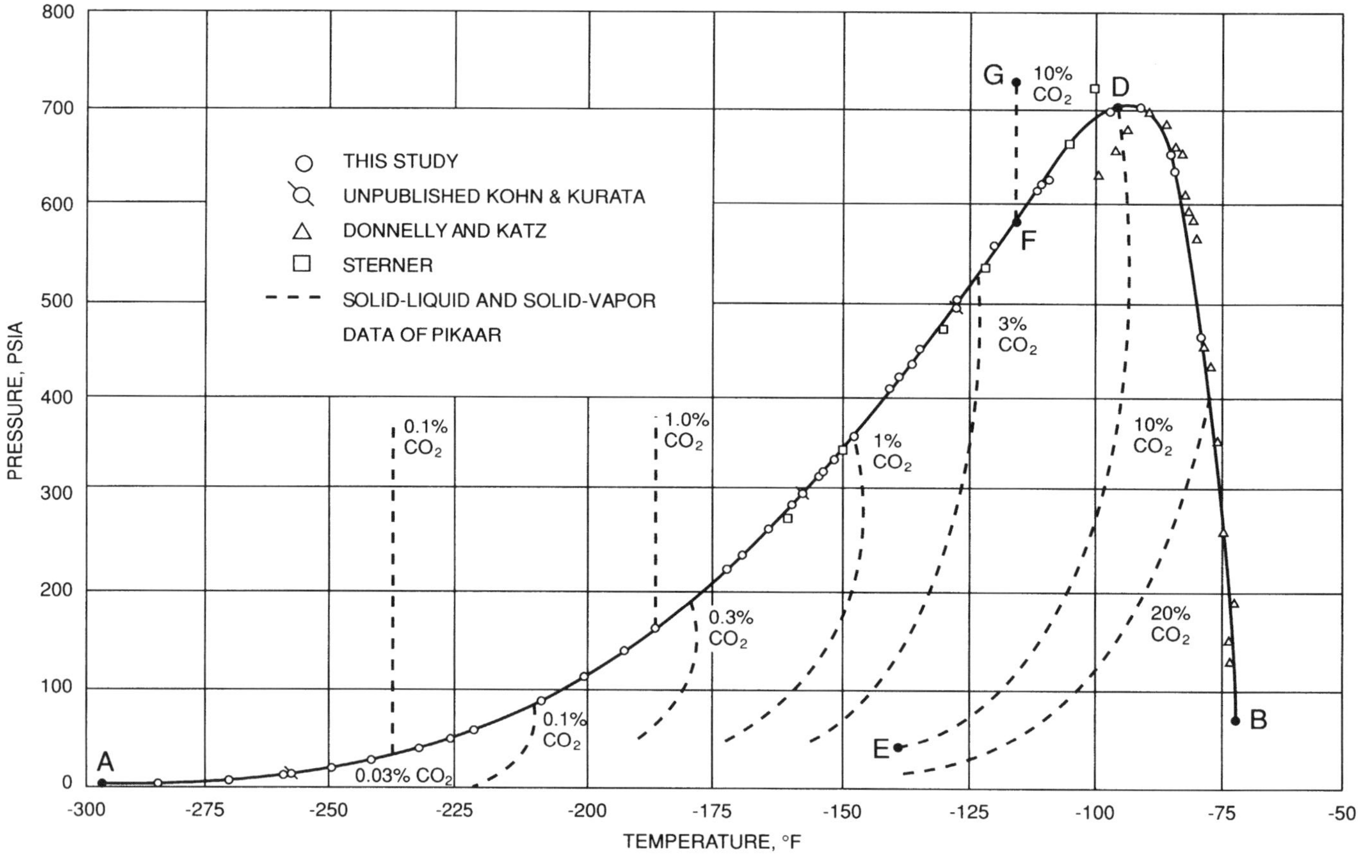

Figure 4–24. L-V-S locus for C1-CO₂ binary (Davis et al., 1962) (Reproduced by permission of the American Institute of Chemical Engineers).

In NGL recovery a key question may well be "if our natural gas contains 10 mol percent CO_2, how far can we cool it without CO_2 frost forming?" Figure 4–25 answers by showing the P-T diagram for a 90 mol percent C1, 10 mol percent CO_2 binary mixture.

The phase rule aids interpretation and understanding of Figure 4–25. First, let four phases coexist—solid CO_2, solid CH_4, liquid mixture, and vapor mixture. For two components (C = 2) and four phases (δ = 4) the phase rule shows zero degrees of freedom. Therefore, regardless of the initial CO_2 concentration (mol percent CO_2 cannot be 0 or 100% because C must be 2) there is only one pressure and one temperature at which four phases can coexist. This so-called *quaternary point*, Q, essentially coincides with the C1 triple point. Figure 4–24 shows that as the temperature approaches $-300°F$ the CO_2 solubility in both the L and V phases becomes very small. Therefore,

at Q the four phases are: pure solid CO_2, pure solid, liquid, and vapor C1.

Figure 4–25 also shows the L-V phase envelope, line FC_MD. As before, the bubble-point line "starts" at the L-V-S locus (point F) and ends at the critical (point CM). The dew-point line also "starts" at the L-V-S locus (point D) and ends at the critical (point C_M). The dew-point line also "starts" at the L-V-S locus (point D) and ends at the critical (point C_M).

As in Figure 4–24, line DE represents the solubility of CO_2 in the gas-phase while line FG represents CO_2 solubility in the liquid. Formation of solid CO_2 or CO_2 frost occurs on and to the left of these two lines (lower temperature). If the mixture is initially vapor at a pressure of about 700 psia or less and is cooled, the mixture will reach line DE and solid CO_2 will drop out of the vapor, as along line WX. If the pressure is higher than about 700 psia, cooling,

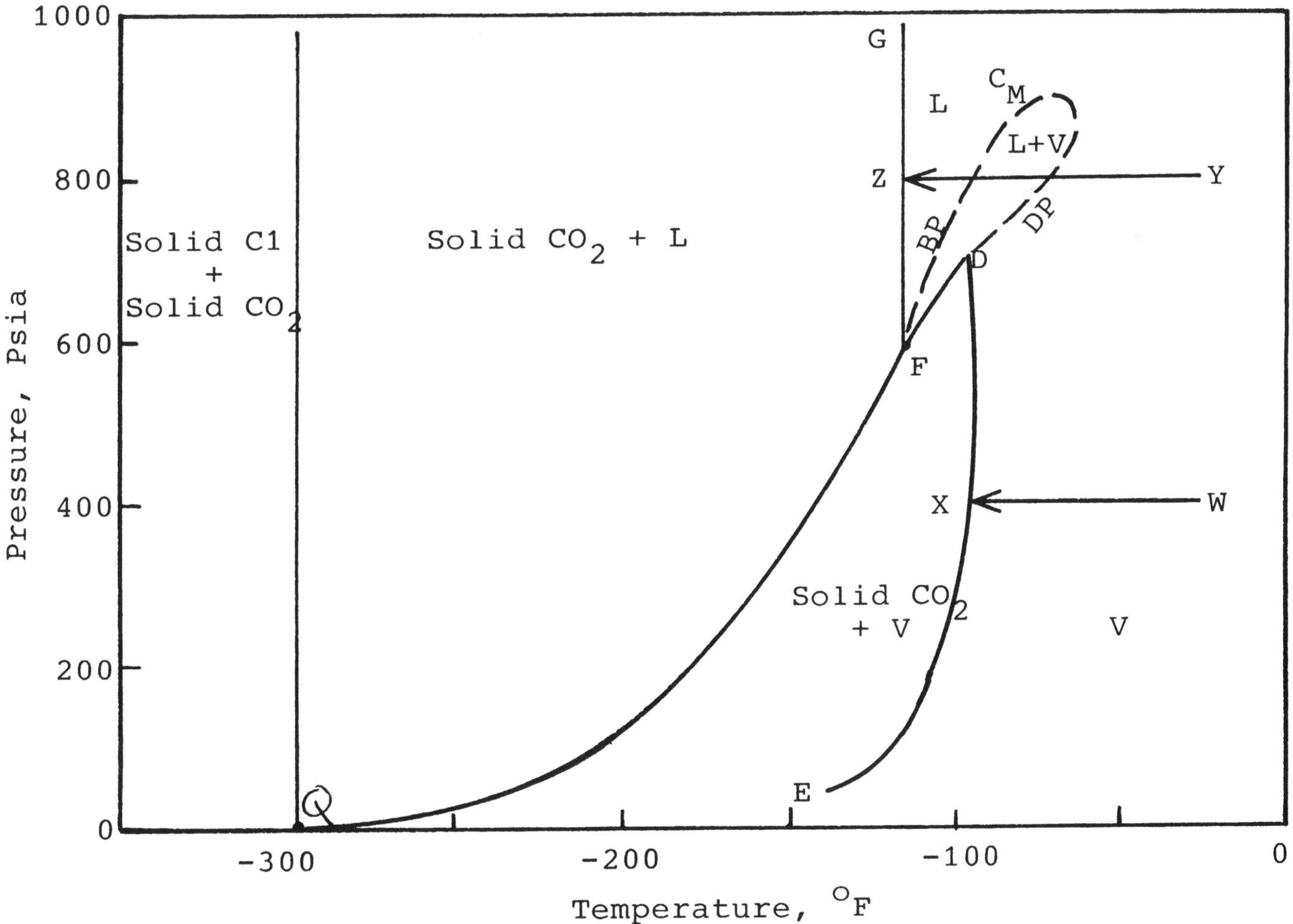

Figure 4–25. Phase diagram for 10 mol% CO_2-90 mol% C1 binary.

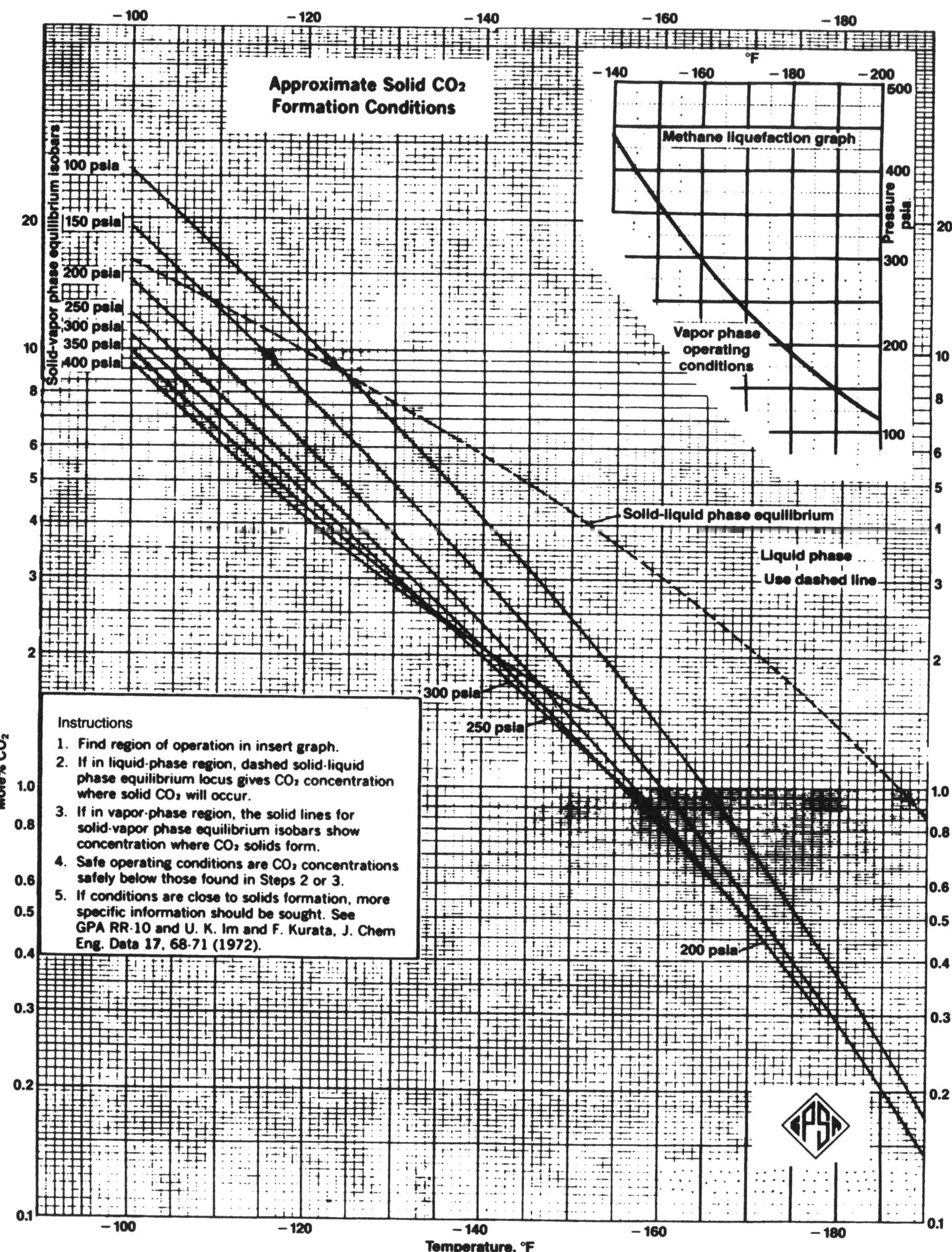

Figure 4–26. Approximate solid CO_2 formation conditions (GPSA 1987, p. 13–43).

as along line YZ, first causes condensation at the dew-point locus, DC_M. Further cooling will cause more liquid to form until the bubble-point locus, FC_M, is reached. Further cooling of the liquid will ultimately produce solid CO_2 formation when line FG is reached.

How does Figure 4–25, which represents the C1–CO_2 *binary*, relate to a *natural gas* containing CO_2? Specification of the complete composition of the natural gas fixes C–1 variables, so f = 3 − ϕ, the same as for the binary. Therefore, a water-free natural gas of known composition will exhibit a P-T diagram quite similar to Figure 4–25. Of course the additional components (N_2, C_2, C_3, etc.) will cause the frost point loci to shift. For lean natural gases the shift is not great. Cryogenic processing is not used for rich gas streams, and gases of intermediate richness will have already undergone chilling and condensate knock-out before cryogenic processing. A single correlation is adequate for hand calculations. For computer calculations, equation-of-state methods can be used as described earlier.

Figure 4–26 (GPSA, 1987, p. 13–43) provides an approximate answer to "when will solid CO_2 form?" for lean gases. Figure 4–26 is explained by relating it to Figures 4–24 and 4–25. Figure 4–25 shows that solid CO_2 will form from vapor as long as the pressure lies below the intersection of the S-V and dew-point lines. Line MN on the insert graph in Figure 4–26 relates to the L-V-S line of Figure 4–24, which represents the upper limit of S-V loci for mixtures with 10% CO_2 or less. Above these pressures, solid CO_2 forms from the liquid phase. Below these pressures solid CO_2 forms from the vapor. The insert graph in Figure 4–26 determines the controlling phase.

Figure 4–24 shows that liquid-phase CO_2 solubility is essentially independent of pressure; the L-S lines are essentially vertical. Therefore, Figure 4–26 needs only one solid-liquid phase equilibrium line to relate the CO_2 frost temperature to the mol percent CO_2 in the natural gas. This single line is the dashed line in the main portion of Figure 4–26.

Conversely, Figure 4–24 shows clearly that CO_2 solubility in the gas phase varies with pressure; the S-V lines are curved and their upper limit depends strongly on the composition. Accordingly, Figure 4–26 presents a series of constant-pressure lines (100 to 400 psia) that relate vapor-phase frost temperature to mol percent CO_2 in the natural gas.

In cryogenic processing (discussed in Chapter 14), CO_2 frost is important primarily in two equipment items, the turboexpander and the demethanizer column. In the case of the turboexpander, frost will be from the gas phase due to the relatively-low pressure at the expander outlet. In the demethanizer column the concern is solid formation

from the liquid near the top of the tower, where the temperatures are coldest.

Example 4–3. Natural gas containing 2 mol percent CO_2 is expanded to 250 psia, −140°F. Is there danger of CO_2 frost?

Solution: The insert graph of Figure 4–26 shows that at −140°F and 250 psia, vapor phase operating conditions apply. Enter the main part of Figure 4–26 at −140°F and go vertically upward to the 250 psia line. Read to the left to obtain 2.1 mol percent CO_2. The given 2.0 mol percent CO_2 is slightly less, so that CO_2 frost will probably not be a problem, but it is on the borderline. HYSIM indicates −138°F as the frost point.

Example 4–4. A demethanizer is to operate at 450 psia. The temperature at the column top is approximately −145°F. What percent CO_2 can be tolerated?

Solution: The inset graph of Figure 4–26 shows liquid-phase operating conditions control. The main graph shows that 5% CO_2 can be tolerated. HYSIM verifies this temperature.

Review Questions

1. In designing oilfield gathering lines, what two predictions must be made with respect to water in natural gas?
2. Name four methods of estimating the saturated water vapor content of a natural gas. Rank your four methods using the following criteria:
   - ease of use
   - applicability
   - information required
   - accuracy.
3. When using a Bureau of Mines dew-point tester:
   a. What precautions should be taken?
   b. How can a water dew point be confirmed?
   c. How do water dew points differ from other dew points?
4. A sample line from a natural-gas stream to a water content analyzer is exposed to ambient air. Does this pose any potential problem?
5. How do acid-gas components affect the saturated water vapor content of a natural gas? When is this acid-gas effect
   a. Of minor significance, i.e., less than 10%?
   b. Of major importance, i.e., greater than 50%?

6. Name an advantage of equation-of-state methods for estimating equilibrium water vapor content of natural gas.

7. What is a natural gas hydrate?
   What components in a natural gas form hydrates?
   Describe in words the structure of gas hydrates.
   How can gas hydrates exist at temperatures much higher than the freezing point of water?

8. Name the three conditions essential for hydrate formation.

9. Identify four methods of estimating the temperature (or pressure) at which natural gas hydrates form. Rank your four methods using the following criteria:
   - ease of use
   - applicability
   - information required
   - accuracy.

10. How do acid gases affect hydrate formation?
    How does gas gravity affect hydrate formation?
    Estimate the magnitudes of these effects.

11. If a natural gas containing water vapor is cooled at constant pressure, which will condense first, hydrocarbon liquid or liquid water? Explain.

12. Under what conditions is CO_2 frost a potential problem in natural-gas processing?

13. Does CO_2 frost form from a CO_2-hydrocarbon vapor or liquid phase? Explain.

Problems

1. Determine the saturated water content of the following gases:

Gas	SG	Temp °F	Pressure Psia	Mol % H_2S	Mol % CO_2
#1	0.8	120	1000	5	15
#2	0.7	100	1000	20	0
#3	0.65	100	1500	20	15
#4	0.7	100	1000	10	0

Use the following methods:
a. McKetta-Wehe
b. Robinson, Wichert, Moore, Heidemann correlation
c. Campbell method
d. A computer simulation method, if available.

2. A sweet, water-saturated natural gas enters a compressor at 50 psia and 100°F and leaves at 200 psia and 300°F.
   a. Is there any chance of water condensing in the compressor? If so, how much? Answer in lb H_2O/ MMscf gas compressed.
   b. Estimate the mole fraction of water vapor in the gas stream leaving the compressor.
   c. The natural gas is then cooled to 190 psia and 100°F before entering the second stage of compression. Should a vapor-liquid scrubber be installed between the intercooler and the second stage of compression? If so, how much water (lb H_2O/MMscf) will be removed? Will any hydrocarbon liquids condense?

3. To confirm a water-dew-point measurement, the dew point of the same sample of gas was measured at 100, 300, 600, and 1000 psia. If the 1000 psia dew-point reading is 32°F:
   a. What is the water content (lbH_2O/MMscf) of the gas?
   b. What should the 100, 300, and 600 psia dew points be?

4. Fifty MMscfd of a sour (35 vol% CO_2, 0 vol% H_2S) natural gas enter a remote amine unit at 1000 psia and 100°F.
   a. Estimate the water content of this gas.
   b. This produced gas (SG = 0.7) leaves the amine contactor at 990 psia and 130°F with 2 vol % CO_2 (assume saturated with water as the amine contactor has a water wash at its top). How much, if any, make-up water (in gpm) is needed? State any assumptions.
   c. It is suggested that this treated gas be fed into a sales pipeline. While flowing in the pipeline the temperature and pressure are expected to drop to 50°F and 500 psia, respectively. Do you approve or oppose this suggestion? Support your answer with appropriate calculations.
   d. What processing and what specifications would you recommend to ensure trouble-free pipelining of the gas?

5. Estimate the hydrate formation temperature at 1000 psia (6.9 MPa) for the following gas:

Component	Mol (Vol) %
N_2	10.0
H_2S	5.0
CO_2	10.0
C1	63.8
C2	5.0
C3	3.0
iC4	.5
nC4	1.5
iC5	.3
nC5	.6
C6	.1
C7+	.2

Use the following methods:
a. GPSA gas gravity curve
b. Katz *et al.* K-values
c. Baillie and Wichert method
d. A computer simulation package, if available.
6. A sweet, natural-gas stream (SG = 0.75) leaves a three-phase (gas-crude oil-brine) separator at 500 psia and 100°F.
   a. What new phases, if any, are formed if the gas is cooled at essentially constant pressure to 30°F?
   b. Estimate the temperature(s) at which the phase(s) are formed.

Nomenclature

C	= Number of components in a mixture
f	= Number of degrees of freedom at equilibrium
K_{V-S}	= Equilibrium constant for gas-hydrate equilibrium
MW	= Molecular weight
W	= Water content of gas (lb H_2O/MMscf)
P	= Pressure
SG	= Specific gravity
T	= Temperature
W_{CO_2}	= Effective water content of CO_2 (lb H_2O/MMscf)
W_{hc}	= Water content of sweet gas (lb H_2O/MMscf)
W_{H_2S}	= Effective water content of H_2S (lb H_2O/MMscf)
y_{CO_2}	= Mol fraction of CO_2 in natural gas
y_{hc}	= Mol fraction of all components in natural gas except H_2S and CO_2
y_{H_2S}	= Mol fraction of H_2S in natural gas
z	= Mol fraction of any component in natural gas
ϕ	= Number of phases in equilibrium

References

Adler, S. B., H. Ozkardesh, and C. F. Spencer (1980), "Water/Hydrocarbon Systems Pose Special Problems," *Oil & Gas J.*, Vol. 78, No. 46, pp. 107–113 (Nov. 17).

API (1983), "Technical Data Book—Petroleum Refining," Vol. 11, Chapter 9, 4th Edition, American Petroleum Institute, 1801 K Street, Washington, DC.

ASTM (1963) Method ANSI/ASTM D1142–63 (Reapproved 1975), "Water Vapor Content of Gaseous Fuels by Measurement of Dew-Point Temperatures," American Society for Testing and Materials, 1916 Race Street, Philadelphia, PA 19103.

ASTM (1988) Method ANSI/ASTM D 4810–88, "Standard Test Method for H_2S in Natural Gas Using Length-of-Stain Detector Tubes," American Society for Testing and Materials, 1916 Race Street, Philadelphia, PA 19103.

Baillie, Chris, and Edward Wichert (1987), "Chart Gives Hydrate Formation Temperature for Natural Gas," *Oil & Gas J.*, Vol. 85, No. 14, pp. 37–39 (April 6).

Barnes, William R. (1988), "Devices for Moisture Measurement in Natural Gas," Proceedings of 63rd International School of Hydrocarbon Measurement, p. 401, University of Oklahoma, Norman, OK (May 24–26).

Bukacek, R. F., (1955), "Equilibrium Moisture Content of Natural Gases," Institute of Gas Technology, Research Bulletin 8, IGTBA, Chicago, IL.

Campbell, John M. (1976), "Gas Conditioning and Processing," Vol. 2, Campbell Petroleum Series, Norman, OK.

Carson, D. B., and D. L. Katz (1942), "Natural Gas Hydrates," *Trans. A.I.M.E.*, Vol. 146, pp. 150–158.

Chandler Engineering (1986), "Instruction Manual for Operating Dew-Point Testers," EG&G Chandler Engineering Co., 7707 E. 38th St., Tulsa, OK (March).

Davis, J. A., N. Rodewald, and F. Kurata (1962), "Solid-Liquid-Vapor Phase Behavior of the Methane-Carbon Dioxide System," *A.I.Ch.E. J.*, Vol. 8, No. 4, pp. 537–539 (Sept.).

Deaton, W. H., and E. M. Frost, Jr. (1938), "Bureau of Mines Apparatus for Determining the Dew Point of Gases Under Pressure," U.S. Bureau of Mines R.I. 3900 (May).

Donnelly, H. G., and D. L. Katz (1954), "Phase Equilibria in the Carbon Dioxide-Methane System," *Ind. Eng. Chem.*, Vol. 46, No. 3, pp. 511–517 (March).

EG&G Chandler Engineering (1985), "Advanced Aluminum Oxide Sensor," 7707 E. 38th St., Tulsa, OK 74145.

GPSA (1987), "Engineering Data Book," 10th Ed., Gas Processors Suppliers Association, Tulsa, OK.

Gibson, P. S. (1980), "A Review of the Bureau of Mines Dew-Point Tester," 28th Annual Gas Measurement Institute, Liberal, KS (Sept. 24–25).

Hammerschmidt, E. G. (1933), "Calculation and Determination of the Moisture Content of Compressed Natural Gas," *Western Gas*, Vol. 9, No. 12, pp. 9–11, 29.

Hammerschmidt, E. G. (1939), "Gas Hydrate Formations," *Western Gas*, Vol. 15, No. 5, pp. 30–34, 94 (May).

Hammerschmidt, E. G. (1939), "Preventing and Removing Gas Hydrate Formations in Natural Gas Pipelines," *Oil & Gas J.*, Vol. 37, No. 52, pp. 66–72 (May 11).

Hammerschmidt, E. G. (1939), "Preventing and Removing Hydrates in Natural Gas Lines," *Gas Age*, Vol. 83, No. 9, pp. 45–49 (April 27).

Hyprotech (1988), HYSIM Process Simulator for PC-compatible computers, Hyprotech Ltd., Calgary, Alberta, Canada.

Katz, D. L. (1944), "Prediction of Conditions for Hydrate Formation in Natural Gas," *Petroleum Technology*, Vol. 7, No. 4, pp. 1–10 (July).

Katz, D. L. (1945), "Prediction of Conditions for Hydrate Formation in Natural Gases," *Trans. A.I.M.E.*, Vol. 160, pp. 140–149.

Loh, James, R. N. Maddox, and J. H. Erbar (1983), "New Hydrate Formation Data Reveal Differences," *Oil & Gas J.*, Vol. 81, No. 20, pp. 96–98 (May 16).

Maddox, R. N., and J. H. Erbar (1982), "Gas Conditioning and Processing," Vol. 3, Chapter 14, Campbell Petroleum Series, Norman, OK.

McCarthy, E. L., W. L. Boyd, and L. S. Reid (1950), "The Water Vapor Content of Essentially Nitrogen-Free Natural Gas Saturated at Various Conditions of Temperature and Pressure," (TP 2914) *Trans. A.I.M.E.*, Vol. 189, pp. 241–242.

McKetta, J. J., Jr., and A. H. Wehe (1958), "Use This Chart for Water Content of Natural Gases," *Petroleum Refiner*, Vol. 37, No. 8, pp. 153–154 (August).

Mayeaux, Donald P. (1987), "Electrolytic Cell and Process for the Operation of Electrolytic Cells for Moisture Analyzers," U.S. Patent 4,842,709, assigned to EG&G Chandler Engineering, Tulsa, OK (Dec. 7).

Ng, H.-J., and D. B. Robinson (1976), "The Measurement and Prediction of Hydrate Formation in Liquid Hydrocarbon-Water Systems," *Ind. Eng. Chem. Fund.*, Vol. 15, No. 4, pp. 293–298 (November).

Noaker, L. J., and D. L. Katz (1954), "Gas Hydrates of Hydrogen Sulfide-Methane Mixtures," *Trans. A.I.M.E.*, Vol. 201, pp. 237–239.

Nolte, Frank W., D. B. Robinson, and H.-J. Ng (1985), "Capabilities of the Current Equi-Phase Programs," *Proc. 64th Annual Conv. GPA*, Houston, TX, pp. 137–146 (March 18–20).

Parrish, W. R., and J. M. Prausnitz (1972), "Dissociation Pressures of Gas Hydrates," *Ind. Eng. Chem. Proc. Des. Dev.*, Vol. 11, No. 1, pp. 26–35 (January).

Poettmann, F. H. (1984), "Here's Butane Hydrates Equilibria," *Hydro. Proc.*, Vol. 63, No. 6, pp. 111–112 (June).

Poettmann, Fred H., E. Dendy Sloan, Susan L. Mann, and Louise M. McClure, (1989), "Vapor-Solid Equilibrium Ratios for Structure I and II Natural Gas Hydrates," *Proc. 68th Ann. GPA Conv.*, San Antonio, TX, (March 13–14).

Ranarex (1988), "Instruction Manual for Portable Moisture Analyzer," EG&G Chandler Engineering, 7707 E. 38th St., Tulsa OK 74145 (March)

Robinson, D. B., and H.-J. Ng (1975), "Improve Hydrate Predictions," *Hydro. Proc.*, Vol. 54, No. 12, pp. 95–96 (Dec.).

Robinson, D. B., H.-J. Ng, and C.-J. Chen (1987), "The Measurement and Prediction of the Formation and Inhibition of Hydrates in Hydrocarbon Systems," *Proc. 66th Ann. GPA Conv.*, Denver, CO, pp. 154–164 (March 16–18).

Robinson, J. N., Edward Wichert, R. Gordon Moore, and Robert A. Heidemann (1978), "Charts Help Estimate H_2O Content of Sour Gases," *Oil & Gas J.*, Vol. 76, No. 6, pp. 76–77 (Feb. 6).

Scelzo, M. J. (1989), "Devices for Moisture Measurement in Natural Gas," *Proc. 64th Int'l. School of Hydrocarbon Measurement*, University of Oklahoma, Norman OK, (May 16–18).

Sensidyne Inc. (1984), "Technical Data Book," 12345 Starkey Rd., Largo, FL 34643.

Sloan, E. Dendy (1984), "Phase Equilibria of Natural Gas Hydrates," *Proc. 63rd Ann. Conv. GPA.*, New Orleans, LA (March 19–21).

Sloan, E. Dendy (1985), "The Colorado School of Mines Hydrate Program," *Proc. 64th Ann. GPA Conv.*, Houston, TX, pp. 125–128 (March 18–20).

UGC Industries (1983), "Operating Instructions for Moisture Titrator," Shreveport, LA (Dec.).

Unruh, C. H., and D. L. Katz (1949), "Gas Hydrates of CO_2-CH_4 Mixtures," *Trans. A.I.M.E.*, Vol. 186, pp. 83–86.

van der Waals, J. H., and J. C. Platteuw (1959), "Advances in Chemical Physics," Vol. 2, Interscience, NY.

Wagner, Jan, Ruth C. Erbar, and Ali I. Majeed (1985), "AQUA*SIM—Phase Equilibria and Hydrate Inhibition Using the PFGC Equation of State," *Proc. 64th Ann. Conv. GPA.*, Houston, TX, pp. 129–136 (March 18–20).

Welker, T. F. (1989), "Devices for Moisture Measurement in Natural Gas," Proc. 64th International School of Hydrocarbon Measurement, University of Oklahoma, Norman OK. (May 16–18)

Wilcox, W. L., D. B. Carson, and D. L. Katz (1941), "Natural Gas Hydrates," *Ind. Eng. Chem.*, Vol. 33, p. 662.

Wilson, A., E. Strange, A. Goethe, and D. G. Elliot (1985), "Hydrate Predictions: Importance of GPA Research," *Proc. 64th Ann. GPA Conv.*, Houston, TX, pp. 155–160 (March 18–20).

Chapter 5

Field Processing of Natural Gas

INTRODUCTION

Extensive processing of the gas produced at the well is sometimes required. In every case, the *specific processing* needed is determined by the flowrate, composition, temperature, and pressure of the produced gas and by the components/impurities that must be removed to meet *delivery specifications*. Processing may vary from simple separation plus dehydration all the way up to compression, sweetening, natural gas liquids (NGL) recovery, and dehydration.

The hydrocarbon portion of the natural gas may be suitable as a sales gas as produced. On the other hand, if the condensate content is high, these liquids must be recovered and either stored or transported. Field location, sales specifications, environmental requirements, and economics determine the extent to which nonhydrocarbon contaminants are removed and the ultimate destination of the various streams to sales, flare, or disposal.

Figure 5–1 is a comprehensive diagram of the various *unit operations*, or *modules*, carried out in field processing. All the various modules shown will not be present in every system. Further, the modules used in a given application may not be arranged in the sequence shown, although in general the sequence is typical. The choice of modules to be used and the sequencing of these modules is determined during the design phase of field development (Meyer and Sharma, 1980).

DESIGN BASIS

As discussed in Chapter 2, well streams must be *sampled* and *analyzed* with great care if the resulting data are to serve as a reliable design basis. Even then, the well-stream composition used in design must be viewed realistically. The likelihood that the actual feed to the processing equipment will be identical to the design basis is remote.

Production is usually from a combination of wells that may have differing compositions. The producing pattern (number of wells and flow rate from each) may be different from that originally envisioned. Also, well-stream compositions change as reservoir pressure declines. *Equipment* must be sufficiently *flexible* and/or adequately sized to operate well in the face of changing feed composition and flow rate.

In addition to well-stream composition, important design data are wellhead pressure, temperature, and flow rate. These variables change with time in a manner often difficult to predict, but nevertheless important. Decline of reservoir pressure with production is a much-studied, but often difficult phenomenon to forecast. Pressure and flow variation in oil formations depends on the reservoir drive mechanism as well as the possible use of secondary and tertiary recovery techniques. Gas-oil ratio often increases with time. This latter factor is important in predicting associated-gas yield.

Pressure and flow-rate changes will affect selection and design of vessel and pipe sizes and of compressor horsepower, or even the necessity for compression. Design of equipment based solely on initial production conditions only can lead to inadequate sizing and/or omission of vital equipment. Generalizations are difficult; each project must be analyzed with consideration of its peculiar characteristics, including capacities both early and late in the life of the project.

GAS PROCESSING

As shown in Figure 5–1, hydrocarbon condensate, liquid water, and solids should be separated from the gas stream as early as practical. Then the individual streams can be treated with less difficulty. Processing of produced water and sand will be discussed in Volume 3. Condensate processing is discussed in Chapter 14. Processing of the gas stream is emphasized here.

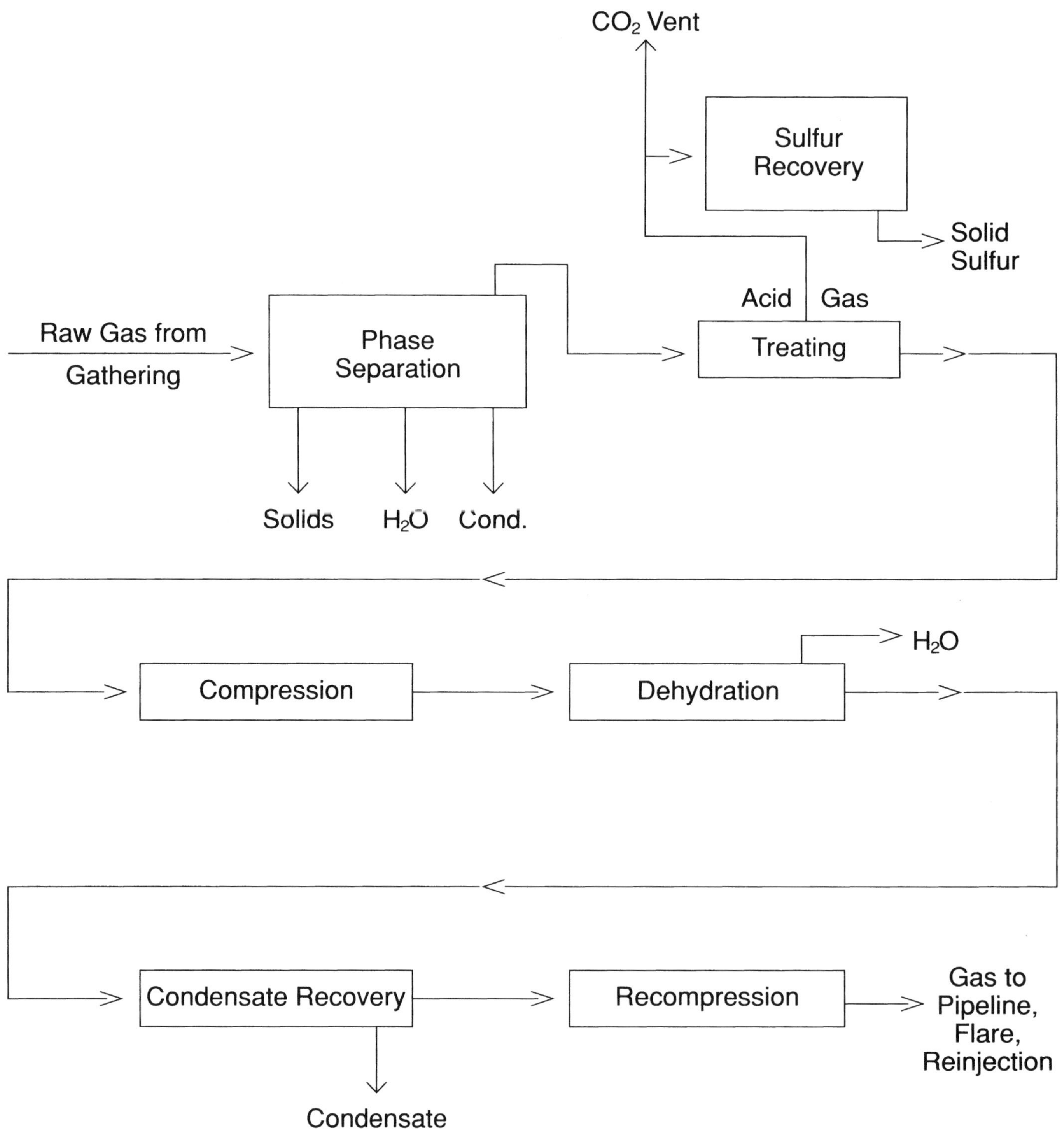

Figure 5–1. General scheme for gas processing.

The individual operations in Figure 5–1 are reviewed briefly below and then in greater detail in subsequent chapters. The discussion follows the gas flow shown in Figure 5–1.

Process Modules

The first unit operation is the physical *separation* of the distinct phases, which are *gas* and possibly *liquid hydrocarbon*, *liquid water*, and/or *solids*. The temperature and pressure of the gas stream dictates whether liquid hydrocarbon and/or water are present. Phase separation usually occurs in a pressure vessel provided for that purpose. Gas leaves from the top of the vessel and liquid from the bottom. If a three-phase separator is needed, special internal elements are required to permit separation of the water and hydrocarbon liquids. Solids fall to the bottom and must be removed by special techniques. Mist extractors—wire mesh, vanes, or baffles—are sometimes used to reduce liquid entrainment in the gas stream. In gas processing, natural gas/liquid phase separation is generally not too difficult, and entrainment of one phase in the other is not excessive. Separators are discussed in great detail in Volume 2. Slug catchers are required in some gathering systems or at the terminus of offshore pipelines where large slugs of liquid may occur at random intervals.

The next step in processing is *treating*, if necessary, to remove the acid-gas components hydrogen sulfide and carbon dioxide; the former is toxic and both are corrosive. Hydrogen sulfide removal must be essentially complete, while the extent of carbon dioxide removal depends on the intended use for the gas. If the gas is to be cooled to cryogenic temperatures (less than, say, $-100°F$), CO_2 removal to a few tenths of a percent may be required to prevent formation of solid CO_2. Also, if the heating value of the gas is too low, partial CO_2 removal may be required. In some cases the gas is compressed prior to treating.

Treating processes are detailed in Chapter 7, but frequently consist of countercurrent contacting of the gas in an absorber with a chemical or physical solvent—most often a water solution of an ethanolamine. Therefore treating precedes dehydration.

The side stream from treating consists mainly of H_2S and/or CO_2. Environmental regulations usually limit venting or flaring H_2S to the surroundings, so conversion to elemental sulfur is necessary. The primary conversion process is the *Claus process*, in which one-third of the H_2S is burned to SO_2 and then reacted with the remaining H_2S to form elemental sulfur and water vapor.

Sulfur recovery is a big business, and many plants recover 500 to 1000 tons a day. Carbon dioxide is usually vented to the atmosphere but sometimes is recovered for CO_2 floods

Dehydration is often necessary to prevent formation of gas hydrates, ice-like substances that may form and plug processing equipment or pipelines at high pressure, even at temperatures considerably higher than $32°F$ ($0°C$). The two principal dehydration methods are glycol contacting and solid-desiccant adsorption. Methanol or glycol also can be injected to prevent hydrate formation. Dehydration is discussed in Chapters 8 and 9, and hydrate inhibition in Chapter 6.

Gas containing considerable amounts of liquefiable HC (ethane, propane and heavier) produces condensate upon cooling or compressing and cooling. If condensation would occur in the transportation, processing, or use of the gas, it may be better to remove the condensate at or near the wellhead.

In some cases the potential *NGL* are sufficiently valuable to justify their *recovery*, quite apart from other considerations. Normally, condensate is fractionated into the NGL products in a central facility rather than in the field. Recovered condensate may have to be stabilized by partial removal of dissolved gaseous components to obtain a liquid product with low-enough vapor pressure to be transported safely. Natural gas processing for liquid recovery is discussed in Chapter 14.

If the wellhead pressure is low, the natural gas must often be *compressed* to permit *pipeline transport*. Transport is commonly done at high pressure levels (700–1000 psig) to reduce the pipeline diameter. Pipelines may operate at very high levels (above 1000 psig) to keep the gas in the dense phase and thus prevent condensation and two-phase flow. Compression also may be required as a necessary step in condensate recovery. Compression is discussed in Chapter 10 and transport in Chapter 13.

The final destination of the finished gas stream depends on the project. Natural gas or associated gas is usually injected into a pipeline for sales. Early in the life of a remote field, before the availability of a gas pipeline, separator gas may be flared temporarily. The ability to flare depends on governmental regulations as well as the field location. More and more, separator gas is being conserved by compression and reinjection into the formation for eventual recovery and sales. Also, in gas condensate reservoirs, the gas is often reinjected, or "cycled," to prevent condensation of valuable liquid hydrocarbons in the reservoir.

Ancillary Equipment

In addition to the main processing modules described above, additional equipment is required for important *ancillary functions*. These are described briefly here. In the remainder of the book, emphasis is primarily on the process steps as such.

The general arrangement of production equipment for "*safe, pollution-free*, and efficient production of oil and gas" is discussed in American Petroleum Institute (API) Recommended Practice (RP) 2G. This RP deals specifically with offshore structures, but many of the guidelines are pertinent for land facilities. The same can be said for API RP–14E (1984), which deals specifically with piping systems.

Test separators with gas and liquid flow meters are often

provided for periodic determination of the production rate of individual wells or groups of wells. Such testing is often a legal requirement (Sivalls, 1983). Appropriate flow lines, valving, and manifolding must be provided to isolate the flow of the various wells through the test or production separators. Gas production is generally measured by orifice meter, while liquid flow may be determined by orifice or turbine meter. Gas metering is discussed in Chapter 11 and liquid metering in Volume 2.

Although not shown in Figure 5–1, *heat exchange* is a common operation in gas processing: e.g.,

1. Heating wellhead gas to prevent hydrate formation in the choke
2. Cooling or refrigeration of a gas stream to condense hydrocarbon liquid
3. Intercooling and aftercooling of compressor outlet gas streams
4. Heat interchange of lean and rich amine solutions in a treating unit
5. Heat interchange of lean and rich glycol in a glycol contacting or injection system
6. Reboiling of lean amine or lean glycol in regenerator towers.

Heat exchange is discussed in Chapter 12.

Automatic safety systems are a necessary and integral part of any processing unit. Protection must be provided against such unsafe conditions as overpressure, liquid overflow, excessive temperature in fired equipment, sources of ignition, etc. Flares or vents, with necessary manifolding, are required for the emergency removal of gas or vapors that cannot be discharged locally from a safety-relief valve. Relief is necessary whenever the inlet gas flow to a processing facility exceeds the outlet flow for short periods of time. Specific treatment of these aspects of field processing are discussed in API Recommended Practices 14C (1986), 520 (1976,1963), and 521 (1982). Controls, relief systems, and flares are discussed in Volume 2.

Additional facilities include such items as electrical power generation, instrument air supply, sewage disposal, as well as fire, drinking, and sanitary water supply.

PROCESSING SCOPE

The important factors in selecting the scope of processing are now discussed. These factors generally include the processing objective, the type or source of the gas, and the location and size of the field.

Compression requirements depend simply upon the produced gas pressure and the pressure required for the particular end use.

Process Objective

Processing of a gas stream may have one of *three basic objectives* (Odello, 1981):

1. To produce a transportable gas stream
2. To produce a salable gas stream
3. To maximize condensate production.

Each of these objectives merits explanation; details concerned with pipeline transport are developed in Chapter 13. The emphasis here is on processing.

Transportable Gas. Production of a *transportable gas* stream implies *minimal processing* in the field and transport through a pipeline to a final processing plant. Three main constituents are of concern: *water, hydrogen sulfide*, and *condensate*.

Pipeline transport occurs at relatively high pressure and low temperature. High pressure is used to increase gas density and thus decrease the pipe diameter. Relatively low temperatures may be encountered because the pipeline is exposed to ambient temperature. Water must be removed to a level that will prevent hydrate formation in the pipeline. Dehydration is required to reduce the water dew point to a temperature lower than the lowest expected temperature (LET) in the pipeline.

If the LET is higher than the hydrate point of the gas, then water removal may not be absolutely necessary. Liquid water can cause corrosion which can be controlled by injecting corrosion inhibitors or using internal pipe coatings. Protection may be necessary in two-phase lines thought to be dry, but into which small amounts of water enter with injected HC condensate.

Hydrogen sulfide is highly corrosive and must be removed if present in sufficiently high concentration. The technical alternative is to use internal pipe coatings and/or corrosion inhibitors. Safety and/or local regulations may require removal or special measures that make removal the economic choice.

Condensate may or may not be recovered prior to pipeline transport. If the hydrocarbon dew point is less than the LET, then no processing is required. If the cricondentherm of the gas is higher than the LET, pipeline pressure may fall into the retrograde zone and condensation will occur.

If HC condensation is likely, there are three options:

1. No condensate removal and two-phase pipeline flow
2. Condensate removal to produce a hydrocarbon dew point less than the LET
3. No condensate removal and dense-phase transport.

If condensate is not removed, the pipeline path passes into the retrograde zone and two phases are produced, as shown by curve 1–2 in Figure 5–2a. As shown, the pipeline path is one of decreasing pressure with temperature falling rapidly to an almost constant level. This occurs because the flowing gas temperature often approaches ambient temperature in a short distance. Two-phase flow requires a larger pipe diameter than single-phase flow. Furthermore, the condensate must be handled in a slug catcher at the pipeline end.

Condensate removal requires cooling the gas to condense liquids, followed by phase separation in a vapor-liquid separator (Fig. 3–9). The phase diagram of the liquids-depleted gas is shown in Figure 5–2b; transport for the path shown results in no condensation. In this alternative a separate means of condensate transport or disposal is required.

In dense-phase transport the gas is maintained at a pressure

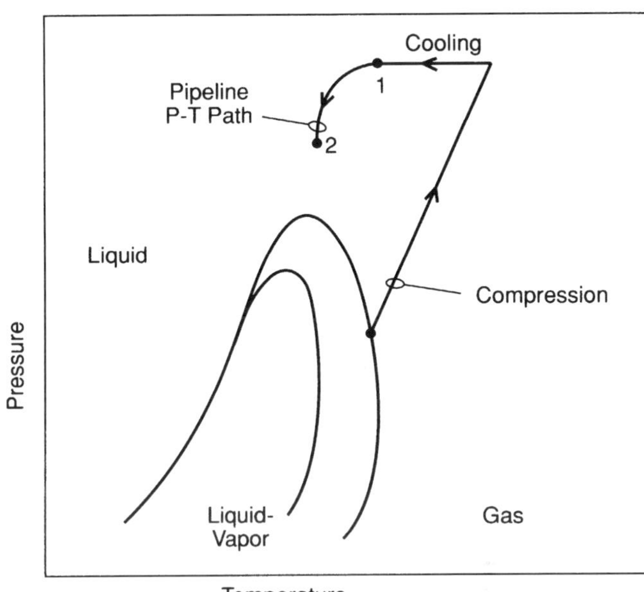

Figure 5–2c.   Phase diagram for dense-phase pipeline gas.

higher than the cricondenbar throughout the line, which prevents the formation of two phases at any point (Figure 5–2c). Expensive compression may be required and pipe with high pressure rating and thickness is required, but the diameter will be smaller because of the increased density of the compressed fluid.

In a given case, any one of the above alternatives might be preferable. Comparison of costs and other constraints must be made.

Salable Gas. Production of *sales-quality gas* requires all processing necessary to meet *specifications* of the type shown in Table 2–8. Hydrogen sulfide and CO_2 must be removed if present in sufficient amount. Dehydration is required also, generally after compression, since free water is condensed out in the compressor aftercooler. Removing free water in a separator reduces the load on and cost of the downstream TEG dehydration unit.

Recovery of condensate depends on the amount present, the heating value requirement of the sales gas, and the hydrocarbon dew-point specification (if any). If the condensate content is low, there may be little need to remove it. If the gas contains a high percentage of nitrogen, it may be necessary to retain the heavier hydrocarbons in order to meet the heating-value specification. If the HC dew point is regulated, condensate recovery may be required to meet the dew-point specification.

Three basic techniques are used for condensate recovery. From the standpoint of the phase diagram, the task is to move the state of the system to the left of the dew-point line of the gas. In other words, cool the gas below its

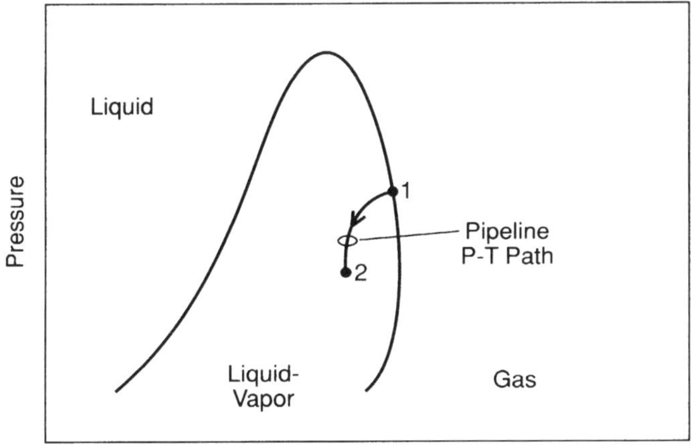

Figure 5–2a.   Phase diagram for pipeline gas with condensation.

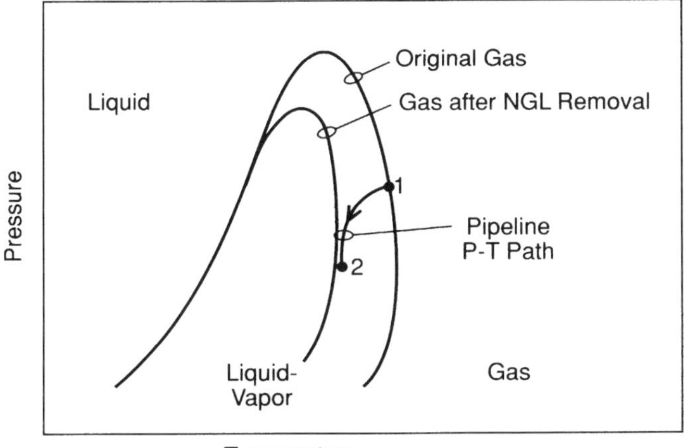

Figure 5–2b.   Phase diagram for pipeline gas after NGL removal.

dew-point temperature. Then the resulting condensate can be separated.

Cooling and/or refrigeration is the most obvious technique, indicated by path ABC in Figure 5–3. Refrigeration requires a vapor-compression refrigeration cycle, as shown in Figure 5–4a. An alternative process is Joule-Thomson (J-T) expansion through a valve to lower pressure, path ABC′ in Figure 5–3. Line AB represents cooling the gas by exchange with the cold outlet gas from the separator, as shown in Figure 5–4b.

Expansion through a turbine to a lower pressure, as in path ABC″ of Figure 5–3, results in about the same condensate recovery at a higher pressure. Figure 5–4c depicts the process. Path AB represents cooling the gas by exchange with the cold outlet gas as before. Joule-Thomson and turbine expansion may require expensive gas recompression. Refrigeration can be combined with J-T or turbine expansion for even greater condensate recovery.

In the alternatives shown in Figure 5–4, condensate "stabilization" refers to removing the gaseous components by heating or stripping, in order to produce a liquid product that will be stable in low-pressure storage or transport.

Field processing of a low-pressure, sour gas to yield specification pipeline gas will require essentially all the process elements shown in Figure 5–1.

Maximum Liquid Production. Three situations motivate maximum condensate recovery. The first is the desire to maximize crude oil production when processing associated gas. Condensate is often more valuable if recovered from the associated gas and injected into the crude oil. The second case is processing retrograde condensate gas; here the objective is sometimes to recover the condensate and reinject the gas to the formation, as described below. Third, in some markets the NGL produced from the condensate may be more valuable as liquid products than as sales-gas components; their recovery will yield a better profit.

Processing for maximum condensate recovery is similar to that shown in Figure 5–3. One or a combination of the processes shown in Figure 5–4 are used with conditions chosen to maximize condensation and removal of condensate.

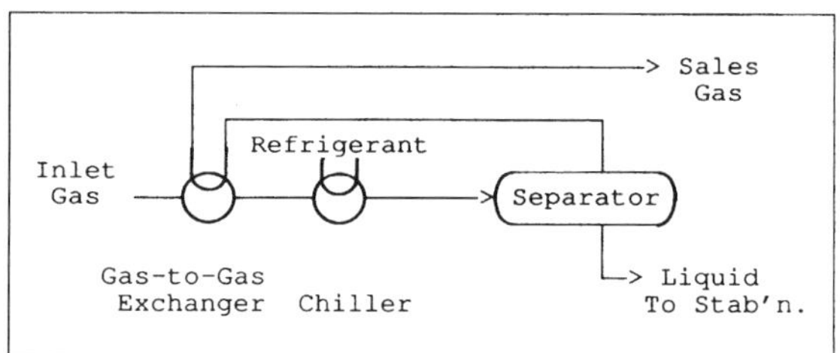

Figure 5–4a. Condensate recovery by refrigeration.

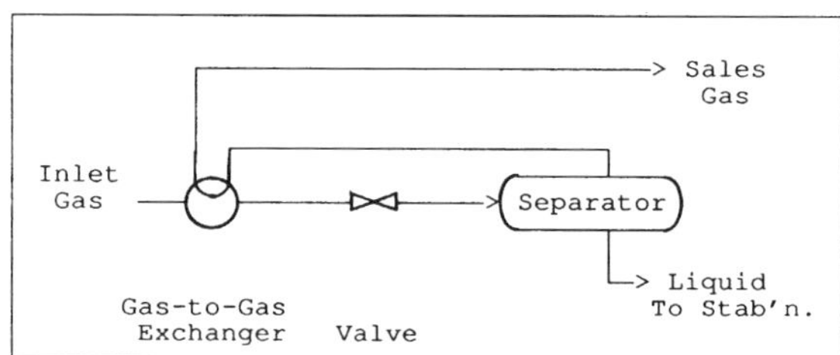

Figure 5–4b. Condensate recovery by Joule-Thomson expansion.

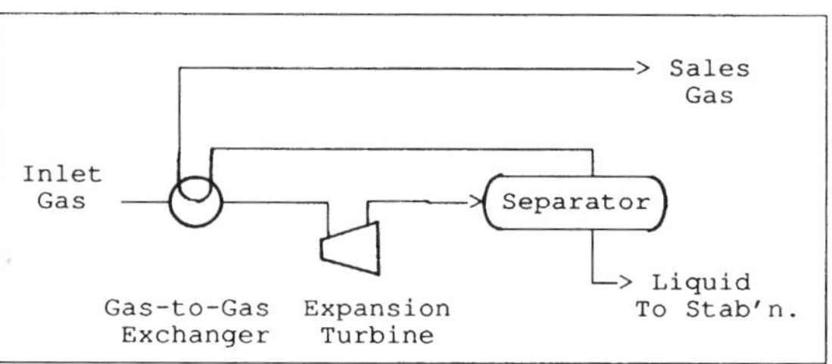

Figure 5–4c. Condensate recovery by expansion-turbine cooling.

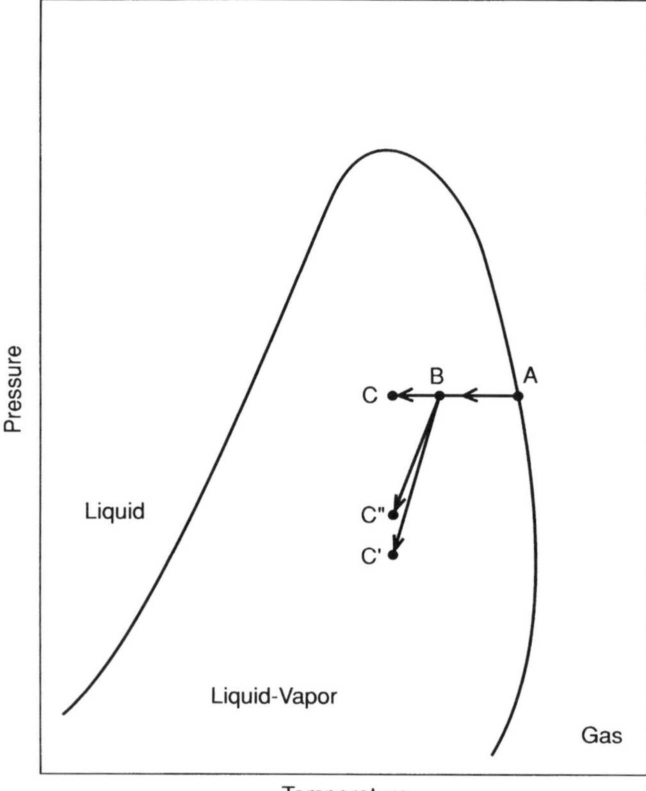

Figure 5–3. Phase-diagram paths for NGL recovery.

Type of Gas

As discussed in Chapter 2, there are three basic types of natural gas: gas-well gas, associated gas, and condensate-reservoir gas. Gas type influences its processing primarily in relation to the amount of liquefiable hydrocarbons contained.

Gas-well gas, whether "wet" or "dry," is composed mainly of methane. The condensate content is not particularly high and may not have to be reduced to produce a salable gas. The only motivation for condensate recovery would be if the liquefiable components were more valuable as liquid products than as sales-gas components. In the U.S., this has often been the case in the past, but more recently liquid-product prices have been low and NGL recovery of little or no economic benefit.

Associated gas is very rich in liquefiable components and typically must undergo condensate recovery to meet hydrocarbon dew-point or maximum heating value requirements.

So-called *gas-condensate* reservoirs contain reservoir fluid having a high percentage of dissolved heavier hydrocarbons. As the pressure in the formation falls, the fluid may reach its dew point in the retrograde region, as along line AB in Figure 5–5. Then liquid hydrocarbon—termed "condensate"—forms as the pressure continues to

decline. Because of its small volume, the condensate is not produced but remains in the reservoir.

The reservoir fluid produced from a gas-condensate formation undergoes a decrease in pressure and temperature as it flows up the well string, as along line AC in Figure 5–5. Condensate is formed and recovered by phase separation at the wellhead. If the residue gas from the wellhead separator is compressed and recycled back into the formation, reservoir pressure will be maintained near point A; the condensate will be retained in the gas phase and can be produced, thus conserving this material.

Gas cycling in condensate reservoirs has been practiced for many years (Thornton, 1946) and continues to be (Huzyk, 1988). Wellhead separator gas is still rich in liquefiable components; condensate is recovered from it in modern plants (Huzyk, 1988, and Petzet, 1985). In other words, after the condensate is removed from the produced gas in the wellhead separator, further processing of the type shown in Figure 5–4 is used to recover additional condensate from the rich separator gas.

Alternative methods of maintaining pressure in gas-condensate reservoirs are partial gas recycling (Sarssam, 1988), nitrogen injection (Hagoort *et al.*, 1988) and, waterflooding (Matthews *et al.*, 1988).

Location and Size of Field

The geographical location of the gas or oil field is an important factor in choosing the processing scheme. There are at least two aspects of location that are important: remoteness and climate. In addition, there is the important factor of size.

Remoteness. In the sense used here, remoteness refers to the distance from population centers and hence to ease of transport of personnel and materiel to and from the production facility, as well as ease of maintenance, availability of support personnel, etc.

Offshore platforms and *outback* are examples of remote locales. Even these locales are not strictly comparable because of the factor of the sea vs. dry land. Platform facilities must be placed on decks that are extremely limited with respect to size and allowable weight; only those operations absolutely needed are performed. Currently, most offshore production is sweet, thus reducing the required processing; however, more and more gas is showing traces of H_2S as sea water is used for waterflooding. A dry-land outback facility has essentially unlimited area available and can support operations not practical or desirable offshore, such as treating or processing involving fire hazards.

Climate. The hostility of the environment is important. Temperature obviously affects the tendency for hydrate

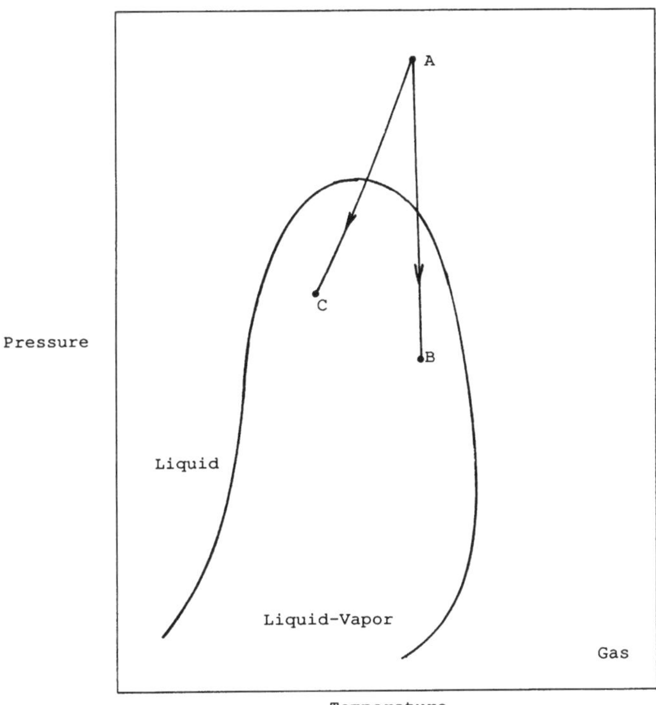

Figure 5–5. Phase-diagram paths for a gas-condensate reservoir fluid.

formation. A tropical location such as offshore Indonesia may pose little problem. The opposite extreme is North Slope Alaska, where process equipment may have to be housed.

Size. Magnitude of production rate is also important. *Small flow* rates justify only *simple* operations; *large rates* can sustain sizable and *complex processing* facilities.

EXAMPLE PROCESSING SCHEMES

Three typical cases illustrate actual processing schemes. Each example is calculated using the OPSIM flowsheet simulator program (Appendix 5) to develop the material and energy balances. The compositions of the initial fluid and the final products are given. Selected intermediate results of the simulation are summarized in the flow diagrams.

The first example is an associated gas produced on an offshore platform. The second is a Gulf Coast dry natural gas produced at a relatively-high pressure. The final example is a mixed low-pressure gas-well/associated gas typical of West Texas.

Example 5–1. Platform Associated Gas

This *offshore case* is a *sweet crude oil* with *dissolved gas* produced to the wellhead at 1000 psig and 120°F. The objective is to produce transportable oil and gas streams which are injected into separate pipelines.

Figure 5–6 shows the schematic diagram and the analysis of the reservoir fluid. The heavy ends of the crude have been split into five boiling-point cuts for simplicity. More cuts, or "pseudocomponents," can be used for better accuracy in the phase equilibrium calculations. The present breakdown is adequate for estimating approximate compressor loads and gas volumes.

The well stream flows into a separator where water and produced gas are separated. Then the oil is passed through a series of three separators at 300 psig, 55 psig, and 2 psig, respectively, to obtain a stabilized oil for transport. The gas from each stage is compressed to the pressure of

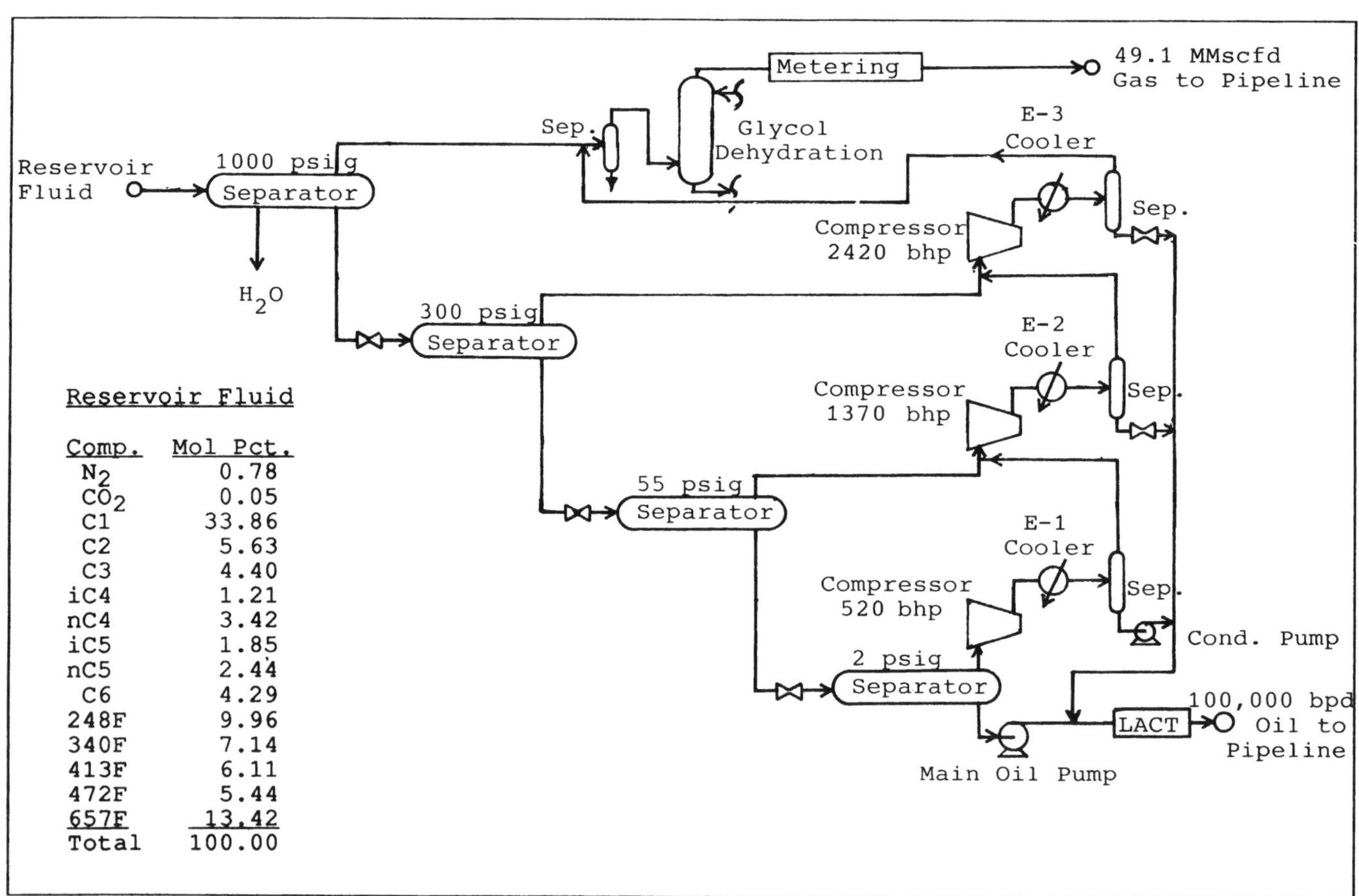

Figure 5–6. Platform processing of associated gas.

the previous stage, cooled, and mixed with the gas from the previous stage. By this means a total of 49.1 MMscfd gas is obtained at 1000 psig. The overall GOR is 491 scf/Bsto.

Pressures for the intermediate separators are normally selected to *maximize oil production or minimize compressor horsepower*, depending on the application. The pressures in Figure 5–6 are typical, not optimal, values.

Figure 5–6 shows that separator liquid is obtained after compressing and cooling the gas from each stage separator, for a total of 1481 bpd of the total oil production of 100000 bpd. The condensate liquid is not stabilized but simply combined with the low-pressure separator oil. The vapor pressure of the combined oil production at 100°F is 28.1 psia. The pipeline must be maintained above this pressure to prevent vapor breaking out in the line. If the crude oil from the last separation stage were stored at the platform before transport, it would be necessary to stabilize it to reduce its vapor pressure to atmospheric pressure or slightly below.

The oil is metered with a so-called *LACT* unit (lease automatic custody transfer) and delivered to the main pipeline oil pumps.

The high-pressure gas must be dehydrated for transport. A glycol contacting unit, the usual choice, is shown.

The combined gas flows into the pipeline for transport to shore. The gas leaving the platform is a dew-point gas. As the temperature and pressure in the gas pipeline decrease, condensate is dropped out. For example, at 500 psia and 44°F (typical North Sea bottom temperature), the condensate amounts to about 16.5 bbl/MMscf.

The present gas-stream processing is relatively simple, though not inexpensive. The total compression requirement of about 4310 bhp is substantial. The only other processing required is cooling, phase separation, and dehydration. Not shown in the diagram, but usually provided in this application, is a sphere-launcher for periodic pigging of the gas pipeline to reduce the liquid holdup, pressure drop, and required compression horsepower.

Recently, some offshore gas contains small (100 ppm)

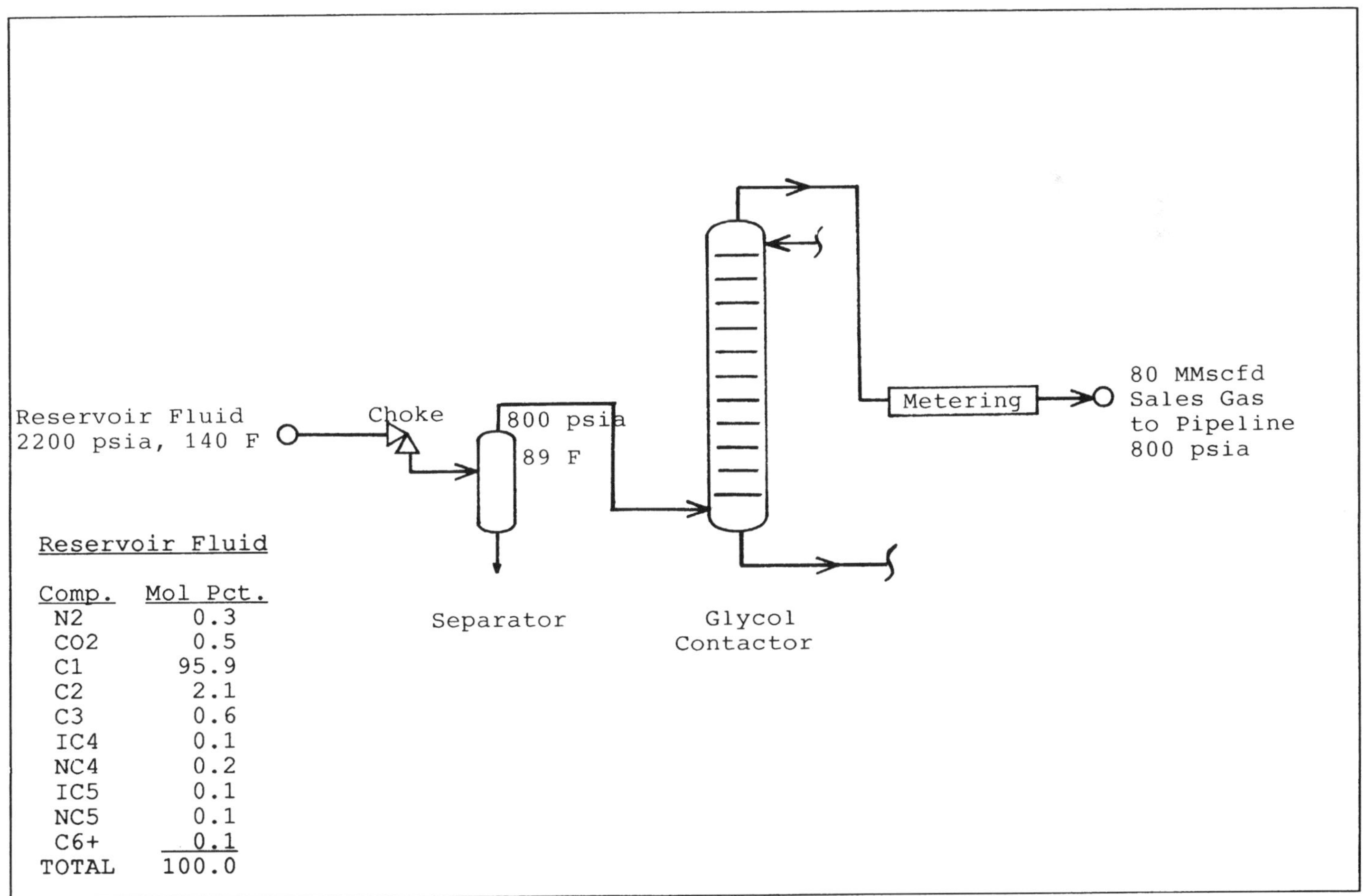

Figure 5–7. High-pressure lean gas processing.

concentrations of H_2S. Special mole sieves can be used to sweeten and dry the gas simultaneously. This method is especially attractive if the quantities of H_2S are low enough to permit flaring of the H_2S evolved when the mole sieves are regenerated. One offshore gas contained 36 vol percent CO_2 and 90 ppmv H_2S. Treating with Selexol produced dry gas with 1.2 vol percent CO_2 and 1.6 ppmv H_2S (see Figure 7–12). Again, the H_2S could be flared.

Example 5–2. High-Pressure Lean Gas

The second example is processing a *high-pressure lean gas* to produce a pipeline-grade sales gas. As shown in Figure 5–7, the gas is passed through a wellhead choke, a separator, and thence through glycol dehydration to lower the water content to pipeline specification. Finally, the dry gas is metered and sent to the pipeline. Because of the high wellhead pressure, no compression is required.

This gas is a sweet, very-lean gas (0.94 gal liquefiable C2+ per Mscf), typical of some Texas Gulf-Coast gases.

The flow rate is quite high, 80 MMscfd. The heating value is 1044 Btu/scf, not much higher than typical minimum values. For this latter reason and because of the low NGL content, it is probably not economical to remove NGL. The gas contains no condensate and has a cricondentherm of about 31°F, so that pipeline condensation should not be encountered.

Example 5–3. Low-Pressure Rich Gas

This *low-pressure, sour, rich gas* is typical of West Texas and other parts of the U.S. in which the small productions from many wells are combined and processed in a single plant. The analysis in Figure 5–8 indicates that the gas is either all or in part associated gas (because of the high content of C2+). Note the very low flow rate of 6.0 MMscfd.

The gas is very rich, containing 5.86 gal liquefiable C2+ per Mscf. Condensate removal is necessary to produce a salable pipeline gas, which is the objective. The selected process is refrigeration, with propane refrigerant being

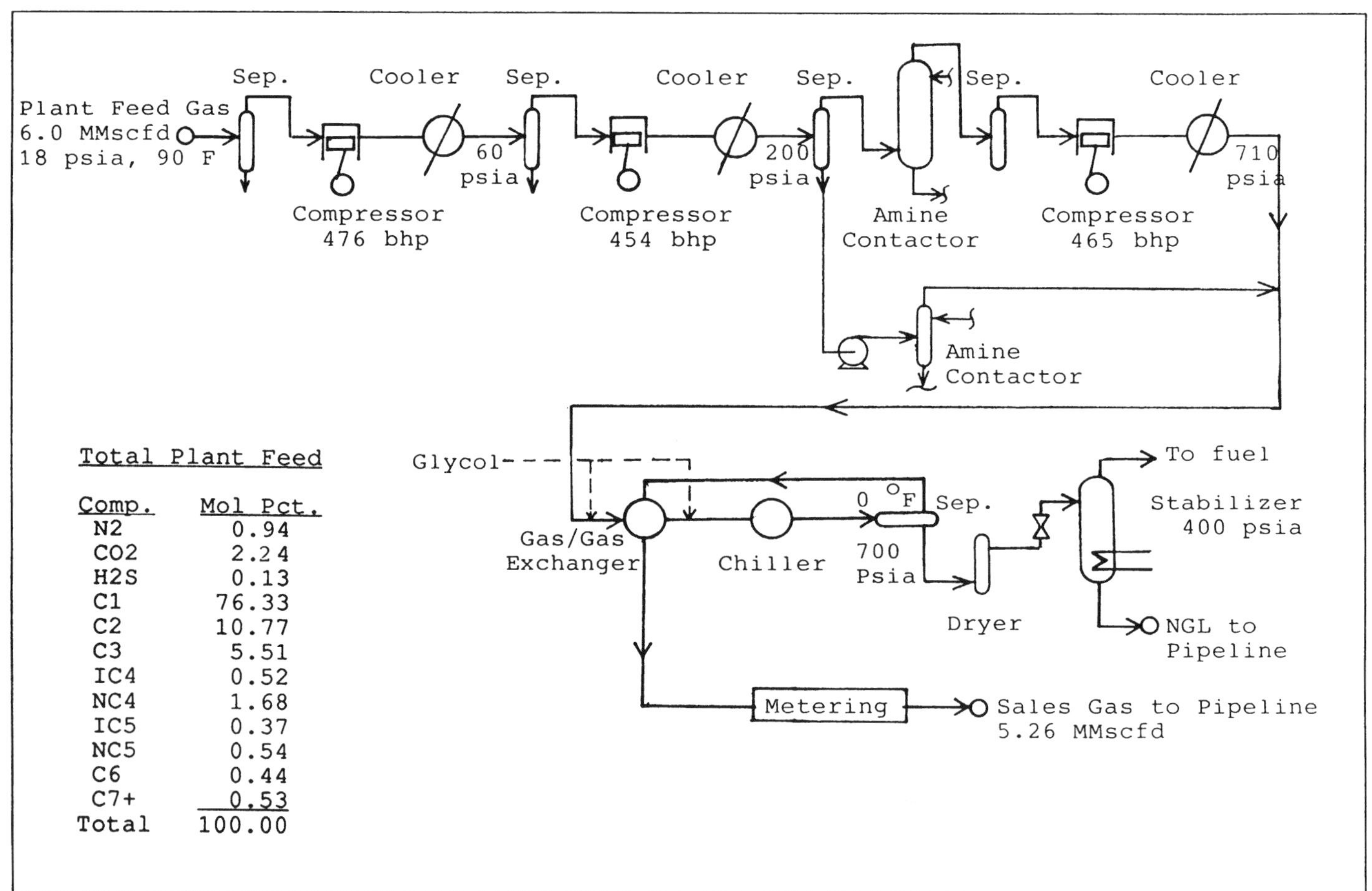

Figure 5–8. Low-pressure rich gas processing.

chosen because of its ready availability. The chiller cools the high-pressure gas to 0°F to produce a sales gas with low hydrocarbon dew-point. Glycol is injected to prevent hydrate formation in the heat exchangers. A beneficial result of glycol injection is that water is absorbed, yielding a gas of water dew point no higher than 0°F. Separate dehydration is not required. Approximately 65 bhp compression duty is required in the refrigeration unit.

The gas contains a finite, though low, H_2S content that must be reduced to pipeline specification of ¼ grain per 100 scf. Amine contacting is chosen.

The gas must be compressed from the low field pressure of 18 psia to a sales-gas pressure of 700 psia. Reciprocating compressors are used due to the small volume of gas handled. Three stages of compression are required to retain reasonable compression ratios and temperatures in each stage. Amine treating is done at the second-stage discharge pressure.

The 700–psia condensate is stabilized using a reboiled tray contactor operating at 400 psia to produce approximately 300 bpd of liquid. This liquid has a vapor pressure of 243 psia at 100°F and is supplied to a pipeline. Fractionation of such a liquid into its individual hydrocarbon products—ethane, propane, butanes, and pentanes plus—is not economical for such a small product rate. It is common in the U.S. to pipeline such products to a large-capacity, central fractionation facility.

The 700–psia residue gas has a heating value of 1119 Btu/scf, and the production rate is 5.26 MMscfd.

Review Questions

1. Identify the individual unit operations or process modules commonly used in field processing of natural gas.
2. Discuss the correct sequence for arranging the modules.
3. State the reasons for removing acid gases from natural gas.
4. Why is water vapor removed from natural gas?
5. Removal of hydrocarbon condensate from natural gas may be desirable. Under what circumstances would this be true?
6. When is compression required in processing natural gas?
7. Name three possible destinations for natural gas (after suitable processing, of course).
8. State three important objectives in gas processing.
9. Name three types of ancillary equipment of systems used in natural gas processing.
10. If a natural gas may exhibit hydrocarbon condensation in the pipeline, what three alternative procedures can be adopted?
11. How does the gas type or source (gas-well gas, retrograde condensate gas, or associated gas) affect its processing?
12. How do production location and quantity of produced gas affect its processing?
13. Name three basic techniques for NGL recovery.

Problems

1. A gas well in Alberta, Canada, produces 30 MMscfd of gas with the following composition:

Comp.	Mol %
N_2	0.4
CO_2	11.6
H_2S	23.0
C1	64.7
C2	0.2
C3	0.1

Wellhead pressure is 835 psia. The gas must be of sales quality:

GHV	980 Btu/scf, dry ideal gas
H_2S	0.25 grain/100scf
H_2O	0°F dew-point

a. Sketch a flow sheet suitable for processing this gas.
b. Label the sketch clearly.
c. Estimate the allowable CO_2 in the product gas.
d. Estimate the MMscfd of sales gas.
Do not simulate the process.

2. An Indonesian natural gas contains 20 vol percent CO_2 but no H_2S. This gas is pipelined 60 miles to a natural gas liquefaction plant where gas sweetening (DEA) and dehydration (silica gel solid desiccant) facilities already exist with sufficient capacity to handle this gas-gathering production (200 Mscfd). The temperature never falls below 70°F in this sea-level on the equator location.
a. Suggest minimum specifications for the gas as it enters the pipeline.
b. Suggest a suitable processing sequence for this "jungle camp" facility.
c. Outline any alternative schemes worthy of consideration.
Do not simulate the processes.

References

API Recommended Practice 2G (1974), "Production Facilities on Offshore Structures," American Petroleum Institute, Division of Production, Dallas, TX, 1st Ed.

API Recommended Practice 14C (1986), "Analysis, Design, Installation and Testing of Basic Surface Safety Systems on Offshore Production Platforms," American Petroleum Institute, Division of Production, Dallas, TX, 4th Ed.

API Recommended Practice 14E (1984), "Design and Installation of Offshore Production Platform Piping Systems," American Petroleum Institute, Division of Production, Dallas, TX, 4th Ed.

API Recommended Practice 520 (1976), "Recommended Practice for Design and Installation of Pressure-Relieving Systems in Refineries," Part I, American Petroleum Institute, Division of Refining, Washington, DC, 4th Ed.

API Recommended Practice 520 (1963), "Recommended Practice for Design and Installation of Pressure-Relieving Systems in Refineries," Part II, American Petroleum Institute, Division of Refining, Washington, DC, 2nd Ed.

API Recommended Practice 521 (1982), "Guide for Pressure Relief and Depressuring Systems," American Petroleum Institute, Division of Refining, Washington, DC, 2nd Ed.

Hagoort, J., J. W. Brinkhorst, and P. H. van der Kleyn (1988), "Development of an Offshore Gas-Condensate Reservoir by Nitrogen Injection vs. Pressure Depletion," Vol. 40, No. 4, pp. 463–469 (April).

Huzyk, S. L. (1988), "Anschutz Ranch Facilities' Expansion Ups Production," *Oil & Gas J.*, Vol. 86, No. 24, pp. 35–38 (June 13).

Matthews, J. D., R. I. Hawes, I. R. Hawkyard, and T. P. Fishlock (1988), "Feasibility Studies of Waterflooding Gas-Condensate Reservoirs," *J. Pet. Tech.*, Vol. 40, No. 8, pp. 1049–1056 (Aug.).

Meyer, Paul E., and Suresh C. Sharma (1980), "Field Production Systems and Oil Processing," *Oil, Gas & Petroleum Eqpt.* (Sept. and Oct.).

Odello, R. (1981), "Systematic Method Aids Choice of Field Gas Treatment," *Oil & Gas J.*, Vol. 79, No. 6, pp. 103–109 (Feb. 9).

Petzet, G. A. (1985), "Gas Cycling to Boost Hatter's Pond Recovery," *Oil & Gas J.*, Vol. 83, No. 17, pp. 56–57 (April 29).

Sarssam, S. (1988), "Condensate Fields—Processing Options," IBC Tech. Serv. Ltd., *Proc. Conf. Development of Condensate Fields*, London, p. 22.

Sivalls, C. Richard (1983), "Fundamentals of Oil Production Processing," *Proc. 30th Annual Southwestern Petrol. Short Course*, Midland, TX (April 27–28).

Thornton, O. F. (1946), "Gas-Condensate Reservoirs—A Review," *API Drill. Prod. Prac.*, pp. 150–159.

Chapter 6

Prevention of Hydrate Formation

INTRODUCTION

Gathering lines moving water-saturated gas from the wellhead to a central treating station can become plugged with hydrates if the gas temperature drops below the hydrate expectancy value (Fig. 4–10). Part of the gas temperature drop occurs at the wellhead due to Joule-Thomson (J-T) cooling as the gas is expanded through a choke to reduce pressure and control the flow rate. Heat loss from the flow line causes additional gas cooling as is discussed in Chapter 13.

Hydrate formation is prevented by drying the natural gas, by heating the flowline, or by chemical addition. *Drying* the natural gas with triethylene glycol (TEG) or solid desiccant is the best protection; but dehydration is generally carried out at a central facility, as discussed in detail in Chapters 8 and 9. Heating or chemicals prevent hydrate formation until the gas can be dried economically. *Flowline heating* is used upstream of chokes and in short gathering lines because: initial investment is modest, heaters operate with a minimum of attention, and fuel is readily available, often at low cost (Harrell, 1988). For long flowlines (over 1 or 2 miles) the gas approaches ambient temperature and chemical injection may be preferred.

Many *chemicals* depress the temperature at which hydrates and/or ice form. While ammonia and brine were used in the past, the current choice is either a glycol or methanol. Triethylene glycol (TEG) and tetraethylene glycol (TREG) are too soluble in liquid hydrocarbons and too viscous for general use; therefore, the most popular inhibitors are ethylene glycol (EG), diethylene glycol (DEG), and methanol (MeOH).

The gas stream leaving a hydrate-inhibition unit should be considered saturated with water. For example, gas leaving the low temperature separator in Figure 6–1 should be considered water saturated at the separator temperature and pressure. Very little, if any, gas-phase dehydration occurs because the liquid-phase glycol concentration is too dilute and the gas-liquid contact is co-current. Conversely, the water dew-point of the gas is reduced at least to the temperature in the separator.

Prevention of hydrate formation is now discussed under the following headings: design considerations, glycol and methanol injection, design procedures, example calculations, and operation.

DESIGN CONSIDERATIONS

Glycol and methanol injection are compared by considering temperature effects, injection techniques, potential downstream problems, and recovery economics.

Methanol may be used at any temperature, but DEG is usually not chosen below 15°F because of its high viscosity and difficulty of separation from liquid hydrocarbons. Above 20° F DEG might be preferred over EG due to lower vaporization losses. While ethylene glycol exhibits a higher depression of the hydrate formation temperature than DEG on a per pound basis, methanol is superior to both EG and TEG on a per pound basis.

To be effective, *glycols must be sprayed* into the wet gas as very small droplets. If intimate mixing of the fine liquid glycol spray and the natural gas is not achieved, then glycol injection may not prevent hydrate formation. With methanol the injection technique is not as critical because all or a considerable fraction of the methanol evaporates into the gas stream, thus providing protection.

Diluted glycols are separated from the gas, reconcentrated, and reused. Methanol recovery may be marginally economical due to its low cost and high vaporization losses. However, if the gas stream is dried downstream in a TEG unit, methanol can be recovered easily and economically from the TEG regenerator overhead. Glycol is usually cheaper where continuous injection is

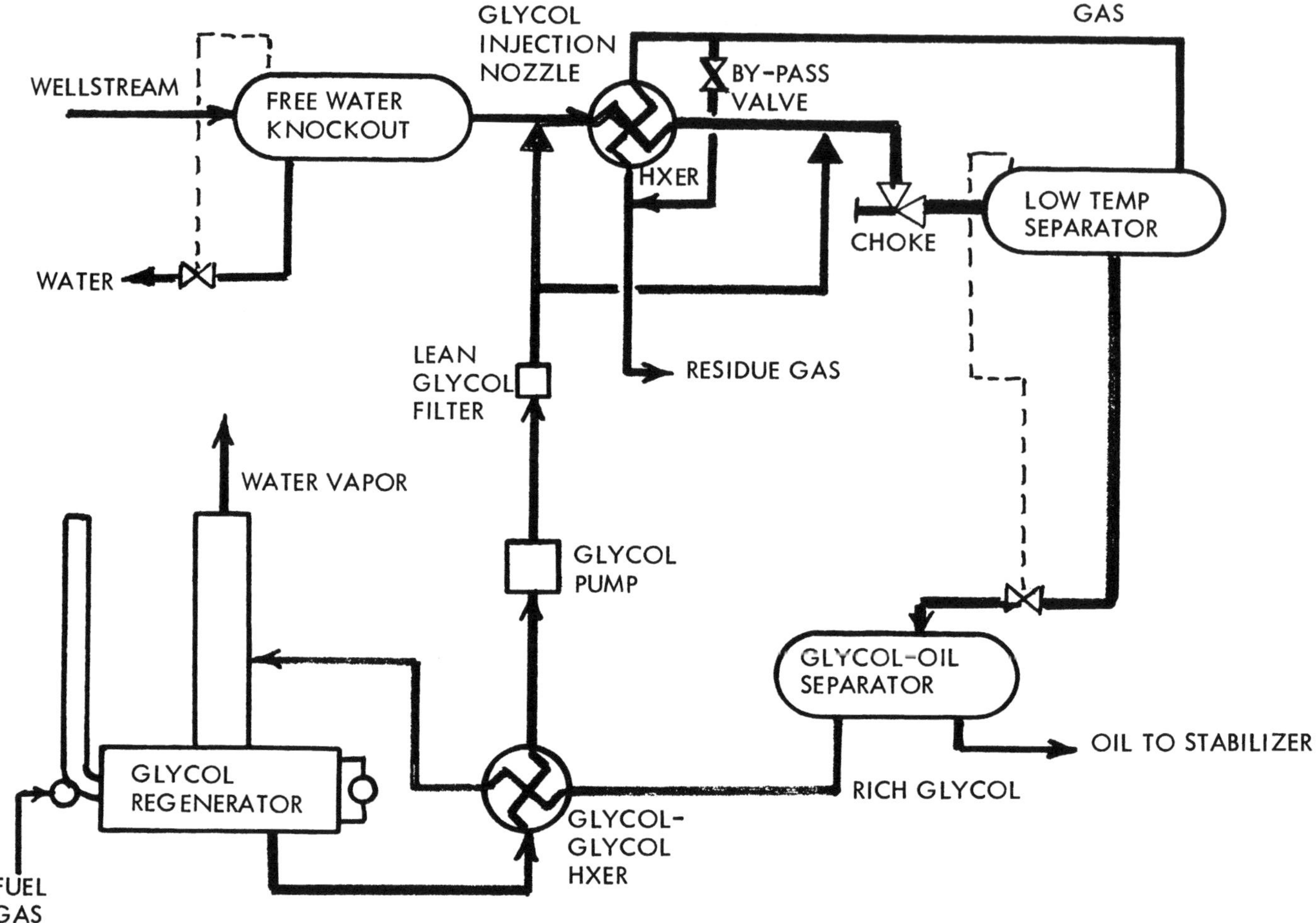

Figure 6–1. Typical glycol injection system.

required and/or high gas volumes are being inhibited. Because of its lower capital investment (no recovery units), methanol is frequently used for low gas volumes, temporary installations, or where hydrate problems are mild, infrequent, or seasonal. Glycol is seldom used in long lines—say a mile or more—because methanol offers much better protection.

Methanol injection can cause the following three problems if the gas is subsequently passed through a glycol dehydration plant (Guenther, 1979):

1. Methanol is co-absorbed with water vapor by the glycol and so increases the glycol regenerator heat load. (Any methanol vented to the atmosphere with the water vapor from the regenerator still is hazardous.)
2. Aqueous methanol can corrode carbon steel in the glycol still and reboiler vapor space.
3. Methanol also can reduce the capacity of solid desiccant

pellets because methanol is readily co-adsorbed and competes with water for desiccant surface.

The relative costs of methanol injection, glycol injection, and flow-line heating can be summarized as follows. *Methanol* injection has a *low investment cost* but a *high operating cost* due to the chemical consumption. Injection of *EG or DEG* requires a *higher initial cost* but a relatively *low operating cost*. Line heaters are characterized by intermediate investment cost; the operating cost depends on the value of the fuel consumed. One rule of thumb is that glycol units are used when the required methanol injection rate exceeds 30 gal/hr.

While glycol can prevent hydrates, it will not attack or dissolve hydrates already present. On the other hand, *methanol can dissolve existing hydrates* to some degree. A pipeline plugged with hydrates can be unplugged by reducing the pipeline pressure—see Figure 4–10 for the relevant equilibrium data. Be sure to *reduce* the *pressure on both sides* of the *plug.* Reducing the pressure on only one side is **EXTREMELY DANGEROUS**—the solid hydrate plug can break loose. Then the pipeline pressure drives the ice-hard hydrate plug towards the lowered pressure side at very high velocity. When the hydrate hits a bend or restriction it can break the pipe and even unearth a buried pipeline. Such accidents have resulted in loss of lives and extensive equipment damage.

INJECTION PROCESSES

Figure 6–1 shows a typical *glycol injection* system. The free-water knockout (FWKO) removes all free water so that it is only necessary to inhibit the water condensing from the gas. The gas-to-gas heat exchanger (HXER) pre-cools the entering gas with cold gas from the low temperature separator (LTS). Because this HXER usually cools the incoming gas below its hydrate formation temperature, lean glycol is injected into the gas ahead of the HXER.

Then the gas is cooled further by a Joule-Thomson expansion across the choke. More glycol is sprayed into the gas ahead of the choke as shown in Figure 6–1. A mechanically refrigerated chiller can be substituted for the choke. Then the dry cold gas is separated from glycol and any hydrocarbon (HC) condensate in the low-temperature separator (LTS). Afterwards, the cold separator gas is used to cool incoming gas in the HXER before being fed to the pipeline, thus providing a lower temperature in the LTS.

The liquids leaving the LTS are flashed to a lower pressure. Then the rich or water-diluted glycol is separated from the HC condensate in the low-pressure, glycol-oil separator. Any low-pressure hydrocarbon gas can be used for plant fuel. Finally, the rich glycol is regenerated, repressured, filtered, and reinjected.

Several variations of the scheme shown in Figure 6–1 are possible.

1. The low-temperature separator and the low-pressure glycol-oil separator may be replaced by a single *three-phase separator* to separate the cold gas, rich glycol, and hydrocarbon condensate. At low temperatures and high pressures glycol can absorb natural gas, and the resulting decrease in aqueous liquid density makes glycol-condensate separation difficult. Long liquid residence times—say 20 to 40 minutes—may be necessary. The resulting glycol losses may be prohibitively expensive (Arnold and Pearce, 1961; Campbell, 1979).

2. Three-phase separation permits *regeneration* of only a *portion* of the *rich glycol* stream. One pump, with a relatively low differential head, recirculates the bulk of the glycol stream. A second pump with a large differential head circulates the relatively small glycol stream that flows through the regenerator and is concentrated to a relatively high concentration. This arrangement *reduces H_2S carry-over* from the regenerator when sour gas is being handled.

3. An *activated carbon filter* is sometimes used, in addition to the particulate filter, to remove absorbed hydrocarbons.

4. *Stabilization by distillation* may be used on the HC liquids from the low temperature separator, or the stabilizer may be replaced by stage separation. In this case, supply gas for fuel and general use would be taken from the second separation stage. The glycol-oil separator would be the third stage with some additional gas vented from it.

If the HXER is a horizontal shell-and-tube or plate type, supplementary glycol injection is required at the inlet end. Glycol tends to separate out by gravity and the upper portions of the HXER may plug. Usually one or more nozzles are placed to spray glycol against the tube sheet. Alternatively use a vertical exchanger. The tube sheet provides distribution; slightly more efficient heat transfer is also obtained.

Boiling point curves for EG and TEG show that the concentration of the lean injection glycol is fixed by the reboiler temperature. Because the concentrations for inhibition are lower than those needed for dehydration, no stripping gas or vacuum is needed for the reboiler or still. In all other respects, regeneration is comparable for injection or for dehydration. Efficient separation of formation water prior to glycol injection minimizes salt in the reboiler. Water condensed from the gas phase is fresh.

Methanol injection is quite different from glycol injection. First, *methanol is often not recovered* and reconcentrated so no regeneration equipment is needed. Second, *methanol does not have to be atomized*—methanol's high vapor pressure insures adequate gas-phase inhibitor concentration. A suitable low-flow metering pump is all that is required. The assurance of protection and the simplicity of injection account for methanol's popularity.

Methanol injection pumps are the positive-displacement type that deliver the same quantity of liquid per stroke whether the discharge pressure is 10 or 1000 psig. The injection rate can be checked by pumping into a measuring vessel for a short period. Observe the following precautions:

(i) Avoid any pump with copper or brass trim, especially with gases containing H_2S or Hg.

(ii) Place a rupture disc in the discharge piping to prevent catastrophic failure if the injection pump is inadvertently started against a closed valve.

(iii) Observe proper safety precautions when handling methanol. **Methanol is toxic,** therefore avoid inhaling the vapors, prolonged contact with skin, and do not swallow. See Curry (1981) for further details.

DESIGN PROCEDURES

Hydrate inhibition involves the transfer of water from the natural gas to the liquid inhibitor. In steady-state operation the mass flow rates of all three components—gas, water, and inhibitor—into the system must equal the corresponding exit flow rates. The required injection rate of inhibitor solution is calculated using 1 MMscf of dry gas as a basis. Therefore all other flow rates are expressed in lb/MMscf of dry gas.

Minimum Liquid-Phase Inhibitor Concentration

The *minimum amount of inhibitor* required may be calculated using two approaches:

1. The semi-empirical correlation of Hammerschmidt (1939),
2. Computer simulations, such as Aqua*Sim (Wagner *et al.*, 1985) or Equi-Phase (Robinson *et al.*, 1987).

Hammerschmidt's equation estimates the minimum inhibitor concentration in the exit liquid stream that is required to prevent hydrate formation.

$$W = \frac{(d)(MW)(100)}{(d)(MW) + K_H} \qquad (6\text{-}1)$$

where W = weight percent pure inhibitor in liquid water phase
     d = desired depression of hydrate formation temperature, °F
  MW = molecular weight of inhibitor
  K_H = Hammerschmidt constant for inhibitor.

The Hammerschmidt constant, K_H, and molecular weight, M, for EG, DEG, and methanol are (GPSA, 1987):

	EG	DEG	Methanol
K_H	4000	4000	2335
MW	62	106	32

The Hammerschmidt equation, Equation 6-1, was developed originally for typical natural gases and dilute solutions (less than 0.2 mole fraction inhibitor). Glycol injection systems operating down to −10°F meet these limitations (Nielsen and Bucklin, 1983). Surprisingly, the Hammerschmidt equation has been used successfully for glycol injection systems as low as −40 to −50°F that require

0.4 mol fraction EG (70 weight percent) (Nielsen and Bucklin, 1983).

For concentrated methanol solutions and temperatures as low as −160°F, Nielsen and Bucklin (1983) recommend:

$$d = -129.6 \ln (1 - x_{MeOH}) \qquad (6\text{-}2)$$

where x_{MeOH} = mol fraction methanol in aqueous phase

Glycol Dilution Restriction

When glycol injection is performed below 20°F, the freezing points of EG and DEG must be considered. Crystallization temperatures for EG and DEG are presented in Figure 6-2. Glycols do not truly freeze solid, they merely become "mushy"; however, this behavior inhibits flow and proper separation. Accordingly, it is common practice to keep the glycol concentrations between 60–70 weight percent (Robirds and Martin, 1962; Kohl and Riesenfeld, 1985).

Inhibitor Required in Liquid Phase

The quantity of inhibitor required may be calculated readily by a material balance:

$$\begin{aligned} \genfrac{}{}{0pt}{}{\text{water removed}}{\text{from gas}} = &\left(\frac{\text{lb pure}}{\text{inhibitor}}\right)\left(\frac{\text{lb water in outlet stream}}{\text{lb inhibitor}}\right. \\ &\left. - \frac{\text{lb water in inlet stream}}{\text{lb inhibitor}}\right) \end{aligned}$$

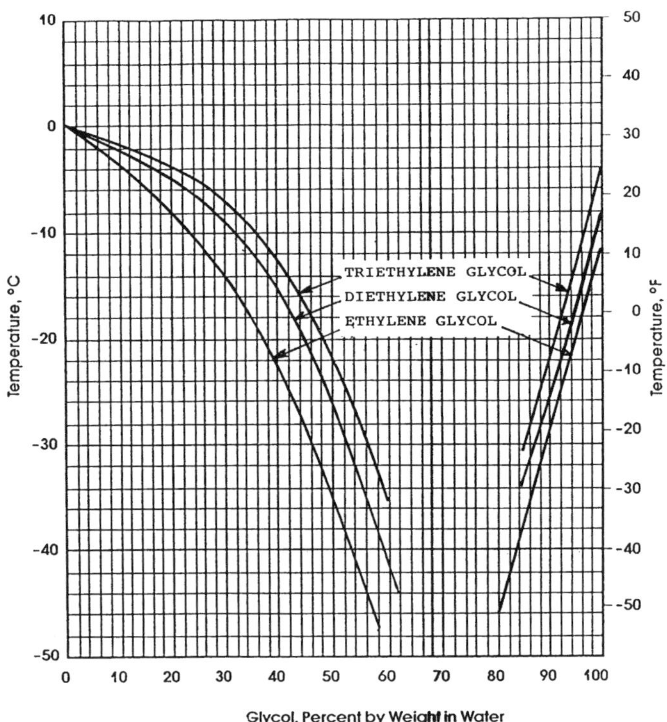

Figure 6-2. Freezing points of aqueous glycol solutions (UOP, 1989).

Stated mathematically, the equation becomes:

$$w_G = I \left[\left(\frac{100 - W_{out}}{W_{out}} \right) - \left(\frac{100 - W_{in}}{W_{in}} \right) \right] \quad (6\text{–}3)$$

where w_G = water removed from gas stream (lb/MMscf)
  I = pure inhibitor required (lb/MMscf)
  W_{in} = concentration of inhibitor in inlet inhibitor stream (weight percent)
  W_{out} = concentration of inhibitor in outlet inhibitor stream (weight percent)

The water removed from the gas stream is also equal to the increase in weight of the liquid-phase inhibitor solution; therefore:

$$w_G = I \left[(100/W_{out}) - (100/W_{in}) \right] \quad (6\text{–}4)$$

If W_{out} is fixed by the Hammerschmidt equation, combine Equation 6–1 and Equation 6–4 to obtain:

$$w_G = I \left(\frac{(d)\,(MW) + K_H}{(d)\,(MW)} \right) - \frac{100}{W_{in}} \quad (6\text{–}5)$$

If, in addition, the inhibitor enters pure—i.e., W_{in} is 100 weight percent—then:

$$w_G = I\, K_H\, /\, (d\, MW) \quad (6\text{–}6)$$

Both the Hammerschmidt and the dilution restrictions must be satisfied. If unknowns exist, the inhibitor should not be diluted over 5–10% by the pipeline water being inhibited. For pipeline protection above 20°F, a greater dilution may be tolerated but should not exceed about 20%. For spot injection, such as in a heat exchanger, where distribution is a problem, dilution may be limited to 5%. It is recommended that the *initial 24-hr minimum injection rate be higher* than the normal rate in order to obtain adequate distribution as glycol wets the steel and forms a film on the pipe walls.

Vapor-Phase Inhibitor Losses

Glycols have very low vapor pressures and accordingly vapor-phase losses should be very small. Methanol, on the other hand, is quite volatile and vapor-phase losses

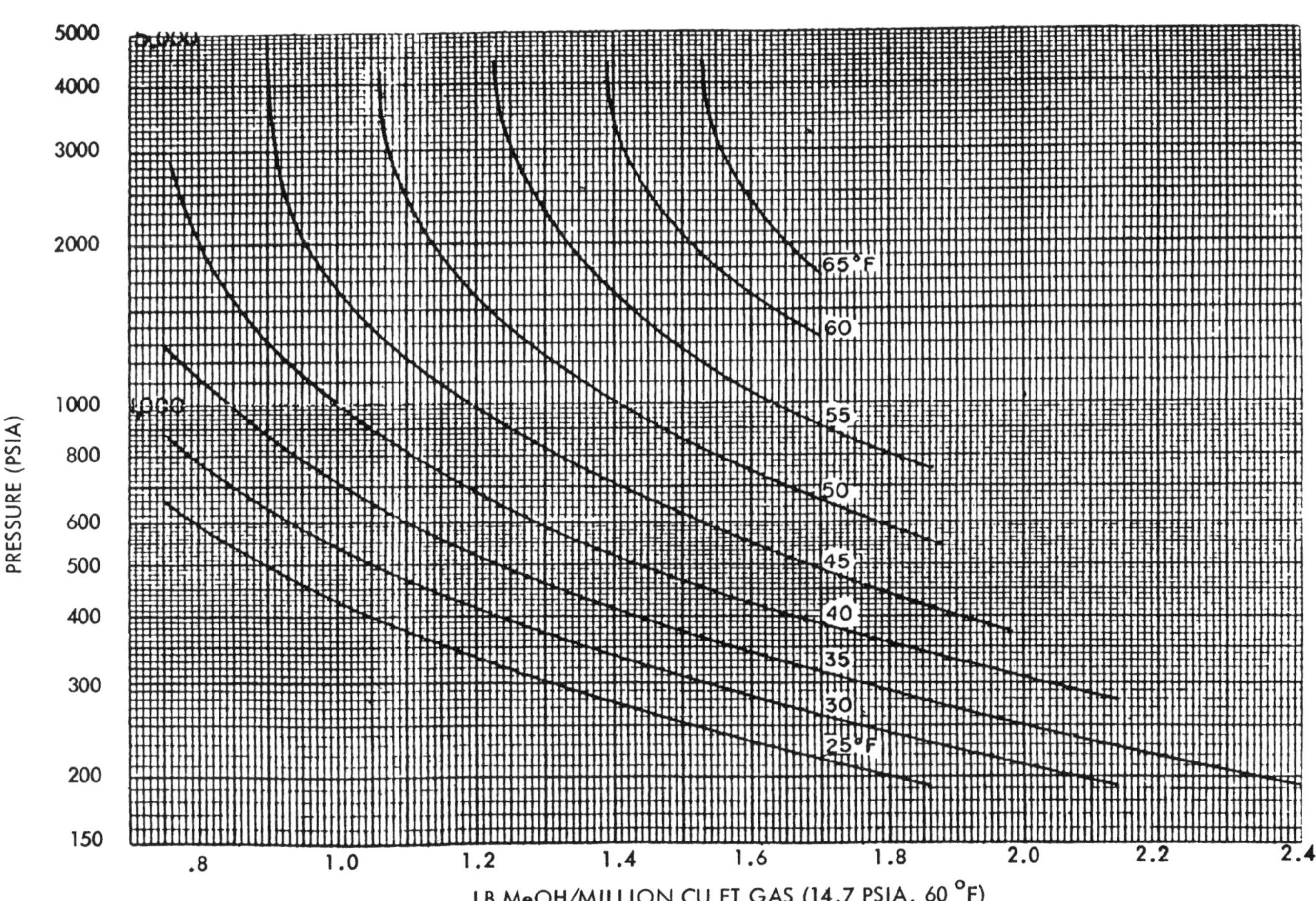

Figure 6–3. Vapor-phase methanol losses (Jacoby, 1955; GPSA, 1987, p. 20–18).

must be included. Jacoby (1955) correlated the vaporization losses in terms of the lowest temperature in the hydrate inhibition system and the corresponding pressure. Figure 6–3 presents the ratio of the gas-phase methanol concentration (lb MeOH/MMscf) to that in the liquid (weight percent MeOH) at the previously-cited minimum temperature and corresponding pressure.

Computer Simulations

Equation-of-state methods of predicting hydrate formation without inhibitor present are summarized in Chapter 4. The same simulation packages—Aqua*Sim (Wagner *et al.*, 1985), CSM Hydrate Program (Sloan, 1985), and Equi*Phase (Robinson *et al.*, 1987)—can be used to predict hydrate formation when inhibitors (EG, DEG, methanol) are present. However, it is difficult to calculate the vapor/hydrocarbon-liquid/aqueous-liquid/solid-hydrate equilibrium for systems containing inhibitor, water, hydrate-forming hydrocarbons, and nonhydrate-forming hydrocarbons. Calculation is much easier when liquid water or ice is not present. Another complicating factor is the lack of reliable experimental data to check and "tune" the computer predictions (Robinson *et al.*, 1987).

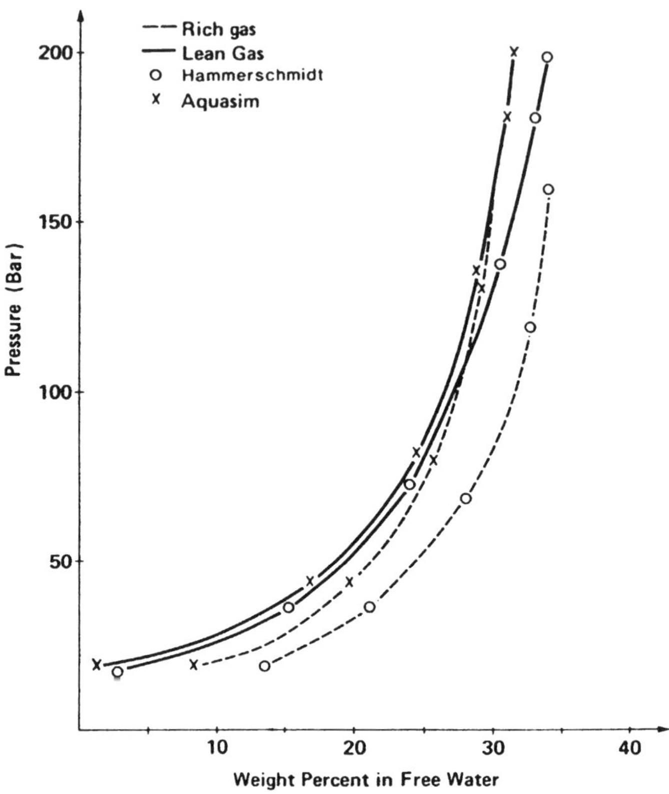

Figure 6–5. Methanol Injection rate for hydrate depression at 0°C (Wilson *et al.*, 1985).

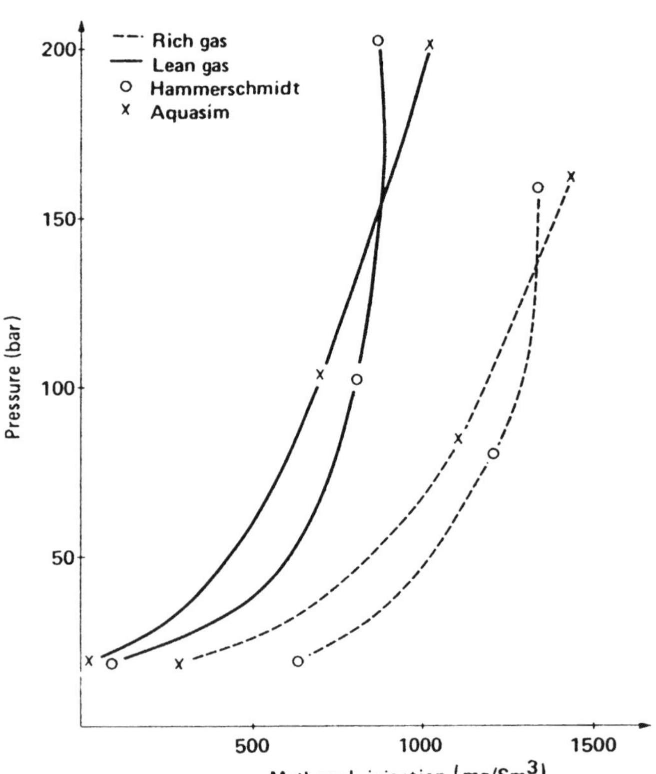

Figure 6–4. Free water-methanol mixture for hydrate boundary at 0°C (Wilson *et al.*, 1985).

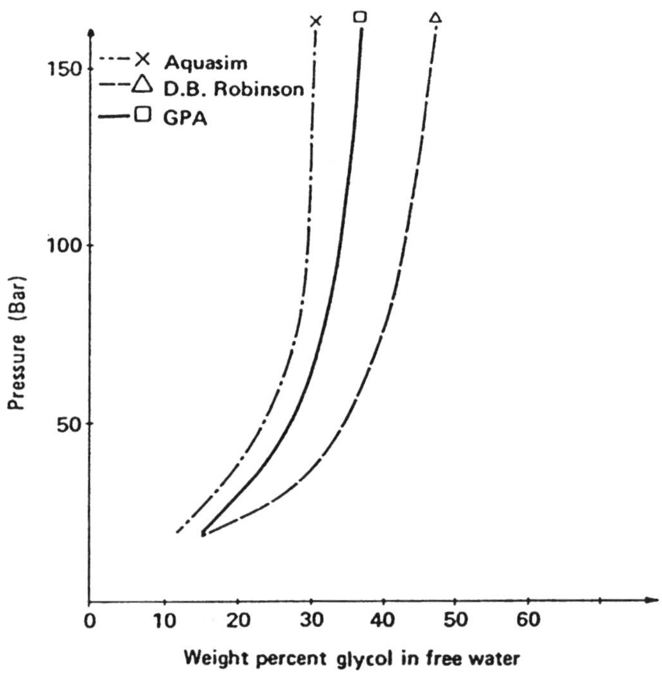

Figure 6–6. Hydrate inhibition using ethylene glycol for a rich gas at 0°C (Wilson *et al.*, 1985).

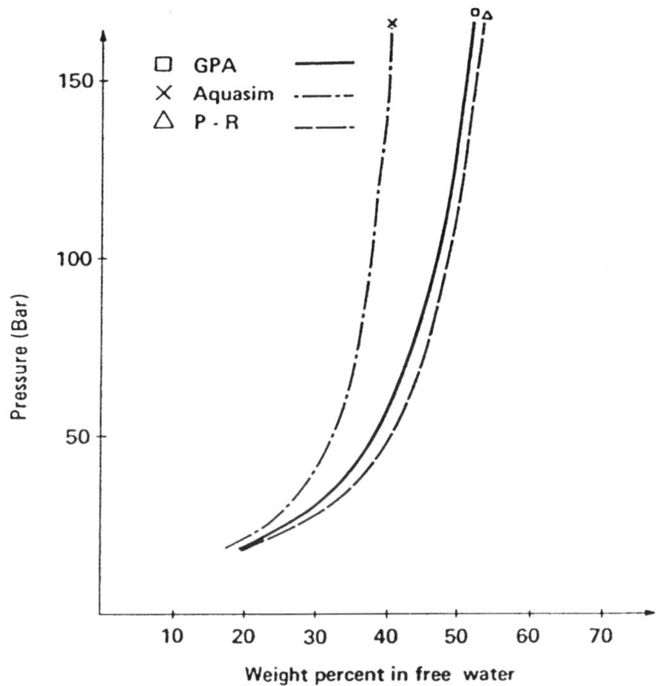

Figure 6–7. Hydrate inhibition using diethylene glycol for a rich gas at 0°C (Wilson *et al.*, 1985).

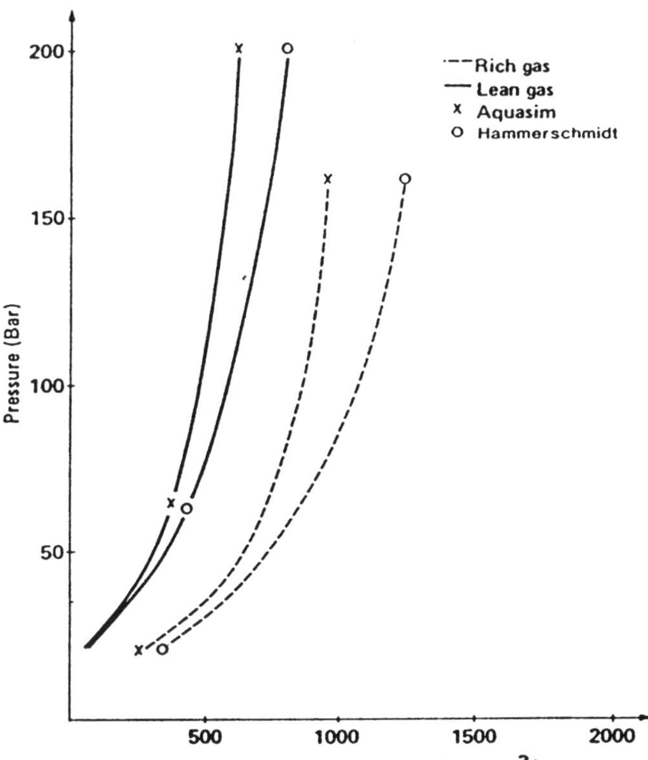

Figure 6–8. Glycol injection rate for hydrate depression at 0°C using ethylene glycol (Wilson *et al.*, 1985).

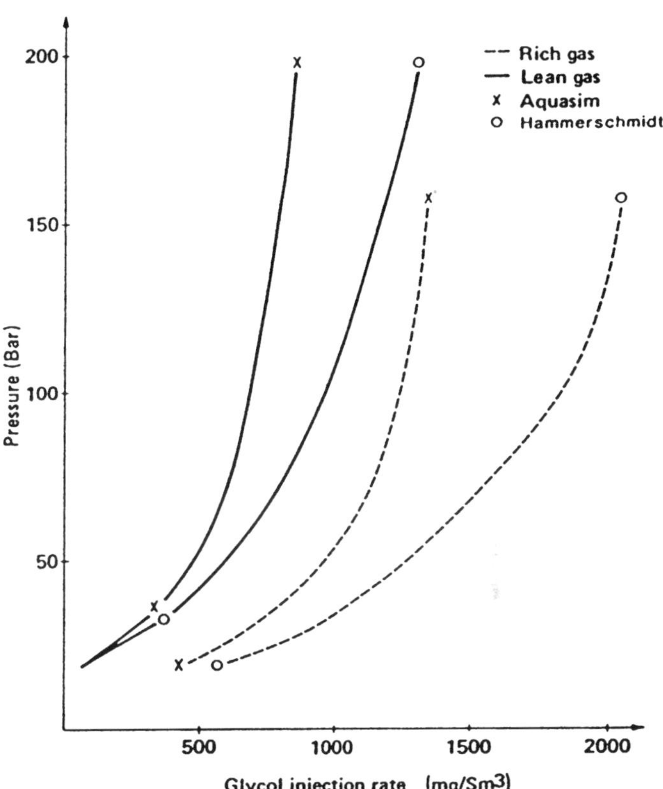

Figure 6–9. Glycol injection rate for hydrate depression to 0°C using diethylene glycol (Wilson *et al.*, 1985).

Comparison of Design Methods.

Wilson *et al.* (1985) compared the Hammerschmidt and Aqua*Sim methods of predicting the minimum methanol liquid-phase concentration (Fig. 6–4) and the required methanol injection rate (Fig. 6–5) for both lean and rich natural gases. Similar comparisons are presented for minimum EG concentration (Fig. 6–6), required EG injection rate (Fig. 6–7), minimum DEG concentration (Fig. 6–8), and required DEG injection rate (Fig. 6–9). The *Hammerschmidt correlation* appears to be considerably more *conservative* than Aqua*Sim both for minimum inhibitor concentration and required injection rate. Both Wilson *et al.* (1985) and Robinson *et al.* (1987) emphasize the need for more reliable data to improve the reliability of the computer predictions.

Stange *et al.* (1989) simulated hydrate inhibition in a 20 km offshore pipeline carrying 280 MMscfd. Injection of 50000 lb/day of methanol is required to protect against hydrate formation down to 32°F (a 9°F safety factor). Their computer simulation, which uses the Parrish and Prausnitz (1972) model, accurately predicted the methanol distribution between the gas, HC liquid, and water phases.

DESIGN EXAMPLES

Example 6–1. Ten MMscfd of a 0.7 SG natural gas cools down to 45°F in a buried pipeline. Minimum pipeline pressure is 900 psia. Concentration of commercially available glycol inhibitor is 75% by weight. What volume of inhibitor solution must be added daily if the gas enters the line saturated at 90°F? Solve for EG, DEG, and MeOH.

Solution: Some *safety factor is warranted* because hydrate formation can cause severe problems—plug the pipeline and so disrupt sales. A safety margin of 5 to 10°F is recommended to allow for operating excursions and uncertainties in data and in J-T cooling effects (e.g., see Figs. 9–17, 9–18). Because both the Hammerschmidt and dilution methods are quite conservative, 5°F is used here.

Normal hydrate formation temperature = 64°F (Fig. 4–10, or GPSA, 1987, Fig. 20–13)

Free water formed (Fig. 4–6, or GPSA, 1987, Fig. 20–3).

Water content of gas at:

90°F, 900 psia	= 48 lb/MMscf
40°F, 900 psia	= 9.6 lb/MMscf
Free water formed	= 48 − 9.6
	= 38.4 lb/MMscf
	= (38.4)(10)
	= 384 lb/day.

<u>EG Injection Rate</u>
1. Estimate minimum EG concentration using Hammerschmidt equation (Eq. 6–1):

$$W = \frac{(d)\,(MW)\,(100)}{(d)\,(MW) + K_H}$$
$$= \frac{(64 - 40)\,(62)\,(100)}{(64) - 40)\,(62) + 4000}$$
$$= 27.1 \text{ weight percent}$$

The minimum EG concentration (27 weight percent) is much less than the EG concentrations required to prevent any freezing of the glycol at any temperature (60 to 80 weight percent). Therefore, the EG injection rate is calculated using Equation 6–4.

2. If the EG enters at 75 weight percent and the dilution is limited to 10 weight percent then W_{out} = 65 weight percent.

The following is obtained by substituting these concentrations into Equation 6–4:

$$w_G = I\,[(100/W_{out}) - (100/W_{in})]$$
$$38.4 = I\,[(100/65 - (100/75)]$$
$$I = 187 \text{ lb EG/MMscf}$$

lb pure EG/day
$$= (\text{lb EG/MMscf})(\text{MMscf/day})$$
$$= (187)\,(10) = 1870 \leftarrow$$

3. If the 75 weight percent inlet EG enters at 90°F, the density of inlet glycol solution is 1.084 g/ml or 9.03 lb/gal (Fig. 6–10). And:—

$$\left(\frac{\text{lb EG}}{\text{day}}\right)\left(\frac{\text{lb soln}}{\text{lb EG}}\right)\left(\frac{\text{gal soln}}{\text{lb soln}}\right)$$
$$= (1870)\quad(100/75)/(9.03/1)$$
$$= 276 \text{ gal soln/day}$$

The recommended glycol flow rate is quite small, approximately 12 gal/hr or 28 gal/MMscf. This latter figure is higher than the typical rule of thumb of 6 to 10 gal/MMscf. Because the lowest temperature is 40°F, the glycol dilution might be increased to 20% thus lowering the injection rate.

<u>DEG Injection Rate</u>
1. Estimate minimum DEG concentration using the Hammerschmidt equation (Eq. 6–1):

$$W = \frac{(d)(MW)(100)}{(d)(MW) + K_H}$$
$$= \frac{(64 - 40)(106)(100)}{(64 - 40)\,(106) + 4000}$$
$$= 38.9 \text{ weight percent}$$

The minimum DEG concentration (38.9 weight percent) is much less than the DEG concentrations required to prevent any freezing of the glycol (65 to 75 weight percent). Therefore, the DEG injection rate is calculated using Equation 6–4.

2. If the DEG enters at 75 weight percent and the dilution is limited to 10 weight percent then W_{out} = 65 weight percent. Substituting into Equation 6–4, one obtains:

$$w_G = I\,[(100/W_{out}) - (100/W_{in})]$$
$$38.4 = I\,[(100/65) - (100/75)]$$
$$I = 187 \text{ lb DEG/ MMscf}$$

lb pure DEG/day
$$= (\text{lb DEG/MMscf})\,(\text{MMscf/day})$$
$$= (187)(10) = 1870$$

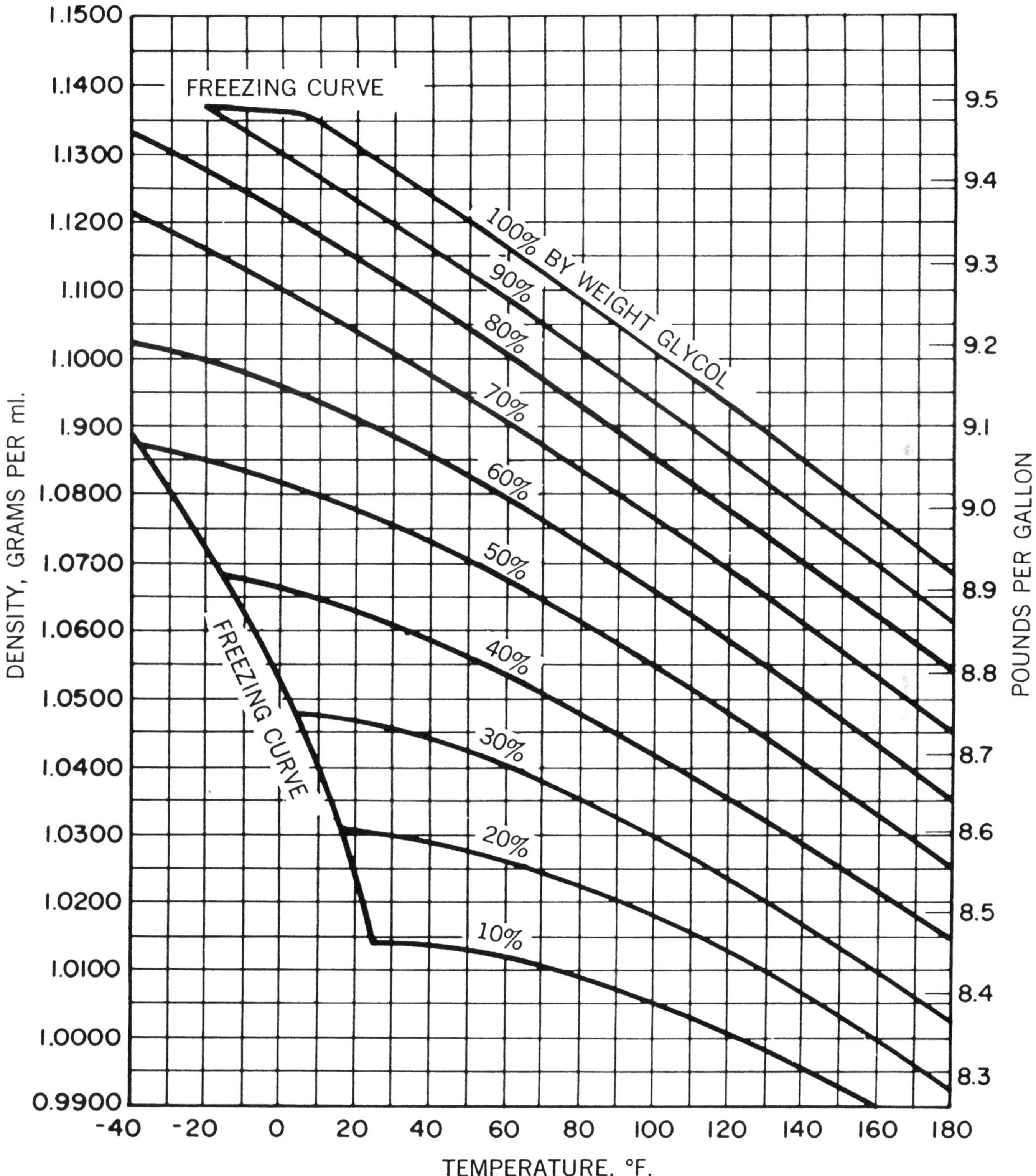

Figure 6–10. Densities of aqueous ethylene glycol solutions (courtesy Dow Chemical Company, 1962, p. 89).

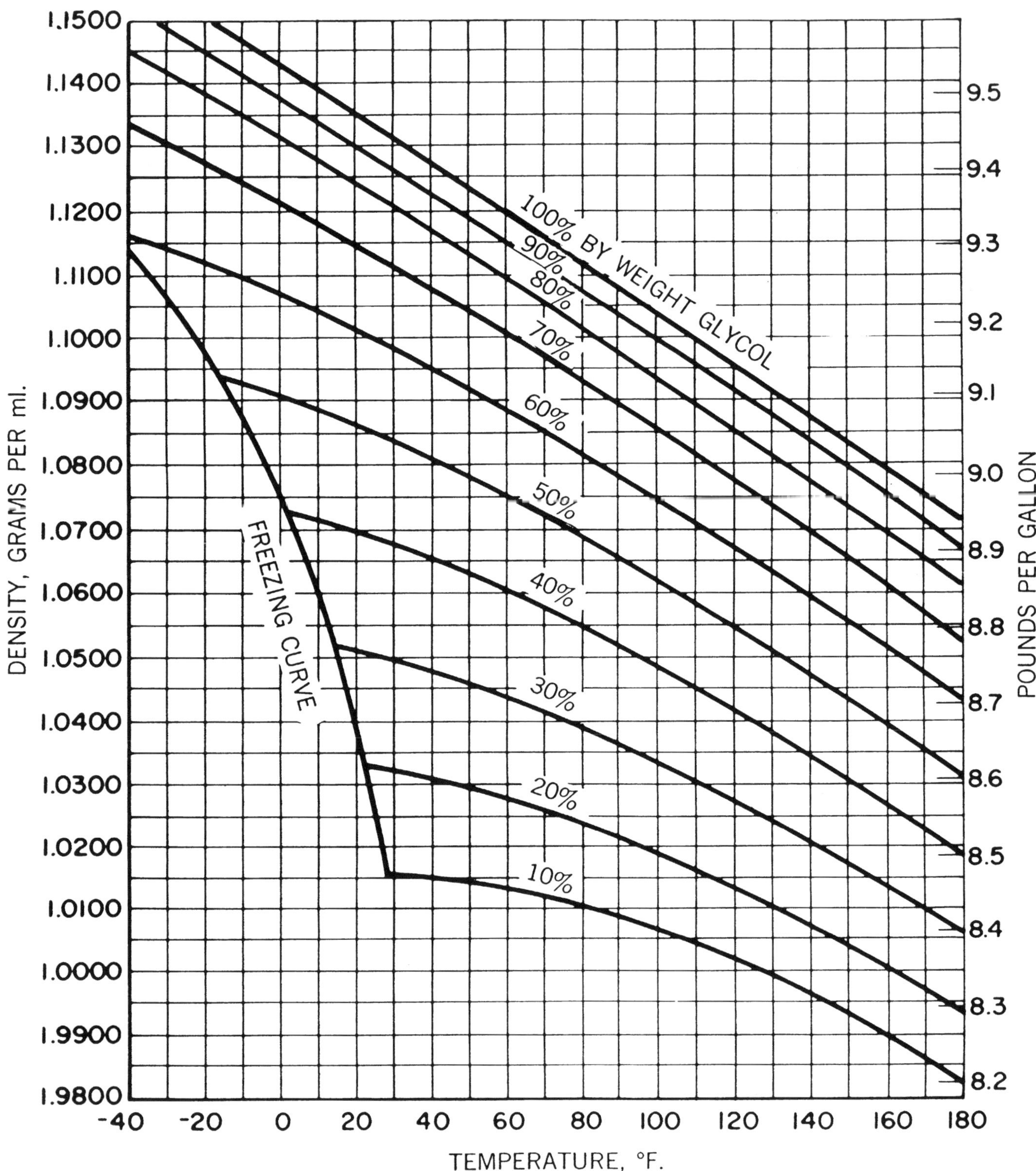

Figure 6–11.   Densities of aqueous diethylene glycol solutions (courtesy Dow Chemical Company, 1962, p. 90).

3. If the 75 weight percent inlet DEG enters at 90°F, the density of inlet glycol solution is 1.093 g/ml or 9.10 lb/gal (Fig. 6–11). And:

$$\left(\frac{lb\ DEG}{day}\right)\left(\frac{lb\ soln}{lb\ DEG}\right)\left(\frac{gas\ soln}{lb\ soln}\right)$$
$$= \quad (1870) \quad (100/75)/(9.10/1)$$
$$= 274\ gal\ soln/day$$

As with the previous EG case, this DEG injection rate could be reduced by increasing the dilution.

MeOH Injection Rate

1. Estimate minimum liquid-phase MeOH concentration using the Hammerschmidt equation.

$$W = \frac{(d)(MW)(100)}{(d)(MW) + K_H}$$
$$= \frac{(24)(32)(100)}{(24)(32) + 2335} = 24.8\ weight\ percent$$

2. If the methanol enters pure, $W_{in} = 100\%$ and is diluted down to the Hammerschmidt minimum, then Equation 6–4 yields:

$$w_G = I\ (K_H)/(d)(MW)$$
$$38.4 = I\ (2335)/(24)(32)$$
$$I = 12.6\ lb\ pure\ liquid\ MeOH/MMscf$$

3. Methanol usage in the gas phase is obtained from Figure 6–3. At 40°F and 900 psia:

$$\frac{(lb\ vapor\ MeOH/MMscf)}{(weight\ percent\ MeOH\ in\ soln)} = 1.05$$
$$MeOH\ used = (1.05)(24.8) = 26.0\ lb\ MeOH/MMscf.$$

4. Total MeOH required
$$= 12.6 + 26.0$$
$$= 38.6\ lb\ pure\ MeOH/MMscf$$
$$= 386\ lb\ pure\ MeOH/day.$$
$$= 386/6.64 = 58.1\ gal\ MeOH/day$$
Note: 1 U.S. gal pure MeOH at 60°F and 1 atm weighs 6.6385 lb (GPSA, 1987, p. 23–3).

OPERATION

Operation of injection plants is reviewed under the following headings: glycol injection, glycol-water-oil separation, glycol losses, corrosion inhibition, and troubleshooting.

Glycol Injection

The *glycol* should be *injected* into the gas stream just *ahead* of *cooling* vessels (HXER, choke, expander, or chiller), thus providing sufficient time for good glycol-gas contact. Ballard (1966) offered the following recommendations:

1. Improved contact may be realized by spraying the glycol directly onto the tube sheets as shown in Figure 6–12. Such injection may supplement or replace injection ahead of the exchanger. The glycol must contact the upper portions of horizontal exchangers, or the glycol may separate by gravity thus allowing the upper tubes to plug. Vertical exchangers may be preferred because the tube sheet distributes the glycol and slightly improved heat transfer may be obtained.

2. The *glycol* should be *sprayed-in* as a *very fine mist*. Ballard (1966) recommends 100 psi spray nozzle pressure and full-cone nozzles over hollow-cones.

3. Just ahead of each nozzle the *glycol* should be *filtered* to remove 5 μm and larger particles. This minimizes clogged nozzles, distorted spray patterns, and erosion of the nozzle openings. A backflow blowdown helps to keep the nozzles working freely.

4. Installation of backup nozzles in parallel will permit nozzle removal, inspection, and repair/replacement without interrupting operation.

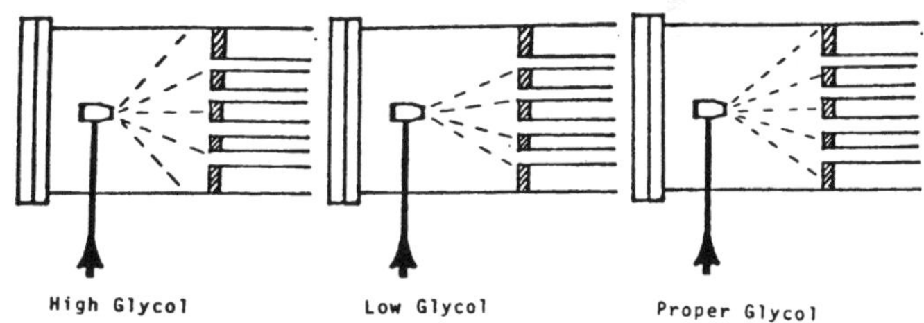

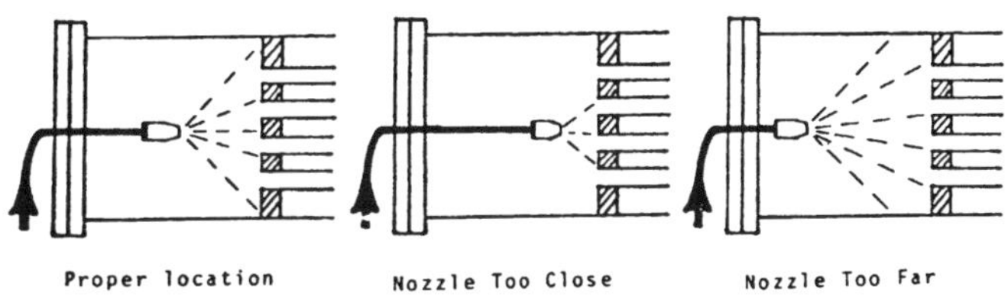

Figure 6–12. Glycol injection practices.

Glycol-Water-Oil Separation

All vessels for *separating glycol-water-oil* mixtures (below 30–40°F) should be designed for a residence time of about 30 minutes for maximum efficiency. This residence time is found by dividing the vessel volume by the volumetric flow rate of the inlet mixture. Above 30–40°F, a residence time of 20 minutes may be adequate if foaming or emulsions are no problem. If possible, it is recommended that separation be carried out at 80–100°F, which lowers the viscosity and promotes settling. At colder temperatures, separation becomes increasingly difficult.

Glycol Losses

Glycol is *lost* in three ways:

1. Solution in liquid hydrocarbons
2. Vaporization and entrainment in the exit gas
3. Decomposition in and carry-over from the regenerator.

Glycol solubility in hydrocarbon liquids increases with:

1. An increase in molecular weight of the glycol (TEG is more soluble than DEG, and DEG than EG)
2. An increase in temperature
3. An increase in weight percent glycol in the water-glycol mixture.

Glycol solubility also depends on the type of hydrocarbon liquid present. Glycols are more soluble in aromatics and naphthenes than in paraffin hydrocarbons.

True glycol solubility in the HC liquids normally occurring at a separation temperature of 60°F and for 50–70 weight percent glycol concentrations, range from 10 to 50 ppm for EG and 20 to 100 ppm for DEG. These losses are approximately 0.3 to 3.0 gal glycol/1000 bbl of condensate. Glycol solubility losses would be lower at lower separation temperatures. If good vapor-liquid separation is not obtained, losses can be substantially higher.

Total losses of 1 gal/100 bbl condensate are not uncommon. Vaporization losses are very low below 100°F and may be estimated as 0.23 lb/MMscf. Carry-over losses from the separator and regenerator vary widely with equipment design and operation but are usually less than 0.2 gal/MMscf (Harrell, 1988).

Corrosion Inhibition

A *full-flow filter* comparable to that in a TEG dehydrator is needed to remove solid particles that promote corrosion and degradation. The pH of glycol solutions does not appear to affect emulsion-breaking time. However, some corrosion inhibitors do have a pronounced effect upon emulsion stability.

Almost all degradation products are acidic. Corrosion is minimized by adding inhibitors to hold the *solution pH* near *7.2*. Common inhibitors and the concentrations at which they are used are: MEA (1–2 weight percent), Nacap (0.3 weight percent), Boraz (0.3 weight percent) and $K_2 HPO_4$.

Troubleshooting

Ballard (1966) provides the following advice.

Prevention of Freeze-Ups. Check the following items:

1. Be sure that the inlet FWKO or scrubber is removing all free water and potential contaminants such as salt and solids.
2. Check gas-glycol contact (is the atomizing nozzle eroded or plugged?). The glycol must be sprayed in a very fine mist to form a glycol film on the inner surfaces of all equipment. Hydrates can form on bare surfaces.
3. Check glycol circulation rate; make sure the pump is working properly.
4. Check the glycol concentration. Hydrometers can provide simple and rapid determination of approximate water content of glycol solutions. However, glycols often become contaminated with salt or other chemicals that alter the specific gravity.

Improvement of Glycol-Condensate Separation. Vapor-liquid and liquid-liquid separators will be discussed in Volume 2, Field Processing of Crude Oil; accordingly, the present treatment is limited to identifying potential troubles. If condensate samples taken downstream of the separator appear cloudy, then the *glycol-condensate separation* is incomplete. Possible causes are:

1. Excessive gas velocities in the separator.
2. Inadequate liquid retention time in the separator.
3. Foaming and/or emulsion forming in the separator.
4. Plugging of the glycol dump valve.
5. Improper operation of the liquid-level controllers.

Defoamers and/or emulsion-breaking chemicals can be added to temporarily control foam or emulsion formation. Permanent resolution requires that the sources of the foam and/or emulsion be determined and eliminated. Remember that glycol-condensate separation is much easier at higher temperatures (60–70°F) and lower pressures.

Review Questions

1. A natural gas properly dried with TEG or solid desiccants will not form hydrates. True or False?

2. If a dry gas will not form hydrates, why not dry the gas and forget about flowline heating and/or injection of EG, DEG, or methanol?

3. Name three methods of preventing hydrates in natural gases. Under what circumstances are these three methods used most most advantageously?

4. Name the three chemicals used most frequently to prevent hydrate formation.

5. Compare the hydrate-prevention characteristics of EG, DEG, and methanol, using the following criteria:
   • Ease of use
   • Capital equipment requirements
   • Operating costs
   • Vapor-phase losses
   • Quality of protection provided.

6. When would you use methanol injection? EG? DEG?

7. How would you modify the flowsheet shown in Figure 6–1
   a. To handle a sour natural gas stream?
   b. To remove absorbed hydrocarbons from the rich glycol?

8. How would you clear a long length (10 miles) of buried pipe line that is plugged with hydrates? What safety precautions must be taken?

Problems

1. A sweet, water-saturated 0.65 SG natural gas will be cooled from 100°F and 1000 psia to 30°.
   a. At what temperature (°F) will liquid water condense? At what temperature (°F) will hydrates form?
   b. How much water (lb/MMscf) will be removed from the gas?
   c. How much DEG (gal pure DEG/MMscf) should be added to prevent hydrate formation? Assume that the DEG is injected as a 75 weight percent solution and is recovered at 65 weight percent.
   d. What is the minimum aqueous-phase methanol concentration (weight percent MeOH) required to prevent hydrates for the gas of Problem 1?
   e. How much methanol (lb/MMscf) should be added to prevent hydrates from forming?

2. Review the following glycol-injection flowsheet. What changes would you recommend? Why?

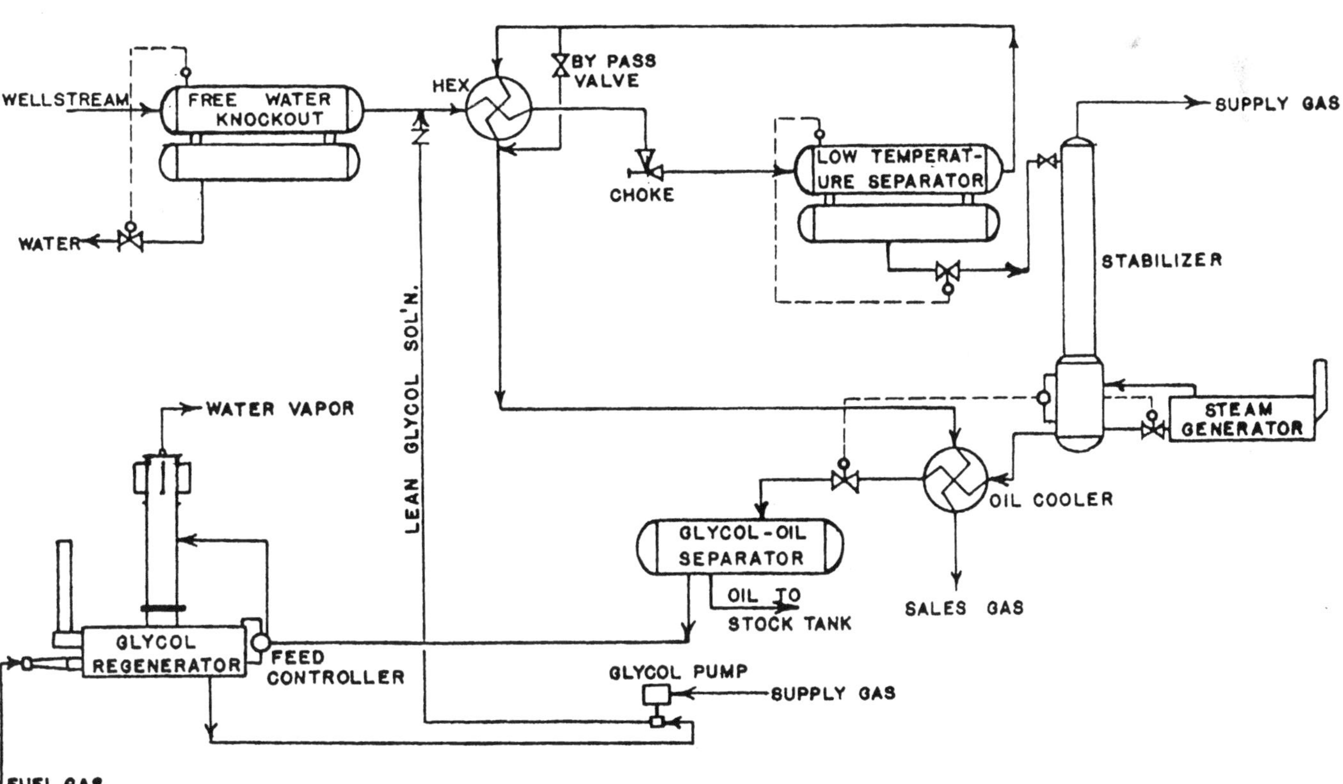

Figure 6–13. Flow sheet for a glycol injection system.

Nomenclature

d = depression of hydrate formation temperature (°F)

I = weight of pure inhibitor required (lb/MMscf)

K_H = Hammerschmidt constant for inhibitor (°F − lb/lb mol)

MW = molecular weight of inhibitor (lb/lb mol)

W = required concentration of inhibitor in water phase in the Hammerschmidt equation (weight percent), Equation 6–1

W_{in} = concentration of inlet inhibitor stream (weight percent)

W_{out} = concentration of exit inhibitor stream (weight percent)

w_G = water removed from gas stream (lb/MMscf)

x_{MeOH} = mol fraction of methanol in aqueous liquid phase

References

Arnold, J. L., and R. I. Pearce (1961), "Let Glycols Help Inhibit Hydrate Formation," *Oil & Gas J.*, Vol. 59, No. 25, pp. 92–95 (June 19); Vol. 59, No. 27, pp. 125–129 (July 3).

Ballard, D. (1966), "How to Operate a Glycol Plant," *Hydro. Proc.*, Vol. 45, No. 6, pp. 171–188 (June).

Campbell, J. M. (1979), "Gas Conditioning and Processing," Campbell Petroleum Series Inc., Vol. 2, p. 159, Norman, OK.

Curry, R. N. (1981), "Fundamentals of Natural Gas Conditioning," Part III Additive Injection, PennWell Pub. Co., P.O. Box 1260, Tulsa, OK 74101.

Dow (1962), Gas Conditioning Fact Book, Dow Chemical Co., Midland, MI.

GPSA (1987), Engineering Data Book, 10th Ed., Gas Processors Suppliers Association, Tulsa, OK.

Guenther, J. D. (1979), "Natural Gas Dehydration," Seminar on Process Equipment and Systems, Teknologist Institut, Taastrup Denmark, April 26.

Hammerschmidt, E. G. (1939), "Preventing and Removing Gas Hydrate Formations in Natural Gas Pipelines," *Oil & Gas J.*, Vol. 37, No. 52, pp. 66–72 (May 11).

Hammerschmidt, E. G. (1939), "Preventing and Removing Hydrates in Natural Gas Lines," *Gas Age*, Vol. 83, No. 9, pp. 45–49 (April 27).

Harrell, B. E. (1989). Personal Communication, NATCO, P. O. Box 1710, Tulsa, OK 74101.

Jacoby, R. H. (1955), "Calculation of Methanol Requirements to Prevent Formation of Gas Hydrates," *Gas*, Vol. 31, No. 2, p. 114 (Feb.).

Kohl, A. L., and F. C. Riesenfeld (1985), "Gas Purification," 4th Ed., Gulf Publishing Co., Houston, TX, pp. 591–599, 609–610.

Nielsen, R. B., and R. W. Bucklin (1983), "Why Not Use Methanol for Hydrate Control?," *Hydro. Proc.*, Vol. 62, No. 4, pp. 71–78 (April).

Parrish, W. R., and J. M. Prausnitz (1972), "Dissociation Pressures of Gas Hydrates," *Ind. Eng. Chem. Proc. Des. Dev.*, Vol. 11, No. 1, pp. 26–35 (Jan.).

Robinson, D. B., H. J. Ng, and C-J Chen (1987), "The Measurement and Prediction of the Formation and Inhibition of Hydrates in Hydrocarbon Systems," *Proc. 66th Ann. GPA Conv.*, Denver, CO, pp. 154–164 (March 16–18).

Robirds, K. D., and J. C. Martin III (1962), "How to Design a Glycol-Injection System," *Oil & Gas J.*, Vol. 60, No. 18, pp. 85–89 (April 30).

Sloan, E. Dendy (1985), "The Colorado School of Mines Hydrate Program," *Proc. 64th Ann. GPA Conv.*, p. 125, Houston, TX (March 18–20).

Stange, E., A. Majeed, and S. Overa (1989), "Experimentations and Modeling of the Multiphase Equilibrium and Inhibition of Hydrates," *Proc. 68th Ann. GPA Conv.*, San Antonio, TX (March 13–14).

Union Carbide Chemical and Plastics Company, Inc. (1989), "Ethylene Glycol, Diethylene Glycol, and Triethylene Glycol," 39 Old Ridgebury Rd., Danbury, CT06817–0001.

Wagner, Jan, Ruth C. Erbar, and Ali I. Majeed (1985), "AQUA*SIM Phase Equilibria and Hydrate Inhibition Using the PFGC Equation of State," *Proc. 64th Ann. GPA Conv.*, Houston, TX, pp. 129–136 (March 18–20).

Wilson, A., E. Stange, A. Goethe, and D. G. Elliot (1985), "Hydrate Prediction: Importance of GPA Research," *Proc. 64th Ann. GPA Conv.*, Houston, TX, pp. 155–160 (March 18–20).

Chapter 7

Gas Sweetening

INTRODUCTION

By far the most stringent specifications for selling natural gas are those for sulfurous gases. In the U.S., hydrogen sulfide, H_2S, is nearly always limited to *0.25 gr/100 scf (4 ppmv)* and the specification can be *as low as 1 ppmv* in some countries. The maximum total sulfur content including mercaptans (RSH), carbonyl sulfide (COS), disulfides (RSSR), etc., is usually to 10 to 20 gr/100 scf. However, some transcontinental pipelines now also require as low as 0.25 gr/100 scf of organic sulfur compounds.

Carbon dioxide, CO_2, is usually included with nitrogen as an inert gas specification of 2 to 3%. So there are occasions where sweetening requires only H_2S removal. When the CO_2 content exceeds the specification and/or cryogenic processing follows, removal to as low as 100 ppmv may be needed.

There are more than *30 natural gas sweetening processes*. For detailed reviews consult Goar (1972), GPSA (1987), Kohl and Riesenfeld (1985), Love (1985), Maddox (1974), and Wall (1986, 1988). These processes can be classified as:

1. *Batch processes*, e.g., iron sponge, Chemsweet, Sulfa-Check, and caustic soda. Because the reactant is discarded, the use is limited to removing small amounts of sulfur, i.e., low gas flow rates and/or small concentrations of hydrogen sulfide.
2. *Aqueous amine solutions*, e.g., monoethanolamine, diethanolamine, diglycolamine, methyldiethanolamine, and proprietary formulations such as Amine Guard, Ucarsol, Flexsorb, and Gas/Spec. These solutions are regenerated and are used to remove large amounts of sulfur, and CO_2 when needed.
3. *Mixed Solutions* (mixtures of an amine, physical solvent and water), e.g., Sulfinol, Ucarsol, Flexsorb, and Optisol. These solutions also absorb organic sulfur and are capable of high acid gas loadings.
4. *Physical Solvents*, e.g., Selexol, Rectisol, Purisol, Fluor Solvent. These can be regenerated without heat and simultaneously dry the gas. They are used for bulk removal of CO_2 frequently offshore.
5. *Hot potassium carbonate solutions*, e.g., Hot Pot, Catacarb, Benfield and Giammarco-Vetrocoke. These are chemical analogs of the physical solvents.
6. *Direct oxidation* to sulfur, e.g., Stretford, Sulferox Lo Cat, etc. These processes practically eliminate H_2S emissions.
7. *Adsorption*, e.g., Linde, Zeochem, and Davison Chemical molecular sieves. The use is limited to low acid gas concentrations, and the gas is simultaneously dried.
8. *Membranes*, e.g. AVIR, Air Products, Cynara (Dow), duPont, Grace, International Permeation, and Monsanto are most suitable for bulk CO_2 separations, especially when the feed-gas concentration is very high.

Figure 7–1 shows when these processes are used (based on Tennyson and Schaaf, 1977). It is a general guide because many other factors must be considered, e.g., detailed gas analysis, operating temperature and pressure, well location, environmental regulations, sales contract, etc.

The batch and amine processes are used for over 90% of all onshore wellhead applications. Amines are preferred when the lower operating cost—the chemical cost for batch processes is prohibitive—justifies the higher equipment cost. The key is the sulfur content of the feed gas. *Below 20 lb/day batch processes* are more economical, and *over 100 lb/day amine* solutions are preferred.

Sulfur content, lb/day = 1.34 (MMscfd)(gr H_2 S/100 scf).

BATCH PROCESSES

Many chemicals absorb acid gases. The trick is to find an *inexpensive, non-hazardous material* that has a *high capacity*

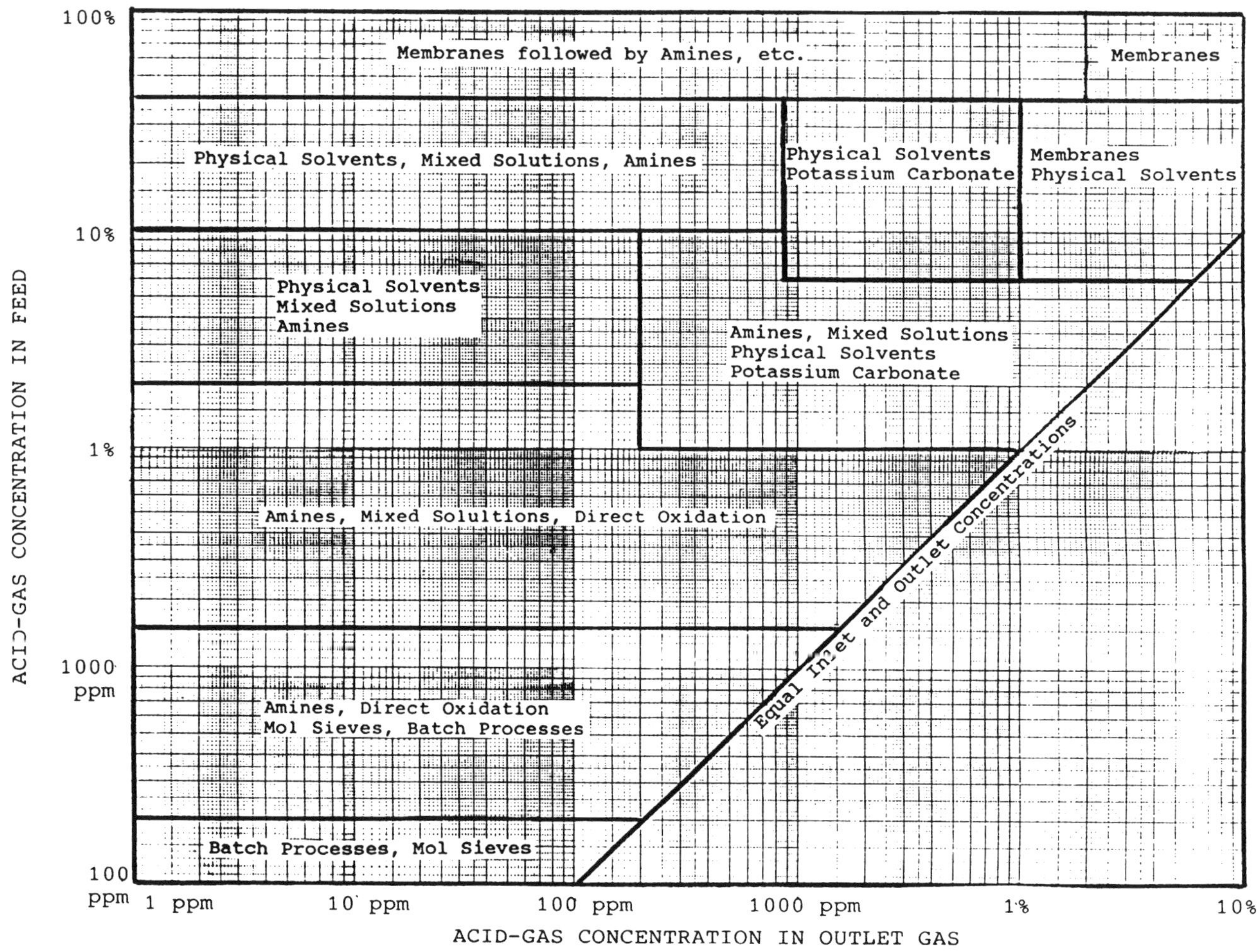

Figure 7–1. Guide to selection of gas sweetening processes (Tennyson and Schaff, 1977; modified).

for H_2S and produces an *environmentally acceptable* waste. Currently marketed processes are iron sponge, Chemsweet, and Sulfa-Check. Caustic soda is used—especially when mercaptans must be removed or when the resulting product can be sold to a paper company. Aqueous slurries of iron oxide particles and mixtures of formaldehyde, methanol, and water are encountered also.

The *advantages* of *batch processes* are:

1. Complete removal of low to medium H_2S concentrations with no consumption of the reactant by CO_2.
   (Caustic soda is the exception.)
2. Relatively low capital investment when compared to regenerative processes.
   (However, large high-pressure vessels are costly.)
3. The affinity for sulfur-containing gases is largely independent of the operating pressure.
4. Often the removal of organic sulfur contaminants such as the lower molecular weight mercaptans is adequate.

The *disadvantages* are:

1. Uninterrupted operation requires two or more contact towers, so that at least one tower is on stream when another is being recharged.
2. The presence of liquids—poor upstream separation or condensation in the tower due to temperature changes— ruins iron sponge chips and makes liquid sweeteners foam.
3. Hydrate formation can occur at higher pressures and lower temperatures.

Table 7–1 compares the iron sponge, Chemsweet, Sulfa-Check, and caustic soda processes.

Process Descriptions

Iron Sponge. *Iron sponge* is the oldest and still the *most widely used* batch process (Duckworth and Geddes, 1965; Anerousis and Whitman, 1985; and Zapffe, 1963). Because

Table 7–1 Comparison of Batch Processes

Process:	Iron Sponge	Chemsweet[1]	Sulfa-Check[2]	Caustic
General:				
Age, years	100	10	4	50
Units in service	Thousands	Hundred	Hundred	Hundred
H_2S selective	Yes	Yes	Yes	No
Vessel:				
Cross-sectional area	Lowest	High	Medium	Medium
Height, ft	10 to 20	20 to 30	20 to 30	10 to 20
Relative cost	Lowest	High	Medium	Medium
Chemical:				
Relative Cost	Low	High	High	Low
lb S/ft^3 bed	7	4	13	varies[3]
Disposal:				
Effort and time needed	High	Low	Low	Low
Environmental approval	Many States	7 States	14 States	Best sold to paper company

Notes:
1. Data are from National Tank Company.
2. Data are from NL Treating Chemicals.
3. Using 20% NaOH the NaSH process absorbs 11 lb S/cu ft.

the bed consists of solid particles, the superficial gas velocity is higher and the tower diameter smaller than for slurries. The chemical cost is low, but a major drawback is the labor needed to empty the tower and dispose of the pyrophoric spent sponge.

Carnell (1986) describes a granular zinc oxide powder that reacts with H_2S to form zinc sulfide. The efficiency is about 50% at ambient temperatures.

Iron sponge consists of *wood chips* or shavings impregnated with *hydrated ferric oxide*, Fe_2O_3, and sodium carbonate to control the pH—major suppliers are Connelly-GPM and Physichem Technologies. The chips are dumped into the vertical contact tower (Figure 7–2) through the manway at the top. (How the chips are loaded is very important and this is discussed in the Operation section.) They rest on a perforated support plate. Process gas enters at the top of the vessel and is distributed by an inlet gas diverter. It passes through and reacts with the sponge and leaves sweetened at the bottom of the vessel. The sweetening reactions form ferric sulfide, Fe_2S_3, and ferric mercaptide.

$$Fe_2O_3 + 3\ H_2S = Fe_2S_3 + 3\ H_2O$$
$$Fe_2O_3 + 6\ RSH = 2\ Fe(RS)_3 + 3\ H_2O$$

Downflow of the process gas is preferred to upflow because this reduces channeling. Also, flow of any water spray or condensate formation is concurrent to the gas flow, and the downflow pressure drop tends to scrub the Fe_2S_3 off the unreacted Fe_2O_3.

Many iron sponge vessels have a water spray nozzle above the bed. The bed support is several layers of wire mesh screen on a structural grate. The mesh serves as a filter to keep the chips from being carried out with the gas. In larger vessels the space below the grate is often filled with coarse packing, e.g., pipe thread protectors for added support. The vessel has loading and unloading manways at the top and bottom of the bed, respectively. The platform and ladder facilitate bed loading and cleanout. Bed cleanout is easier if a hose and a supply of high-pressure water are available. Steel chains attached to the top and side of the vessel or coiled in the bed help dislodge the spent sponge.

The *regeneration* or revivification reactions are:

$$Fe_2S_3 + 3\ O = Fe_2O_3 + 3\ S$$
$$2\ Fe(RS)_3 + 3\ O = Fe_2O_3 + 3\ RS{:}SR$$

For continuous revivification a small amount of air or oxygen is added to the inlet sour gas stream to oxidize the Fe_2S_3 back to Fe_2O_3 as soon as the H_2S is absorbed. Periodic revivification entails tower shutdown, depressurization, and circulation of the remaining gas with incremental additions of oxygen. Both revivification methods are dangerous due to the heat of the oxidation reactions (85,000 Btu/lb mol sulfur). Precise air additions are difficult and the injection equipment is expensive. Also, oxygen promotes corrosion, especially in the presence of CO_2, and caking of the bed is increased by the melting and solidification of the sulfur.

Oxidants such as hydrogen peroxide, H_2O_2, also can be used to regenerate spent sponge (Frech and Tazuma, 1982).

Iron sponge unit design

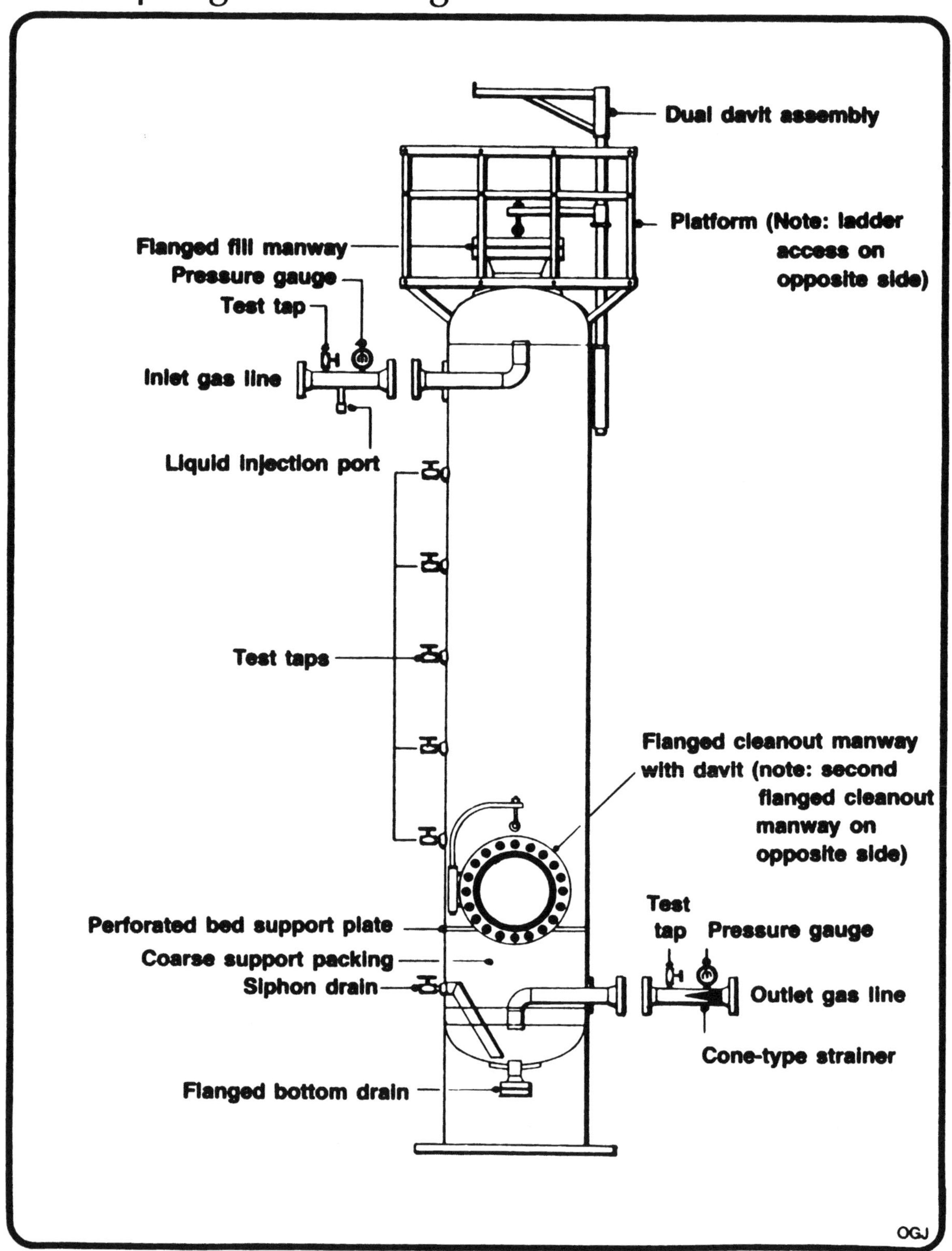

Figure 7–2.   Iron sponge unit (Anerousis and Whitman, 1984, copyright 1984, Society of Petroleum Engineers).

Field experience indicates that bed life can be extended three to five times and the spent sponge rendered more manageable, i.e., reduced tendency to burn and cake. But handling H_2O_2 is difficult and requires a degree of cleanliness not usually found in the oil patch. Also, the cost of H_2O_2 can be justified only by decreased downtime, i.e., one tower operation.

The iron sponge process has been used to *sweeten natural gas liquids* (Davis *et al.*, 1985). The liquid flow is usually upward at less than 3 gal/min ft^2.

Liquid and Slurry Processes. These were developed as alternatives to iron sponge. Occasionally chemicals other than those described below are encountered, but their use is limited by significant shortcomings. Inhibit, Scavinox, and Di-Chem are mixtures of *formaldehyde, methanol*, and *water*. They react with H_2S to form cyclic carbon-sulfur compounds known as formthionals. Use is limited in the USA because formaldehyde is classified by the EPA as a carcinogen. However, in Canada the spent material can be injected into disposal wells (Schaak and Chan, 1989a, 1989b).

Slurries of *iron oxide* have been used to selectively absorb H_2S (Fox, 1981; and Kattner *et al.*, 1986). The results have been inconsistent; problems include obtaining stoichiometric use, premature H_2S breakthrough, and foaming. Also, high corrosion rates have been reported. This has been circumvented by coating the contact vessels with phenolic or epoxy resins (Samuels, 1988).

The chemical cost is higher than that for iron sponge. But this is partially offset by the ease and lower cost with which the contact tower can be cleaned out and recharged. Also, liquids freeze and foam—so antifreeze must be added in winter—and an antifoamant should be available. Finally, obtaining approval to dispose of the spent chemicals, even if they are non-hazardous, is time consuming.

Chemsweet is a NATCO process (Manning, 1979; and Manning *et al.*, 1981). The white powder product is a mixture of *zinc oxide, zinc acetate*, and a *dispersant* to keep the zinc oxide particles in suspension. When one part is mixed with five parts of water the acetate dissolves and provides a controlled source of zinc ions that react instantaneously with the bisulfide and sulfide ions that are formed when H_2S sulfide dissolves in water. The zinc oxide replenishes the zinc acetate.

Sweetening	$ZnAc_2 + H_2S$	$= ZnS + 2\ HAc$
Regeneration	$ZnO + 2\ HAc$	$= ZnAc_2 + H_2O$
Overall	$ZnO + H_2S$	$= ZnS + H_2O$

The presence of CO_2 in the natural gas does not matter. The pH of the Chemsweet slurry is low enough to prevent significant absorption of CO_2, even when the ratio of CO_2 to H_2S is very high.

Sulfa-Check is NL Treating Chemicals' contribution (Bhatia and Allford, 1986; Dobbs, 1986). It is an aqueous solution of *sodium nitrite buffered* to stabilize the pH above 8. Also, there is enough strong base to raise the pH of the fresh material to 12.5. The claimed reaction with H_2S forms elemental sulfur, ammonia, and caustic soda.

$$NaNO_2 + 3\ H_2S = NaOH + NH_3\ + 3\ S + H_2O$$

But the process patent indicates that other reactions forming the oxides of nitrogen do occur (Burnes and Bhatia, 1985). And, CO_2 in the gas reacts with the sodium hydroxide to form the carbonate and bicarbonate. The spent solution is a slurry of fine sulfur particles in a solution of sodium and ammonium salts.

Caustic soda, NaOH, is the standard base and acid gas reaction that proceeds in two steps:

$$NaOH + H_2S = NaHS + H_2O$$
$$NaOH + NaHS = Na_2S + H_2O$$

Similar though slower reactions occur with CO_2 and RSH. Indeed, a NaOH wash is often used to specifically remove RSH from natural-gas streams after the H_2S and CO_2 have been removed. The rate of the reaction with RSH can be increased using a catalyst, e.g., Merox. The RSH rich caustic is regenerated using air.

$$NaOH + CO_2 = NaHCO_3$$
$$NaOH + NaHCO_3 = Na_2CO_3 + H_2O$$
$$NaOH + RSH = NaSR + H_2O$$
$$2NaSR + H_2O + O = 2NaOH + RSSR$$

Limiting the reaction time so that only the bisulfide is formed is the crux of the *NaSH process*. The reaction time between the NaOH and sour gas is controlled with a static mixer, and the solution properties—pH, concentration, and temperature—are controlled by recirculation of the product solution to dilute the NaOH feed.

By keeping the pH at 9 to 11 and the contact time short, over 90% of the H_2S can be removed in a single stage while 1 to 5% of the CO_2 is absorbed (Hohlfeld, 1980; Kent and Abib, 1985). Two or three stages produce pipeline quality gas. The product solution cannot be dumped indiscriminately, and injection into deep waste water wells is no longer condoned. It is best sold to a nearby paper company for bleaching wood pulp.

A NaSH plant is far more complex than the one or two contact towers that suffice for batch processes, especially if the sweetened gas is washed with water. This is recommended to prevent entrainment of caustic soda. Typically the unit includes storage tanks for 50% NaOH and water; metering, mixing, and a storage tank for the 20% NaOH feed; sour gas flow control; mixing tee or contactor; gas-liquid separator; NaSH recirculation and storage tank; and a control system.

Figure 7–3 shows a *typical contact tower* for liquids and slurries. The vessel is cylindrical with elliptical heads. It is equipped with inlet and outlet gas nozzles, slurry feed and overflow nozzles, a drain, a manway to meet code requirements, a mist eliminator, and a gas distributor. The sour gas enters at the bottom and rises to the distributor plate where it is dispersed into small gas bubbles. These rise through the slurry and provide the agitation needed to keep the slurry mixed. The sweetened gas flows through the mist eliminator and leaves at the top.

Bubbling the gas into the slurry results in an expansion of the slurry equal to the volume of all the gas bubbles rising through the slurry at any given time. Consequently, the depth of the slurry increases from the static value to the dynamic value.

Vessel height must be sufficient to maintain an adequate disengaging space, i.e., the distance between the top of the expanded slurry and the mist eliminator.

Vessel diameter is a function of the actual gas flow rate at operating conditions. A minimum superficial gas velocity (i.e., actual gas flow rate divided by the cross-sectional area) is needed to provide the agitation that keeps the slurry mixed. Too high a superficial gas velocity results in excessive expansion and entrainment of the slurry.

Vessel height is dictated by the chemical requirement. This requirement is determined by the bed life desired between charges. *Normal bed life is 30 to 90 days.*

Design

Table 7–2 summarizes *five design procedures* for iron sponge units. Agreement is excellent. Operation at a superficial velocity of 10 ft/min results in lower loadings, e.g., 0.25 lb S/lb sponge. Higher loadings correspond to lower space velocities. The iron oxide must be hydrated, i.e., $Fe_2 O_3.x\ H_2O$ with a moisture content of at least 20%. A basic environment is assured by including about one pound of sodium carbonate per bushel.

When gas bubbles flow through a liquid or slurry they quickly reach a constant terminal velocity of about 48 ft/min. The volume of the liquid expands by the volume of gas in the liquid. The *expansion factor*, f, is the ratio of the dynamic volume—or liquid depth in a vessel of constant diameter—to the static volume or depth. The expansion factor should not be less than 1.03 to provide mixing of the solution or above 1.20 where liquid entrainment occurs.

The following design procedure is recommended:
Required Data:

1. Gas flow rate, Q, MMscfd
2. Operating pressure, P, psig
3. Operating temperature, T, °F
4. H_2S inlet concentration, (ppmv)
   (If the H_2S concentration is in grains H_2S/100 scf, multiply by 15.9 to get ppmv.)
5. Compressibility factor, Z—Use Figures 3–16 through 3–21 (If the gas gravity is not known use 0.7.)
6. Expansion factor, f (Use 1.10 to 1.20.)

When the bubbles rise at 48 ft/min f is related to the superficial velocity V_s in ft/min

$$V_s = 48\ (f - 1)/f.$$

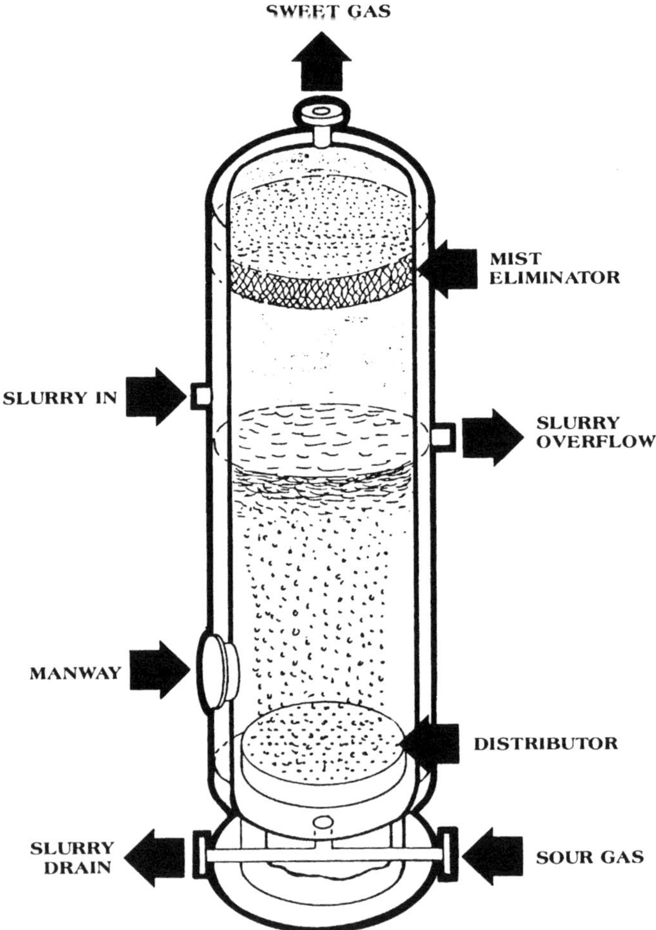

Figure 7–3. Contact tower for Liquid Processes (Manning, 1979).

Table 7–2 Design Criteria for Iron Sponge Towers

	Taylor (1956)	Perry (1970)	Campbell (1976)	Maddox (1974)	Anerousis Whitman (1985)
Bed Height, ft	10 (min.)	10 to 20 (min.)	10 (min.)	4 to 10 (min.)	10 (min.) 20 with RSH
Superficial Velocity, ft/min		10 (max.)			2 to 10
Max. Mass Velocity lb/min sq ft	$78\,\rho_g^{0.5}$		$78\,\rho_g^{0.5}$		
Max. Space Velocity, acfh/cu ft bed	180 90 (50 gr H_2S/100scf)	60	50 to 180	55 to 180	
Contact Time, sec					60 (min.)
Loading, lb S/lb sponge	0.60	0.45	0.51	0.51	0.25 to 0.35
Heat Effect, Sulfur Deposition; gr/min/ sq ft. of bed	15	—	15	15	15
Max. Gas Temp, °F dry water spray	100 120		113	110 —	— —
Minimum Gas Pressure, psig	20			—	20

Calculation Procedure:

1. Calculate the Gas *actual flow rate*, Q_a, at T & P

$$Q_a = \frac{19.63\,(Q)(T + 460)\,(Z)}{(P + 14.7)}\ \text{acfm}$$

where 14.7 = atmospheric pressure at sea level

2. Calculate the internal *cross-sectional area*, A, and the internal vessel diameter, ID.

Process	Superficial Velocity ft/min	Cross Sectional Area sq ft
Iron Sponge	10.0	$0.10\,Q_a$
Chemsweet	6.5	$0.16\,Q_a$
Sulfa-Check	8.0	$0.11\,Q_a$
Caustic Soda	8.0	$0.13\,Q_a$

$$ID = [A/0.7854]^{0.5}\ \text{ft}$$

Select a vessel OD that provides at least this internal diameter after considering the wall thickness. Calculate the actual ID (ft) and use this value.

3. Calculate the *daily chemical requirement*.

Iron Sponge	IS = 0.0133 (Q)(ppmv H_2S)	ft³/day	
Chemsweet	CS = 0.248 (Q)(ppmv H_2S)	lb/day	
Sulfa-Check	SC = 0.0474 (Q)(ppmv H_2S)	gal/day	

Caustic Soda   Depends on the CO_2 content
NaSH   NS = 0.109 (Q)(ppmv H_2S)   gal 20% NaOH/day

4. Select a *static bed height*, L—typically 10 to 20 ft. The volume, RV, of reactant is:

$$RV = 0.7854\,(ID)^2(L)\ \text{ft}^3$$

Note that bubbling gas into a liquid increases the volume (and the height in the tower) by the volume of the gas bubbles in the liquid at any given time. This increase in bath height—up to 20%—must be considered in selecting the shell length of the tower.

5. The *bed life*, BL, is:

Iron Sponge	BL =	RV/IS days
Chemsweet	BL = 11.7	RV/CS days
Sulfa-Check	BL = 7.5	RV/SC days

6. The *charge requirements* are:

Iron Sponge:	RV	bushels
Chemsweet:	11.7 RV	lb solids
	7.0 RV	gal water
Sulfa-Check:	7.5 RV	gallons

The above correspond to a minimum vessel diameters, reactant volumes, and bed lives. To increase the bed life, the tower diameter and height can be increased.

However, reducing the superficial gas velocity below 2 ft/min causes channeling in iron sponge towers. For slurries, operation velocities below 1 ft/min are not recommended because of poor mixing.

7. For *iron sponge* units the following *checks* (Table 7–2) are advisable even though they seldom invalidate a design. Make sure the *mass velocity* is less than 78 $(\rho_g)^{0.5}$ lb/min sq ft. This occurs only with liquid hydrocarbons and a density over 30 lb/cu ft.

8. Check the *contact time* by determining the space velocity acfh/ft^3 of bed. This should be less than 180 acfh/ft^3 for low H_2S contents and 90 acfh/ft^3 for over 50 gr/100 scf gas.

9. Check the *heat effect*. The sulfur deposition or absorption should be less than 15 gr/min ft^2 of bed area.

Example 7–1. Select and size a sweetening process for the following gas stream: 100 Mscfd, 100 ppm H_2S, 1% CO_2, 200 psig, 80°F, SG = 0.7, elevation = 1000 ft.

First estimate sulfur content:

$$\frac{\text{lb S}}{\text{day}} = \left(\frac{\text{scf}}{\text{day}}\right)\left(\frac{\text{lb mol gas}}{\text{scf}}\right)\left(\frac{\text{mol } H_2S}{\text{mol gas}}\right)$$

$$\left(\frac{\text{lb S}}{\text{mol } H_2S}\right)$$

$$= \frac{100,000}{1}\,\frac{1}{379.5}\,\frac{100}{1,000,000}\,\frac{32}{1}$$

$$= 0.85$$

Use a batch process—select iron sponge. Design steps are:

1. At actual gas flowing conditions
   Z = 0.955 (Fig. 3–19).

   P_{atm} = 14.2 at 1000 ft elevation
   (Appendix 6, Table A6–2)

   $$Q_a = \frac{19.63\ (0.1)(80 + 459.7)(0.955)}{(200 + 14.2)}$$
   $$= 4.72\ \text{acfm}.$$

2. Assume superficial velocity is 2.5 ft/min.

   Bed cross-sectional area = 0.4 Q_a
   $$= 1.89\ \text{ft}^2$$
   Bed diameter = 1.5 ft ID

3. Daily chemical requirement

   IS = (0.0133)(.1)(100) = 0.133 ft^3/day

4. Select bed height 5 ft, D = 1.5 ft

   RV = (.7854)(1.5)^2(5) = 8.84 ft^3

5. Bed life = RV/IS

   $$= 8.84/0.133 = 66\ \text{days}$$

6. Check contact time
   Space velocity = acfh/RV
   $$= (4.72)(60)/(8.84)$$
   $$= 32\ \text{acfh/ft}^3$$

   Okay because < 180.

7. Check heat effect

   100 ppmv = 100/15.9
   $$= 6.3\ \text{gr } H_2S/100\ \text{scf}$$
   S deposition
   = (scf/min)(gr H_2S/scf)/
   (bed area)
   = (100,000/24 × 60)(6.3/100)/(1.9)
   = 2.3 gr H_2S/min ft^2 bed

   Okay because less than 15 gr H_2S/min ft^2 bed.

Operation

The keys to *trouble-free operation* are *proper installation* of the tower and *preparation* of the *reactant*. Be sure to:

1. Check the *design parameters*, especially the superficial velocity of the sour gas. For iron sponge units, too low a velocity results in poor gas distribution, gas channeling, and loss of absorption efficiency. Too high a velocity compacts the bed and increases the pressure drop. For slurries, too low a velocity means poor mixing and settling of the solids. Too high a velocity promotes foaming, liquid entrainment, and carryover of the bath.

2. Anticipate *condensation* and *hydrate formation* during cold weather. A high pressure drop across an iron sponge bed can produce sufficient cooling for these phenomena.

3. Include an *inlet separator*. Liquids—water or hydrocarbons—quickly ruin an iron sponge bed by washing the iron oxide off the wood chips and swelling the wood. The result is a rapid pressure build-up across the bed— as much as 200 psi in one day. For slurries the results are equally disastrous: increased bath volume, formation of water/hydrocarbon emulsions, foaming, and carryover.

4. *Pressurize the tower slowly* and, if possible, from both the top and bottom. Be careful: too rapid a pressurization from the top compacts the iron sponge chips and strains the bed support. From the bottom too rapid a pressurization either lifts the chips and settles them unevenly or bumps liquids into the mist eliminator. Start and increase the gas flow slowly.

5. Unless impractical (e.g., remote areas), *check the operation daily*. Measure the gas flow, temperature, and

pressure; the inlet and outlet H_2S and RSH concentrations; the liquid height for slurries, the pH of the bath, or any water that collects in the bottom of sponge units. Larger installations usually have continuous H_2S monitors.

Iron Sponge.

1. For medium and large vessels, coarse packing can be placed below the bed support to prevent collapse of the support structure and possible damage to the vessel shell.
2. Make sure there is a *strainer* in the exit gas line to remove entrained particles of iron sponge.
3. Install a *differential pressure controller* across the tower. This can be used for shutdown or to bypass some of the sour gas when a predetermined pressure drop is reached.
4. *Check* the condition of the *iron sponge before loading*. The usual containers are polyethylene-lined burlap sacks. If possible use only unbroken sacks. Iron sponge does not need any special storage. But open sacks result in dehydration of the sponge in hot dry weather and washing of the iron oxide from the wood chips when rained on.
5. *Load the sponge* into the vessel in incremental layers of one foot. Rake each layer level before adding the next layer. Do not drop the sponge continuously in one spot (e.g., the center). This produces a cone of more densely packed sponge and promotes gas channeling. Do not compress the sponge. Some operators prefer to add a little water. Do not soak the sponge with a hose or fill the bed with water.

Slurries.

1. *Check the water supply.* Usually drinking water is satisfactory. Do not use produced water, pond water, etc., because the more impurities water contains the more likely it is to foam. If antifreeze is added, use pure ethylene glycol and not commercial antifreeze.
2. *Check the pH* of the batch. Typical values are over 11 for Sulfa-Check and 7 for Chemsweet.
3. *Perform a foam test* before starting the unit. For example, bubble a small stream of the process gas into two gallons of the batch in a five-gallon bucket. If foaming occurs note the stability and what antifoamant is most effective.
4. Make sure that the sour gas is saturated with water. If not, add water periodically to *restore the slurry level*.
5. *Include a downstream separator* with a sight glass. This gives immediate warning of batch carryover and protects downstream equipment.

Trouble-shooting

Even with well-trained, experienced operators, bed life may be inconsistent. The *major problems* are: premature H_2S breakthrough; rapid buildup of the pressure drop across iron sponge beds; and foaming with bath carryover in slurries. The usual cause of these problems is the condition of the inlet sour gas. If the gas is too close to or at its HC dew point, any cooling condenses hydrocarbons, and if it is unsaturated with respect to water, it dries the charge. Keep the gas a few degrees above its HC dew point and nearly saturated with water.

Iron Sponge.

1. *Check the inlet gas temperature*. Above 120°F the sponge dehydrates and loses its reactivity. A water spray can be used to cool the bed and maintain the sponge activity, but too much water leaches the sodium carbonate and washes the oxide off the chips.
2. *Check for channeling* which usually occurs at very low gas superficial velocities, i.e., below 2 ft/min. Often H_2S measurements and/or pressure readings at the sample taps around the tower can detect channeling.
3. *Check for hydrate formation* and *condensation*.
4. *Check the alkalinity* of the water in the bottom of the tower. If the pH is below 8.0, add soda ash.
5. *Check the pressure drop across the tower*. If it is very high, e.g., 100 to 200 psi, condensed liquids may have already swelled the chips and ruined the charge.
6. *Find the cause of the liquids*. It is generally poor upstream separation or condensation in the tower.
7. As a last resort back flow the bed to lift and repack the chips. This is often unsuccessful.

Slurries.

1. *Check the sour gas velocity* in the contactor. Above 10 ft/min the bath is carrying over. Below 1 ft/min the mixing is inadequate.
2. *Check the inlet gas temperature*. It should be 50 to 98°F. Too low explains condensation or hydrate formation and too high removes water.
3. *Check the bath level*. If it is foaming and too high, add antifoamant; if it is too low, add water.
4. *Check the outlet H_2S concentration*. If it is rising rapidly, add zinc acetate to Chemsweet towers and more buffer to Sulfa-Check units.
5. *Check the pH of the slurry*. If it is too low it confirms why the exit H_2S concentration is rising.
6. *Check for hydrocarbon condensation* because this fouls the reactive solid particles in Chemsweet. Remove a sample of the slurry from the top of the bath. Pour

100 ml into a beaker and warm to 140°F. The appearance of an oil layer indicates hydrocarbon contamination.

7. *Check for settling of the slurry*, especially if the gas flow has been very low or interrupted. Recirculate and remix the slurry. To do this release the pressure in the contact tower, and use the charge pump to circulate the slurry. Repressure the tower and observe the performance.

8. *Check the inlet gas for RSH, RSSR, and COS.* If these exceed 25% of the H_2S concentration, foaming is likely. Add more antifoamant.

9. *Check the distributor plate* or sparger. After a batch with an unsatisfactory bed life, remove the manway cover and check for plugged holes.

There have been three recent comparisons of the operating costs of batch processes. Harrell and Manning (1986) and Samuels (1988) used manufacturers' design procedures to size the contact vessels and determine the bed lives. Schaak and Chan (1989a, 1989b) used equal bed lives for all processes. Their results show that the relative costs are determined by vessel size, bed life, downtime for change out, labor costs, and equipment rental.

AMINE AND MIXED SOLUTIONS

Most continuous gas sweetening processes use aqueous solutions of *alkanolamines*. A physical solvent is added to enhance the performance of the amine in special situations; e.g. recent gas contracts with very low total sulfur limits or gas streams with high concentrations of acid gases and/or RSH where the alkanolamine alone is either inadequate or not the most economic approach. The *advantages* of these processes where the treating solution is regenerated and recirculated are:

1. *Complete removal* of medium to high concentrations of *acid gases* even at high gas flow rates with negligible consumption of the reactant.
2. Relatively *low operating cost* per pound of sulfur removed as compared with batch processes.
3. The solution composition can be tailored to the sour gas composition.
4. Large amounts of *organic sulfur compounds* can also be *removed* when a physical solvent is added to the amine solution.

The *disadvantages* are:

1. *High capital investment* compared with batch processes.
2. The *operational* and *maintenance costs* are significant.
3. Some of the processes, e.g., Sulfinol and Flexsorb require license or royalty fees.

Chemical Reactions

The reactions between amines and acid gases have been examined extensively (Dankwerts, 1970; Deshmukh and Mather, 1981; Kent and Eisenberg, 1976; Moshfeghian *et al.*, 1977).

1. *Ionization* and *hydrolysis* of H_2S and CO_2

$$H_2S = H^+ + HS^- = 2\,H^+ + S^{--}$$
$$CO_2 + H_2O = H^+ + HCO_3^- = 2H^+ + CO_3^{--}$$

Note that in basic solutions, bisulfide and bicarbonate ions may also be produced by the reactions:

$$H_2S + OH^- = HS^- + H_2O$$
$$CO_2 + OH^- = HCO_3^-$$

The dissociation constants for H_2S, HS^-, CO_2, and HCO_3^- are approximately $3(10^{-7})$, 10^{-13}, $5(10^{-7})$, and $6(10^{-13})$, respectively at 140°F (Kent and Eisenberg, 1976). So at a pH of 11 to 12 there is substantial ionization into the bisulfide and bicarbonate ions.

2. *Protonation of the amine.* RNH_2, $RR'NH$, and $RR'R''N$ denote primary, secondary, and tertiary amines, respectively—and the substitution of one, two, and all three of the hydrogen atoms of the parent ammonia molecule, NH_3, with alkanol groups, for example, with a primary amine.

$$RNH_2 + H^+ = RNH_3^+$$

The equilibrium constants for the simpler alkanolamines are high, e.g., 10^8 (Kent and Eisenberg, 1976). But a high pH limits the RNH_3^+ concentration; so reactions with both amine species are likely.

3. *Reactions with H_2S*, e.g., $RR'NH$. The addition compound between the protonated amine and bisulfide ion can be formed by two parallel mechanisms:

$$RR'NH_2^+ + HS^- = RR'NH_2.SH$$
$$RR'NH + H_2S = RR'NH_2.SH$$

The first reaction is fast, and together they limit the formation of the sulfide addition compound:

$$2RR'NH_2^+ + S^{--} = (RR'NH_2).S$$

4. *Reactions with CO_2*, e.g. $RR'R:N$. The bicarbonate addition compound is formed similarly to the bisulfide.

$$RR'R''NH^+ + HCO_3^- = RR'R''NH.HCO_3$$
$$RR'R''N + H_2O + CO_2 = RR'R''NH.HCO_3$$

However, primary and secondary amines—but not tertiary—can form carbamates also (Dankwerts, 1970; Deshmukh and Mather, 1981).

$$RR'NH + CO_2 = RR'NCOOH$$
$$RR'NCOOH + RR'NH = (RR'NCOO^- + RR'NH^+)$$

The stoichiometry is 0.5 mol CO_2 per mol of amine. The instability of the carbamate anion is its tendency to hydrolyze to the amine and bicarbonate ion.

$$2RR'NH + CO_2 = RR'NH_2^+ + RR'NCOO^-$$
$$RR'NCOO^- + H_2O = RR'NH + HCO_3^-$$

For the simpler alkanolamines the formation of stable carbamates is faster than the bicarbonate addition compounds, and it is the dominant reaction.

5. *Other sulfur gases.* The amine bases are too weak to form mercaptides. And COS does not hydrolyze significantly to H_2S and CO_2 at ambient temperatures. So removal of RSH, COS, and RSSR is by absorption at high pressure, and the solvent regeneration is desorption at low pressure and high temperature. A physical solvent is added to the amine solution when the affinity for sulfur gases is insufficient.

Comparison of Amine Solutions

Currently there are about *700 amine units* operating in the United States. Table 7–3 compares the principal properties and lists the typical operating conditions for the commonly used amine solutions. For more data consult Maddox, 1974; Richardson and O'Connell (1975); Bucklin, 1982; Dingman *et al.*, 1983; Kohl and Riesenfeld, 1985; GPSA 1987; and Leci, 1988). (Selexol and hot potassium carbonate are included also.) Figure 7–4 shows the molecular structures of the alkanolamines, some typical hindered amines, and some physical solvents.

Monoethanolamine, MEA. *MEA* was the *first alkanolamine* to be used and the process remains essentially unchanged after 50 years. While the predominant use has declined recently, it is often used when the partial pressure of the acid gases is low, e.g., low pressures and/or low acid gas concentrations. MEA is a primary amine with the lowest molecular weight. Accordingly, it is the most reactive, volatile, and corrosive. This is why it is used in relatively dilute solutions, has the highest vaporization losses, requires more heat for regeneration, and has the lowest hydrocarbon pickup (Perry, 1978).

Primary amines form stronger bonds with acid-gas anions than secondary and tertiary amines. As such, regeneration—

Table 7–3 Comparison of Gas Sweetening Solutions

Process	MEA	DEA	DGA	DIPA	MDEA	Mixed Solution	Selexol	K_2CO_3
Amine Type	Primary	Secondary	Primary	Secondary	Tertiary	Tertiary	—	—
Reactivity	High	Moderate	Moderate	Moderate	Moderate	Moderate		
Stability	Fair	Good	Fair	Good	Good	Good	Good	Excellent
HC Absorptivity	Low	Moderate	High	Moderate	High	High	High	None
Vaporization								
Losses	High	Moderate	High	Moderate	Low	Low	Low	None
H_2S Selective	No	No	No	No	Yes	Yes	Yes	No
Organic S								
Removal	Low	Low	Moderate	Low	Low	High	High	No
Corrosivity	High	Moderate	Moderate	Low	Low	Low	Low	Low
Tendency to Foam	Low	Low	Low	Moderate	High	High		
Cost	Low	Low	Moderate	Low	Moderate	High	High	Low
Degradability,								
H_2S	None	None	None	None	None	None	None	None
CO_2	Some	Low	Some	Some	Low	None	None	None
COS	Yes	Minor	Some	Severe	Minor	None	None	None
Solvent Conc.,								
wt %	15–20	20–35	45–65	30–40	40–55	50–80	100	25–35
AG mol/mol	0.3–0.4	0.5–0.6	0.3–0.4	0.3–0.4	0.3–0.45	0.3–0.4		
Loading scf/gal						3–6	4–6	4–8
Circulation,								
gal/mol AG	100–165	60–125	50–75		65–110	65–110		
Steam Rate,								
lb/gal	1.0–1.2	0.9–1.1	1.1–1.3		0.9–1.1	0.8–1.1	None	0.6–0.8
Reboiler Temp, °F	240	245	260	255	250	250	80	
Freezing Pt., °F	15	20	−40	16	−25	−20		
Heat of Reaction								
Btu/lb AG, H_2S	620	550	675		500	500		
CO_2	660	630	850		600	600		

Note: AG is "acid gas."

ALKANOLAMINES

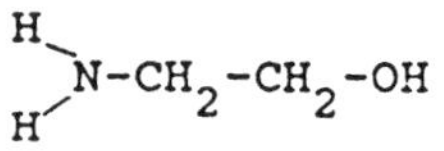

Monoethanolamine

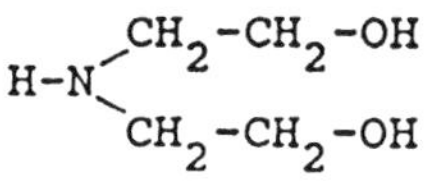

Diethanolamine

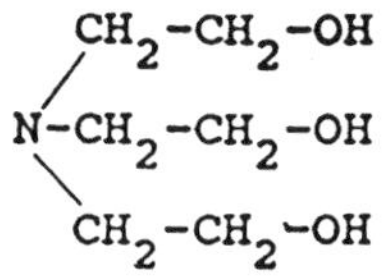

Triethanolamine

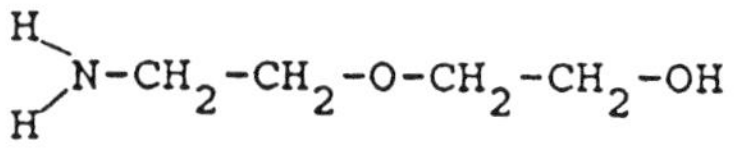

Diglycolamine

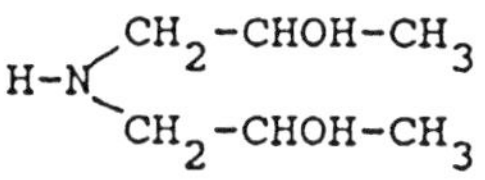

Diisopropanolamine

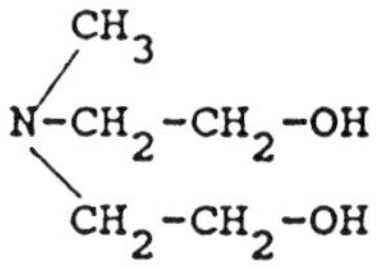

Methyldiethanolamine

HINDERED AMINES

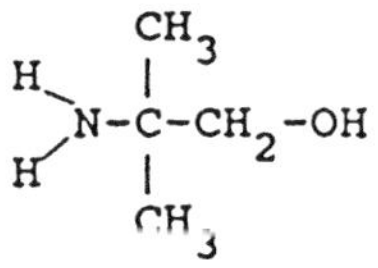

2-amino-2-methyl-1-propanol

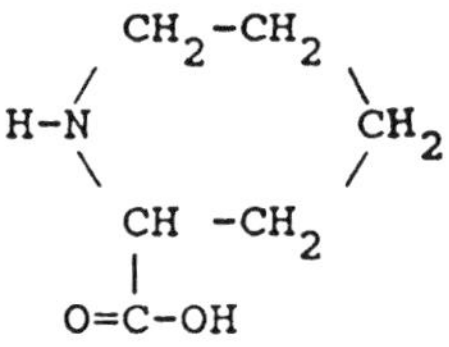

Pipecolinic acid

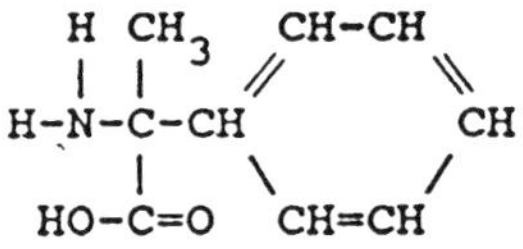

2-amino-2-phenylpropionic acid

PHYSICAL SOLVENTS

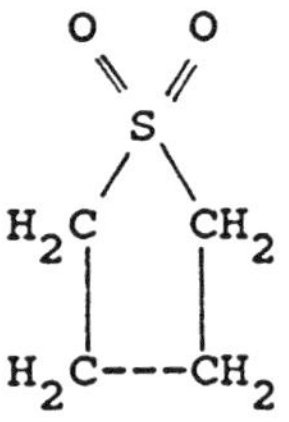

Sulfolane

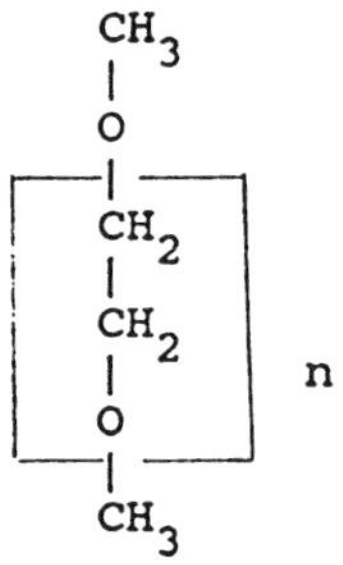

Selexol

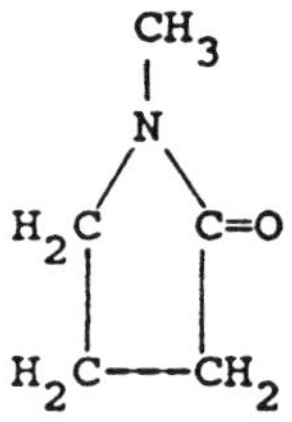

1-methyl-2-pyrrolidone
(M Pyrol or NMP)

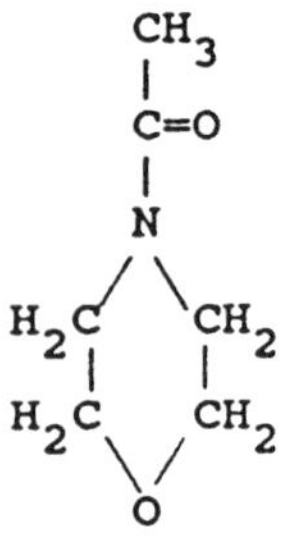

1-acetylmorpholine

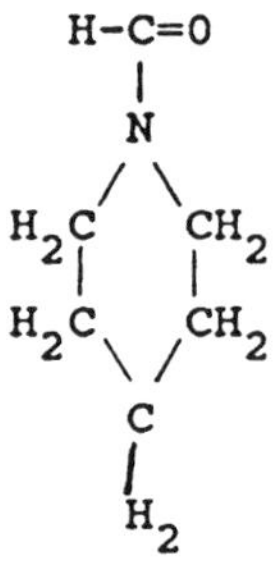

1-formylpiperdine

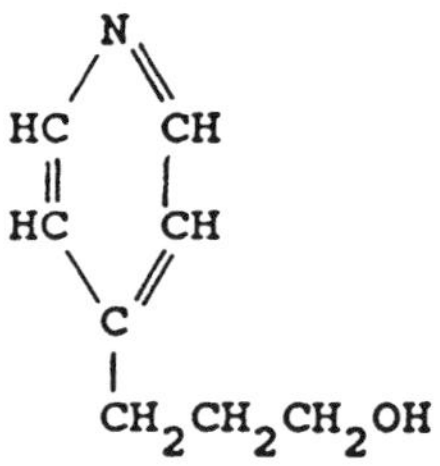

4-propanolpyridine

Figure 7–4.   Molecular structures of gas sweetening liquids.

i.e., the decomposition of the salts formed during absorption of acid gases—is more difficult. A reclaimer operating at a higher temperature than the reboiler is usually used to augment regeneration.

Diethanolamine, DEA. *DEA* is currently the *most widely used* gas sweetening solvent. Compared with MEA, it has lower heats of reaction with H_2S and CO_2, is less corrosive, and can be used in higher concentrations with greater acid gas loadings. This results in a reduced circulation rate and in lower capital and operating costs. DEA also is very resistant to degradation from RSH and COS. This was very significant in the early commercial successes in Canada (Goar, 1972). Major disadvantages are the inability to slip CO_2 and that some newer processes are more economical for specific situations.

There is an adaptation—*SNPA*, developed and licensed by Société Nationale Elf-Aquitaine—that achieves very high acid gas loadings—as much as 0.7 mol AG/mol DEA—without excessive corrosion. However, to achieve this (and the reduction in solvent circulation and regeneration cost) high inlet acid gas partial pressures are needed, e.g., 4 atmospheres (Dailey, 1970). The process is used extensively in Canada.

Diisopropanolamine, DIPA. *DIPA* is the amine used most frequently in the ADIP process licensed by Shell—MDEA is occasionally used. Klein (1970) reports low steam requirements, low corrosion rates, and suitability for gas streams containing COS. However, Bucklin (1982) rates the irreversible degradation from CO_2 and COS higher than that for MEA, DEA, or DGA. It is used to remove H_2S and COS from LPG also (Ouwerkerk, 1978).

Diglycolamine, DGA. *DGA* is a primary amine with the same molecular weight as DEA. As practiced in the Econamine process which is licensed by Fluor, a 50–70% solution can be loaded to 0.4 mol AG (acid gas)/mol DGA. This reduces the circulation rate substantially below that for DEA (Bucklin, 1982; Moore *et al.*, 1984).

A solution containing 65% DGA freezes at $-40°F$. This makes *DGA very suitable for cold climates*. Note that 20% MEA and 30% DEA solutions freeze at 15°F and 20°F, respectively. Similarly to MEA, it is suitable for treating gas streams with low partial pressures of acid gases, and a reclaimer is needed to complete the regeneration. DGA also exhibits a significant affinity for organic sulfur compounds.

Methyldiethanolamine, MDEA. *MDEA* has gained a greater share of the gas treating market in recent years. This success is due in part to the fact that tertiary amines such as MDEA exhibit a selectivity for H_2S over CO_2 when contacting gas streams containing both acid gases. This selective property is useful in upgrading the H_2S content of sulfur plant feed gas (Blanc *et al.*, 1981) and removing H_2S from CO_2 used for flooding projects (Laengrich *et al.*, 1982).

MDEA solutions have the *lowest heat requirements* for regeneration because they can be used at 50% strength and acid gas loadings of 0.4 mol/mol, have the lowest heats of reaction with H_2S and CO_2, and the lowest specific heat (Pearce, 1978 and Daviet *et al.*, 1984). Solvent losses are very low and the freezing point is about $-25°F$. Also MDEA is used in many of the special solvent formulations.

Special Solvents. Several *new alkanolamine-based solvents* have been developed recently. Some of Dow's Gas/Spec and Union Carbide's Amine Guard Solvents are MEA and DEA formulations with corrosion inhibitors. These permit the use of substantially higher concentrations, e.g., 30% for MEA and 50% for DEA. The results are significant reductions in the amine circulation rate and the heat requirements for regeneration (Kosseim *et al.*, 1984; and Pearce and Wolcott, 1986).

The most significant penetrations into the gas sweetening market—currently there are over 100 units—are Dow Chemical's *Gas/Spec* and Union Carbide's *Ucarsol* solvents. Some of these are *MDEA*-based with other *amines, buffers, promoters, corrosion inhibitors, and antifoamants* added. These additives control both the extent and rate of the reactions with H_2S and CO_2. The proprietary solvents are blended for specific applications; e.g. selective H_2S removal, partial or bulk CO_2 removal with or without high H_2S removal, high acid gas loadings, COS removal, etc. (Bacon and Pearce, 1985; Robinson *et al*, 1988; Thomas, 1988; and Dwyer,1989)

Both Dow Chemical and Union Carbide have extensive research and development programs, and it is very likely that new formulations with even better performance will be developed. Vickery *et al.* (1988) show that the addition of relatively small amounts of MEA or DEA to MDEA can be used to control the selectivity and acid gas loading.

Exxon's *Flexsorb* solvents use *steric hindrance*—i.e., the presence of a large group next to the nitrogen atom—(Fig. 7–4)—and basicity to control the CO_2/amine reaction (Sartori and Savage, 1983; and Goldstein *et al*, 1984). There are Flexsorb solvents for selective H_2S removal and bulk CO_2 removal (Clem *et al*, 1985).

Mixed Solutions. Shell's *Sulfinol* was the first of these processes. The solution uses *sulfolane* as the *physical solvent* and either *DIPA or MDEA* as the *amine* (Wansink, 1985). In addition to absorbing organic sulfur compounds,

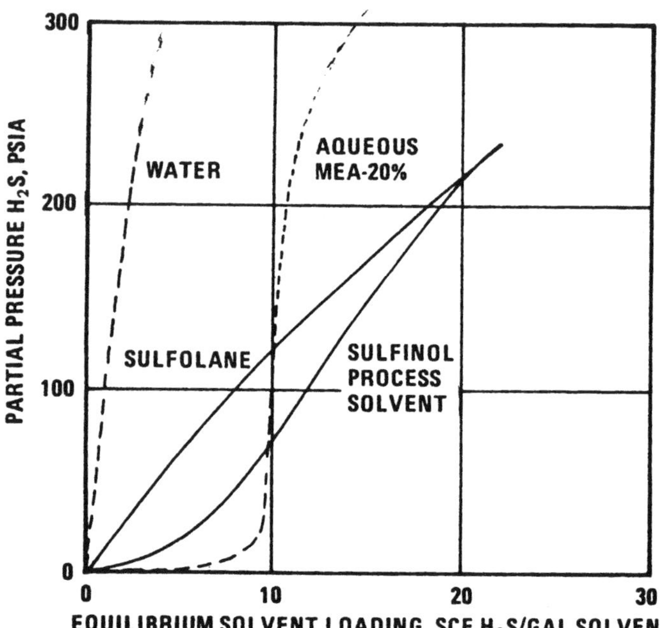

Figure 7-5. Equilibrium solvent loadings (Kohl and Riesenfeld, 1985, p. 8681).

Sulfinol's capacity for acid gases increases with the partial pressure of the acid gases. This is a Henry's Law phenomenon with no limiting chemical stoichiometry and a much lower heat release. Figure 7–5 compares the equilibrium solvent loadings for MEA, Sulfolane, and Sulfinol (Kohl and Riesenfeld, 1985). The solvent capacity of Sulfinol is less than MEA at low H_2S pressures. However, as the H_2S pressure increases the Sulfinol capacity continues to increase while the MEA loading levels off at the stoichiometric proportion. Accordingly, a major application has been treating gas streams with high acid gas contents.

Like most *physical solvents, Sulfinol* has a *significant affinity for hydrocarbons* and especially aromatics. Therefore, pretreatment facilities should be installed ahead of the acid gas if it is fed to a sulfur recovery plant. When MDEA is used the solution can selectively remove H_2S and leave up to 50% of the CO_2 originally in the gas. Several similar solvents have been developed recently; e.g., Ucarsol LE, Flexsorb PS, and Optisol (Byeseda *et al.*, 1985). Flexsorb SE units perform significantly better than plain MDEA units (Exxon, 1989) These formulations are proprietary and cost more than the alkanolamines. Ferrin

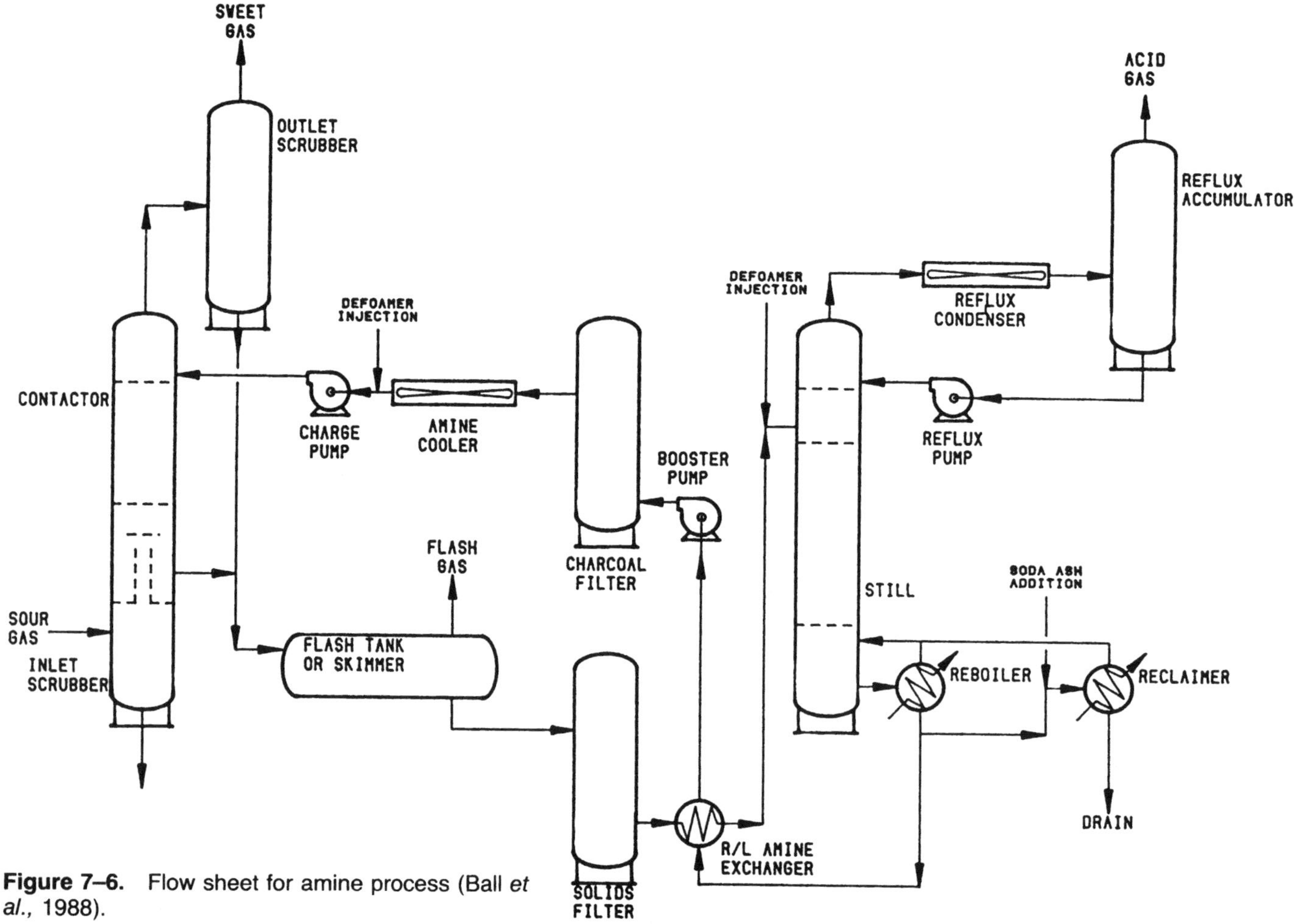

Figure 7–6. Flow sheet for amine process (Ball *et al.*, 1988).

and Manning (1982 and 1984) report the use of alkylated pyridines and pyrrolidones as absorbants for RSH and other organic sulfur compounds.

Physical solvents in mixed solutions are the most effective way to *remove COS*; COS absorption is complicated by reversible hydrolysis to H_2S and CO_2. Some absorption is obtained with DEA, DGA, DIPA, and the BASF MDEA. MEA should not be used because it reacts directly and nearly irreversibly to form a stable thiocarbamate.

Process Flow Sheet

Figure 7–6 shows the arrangement of the equipment needed for the *amine process*. There are two basic parts: the high pressure process gas contactor and scrubbers, the sizes of which depend on the sour gas flow rate and pressure, and the low pressure regeneration system that is sized primarily according to the amine circulation rate.

Insurance codes and *safety requirements* often require that the absorber tower and regeneration system with the direct-fired reboiler be on separate foundations 50 to 150 ft apart. For smaller units all of the regeneration equipment can be mounted on one skid. For larger units the reboiler

and aerial cooler are installed on adjacent concrete slabs.

The *simple amine flow sheet* can be *modified to improve performance*, usually for a specific purpose or application. For example, multiple amine feed and withdrawal ports on the contact tower improve the selectivity of H_2S absorption as shown in Figure 7–7 (Sigmund *et al.*, 1981). *Contacting the sour gas successively* with *lean MDEA* exploits the fact that the bisulfide reaction is fast and controlled by the rate at which the H_2S dissolves in the solution. The bicarbonate reaction—there is no carbamate reaction with a tertiary amine—is slower than the dissolution rate.

Another example is the *split stream* BASF Activated MDEA process (Fig. 7–8). The sour gas is treated in two stages: first with a semi-lean or partially regenerated solution and second with the lean or fully regenerated amine. The rich solution is flashed at an intermediate pressure and again at low pressure where it is also stripped with the overhead vapors from the regenerator. Most of this semi-lean solution is fed to the lower part of the contactor. The rest is regenerated and fed to the top of the contactor. The advantages are increased plant capacity and less energy for regeneration (Stanbridge *et al.*, 1984).

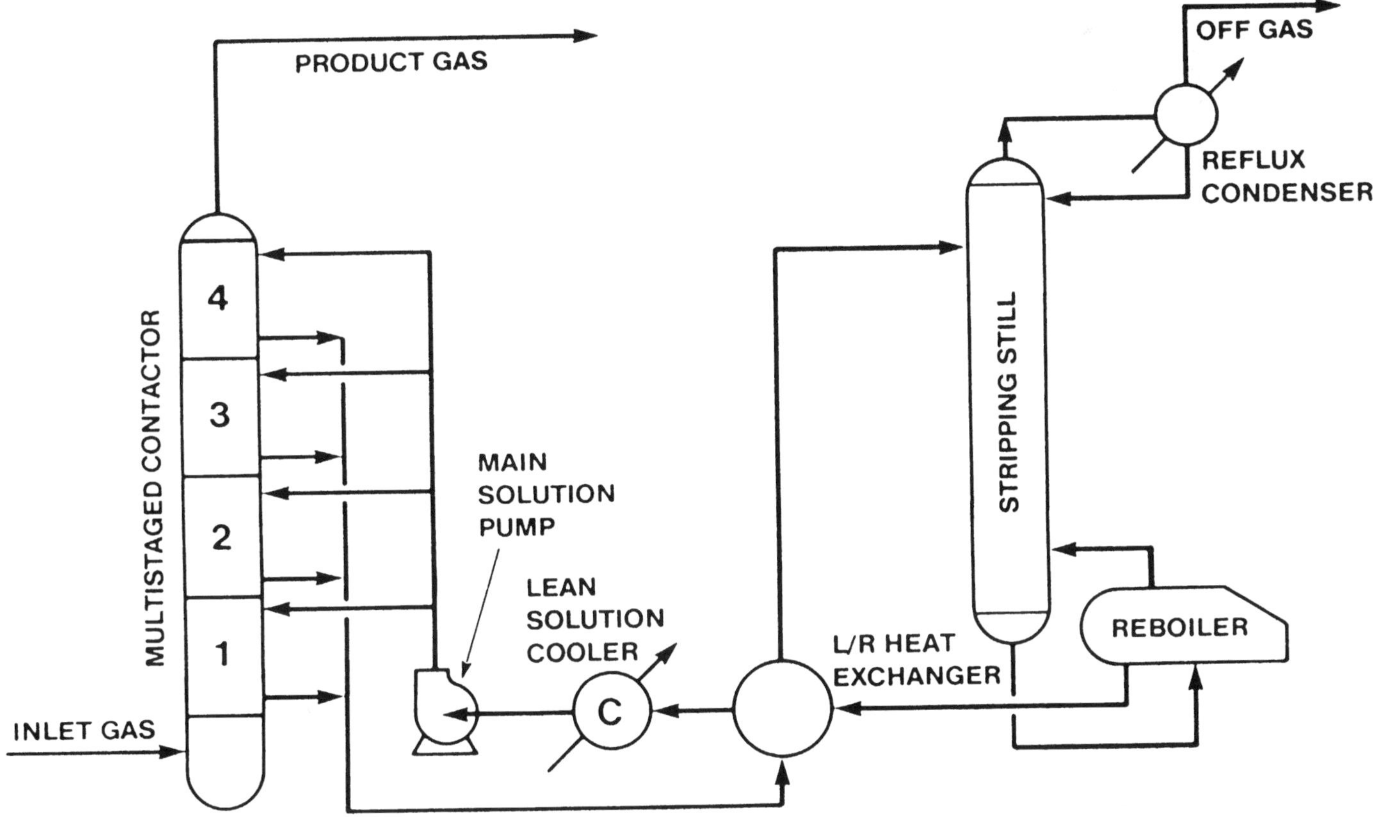

Figure 7–7.   HS process flow sheet (Sigmund *et al.*, 1981).

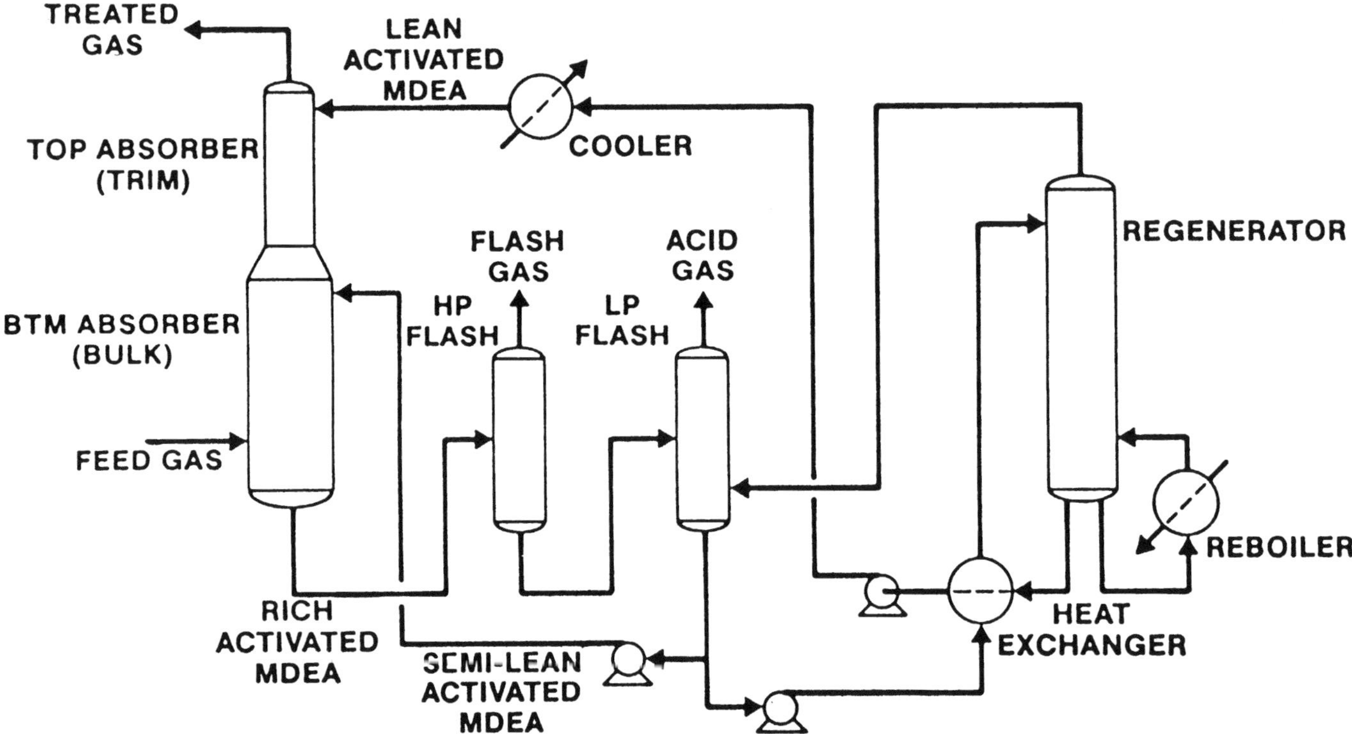

Figure 7–8. Split stream flow sheet (Stanbridge, 1984, courtesy Parsons).

Design

Input Data. The following information is required:

Technical Data:
Gas flow rate   MMscfd; nm^3/hr
H_2S content   mol percent; ppmv; gr/100 scf
CO_2 content   mol percent; ppmv
RSH and total S content ppmv; gr/100 scf
Gas composition, vol or mol percent, gravity (reference
   air = 1.0)
Gas pressure & temperature psig; kPa; kg/cm^2. °F; °C
Site elevation ft; m
Ambient temperature °F; °C (Min. and Max.)
Treated gas specifications H_2S, CO_2, RSH, "S" content
EPA and local regulations for disposal and emissions

Customer's Requirements:
Budget allocation
Time limitation
Use and/or retrofit existing equipment

Design specifications:
Codes and standards
Corrosion allowances
Welding procedures
Coating and paints, etc.

Ancillary equipment
Flash tank
Reclaimer
H_2S monitor, etc.

Computer Programs. Several *computer programs* are available for simulating amine plants. Probably the most accurate and detailed are those used by Dow Chemical for the Gas/Spec solvents and Union Carbide for the Ucarsol solutions. These programs simulate the mass transfer rates in the absorber and stripper and are based on the reaction stoichiometry, partial pressure of the acid gases, chemical kinetics, etc. The results are compared with operating data. The printout is a complete process flow sheet with tray by tray information (Katti and Langfitt, 1986 and Vickery *et al*, 1988).

Commercially available programs include:

AMSIM	—D. B. Robinson & Associates, Ltd., Alberta, Canada (Tomcej *et al.*, 1987)
APM	—Oklahoma State University, Stillwater, OK (Vaz *et al.*, 1981)
Chemcalc II	—Gulf Publishing Co., Houston, TX
Gas-Plant	—Taylor, Weiland & Associates, Pottsdam, NY

T-Sweet   —Bryan Research & Engineering, Bryan, TX
(Bullin and Polasek, 1982; King *et al.*,
1985)

Typically, these programs require considerable input. For example:

1. The type and concentration of the amine solution
2. The lean amine loading or the amine circulation rate
3. The residence time of the amine solution in the contactor or the percent approach to equilibrium loading of acid gas
4. The number of theoretical stages in the contactor and still
5. The reflux ratio or the steam required per gallon of circulation.

Often some of the input data must be calculated, e.g., acid gas loadings. Table 7–3 (Bucklin, 1982; Dingman *et al.*, 1983; GPSA Eng. Data Book, 1987; Kohl and Riesenfeld, 1985; and Maddox, 1974) provide accurate data as long as the sour gas is at over 400 psig and under 120°F. For applications out of this range use the partial pressures of the acid gas to obtain the loadings.

The programs allow the user to:

1. Compare amine solutions—type and concentration—quickly.
2. Optimize the number of theoretical stages in the contactor and still.
3. Obtain the heat duties and size equipment.
4. Get the physical properties of the amine solution.
5. Rerate an existing plant for a different amine solution.

Amine Flow Rate. Select the *appropriate sweetening solution* and the concentrations of amine and physical solvent. For example, MEA or DGA for low gas pressure streams, DEA for most applications, DGA for cold climates, MDEA for selective removal of H_2S, and physical solvent process for organic sulfur removal.

Calculate the *partial pressures of H_2S, CO_2, RSH* and other organic sulfur gases using the tower pressure and the inlet gas composition.

Estimate the *temperature of the rich amine* solution at the bottom of the contact tower. This is usually 20 to 40°F hotter than the inlet gas.

Calculate the *equilibrium loadings for H_2S, CO_2, RSH*, etc., in the rich amine solution. This requires solubility data from sources such as Butwell and Kroop (1986), Dow Fact Book (1962), Jou *et al.* (1982), Kohl and Riesenfeld (1985), GPSA (1987), Maddox (1974), Kent and Eisenberg (1976). These cover most amine solutions and account for the interactive effect of H_2S and CO_2 absorption and vice

versa. Nearly all data on physical solvents are proprietary. For temperature interpolation of equilibrium data consult the GPSA (1987).

Assume an *approach to equilibrium* and *determine the pickups for H_2S, CO_2, RSH*, etc. Mixed solutions are more complex because both amine and physical solvent must be considered separately, and also the interactions between these absorbents. The typical rich loading is about 80% of the equilibrium concentration and the lean (regenerated) loading (mol AG/mol amine) is 0.05 to 0.10 for primary amines, 0.03 to 0.05 for secondary amines, and as low as 0.005 for MDEA-based special solvents (Dwyer, 1989).

Calculate the *amine circulation rate* from the pickup and the acid gas content of the sour gas.

Calculate the *heats of reaction and solution* for all of the gas pickups. Use the data in Table 7–3 and the GPSA (1987) Engineering Data Book.

Estimate the *lean amine feed temperature*—100 to 130°F—and the sweet gas outlet temperature—15 to 30°F hotter than the inlet gas and/or 0 to 15°F above the lean amine.

Make a *heat balance* around the *contact tower*. Verify all assumptions. If not, another iteration is required. A sample calculation for H_2S and CO_2 absorption in DEA is presented by Butwell and Kroop (1986).

A quick estimate of the *flow rate* for typical MEA, DEA, DGA and MDEA solutions can be obtained from the following equation (based on Khan and Manning, 1985):

Circulation rate, gpm =
$$K\ (MMscfd)(Mol\ percent\ AG\ removed)$$

Amine	Soln wt%	mol AG/mol Amine	K
MEA	20	0.35	2.05
DEA conventional	30	0.50	1.45
high load	35	0.70	0.95
DGA	60	0.30	1.28
MDEA	50	0.40	1.25
Mixed solvent	75	0.40	varies

Note that K is inversely proportional to both the acid gas pickup and amine concentration. (These K values are for contact tower pressures above 400 psig and temperatures below 120°F.)

Absorption System. The high-pressure process gas system consists of an inlet scrubber, contact tower, and outlet separator.

The *inlet scrubber* removes slugs and drops of condensed hydrocarbons, produced water, corrosion inhibitors, and well treating chemicals from the inlet sour gas stream. It should be located as close as possible to the amine contactor. It can be complemented, but not replaced, by an integral scrubber in the contact tower. Removal of contaminants from the inlet gas is enhanced by washing the gas with water. However, this is an additional processing step and is needed only when the sour gas contains drops or particles too small to be effectively removed by conventional separators. Entrained liquids, most commonly slugs of produced water (often saline), must be removed because they quickly contaminate the amine solution and cause problems such as amine foaming equipment fouling, high corrosion rates, solvent loss, and sour effluent gas. Entrained liquids are the most frequent contamination source for all sweetening processes.

Most *absorber towers* use trays to contact the sour gas with the amine solution. However, packed columns are sometimes used for small applications. Twenty valve-type trays are the standard design, as is the inclusion of a mist pad above the top tray. The typical tray spacing is 18 to 24 inches and the distance between the top tray and the mist pad is 3 to 4 ft.

Some contactors have a *water wash* consisting of two to five trays above the amine feed tray. This eliminates amine carryover, conveniently adds make-up water, and is often used in low pressure MEA units.

The best way to monitor the performance of the absorber and to detect foaming is with a differential pressure cell connected to the gas inlet and outlet. Accordingly, connections for a *DP cell* should be included.

The *cross-sectional area* of the contactor is sized for the gas and amine flow rates. The maximum gas superficial velocity is obtained from the Souders-Brown equation.

$$V_{gas} = 0.25 \ [(\rho_{amine} - \rho_{gas})/\rho_{gas}]^{0.5} \ ft/sec.$$

Reduce the gas velocity by 25 to 35% to avoid jet flooding and by 15% to allow for foaming. Use an amine velocity of 0.25 ft/sec in the downcomer. Figure 7–9 estimates the internal diameter in terms of the sour gas flow rate and pressure (Khan and Manning, 1985).

Check the hydrocarbon dew point in the contactor at both the inlet and outlet conditions. Remember that the dew point increases as the acid gases are removed, and hydrocarbon condensation is one of the most common causes for excessive foaming.

Check for retrograde condensation in the absorber tower. (Hydrocarbon condensation can result from a temperature rise or a pressure drop in the contactor.)

The outlet gas separator removes any liquid carryover from the sweet gas and prevents contamination of downstream equipment. It also signals excessive foaming in the contactor.

Regeneration System. The *regeneration system* consists of a flash tank, rich/lean amine heat exchanger, stripping still, reboiler filters, and aerial cooler. There also are reflux, booster, and circulation pumps. And, primary amines require a reclaimer.

The amine solution absorbs hydrocarbons (HC) as well as acid gases in the contact tower. These flash out when the pressure is reduced. The flash tank provides the residence time for this degassing of the rich amine solution. HC absorption depends on the inlet gas composition and pressure. For lean sour gas the rule of thumb is 2 scf/gal. Recommended residence times are 10–15 minutes for two-phase units and 20–30 minutes for three-phase vessels. When flash gas containing H_2S is used as fuel, treat it with the lean amine. A small packed tower installed on top of the flash tank suffices.

The flash tank also improves the condition of the amine solution and the amine sweetening system. HC contamination in aqueous amine solutions often promotes foaming. Equipment fouling is more severe and occurs at a faster rate in the absence of a flash separator. Sulfur plant operations are hindered if HC are volatilized in the amine regenerator.

The *lean/rich amine heat exchanger* preheats the rich amine solution and reduces the duty of the reboiler. It also cools the lean amine and reduces the duty of the aerial cooler. Both shell and tube and plate and frame exchangers are used. The rich solution is usually tube side with a low inlet velocity—2 to 4 ft/sec—to reduce corrosion. Typically the temperature change for both streams is 70 to 100°F and the pressure drops 2 to 5 psi. Using two or more exchangers in series is common practice to accomodate a temperature cross; for example, the exit rich amine stream is hotter than the exit lean amine stream.

Like the absorber, the *still* or *stripping tower* is either trayed or packed, and the same sizing procedure can be used. A standard design consists of 20 V-grid trays spaced 24 inches apart. This is equivalent to 7 to 10 theoretical plates when the tray efficiency is 40% to 50%. The trays are designed for liquid and jet flood rates of 65 to 75% with a foam factor of 0.75.

The *reflux condenser* and the *amine cooler* are air-cooled, forced-draft heat exchangers with automatic louvers for temperature control. Add a 10% safety factor to the heat duty. With up to a 200 gpm amine flow rate, both services can be combined on one structure with a common fan. Air recirculation is advisable when the ambient air temperature is expected to be below 10°F. Often a trim cooler using water is needed for the lean amine stream.

The *reflux accumulator* separates the reflux or condensed water from the water-saturated acid gases. The water is returned to the still and the acid gases flared or sent to a

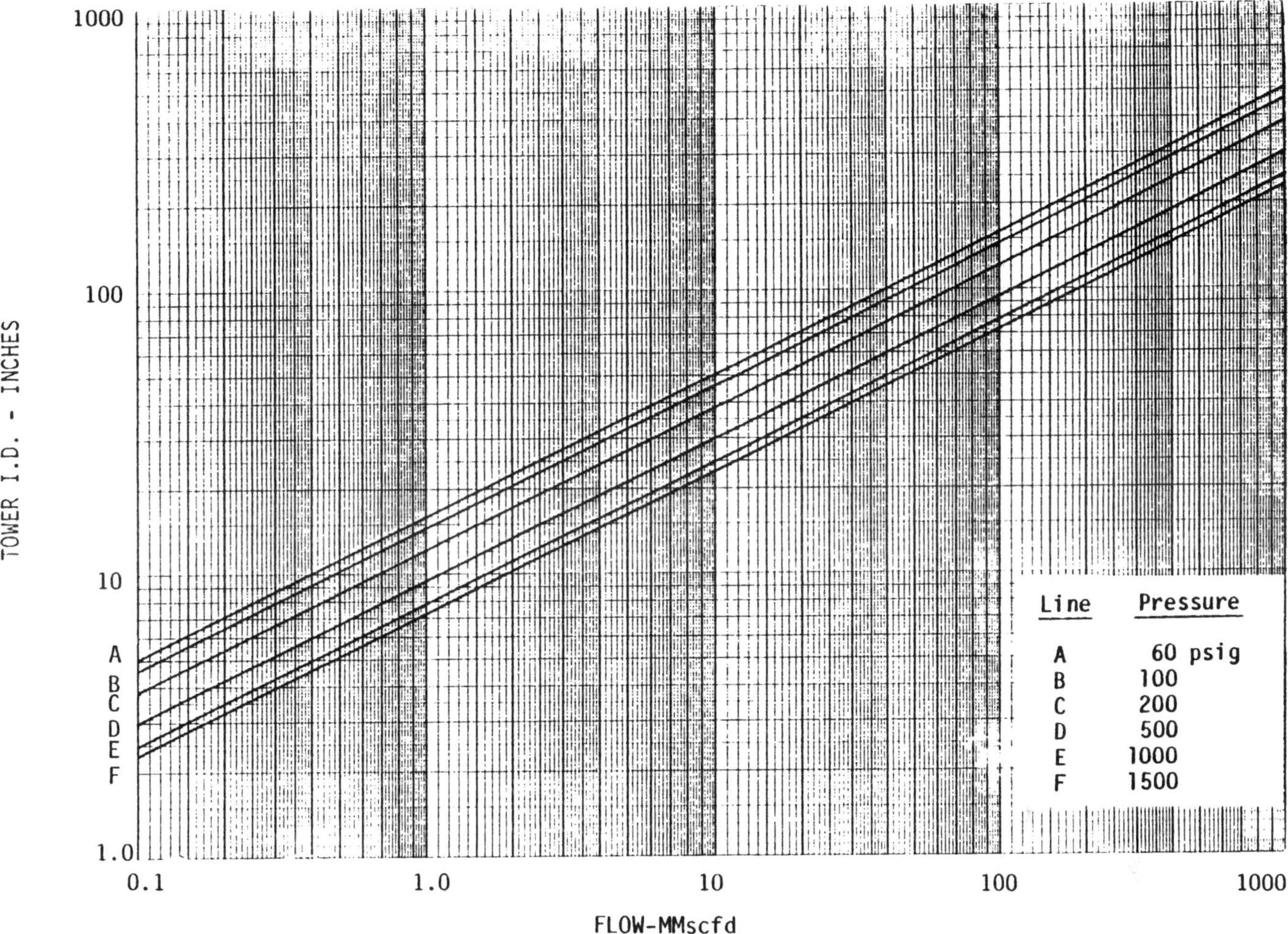

Figure 7–9. Contact tower diameter (Khan and Manning, 1985).

sulfur recovery unit. A mist pad is usually included, and a means of measuring the reflux rate is most helpful.

Establish the reflux ratio; i.e., mols of water returned to the still to mols of acid gases leaving the reflux condenser. This varies from 3.0 for primary amines and low CO_2/H_2S ratios to 1.2 for tertiary amines and high CO_2/H_2S ratios.

Smaller *amine reboilers* are direct fired heaters that use natural gas. The firetube surface is based on a heat flux of 6500 to 8500 Btu/hr ft^2 to keep the tube-wall temperature below 300°F. Larger reboilers use a steam or hot oil bundle. For design guidelines consult Ballard (1966). There is usually more corrosion with thermosyphon than with kettle designs.

A *surge tank* holding about 10–15 minutes of flow is needed. This is usually an extension of the firetube-type reboiler for smaller units and a separate vessel for larger systems.

The *reboiler duty* is: (1) the heat to bring the acid amine solution to the boiling point, (2) the heat to break the chemical bonds between the amine and acid gases, (3) the heat to vaporize the reflux, (4) the heat load for the makeup water, and (5) the heat losses from the reboiler and still. A safety factor of 15 to 20% is needed for start up. Also, meter the heat input.

The heat duties and transfer areas of the regeneration equipment can be estimated from the amine circulation rate (Jones and Perry, 1973).

	Duty—Btu/hr	Area—sq ft
Reboiler	72,000 × gpm	11.30 × gpm
Solution Exchangers	45,000 × gpm	11.25 × gpm
Aerial Cooler	15,000 × gpm	10.20 × gpm
Reflux Condenser	30,000 × gpm	5.20 × gpm

Reclaimer. Primary amines require additional regeneration, i.e., semi-continuous distillation in a *reclaimer* (Fig. 7–6). Initially the reclaimer is filled two-thirds full with lean amine and one-third full with a strong base such as 10 weight percent sodium hydroxide or carbonate. Heat distills the amine and water leaving the contaminants in the reclaimer. A slipstream of 1 to 3% of the lean amine flow rate is fed continuously. Add base as required to keep the pH at 12. The reclaimer temperature rises as the contaminants accumulate and operation is stopped at a predetermined temperature—300°F for MEA—to limit thermal degradation of the amine.

The *caustic soda or soda ash* releases the amine from salts formed with strong acids, e.g., formic, acetic, and sulfuric. Other contaminants are MEA degradation products, iron sulfide, and high boiling hydrocarbons. These are removed as a sludge at the end of each cycle.

Filters. Both mechanical filters for particulates and activated carbon filters for heavy hydrocarbons and amine degradation products are needed. Installing filters in both the rich and lean amine lines is the preferred but most costly approach. A *mechanical filter* in the rich amine line just after the flash tank prevents plugging in the lean/rich heat exchanger and still. Including a *carbon filter* reduces the tendency to foam in the still, but this adsorbs hydrocarbons that would otherwise be removed in the still. The carbon filter is often installed in the lean line before the aerial cooler to improve the quality of the amine feed stream.

The *mechanical filter* should be *full flow* and remove 95% of 10 microns or larger particles. For large units with high circulation rates, a slip stream may be more practical. The *carbon filter* is usually sized for *10 to 25 percent* of the circulation rate. Use hard coal-based charcoal with 5×7 mesh size. For very rich gases a carbon filter in both the rich and lean lines is advisable.

Both the particulate and carbon filters are more effective when installed in the rich amine stream. But, it is harder and more hazardous to change cartridges and carbon when the amine contains a high H_2S loading. Therefore, it is sometimes best to install both systems on the lean side.

Pumps. The *amine booster* and *reflux pumps* are centrifugal, preferably in-line or horizontal, with TEFC motors. *One hundred percent standby capacity* is recommended.

Selection of the circulation pump depends on the contactor pressure and the amine circulation rate. Include 50% or 100% standby capacity. Normally, reciprocating pumps are preferred. However, centrifugal pumps are used for low pressures (e.g., 100 psig) and multistage horizontal centrifugal pumps for high pressures (e.g., 700 psig)—or high circulation rates (e.g., 300 gpm).

In sizing and rating pumps use a low positive suction pressure of 3–10 psig. Motor sizes can be estimated from the amine circulation rate (Jones and Perry, 1973).

Motor	Horsepower
Circulation Pumps	gpm × psig × 0.00065
Booster Pumps	gpm × 0.06
Reflux Pumps	gpm × 0.06
Aerial Cooler	gpm × 0.36

Control Panel. The *alarm* and *shutdown panel* can be electronic or pneumatic. In either case a process or safety malfunction shuts the unit down, sounds the alarm horn, and indicates the condition responsible.

As a minimum, the following should trigger a *plant shutdown*:

Emergency Manual Shutdown

Treated Gas	— Off Spec
Inlet Scrubber	— High Level
Amine Cooler	— High Temperature
Reflux Condenser	— High Temperature
Surge Tank	— Low Level
Amine Reboiler	— Low Level
Amine Reboiler	— High Pressure
Amine Reboiler	— High Temperature
Amine Reboiler	— Flame Failure
Cooler Fan(s)	— High Vibration

Construction Materials. Pearce and DuPart (1985) and Richert *et al.* (1989) reviewed the *types of corrosion that occur in amine plants*: general, galvanic, crevice, pitting, intergranular, selective leaching, erosion, and stress corrosion cracking. The last is important because gas streams containing *hydrogen sulfide* cause metals to "*sulfide stress crack.*" Carbon steels become susceptible when the H_2S partial pressure exceeds 0.05 psia at system pressures above 65 psia. Corrosion problems are more severe in MEA units. For a general review of government regulations and design procedures consult Sivalls (1985), and for guidelines on selecting construction materials refer to Arnoldi *et al* (1987).

In considering the susceptibility of steel to stress corrosion cracking the key property is "*hardness.*" *N.A.C.E. publication MR–01–75* is the leading guideline for specifying the appropriate metal hardness. In addition to H_2S corrosion, the following factors must be considered:

1. *CO_2 corrosion*, especially when the rich amine temperature exceeds 150°F. Limiting the amine

concentration and using corrosion inhibitors are recommended (Hawkes and Mago, 1971).

2. *Electrolytic corrosion* due to the presence of unlike metals in aqueous environments.

3. *Oxidation* corrosion due to free oxygen in the makeup water and feed gas streams.

For most applications the *following guidelines* for *material selection* will provide adequate life and good corrosion resistance.

1. *Stress relieve* all pressure vessels, heat exchangers if shell-and-tube, reboilers, coolers, headers and piping in sour gas or rich amine service.

2. Include a *corrosion allowance*: 0.0625 inch for ASME code pressure vessels and 0.05 inch for piping. Stainless steel equipment requires no corrosion allowance.

3. *Shells, heads, and externals*: Use carbon steel with an "as welded condition" *maximum hardness RC–22*. Steels of fine grain manufacture are least susceptible to corrosion and should be specified for all pressure containing parts.

4. Internals such as *baffles, lugs, and clips*: Use carbon steel compatible with the shell and head material.

5. Internals such as *demisters, bubble caps, valves, bolts,* and parts subject to high velocities: Use 300 or 400 series stainless steel of low carbon type when attached by welding or are otherwise not removable.

6. *Tubes and tube sheets*: Use carbon steels below 150°F; consider 300 series stainless steels above 150°F.

7. Use *carbon steel* for firetubes in direct-fired reboilers.

8. *High pressure pumps*—above 200 psig: Use carbon steel or ductile iron case or fluid ends with 300 and 400 series stainless steel internals.

9. *Low pressure pumps*—below 200 psig service: Use ductile iron case or fluid ends with 300 and 400 series stainless steel internals.

10. Ductile iron *impellers* in centrifugal pumps provide satisfactory service. Consider 300 series stainless steel impellers for pumping temperatures above 150°F or for rich amines.

11. *Valves and piping*: Use carbon steel. Avoid ductile or cast iron valves because of the lack of impact resistance and tensile strength. When removing only CO_2 consider 300 series stainless steels for service above 150° F where the protection afforded by amines is not significant. Use stainless steel for the feed line to the still, the overhead line from the still, and the reflux condenser tubes, and upper still trays.

12. Size all *amine piping* for low velocities, e.g., 2 to 5 ft/s. Avoid using slip-on flanges and other fabrication techniques that provide cavities which collect the corrosive amine solution. Use seal welds to fill in all gaps, e.g., tray support rings.

13. Differential-pressure (DP) cells are recommended for monitoring amine flow.

Winterization. Amine units should be *winterized* for cold climates. This includes:

1. Extra insulation for already insulated equipment and insulation of equipment not normally insulated, i.e., the contactor tower.

2. Air recirculation louvers on air-cooled heat exchangers.

3. Heat tracing of the amine piping and stagnant ports; e.g., level gages, external float cages, switches and controllers. The heat can be electric, steam, hot oil, or hot glycol.

4. The foundation design should allow for frost protection to prevent movement.

5. Some manufacturers prefer to erect a building around the equipment. Then only the uncovered equipment needs special protection.

Example 7–2. Select and size a sweetening process for the following gas stream
5 MMscfd, 5% H_2S, 2% CO_2, 110°F, 700 psig.
First estimate sulfur content.

$$\frac{lb\ S}{day} = \frac{scf}{day}\ \frac{lb\ mol\ gas}{scf}\ \frac{mol\ H_2S}{mol\ gas}\ \frac{lb\ S}{mol\ H_2S}$$

$$= \frac{5,000,000}{1}\ \frac{1}{379.5}\ \frac{5}{100}\ \frac{32}{1} = 21,000$$

This is much higher than 100 lb S/day, so use an amine process—like DEA.
Circulation
gpm = K (MMscfd) (mol percent AG removed)
= (1.45) (5) (5 + 2) = 50.8 gpm
Tower ID = 19 ID, use 24 in OD.
Reboiler duty = (72000)(gpm)
= (72000)(50.8)
= 3.7 MMbtu/hr.
Circulation pump
hp = (gpm)(psig) (0.00065)
= (50.8)(700) (0.00065)
= 23 hp

Operation

Amine plants are far more difficult to operate than batch units. Accordingly, it is important to follow the assembly, start up, and operation procedures. These should include:

Plant Check Out.

1. Check that all materials have been received and order replacements for damaged items.
2. Install blinds on the inlet gas line to prevent accidental and premature pressurization of plant.
3. Check assembly of plant. Use the Process and Instrument flow sheets to be sure that all vessels, valves, instruments, etc., are tagged; interconnecting pipes, tubing, and wiring are clean and correctly installed; and all leads to the control panel are functional.
4. Check mechanical equipment for lubrication, alignment, rotation, etc. Pack pumps and put filter elements into filters.
5. Check calibration of pressure, flow, and temperature and level instruments.
6. Pressure test vessels, pipes, etc., using water or clean gas. Repair leaks.
7. Check the alarm and safety devices. Be sure settings conform to process design and that qualified personnel were used in manufacturing.
8. Remove blinds and establish fuel gas, instrument air, and power supply.

Clean Out of Amine Unit. Cleaning removes protective oils, valve grease, mill scale, rust, and other contaminants which foul amine solutions. The wash-out procedure is very similar to actual start up and operation. New units require a soda ash or caustic wash and deionized water rinse. Used units also need an acid wash to remove rust and sulfide scale.

1. Charge the unit with 2% Na_2CO_3. Fire the reboiler and heat to 200°F.
2. Check the circulation pump. Note that plunger pumps leak lubrication slowly. Piston pumps should not leak.
3. Start the reflux condenser fan. Increase the firing rate to establish reflux.
4. Pressurize contact tower slowly. Bypass all filters. Start amine cooler fan and booster and circulation pumps.
5. Vent trapped air from valves at high points.
6. Establish an operating level in the contactor.
7. Circulate for six hours. Periodically, open vessel bottom and piping drains and flush for one minute. Add soda ash solution as required.
8. Shut off the boiler and the pumps and fans when the system has cooled to 180°F.
9. Drain the entire system. Use blanket gas at 2 psig to keep air out.
10. Refill the amine system with deionized water.
11. Repeat the wash-out procedure. Circulate for about eight hours, flushing low points every two hours and observe the cleanliness.

12. If clean, drain sufficient water to add enough amine to obtain the desired solution strength. If the rinse water appears dirty, drain completely and fill with proper strength amine solution.
13. Circulate for two hours to mix thoroughly.
14. Open the block valves on filters, close bypass valves.

The amine plant is now ready for start up. Remove blinds to process and sales gas lines.

Start Up.

1. Close all block valves on process gas flow lines.
2. Turn on power to electrical panels. Make sure all alarm and shutdown signals are operational.
3. Start reboiler, fans, reflux pump, booster and circulation pumps as described in wash-out procedure.
4. Slowly open the inlet gas valve to pressurize the contactor. Leave the sales line valve closed.
5. Check amine solution concentration.
6. Circulate until the still overhead temperature at the condenser inlet reaches the minimum operating temperature of 190°F.
7. Slowly open the sales valve and increase the gas flow to the desired operating rate.

So far the operation may have been manual. If so, switch to the fully automatic mode. None of the safety shutdowns should activate.

Performance. The keys to successful operation are *common sense, knowledge of the plant operation*, and *good record keeping*. The latter should include:

1. *Inlet gas conditions*, i.e., temperature, pressure, and flow rate—at least once daily.
2. The *acid gas contents* of the feed and sales gas daily during the first two weeks and thereafter weekly. Also, an amine analysis weekly.
3. *Pressures and temperatures daily* of the amine contactor, amine still, reflux accumulator, and amine reboiler.
4. *Filter pressure drop daily*.

Typical operating ranges follow:
Rich amine from contactor (taken downstream of the level control valve)

$$Pressure = 40\text{–}80 \text{ psig}$$
$$Temperature = 100\text{–}180°F$$

Amine Exchanger
Rich Side

$$Temperature \text{ in} = 100\text{–}180°F$$
$$Temperature \text{ out} = 190\text{–}220°F$$

Lean Side

> Temperature in = 240–250°F
> Temperature out = 170–190°F

Amine Cooler

> Temperature in = 170–190°F
> Temperature out = 100–130°F

Overhead Condenser

> Temperature in = 190–225°F
> Temperature out = 100–130°F

Reflux Pumps

> Pressure Suction = 3–6 psig
> Pressure Discharge = 30–40 psig

Booster Pumps

> Pressure Suction = 3–6 psig
> Pressure Discharge = 50–65 psig

Circulation Pumps

> Pressure Suction = 0–40 psig
> Pressure Discharge = 50 psig over contactor
> operating pressure

Amine Analyses. The *color* of the *regenerated solution* can be most informative (Pearce *et al.*, 1980).

1. Straw or light brown: good condition and well regenerated.
2. Green: very fine FeS particles (less than 1 micron). Larger FeS particles are black and will settle.
3. Blue, green: copper or nickel. Copper oxide reacts with MEA to form a royal-blue complex.
4. Amber, dark red: iron compounds complexed with the amine.
5. Red, brown: either oxidation products or thermal degradation. Primary amines may smell like ammonia.

The following analyses may be used to determine the condition of the amine. For procedures and interpretation, consult Pearce *et al.* (1980) and the Dow Gas Conditioning Fact Book (1962).

1. Alkalinity titration: amine concentration
2. Gas chromatography: free amine, degradation products, acid gas
3. Total and primary nitrogen
4. Acid gas loadings: H_2S, CO_2
5. Karl Fischer: water content
6. Elemental analyses: Fe, Cl, Al, Cu, Na, etc.
7. Foam: extent and stability

Amine Losses. There are two types of day-to-day losses: those that are an integral part of the process and those due to *leaks, spills* and *mishandling*. Unfortunately, the latter are often greater than the process losses. There are also upsets (e.g., foaming in the absorber) and contamination (e.g., inlet separator dumps produced water into the absorber). These losses can be considerable, especially when the amine solution has to be replaced.

Process losses (Veldman, 1988) include:

1. Sweet gas from the absorber: MEA is volatile enough to result in losses of 0.45 lb/MM scf of processes gas. DEA, DGA, DIPA, and MDEA are much less volatile, and the losses are 0.02 to 0.03 lb/MM scf.
2. Entrainment from the absorber: This averages 0.5 to 3.0 lb/MM scf. This loss can be reduced by operating the absorber at less than 70% of the flooding velocity and installing a demister in the top of the tower. Two water wash trays above the amine trays are very effective.
3. Gas from the flash tank and liquids from a three-phase tank: These losses are usually small.
4. Reflux overhead gas: These are minor because the vapors and liquid do not contain much amine.
5. Reclaiming. This loss depends on how often the primary amine is reclaimed.

Amine consumption for a well designed system with few upsets averages 2 lb/MMscf for DEA, DIPA, and MDEA solutions. Losses for MDEA are about 3 lb/MMscf because of the volatility and reclaiming. DGA losses are intermediate.

Makeup Water. This replaces operating losses from the absorber—the sweet gas is hotter and therefore more humid than the sour gas—and the off gas from the reflux condenser. Periodic additions suffice for smaller units. Continuous addition is recommended for larger units especially when the acid gas rate is high. This makeup can be combined with the reflux to the stripper or fed to wash trays in the top of the absorber.

Makeup water can contain contaminants that accumulate in the unit and so cause foaming, corrosion, plugging, and fouling. Makeup water should meet the following specifications: 20 ppm chloride, 100 ppm TDS, 50 ppm total hardness, 10 ppm sodium/potassium, and 10 ppm iron (Dwyer, 1989).

Optimization. Ballard (1986) gives an excellent review of how to *cut fuel costs, use less amine* and *reduce corrosion*. Note that the larger the unit the greater is the need for optimization.

Gas fired reboilers should be equipped with a secondary air controller and a firetube turbulator. The thermostat should be throttling and not on-off. Obtain the thermal efficiency from the oxygen content and temperature of the stack gas. Use this and the fuel consumption to check that the steam rate is 0.9 to 1.2 lb per gal of amine circulated.

The weekly amine analysis is most informative. Check the amine concentration—too high increases the corrosion rate, and too low increases the circulation rate and wastes fuel. Check the acid gas loadings—rich stream too high increases the corrosion rate, and too low indicates inefficient use of the amine and too high a circulation rate; lean stream too high indicates poor regeneration and too high a circulation rate, and too low means unnecessary regeneration and fuel usage. Check the pH—too low add soda ash. Check the makeup water quality regularly.

The reflux ratio is usually not measured. However, it can be obtained from Figure 7–10 (Manning, 1989). The optimum value depends on the amine system. Keeping the overhead still temperatures below 225°F and the reflux condenser effluent near 130°F is recommended.

Check the temperature approach of the lean/rich amine

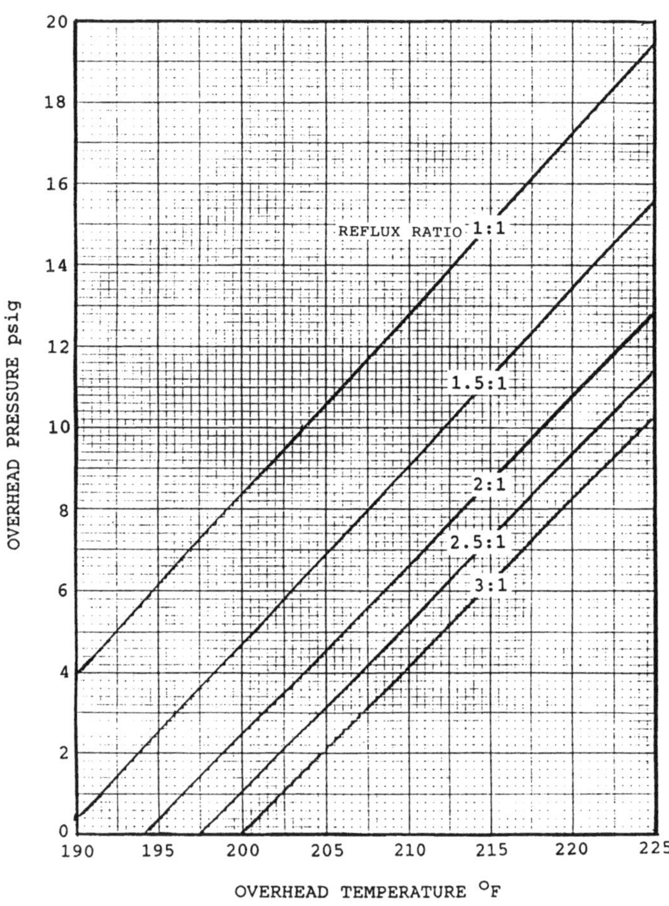

Figure 7–10. Reflux ratio as a function of overhead pressure and temperature (Manning, 1989).

heat exchanger. Departure from design and/or start up conditions denotes fouling.

Troubleshooting

Consult the daily record sheets. Sudden changes in operating conditions are the best clues for malfunctions. A brief review of the more *frequent problems* and *corrective procedures* follow.

1. Gas is not "Sweet"
   Check amine solution concentration: too low—check makeup water addition.
   Check amine flow rate: too low—open bypass valve.
   Check amine regeneration: increase firing rate.
   Check reflux rate and temperature: probably too low—increase firing rate.
   Check stripping column pressure: It may be too low.
   Check for foaming: carryover into outlet separator and/or pressure fluctuations across absorber.
2. Amine solution not regenerated
   Check reboiler temperature and pressure and the reflux rate.
   Check for leaks in lean/rich amine heat exchanger.
   Check the reclaimer for primary amines.
   Check for foaming in stripper: pressure fluctuations.
3. Dirty, degraded amine
   Gas contains oxygen.
   Storage-tank blanket-gas valve is not functioning: repair.
   Makeup water contains free oxygen: add oxygen scavenger.
   Sparge amine with sweet gas to strip oxygen.
   (Oxygen can enter when the feed gas contains vapor recovered from stock tanks, and can also oxidize H_2S to elemental sulfur.)
4. Excessive corrosion
   Amine concentration is too high: add makeup water.
   Amine is highly degraded: replace.
   Makeup water is high in dissolved solids: treat makeup water or use deionized water.
   Insufficient amine regeneration.
   Insufficient amine filtration: increase filter rate or change filter elements more frequently.
   Oxygen is entering system: eliminate.
   Velocities too high: reduce temperature to stripper.
5. Foaming of amine solution
   Foaming is a very unpredictable phenomenon. It can be caused by any or a combination of the following conditions:
   Dirty amine (solids)—check filter elements.
   Degraded amine.
   Liquid hydrocarbons in gas or amine.

Inlet gas temperature too low—hydrocarbon condensation.

Inlet amine temperature too low, at least 10°F above inlet gas temperature.

Wrong or off-spec. chemicals.

Well treating chemicals.

Surfactants.

Corrosion inhibitors.

Very fine particles; e.g., iron sulfide, in sour gas.

Inadequate cleaning of amine plant before startup.

Misuse or abuse of antifoam chemicals in amine units, too much or too little defoamer.

Incoming gas is not adequately scrubbed and contains salt water or produced water.

Makeup water contains iron, sulfides, chlorides, etc. (use deionized or demineralized water).

Cotton-sizing chemicals being washed off a newly installed cloth filter.

Tray downcomers are plugged, causing amine to stack up in the trays. (This is really not a foaming problem but behaves so; usually with older plants.)

Always add antifoamants downstream of the carbon filter. The following antifoamants are recommended. Dilute with 50% isopropyl alcohol and use in concentrations of 5 to 50 ppmw. Consult a reputable chemical dealer listed below.

• Dow-Corning DB–100 Antifoam Compound
• Dow-Corning DB–31 Antifoam Emulsion
• Tretolite VEZ D–83
• Tretolite D–95
• Natco DF–971
• Exxon Corrects-It for chlorides
• Union Carbide SAG Antifoamants (GT–101, 102, 301 or 302)

OTHER PROCESSES

These include regenerative processes (other than the alkanolamine and mixed solutions), adsorption, and membranes. Unique properties of the solvents and membranes offer substantial advantages—often for special situations with specific requirements (e.g., offshore). A convenient classification follows:

1. *Physical solvents*; e.g., Selexol, Rectisol, Purisol, and Fluor Solvent.
2. *Hot potassium carbonate* solutions; e.g., Hot Pot, Catacarb, Benfield, and Giammarco-Vetrocoke.
3. *Direct oxidation* to sulfur; e.g., Stretford, Sulferox and Lo-Cat.
4. *Adsorption*; e.g., molecular sieves.
5. *Membranes*; e.g., AVIR, Separex, Cyrano (Dow), Grace Membranes International Permeation, and Monsanto (Spillman, 1989).

The *advantages* include:

1. Reduced equipment size and cost—general.
2. Very high acid gas loadings; i.e., 10–12 scf/gal at high acid-gas partial pressures, and reduced circulation rates; e.g., physical solvents.
3. Regeneration without heat; e.g., physical solvents or no regeneration; e.g., membranes.
4. Reduced heat requirements and heat exchange; e.g., potassium carbonate solutions.
5. Elimination of H_2S emission; e.g., direct oxidation processes.
6. Ability to process gas streams with very high CO_2 contents (i.e., over 50%, economically), e.g., membranes.
7. No moving parts (e.g., molecular sieves), or only compression (e.g., membranes).
8. Simultaneous dehydration of the gas; e.g., physical solvents and molecular sieves.

Figure 7–1 shows when these processes are usually used depending on whether H_2S, CO_2, H_2S and CO_2, or selective H_2S removal is desired.

The *disadvantages* include:

1. More complex process designs often limited to a specific application. Higher engineering costs.
2. Some physical solvents absorb significant amounts of the heavier hydrocarbons. These are sometimes lost.
3. Some physical solvents are expensive and very corrosive to the elastomers used as seats and seals in valves and instruments.
4. Many processes have license and/or royalty fees; e.g., Selexol, Rectisol, Purisol, Fluor Solvent, Catacarb, Benfield, Stretford, and Lo-Cat.
5. Membrane separators are expensive.

Physical Solvents

These are *organic liquids* that absorb CO_2 and H_2S at high pressures and ambient or low temperatures. Regeneration is by flashing to atmospheric pressure and sometimes with vacuum—usually with no heat.

The basic flow sheets are similar to alkanolamine processing, and the options of single-stage absorption, split streams, and two-stage absorption are available.

The physical solvent should be low-melting of low viscosity, chemically stable, nontoxic, non-corrosive, selective for the contaminant gas, and available. The gas pickup is proportional to the acid-gas partial pressure—i.e., Henry's law applies.

In the following process descriptions, Selexol is emphasized because the primary use is for natural gas streams and not synthetic gases.

Selexol. Selexol is used in over 50 installations worldwide for the bulk removal of CO_2 and also more recently for simultaneous H_2S removal. It is a mixture of dimethyl ethers of polyethylene glycols (Fig. 7–4)—mostly the trimer through hexamer. It is nontoxic, high boiling, can be used in carbon steel equipment, and is an excellent solvent for acid gases, other sulfurous gases, heavier hydrocarbons, and aromatics (Raney, 1977).

There are several process variations. Figure 7–11 shows the most basic, which is very suitable for bulk CO_2 removal. The rich solvent is regenerated by flashing to a vacuum of about 20 in. Hg (Sweny, 1980).

The inlet gas is mixed with the rich solvent from the contact tower, cooled, and separated before entering the contactor. Following the separator there are four flashes: the first is a high-pressure flash, needed to recover co-absorbed methane; the second is an intermediate-pressure flash, from which CO_2 is released at elevated pressure; the third is an atmospheric flash; and the fourth is a vacuum flash. Two of these flashes are not essential, but they improve

the economics significantly. The intermediate flash yields CO_2 at elevated pressure, from which it can be expanded to yield power or refrigeration, or from which it can be recompressed for injection to the reservoir. The atmospheric flash serves to reduce the load on the vacuum flash.

With absorption at 1000 psia and vacuum flashing at 5 psia, the CO_2 level in the product is 1%, a suitable level for most natural gas purification. Higher levels such as 2%, 3%, or 5% are obtained by flashing at higher pressure. Atmospheric pressure results in a CO_2 level of about 3.5%.

Removing and simultaneously expanding the CO_2 produces significant cooling. A 30% CO_2 content in the feed gas at 1000 psia usually supplies enough cooling to offset the heat for pumping and pressure letdown, compression, the sun's radiation, and the temperature difference between feed and product. Refrigeration is needed if a sizeable quantity of the CO_2 is not flashed—e.g., injected back into the reservoir. The amount of CO_2 removed at any of the intermediate pressures is an economic decision—e.g., the cost of lost cooling versus the compression cost.

Note that *Selexol* also *absorbs* the *heavier hydrocarbons*. For rich gas streams buildup is prevented by mixing with water to form a hydrocarbon phase. Then the water must be stripped off which requires heat.

Another process variation is the "split cycle," which is very effective in removing H_2S (Fig. 7–12). Because Selexol absorbs H_2S preferentially, the regenerated solvent can produce 0.25 gr H_2S/100 scf once most of the CO_2 has been removed.

In the split-cycle process, the bulk of the Selexol flows through the lower section of the absorber tower and is regenerated economically by flashing. Typically, the gas

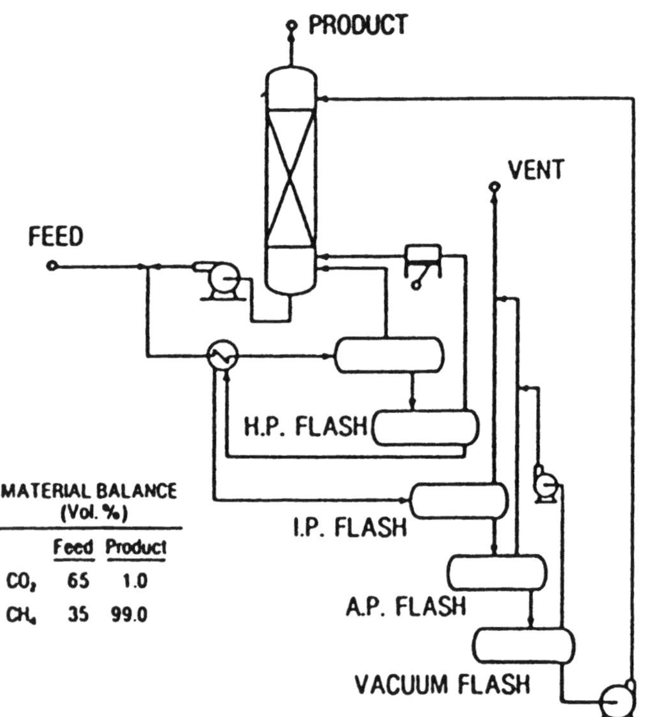

Figure 7–11. Basic Selexol Process (Sweny, 1980).

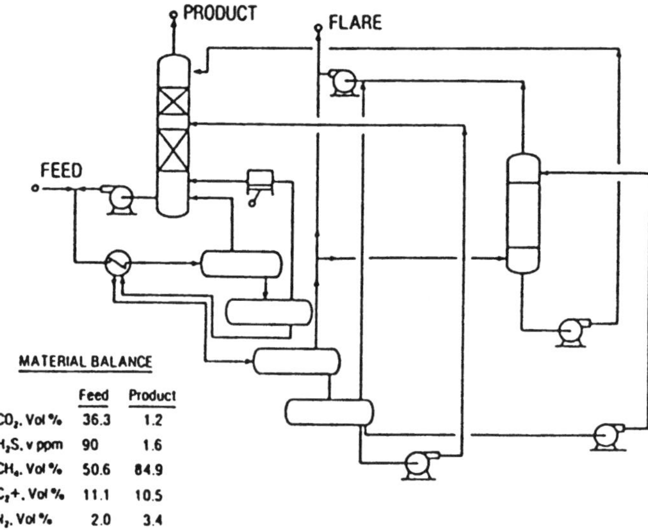

Figure 7–12. Split cycle Selexol Process (Sweny, 1980).

leaving the lower section contains less than 1% CO_2. The smaller part of the Selexol stream is further regenerated by stripping with flash gas. Then it can be used in the upper section of the contactor to remove the H_2S. This stripping arrangement offers regeneration without heating, a major advantage on a offshore platform.

Another approach for removing H_2S is to first treat the sour gas with Selexol that has been saturated with CO_2 and cooled to remove the heat of absorption. The Selexol is flashed to remove most of the CO_2 and the flash gases returned to the effluent from the contact tower. The CO_2 is recovered in a second Selexol unit.

The flashed Selexol from the first tower is stripped with steam to remove the H_2S, which is fed to a sulfur recovery unit. The stripped Selexol is cooled, saturated with CO_2, and returned to the first contactor. Mortko (1984) discusses process variations for a gas stream containing 65% CO_2 and 5% H_2S. Approximately 90% of the CO_2 was recovered for EOR, and the sales gas was pipeline quality (Hunter and Bryan, 1987).

Rectisol. *Methanol* is the Rectisol solvent, and the high volatility requires that the contact tower be at very low temperatures, e.g., 0 to –70°F. This limits the applicability to very lean natural-gas streams. It is mainly used for treating synthesis gas in Europe. The ITR process variation is similar (Shah, 1989). Additional processing is included to recover hydrocarbons from the H_2S and CO_2 streams.

Purisol. Lurgi Purisol solvent is n-methyl–2-pyrrolidone—also known as NMP or M Pyrol. It is an excellent solvent of H_2S, CO_2 , H_2O, RSH, hydrocarbons, and unfortunately many elastomers. Also, it is very selective towards H_2S. But the boiling point of 396°F, although suitable for the Purisol process, is too low for use as an additive to alkanolamine solutions. As with Rectisol most Purisol applications have been used in Europe for synthetic gases.

Figure 7–13 shows a typical flow sheet. Regeneration is accomplished by a two-stage flash to atmospheric pressure. Strip gas, heat with reflux, or a combination of these are used. The reabsorber is the source of fuel gas. High pressure in the absorption tower improves the performance.

Fluor Solvent. This process uses *propylene carbonate* to remove H_2S, CO_2 , COS, and RSH from natural gas streams. It is not selective towards H_2S. All types of sulfur compounds can be reduced to 4 ppm or less. Nevertheless the major use has been as an alternative to Selexol for bulk CO_2 removal (Bucklin and Schendel, 1984).

Figure 7–14 shows a typical flow sheet with regeneration

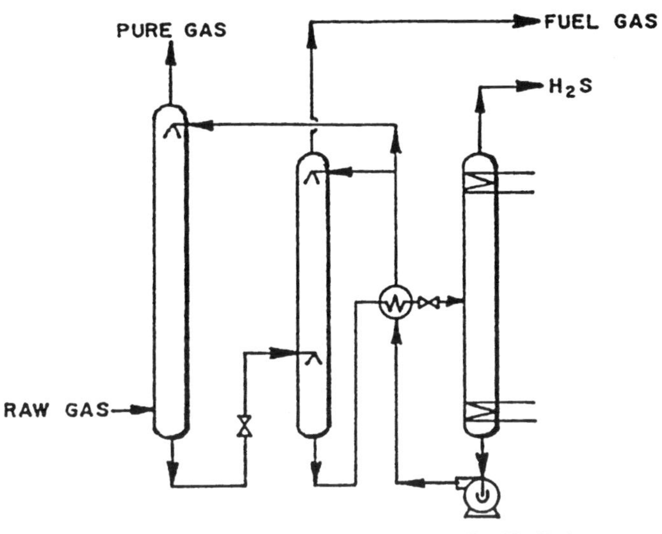

Figure 7–13.   Purisol flow sheet.

consisting of two stages of flashing. The first flash gas containing most of the hydrocarbons is compressed and recycled. The second flash drives the expansion turbine.

The best applications are lean gas streams with an acid gas partial pressure over 60 psi. Solvent temperatures below ambient reduce the circulation rate and the equipment size. Expanding the flashed CO_2 provides refrigeration.

Comparison. Table 7–4 gives the solubility of H_2S, CO_2, COS, and propane in the Selexol, Purisol, and Fluor solvents (Ferrin and Manning, 1984). Purisol has the greatest affinity for the acid gases, and it is the most selective; however, it is the most volatile. Selexol is more selective than the Fluor solvent, but it dissolves the most propane.

Table 7–4 also shows the solubilities of the gases that occur most frequently in the process gas. All of the solvents exhibit significant affinity for the heavier paraffins (C6 and

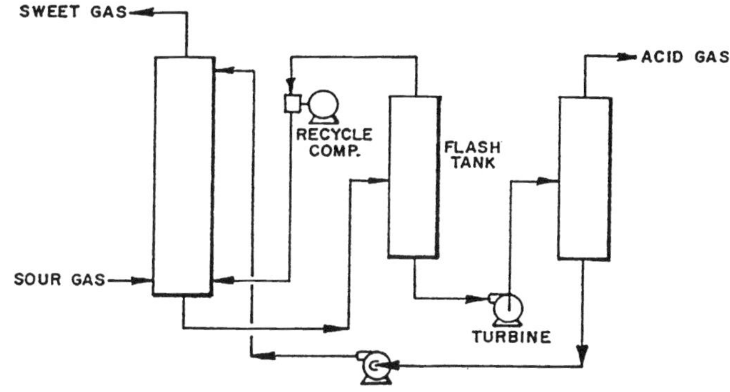

Figure 7–14.   Fluor solvent flow sheet.

Table 7–4 "Solubilities of Gases in Physical Solvents"
Data at 75°F

Gas Solubility (cc gas at 1 atm, 75°F/cc solvent) (Ferrin and Manning, 1984)			
Gas	Selexol	Purisol	Fluor Solvent
H_2S	25.5	43.3	13.3
CO_2	3.6	3.8	3.3
COS	9.8	10.6	6.0
C3	4.6	3.5	2.1

Gas Solubilities Relative to CO_2 (Bucklin and Schendel, 1984)			
Gas	Selexol	Purisol	Fluor Solvent
H_2	0.013	0.006	0.008
CO	0.028	0.021	0.021
C1	0.067	0.072	0.038
C2	0.42	0.38	0.17
CO_2	1.0	1.0	1.0
C3	1.02	1.07	0.51
nC4	2.33	3.48	1.75
COS	2.33	2.72	1.88
H_2S	8.93	10.2	3.28
nC6	11.0	42.7	13.5
CH_3SH	22.7	34.0	27.2
C_6H_6	253.0	—	200.0
H_2O	733.0	4000.0	300.0

heavier), high solubility for aromatics, and very high affinity for water. The latter makes the solvents good desiccants, and the processed gas is usually very dry. Sweeney *et al.* (1988) and Bucklin and Schendel (1984) compare physical solvents and discuss design methods.

Hot Potassium Carbonate Solutions

Benson and Field developed the original Hot Pot process at the U.S. Bureau of Mines in the 1950s. Improvements followed—e.g., catalysts to increase the reaction rates, corrosion inhibitors, etc.—and resulted in proprietary processes (Catacarb and Benfield). However, most of the applications are for synthesis gases rather than natural gases. The Alkacid, Seabord, and Vacuum Carbonate processes are essentially defunct (Goar, 1972). The Giammarco-Vetrocoke process has environmental problems because of the arsenic catalyst. It has been used largely in Europe. These processes use aqueous inorganic solutions containing 25 to 35 weight percent of K_2CO_3. The absorption is chemical, and not physical. The reactions are:

$$K_2CO_3 + CO_2 + H_2O = 2\ KHCO_3 + heat$$
$$K_2CO_3 + H_2S = KHCO_3 + KHS + heat$$

The *Catacarb* solution contains alkali metal borates and a corrosion inhibitor. The *Benfield*—named after Benson and Field—has a vanadium oxide additive. Both produce higher acid gas loadings, lower circulation rates, smaller equipment and less maintenance (Bartoo, 1984; Eickmeyer, 1976; and Strehzoff, 1975).

Figure 7–15 shows the available process flow sheets. The basic process—no flash tank is shown—has no heat exchange between the rich and lean streams. The contactor is hot, 220 to 400°F depending on the pressure. Variations such as split flow and two-stage regeneration improve the quality of the treated gas and reduce heat requirements (Fig. 7–15).

The *salient features of these processes* are:

1. The chemical reactions are specific for CO_2 and H_2S, and at the contact temperature the solubility of other gases is negligible. So the loss of valuable process gases is insignificant.
2. The heats of the reactions that absorb CO_2 and H_2S are about half those for alkanolamines, and this reduces the regeneration heat requirements comparably—e.g., 0.6 to 0.8 lb steam per gallon of solution circulated.
3. Complete H_2S removal requires the presence of CO_2. COS and CS_2 are removed by hydrolysis to H_2S and CO_2. Mercaptans are difficult to remove. Earlier process designs recommended an acid gas content of 5 to 8% and a contactor pressure over 306 psig (Swaim, 1970). Indeed, additional treatment with MEA has been suggested (Goar, 1972).
4. Acid-gas pickups of 4 to 8 scf/gal are achievable. Higher loadings are unobtainable because the absorption is chemical as opposed to physical.
5. The precipitation of potassium bicarbonate is prevented by limiting the carbonate concentration to 35%.
6. Like other liquid absorbents contaminants such as suspended solids and heavy hydrocarbons cause foaming and solvent carryovers.

Direct Conversion Processes

Only processes used to remove H_2S from natural gas are considered. (This eliminates so-called tail gas cleanup, where two mols of H_2S are reacted with one mol of SO_2.)

The basic schemes use *oxidation-reduction reactions.* Typically the H_2S is absorbed in an alkaline solution containing oxygen carriers and converted to elemental sulfur. The solution is regenerated by air which also acts as a flotation agent for the sulfur formed. The *sulfur sludge* has often been the Achilles heel of these processes. The collection and removal of the sulfur are difficult, and line plugging is a common cause of shutdowns.

The *Stretford process* (Fig. 7–16) has been used widely. The absorbing solution is a dilute solution of Na_2CO_3,

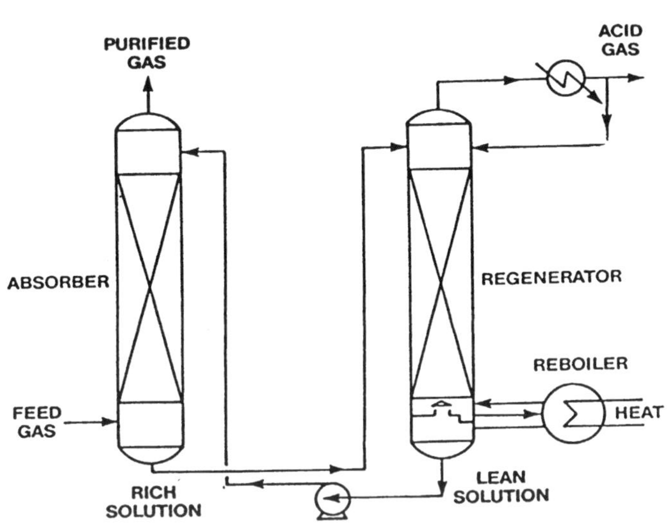

Basic Process

Two-Stage Regeneration

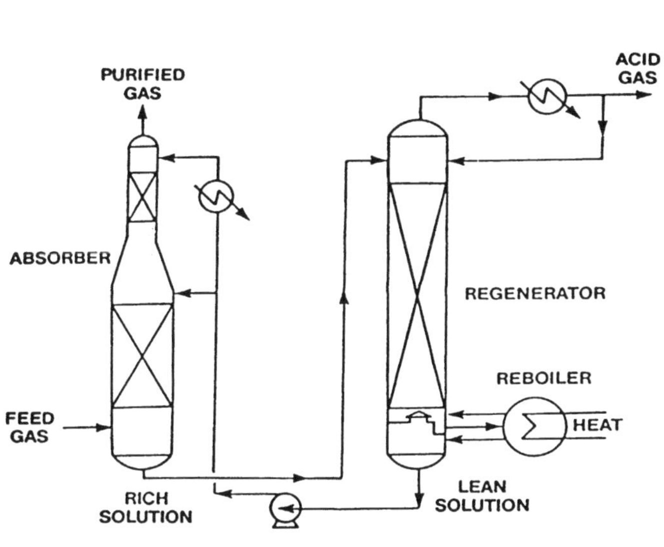
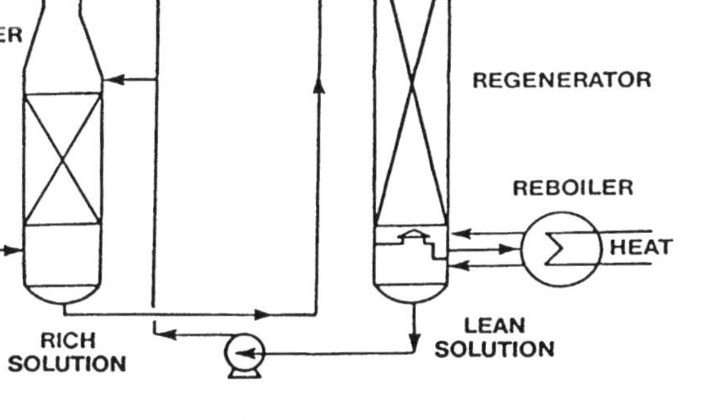

Split Flow

Comparison of Performance

Figure 7–15. Hot K_2CO_3 flow sheets (Bartov, 1984).

$NaVO_3$, and anthraquinone disulfonic acid (ADA). The reactions for the four step process are:

$$H_2S + Na_2CO_3 = NaHS + NaHCO_3$$
$$4\ NaVO_3 + 2\ NaHS + H_2O = Na_2V_4O_9 + 4\ NaOH + 2\ S$$
$$Na_2V_4O_9 + H_2O + 2\ ADA\ (quinone) = 4\ NaVO_3 + 2\ ADA\ (hydroquinone)$$
$$ADA\ (hydroquinone) + O = ADA\ (quinone)$$

These reactions are unaffected by pressure and tolerate temperatures up to 120°F. The solution is not corrosive.

The *Lo Cat process* uses an extremely dilute solution, 0.15 to 0.25 weight percent of iron chelates. The absorption and regeneration reactions may be summarized as:

$$H_2S + 2\ Fe^{+++} = 2\ H^+ + S + 2\ Fe^{++}$$
$$2\ Fe^{++} + H_2O + O = 2\ OH^- + 2\ Fe^{+++}$$

The solution is buffered with $KHCO_3$ formed by reaction

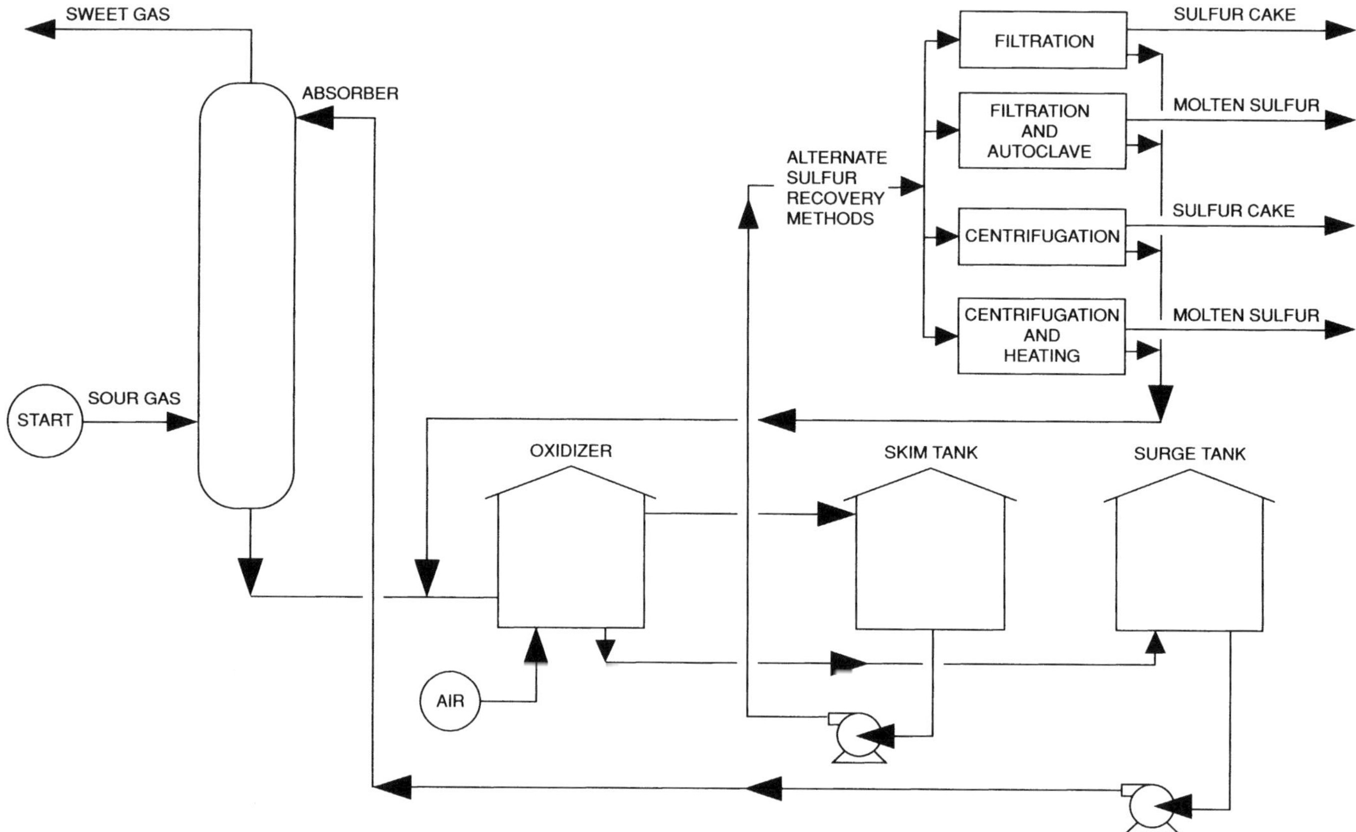

Figure 7–16.   Stretford flow sheet.

of the CO_2 in the gas stream with the KOH in the formulation.

To overcome the very dilute solutions, the sweetening and regeneration are sometimes accomplished in one large vessel, using circulation in the tank rather than pumps. Then the process gas must be compressed for sale.

Sulferox is another iron chelate process (Fong *et al.*, 1987). Difficulties with soluble iron scavengers include:

1. Precipitation of ferric hydroxide under alkaline conditions.
2. Precipitation of ferrous sulfide from acidic solution, so pH control is critical.
3. H_2S loading, ''oversulfiding'' produces thiosulfates in the regeneration step.
4. Corrosion: chelates corrode carbon steel; stainless steel is required.

Buenger *et al.* (1988) review ligand chemistry in detail.

Acceptance of these processes and others—e.g., Lacy Keller, Konox, Takahax, Thylox—has been limited. However, there are occasional glowing reports, e.g., Ellwood (1964) and King *et al.* (1986).

Adsorption Processes

Adsorption is the process in which gas or liquid molecules are held on the surface of solids. This retention may be a chemical reaction, capillary condensation, intermolecular forces, or a combination of these. A major application is the dehydration of natural gas, especially as the first step in the cryogenic recovery of natural gas liquids as is discussed in Chapter 9.

Zeolite molecular sieves are the most suitable adsorbents for H_2S, CO_2, RSH, etc. They are nontoxic, non-corrosive, and available in pore sizes ranging from 3 to 10 angstroms. The smaller sizes adsorb H_2S but are too small for the heavier mercaptans. The larger sizes are often used for total sulfur removal.

Molecular sieves adsorb only polar molecules. Among the common contaminants in natural gas, H_2O is the most strongly adsorbed followed by mercaptans, H_2S, and CO_2 in that order. Because components with a stronger affinity for adsorption will displace less tightly-held components, selective removal of sulfur contaminants is obtainable, as are very low effluent concentrations, e.g., 0.01 gr/100 scf.

Figure 7–17 illustrates what happens in a molecular sieve

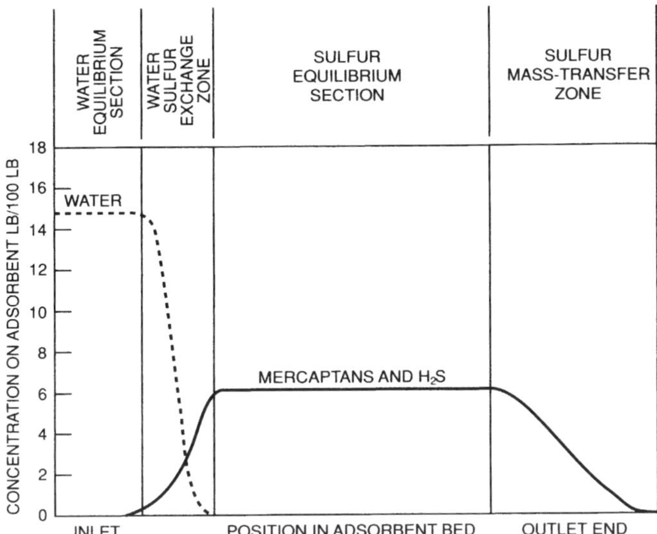

Figure 7–17. Adsorption zone in molecular sieve bed (Kohl and Riesenfeld, 1985).

bed. There are a number of adsorption zones as the different contaminants are adsorbed and displaced. Water occupies the position closest to the inlet followed by RSH, H_2S, and CO_2. During operation, these zones progress towards the outlet, and when the key contaminant—usually H_2S—reaches the end, the bed must be regenerated.

Multicomponent adsorption in fixed beds is discussed in more detail in Chapter 9.

Commercial installations require at least two beds so that one is always on line while the other is being regenerated. Disposal of the sour regeneration gas is a problem that is solved either by using an amine plant or flaring. Figure 7–18 shows a process scheme where two towers are used in series and the sour vent gas is flared (Thomas, 1967), simultaneously drying and sweetening the gas.

Chi and Lee (1973) provide a detailed design procedure for H_2S and water removal from natural gas using type 5A molecular sieves. Experimental data confirm that the Chi-Lee method adequately predicts H_2S break time. Maddox and Erbar (1982) also recommend a design method for sieves removing H_2S, CO_2 and water.

Membranes

Membrane separation is the most recent development, and in spite of the economic downturn, use is steadily increasing. Although the major service has been H_2 separation from NH_3 and refinery gases, CO_2 separation is significant.

Gas Separation Elements. Gas from gas separation occurs on a molecular basis, therefore thin films with no

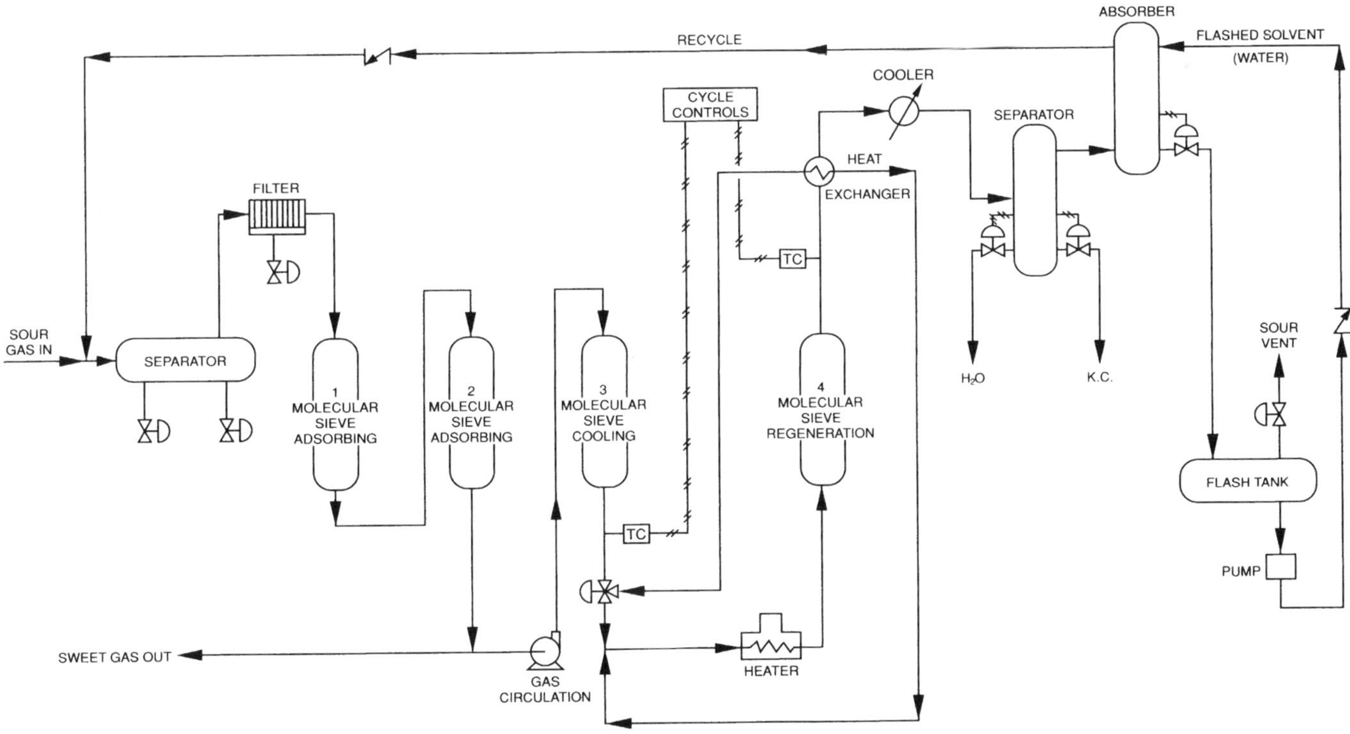

Figure 7–18. Closed cycle sweetening process (Thomas, 1967).

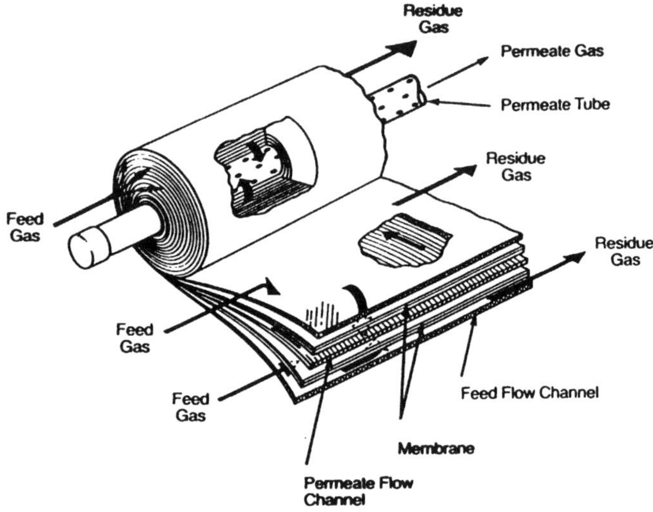

Spiral-Wound Element

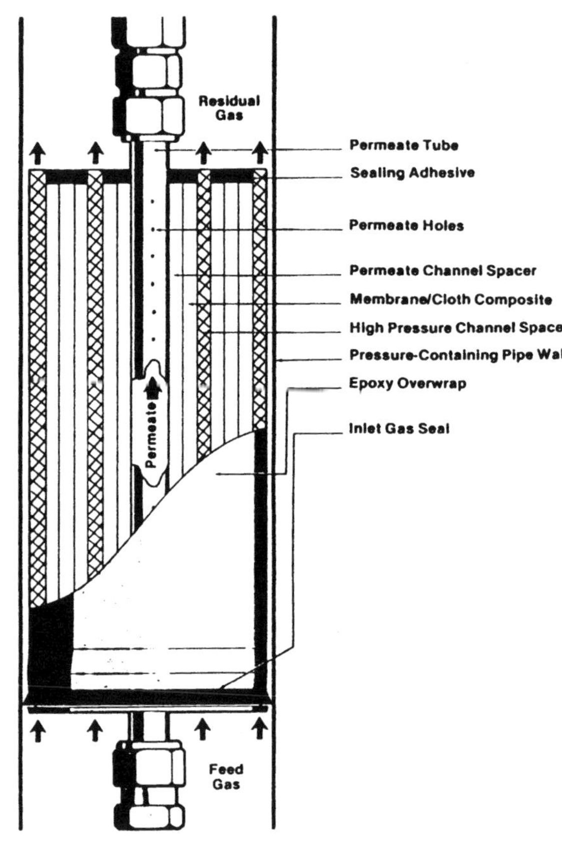

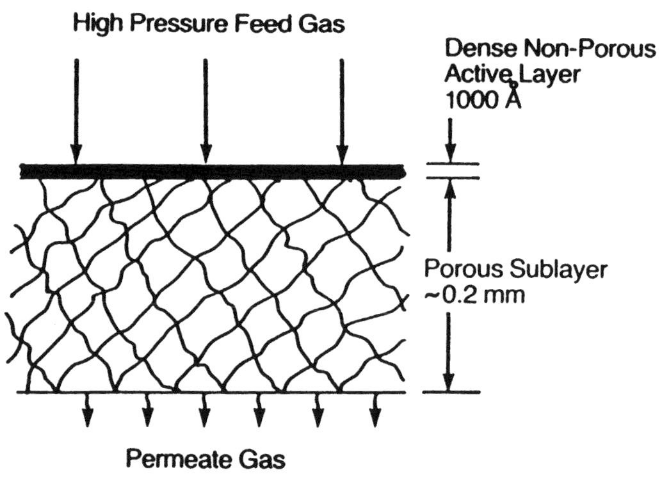

Asymmetric Membrane for Gas Separation   **Spiral-Wound Element Construction**

Figure 7–19.   Construction of a spiral-wound separator.

pores must be used. (Permeation across the film is not filtration; it is sorption on the high pressure side, diffusion across the film, and desorption or the low pressure side.) The membranes consist of an ultra-thin polymer film on top of a thin porous substrate. Depending on the design, either the coating or the substrate controls the permeation rate of the composite film.

There are two basic configurations: *spiral-wound* (e.g., Separex and Grace) and *hollow fibers* (e.g., Prism and duPont). Figure 7–19 illustrates the construction of a spiral-wound separator (Mazur and Chan, 1982). It consists of

successive layers of feed gas flow channel, separation membrane, permeate flow channel, and separation membrane wound around an axial permeate tube. The package is fitted into a cylindrical pressure vessel to form an element which is typically 4–8 in. O.D. by 4 to 5 ft long. Elements are combined in parallel and/or series to form the separator package. For these membranes, the dense nonporous cellulose acetate coating is typically the active or permeation-controlling layer.

Figure 7–20 shows a hollow-fiber separator. The fibers are hollow cylinders—less than 1 mm O.D.—spun from

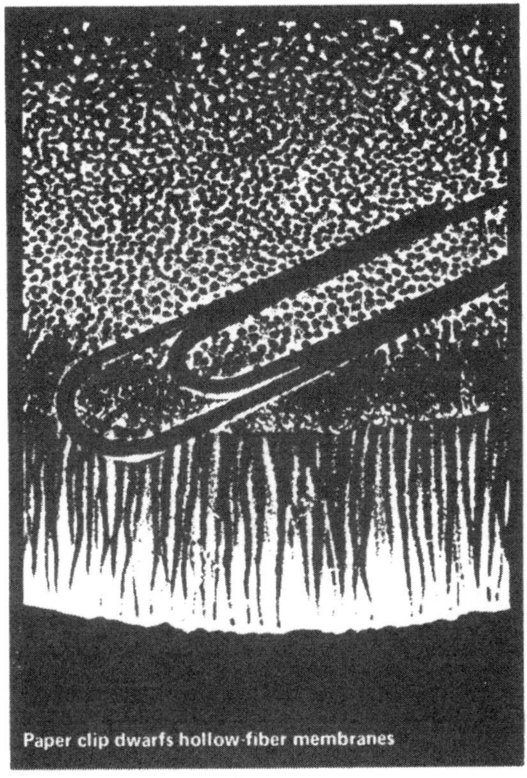

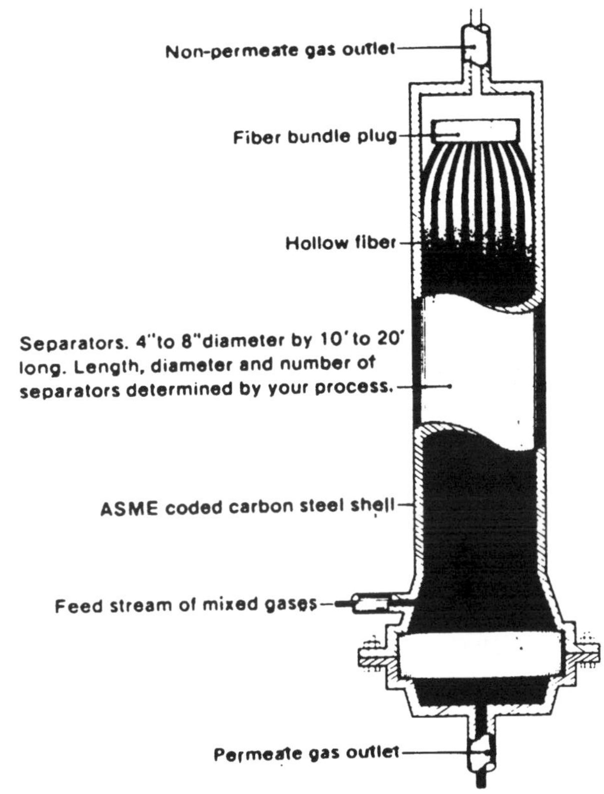

Cutaway of Prism Separator Module

Figure 7–20. Hollow fiber separator.

the separation-control material, usually a polysulfone coated with a silicone elastomer layer for protection (Bollinger *et al.*, 1982). Typical wall dimensions are up to 300 microns substrate—the separating layer is only 500 to 1000 angstroms of this—1 to 2 microns coating. The bundle of hollow fibers, which is sealed on one end, fits into the steel shell. The unit resembles a shell-and-tube heat exchanger. The feed and nonpermeate outlet gas are on the "shell" side, and the permeate gas is on the "tube" side. Typical dimensions are 4 to 8 in. diameter and 10 to 20 ft long.

Membrane Characteristics. Membranes separate gases by the difference in the rates at which the gases diffuse across the film. "Fast" gases collect in the permeate stream, and "slow" gases remain in the nonpermeate stream. Table 7–5 shows that the relative permeation rates for three commercial membranes are remarkably similar.

The permeation flow rate of any gas is given by:

$$\begin{pmatrix}\text{Permeation}\\ \text{Rate}\end{pmatrix} = \begin{pmatrix}\text{Permeability}\\ \text{Coefficient}\end{pmatrix}\begin{pmatrix}\text{Membrane}\\ \text{Area}\end{pmatrix}\begin{pmatrix}\text{Partial Pressure}\\ \text{Across Membrane}\end{pmatrix}$$

When the difference in the permeation coefficients is large,

Table 7–5 Gas Permeation Rates

Gas	Membrane	
	Spiral Wound	Hollow Fibers
H_2	100.0	fast
He	15.0	fast
H_2O	12.0	fast
H_2S	10.0	medium
CO_2	6.0	medium
O_2	1.0	medium
Ar	—	slow
CO	0.3	slow
CH_4	0.2	slow
N_2	0.18	slow
C_2H_6	0.1	

good separation of the gases is possible—e.g., H_2 from NH_3 or refinery gases. But when the difference is small—e.g., H_2S and CH_4, separation is difficult and expensive because multiple stages are needed. In any case the separation is not complete: the permeate gas contains some of the residual gas and vice versa.

Sweet gas corresponds to 4 ppm of H_2S, an impractical result considering the coefficients of the two gases. Dry gas (7 lb H_2O/MMscf) corresponds to 147 ppm. This is achievable, but the CH_4 ''lost'' with the water is too valuable. Recovery involves chilling to condense most of the water, compression to the feed pressure, and recirculation. This is usually too much equipment to compete economically with glycol dehydration. Separation of CO_2 is far more practical. The residual gas can contain 2 to 3% CO_2, and CH_4 in the CO_2 is generally not a problem.

Process Flow Sheets. Figure 7–21 shows single- and double-stage separations. Note that the gas is first dried, and that there is a filter separator just ahead of the membrane separator. These steps are important because the membrane

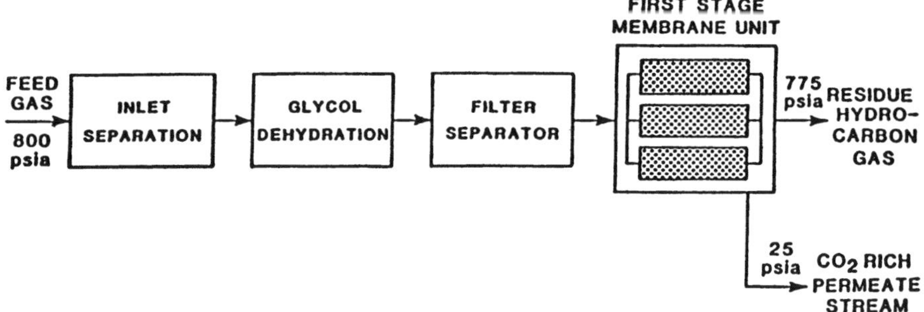

Single Stage Separation (Mazur and Chan, 1982)

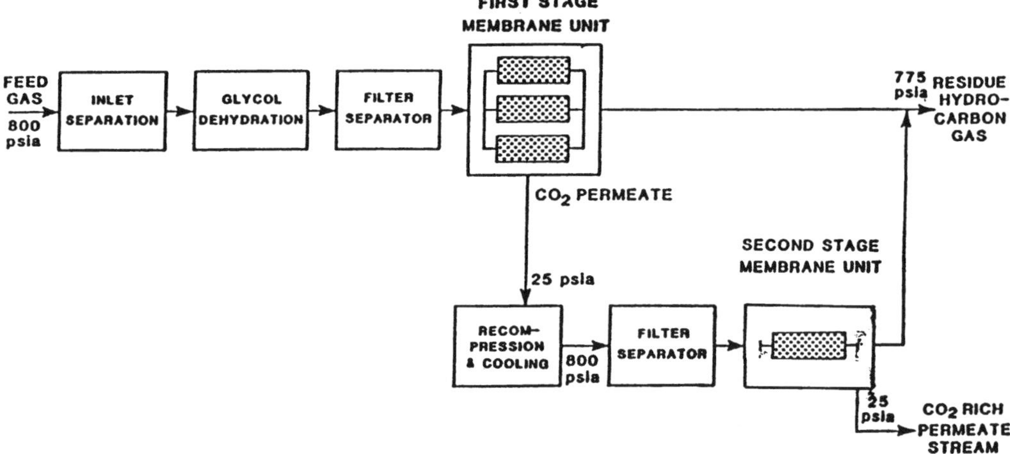

Figure 7–21. Double stage separation (Mazur and Chan, 1982).

Table 7–6 CO_2 Separation for EOR

	Feed	Permeate	Residual
Flow, MMscfd	50.0	35.8	14.2
Pressure, psig	1000.0	390.0	985.0
CO_2, %	75.0	95.0	42.0
CH_4, %	25.0	5.0	58.0
CO_2 Recovery 85%	CH_4 Recovery 89%		

films are easily fouled by dirt and liquids. (Indeed, even with clean dry gas there is a loss in the permeation efficiency of about 20% in 1000 days.) Cleaning the membranes is very difficult and often the fouled elements must be discarded.

As the volume of residual or nonpermeate gas decreases, fewer elements are needed. So a tapered arrangement is used when the feed gas contains a significant amount of the permeate gas. For example, six, four, and two elements can be used for the three steps of a separator taper configuration.

Applications. Table 7–6 summarizes the performance of an EOR plant that produces 95% CO_2 for reinjection (Schell and Houston, 1982). The results are typical of single-stage operation. The residual stream requires further processing—for example, another membrane unit or hot potassium carbonate to remove the CO_2. But this performance is satisfactory for EOR (Parro, 1984).

Several studies show both lower capital and operating costs for membrane systems when the CO_2 content of the gas is high, e.g., over 40% (Goddin, 1982; Markiewicz, 1988; and Spillman, 1989). The last reference contains detailed comparisons with other processes.

Review Questions

1. List the reasons for sweetening natural gas.
2. Classify the processes used to sweeten natural gas.
3. Summarize the areas of application for the common gas-sweetening processes. What factors should be considered?
4. Match the following gas stream flow rates and compositions (alphas) with the most appropriate gas sweetening process (numbers).

   A   100 M scfd, 200 ppm H_2S, 1% CO_2
   B   5 MM scfd, 1% H_2S, 4% CO_2
   C   10 MM scfd, 1% H_2S, 2000 ppm RSH, 3% CO_2
   D   5 MM scfd, 70% CO_2
   E   5 MM scfd, 1% H_2S, 4% CO_2, very cold climate
   F   10 MM scfd, 500 ppm H_2S, 5% CO_2 and dry the gas

   1 DGA
   2 DEA
   3 Selexol
   4 membranes
   5 batch (Fe sponge, Chemsweet, Sulfa-Check)
   6 Sulfinol

5. List the advantages and disadvantages of batch processes.
6. Describe a typical iron sponge unit.
7. The pressure drop across your iron sponge tower increases daily. Also the rate of increase is greatest at night. The back pressure valve in the exit line is no longer controlling the pressure in the tower properly. What is wrong?
8. List the advantages and disadvantages of amine units.
9. Follow the path of amine through a typical amine unit and discuss the function of the different components.
10. What purpose does the inlet separator serve in an amine unit?
11. Where do amine losses occur?
    How can such losses be minimized?
    What loss level is acceptable?
12. How can energy consumption be minimized in an amine unit?
13. What instrumentation would you recommend for an amine unit?
14. What malfunctions should trigger an alarm/shutdown of an amine unit?
15. Discuss material selection for an amine plant.
16. What modifications/precautions would you recommend for operating amine plants in cold climates?
17. Describe the plant check out, clean out, and start up of an amine plant.
18. What are the recommended temperatures for the amine in the various units of an amine plant? What are the consequences of deviating from these recommended temperatures?
19. The particulate filter downstream of the rich-amine flash tank is plugged with fine particles of elemental sulfur. Explain.
20. What is the best way to monitor the flow in both the contact tower and the stripping column of an amine plant?
21. What contaminant in the inlet gas ruins the performance of all gas-sweetening processes most frequently?
22. Under what conditions can retrograde condensation occur in the contact tower of an amine plant?
23. Why is it important not to add too much antifoam to an amine solution?

24. Your amine solution foams shortly after every change of the elements in the particulate filter. Explain.
25. What is the most troublesome stage in the operation of Stretford and Low Cat processes? How does this relate to the Sulfa-Check process?
26. Why is it critical to keep liquid and solid contaminants out of membrane elements?
27. Why do hot potassium carbonate systems not have an inlet gas/outlet gas heat exchanger?
28. The Rectisol process is not used in the gas patch. Why?

Problems

1. Convert 0.25 gr/100 scf of H_2S to ppmv and ppmw. State any assumptions.
2. Select a suitable amine, and size the plant to treat the following gas streams:
   a. 5 MMscfd, 1% H_2S, 4% CO_2, 600 psig, 100°F.
   b. 5 MMscfd, 2% H_2S, 5% CO_2, 800 psig, 90°F, very cold climate.
3. Figure 1–12 provides the following compositions for the feed (sour) and product (sweetened) gas streams in a Selexol plant.

Component	Feed	Product
CO_2 (vol %)	36.3	1.2
H_2S (ppmv)	90.0	1.6
C1 (vol %)	50.6	84.9
C2+ (vol %)	11.1	10.5
N_2 (vol %)	2.0	3.4

Calculate:
   a. The percentage of the inlet H_2S removed.
   b. The percentage of the inlet C2+ lost.
      For simplicity assume that no nitrogen is lost.
   c. Are the gas analyses complete or are any components missing?

Nomenclature

A = cross-sectional area of tower (ft^2)
AG = acid gas (H_2S and/or CO_2)
acfh = actual cubic feet per hour
acfm = actual cubic feet per minute
BL = bed life (days)
CS = Chemsweet chemical requirement (lb/day)
EOR = enhanced oil recovery
f = expansion factor
gpm = U.S. gallons per minute
gr = grains
ID = internal diameter (ft)
IS = iron sponge chemical requirement (bushels/day)
K = amine circulation factor (gpm)/(MMscfd)(mol% AG)
L = bed height (ft)
NS = NaSH chemical requirement (gal 20% NaOH/day)
OD = outside diameter (ft)
P = pressure (psia, psig)
ppmv = parts per million by volume
ppmw = parts per million by weight
Q = gas flowrate, MMscfd
Q_a = actual gas flowrate at contactor T & P (acfm)
RV = volume of reactant in tower (ft^3)
SC = Sulfa-Check chemical requirement (gal/day)
T = temperature (°F, °R)
V_s = superficial gas velocity (ft/min)
Z = compressibility factor

Greek

ρ = density (lb/ft^3)

Abbreviations

DEA = diethanolamine
DGA = diglycolamine
DIPA = diisopropanolamine
MDEA = methyldiethanolamine
MEA = monoethanolamine
SNPA = Societé National Elf–Aquitinè DEA
MMSCFD = million standard cubic feet per day

References

Anerousis, J. P., and Whitman, S. K. (1984), ''An Updated Examination of Gas Sweetening by the Iron Sponge Process,'' SPE 13280, Houston, TX (Sept.).

Arnoldi, J. C., Dempsey, P. E., and Morgan, D. J. (1987), ''Selection of Materials for Sour Gas Processing,'' Laurance Reid Gas Conditioning Conference, pp. A1–17, Norman, OK (March).

Bacon, T. R., and Pearce, R. L. (1985), ''Energy Conservation and H_2S Selectivity with COS Removal,'' Petroenergy, Houston, TX (Sept.).

Ballard, D. (1966), ''How to Operate an Amine Plant,'' *Hydrocarbon Processing*, Vol. 45, No. 4, p. 137 (April).

Ballard, D. (1986), ''Cut Energy, Chemical and Corrosion Costs in Amine Units,'' *Energy Progress*, Vol. 6, No. 2, pp. 112–124 (June).

Bartoo, R. K. (1984), ''Benfield Process for Acid Gas Removal,'' AIChE National Meeting, Atlanta, GA (March 11–14).

Bhatia, K., and K. T. Allford (1986), ''One-step Process

Takes H2S From Gas Stream," *Oil & Gas J.*, Vol. 84, No. 42, p. 44 (Oct. 20).

Blanc, C. J., J. Elgue, and F. Lallemand (1981), "MDEA Process Selectively Extracts H$_2$S," *Hydrocarbon Processing*, Vol. 60, No. 8, pp. 111–116 (Aug.).

Bollinger, W. A., D. L. MacLean, and R. S. Narayan (1982), "Separation Systems for Oil Refining and Production," *Chem. Eng. Prog.*, Vol. 78, No. 10, pp. 27–32 (Oct.).

Bucklin, R. W. (1982), "DGA—A Workhorse for Gas Sweetening," *Oil & Gas J.*, Vol. 80, No. 45, pp. 204–210 (Nov. 8).

Bucklin, R. W. and R. L. Schendel (1984), "Comparison of Fluor Solvent and Selexol Processes," AIChE Symposium, Atlanta, GA (March 11–14).

Buenger, C. W., S. A. Bedell, and K. H. Kirby (1988), "Chelates in H$_2$S Removal," Laurance Reid Gas Conditioning Conference, Norman, OK (March).

Bullin, J. A., and J. C. Polasek, (1982), "Selective Absorption With Amines," 61st Annual GPA Convention, pp. 86–90, Dallas, TX (March 15–17).

Bullin, J. A., J. C. Polasek, and J. W. Holmes (1981), "Optimization of New and Existing Amine Gas Sweetening Plants by Computer Simulation," Proceedings 60th Annual GPA Convention, p. 142 (March).

Burnes, E. E., and K. Bhatia (1985), "Process for Removing Hydrogen Sulfide From Gas Mixtures," U.S. Patent 4,515,759 (May 7).

Butwell, K. F., and L. Kroop (1986), "Fundamentals of Gas Sweetening," Laurance Reid Gas Conditioning Conference, University of Oklahoma, Norman, OK.

Byeseda, J. J, J. A. Deetz, and W. P. Manning (1985), "Optisol—A New Gas Sweetening Solvent," *Oil & Gas J.*, Vol. 83, No. 23, pp. 144–146, (June 10).

Carnell, P. J. H. (1986), "New Fixed-Bed Absorbent for Gas Sweetening," *Oil & Gas J.*, Vol 84, No. 33, pp. 59–62 (Aug. 18)

Chi, Chang W., and H. Lee (1973), "Gas Purification by Adsorption," AIChE Symposium Series, Vol. 69, No. 134, pp. 96–101.

Clem, K. R., J. L. Kaufman, M. M. Lambert, and D. L. Brown (1985), "Gas Treating Advances With Flexsorb Technology," Laurance Reid Gas Conditioning Conference, Norman, OK (March 4).

Dailey, L. D. (1970), "Status of SNPA-DEA," *Oil & Gas J.*, Vol. 68, pp. 136–141 (May 4).

Dankwerts, P. V. (1970), "Gas-Liquid Reactions," McGraw-Hill, NY.

Daviet, G. R., R. Sundermann, S. T. Donnelly, and J. A. Bullin (1984), "Dome's NC Plant Successful Conversion to MDEA," 63rd Annual GPA Convention, New Orleans, LA (March 20).

Davis, B. J., J. P. Anerousis, and O. D. Handley (1985), "Iron Sponge is Economical and Effective for NGL Sweetening," *Oil & Gas J.*, Vol. 83, No. 38, pp. 122–125 (Sept. 23).

Deshmukh, R. D., and A. E. Mathes (1981), "A Mathematical Model for Equilibrium Solubility of H$_2$S and CO$_2$ in Alkanolamine Solutions," *Chem. Eng. Science*, Vol. 36, pp. 355–362.

Dingman, J. C., J. L. Jackson, T. F. Moore, and J. A. Branson (1983), "Equilibrium Data for the H$_2$S-CO$_2$-DGA-System," 62nd GPA Annual Convention, pp. 256–268, San Francisco, CA (March 14).

Dobbs, J. B. (1986), "One Step Process," Laurance Reid Gas Conditioning Conference, University of Oklahoma, Norman, OK (March).

Dow Chemical Company (1962), "Gas Conditioning Fact Book," pp. 145–234, Midland MI.

Duckworth, G. L., and J. H. Geddes (1965), "Natural Gas Desulfurization by the Iron Sponge Process," *Oil & Gas J.*, Vol. 63, No. 37, pp. 94–96 (Sept. 13).

Dwyer, D. L. (1989), "Operational Experience with Formulated Solvents," Presented Venezuela GPA Meeting, Maracaibo (Feb. 15)

Eickmeyer, A. G. (1976), "Methods and Compositions for Removing Acid Gases From Gaseous Mixtures," U.S. Patent 3,932,582 (Nov. 14).

Ellwood, P. (1964), "Meta-Vanadates Scrub Manufactured Gas," *Chemical Engineering*, Vol. 71, No. 15, p. 128 (July 20).

Exxon (1989) "Flexsorb SE and Flexsorb SE Plus," Exxon Research and Engineering Co., Florham Park, NJ.

Ferrin, C. R., and W. P. Manning (1982), "Physical Solvents for Gas Sweetening," U.S. Patent 4,360,363 (Nov. 23).

Ferrin, C. R., and W. P. Manning (1984), "Physical Solvents for Gas Sweetening," U.S. Patent 4,784,934 (Nov. 27).

Fong, H. L., D. S. Kushner, and R. T. Scott (1987), "Gas Desulfurization Using Sulferox," Laurance Reid Gas Conditioning Conference, Norman, OK (March).

Fox, I. (1981), "Process for Scavenging Hydrogen Sulfide From Hydrocarbon Gases," U.S. Patent 4,246,274 (Jan. 20).

Frech, K. J., and J. J. Tazuma (1982), "Method for Removal of Sulfur Compounds From a Gas Stream," U.S. Patent 4,311,680 (Jan. 19).

Gas Processors Suppliers Association (1987), "Engineering Data Book," 10th Ed., Vol. II, Sec. 21 "Hydrocarbon Treating."

Goar, B. G. (1972), "Today's Gas Treating Processes," N.G.P.A. Permian Basin Regional Meeting, Odessa, TX (May 4).

Goddin, C. S. (1982), "Pick Treatment for High CO_2 Removal," *Hydrocarbon Processing*, Vol. 61, No. 5, pp. 125–130 (May).

Goldstein, A. M., A. M. Edelman, W. D. Beisner, and P. A. Ruziska (1984), "Hindered Amines Yield Improve Gas Treating," *Oil & Gas J.*, Vol. 82, No. 29, p. 70 (July 16).

Harrell, B. E., and Manning, W. P. (1986), "Review of Batch Gas Sweetening Processes," International Technological Exchange, Kern County, CA (Nov.).

Hawkes, E. N., and B. F. Mago (1971), "Stop MEA CO_2 unit corrosion," *Hydrocarbon Processing*, Vol. 50, No. 8, pp. 109–112 (Aug.).

Hegerty, M. J., R. J. Wissbaum, and J. E. Johnson (1988), "Equilibrium of $H_2S + CO_2 = COS + H_2O$: This Reaction Can Upset Your Sour Gas Plant," AIChE Summer Nat. Mtg., Denver, CO (Aug. 21–24)

Hohlfeld, R. W. (1979), "Selective Absorption of H_2S from Sour Gas," S.P.E. 7972, Ventura, CA (April 18–20).

Hunter, J. K., and L. A. Bryan (1987), "LaBarge Project Availability of CO2 for Tertiary Projects," *J. Petroleum Technology*, Vol. 39, No. 11, pp. 1410–1410 (Nov.).

Johnson, J. T. (1985), "Natural Gas Odorants and Their Components," Proceedings 60th International School of Hydrocarbon Measurement, University of Oklahoma, Norman, OK (April 16–18).

Jones, V. W., and C. R. Perry (1973), "Fundamentals of Gas Treating," Gas Conditioning Conference, Norman, OK.

Jou, Fang-yuan, A. E. Mather, and F. Otto (1982) "Solubility of H_2S and CO_2 in Aqueous MDEA Solutions," *Ind. Eng. Chem. Proc. Des. Dev.*, Vol. 21, No. 4, pp. 539–544.

Katti, S. S., and B. D. Langfitt (1986), "Simulator Development for Commercial Absorbers," 65th Annual GPA Convention, pp. 158–163, San Antonio, TX (March).

Kattner, J. E., A. Samuels, and R. P. Wendt (1988), "Iron Oxide Slurry Process for Removing Hydrogen Sulfide," *J. Petroleum Technology*, Vol. 40, No. 9, p. 1237 (Sept.).

Kent, R. L., and B. Eisenberg (1976), "Better Data for Amine Treating," *Hydrocarbon Processing*, Vol. 55, No. 12, p. 87 (Feb.).

Kent, V. A., and R. A. Abib (1985), "Selective H_2S Caustic Scrubber," Laurance Reid Gas Conditioning Conference, Norman, OK (March).

Khan, M., and W. P. Manning (1985), "Practical Designs for Amine Plants," Petroenergy Workshop, Houston, TX (Sept.).

King, J. C., D. W. Stanbridge, Y. Ide, T. A. Trinker, and S. R. Gupta (1986), "Process Selection for an Oil Associated Sour Gas Plant," Laurance Reid Gas Conditioning Conference, Norman, OK (March).

King, R. L., M. L. Spears, and J. A. Bullin (1985), "Design Alternatives for Sweetening Liquids With Amines," Petroenergy Conference, Houston, TX (Sept. 17).

Klein, J. P. (1970), "Developments in Sulfinol and Adip processes increases uses," *Oil and Gas International*, Vol. 10, No. 9, pp. 109–112 (Sept. 10).

Kohl, A. L., and F. C. Riesenfeld (1985), "Gas Purification," 4th Ed., Gulf Pub. Co., Houston, TX.

Kosseim, A. J., J. G. McCullough, and K. F. Butwell (1984), "Corrosion-Inhibited Amine Guard ST Process," *Chemical Engineering Progress*, Vol. 80, No. 10, pp. 64–71 (Oct.).

Laengrich, A., C. Harmon, and K. Carlisle (1982), "Selective H2S Removal in CO2 Floods Requires Careful Study," *Oil & Gas J.*, Vol. 80, No. 40, pp. 90–93, (Oct. 4).

Leci, C. L. (1988), "Acid Gas Removal from Condensate Fields," *IBC Tech. Serv. Ltd. Develop. of Condensate Fields Conf. (London, England) Proc.* (Sept. 20–21).

Love, D. E. (1985), "Review of H2S and CO2 Removal Processes," presented at Engineer's Society of Tulsa Meeting, Tulsa, OK (March 12).

Maddox, R. N. (1974), "Gas and Liquid Sweetening," Campbell Petroleum Series, Norman, OK.

Maddox, R. N., and J. H. Erbar (1982), "*Gas Conditioning and Processing*," Vol. 3, Chapter 15, Campbell Petroleum Series, Norman, OK.

Manning, W. P. (1979), "Chemsweet, A New Process for Sweetening Low-Value Sour Gas," *Oil & Gas J.*, Vol. 77, No. 42, p. 122–124 (Oct. 15).

Manning, W. P. (1989), "Guidelines for Designing Amine Plants," Petroenergy, Houston, TX (Oct.).

Manning, W. P., S. J. Rehm, and J. L. Schmuhl (1981), "Method of Removing Hydrogen Sulfide From Gas Mixtures," U.S. Patent 4,276,271 (June 30).

Markiewicz, G. (1988), "Membrane System Lowers Treating Costs at Gas Plant," *Oil & Gas J.*, Vol. 86, No. 44, p. 71–73 and 76 (Oct. 31).

Mazur, W. H., and M. C. Chan (1982), "Membranes for Natural Gas Sweetening and CO_2 Enrichment," *Chem. Eng. Prog.*, p. 38 (Oct.).

Moore, T. F., J. C. Dingman, and F. L. Johnson, Jr. (1984), "Review of Current Diglycolamine Agent Gas Treating Applications," AIChE Meeting, Atlanta, GA (March 11–15).

Mortko, R. A. (1984), "Remove H_2S Selectively," *Hydrocarbon Processing*, Vol. 43, No. 6, p. 78 (June).

Moshfeghian, M., K. J. Bell, and R. N. Maddox (1977),

''Reaction Equilibria for Acid Gas Systems,'' Laurance Reid Gas Conditioning Conference, Norman, OK (March).

Ouwerkerk, C. (1978), ''Design for Selective H_2S Absorption,'' *Hydrocarbon Processing*, Vol. 57, No. 4, p. 89 (April).

Parro, D. (1984), ''Membrane CO_2 Separation Proves Out at Sacroc Tertiary Recovery Project,'' *Oil & Gas J.*, Vol. 82, No. 38, p. 85 (Sept. 24).

Pauley, C. R., R. Hashemi, and S. Caothien (1988), ''Analysis of Foaming Mechanism in Amine Plants,'' Presented AIChE National Summer Meeting, Denver, CO (Aug. 21–24).

Pearce, R. L. (1978), ''H_2S Removal With MDEA,'' 57th Annual GPA Convention, pp. 139–144, New Orleans, LA (March 22).

Pearce, R. L., S. Grosso, and D. C. Cringle (1980), ''Amine Gas Treating Solution Analysis,'' Proceedings 59th Annual GPA Convention, pp. 150–162, Houston, TX (March 17–19).

Pearce, R. L., and M. S. DuPart (1985), ''Amine Inhibiting,'' *Hydrocarbon Processing*, Vol. 64, No. 5, p. 70 (May).

Pearce, R. L., and Wolcott (1986), ''Basic Considerations in Acid Gas Removal,'' AIChE Annual Meeting, New Orleans, LA (April 6–10).

Perry, C. R. (1970), ''A New Look at Iron Sponge Treatment of Sour Gas,'' Gas Conditioning Conference, Norman, OK (April).

Perry, C. R. (1978), ''Hydrogen Sulfide Removal With MEA and DEA,'' GPA Annual Meeting (Nov. 21).

Raney, D. R. (1977), ''Bulk Removal of Carbon Dioxide With Selexol Solvent at Pikes Peak Plant,'' Gas Conditioning Conference.

Richardson, I. M. J., and J. P. O'Connell (1975), ''Some Generalizations about Processes to Absorb Acid Gases and Mercaptans,'' *Ind. Eng. Chem.*, Vol. 14, No. 4, p. 457.

Richert, J. P., A. J. Bagdasarian, and C. A. Shargay (1989), ''Extent of stress corrosion cracking in amine plants revealed by survey,'' *Oil & Gas J.*, Vol. 87, No. 23, p. 45 (June 5).

Robinson, A., V. Bradley, and J. Daughtry (1988), ''Formulated Amine Solvent Improves Natural Gas Treating Process,'' Laurance Reid Gas Conditioning Conference, Norman, OK (March).

Rosenzweig (1981), ''Unique Membrane System Spurs Gas Separations,'' *Chem. Eng.*, Vol. 88, No. 24, pp. 62–66 (Nov. 30).

Samuels, A. (1988), Gas Sweetener Associates' Technical Manual 3–88, Metairie, LA.

Sartori, G., and F. Leder (1978), ''Process for Removing Carbon Dioxide Containing Acid Gases from Gaseous Mixtures Using a Basic Salt Activated with a Hindered Amine,'' U.S. Patent 4,112,050; ''Process and Amine-Solvent Absorbent for Removing Acidic Gases from Gaseous Mixtures,'' U.S. Patent 4,112,051; and ''Process for Removing Carbon Dioxide Containing Acid Gases from Gaseous Mixtures Using Aqueous Amine Scrubbing Solutions,'' U.S. Patent 4,112,052 (Sept. 5).

Sartori, G., and D. W. Savage (1983), ''Sterically Hindered Amines for CO_2 Removal From Gases,'' *I&EC Fundamentals*, Vol. 22, No. 2, p. 239.

Schaak, J. P., and F. Chan (1989a), ''H_2S Scavenging—1,'' *Oil & Gas J.*, Vol. 87, No. 4, p. 51 (Jan. 23).

Schaak, J. P. and F. Chan (1989b), ''H_2S Scavenging—2,'' *Oil & Gas J.*, Vol. 87, No. 5, p. 81 (Jan. 30).

Schell, W. J., and C. D. Houston (1982), ''Process Gas With Selective Membranes,'' *Hydrocarbon Processing*, Vol. 61, No. 9, pp. 249–252 (Sept.).

Schell, William J., C. Douglas Houston, and William L. Hopper (1983), ''Membranes Can Efficiently Separate CO_2 From Mixtures,'' *Oil & Gas J.*, Vol. 81, No. 33, pp. 52–56 (Aug. 15).

Shah, V. A. (1989), ''Integrated Gas Treating and Hydrocarbon Recovery Process Using Selexol Solvent Technology,'' Proc. 69th Annual GPA Convention, San Antonio, TX (March).

Sigmund, P. W., K. F. Butwell, and A. J. Wussler (1981), ''HS Process Removes H_2S Selectively,'' *Hydrocarbon Processing*, Vol. 60, No. 5, p. 118 (May).

Sivalls, C. R. (1985), ''Hydrogen Sulfide Service Dictates Special Equipment Design Above Requirements,'' *Oil & Gas J.*, Vol. 83, No. 27, p. 56 (July 8).

Spillman, R. W. (1989), ''Economics of Gas Separation Membranes,'' *Chem. Eng. Progress*, Vol. 85, No. 1, pp. 41–62 (Jan.).

Stanbridge, D. W., Y. Ide, and W. Hefner (1984), ''Basic Developments in the BASF Activated MDEA Process,'' AIChE Meeting, Anaheim, CA (May).

Strelzoff, S. (1975), ''Choosing the Optimum CO_2 Removal System,'' *Chem. Eng.*, Vol. 82, No. 19, p. 115 (Sept. 15).

Swaim, Jr. C. D. (1970), ''Gas Sweetening Processes of the 1960's,'' *Hydrocarbon Processing*, Vol. 49, No. 3, p. 127 (March).

Sweeney, C. W., T. J. Ritter, and McGintey (1988), ''A Strategy for Screening Physical Solvents,'' *Chem. Eng.*, Vol. 95, No. 9, p. 119 (June 20).

Sweny, W. (1980), ''High CO_2—High H_2S Removal With Selexol Solvent,'' Proceedings 59th GPA Annual Meeting, Houston, TX (March).

Taylor, D. K. (1956), ''How to Desulfurize Natural Gas,'' *Oil & Gas J.*, Vol. 54, No. 79, p. 125 (Nov. 5).

Taylor, D. K. (1956), "Natural-Gas Desulfurization—2," *Oil & Gas J.*, Vol. 54, No. 81, p. 260 (Nov. 19).

Taylor, D. K. (1956), "Natural-Gas Desulfurization—3," *Oil & Gas J.*, Vol. 54, No. 83, p. 139 (Dec. 3).

Taylor, D. K. (1956), "Natural-Gas Desulfurization—4," *Oil & Gas J.*, Vol. 54, No. 84, p. 147 (Dec. 10).

Tennyson, R. N., and R. P. Schaaf (1977), "Guidelines Can Help Choose Proper Process for Gas-Treating Plants," *Oil & Gas J.*, Vol. 75, No. 2, pp. 78–86 (Jan. 10).

Thomas, T. L., and E. L. Clark (1967), "Molecular Sieves Reduce Problems," *Oil & Gas J.*, Vol. 65, No. 12, p. 112 (March 20).

Thomas, J. C. (1988), "Improved Selectivity Achieved with Ucarsol Solvent," Laurance Reid Gas Conditioning Conference, Norman, OK (March).

Tomcej, R. A., F. D. Otto, H. A. Rangala, and B. R. Morrell (1987), "Tray Design for Selective Absorption," Laurance Reid Gas Conditioning Conference, Norman, OK (March 2).

Vaz, R. N., G. J. Mains, and R. N. Maddox (1981), "Ethanolamine Process Simulated by Rigorous Calculation," *Hydrocarbon Processing*, Vol. 60, No. 4, pp. 139–142 (April).

Veldman, R. (1988), "How to Reduce Amine Losses," Petroenergy, Houston, TX (Oct.).

Vickery, D. J., S. W. Campbell, and R. H. Weiland (1988), "Gas Treating with Promoted Amines," Laurance Reid Gas Conditioning Conference, Norman, OK (March).

Wall, J. (Editor) (1986), "1986 Gas Process Handbook," *Hydrocarbon Processing*, Vol. 65, No. 4, p. 75–106 (April).

Wall, J. (Editor) (1988), "1988 Gas Process Handbook," *Hydrocarbon Processing*, Vol. 67, No. 4, pp. 51–80 (April).

Wansink, D. H. N. (1985), "The Shell Sulfinol-M Process," Gas Sweetening and Sulfur Recovery Seminar, Amsterdam (Nov.).

Zapffe, F. (1963), "Iron Sponge Process Removes Mercaptans," *Oil & Gas J.*, Vol. 61, No. 33, pp. 103–104 (Aug. 19).

Chapter 8

Gas Dehydration Using Glycol

INTRODUCTION

At the wellhead, reservoir fluids almost invariably contain water and, except for a few shallow wells, natural gas is produced saturated with water. The major *reasons* for *dehydrating* natural gas are:

1. Natural gas can combine with liquid or free water to form *solid hydrates* that can *plug* valves, fittings or even pipelines.
2. If not separated from the produced water, the natural gas is *corrosive*, especially when CO_2 *and/or* H_2S are also present.
3. Water can condense in the pipeline causing slug flow and possible *erosion and corrosion*.
4. Water vapor increases the volume and decreases the heating value of the gas.
5. *Sales* gas *contracts* and/or *pipeline specifications* have a maximum water content—usually 7 lb H_2O per MMscf—or dew point.
6. Dehydrating allows operation of cryogenic and refrigerated absorption plants without *freeze-ups*.

For convenience, the broad subject of dehydration will be discussed in both Chapters 8 and 9. The present chapter begins with an overview of dehydration methods, followed by a detailed discussion of glycol contacting. Chapter 9 describes dehydration using solid desiccants, expansion refrigeration, and calcium chloride, and finally compares all four dehydration methods.

Methods of Dehydration

Natural gas can be dehydrated to pipeline specifications by several processes, including liquid desiccants (glycols), solid desiccants (alumina, silica gel, molecular sieves), expansion refrigeration, and calcium chloride.

In the liquid state, water molecules are highly associated because of hydrogen bonding. The hydroxyl and ether groups in glycols form similar associations with water molecules. This liquid-phase hydrogen bonding explains why *glycols* have such a high affinity for water and why the equilibrium partial pressure of water vapor over a glycol water solution is far less than that predicted by Raoult's law.

Solid desiccants are characterized by an internal porous structure that contains very large internal surface areas (e.g. 200–800 m^2/g) with very small radii of curvature (0.001–0.2 μm). The equilibrium partial pressure of water vapor above such concave surfaces is much less than that above plane surfaces and so solid desiccants exhibit a very great affinity for water. These desiccants exhibit capacities from 5 to 15% on a weight basis and can dry natural gas to less than 0.1 ppm of water or a dew point of $-150°F$.

Expansion refrigeration processes are known also as low-temperature separation (LTS) units. Examples are BS&B's LTS with ethylene glycol, or NATCO's LTX with no inhibitors, or Hot Bottom Units (HBU). They employ the Joule-Thomson (J-T) effect to both dry the gas and recover natural gas liquids. The J-T expansion requires a large pressure drop to obtain all of the cooling needed to dehydrate the gas. Therefore, expansion refrigeration, with or without hydrate inhibition, is normally used when the prime objective is hydrocarbon recovery.

Anhydrous calcium chloride absorbs 1 lb H_2O per lb $CaCl_2$ before forming brine. Dilution of the brine increases the capacity to 3.5 lb of H_2O per lb $CaCl_2$. The brine is not reusable, and dew-point depressions range between 50–80°F. Calcium chloride is suited best for modest dew-point depressions of small gas volumes in remote locations.

Usually the choice of dehydration method is between glycol and solid desiccants. *Glycol* dehydration possesses the following *advantages* over solid desiccants:

1. *Lower installed costs*—Kohl and Riesenfeld (1979) report that solid desiccant plants cost 50% more at 10 MMscfd and 33% more at 50 MMscfd.

2. *Lower pressure drop*—5 to 10 psi vs. 10–50 psi for dry desiccant units.
3. Glycol dehydration is continuous rather than batch.
4. Glycol makeup is easily accomplished. Recharging solid desiccant towers is time-consuming and sometimes requires interruption of gas sales.
5. Glycol units *require less regeneration heat* per pound of water removed thus lowering utility costs.
6. Glycol systems will operate in the presence of materials that would foul solid desiccants.
7. Glycol units can dehydrate natural gas to 0.5 lb H_2O/ MMscf. Indeed, 0.25 lb H_2O/MMscf has been achieved on the North Slope with stripping gas and a Stahl column.

However, *glycol* dehydration has the following *disadvantages*:

1. Water dew points below $-25°F$ require stripping gas and a Stahl column.
2. Glycol is susceptible to contamination.
3. Glycol is corrosive when contaminated or decomposed.

The *advantages* of *solid desiccants* are:

1. Dew points as low as $-150°F$ (1 ppmv of H_2O) are obtainable.
2. They are less affected by small changes in gas pressure, temperature, or flow rate.
3. They are less susceptible to corrosion or foaming.

The *disadvantages* of *solid desiccants* are:

1. Higher capital costs and higher pressure drops.
2. Desiccant poisoning by heavy hydrocarbons, H_2S, CO_2, etc.
3. Mechanical breaking of desiccant particles.
4. High space and weight requirements.
5. High regeneration heat requirements and high utility costs.

The above summary indicates why glycol dehydration is by far the most commonly used process.

Choice of Glycol

Four glycols have been successfully used to dry natural gas: ethylene glycol (EG), diethylene glycol (DEG), triethylene glycol (TEG), and tetraethylene glycol (TREG). Grosso (1978) discusses the relative merits of EG, DEG, TEG, and TREG in detail. *TEG* has gained almost *universal acceptance* as the most cost-effective choice because:

1. *TEG is more easily regenerated* to a concentration of 98–99.95% in an atmospheric stripper because of its high boiling point and decomposition temperature.

This permits higher dew-point depressions of natural gas in the range of 80–150°F.
2. TEG has an initial theoretical *decomposition temperature of 404°F*, while that of diethylene glycol is only 328°F (Ballard, 1966).
3. Vaporization losses are lower than EG or DEG.
4. *Capital and operating costs are lower.*
5. TEG is not too viscous above 70°F.

For more than 40 years sweet and sour natural gases have been dehydrated using TEG. Dew-point depressions range from 40–150°F while inlet gas pressures and temperatures vary from 75 to 2500 psig and from 55 to 160°F, respectively (Worley, 1967).

Natural gas dehydration with TEG is discussed under the following topics: process description, operation, design, and troubleshooting.

PROCESS DESCRIPTION

Dehydration of natural gas by TEG is first outlined by summarizing the flowpaths of the natural gas and glycol. Then the individual components of a *typical TEG unit* are described in detail.

As shown in Figure 8–1, wet natural gas first flows through an inlet separator or scrubber to remove all liquid and solid impurities. Then the gas flows into and upward through the absorber or contactor where it is contacted countercurrently and dried by the glycol. Finally, the dried gas passes through a gas/glycol heat exchanger, and then into the sales line.

Reconcentrated or "lean" glycol enters the top of the contactor where it flows downward from tray to tray and absorbs water from the rising natural gas. The wet or "rich" glycol leaves the absorber and flows through a coil in the accumulator where it is preheated by hot lean glycol. After the glycol-glycol heat exchanger, the rich glycol enters the stripping column and flows down the packed bed section into the reboiler. Steam generated in the reboiler strips water from the liquid glycol as it rises up the packed bed. The water vapor and desorbed natural gas are vented from the top of the stripper.

The hot reconcentrated glycol flows out of the reboiler into the accumulator where it is cooled by heat exchange with rich glycol. Finally, the lean glycol flows through the glycol/gas exchanger and is pumped back into the top of the absorber.

The simple flow diagram shown in Figure 8–1 is typical of small wellhead units where unattended operation is the prime concern. *Larger* and *offshore units* are monitored

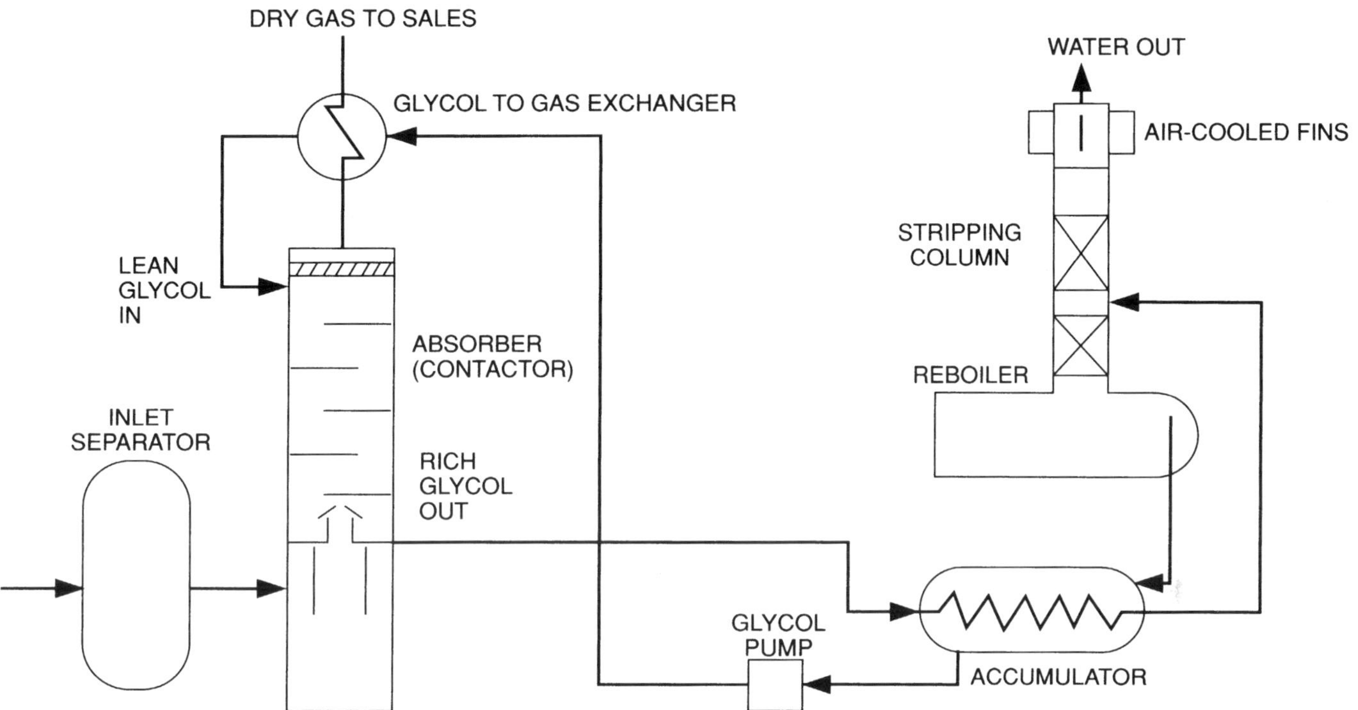

Figure 8–1. Simplified flow diagram for TEG dehydration.

daily, and the efficiency and operational cost are improved by the additional equipment shown in Figure 8–2.

1. Rich glycol leaves the absorber and enters a *cooling coil* that controls the water reflux rate at the top of the stripper. This temperature control insures that the water vapor leaving the still does not carry over excess glycol.
2. *Heat exchange* between the cool, rich glycol and the hot, lean glycol is improved by using two or more shell-and-tube heat exchangers in series. The increased heat recovery reduces fuel consumption in the reboiler and protects the glycol circulation pump from being overheated; it also allows the flash tank and filter to operate at approximately 150°F.
3. Rich glycol is *flashed* to remove dissolved hydrocarbons. The latter can be used for fuel and/or stripping gas.
4. The rich glycol is *filtered* before being heated in the reconcentrator. This prevents impurities such as solids and heavy hydrocarbons from plugging the packed column and fouling the reboiler fire tube.
5. The pump is protected by filtering the lean glycol as it leaves the accumulator.

Inlet Scrubber

About half of all gas dehydration problems are caused by inadequate scrubbing of the inlet gas. Five of the more common *contaminants* that impair the performance of the TEG are (Simmons, 1981a):

1. *Entrained* or *free water*. This water increases the glycol recirculation, reboiler heat duty, and fuel costs. If the unit becomes overloaded, glycol can be carried over from the contactor and/or still.
2. *Oils* or hydrocarbons: Dissolved oils (aromatic or asphaltic) reduce the drying capacity of the glycol and, with water, cause foaming. Undissolved oils can plug the absorber trays. Undissolved oils coke on the heat transfer surfaces in the reboiler and increase the viscosity of the glycol.
3. *Entrained brine*: These salts dissolve in the glycol, are corrosive to steels—especially stainless—and can deposit on the fire tubes in the reboiler, causing hot spots and fire-tube burnout.
4. *Down-hole additives* such as corrosion inhibitors, acidizing and fracturing fluids: These materials cause foaming, corrosion, and hot spots if they deposit on the fire tubes.
5. *Solids* (such as sand and corrosion products, e.g., rust or FeS): Solids promote foaming, erode valves and pumps, and eventually plug trays and packing.

Complete scrubbing of the incoming wet gas is absolutely essential for satisfactory performance. Many large units

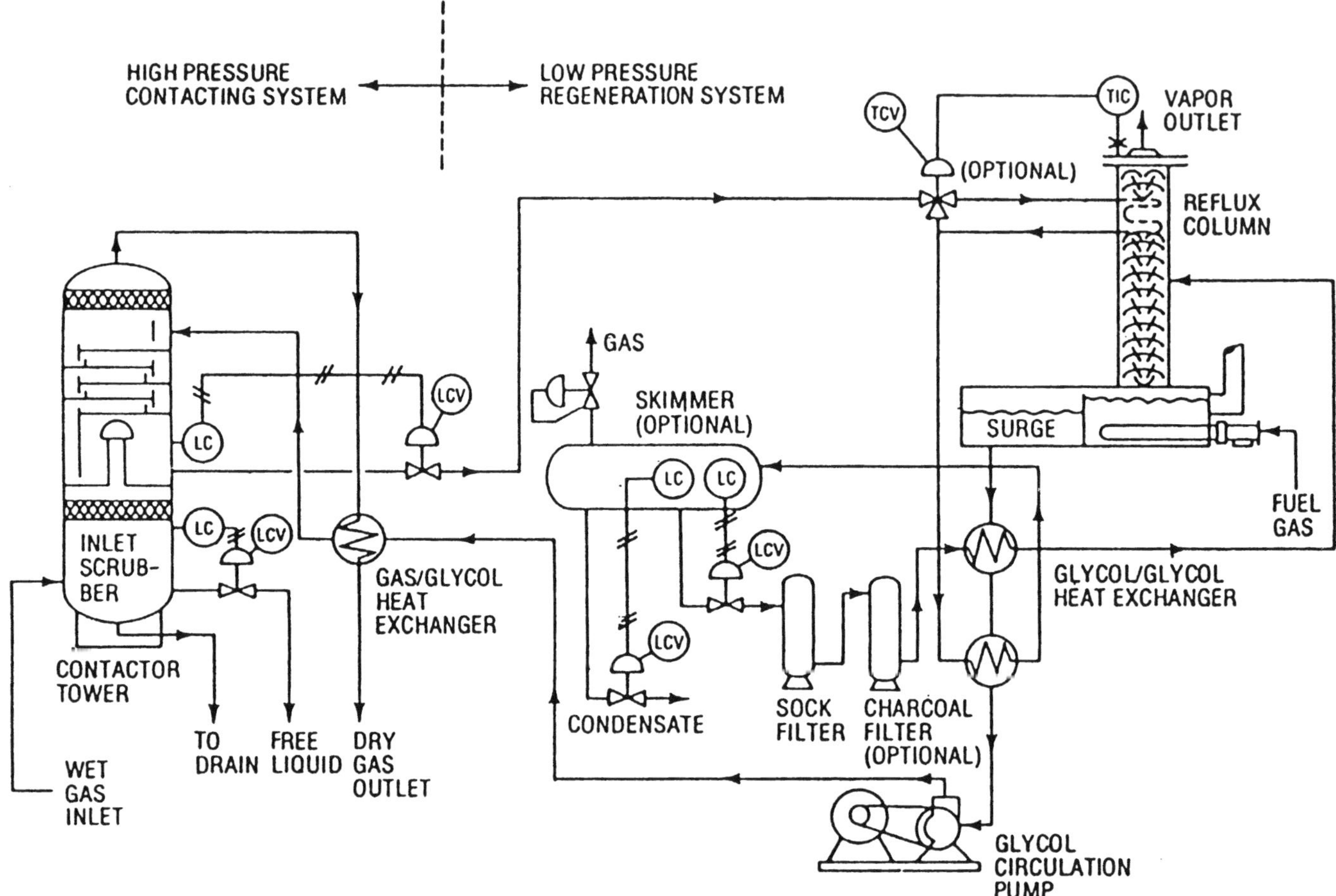

Figure 8–2. Typical flow diagram glycol dehydration unit (NATCO, 1984).

place a scrubber or separator upstream of the contactor, and possibly a filter-coalescer. The scrubber should be sized to handle 125% of the maximum gas flow rate and have a high-liquid-level shutdown.

Absorber

As shown in Figure 8–2 and in further detail in Figure 8–3, the *absorber* (or contactor) consists of an integral scrubber at the bottom, a mass transfer or drying section in the middle, and a glycol cooler and mist extractor at the top.

Wet natural gas enters the *integral scrubber* tangentially, and then passes through a wire-mesh mist extractor that removes most of the remaining entrained liquid droplets. The two stages of scrubbing and the mist extractor minimize contamination of the glycol and prevent free water from overloading the unit. The integral scrubber complements the inlet separator; it should not be used as a substitute.

In the *drying section*, the gas flows upward and is contacted intimately by the descending glycol solution. This counter-current contact usually employs 4 to 12 bubble cap or valve trays. Even though valve trays may be more efficient—33 vs. 25% for bubble caps (Sivalls, 1976)—bubble caps are usually used. Bubble caps do not weep at low gas flow rates and do not drain the trays. They are suitable for viscous liquids also. Turndown to as low as 16 to 20% of design flow is achievable.

Because glycols tend to foam, the trays should be spaced at least 18 in. and preferably 24–30 in. apart. The absorber column must be vertical. If not, the unequal liquid level on the trays seriously impairs the gas/liquid contact.

TEG circulation rates vary from 1.5 to 4 gal TEG per lb water removed. Smaller units with 4 to 6 trays are often operated at 3 gal TEG per lb water. In larger units of eight or more trays, the TEG circulation rate is reduced to 2 gal TEG per lb water removed with simultaneous reduction in reboiler duty. Good glycol-gas contact on the trays is difficult with less than 2 gal TEG per lb water removed.

A mesh or woven *mist extractor* at the top of the absorber

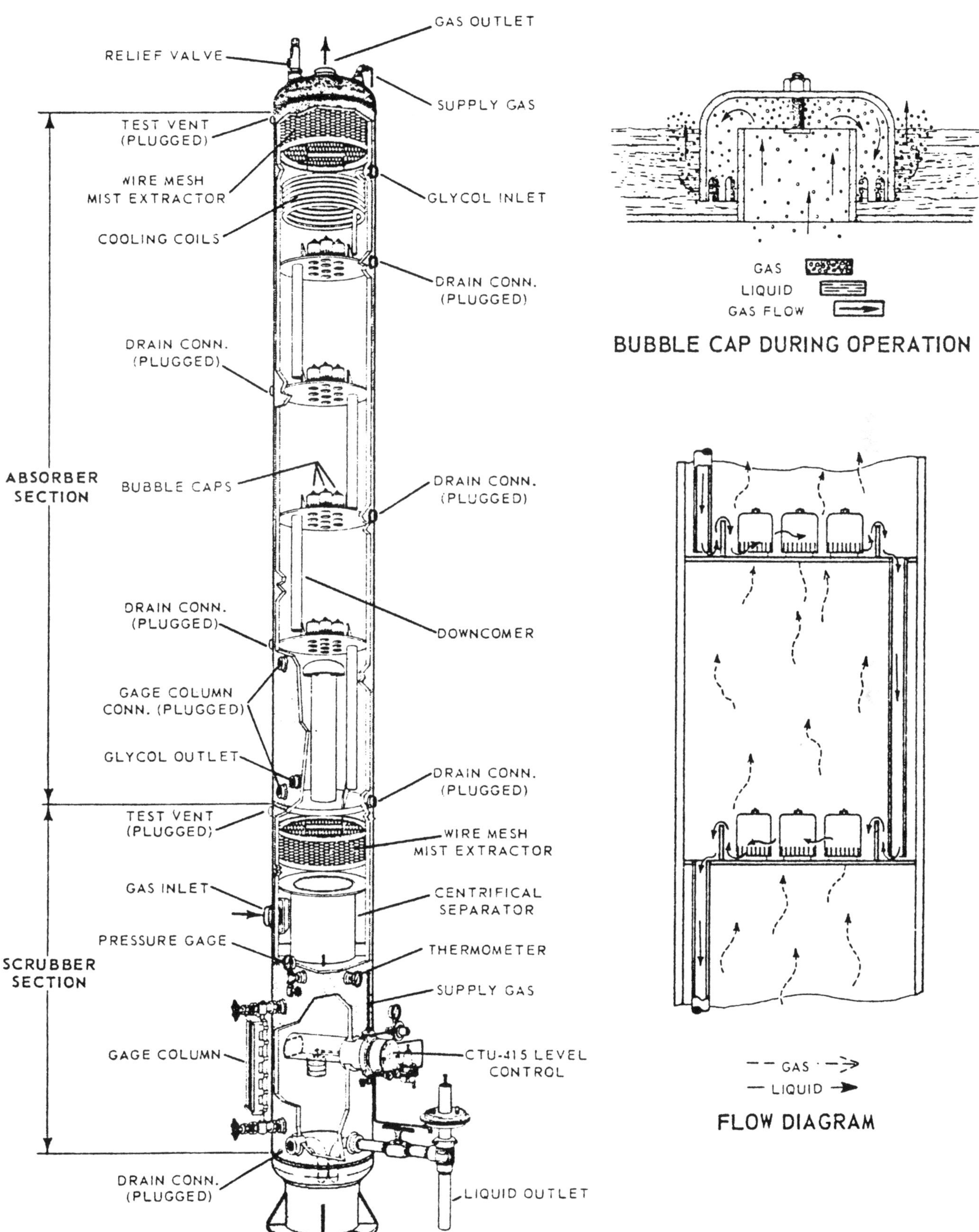

Figure 8–3.   Typical absorber tower with integral scrubber (Caldwell, NATCO, 1976).

reduces liquid glycol carryover to less than 1 lb per MMscf (Sarma, 1981). Combination pads consisting of 4 to 8 in. of 304 stainless steel and 4 in. of dacron are recommended. Mapes (1960) emphasizes the importance of allowing adequate disengagement space between the top tray and the demister. The minimum disengaging height should be 1.5 times the tray spacing (NATCO, 1984).

Flash Tank

While TEG absorbs 1 scf of natural gas per gal at 1000 psig and 100 °F (Guenther, 1979; GPSA, 1987, p. 20–18), heavier hydrocarbons in the natural gas, including aromatic compounds, are much more soluble. A *two-phase separator* with a 5–10 min liquid retention time prevents excess gaseous hydrocarbons from entering the still column where they will flash, increase glycol losses, and possibly break the ceramic packing (Ballard, 1977). A *three-phase separator* with a 20–30 minute liquid retention time is preferred when the gas gravity is high and the glycol adsorbs significant quantities of heavier hydrocarbons. The gas-condensate-glycol separation is best performed at 100–150°F and at 50–75 psig (PETEX, 1973c). These conditions permit the flashed gas to be used without compression for fuel or stripping gas.

When *oil removal* is desired, the rich glycol can be preheated in the glycol-glycol exchanger before flashing. Heating reduces the viscosity and accelerates glycol-liquid hydrocarbon separation. However, increased temperatures increase the solubility of liquid hydrocarbons in the glycol.

The pressure in the *flash separator* must be sufficient to permit the exit glycol stream to flow through all downstream equipment, i.e., the heat exchangers and the filters.

Better condensate-glycol separation is obtained with horizontal flash tanks; however, vertical separators require less platform area.

Filters

The *solids content* in the glycol should be kept below 0.01 weight percent (100 ppmv) to prevent pump wear, plugging of heat exchangers, foaming, fouling of contactor trays and still packing, cell corrosion (where the solids settle on metal surfaces), and hot spots on the fire tubes (Ballard, 1977).

Sock filters, made from cloth fabrics, paper, or fiberglass are designed to remove 5-μm and larger particles from the glycol before it enters the still. Placement of the filter after the glycol-glycol heat exchanger exploits the reduced viscosity of the rich glycol. Sock filters are designed for initial pressure drops of 3 to 6 psi and should be changed when the pressure differential reaches 15 to 25 psi.

Used filter elements should be cut to the core and inspected. If the element is still clean on the inside, problems such as incorrect selection of element type or paraffin plugging have occurred (Ballard, 1977, 1979).

Rigid filter cartridges last longer and do not collapse and dump filtered solids back into the glycol. Even though these cartridges cost more, the total filtration cost—cartridges, labor, glycol loss, and disposal—is usually less, (Ballard and von Puhl, 1988).

Activated carbon filters are used to remove liquid hydrocarbons, surfactants, well-treating chemicals, compressor lube oils, etc., from a glycol side stream. Cannister filters, which seldom rupture or bleed, are recommended. Hard, dense, coal-based carbons are preferred over wood-based carbons that are light, soft, and create dusting problems (Ballard, 1977). Allowable flow rates are low, 1–2 gpm per sq ft bed area.

Alternatively, a carbon filter vessel operating on the full glycol stream can be used. Granular carbon in the range of 4 to 30 mesh, a flow rate of 2 to 10 gal per min per sq ft of bed area, and a retention time of 5 to 10 minutes is recommended.

Cramer and Cook (1981b) and others recommend a low-pressure, lean-glycol, sock filter to reduce pump wear by removing particulates such as cracked hydrocarbons, resins, free salt, and metallic compounds. However, this practice can cause pump cavitation. Some operators solve this dilemma by placing the carbon filter downstream of the pump, while others place a Y-type strainer in the pump suction line.

Glycol Pump

The *glycol recirculation pump*, which contains the only moving parts in the entire unit, returns the low-pressure lean glycol to the high-pressure contactor. Three types are used:

1. High-pressure-gas operated (e.g., Texsteam)
2. High-pressure-liquid operated (e.g., Kimray)
3. Electric-motor driven.

Because dehydration ceases without glycol circulation, two glycol pumps, each capable of providing the full circulation, are recommended. In larger units an *electric-motor driven*, horizontally-mounted, multiple-cylinder positive displacement pump is preferred, and a high-pressure-liquid standby pump can be used.

For smaller units and in remote areas, a *high-pressure-liquid* or *gas-driven glycol pump* is often selected. High-pressure, rich glycol is taken from the bottom of the chimney tray of the absorber and used to supply part of the power for this double-acting pump. High-pressure gas from the absorber supplies the remainder of the power. This additional gas consumption increases with increasing absorber pressure. Typical rates are 1.7 scf per gal TEG at 300 psig and 5 to 6 scf per gal TEG at 1000 psig (Sivalls, 1976). This pump features serviceable controls and, if properly adjusted, gives dependable, trouble-free operation. Unfortunately, and especially at higher absorber pressures, the gas consumption exceeds the demand for fuel and stripping gas.

Surge Tank

The *surge tank* must be big enough to handle a complete drain-down of TEG from the absorber-tower trays. During normal operation, the surge tank should be half full. A gas blanket is recommended to prevent air contamination (oxygen, actually).

Heat Exchangers

The glycol-glycol heat exchanger extracts heat from the hot, lean glycol returning to the absorber and delivers it to the rich glycol going to the still, thus saving energy. However, a glycol unit responds very sluggishly to surges in the gas flow rate when the reboiler carries less than 60% of the total heat load. If changes in gas flowrate are expected, it is best to have additional heat readily available for the reboiler—a higher firing rate or increased flow of heat-transfer fluid—than to reduce the thermal efficiency by decreasing the heat exchanger duty.

The *heat exchangers* in a glycol unit should be designed to accomplish the following:

1. Supply the lean glycol to the absorber 5–15°F warmer than the dried gas leaving the absorber. This aim is achieved by locating a cooler down stream of the rich/lean-glycol exchanger. In smaller units, the glycol is cooled in a coil located in the contactor just below the mist extractor. In larger units, a separate, external gas-glycol exchanger or an air-cooled, finned-tube exchanger is used. Failure to cool the glycol makes the top tray act as a heat exchanger. The temperature of the glycol on this tray is higher, which increases the partial pressure of the water vapor and thus reduces the drying efficiency.
2. Maintain the top of the stripping still at 210°F (at sea level). The cold rich glycol can be used as coolant for the reflux coil.
3. Control the preheat of the rich glycol entering the stripping still to a maximum (and thus the maximum heat recovery from the lean glycol leaving the reboiler).

Too high a temperature flashes water vapor and produces two-phase flow. An approach of 60–90°F at the hot end is desirable in the rich/lean-glycol heat exchanger. This exchanger has a large temperature cross; that is, the temperature of the exit lean glycol is much lower than that of the exit rich glycol. This type of service requires true counter-current flow. Double-pipe (finned-tube inside) or plate-and-frame heat exchangers are used. Two or more exchangers in series are often used in larger units to control the temperature at which the rich glycol enters the flash tank and filters.

Still Columns

Reconcentration occurs in both the vertical stripper, or still column, and the reboiler. In smaller units, the still is frequently mounted on top of the reboiler. The *still column* contains a stripping section usually filled with 4 to 8 ft of ceramic packing (1– to 1.5–in. Intalox saddles or Pall rings). Trays are sometimes used in very large units. The water in the rich glycol leaves the still as steam. The cooling coil in the top of the still condenses some of the exit steam to provide reflux for the column. This arrangement controls condensation and reduces glycol losses. The air-cooled, finned condensers used in many older units provided excessive ambient cooling in cold weather and inadequate cooling in hot weather.

Reboiler

The *reboiler* provides the heat necessary to boil the water out of the glycol. Direct-fired heaters are often used, but they constitute an open flame hazard. In some locations, such as offshore platforms, indirect heating with oil or steam is required by fire code and prudent practice. Use of exhaust gases from gas turbines and engines as the heating medium (when available) achieves substantial energy savings.

TEG does not undergo significant thermal decomposition if the bulk glycol temperature in the reboiler is kept *below 400°F* and the maximum outside fire-tube skin temperature does not exceed 430°F (Guenther, 1979). Accordingly, the U-shaped fire tube should be sized so that the average heat flux is 6000–8000 Btu/hr-ft^2. The heat flux is not uniform throughout the fire tube but is highest near the flame in the bottom tube. A laminar-flow, long, rolling, yellow-tipped flame produces a more uniform heat flux, (Ballard, 1966; Simmons, 1981a). Simmons (1981a) recommends a vaned air register which causes spiraling of the flame and combustion gases, lower fire-tube skin temperatures, and reduced fouling. Installation of a *Smick Turbulator* in the return leg of the U-shaped fire tube improves the heat flux distribution, lowers the stack-gas temperature, reduces fuel consumption by at least 8%, extends fire-tube life, and minimizes glycol degradation (Ballard, 1979).

Instrumentation

Governmental regulations such as OCS Order No. 5, fire and safety regulations issued by insurance companies, and company specifications must be followed. Industry recommendations, such as API Recommended Practice 14C (1978) and 14E (1984), are not mandatory but should be considered very carefully.

Landes (1983) describes two extremes of engineering

Table 8–1 Recommended Instrumentation for Lean Design (Landes 1983)

Item	Controls
Contactor	PC on exit gas line
	PI on contactor
	TI on contactor
	LC on contactor
Reconcentrator	PSV on reboiler shell
	TSH on glycol in reboiler (to shutdown panel)
	TI on glycol in reboiler
	TIC on glycol in reboiler connected to
	TCV on fuel gas to main burner
	TSH on stack gas temperature (to shutdown panel)
	BSL flame sensor on burner (to shutdown panel)
	PI on fuel line to main burner
	PCV on fuel line to main burner
	SDV on fuel line to main burner (activated by shutdown panel)
	SDV on pilot fuel line (activated by shutdown panel)
Shutdown Panel	LAH on glycol level in glycol flash tank
	LAL on glycol level in glycol flash tank
	BAL on flame in main burner
	TAH on glycol temperature in reboiler
	or TAH on stack gas temperature
	LAH on integral scrubber in contactor
Legend	BAL = Low burner flame alarm
	BSL = Burner flame sensor
	LC = Level control
	LAH = Low liquid-level alarm
	LAL = High liquid-level alarm
	PC = Pressure control
	PCV = Pressure control valve
	PI = Pressure indicator
	PSV = Pressure safety valve (relief valve)
	SDV = Shut-down valve
	TAH = High-level temperature alarm
	TCV = Temperature control valve
	TI = Temperature indicator
	TIC = Temperature-indicating controller
	TSH = High-temperature shutdown

design: "lean" and "fat." Over 10000 off-the-shelf, "lean" units are in operation worldwide. Landes characterizes these lean designs as safe, sane, simple, and suitable for unattended operation; they require minimum maintenance, minimum operator training, minimum operator skill level, and provide an onstream capability of over 98%. Table 8–1 summarizes the instrumentation for a typical "lean" design.

Over 1000 "fat" units are currently in operation. These units follow all API RP 14C (1983) recommendations and are (according to Landes) safe, insane, complicated, operate semi-unattended, require maximum maintenance, maximum operator training, highly-technical electronic skill, and have an onstream capability of over 98%. Further comparison of the "lean" and "fat" approaches follows:

	Number of Instruments Used	
Component	"Lean" Design	"Fat" Design
Contactor	4	24
Reconcentrator	11	over 29
Shutdown panel	5	35

The obvious and platitudinous recommendation is to keep the instrumentation as simple as government regulations, insurance requirements, and company specifications will allow.

PROCESS OPERATION

A correctly designed and properly operated TEG unit will dehydrate natural gas with only minor difficulties and require modest maintenance. Table 8–2 summarizes preferred operating temperatures. The potential ease of operation often leads to the glycol unit being virtually ignored. Thousands of dollars can be wasted annually because of high glycol losses, excessive TEG recirculation rates, improperly operating pumps, needless energy consumption, frequent plant shutdowns, and excessive equipment replacement. Operating procedures are now recommended under the following topics: absorber, reconcentrator, glycol care, pump, start up, shutdown, and preventive maintenance.

Contactor or Absorber

The operating efficiency of the glycol-gas contactor depends on the inlet gas flowrate, temperature, and pressure and also on the lean glycol concentration, temperature, and circulation rate. The effects of these variables are now summarized.

Table 8–2 Recommended Glycol Unit Operating Temperatures

Process Location	Temperature or Temperature Range, °F
Inlet gas	80–100
Glycol into absorber	5–15 warmer than gas
Glycol into flash separator or skimmer	100–150 (prefer 150)
Glycol into filters	100–150 (prefer 150)
Glycol into still	300–350
Top of still	210 / 190 with stripping gas
Reboiler	380–400 (prefer 380) / 350 yields 98.5 wt% TEG / 400 yields 99.0 wt% TEG
TEG entering pump	<200 (prefer 180)

Inlet Gas Flowrate. Obviously the *load* (lb water to be removed/hr) varies directly with the feed gas flowrate. Normally the trays (bubble cap usually and valve sometimes) are operating in a severe spray regime—very little liquid glycol compared to the gas flowrate. Increases in feed gas flowrate can exacerbate this delicate condition of "blowing flood" and can be very detrimental to contactor performance. Random or structured packing is not susceptible to this "blowing flood" because the liquid flows as a wetted film on the packing surface. The glycol circulation rate should be adjusted to match variations in feed gas flowrate.

Most contactors have been designed conservatively and so can handle flowrates 5 to 10%, and perhaps even 20% above capacity. Of course the capabilities of the other units—pumps and reboilers—must be considered. The lower flow limit is determined by the aforementioned "overdesign" and the approximate 5 to 1 turndown ratio of the bubble caps.

Inlet Gas Temperature and Pressure. The inlet gas may be assumed to enter the absorber saturated with water vapor; therefore, its water content and the quantity of water to be picked up by the glycol depend on the inlet gas temperature and pressure. McKetta and Wehe's correlation (Fig. 4–6) shows that at 1000 psia the water content increases from 33 to 62 to 102 lb H_2O/MMscf as the temperature increases from 80 to 100 to 120°F. The pressure effect, while very significant, is not as severe: at 100°F the water content is 62, 72, and 87 lb H_2O/MMscf at 1000, 800, and 600 psia, respectively. An increase in the inlet-gas

temperature or decrease in inlet-gas pressure increases the load on the unit. Sudden changes in pressure and temperature also can affect glycol flow in the absorber and break downcomer seals.

Entering TEG Temperature and Concentration. The drying ability of TEG is limited by the vapor-liquid equilibrium (VLE) of water between the gas phase and the liquid TEG phase. Traditionally, the VLE data are presented as a plot of the water dew point of the natural gas vs. the contact temperature and the water content of the liquid TEG, as shown in Figure 8–4 (Parrish *et al.*, 1986). The drying ability of TEG increases dramatically with concentration; for example, at 100°F the equilibrium water dew points of the gas in contact with 98, 99, 99.5, and 99.7 weight percent TEG are 32, 16, 0, and −16°F, respectively. Also, the equilibrium water dew point decreases with decreasing temperature, but cooling the glycol increases its viscosity (Dow, 1962, p. 95; Union Carbide, 1989) and its tendency to foam.

Glycol Circulation Rate. The water picked up by the glycol increases with increasing inlet-glycol concentration, decreasing glycol temperature, higher circulation rates, and the number of contactor trays. A glycol circulation rate of 3 gal per lb water removed is conservative and was commonly used in the past. However, recent energy conservation practices have lowered the circulation to 2 gal per lb water.

Dehydration Temperature. While TEG can dehydrate natural gas at operating temperatures from 50 to 130°F, the preferred temperature range is 80–100°F. Below 70°F the glycol is too viscous. This reduces tray efficiency, promotes foaming, and increases glycol losses. Below 50°F the drop in dehydration efficiency is very pronounced (BS&B, 1960). Above 110°F the inlet gas contains too much water, and the drying ability of the glycol is reduced.

The inlet glycol temperature should be 5–15°F higher than the inlet gas temperature. If the glycol enters cooler than the gas, the resulting chilling condenses hydrocarbons which, in turn, promote foaming. If the glycol enters more than 15°F above the effluent gas temperature, TEG vaporization losses and sales gas dew point are increased unnecessarily.

Reconcentrator

While the degree of glycol reconcentration depends mainly on the reboiler temperature, additional reconcentration is readily obtainable by use of stripping gas. Glycol losses from the top of the stripping column can be minimized

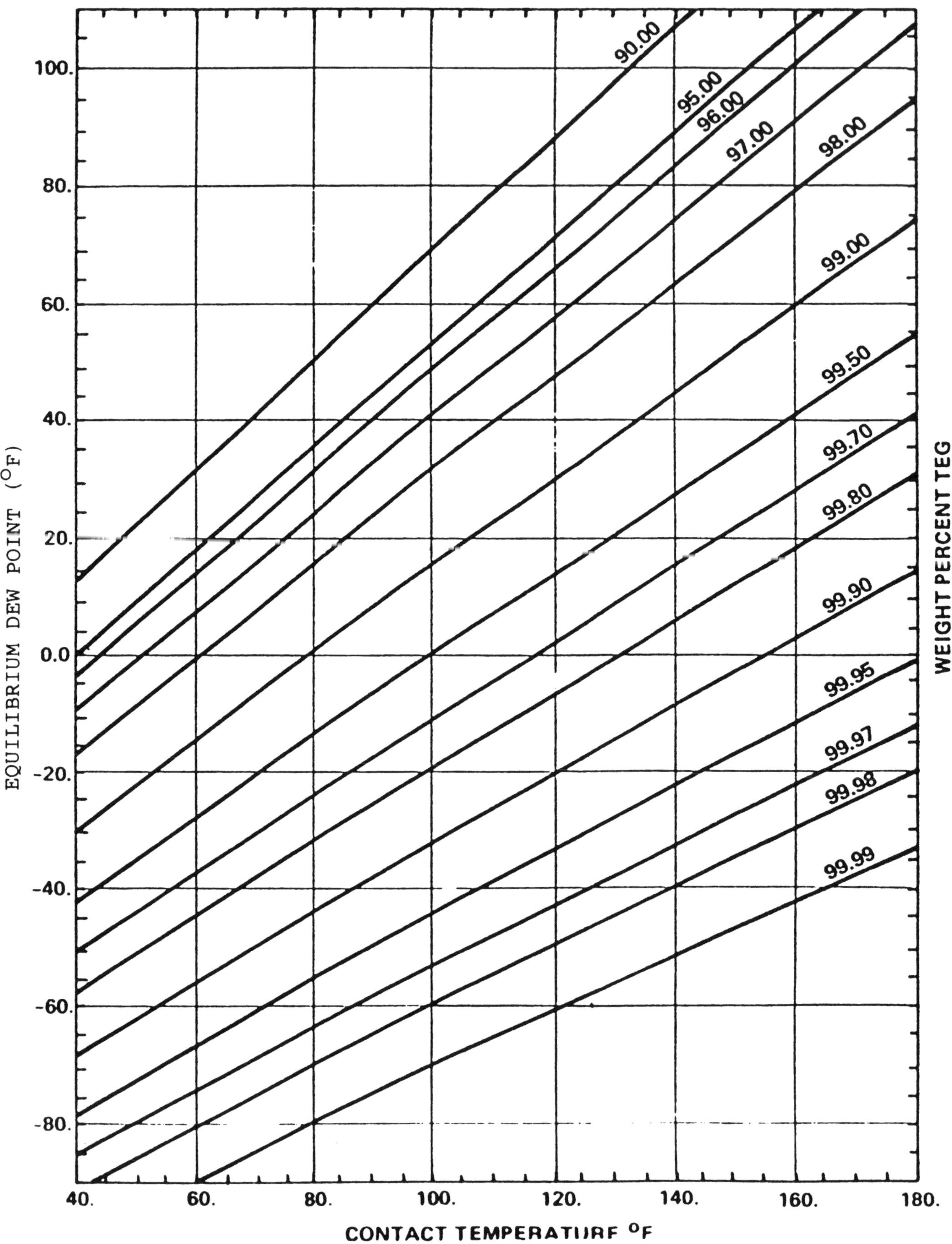

Figure 8–4. Dew-point chart for the TEG-water system (Parrish, Won and Baltatu, 1986, GPA Proceedings).

by temperature control. The effects of these variables are now summarized.

Reboiler Temperature. The concentration of water in the lean glycol leaving the reboiler varies with the reboiler temperature and pressure. Because the reboiler is usually operated at atmospheric pressure, Figure 8–5 shows how the TEG concentration increases with temperature (Ballard, 1979, p. B–33). Reboiler temperatures range from 350 to 400°F because the rate of glycol decomposition increases significantly above 400°F, as mentioned previously. Adjustments of reboiler temperature should be made in 5°F steps to avoid plant upsets.

Stripping Gas. Glycol concentrations up to 99.6 weight percent can be achieved by *sparging stripping gas* directly into the reboiler as shown in Figure 8–6 (Parrish *et al.*, 1986). Figure 8–6 compares Worley's correlation (1967), the GPSA correlation (1981, p. 15–13), and the Parrish *et al.* (1986) rigorous tray-by-tray calculations using VLE and enthalpy data.

In the *Stahl stripping method*, a 2- to 4-ft downcomer between the reboiler and surge tank is packed with Pall rings or Intalox saddles, and the stripping gas is injected at the bottom of the downcomer as shown in Figure 8–7. This more-efficient, countercurrent contact reduces the quantity of stripping gas required and produces glycol concentrations above 99.9 weight percent (Fig. 8–6). Again, Figure 8–6 compares Worley's correlation and the Parrish *et al.* simulations.

Alternative, more complicated concentration methods are (1) reducing the reboiler pressure with an ejector or vacuum pump (Fig. 8–8) and (2) employing azeotropic distillation with isooctane (Fig. 8–9) (Drizo process, Smith, 1990).

Stripping Column Temperature. The temperature at the top of the stripping column should be kept at approximately 210°F, preferably by varying the flow of rich inlet glycol to a reflux-generating, cooling coil located in the top of the still (Fig. 8–2). If the top temperature drops much below 200°F, water-vapor condensation becomes excessive; this may flood the column, overfill the reboiler, increase column pressure, and blow liquids out the top vent (Ballard, 1966; Cramer and Cook, 1981b, p. 76).

Glycol vapor may leave the stripper if the top temperature exceeds 220°F. When stripping gas is used, the top temperature can be reduced to as low as 190°F. An ambient-air-cooled, finned condenser (as shown in Fig. 8–1) cannot provide consistent temperature control. There is insufficient reflux on hot sunny afternoons and overcooling with sudden thundershowers or extremely cold weather.

Glycol Care

The key to avoiding many operating and corrosion problems is to *keep the glycol clean*. Ballard (1966, 1979) identified the following major contamination problems: low pH, oxidation, thermal decomposition, salt contamination, hydrocarbon condensation, sludge accumulation, and foaming. The causes and cures of these problems are now reviewed.

Oxygen in the System. Sources of oxygen include storage tanks without inert-gas blankets, leaking pumps, and—rarely—the inlet gas. A gas blanket consisting of a very small flow of fuel gas into the top of liquid storage tanks will keep air from entering the vessel. Excessively-leaking pump packing glands are another source and should be repaired promptly.

Thermal Decomposition. Thermal degradation occurs when glycol is overheated. Overheating can be avoided by keeping the reboiler temperature below 400°F, using reboiler heat fluxes less than 8000 Btu/hr–ft^2, and by checking the fire tube regularly for hot spots caused by deposits of tar and/or salt. Hot spots are best observed at night; if the flame is shut off, they continue to glow bright red.

Low pH. Solutions in a glycol unit that is not downstream of an amine unit become acidic and corrosive, especially when the inlet gas contains H_2S or CO_2. The optimum glycol pH is 7.0–8.5; a pH above 9 promotes foaming and emulsion formation.

Corrosive acids formed by glycol decomposition or oxidation and dissolved H_2S or CO_2 can be neutralized by bases such as borax, triethanolamine, NACAP, or Coastal 1750C. These bases should be added slowly and in small amounts—e.g., 1 lb of amine per 400 gallons of glycol—because excess base can precipitate sludge. Frequent filter replacement is required while the glycol is being neutralized. The glycol pH should be checked weekly using narrow range pH test paper or a pH meter.

Salt Contamination. Salt deposits on the reboiler fire tube cause hot spots and fire-tube failures. The glycol should be drained and the unit cleaned when the salt level exceeds 2500 ppmw. Special equipment, such as a separate glycol reclaimer or ion-exchange resin bed, is required for removing salt from TEG. Produced water should be prevented from entering the absorber by use of an upstream filter separator.

Liquid Hydrocarbons. Liquid hydrocarbons are either entrained with the inlet gas, are condensed by the lean

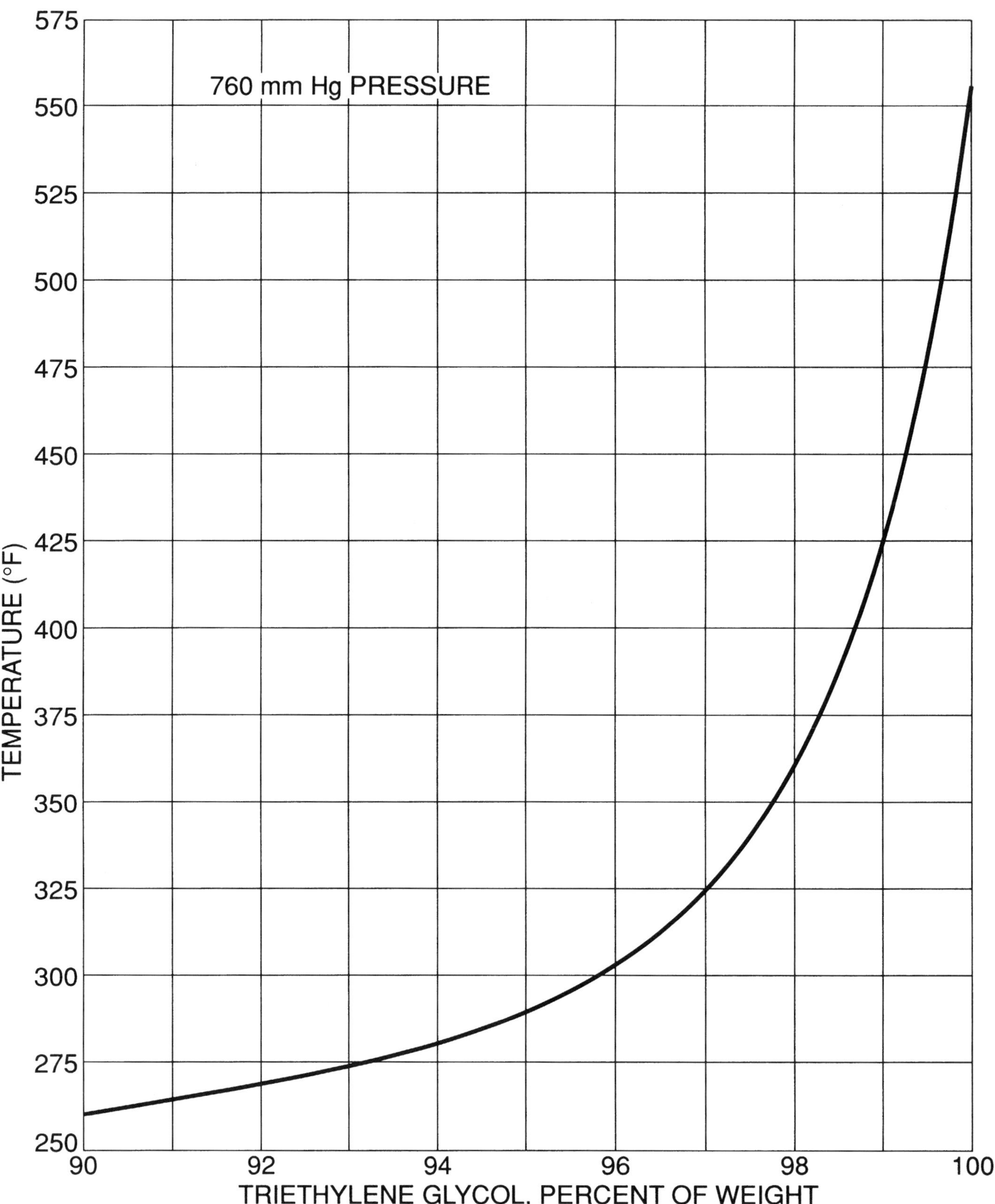

Figure 8–5.   Boiling points of aqueous commercial grade triethylene glycol solutions (Ballard, 1979, p. B-33).

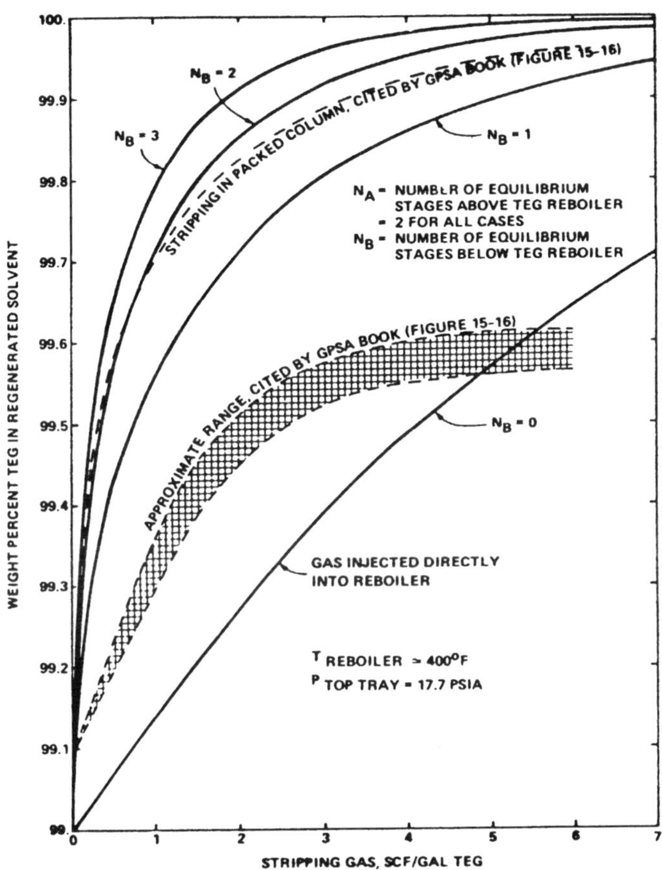

Figure 8–6. Effect of stripping gas rate on the purity of regenerated TEG (Parrish, Won and Baltatu, 1986).

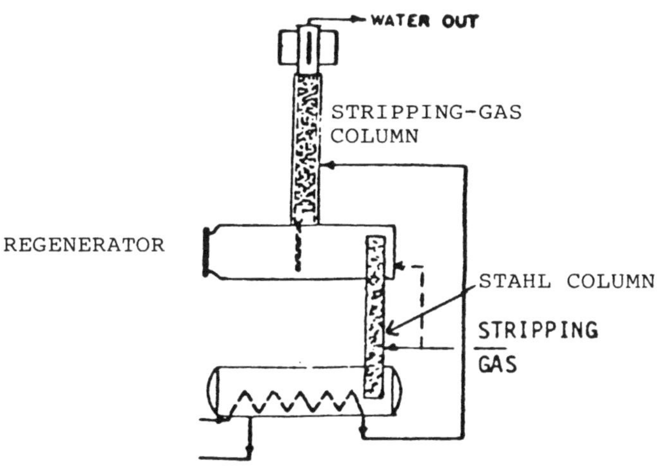

Figure 8–7. Stahl or gas-stripping column.

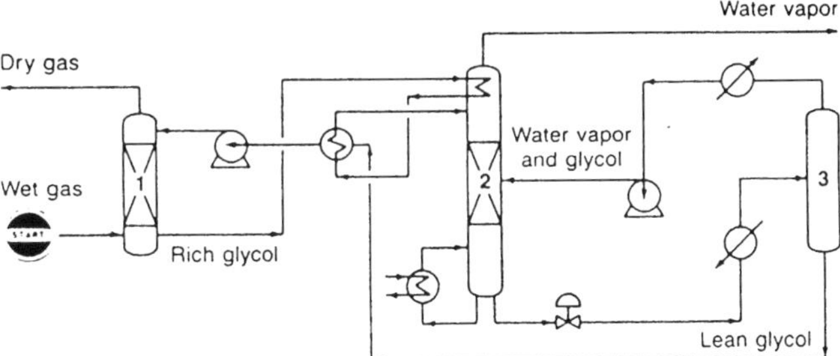

Figure 8–8. Dehydrate (conc glycol) (BSB, 1990).

glycol entering the absorber colder than the sales gas, or simply absorbed from the gas by the glycol. Hydrocarbon contamination should be minimized by scrubbing the inlet gas, keeping the lean glycol 10°F warmer than the sales gas, adequately sizing the three-phase flash separator, and using carbon purification. Liquid hydrocarbons, if permitted to enter the stripper, flow down the column and flash rapidly in the reboiler. This can blow liquid out of the stripper.

In very cold climates, heat losses through the contactor walls can lead to HC condensation. Insulating and/or housing the contactor may be justified.

Sludge Accumulation. Dust, sand, pipeline scale, reservoir fines, and corrosion products such as iron sulfide and rust are picked up by the glycol if not removed by the inlet separator or the integral scrubber. These solids, together with tarry hydrocarbons, eventually settle out and form an abrasive, sticky, black gum which can erode the glycol pump and other equipment, plug the contactor trays and the packing in the stripper, and deposit on the fire tubes. Proper filtration of both the rich and lean glycol streams should keep the solids concentration below 0.01 weight percent.

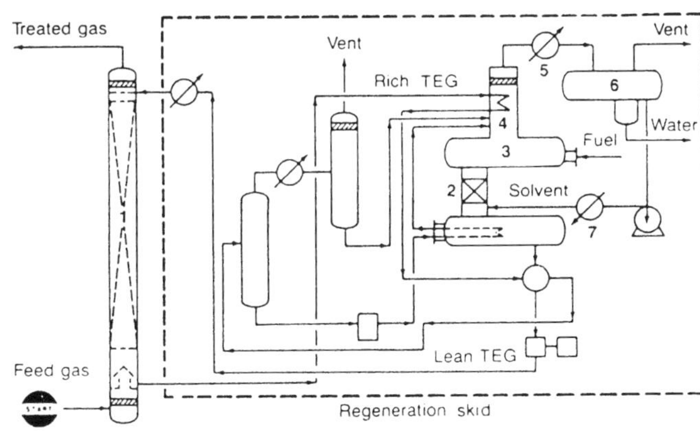

Figure 8–9. Drizo Process (OPC Engineering, Inc., 1990).

Foaming. Foaming may be mechanical or chemical. *Mechanical foaming* is caused by excessively-high gas flow rates in the absorber. Contaminants such as solid particles, salts, corrosion inhibitors, and liquid hydrocarbons cause *chemical foaming*.

Foaming is best detected by monitoring the pressure drop across the contactor. Erratic readings followed by a rapid increase of 2 to 5 psi are symptoms of foaming. Chemical foaming can be detected by bubbling air through a sample of the glycol for five minutes and observing the resulting foam height and stability (Smith, 1979). If the pressure drop across the absorber suddenly increases by 5 to 10 psi and the chemical foam test is negative, then suspect mechanical foaming or some other problem.

Foaming should be prevented by good front-end separation of contaminants, proper solution filtration, and by raising the temperature of the gas and glycol above the dew point of the condensed hydrocarbons. Smith (1979) recommends silicone emulsion breakers, higher molecular weight alcohols, and block polymers of ethylene and propylene as the most effective defoamers. Pape (1983) recommends silicone antifoam in emulsion rather than fluid form. Use of defoamers merely treats the symptom; eliminating the cause is far better. Ballard (1977) observed that silicones are expensive and can decompose in the reboiler and so become foam promoters.

Meusburger and Segebrecht (1980) discussed *defoamers* in detail and recommended the following *precautions*:

1. Clean equipment thoroughly before start up, using chemicals.
2. Establish raw material standards and test methods.
3. Maintain a clean makeup sump and keep it covered when not in use.
4. Prevent surface water contamination and runoff into the sump.
5. Install and maintain carbon and particulate filters to trap surfactants and prevent fine solids from building up in the system.
6. Use an external carbon filter to treat the total solution inventory if a persistent foamy solution develops during operation and conventional treatments fail.
7. Investigate the characteristics of alternative lubricants for triplex pumps and lubricated valves.
8. Keep defoamers on hand at all times—an organic for preventive measures and a silicone for emergencies.

Careful monitoring, elimination of contaminants, carbon filtration, use of a three-phase flash separator, and maintenance of pH between 7 and 8 can extend glycol life to 5–10 years (Harrell, 1989).

Vacuum distillation can recover up to 90% of the remaining glycol from contaminated glycol. One approach is to use a portable unit on a 1 to 3 gpm sidestream.

Glycol Pump

Proper pump care is essential; when the pump fails there is no glycol circulation and no dehydration. A *standby pump* should be included, and small replacement parts should be available. The pump packing gland should be replaced when it leaks glycol excessively. The manufacturer's instructions should be followed when installing new packing. Packing is dry when new and will probably absorb glycol and expand. If installed too tightly, the packing will not receive enough lubrication from the glycol and may burn and score the plungers.

In glycol-powered pumps, the high-pressure, rich glycol can leak through the internal seal rings and dilute the low-pressure, reconcentrated glycol. Such internal leakage, which is detected by sampling the lean glycol as it enters and leaves the pump, can become excessive as the pump seals wear (BS&B, 1960).

Carbon Dioxide Rich Gases

TEG is being used successfully to dry both natural gases containing H_2S and/or CO_2 and pure CO_2 streams (Kemp, 1976; Glaves *et al.*, 1983; Von Puhl, 1986). The presence of H_2S and/or CO_2 requires:

1. The water content of the incoming sour natural gas stream must be carefully determined (see Chapter 4) because sour gases have an increased capacity for water.
2. Both CO_2 and H_2S are more soluble than CH_4 in glycol (Kemp, 1976), and the absorbed acid gases should be removed from the glycol. The glycol-gas flash separator suffices when the acid gas concentrations are not too high. Better stripping of the acid gas from the sour glycol may be required to control reboiler corrosion and/or environmental pollution.
3. The pH of the TEG solution must be monitored closely. Frequent addition of bases (e.g., amine or Nacap) may be required.

Start Up

Start-up procedures are recommended by Caldwell (1976) and PETEX (1973b).

1. Purge system.
2. Pressure absorber slowly.
3. Start pump and establish glycol circulation through the absorber and reconcentrator.

4. Establish fire in the reboiler and bring bath temperature up slowly.
5. Establish gas flow through contactor slowly, taking care not to blow the glycol off the trays.
6. Adjust controls for desired operating conditions.

In cold climates the TEG must be heated to 100°F before starting circulation—glycol at 0°F cannot be pumped. A *cold start* can be expedited by bypassing the glycol-glycol heat exchanger until the glycol warms up.

Shutdown

Localized overheating and attendant decomposition of the glycol is avoided by the following:

1. Shut off the gas flow and close inlet and outlet valves to the contactor.
2. Shut off the heat to reboiler and close burner manifold valves.
3. Circulate glycol until reboiler temperature drops to 175°F.
4. Stop glycol pumps and close appropriate valves.
5. Bleed off pressure in the contactor slowly.

During extended downtimes, perform the following maintenance (Caldwell, 1976):

1. Flush absorber and reconcentrator with water and drain.
2. Clean pumps.
3. Remove packing from the pumps and coat plungers with rust inhibitor.

Preventive Maintenance

Preventive maintenance reduces operating costs. The goal is to repair equipment only when needed; early diagnosis of troubles is the best approach. Such diagnoses require thorough online inspections, careful downtime inspections and servicing, and diligent record keeping.

Daily inspections should include:

1. Check controls for proper operation.
2. Replace broken thermometers and pressure gauges.
3. Measure and record:
   - Sales gas flowrate and dew point
   - Contactor temperature and pressure
   - Glycol circulation rate (pump strokes/min)
   - Glycol temperature at top of contactor
   - Glycol temperature entering pump
   - Reboiler bath temperature.

In addition, if required:

1. Add glycol to surge tank. Note time and amount added.
2. Replace filters when pressure differential reaches 10 psi for soft cartridges and 25 psi for rigid cartridges.
3. Replace strainers when plugged.
4. Repair leaks.

At least *once a week* fill a quart jar with glycol from the drain tap on the storage tank. Hold the jar up to a light and check carefully for the following:

1. Fine black particles which settle when shaken—this FeS and/or Fe_3O_4 precipitate indicates ongoing corrosion and potential failure of the filters.
2. Sniff sample carefully. A sweet, aromatic odor like burnt sugar or overripe bananas suggests thermal degradation.
3. Check the flow characteristics of the sample. A black, thick, viscous solution indicates that heavy hydrocarbons are being absorbed from the gas.
4. Check for an oily layer floating on the glycol—which also indicates that heavy, nonvolatile hydrocarbons are being condensed.

Such evidence of glycol contamination should be confirmed by a complete glycol analysis.

Every month—or sooner if contamination is suspected—the glycol should be sampled and analyzed for hydrocarbons, water, solids, iron, chlorides, specific gravity, pH, and, if deemed appropriate, glycol component analysis (Caldwell, 1976). The following conclusions may arise from such a glycol analysis:

Low pH:	Glycol decomposition, corrosion
Hydrocarbons present:	Flash tank, filter problem
Low SG, H_2O present:	Reconcentrator problems
High solids:	Corrosion, glycol decomposition, filter problem, carryover from inlet separator
High iron:	Corrosion
High chloride:	Brine carryover from separator, scrubber.

Sample both the lean and rich glycol streams and compare with the optimum values shown in Table 8–3 (Fremin, 1988). A check on surge-tank and makeup glycol is worthwhile also. Reclaimed glycol should be checked thoroughly before reuse. Virgin glycol is recommended for makeup—+99 weight percent TEG with <1 weight percent EG, DEG, TetraEG; and buffered with pH 7–8.

Contaminated glycol should be checked for foaming and a suitable defoamer found. The defoamer should be added until the source of the contamination has been identified and removed.

Table 8–3   Optimum Values for Glycol Analysis (Fremin, 1988)

Parameter	Glycol	
	Rich	Lean
pH	7.0–8.6	7.0–8.6
Chlorides (mg/L)	<600	<600
Hydrocarbon (wt%)	<0.3	<0.3
Iron (mg/L)	<15	<15
Water (wt%)	3.5–7.5	<1.5
Suspended solids (mg/L)	<200	<200
Foaming tendency	Foam height, 10–20 mL Break time, 5 sec	
Color and appearance	Clear and light-colored to medium amber	

Remarks:

pH usually lower on rich glycol due to acid-gas content.

Hydrocarbon, iron, and suspended solids may be different for lean and rich glycols, depending on filter placement.

Difference between lean and rich glycol water contents should be 2–6%.

Pearce and Sivalls (1984) discuss the analysis of TEG field samples for water, pH (1:1 dilution), emulsion formation, glycol (acid gases and hydrocarbons by gas chromatography), hydrocarbon by distillation, foaming tendency, chlorides, iron, and other metals (Na, K).

The ideal time for *cleaning* is when the unit is shutdown for repairs. A good cleaning job can improve plant operation greatly. However, a poor job can severely contaminate the glycol for a long time. The best approach is to sample the contaminant(s) and see what will dissolve the deposits. Recommended solutions are:

Salt	Hot water
Rust & FeS	Acids, followed by neutralization with caustic, or sodium carbonate in hot water
Tarry deposits	Hydrocarbon solvents or acetone

Temperatures of 140–150°F and high circulation rates improve the cleaning process. Ballard (1979) enumerates the pitfalls associated with high-detergent soaps, steam cleaning, and acid cleaning.

Hot spots and subsequent burnout of reboiler fire tubes are caused by improper firing and/or deposits on the outside of the tubes (often salt). Adjust burner controls monthly and clean the reboiler thoroughly once a year.

DESIGN METHODS

Standard units can be sized readily from manufacturers' published graphs and tables, e.g., NATCO (Caldwell, 1976) or Sivalls (1976). Several rules of thumb can be used for preliminary sizing, while computer programs permit more rigorous designs. No matter which design approach is selected, the following sequence is recommended:

1. Obtain design information.
2. Select an appropriate combination of lean glycol concentration and circulation rate, and absorber trays.
3. Establish the required material and energy balances.
4. Size the equipment.

Required Information

The inlet gas stream condition and the degree of dehydration are based on the following:

1. Inlet gas flow rate, pressure, and temperature.
2. Required water dew point or water content of exit gas.
3. Inlet gas analysis or inlet gas gravity and acid gas (H_2S, CO_2) content.

Other important considerations include:

1. Available utilities.
2. Safety and environmental regulations for discharging stripper overhead.

Required TEG Reconcentration

Parrish *et al.* (1986) compared the existing VLE data for TEG-water-natural gas—Worley (1967); Dingman and LeBas (1964), Scauzillo (1961); Townsend (1953); Wise *et al.* (1950), Rossman (1973), and Herskowitz and Gottlieb (1984)—with their measurements. There is considerable disagreement, as is illustrated by Figure 8–10 which compares two popular correlations—Worley (1967), as cited by GPSA (1987) and Pearce and Sivalls (1984) with Parrish *et al.* (1986). Obviously, any uncertainty in the VLE data produces an equal uncertainty in the computer simulation. However, the detailed Parrish *et al.* study has reduced this difficulty for sweet natural gas.

The dehydrated natural gas leaving the absorber cannot contain less water than that which would be in equilibrium with the entering lean reconcentrated glycol. Equilibrium is never completely achieved, and the degree to which equilibrium is approached depends on the gal TEG recirculated per lb H_2O and the number of trays in the contactor. In practice, the water dew point of the dried gas leaving the contactor is 5–10°F higher than the equilibrium dew-point. This fact and the rule of thumb that the dew-point depression is 60°F for the first four trays, plus 7°F for each additional tray, are good first estimates.

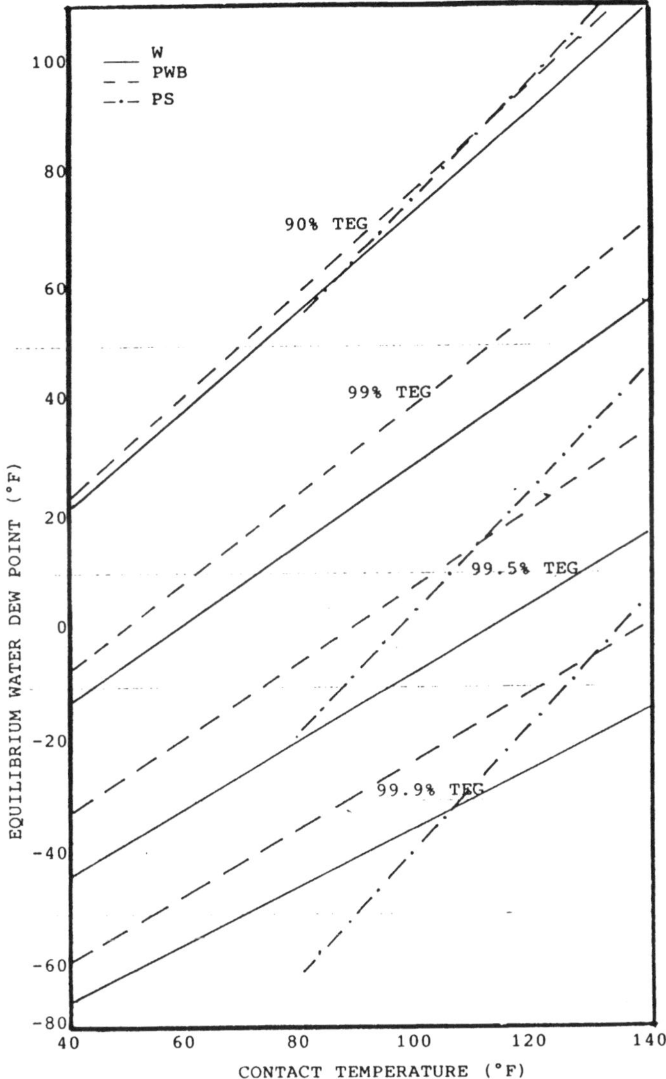

Figure 8–10. Comparison of the Worley, Pearce-Sivalls, and Parrish-Won-Baltatu correlations of TEG-H_2O VLE.

Absorber

Random or structured packing can and is used in place of the traditional, bubble-cap trayed towers. Packing can result in a 25% shorter contactor tower and the resulting reduction in tower size and weight is very welcome on offshore platforms.

Sizing the absorber involves specifying:

1. The type and number of trays
2. The TEG circulation rate
3. The column diameter.

Absorbers may be sized using well-publicized charts, e.g., Sivalls (1976) or Worley (GPSA, 1987, p. 20–21).

However, these charts are based on questionable equilibrium data (Parrish *et al.*, 1986) and do not address the effects of contact pressure and temperature. Olbrich and Manning (Olbrich, 1988) developed charts, shown in Figures 8–11 through 8–20, that predict dew-point depression over the following contactor conditions:

Actual trays	4–12
Lean glycol concentration, wt%	98.5–99.9
Circulation rate, gal TEG/lb H_2O	1.5–6
Temperature, °F	80 and 100
Pressure, psia	300–1400

With all other variables constant, dew-point depressions increase 0.9°F for every 100 psi increase in contactor pressure. Dew-point depressions vary little with temperature, and linear interpolation is recommended

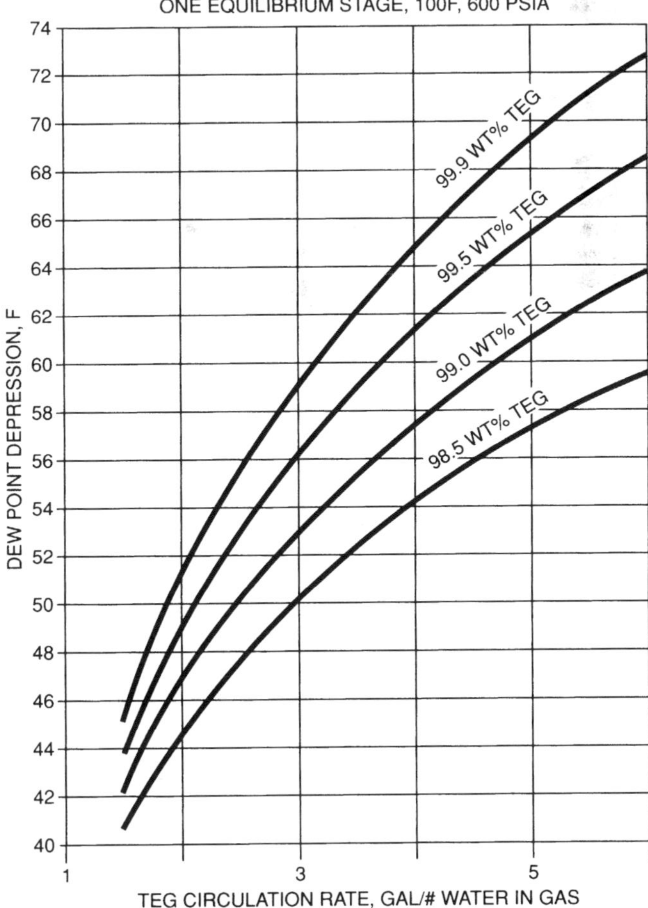

Figure 8–11. Predicted dew-point depression for 1 equilibrium stage (4 trays) at 100°F and 600 psia (Olbrich, 1988).

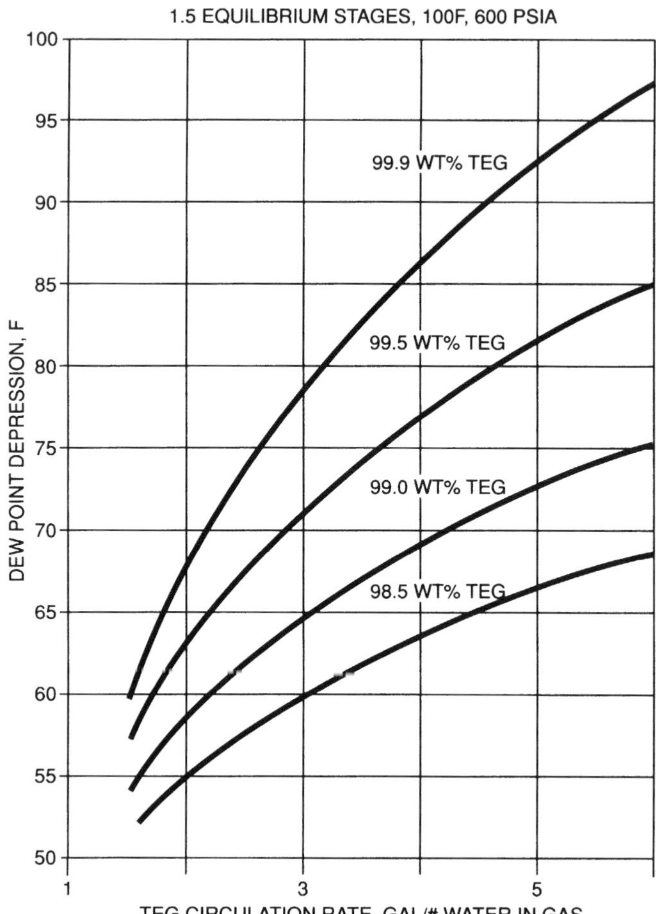

Figure 8–12. Predicted dew-point depression for 1.5 equilibrium stages (6 actual trays) at 100°F and 600 psia (Olbrich, 1988).

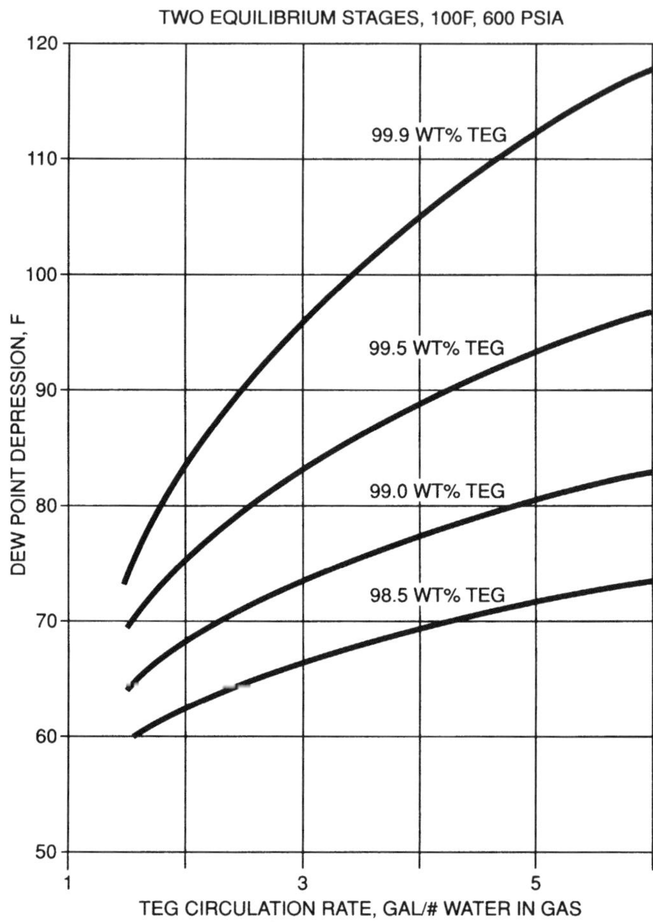

Figure 8–13. Predicted dew-point depression for 2 equilibrium stages (8 actual trays) at 100°F and 600 psia (Olbrich, 1988).

between 80 and 100°F. Olbrich's dew-point depressions are 2–15°F more conservative than those published in the GPSA Engineering Data Book (1987, p. 20–21).

Sivalls (1976) uses a *modified McCabe-Thiele diagram* to estimate the TEG circulation for a given or assumed number of trays and vice versa. Typically, water concentration is quite low in both the natural gas and liquid glycol phases. Therefore, the McCabe-Thiele diagram is characterized by nearly linear equilibrium and operating lines, and the absorber can be modeled by the Kremser-Brown method (Scauzillo, 1961). Such a shortcut approach is especially amenable to computer solution.

The diameter of the contactor can be conservatively estimated either from manufacturers' charts (NATCO-Caldwell, 1976 or Sivalls, 1976) or from the Souders-Brown correlation.

$$V_{max} = C_{SB}[(\rho_L - \rho_v)/\rho_v]^{1/2} \qquad (8–1)$$

where V_{max} = maximum gas superficial velocity (ft/hr)
$\quad C_{SB}$ = Souders-Brown coefficient (ft/hr)
$\qquad$ = 660 ft/hr for towers 30 in. or larger
$\qquad$ in diameter with 18-in. tray spacing
$\quad \rho_L$ = glycol density (lb/ft^3)
$\quad \rho_v$ = gas density at column condition (lb/ft^3)

Pump

A reciprocating pump is sized using manufacturers' catalogs or by the standard mechanical energy balance and an assumed pump efficiency of 70–80%. The temperature rise through the pump may be estimated by increasing the glycol enthalpy by the pump work.

The BS&B Co. (1960) recommends the following quick estimates based on 80% pump efficiency and 90% motor efficiency:

$$\text{Pump BHP} = (.000012)(gph)(psig) \qquad (8–2)$$

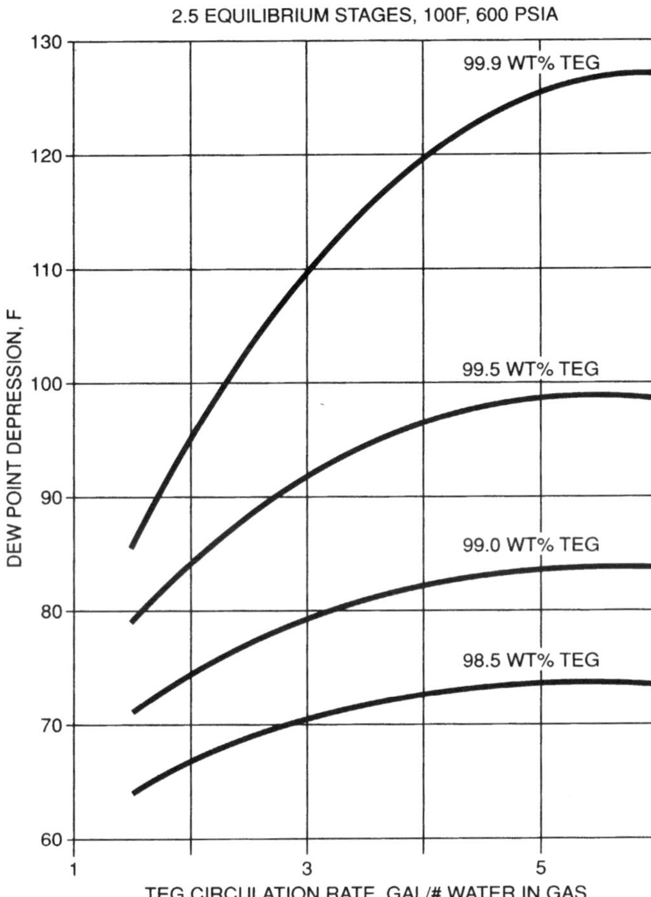

Figure 8–14. Predicted dew-point depression for 2.5 equilibrium stages (10 actual trays) at 100°F and 600 psia (Olbrich, 1988).

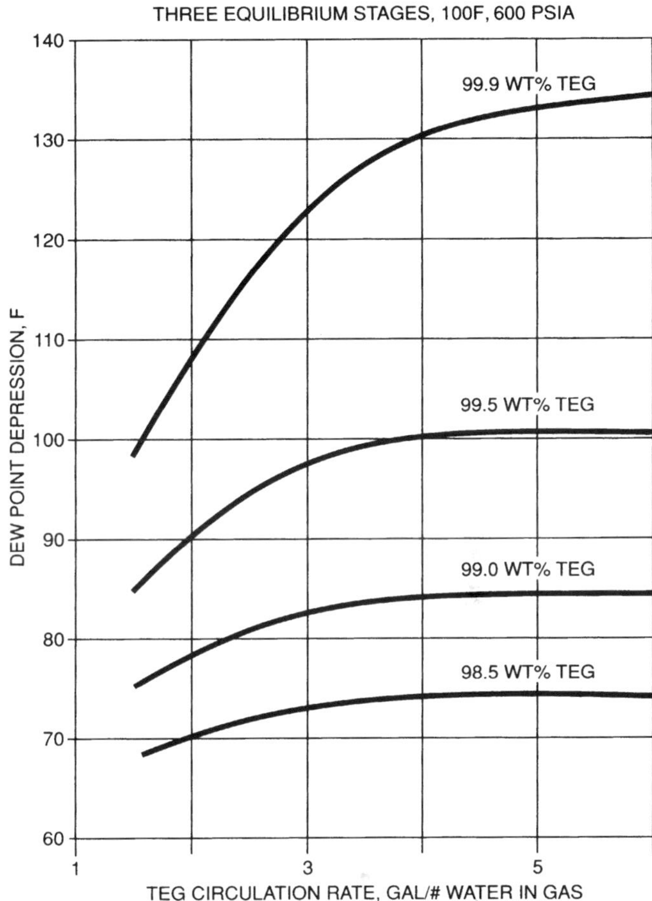

Figure 8–15. Predicted dew-point depression for 3 equilibrium stages (12 actual trays) at 100°F and 600 psia (Olbrich, 1988).

$$\text{Electrical kW} = (.000011)(\text{gph})(\text{psig}) \qquad (8\text{--}3)$$

where gph = gal TEG circulated per hour

Use the manufacturer's catalog for a Kimray Pump.

Glycol Flash Separator

The wet or rich glycol is flashed at 50–100 psia and 100–150°F. Recommended liquid retention times are 5–10 minutes for two-phase (gas-glycol) and 20–30 minutes for three-phase (gas-liquid HC-glycol) separation (Ballard, 1979). Other guidelines are:

1. Vertical Separator

$$\text{Height (ft)} = 3.4 + 0.4 \text{ gpm} \qquad (8\text{--}4)$$

where gpm = gal TEG circulated per min
and minimum height = 4 ft
maximum height = 10 ft
minimum diameter = 1.5 ft

2. Horizontal Separator

$$\text{length/diameter ratio} = 3 \qquad (8\text{--}5)$$
$$\text{minimum length} = 3 \text{ ft}$$
$$\text{minimum diameter} = 2 \text{ ft}$$

Stripping Still

Computer programs usually consider the stripping column as three theoretical trays: one for the reboiler, one for the packed stripping column, and one for the reflux condenser. The diameter of the stripping column is based on the required vapor and liquid loads at the base of the column. Manufacturers' charts or standard sizes based on the required reboiler heat load may be used to determine the column diameter (Caldwell, 1976; Sivalls, 1976). Alternatively use the following approximate equation:

$$\text{Diameter (in.)} = 9 \, (\text{glycol circulation, gpm})^{0.5} \qquad (8\text{--}6)$$

Conservative design and field test data dictate that the packed section should be at least 4 ft high, and that this

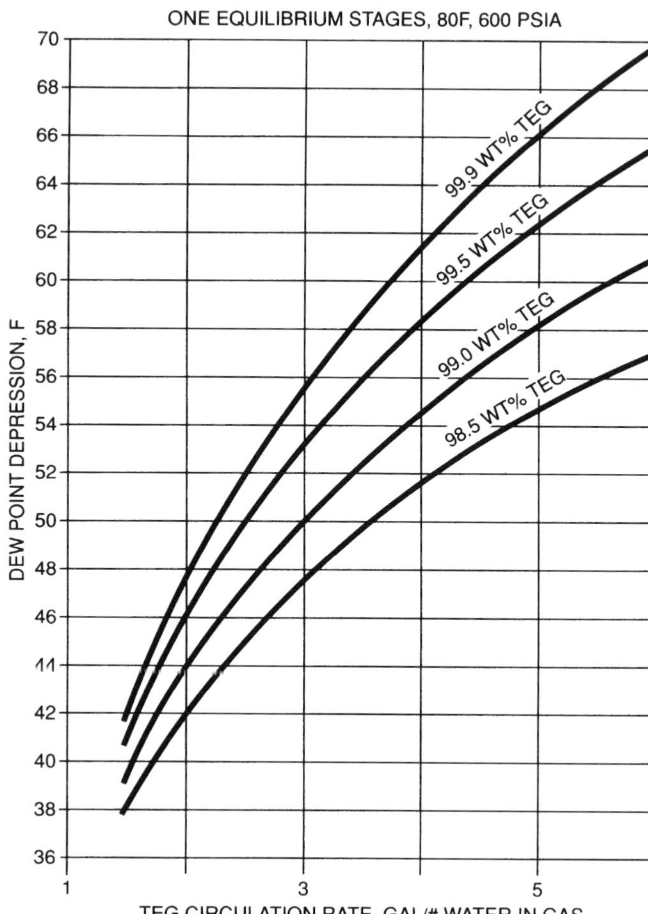

Figure 8–16. Predicted dew-point depression for 1 equilibrium stage (4 trays) at 80°F and 600 psia (Olbrich, 1988).

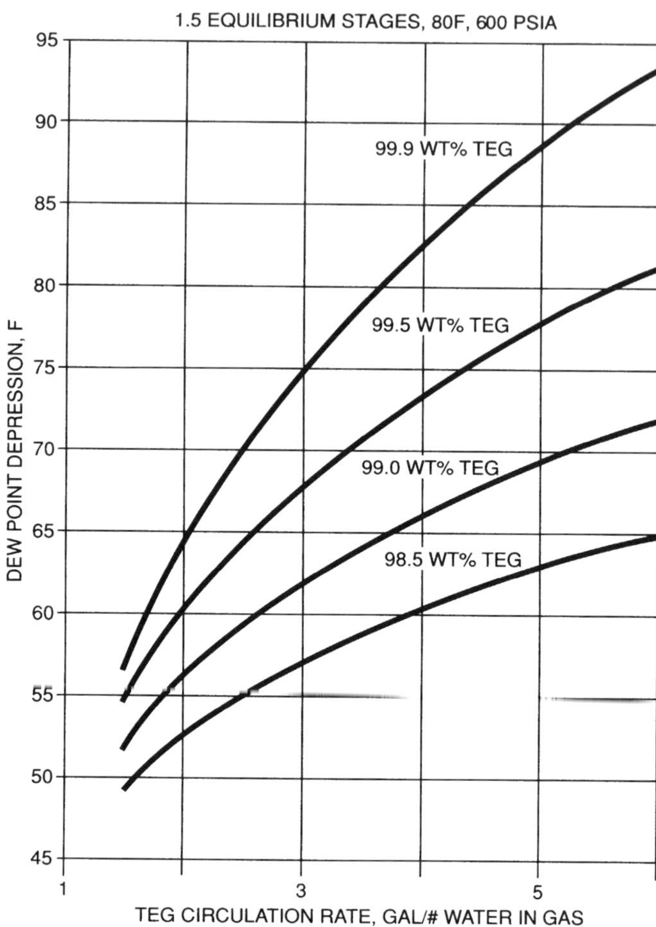

Figure 8–17. Predicted dew-point depression for 1.5 equilibrium stages (6 actual trays) at 80°F and 600 psia (Olbrich, 1988).

height be increased to 8 ft for a 1 MMBtu/hr unit (Sivalls, 1976).

Reboiler

Reboiler duty can be estimated by the following equation:

$$Q_R = 900 + 966 \, m \tag{8-7}$$

where Q_R = regenerator duty Btu/lb H_2O removed
m = gal TEG/lb H_2O removed.

This estimate does not include stripping gas and makes no allowance for combustion efficiency.

A more detailed procedure (patterned after BS&B, 1960, and Sivalls, 1976) is illustrated in the design example below.

Use the following guidelines to size the reboiler.

1. Design duty is calculated requirement duty plus 5% of the condenser and glycol exchanger duties.
2. Vapor disengagement area is based on 14,000 Btu/hr-ft^2 heat flux across the vapor liquid interface.
3. Reboiler shell L/D ratio is 5.
4. Minimum D is 1.5 ft, minimum L = 3.5 ft.

Standard units are available (Caldwell, 1976; Sivalls, 1976).

Heat Exchangers

Typically, *three heat exchangers* are used: the reflux condenser at the top of the stripping still, the glycol-glycol exchanger, and a lean glycol-dry gas or lean glycol-air-fin exchanger. These units can be sized using conventional design procedures. Typical guidelines are:

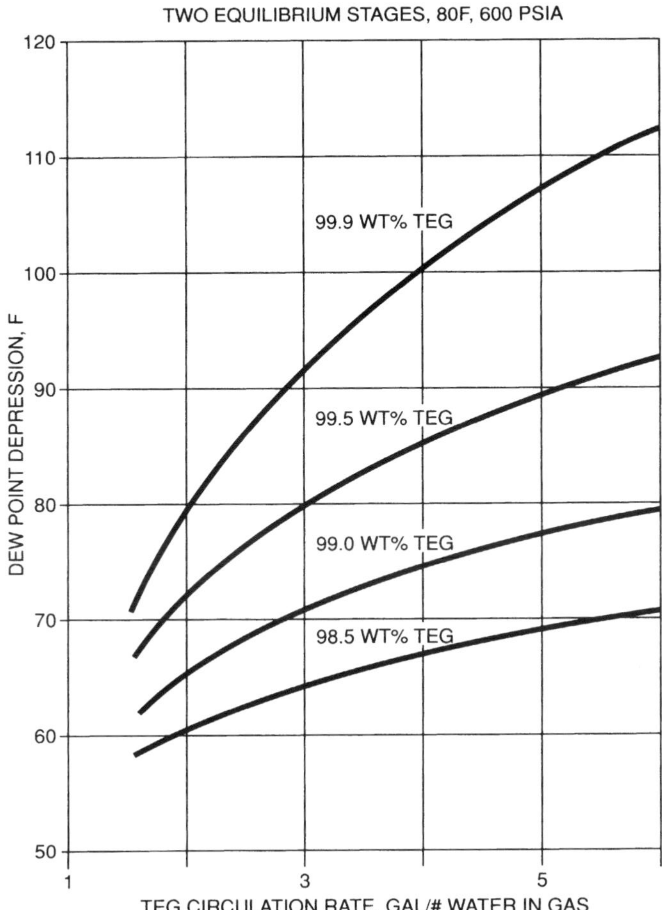

Figure 8–18. Predicted dew-point depression for 2 equilibrium stages (8 actual trays) at 80°F and 600 psia (Olbrich, 1988).

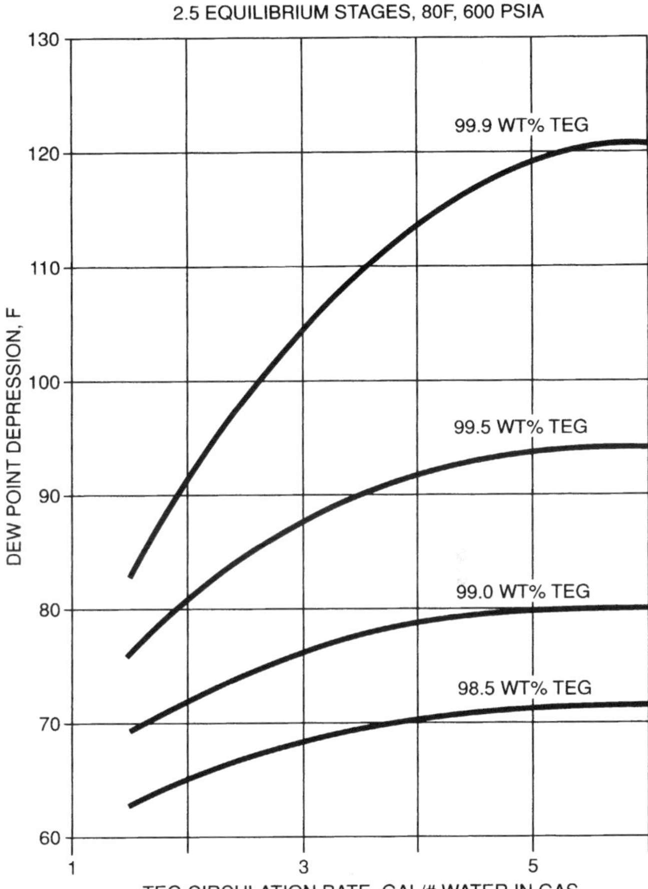

Figure 8–19. Predicted dew-point depression for 2.5 equilibrium stages (10 actual trays) at 80°F and 600 psia (Olbrich, 1988).

1. Reflux Condenser
   a. Design duty is calculated as design requirement plus 5% for fouling and flow variations.
   b. Seider-Tate correlation used for the heat transfer coefficients.
2. Glycol-Glycol Exchanger
   a. Design duty is calculated as design requirement plus 5% for fouling and flow variations.
   b. The entering temperatures of the lean and rich glycol streams are known; a hot end (lean glycol in-rich glycol out) temperature approach of 60°F maximizes the preheat of the rich glycol.
   c. Two or more heat exchangers should be placed in series to avoid any temperature cross.

In smaller units, the exchanger may be replaced by a surge tank and heat-transfer coil. The shell volume is based on a 30-min retention time, an L/D ratio of 4, and a minimum size of D = 1.5 ft, L = 3.5 ft.

3. Lean Glycol Cooler
   a. The lean glycol outlet temperature should be 5–10°F hotter than the inlet gas temperature to the absorber. Therefore, the lean glycol is cooled from 180–200°F to 110–120°F. This may be accomplished in: 1) A double pipe exchanger for smaller units (less than 25 MMscfd), and 2) An aerial, fin-fan exchanger or a water-cooled, shell-and-tube exchanger for larger units (greater than 25 MMscfd).

Again, the design heat duty must provide for a 5 to 10% allowance for fouling and flow variations.

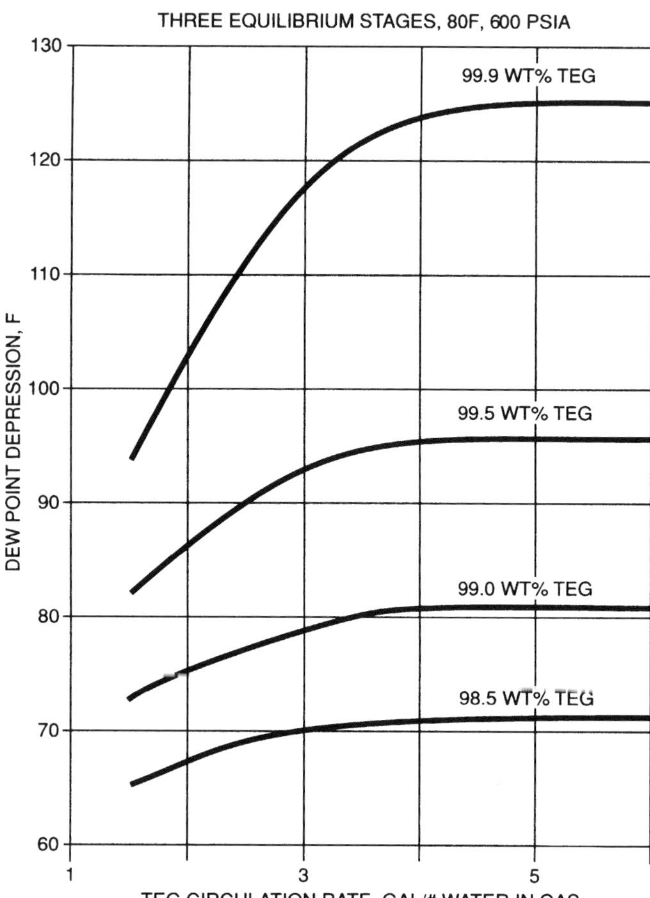

Figure 8–20. Predicted dew-point depression for 3 equilibrium stages (12 actual trays) at 80°F and 600 psia (Olbrich, 1988).

DESIGN EXAMPLES

Example 8–1. (Chapter 5, Example No. 2, High-Pressure, Lean Gas)

A.   Design Data

1. Gas Flow Rate	80 MMscfd at 14.696 psia, 60°F
2. Gas Spec. Grav	0.59
Molecular Weight	$(0.59)(28.96) = 17.0$ lb/lb mole
3. Contactor Pressure	800 psia = 785.8 psig if elevation is 1000 ft and $P_{atm} = 14.2$ psia
4. Gas Inlet Temp	89°F
5. Gas Inlet Humidity	Saturated
6. Outlet Gas H_2O Dew Pt	32°F

B.   Design Criteria

1. Lean, reconcentrated TEG concentration = 98.9 weight percent because:
   a. Can operate reboiler at 390°F (Fig. 8–6).
   b. Do not need stripping gas.
   c. Equilibrium dew point is 10°F (Fig. 8–4).
   d. Dew-point approach of $32 - 10 = 22°F$ is attainable.
2. TEG circulation rate is 2 gal TEG/lb H_2O entering because:
   a. Flow rate on absorber trays is satisfactory.
   b. Flow rate is low enough to conserve energy.
3. Absorber has 7 bubble cap trays (25% efficiency) because:
   a. Required dew-point depression = $89 - 32 = 57°F$.
   b. Dew-point depression predicted with 1.5 equilibrium stages (6 trays with 25% efficiency) at 600 psia is:
   57.5°F at 100°F (Fig. 8–12)
   55.5°F at   80°F (Fig. 8–17)
   56.5°F at   89°F by approximate interpolation

Dew-pt dep = Dew-pt dep +
(0.009)(800 − 600)
800 psia   600 psia
= 56.5 + (0.9)(800 − 600) = 58.3°F

One tray is added for safety so that the common specification of 7 lb H_2O/MMscf (28°F dew point at 800 psia, Fig. 4–6) can be met.

C.   Material Balances

1. Inlet gas contains 52 lb H_2O/MMscf (Fig. 4–6)
2. Exit gas contains 7.0 lb H_2O/MMscf (Fig. 4–6)
3. Water removed = $52 - 7 = 45$ lb H_2O/MMscf
4. lb H_2O entering/hr

$$= \left(\frac{\text{lb } H_2O \text{ entering}}{\text{MMscf}}\right)\left(\frac{\text{MMscf}}{\text{day}}\right)\left(\frac{\text{day}}{\text{hr}}\right)$$

$= (52)(80)(1/24)$

$= 173.3$ ⟵

5. gal TEG circulated/hr

$$= \left(\frac{\text{gal TEG}}{\text{lb } H_2O \text{ entering}}\right)\left(\frac{\text{lb } H_2O \text{ entering}}{\text{hr}}\right)$$

$$= (2.0)(173.3)$$
$$= 346.7 \quad \longleftarrow$$

6. $\dfrac{\text{lb 99 wt\% TEG}}{\text{hr}} = \dfrac{\text{gal}}{\text{hr}} \times \dfrac{\text{lb TEG}}{\text{gal}}$

$$= (346.7)(8.33 \times 1.11)$$
$$= 3205.4 \quad \longleftarrow$$

Densities of TEG solutions are given in Figure 8–21 (Union Carbide, 1989, p. 11, Fig. 6) and are available also from Dow

Chemical Co. (1962, p. 91); and GPSA (1981, p. 15–26), for 60°F only.

7. $\dfrac{\text{lb H}_2\text{O removed}}{\text{hr}} = \dfrac{\text{lb H}_2\text{O removed}}{\text{MMscf}}$

$$\times \dfrac{\text{MMscf}}{\text{day}} \times \dfrac{\text{day}}{\text{hr}}$$
$$= (45)(80)(1/24)$$
$$= 150$$

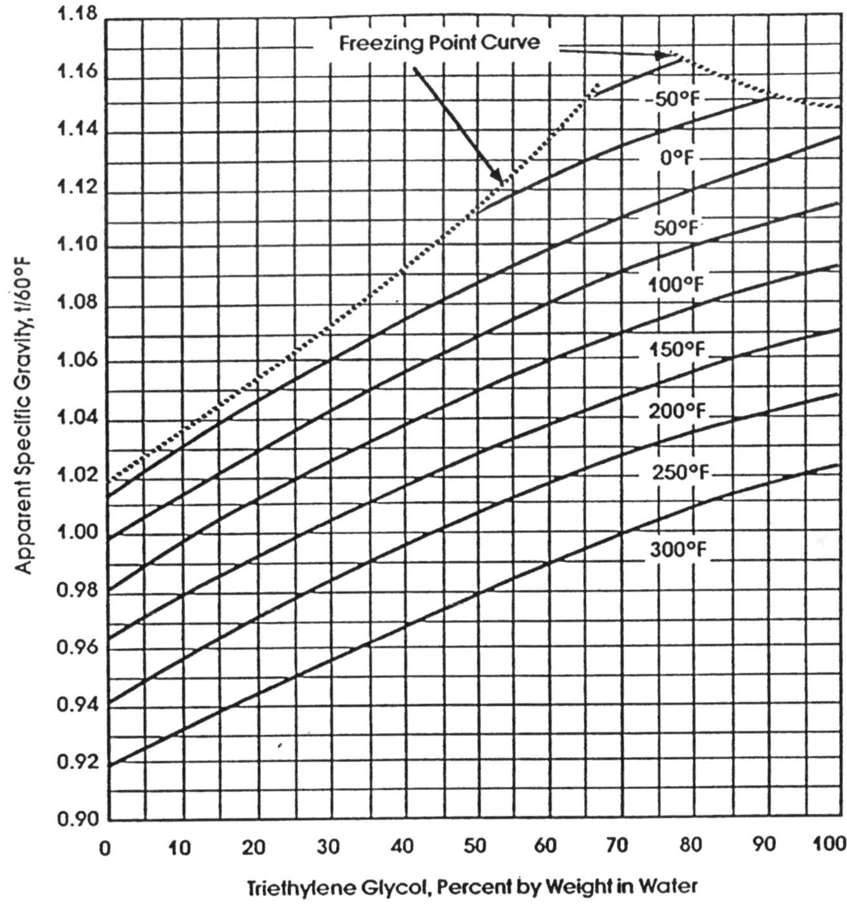

| Specific Gravity at t/60°F = A + Bx + Cx², where x = TEG wt% | | | |
t°F	A	B	C
-50	1.0502	1.8268E-3	-5.2009E-6
0	1.0319	1.7466E-3	-4.8304E-6
50	1.0121	1.5247E-3	-2.8794E-6
100	0.9920	1.7518E-3	-5.4955E-6
150	0.9804	1.5410E-3	-4.3884E-6
200	0.9627	1.4068E-3	-3.5089E-6
250	0.9413	1.3205E-3	-2.7991E-6
300	0.9177	1.2511E-3	-2.0848E-6

Figure 8–21. Specific gravities of aqueous triethylene glycol solutions (Union Carbide, 1989, p. 11, Fig. 6).

8. Basis 1 hour of operation:
Lean glycol stream (98.9 weight percent TEG) contains

$$(3205.4)(0.989) = 3170.1 \text{ lb pure TEG}$$
$$(3205.4)(0.011) = 35.3 \text{ lb } H_2O$$

9. Rich glycol stream contains:

$$3170.1 = 3170.1 \text{ lb pure TEG}$$
$$35.3 + 150 = 185.3 \text{ lb } H_2O$$
$$3170.1 + 185.3 = 3355.4 \text{ lb solution}$$

Rich glycol stream is $100(3170.1/3355.4) = 94.5$ weight percent TEG

10. Glycol pump should handle 360 gal/hr or 6 gpm. Select a pump capable of 6–9 gpm.

D.   Absorber

1. Diameter may be estimated by cross-sectional area.

$$\text{Cross-sectional area} = \frac{\text{gas flow rate}}{\text{superficial velocity}}$$

a. Superficial gas velocity is estimated by Eq. 8–1:

$$V_{max} = C_{SB}\,[(\rho_L - \rho_V)/\rho_V]^{1/2}$$
$$\rho_L = (62.4)(1.11) = 69.3 \text{ lb/ft}^3$$
$$\rho_V = (P)\,(MW)\,/\,(Z)\,(R)\,(T)$$
$$= (800)(17.0)/$$
$$(.89)(10.732)(459.7 + 89)$$
$$= 2.60 \text{ lb/ft}^3$$

Note: $R = 10.732\dfrac{(\text{psia})(\text{ft}^3)}{(\text{lb mole})(^\circ R)}$
Z obtained from Figure 3–17

$$V_{max} = 660\,[(69.3 - 2.60)/(2.60)]^{1/2}$$
$$= 3343 \text{ ft/hr} = 55.7 \text{ ft/min}$$

b. Gas volumetric flow rate of 80 MMscfd at contactor P & T

$$v = n\,Z\,R\,T\,/\,P$$
$$\text{flow rate} = \frac{\text{scf}}{\text{day}} \times \frac{\text{lbmole}}{\text{scf}} \times (Z)$$
$$\times \frac{\text{psia-ft}^3}{\text{lbmole-}^\circ R} \times \frac{^\circ R}{\text{psia}}$$
$$= \frac{(80 \times 10^6)(0.89)(10.732)(548.7)}{(379.5)(800)}$$
$$= 1{,}381{,}000 \text{ ft}^3/\text{day} = 959 \text{ ft}^3/\text{min}$$

c. Cross-sectional area of trayed glycol contactor

$$\text{area} = (\text{gas flow rate})/(\text{superficial gas velocity})$$
$$\text{ft}^2 = (\text{ft}^3/\text{min})/(\text{ft/min})$$
$$\text{area} = 959/55.7 = 17.2 \text{ ft}^2 \longleftarrow$$

d. Absorber diameter

$$\text{area} = (\pi/4)D^2 \text{ and so } D = 4.68 \text{ ft}$$
Use ID = 5 ft as next largest standard
size $\longleftarrow$

2. Absorber Height

Diameter = 60 in. ID
Height for 7 trays (2 ft/tray) = 14 ft
Height for internal scrubber
$= 5$ ft (1 diameter)
Height for glycol inlet, mist
extractor at top = 5 ft (1 diameter)
Total column height = 24 ft $\longleftarrow$

E.   Reboiler

1. Required Heat Load
a. Following Sivalls (1976), p. 8, the required heat load consists of:
Heat glycol from 290 to 390°F.

$$Q_{R1} = M_{lean}\,C_p\,\Delta T$$
$$\frac{\text{Btu}}{\text{hr}} = \frac{\text{lb TEG}}{\text{hr}} \times \frac{\text{Btu}}{\text{lb }^\circ F} \times {}^\circ F$$
$$= (3205)(.72)(390 - 290)$$
$$= 231{,}000$$

C_p for TEG solution (99 weight percent) obtained from Figure 8–22 (Union Carbide, 1989, p. 13, Fig. 81) at average temperature (340°F).
Heat to vaporize water picked up in absorber.

$$Q_{R2} = M_{H_2O}\,h_{f_g}$$
$$\frac{\text{Btu}}{\text{hr}} = \frac{\text{lb }H_2O\text{ removed}}{\text{hr}} \times \frac{\text{Btu}}{\text{lb }H_2O}$$
$$= (150.2)(970.3) = 146{,}000$$

Heat to vaporize reflux water in still—25% is returned.

$$Q_{R3} = 0.25 \times Q_{R2}$$
$$= (.25)(146{,}000) = 36{,}000 \text{ Btu/hr}$$

Heat losses from still.

$$Q_{R4} = 5000 \text{ to } 20{,}000 \text{ Btu/hr depending on size}$$

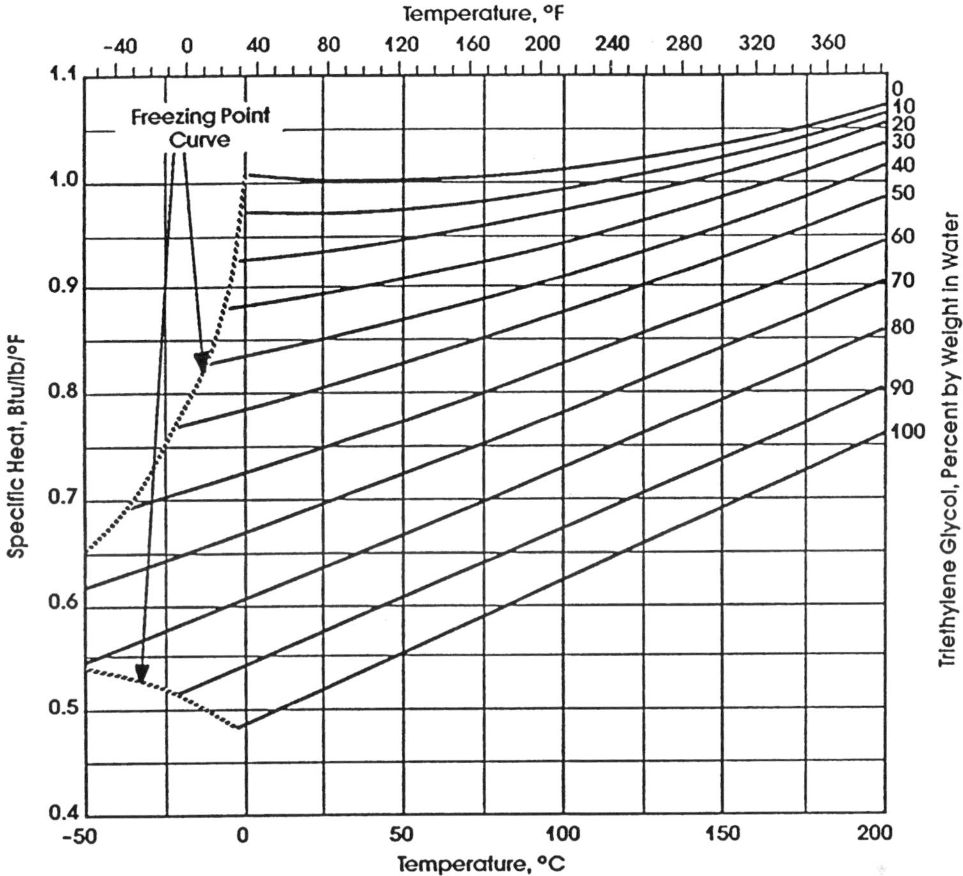

Specific Heat $= A + Bt + Ct^2$, where t = Temperature, °C			
TEG wt%	A	B	C
0	1.00540	-2.7286E-4	2.9143E-6
10	0.96705	-2.7144E-5	2.4952E-6
20	0.92490	2.0429E-4	2.4524E-6
30	0.88012	4.3000E-4	1.6952E-6
40	0.83229	6.2286E-4	1.3714E-6
50	0.78229	7.9286E-4	1.0857E-6
60	0.72200	9.4000E-4	8.0000E-7
70	0.66688	1.0871E-3	4.7620E-7
80	0.60393	1.2043E-3	2.8571E-7
90	0.53888	1.2800E-3	1.9048E-7
100	0.48614	1.3929E-3	-5.7140E-8

Figure 8–22. Specific heat for aqueous triethylene glycol solutions (Union Carbide, 1989, p. 13, Fig. 8).

Total heat load,
$$Q_{RT}' = Q_{R1} + Q_{R2} + Q_{R3} + Q_{R4}$$
$$= 415{,}000 \text{ Btu/hr} \longleftarrow$$

Note that there is no allowance for combustion efficiency. At 70% fuel efficiency:

fuel load = 611 MBtu/hr = 611 scf/hr.

b. Using Rule of Thumb, Equation 8–7.

Regenerator duty $= 900 + 966 \, (m)$
$= 900 + 966 \, (2)$
$= 2832 \text{ Btu/lb } H_2O$ removed

$$\text{Reboiler Duty} = \frac{\text{Btu}}{\text{lb } H_2O \text{ removed}} \times \frac{\text{lb } H_2O \text{ removed}}{\text{hr}}$$

$$= (2832)(150.2)$$
$$= 424,000 \text{ Btu/hr} \longleftarrow$$

2. Required Surface Area of Fire tubes

fire tube area = heat load/heat flux
$$\text{ft}^2 = (\text{Btu/hr})/(\text{Btu/ft}^2\text{-hr})$$
$$= (450000)/(6000)$$
$$= 75 \text{ ft}^2$$

F.   Glycol-Glycol Heat Exchanger
1. Estimate temperature profiles.
Lean glycol enters at 390°F, leaves at 190°F
Rich glycol enters at 90°F, leaves at T°F
2. Lean glycol duty = $M_{lean} C_p \Delta T$

$$= \frac{\text{lb TEG}}{\text{hr}} \times \frac{\text{Btu}}{\text{lb TEG °F}} \times \text{°F}$$
$$= (3205)(0.64)(390 - 190)$$
$$= 436,000 \text{ Btu/hr} \longleftarrow$$

C_p for lean glycol solution (99 weight percent) obtain from Figure 8–22 at average temperature (290°F).

3. Calculate exit temperature (T) of rich glycol.
Assume T = 280°F
Avg C_p = 0.62 at assumed average temp of 190°F and 94.5 weight percent (Fig. 8–22)
Calculate T by heat balance.

$$Q = 436,000 = M_{rich} C_p (T - 90)$$

$$= \frac{\text{lb TEG}}{\text{hr}} \times \frac{\text{Btu}}{\text{lb TEG °F}} \times \text{°F}$$
$$= (3355)(0.62)(T - 90)$$
$$T = 296°F \text{ Close enough} \longleftarrow$$

G.   Electric Pump
Electric pump must handle 9 gpm
Pressure increase (ΔP) = 800 − 14.2
= 786 psi

bhp = Q ΔP/1715 E
$$= (9)(786)/(1715)(.75) = 5.5$$
Use a 7.5 hp pump. $\longleftarrow$

Temperature increase (ΔT) of TEG across pump

$$W_{net} = M_{lean} C_p \Delta T$$

$$hp \times \frac{\text{Btu}}{\text{hp hr}} = \frac{\text{lb TEG}}{\text{hr}} \times \frac{\text{Btu}}{\text{lb TEG °F}} \times \text{°F}$$
$$(5.5)(2545) = (3205)(.58)(\Delta T)$$
$$\Delta T = 8°F \longleftarrow$$

H.   Gas-Glycol Exchanger

Heat load = $M_{lean \, Cp} \Delta T$

$$\text{Btu/hr} = \frac{\text{lb TEG}}{\text{hr}} \times \frac{\text{Btu}}{\text{lb TEG °F}} \times \text{°F}$$
Heat load = (3205)(.56)(190 − 100)
$$= 164,000 \text{ Btu/hr} \longleftarrow$$

Gas temperature rise is very small, as is now shown.
$$Q = M_{gas} C_p \Delta T$$

$$\frac{\text{Btu}}{\text{hr}} = \frac{\text{scf}}{\text{day}} \times \frac{\text{day}}{\text{hr}}$$

$$\times \frac{\text{lb mole}}{\text{scf}} \times \frac{\text{Btu}}{\text{lb mole °F}} \times \text{°F}$$

$$164,000 = (80 \times 10^6) \frac{1}{24} \frac{1}{379.5} (9)\Delta T_{gas}$$

$$\Delta T_{gas} = 2.1°F \longleftarrow$$

Example 8–2. Calculation of Number of Absorber (Contactor) Trays Using McCabe-Thiele Method.

A.   Design Data.   (same as Example 8–1)

B.   Design Criteria.   (same as Example 8–1)

C.   Material Balances.   (same as Example 8–1)

D.   Calculation of Number of Absorber Trays.
1. Equilibrium line at 89°F:

TEG Concn Wt % TEG	Equil Gas Dew-Point Temp °F	Water Content Of Gas At Dew-Pt Temp And Contactor Pres. lb H_2O/MMscf
100*	−	0
99.5	−7.0	1.4
99	7.5	3.0
98	23.0	5.8
97	32.0	8.0
96	39.0	10.0
95	43.5	10.8
90	59.0	20.0

The above equilibrium values are obtained by:

(i) Reading the water dew point in equilibrium with the assumed weight percent TEG at the given contact temperature of 89°F (Fig. 8–4).

(ii) Reading the water content from Figure 4–6 using the 89°F contact temperature, the 800 psia contactor pressure, and the previously obtained equilibrium water dew points.

Note that 100 weight percent implies no water in the liquid phase and this, in turn, means no water in the gas phase.

2. Operating Line

By definition the operating line has the following coordinates:

y axis: water content of gas phase; for convenience in lb H_2O/MMscf

x axis: water content in liquid phase; which, for convenience, is expressed in weight percent TEG.

The operating line represents a water mass balance, which states that the water lost by the natural gas equals the water picked up by the glycol. Because both streams are dilute—the water content is never higher than 7 weight percent—weight percents rather than mass ratios (lb H_2O/lb pure TEG) are used. Any resulting curvature of the operating line may be neglected. Therefore the operating line is defined by:

At top of absorber

y = water content of exit gas
  = 7.0 lb H_2O/MMscf
x = lean glycol concentration
  = 98.9 wt % TEG

At bottom of absorber

y = water content of entering gas
  = 52 lb H_2O/MMscf
x = rich glycol concentration
  = 94.5 weight percent TEG.

3. Plot McCabe-Thiele diagram (Fig. 8–23).

The previously calculated equilibrium data are plotted as solid circles to get slightly curved line DB. The operating line is plotted using the previously calculated top (point F) and bottom (point A) conditions (open circles) as straight line AF.

4. Calculate number of equilibrium stages. Equilibrium stages are drawn in as triangles:

First equilibrium stage = ABC
Fractional equilibrium stage = CE

Calculating the fractional stage requires enlarging the McCabe-Thiele diagram as shown in insert at top of Figure 8–23.

Fraction = length
CE/CD = (12.3 − 7)/(12.3 − 4.8)
  = 0.76
Number of
  equilibrium
  stages   = 1.76

5. Calculate number of actual trays. By definition of tray efficiency:

$$\text{Number of Actual Trays} = \frac{\text{Number of Equilibrium Stages}}{\text{Tray Efficiency}}$$

  = 1.76/.25 = 7.04

Therefore, use 7 actual trays.

TROUBLESHOOTING

Proper preventive maintenance minimizes upsets and shutdowns. Troubleshooting is greatly simplified if the glycol unit is appropriately instrumented and if operating records, glycol analyses, process flow sheets, and drawings of vessel internals are available (Ballard, 1979). Interpreting the records requires knowledge of the design conditions for the plant. Major changes in key variables such as inlet gas flow rate, temperature, or pressure require determination of a new set of optimum operating conditions.

Troubleshooting is described by suggesting possible causes of the more common problems and discussing corrective measures.

High Exit Gas Dew Point
• Change in gas flow rate, temperature, or pressure
• Insufficient glycol circulation (should be 1.5 to 3 gal TEG/lb water removed)
• Poor glycol reconcentration (exit gas dew point is 5–15°F

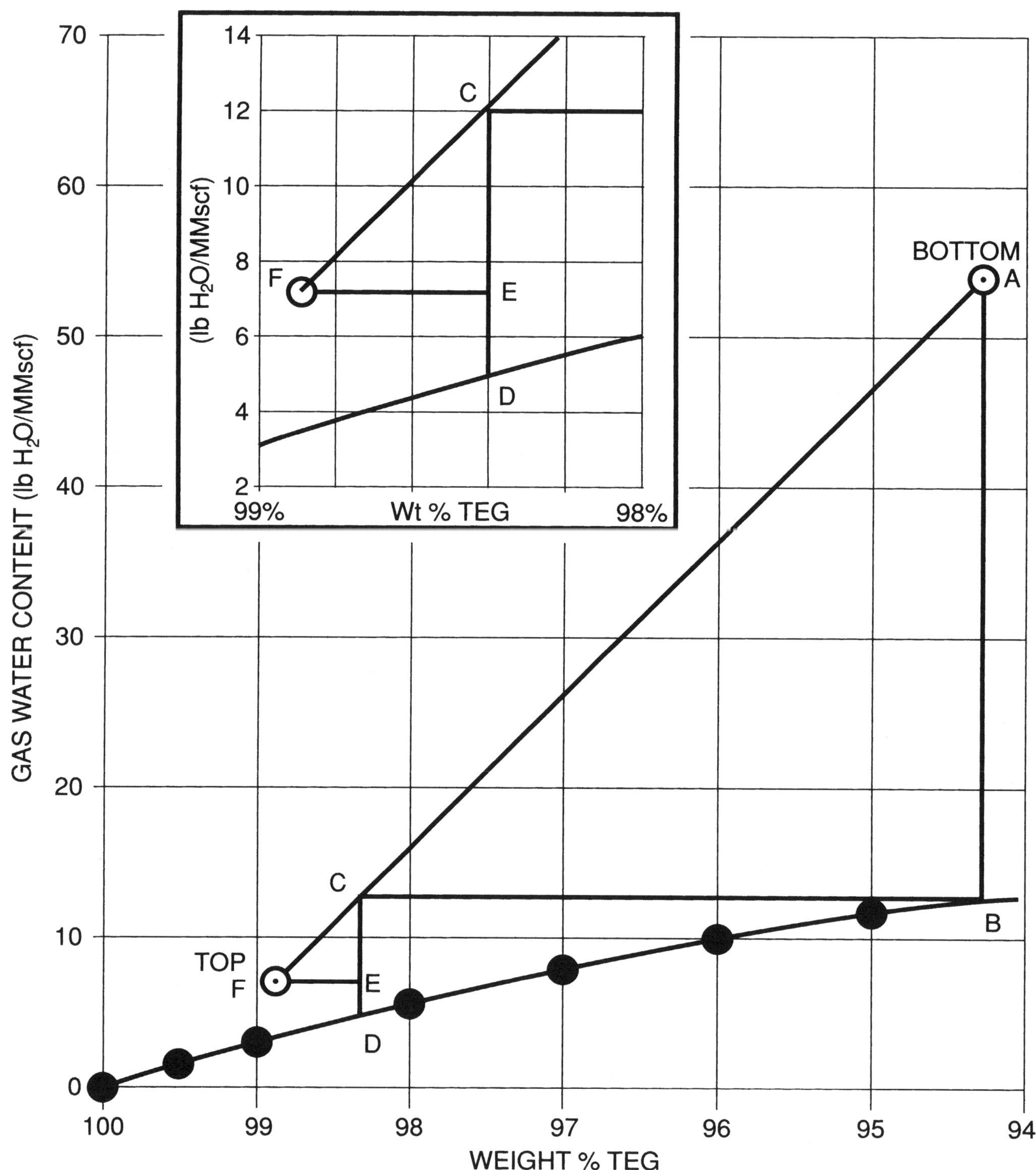

Figure 8–23. McCabe-Thiele diagram.

higher than dew point in equilibrium with lean glycol concentration)
- Current operating conditions different from design
- Malfunction of inlet separator

High Glycol Losses
First determine where loss is occurring: from contactor (i.e., downpipe line), out of still column, or leaking from pump.

Loss from contactor:
- Inlet separator passing liquids
- Carryover due to excessive foaming
- Mist extractor plugged or missing
- Lean glycol entering contactor too hot
- Plugged trays in contactor
- Excessive gas velocity in contactor
- Tray spacing less than 24 in.
- Mist pad too close to top tray

Loss from still:
- Excessive stripping gas
- Flash separator passing condensate
- Packing in still column is broken, dirty, plugged
- Insufficient reflux cooling at top of still
- Too much cooling at top of still
- Temperature at top of still is too high

Loss from separator:
- Glycol dumped with hydrocarbons

Leaks, spills, etc.:
- Check piping, fittings, valves, gaskets
- Check pumps, especially packings
- Pilferage

Many manufacturers will estimate maximum glycol losses to be 1 lb or 0.1 gal per MMscf. In large, well-monitored units with:

Upstream filter-coalescer
Three-phase flash separator
Full-stream activated carbon filter located downstream of glycol-glycol exchanger and operated at 150°F, pH at 7–8

TEG can last 8–10 years without dumping. Losses can be as low as 0.05 gal/MMscf, except immediately downstream from compressors, where 0.1 gal/MMscf loss is good. If no upstream filter-coalescer is present, it will probably be necessary to dump the glycol and clean the unit annually.

Contaminated TEG can be reclaimed by vacuum distillation with about 90% recovery. Portable vacuum distillation units can work successfully on a 1–3 gpm sidestream, but makeup glycol must be added.

Improper operation can increase the actual glycol losses to 1, 10, or even 100 gallons per MMscf. One leaking pump can waste 35 gallons per day (Ballard, 1977).

Glycol Contamination
- Carryover of oils (e.g., compressor lube oils), brine, corrosion inhibitors, well treating chemicals, sand, corrosion scales, etc., from inlet separator
- Oxygen leaks into glycol storage tanks, etc.
- Inadequate pH control (low pH) increases corrosion
- Overheating of glycol in reboiler due to excessive temperature or hot spots on fire tube
- Improper filtration, plugged filters, bypassing of filters
- Improper cleaning of glycol unit, use of soaps or acids (Ballard, 1979)

Poor Glycol Reconcentration
- Low reboiler temperature
- Insufficient stripping gas
- Rich glycol leaking into lean glycol in glycol heat exchanger or in the glycol pump
- Overloading capacity of reboiler
- Fouling of fire tubes in reboiler
- Low fuel rate or low Btu content of fuel
- Glycol foaming in still column
- Dirty and/or broken packing in still column
- Flooding of still column
- High still pressure

Low Glycol Circulation—Glycol Pump
- Check pump operation (Caldwell, 1976). If glycol-powered pump, close lean glycol discharge valve; if pump continues running, it needs repair. If gas or electric pump, check circulation by stopping glycol discharge from contactor and timing fill rate of gauge column on chimney tray section.
- Check pump valves to see if worn or broken
- Plugged strainer, lines, filters
- Vapor lock in lines or pump
- Low level in accumulator
- Excessive packing gland leakage
- Contactor pressure too high

High Pressure Drop Across Contactor
- Excessive gas flow rate
- Operating at pressures far below design
- Plugged trays and/or demister pads
- Glycol foaming

High Stripping Still Temperature
- Inadequate reflux
- Still column flooded
- Glycol foaming
- Carryover of light HC in rich glycol
- Leaking reflux coil

High Reboiler Pressure
- Packing in still column is broken and/or plugged with tar, dirt, etc.
- Restricted vent line, not sloped
- Still column is flooded by excessive boil up rates and/or excessive reflux cooling
- Slug of liquid HC enters stripper, vaporizes on reaching reboiler, and blows liquids out of still

Fire Tube Fouling/Hot Spots/Burn Out
- Buildup of salt, dust, scales, etc., on fire tube—check inlet separator
- Deposition of coke, tar formed by glycol overheating and/or hydrocarbon decomposition
- Liquid glycol level drops exposing fire tubes—low level shutdown

Low Reboiler Temperature
- Inadequately sized fire tube and/or burner
- Setting temperature controller too low
- More water in inlet gas because pressure is low or temperature high
- Temperature controller not operating correctly
- Carryover of water from inlet separator
- Inaccurate reboiler thermometer

Flash Separator Failure
- Check level controllers
- Check dump valves
- Excessive circulation

Review Questions

1. List the reasons for drying natural gas.
2. What factors determine a water dew point specification?
3. List and compare four methods of drying natural gas.
4. Why is TEG used so extensively?
5. Follow the path of glycol through a typical TEG dehydrator, and discuss the function of the different components.
6. What purpose does an inlet separator serve in a TEG dehy unit?
7. Where do glycol losses occur in a TEG unit? How can such losses be reduced? What loss level is acceptable?
8. How can energy consumption be minimized in a TEG unit?
9. What instrumentation would you recommend for a TEG unit? What factors and/or guidelines would you consider?
10. What causes foaming in a TEG unit? How can foaming be detected? How can foaming be prevented?
11. Describe what is meant by good glycol care? What factors determine the useful life of the glycol? How would you conduct a glycol analysis program? What conclusions can be obtained from analyzing the glycol?
12. Describe the procedures for the start up and shutdown of TEG unit. What are the procedures for a cold weather start up?
13. If you were in charge of a TEG unit, what maintenance procedures would you use?
14. How would you clean the interior of TEG contactor towers and/or reboilers?
15. How would you collect and dispose of the water condensed from the reboiler still column?
16. What are the consequences of varying the load to a TEG unit? Discuss the consequences of varying the inlet gas flow rate, temperature, and pressure.
17. What are the recommended temperatures for TEG in the various units of a glycol dehy unit? What are the consequences of deviating from these recommended temperatures?
18. Describe the appropriate troubleshooting procedures for the following problems: high exit-gas dew point, high glycol losses, glycol contamination, insufficient glycol regeneration, low glycol circulation, excessive pressure drop across the contactor, salt build-up in the reboiler, firetube burnout, and flash separator failure.

Problems

1. Dew-point depression vs. lb H_2O removed/MMscf.
   a. A North Sea natural gas enters a TEG dehydrator at 985 psig and 90°F and leaves at 972 psig, 91°F, with a water dew point of -10°F. Estimate the water dew-point depression and the lb water removed in lb H_2O/MMscf dehydrated.
   b. An Indonesian natural gas (20 vol percent CO_2) enters a TEG dehydrator at 650 psig and 130°F and leaves with a water dew point at 70°F. Estimate the dew-point depression and the lb water removed in lb H_2O/MMscf treated.
   c. If the sour (20 vol percent CO_2) Indonesian gas was cooled before dehydration, could the load on the TEG dehydrator units be reduced? If so, by how much? List any concerns.
2. Natural gas enters a TEG dehydrator at 985 psig and 100°F. An exit water dew point of -10°F is required. Design a suitable TEG dehydration unit by recommending:

a. A suitable lean TEG concentration
b. A suitable TEG circulation rate
c. Number of trays in the absorber
d. The rich TEG concentration.
   Using the above values
e. Size the absorber (contactor) column for 100 MMscf/d of a pipeline quality (SG = .6) natural gas
f. Determine the reboiler heat duty.

Nomenclature

C_{SB} = Souders-Brown coefficient (ft/hr) (Eq. 8–1)
C_p = specific heat (Btu/lb °F)
D = internal diameter of absorber (ft)
G = gas specific gravity relative to air
gph = U.S. gallons per hr
gpm = U.S. gallons per min
h_{f_g} = latent heat of water (Btu/lb)
L = length (ft)
M = flowrate (lb/hr)
m = gal TEG circulated/lb H_2O removed (Eq. 8–7)
n = gas flowrate (lb mol/day)
P = pressure (psig or psia)
Q = gas flowrate (MMscfd)
Q_R = reboiler duty (Btu/lb H_2O removed)
Q_{RT} = total reboiler duty (Btu/hr)
Q_{R1} = reboiler duty to heat glycol (Btu/hr)
Q_{R2} = reboiler duty to vaporize water (Btu/hr)
Q_{R3} = reboiler duty to vaporize reflux water (Btu/hr)
Q_{R4} = heat losses from still (Btu/hr)
R = gas constant 10.732 psia ft^3/lb mol °R
T = temperature (°F or °R)
V_{max} = maximum superficial velocity (ft/hr)
v = gas flowrate at contactor T & P (ft^3/min)
W = pump power (hp, kW)
Z = compressibility factor

Greek

ρ = density (lb/ft^3)

Subscripts

L = denotes liquid (glycol)
v = denotes vapor

References

API RP 14C (1978), "Recommended Practice for Analysis, Design, Installation, and Testing of Basic Surface Safety Systems on Offshore Production Platforms," 2nd Ed. (Jan.).

API RP 14E (1984), "Recommended Practice for Design and Installation of Offshore Production Platform Piping Systems," 4th Ed. (April 15).

Ballard, D. (1966), "How to Operate a Glycol Plant," *Hydrocarbon Processing*, Vol. 45, No. 6, pp. 171–188 (June).

Ballard, D. (1977), "Operator Talk . . . Glycol Dehydration," *Hydrocarbon Processing*, Vol. 56, No. 4, pp. 111–118 (April).

Ballard, D. (1979), "The Fundamentals of Gas Dehydration," 1979 Gas Conditioning Conference, University of Oklahoma, Norman, OK.

Ballard, D., and S.A. von Puhl (1988), "Improved Filtration Lowers Cost 80%," Petroenergy, Coastal Chemical, Houston, TX.

Behr, W. R. (1983), "Correlation Eases Absorber-Equilibrium Line Calculations for TEG-Natural Gas Dehydration," *Oil & Gas J.*, Vol. 81, No. 45, pp. 96–98 (Nov. 7).

BS&B (1960), "Product Section 34:00 Dehydrators," Black, Sivalls & Bryson, Inc., Oklahoma City, OK (Jan. 15).

BS&B Engineering Company, Inc. (1990), "Dehydrate (conc, glycol)," *Hydrocarbon Processing*, Vol. 69, No. 4, p. 76 (April).

Caldwell, R. E. (1976), "Glycol Dehydration Manual," NATCO, Tulsa, OK (Jan. 30).

Cramer, D. L., and W. R. Cook (1981a), "Gas Dehydration: Fine-tuning Existing Field Installations, Part 1—Determining Glycol Requirements and Circulation Rates," *World Oil*, Vol. 192, No. 1, pp. 119–125 (Jan.).

Cramer, D. L., and W. R. Cook (1981b), "Gas Dehydration: Fine-tuning Existing Field Installations, Part 2—How Operating Variables Affect Treating Efficiency," *World Oil*, Vol. 192, No. 2, pp. 75–84 (Feb. 1).

Dingham, J. C., and C. L. LeBas (1964), "Now—New Dew Point Data for Triethylene Glycol Solutions," *Oil & Gas J.*, Vol. 62, No. 5, pp. 75–77 (Feb. 3).

Dow (1962), "Gas Conditioning Fact Book," Dow Chemical Co., Midland, MI.

Fremin, Larry (1988), "Operating, Troubleshooting, and Cleaning Glycol Units," Petroenergy 88, Coastal Chemical Co., Abbeville, LA (Oct. 3–7).

Gas Processors Suppliers Association (1987), "Engineering Data Book," 10th Ed., Tulsa, OK.

Glaves, P. S., R. L. McKee, W. W. Kensell, and R. Kobayashi (1983), "Pick Your Data for CO2 Dehydration Carefully," *Hydrocarbon Processing*, Vol. 62, No. 11, pp. 213–215 (Nov.).

Grosso, S. (1978), "Glycol Choice for Gas Dehydration

Merits Close Study," *Oil & Gas J.*, Vol. 76, No.7, pp. 106–110 (Feb. 13).

Grosso, S., R. L. Pearce, and P. D. Hall (1979a), "Glycol Analysis: Dehydrator Problem Solving," Proc. Gas Conditioning Conf., University of Oklahoma, Norman, OK (March 5–7).

Grosso, S., R. L. Pearce, and P. D. Hall (1979b), "Dehydrator Problem Solving–I: Analytical Techniques Can Pinpoint Glycol Problems," *Oil & Gas J.*, Vol. 77, No. 39, pp. 176, 178, 180, 183, 186, 188 (Sept. 24).

Grosso, S., R. L. Pearce, and P. D. Hall (1979c), "Dehydrator Problem Solving-Conclusion: Glycol Analysis Solves Problems," *Oil & Gas J.*, Vol. 77, No. 40, pp. 86, 89, 90, 92, 94 (Oct. 1).

Guenther, J. D. (1979), "Natural Gas Dehydration," Seminar at Teknologisk Institut, Taastrup, Denmark; Crest Engineering, Inc., Houston, TX (April 26).

Halbrook, A. H. (1974), "Sivalls Tanks, Inc. Field Dehydration Operation and Problems," presented at Gas Measurement Institute, University of Kansas, Liberal, KS (Oct. 22 23)

Harrell, B.E. (1989), Personal Communication, NATCO, Tulsa, OK.

Herrin, J. P. (1980), "Fundamentals of Gas Dehydration," Proc. Gas Conditioning Conference, University of Oklahoma, Norman, OK.

Kemp, A. H. (1976), "Glycol Dehydration of Natural Gas Containing H_2S and CO_2," Gas Conditioning Conference, University of Oklahoma, Norman, OK.

Kohl, A. L., and F. C. Riesenfeld (1979), "Gas Purification," 3rd Ed., Gulf Pub. Co., Houston, TX.

Landes, S. H. (1983), "Fat vs Lean Design 1: Production System Designs Must Rediscover Simple Rules," *Oil & Gas J.*, Vol. 81, No. 46, pp. 165–170 (Nov. 14).

Mapes, G. J. (1960), "The Glycol Dehydrator," *World Oil*, Vol. 150, No. 2, pp. 51–58 (Jan.–April).

Meusburger, K. E., and E. W. Segebrecht (1980), "Foam Depressants for Gas Processing Systems," Gas Conditioning Conference, University of Oklahoma, Norman, OK (March 3–5).

NATCO (1984), "Glycol Dehydration Systems," TSL Catalog, p. 511-A2, Tulsa, OK

OCS Order No. 5 (1980), "Production Safety System," U.S. Dept. of Interior, Geological Survey, Conservation Division, Gulf of Mexico Region (Jan. 1).

OPC Engineering Inc. (1990), "Drizo Process," *Hydrocarbon Processing*, Vol. 69, No. 4, p. 76 (April).

Pape, Peter G. (1983), "Silicones: Unique Chemicals for Petroleum Processing," *J. Petroleum Technology*, Vol. 35, No. 7, pp. 1197–1204 (June).

Parrish, W. R., K. W. Won, and M. E. Baltatu (1986),

"Phase Behavior of the TEG-Water System and Dehydration/Regeneration Design for Extremely Low Dew Point Requirements," 65th Annual GPA Convention, San Antonio, TX (March 10–12).

Pearce, R. L. (1973), "Fundamentals of Gas Dehydration with Glycol Solutions," Gas Conditioning Conference, University of Oklahoma, Norman, OK.

Pearce, R. L., and C. R. Sivalls (1984), "Fundamentals of Gas Dehydration, Design and Operation with Glycol Solutions," Gas Conditioning Conference, University of Oklahoma, Norman, OK.

PETEX (1973a), "Glycol Dehydration: Maintenance, Care, and Troubleshooting," Slide-tape Workbook 15–1174, Petroleum Extension Service, University of Texas at Austin, Austin, TX.

PETEX (1973b), "Glycol Dehydration: Unit Startup and Shutdown," Slide-tape Workbook 15–1173, Petroleum Extension Service, University of Texas at Austin, Austin, TX.

PETEX (1973c), "Glycol Dehydration: Operating Conditions and Limits," Slide-tape Workbook 15–1172, Petroleum Extension Service, University of Texas at Austin, Austin, TX.

PETEX (1974), "Plant Processing of Natural Gas," Petroleum Extensive Service, University of Texas at Austin, Austin, TX.

Reid, L. S. (1969), "Predicting Performance Capabilities of Glycol-Gas Dehydration Processes from Design Data," SPE 2598, 44th Annual Fall Meeting of SPE, Denver, CO (Sept. 28–Oct. 1).

Sarma, H. (1981), "How to Size Gas Scrubbers," *Hydrocarbon Processing*, Vol. 60, No. 9, pp. 251–255 (Sept.).

Scauzillo, F. R. (1961), "Equilibrium Ratios of Water in Water-Triethylene Glycol-Natural Gas System," *J. Petroleum Technology*, Vol. 13, No. 7, pp. 697–702 (July).

Schroeder, A. J. (1978), "Studies of the Performance of Dehydration Systems Pay Off," *Oil & Gas J.*, Vol. 76, No. 38, pp. 168, 170, 175, 176 (Sept. 18).

Sivalls, C. R. (1976), "Glycol Dehydration Design Manual," Sivalls, Inc., Odessa, TX (June 1).

Simmons Jr., C. V. (1981a), "Avoiding Excessive Glycol Costs in Operation of Gas Dehydrators," *Oil & Gas J.*, Vol. 79, No. 38, pp. 121–124 (Sept. 21).

Simmons Jr., C. V. (1981b), "Temp Control and Heat Conservation Need Attention in Glycol-Based Dehydration Plants," *Oil & Gas J.*, Vol. 79, No. 39, pp. 313, 315, 316, 321 (Sept. 28).

Smith, R. L. (1979), "Foam Problems and Remedies for in Gas Processing Solutions," Gas Conditioning Conference, University of Oklahoma, Norman, OK.

Smith, R. S. (1990), "Gas dehydration process upgraded," *Hydrocarbon Processing*, Vol. 69, No. 2, pp. 75–77 (Feb.).

Swerdloff, W. (1957), "What We've Learned in 20 Years About Gas Dehydrators," *Oil & Gas J.*, Vol. 55, No. 17, pp. 122–129 (July 29).

Union Carbide Chemicals & Plastics Company Inc. (1989), "Triethylene Glycol," 39 Old Ridgebury Rd., Danbury, CT 06817.

Von Puhl, S. A. (1986), "Problems Associated with Chemical Dehydration of Naturally Produced CO_2," Proceedings 33rd Annual Southwestern Petroleum Short Course, pp. 360–368, Texas Tech University, Lubbock, TX (April 23–24).

Worley, S. (1967), "Super-Dehydration with Glycols," Proc. Gas Conditioning Conf., University of Oklahoma, Norman, OK.

Chapter 9

Gas Dehydration Using Solid Desiccants, Refrigeration and Brines

Chapter 8 outlined the commonly used dehydration processes and described TEG in detail. Chapter 9 discusses solid desiccants in depth, reviews refrigeration methods, summarizes calcium chloride, and finally proposes some process selection guidelines.

INTRODUCTION TO SOLID DESICCANTS

Dehydration with TEG occurs by absorption while solid desiccants dry by adsorption. Gas *absorption* is a *volume* or bulk-concentration phenomenon during which the absorbed-gas molecules condense at the gas-liquid interface, and then migrate into the bulk of the liquid phase. In contrast, *adsorption* is a *surface* phenomenon. All solid surfaces have some ability to adsorb, i.e., capture and hold vapors and liquids on their surfaces. Two types of gas-solid adsorption occur:

1. *Chemisorption* or chemical adsorption that involves specific chemical (or electrovalent) bonding of the gas molecules (or adsorbate) in a monolayer onto the solid-surface atoms.
2. *Physisorption* or physical adsorption that is caused by van der Waals (or non-specific, long-range) forces between the gas molecules and the surface thus forming multilayers of adsorbate on the surface.

Perhaps simplistically chemisorption resembles a chemical reaction between a particular gas and a specific surface. In contrast, physisorption is more like general condensation of gas on any surface. The term "sorption" is used for both chemi- and physi-sorption.

The *adsorption capacity* (lb adsorbate per 100 lb solid adsorbent) is determined by the particular gas being adsorbed, the nature of the solid desiccant, its effective surface area and internal porosity, the temperature, and the partial pressure of the *adsorbate gas*. Because the strength of the adsorption bond depends primarily on the natures of the gas and solid surfaces, adsorbents can separate mixtures by preferentially adsorbing specific components.

Solid desiccants are carefully manufactured to maximize adsorption. Solid-desiccant pellets are characterized by an internal porous structure which generates very large (e.g., 500 m^2/g) internal surface areas with very small radii of curvature (0.001–0.2 μ m). The equilibrium partial pressure of water vapor above such concave curved surfaces is much less than that above plane surfaces and therefore, dry desiccants exhibit a very great affinity for water. In fact, dry desiccants achieve capacities from 5 to 40% and can dry natural gas to <0.01 ppm of water or water dew points below $-150°F$.

Natural-gas dehydration using solid desiccants is described first. Then operating parameters, design methods, and troubleshooting techniques are discussed.

Description of Solid-Desiccant Process

A *typical two-tower* solid-desiccant dehydration unit is described and the influence of choosing "wet" inlet or "dry" outlet gas for desiccant regeneration is discussed. Then the characteristics of solid desiccants are summarized. Next the adsorption of water on desiccant particles and in fixed beds is reviewed. Then multicomponent adsorption in a fixed bed is treated very briefly as this topic is far more relevant for hydrocarbon recovery or "quick cycles." Finally, the individual components of a solid-desiccant dehydration unit are described.

Typical Two-Tower Units

Figure 9–1 shows a typical solid desiccant unit using "wet" or inlet gas for regeneration. To remove all liquid and solid impurities the feed stream first flows through an inlet separator or scrubber (not shown in Figure 9–1). The feed gas leaves the inlet scrubber (point A), flows through the regeneration flow control valve (B), and into the inlet header (C). Then an on-off switching valve (D) directs the wet gas downward through the adsorbing or "on line" tower (E). As wet gas contacts the solid-desiccant bed, water vapor is adsorbed until equilibrium is established between the water contents in the gas stream and on the solid desiccant particles. Generally, only a few seconds are required for complete adsorption or equilibrium to be attained. Dried natural gas leaves the bed, flows through the exit switching valve (F), and finally leaves the solid dehy unit via the dry gas outlet header (G).

If wet gas continues to flow into the adsorption bed, all of the solid desiccant eventually becomes saturated with water, and the wet gas can no longer be dried. Before bed saturation occurs, switching valves (D) and (F) must be closed and another bed used to dry the inlet gas. Therefore, while one bed (E) is "on line" or drying gas, the other bed (H) must be regenerated.

As shown in Figure 9–1, a 5 to 10% side stream (I) of feed gas is heated to 400–550°F in the regeneration heater

(J) and then directed into bed (H) via the switching valve (K). This hot regeneration gas heats up the bed, drives the water off the desiccant particles, and carries the resulting water vapor out of the bed, through switching valve (L), and into the regeneration cooler (M). When this regeneration gas is cooled, most of the water vapor is condensed out and removed in the regeneration separator (N). Finally, the regeneration gas is directed back to be dried. The hot regenerated bed must be cooled before it can be switched back to dry gas. Accordingly, cool "wet" inlet gas (C) is fed via switching valve (Q) into the hot regenerated bed until the bed is cooled.

Alternatively, *dry product gas* can be used for *regeneration* as shown in Figure 9–2. A compressor supplies the pressure boost (typically 40–50 psi) to recycle the dry regeneration gas. In this case the hot dry regeneration gas flows upward through the bed thus desorbed (or removed) water and any liquid contaminants never contact (or saturate) the solid desiccant particles at the bottom of the bed.

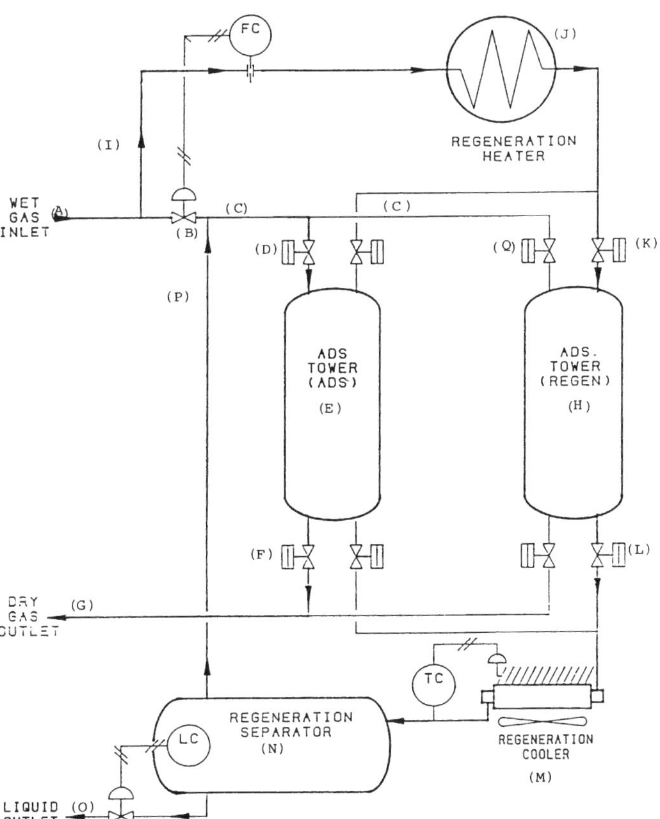

Figure 9–1. Conventional solid desiccant unit with wet gas regeneration (NATCO, 1987).

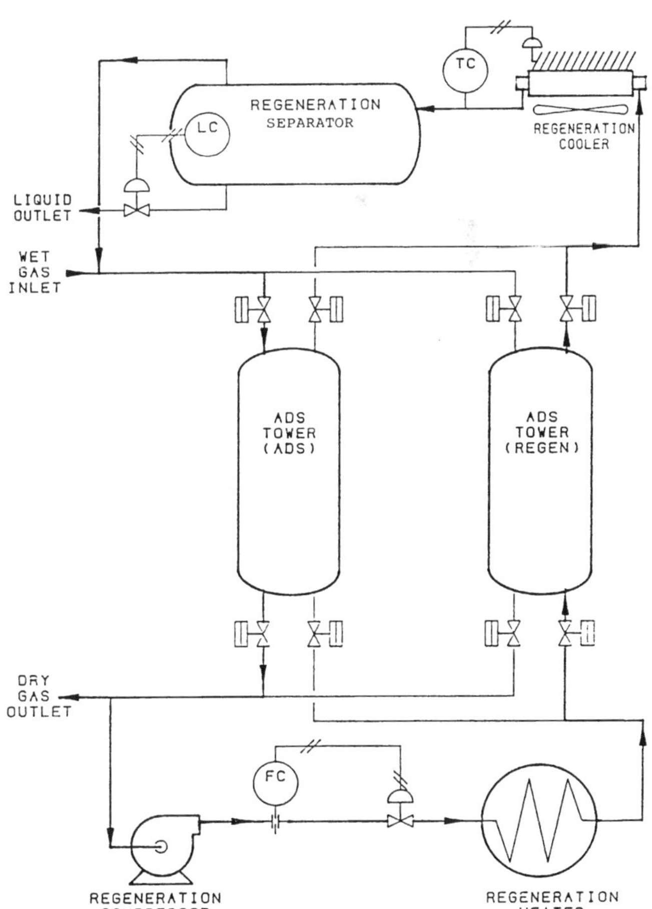

Figure 9–2. Low dew-point gas dehydrator with "dry" gas regeneration (NATCO, 1987).

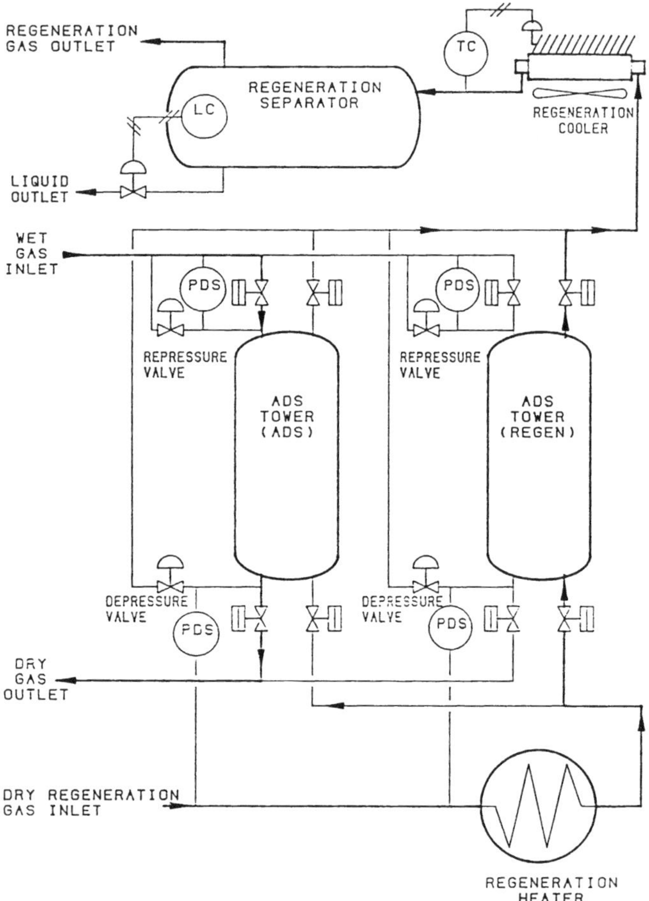

Figure 9–3. Low dew-point gas dehydrator with low-pressure dry gas regeneration (NATCO, 1987).

A separate source of dry gas such as *demethanizer overhead* gas also can be used for *regeneration*. Such regeneration gas is usually available at a lower pressure than the main "wet" gas pressure (typically 500–1200 psia). The required changes (or swings) in bed pressure are achieved using blowdown and repress valves as shown in Figure 9–3.

Solid Desiccants

The adsorbents most commonly used for drying oilfield fluids are silica gel, silica-based beads, activated alumina, activated bauxite, and molecular sieves. The term "gel" indicates that manufacture involves formation of a gelatinous precipitate that is then coagulated, washed, and dried to form solid, hard particles. These desiccants are now described briefly.

Silica Gel is essentially $SiO_2 \cdot nH_2O$—Davison type 03 is 99.71 weight percent SiO_2 with traces of Al_2O_3, CaO,

Na_2O, ZrO_2, TiO_2 and Fe_2O_3. Silica gel is a hard, rugged material with good attrition resistance characteristics and is available commercially as powdered, granular, and spherical beads of various sizes. The bead type is more susceptible than the granular form to breakage caused mostly by entrained liquid drops in the "wet" gas.

Silica-Base Beads (e.g., Mobil's Sorbead) consist of essentially 97% silica (SiO_2) and 3% alumina (Al_2O_3). Adsorption capacity is essentially the same as conventional silica gel, however, the bulk density and hence capacity per unit volume are somewhat greater.

Activated Alumina is a porous, amorphous, partially hydrated alumina (94% $Al_2 O_3$, 5.5% H_2O, .3% Na_2O, and .02% Fe_2O_3 for Alcoa's grade F–200).

Molecular Sieves are crystalline zeolites or metal aluminosilicates having a uniform, three-dimensional interconnecting structure of alumina and silica tetrahedra. These synthetic zeolite crystals are manufactured to contain interconnecting cavities of uniform size, separated by equally uniform pores or narrower openings. The silica and alumina tetrahedra form a sodalite cage or truncated octahedron having a silica or alumina tetrahedron at each point. Different crystal structures are obtained by varying the metal cation (Na, K, or Ca) and the alumina to silica ratio. Two types of structures are commonly used in gas processing.

In type A the sodalite cages form a simple cubic structure. When the metal cation is sodium the unit cell formula is $Na_{12}[(AlO_2)_{12} (SiO_2)_{12}] \cdot x\ H_2O$, the pores are 4.2 angstroms (A°) in diameter, and the mole sieve is called *4A*. Replacing Na with K results in type *3A* (pore size about 3 A°) and substituting Ca for Na yields type *5A* (or 5 A° pores).

In type X the sodalite cages are arranged in a tetrahedral structure. The sodium form, type *13X*, has a unit cell formula of $Na_{86}[(AlO_2)_{86} (SiO_2)_{106}] \cdot y\ H_2O$, and a pore size of 10 A°.

Table 9–1 summarizes some properties of these solid desiccants. These data are for information only and are <u>NOT</u> design recommendations. Consult the desiccant vendor or manufacturer for all design information.

Adsorption

When wet gas contacts solid desiccant particles, water is adsorbed until equilibrium is achieved. This equilibrium is usually described in terms of three variables: the *contact temperature* (°F, °C), the desiccant water content or *static* capacity (wt H_2O/wt dry desiccant), and the *water content* of the gas (water partial pressure—mmHg or water dew point—°F, °C). Common graphical presentations are those of constant temperature (or *isotherms*—Fig. 9–4), constant

Table 9–1 Properties of Solid Desiccants

Desiccant	Silica Gel	Activated Alumina	Sorbead	Molecular Sieves
Reference	Davidson (03)	Alcoa (F-200)	Kali-Chemie	Zeochem (1989)
Pore diam. (Angstroms)	10 to 90	15	20 to 25	3, 4, 5, 10
Bulk density (lb/ft^3)	45	44 to 48	40 to 49	43 to 47
Heat capacity (Btu/lb °F)	0.22	0.24	0.25	0.23
Min. dewpoint (°F)	−60 to −90	−60 to −90	−60 to −90	−100 to −300
Design capacity (wt %)	4 to 20	11 to 15	12 to 15	8 to 16
Regeneration stream temp (°F)	300 to 500	350 to 500	300 to 450	425 to 550
Heat of adsorption Btu/lb			1200	1800

water partial pressure (or *isobars*—Fig. 9–5) and constant desiccant water content (or *isosteres*—Fig. 9–6).

When wet gas flows into a fixed bed of regenerated solid desiccant particles, the result is *dynamic adsorption* (Fig. 9–7). Particles nearest the gas inlet or top soon become saturated with water—the desiccant water content is in equilibrium with the "wet" inlet gas. No more adsorption can occur in this saturated zone. All water is now adsorbed in the zone of particles below the top saturated zone. In this *mass transfer zone* (MTZ), the gas water content is reduced from the inlet concentration, C_o, to the value in equilibrium with the regenerated desiccant, C_s. Below this MTZ the desiccant "sees" only dried gas and therefore remains regenerated. Figure 9–7 shows how the MTZ moves down the bed and eventually reaches the bottom of the bed. Figure 9–7 also shows how the exit-gas water content remains constant at C_s until the MTZ reaches the bottom, and then "break through" occurs. If wet gas continues to

flow, the exit-gas water content continues to increase until it reaches the inlet value, C_o. At this point the entire bed is completely saturated with water and there is no MTZ in the bed.

Multicomponent Adsorption

Alumina, silica gel, and some molecular sieves adsorb other compounds in addition to water. Water is adsorbed preferentially, and the *order of preference* (in decreasing adsorbability) is water, methanol, hydrogen sulfide and mercaptans, carbon dioxide, hexanes and heavier hydrocarbons, pentanes, butanes, propane, ethane, and finally methane. Therefore, when wet inlet gas flows downward through a bed of regenerated desiccant, all adsorbable components are adsorbed at different rates and with varying degrees of intensity. Water is always adsorbed very rapidly by the top layers in the bed. The remaining

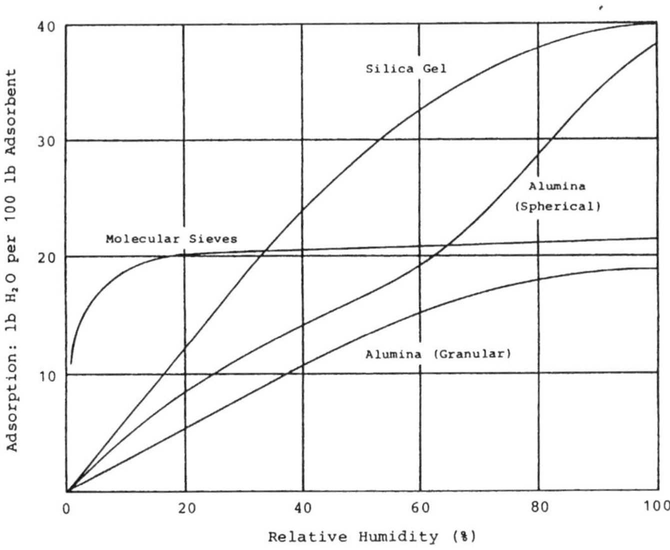

Figure 9–4. Water vapor isotherms (Kali-Chemie, 1987).

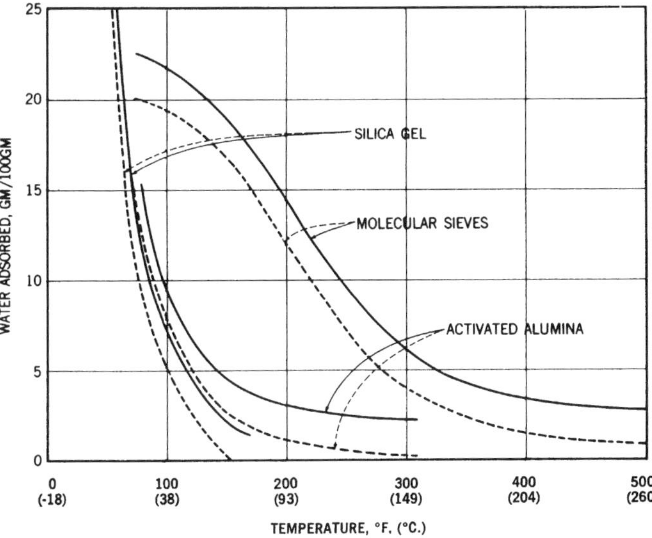

Figure 9–5. Water vapor adsorption isobars (UOP, 1979).

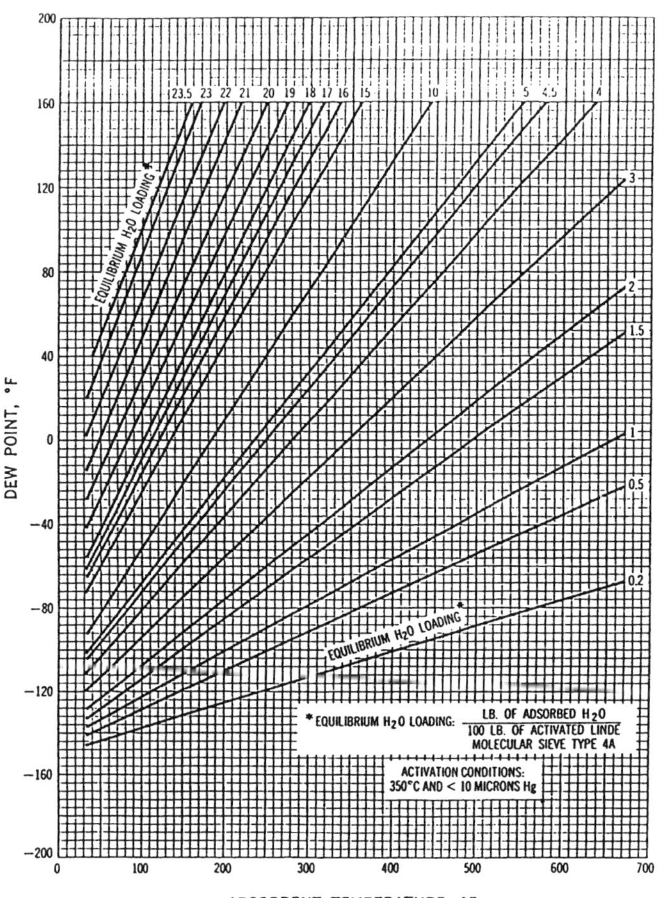

Figure 9–6. Molsiv® Adsorbent, type 4A water adsorption isostere (UOP, 1979).

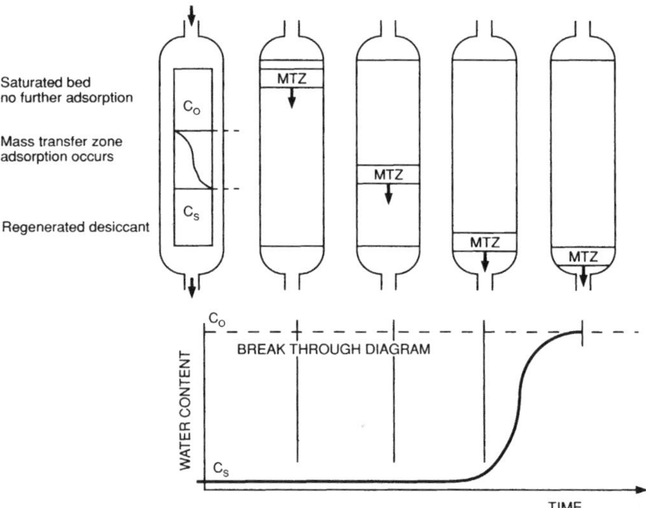

Figure 9–7. Adsorption of water in a fixed bed.

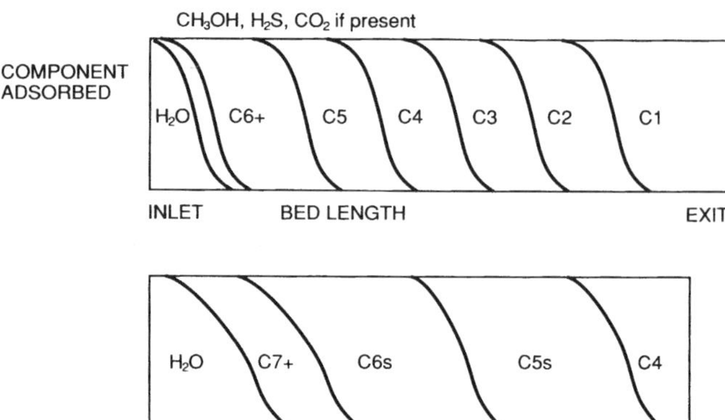

Figure 9–8. Multicomponent adsorption zones.

natural gas components (H_2S, CO_2, C6+, C5s, C4s, C3, C2, and C1) are adsorbed in lower layers of the bed as is shown in Figure 9–8. Another viewpoint is that the desiccant bed is essentially saturated with methane initially. Then the other less prevalent but more intensely adsorbed natural-gas components displace methane (and other less adsorbable components), thus forming the component zones shown in Figure 9–8a. If the adsorption cycle is stopped after 20 to 30 minutes, then the heavier hydrocarbons (C4s, C5s, C6+) can be recovered (Fig. 9–8b). In dehydration cycles of eight hours or longer, water essentially replaces the other natural-gas components. However, molecular sieve type 13X is used to simultaneously sweeten and dehydrate natural gas, i.e., remove H_2S, CO_2, and H_2O.

Individual Equipment Components

All fabrication should meet *ASME Code for Pressure Vessels* (Section 8, Division 1) and be stamped appropriately. If the feed gas contains sufficient H_2S then *NACE MR–01–75 specifications* should be followed.

Inlet Separator. A *separator is absolutely necessary* to protect the desiccant from impurities such as free water, salt water, compressor oils, absorption oil, liquid hydrocarbons, paraffins, corrosion inhibitors, glycol, amines, pipeline rouges, iron sulfide, iron oxide, frac sands, drilling muds, pipeline scale, elemental sulfur, etc. Water and other free liquids will ''hammer'' the desiccant and cause breaking and powdering. Non-volatile liquids coat the desiccant and/or block pores, thus reducing capacity. Solids can plug the bed, and the resulting increased pressure drop can crush the desiccant particles.

The separator should be large enough to handle the most demanding conditions and should have adequate surge capacity for free liquids and solids. A high-liquid-level alarm and a high-liquid-level shutdown are desirable

(Ballard, 1983). A filter-type separator or coalescer should be used when downstream of an amine unit or compressor. All droplets larger than 10 μm and 99% of droplets .5 to 10 μm should be removed (Barrow, 1982).

Adsorption Tower. As shown in Figure 9–9, the adsorption tower consists of a bed support, gas stream distributors at both ends, adsorbent loading and removal connections, and a sample/moisture probe. The bed support must withstand both dead load (weight of desiccant) and live load (flowing pressure drop). One type of bed support is a stainless steel or monel screen (with openings 10 meshes smaller than the desiccant particles) supported horizontally by I-beams and a welded ring. The screen should be securely fastened while allowing for thermal expansion and

contraction (Ballard, 1983). The annular space between screen and vessel wall should be sealed to prevent desiccant loss. Bed support balls normally consist of a 2– to 3–in. layer of ¼–in. balls on top of a 2– to 3–in. lower layer of ½–in. balls.

At the tower top the inlet gas should be introduced radially at low velocity. A screen-wrapped slotted pipe or perforated basket is recommended (Ballard, 1983). Also at the top, the desiccant bed should be protected by a 4– to 6–in. layer of ½– to 2–in. balls resting on a retainer screen with ½–in. openings. This improves distribution of the inlet gas and prevents desiccant damage due to swirling.

The adsorption tower is insulated either externally or internally. External insulation is much easier, but internal insulation can save up to 30% of the total regeneration heat load. Properly applied and cured castable refractory lining can be used internally. However, any liner cracks will allow wet inlet gas to bypass the bed.

Regeneration Heater. Any type of heating (salt bath, direct fired, hot oil, steam, compressor exhaust gases) is acceptable. Small units (8 MMBtu/hr or less) and/or unattended facilities tend to use indirect-fired, salt-bath heaters for safety reasons. During the cooling cycle, a three-way valve must be used to direct the regeneration gas around the heater. Large and/or attended units often use direct fired heaters. Considerable damage can result when a tube fails because the regeneration gas in the tube can be at full line pressure. Undersizing the heater results in incomplete regeneration and thus reduces the performance of the entire unit.

Regeneration Cooler. While air, water, or natural gas can be used, ambient air is generally used to cool the regeneration stream to within 15–20°F of the air temperature. In cold climates overcooling of the regeneration gas should be avoided because free liquids (water and/or HC) may be formed when overcooled regeneration gas is recycled and mixed with the wet feed gas. Locate the cooler away from lights and dirt roads to prevent insects and dust from coating the outside of the tubes (Ballard, 1983). The cooler should have an even number of passes so the inlet and exit connections are at the same end, and the return end is free for expansion.

Regeneration Separator. A horizontal three-phase separator suitably sized to accommodate any surges is recommened. By rule of thumb the dump valve should be sized for four to five times the average liquid-removal rate. Desiccant dust and/or liquid hydrocarbons can plug the dump line. Frequent pH tests on the dumped water help to identify potential corrosion.

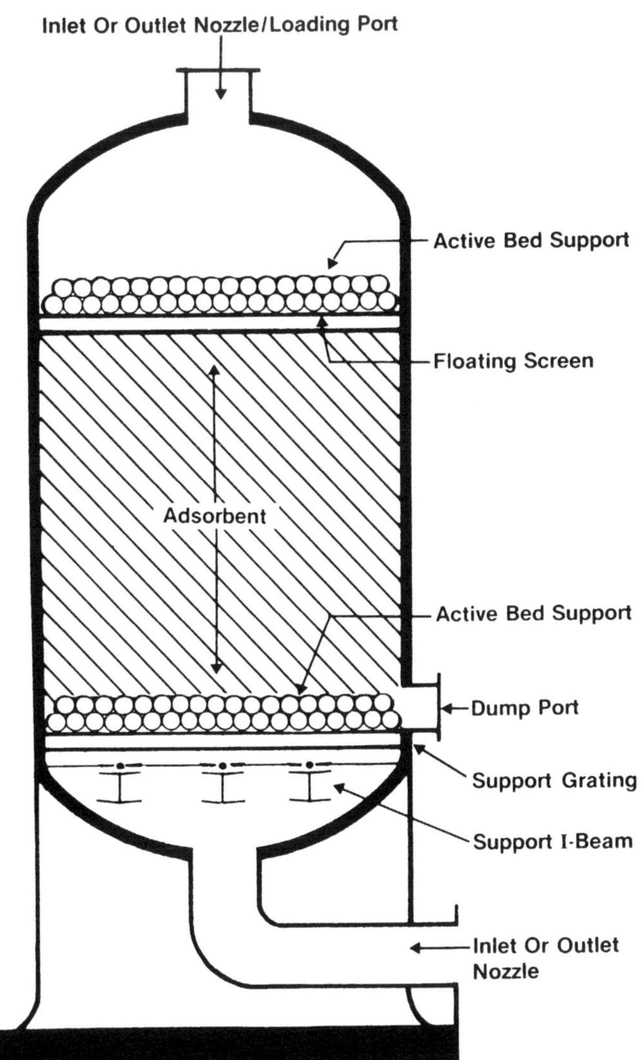

Figure 9–9.   Adsorption tower (Alcoa, 1986).

Switching Valves. Good quality valves will minimize operating problems. The most difficult service occurs where the valves "see" hot (600°F) regeneration gas on one side and ambient (100°F) inlet gas on the other. Usually three-way valves have more problems than two-way valves (Ballard, 1983). Because low leakage is mandatory, ball valves with metal seats are recommended (NATCO, 1987).

Piping. Regeneration piping is subjected to severe temperature cycle service. Follow ANSI B31.3 Piping Code when designing expansion loops and anchors (fixed and "floating"). Insulate any hot piping for personnel protection and heat conservation.

Instrumentation. Switching the solid-desiccant bed from adsorption to regeneration, and vice versa, can be controlled using adsorption time, regeneration heating and cooling times, gas moisture content, regeneration temperature, or any combination. The switching valves can be operated manually or automatically.

The following data should be recorded regularly:

- Inlet-gas flow rate, temperature, pressure
- Exit-gas water content
- Regeneration gas flow rate, pressure
- Regeneration tower inlet and outlet temperature
- Regeneration cooler outlet temperature

Operation of Solid-Desiccant Units

Factors influencing choice of desiccant are presented first. Then the adsorption or dehydration process is described with emphasis on wet-gas flow direction, flow velocity, and resulting pressure drop. Next, desiccant regeneration is discussed, and the influences of regeneration-gas flow direction, flow rate, and water content are noted. Finally, operating topics such as desiccant installation, start up, bed switching, operating data, and energy conservation are summarized.

Desiccant Choice

Desiccant choice is far from routine. Careful consideration of the following factors is required:

1. Inlet-gas pressure, temperature, and composition
2. Required outlet water dew point
3. Hydrocarbon recovery requirement
4. Capital and operating costs

The most relevant parameters are reviewed first and then the more common desiccant choices are summarized.

Desiccant Cost. Currently, silica gel and molecular sieves are about equally expensive while aluminas cost approximately half as much.

Desiccant Capacities vary drastically with temperature and gas water content as is shown in Figure 9–4, 9–5, 9–6, 9–10, and Table 9–2.

Exit Gas Water Dew Point. Activated alumina and silica gel can achieve water dew points of −60 to −90°F, while mole sieves are capable of −150 to −300°F.

Desiccant Life varies from six months to over four years. Mole sieves are more susceptible to liquid water, corrosion inhibitors, and heavy hydrocarbons than some of the other desiccants. Activated alumina is less susceptible to breakage than silica gel.

Regeneration. Mole sieves require up to 16% more total heat load per pound water removed for effective regeneration. Both activated alumina and silica gel adsorb heavier hydrocarbons from natural gas, but it is more difficult to regenerate these hydrocarbons from activated alumina than from silica gel.

Silica Gel is the most widely used solid desiccant for normal dehydration of natural gas to pipeline specifications—1 to 7 lb H_2O/MMscf. It also is excellent for hydrocarbon recovery units (or quick cycle units).

Activated Alumina is most economical within its range of application.

Molecular Sieves are the overwhelming choice for natural-gas dehydration to cryogenic processing standards—<1 ppm water or −150°F dew point. Molecular sieves also are excellent for simultaneous H_2S and/or CO_2 removal and for dehydration of CO_2 streams and hydrocarbon liquids.

Adsorption or Dehydration Step

Downward flow is recommended for gas dehydration and upward flow for hydrocarbon-liquid dehydration. *Downward gas flow* is preferred because (NATCO, 1987):

1. Increasing the upward flow velocity will eventually expand and then fluidize the bed. Any unsettling or movement of the desiccant particles can erode and/or crack then.
2. Downward flow permits higher velocities before the resulting pressure drop crushes the desiccant particles. Of course higher velocities also allow a small diameter, cheaper vessel.
3. Upstream removal of entrained water and/or free liquids is not always perfect, so sooner or later liquid contaminants will contact the desiccant causing oversaturation, cracking, and breakage. It is better to

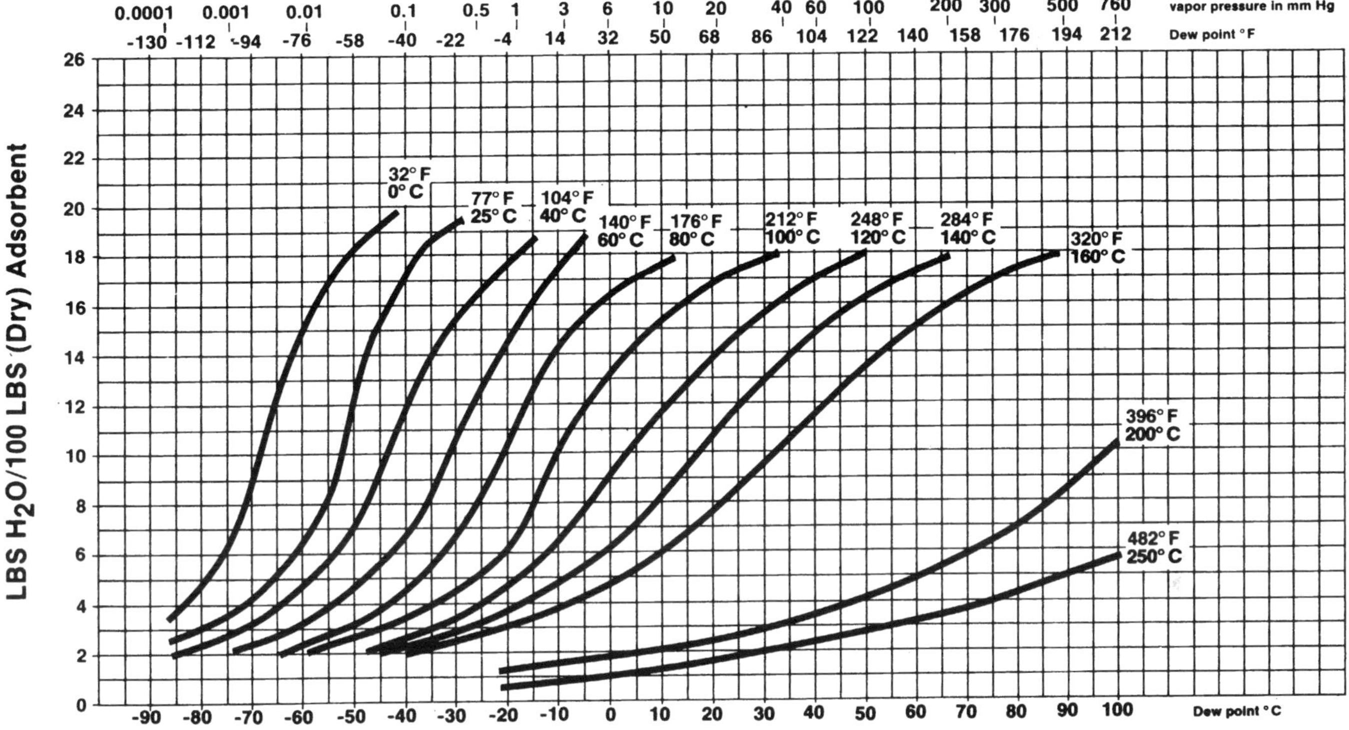

Figure 9–10. Water vapor isotherms of Zeochem molecular sieve type 4A (Zeochem, 1983).

have the top pellets lose mechanical strength because desiccant at the bottom carries the entire bed weight.

Hydrocarbon liquids frequently contain some noncondensed (gaseous) components. *Upward liquid flow* allows any gas bubbles to pass through the dehydration bed. Downward liquid flow could result in accumulation of gas at the top of the tower, thus progressively reducing the amount of desiccant exposed to liquid (NATCO, 1987).

Optimum gas dehydration occurs when the downward gas velocity is large enough to prevent channeling and low enough not to damage the desiccant particles. The pressure drop through a desiccant bed may be estimated by (Grace, 1988):

$$\Delta P/L = B \, \mu \, V + C \, \rho \, V^2 \tag{9–1}$$

where ΔP = pressure drop across bed (psi)
L = bed depth or height (ft)
V = superficial gas velocity (ft/min)
μ = gas viscosity (cp)
ρ = fluid density (lb/ft^3)
B,C = experimentally determined constants given in Table 9–3.

The following rule of thumb may also be used to estimate the maximum gas velocity, V_{max}

$$V_{max} = A/(\rho)^{0.5} \tag{9–2}$$

where A = 55 for ⅛″ pellets or 4–8 mesh
= 40 for ¹⁄₁₆″ pellets or 8–12 mesh.

Table 9–2 Capacities of Solid Desiccants (NATCO, 1987)

Relative Humidity at 25°C (%)	Adsorptive Capacity (lb H₂O/100 lb desiccant)		
	Alumina	Silica Gel	Mol Sieve
2			15
10	7	6	20
20	11	10	22
40	16	22	22
60	22	35	23
80	37	40	23
100	42	42	23

Table 9–3 Pressure-Drop Parameters. (Grace, 1988)

Particle Type	B (psi-min/cp ft²)	C (psi-min²/lbm)
⅛″ or 4–8 mesh	0.0560	0.0000889
⅛″ pellet (extrudate)	0.0722	0.0001240
¹⁄₁₆″ or 8–12 mesh	0.152	0.0001360
¹⁄₁₆″ pellet (extrudate)	0.238	0.0002100

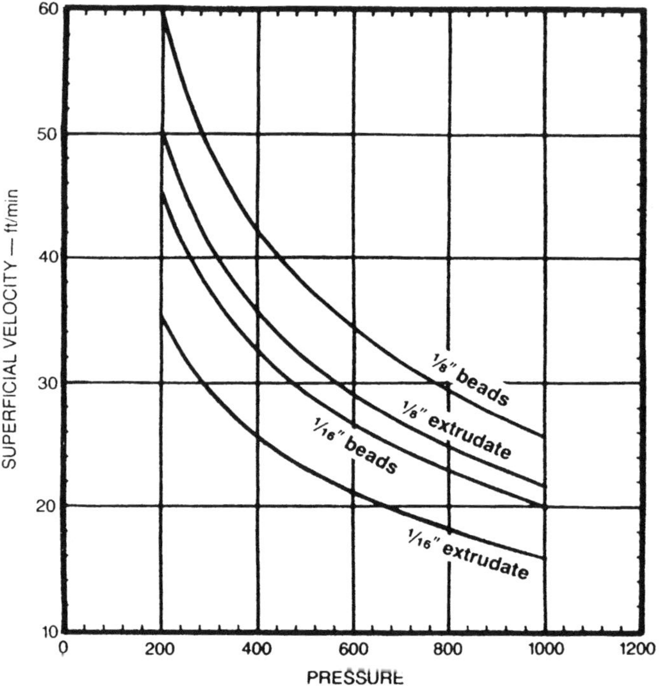

Figure 9–11.   Allowable velocity for mole sieve dehydration (GPSA, 1987, p. 20–22).

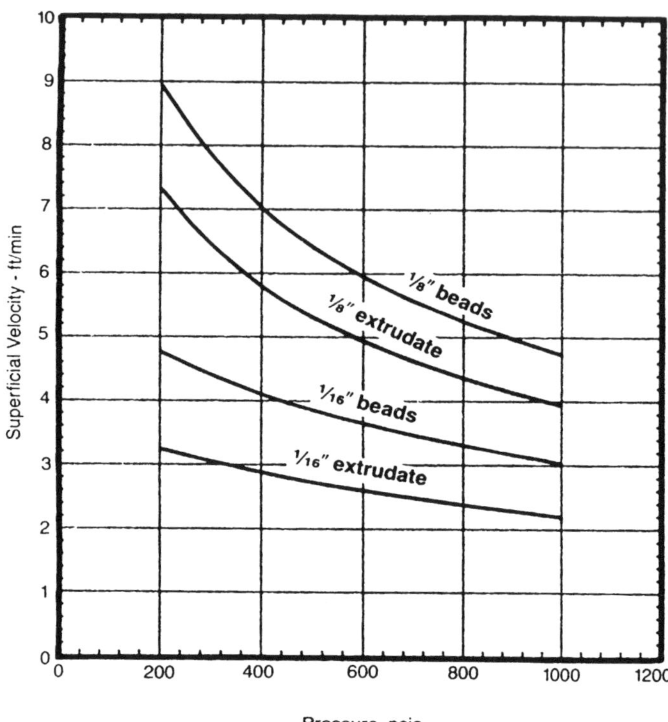

Figure 9–12.   Minimum regeneration velocity for mole sieve dehydration (GPSA, 1987, p. 20–24).

Figure 9–11 (GPSA, 1987) recommends gas velocities that allow an economically-small bed size at a reasonable pressure drop of 0.33 psi/ft. The design pressure drop across the entire bed should be about 5 psi (GPSA, 1987); and values higher than 8 psi are not recommended (Barrow, 1982; GPSA, 1987).

Regeneration

Figure 9–12 recommends minimum regeneration hot-gas velocities that will inhibit channeling. An equivalent criterion is that the hot regeneration gas must generate a pressure drop of 0.01 psi/ft (Barrow, 1982).

Figure 9–13 shows the temperature profiles typically found in a regeneration cycle. At time, t_o, regeneration gas starts to flow through the heater. The heating cycle ends at t_1 when the regeneration gas leaving the tower has reached the desired temperature, T_3. Now, the cooling cycle starts at t_1 and stops at t_2 when the regeneration gas exiting tower is sufficently cool, T_1. Normally T_4 is 50–80°F hotter than T_3, and T_1 is 20–50°F higher than T_o (NATCO, 1987). Usually, the heating time $(t_1 - t_o)$ is 55 to 65% of the entire regeneration cycle $(t_2 - t_o)$. Figure 9–13 also shows typical temperatures and times for regenerating molecular sieves on an eight–hour cycle.

The *regeneration stream* is generally 5–10% of the feed stream. The flow direction of the hot regeneration gas influences the required flow rate, the dew point of the effluent gas, and desiccant life. If the heating gas flow is downward (concurrent with gas flow during dehydration), then all the water and other contaminants must be driven through the entire bed. This can require longer regeneration times and cause additional desiccant contamination, especially at the bottom of the bed.

Therefore, *upward flow* (countercurrent to the downward gas flow during the drying cycle) is preferred. During such upward regeneration, steam, formed from water adsorbed in the lower portions of the bed, can help strip contaminants (e.g. glycol, amine, lube oil, etc.) from the top of the bed.

The flow direction of the cooling gas depends on the water content of the gas. If inlet "wet" gas is used, the flow direction is downward, thus resulting in some "pre-loading" of the desiccant bed with water, especially at the top. More complete desiccant regeneration is obtained using upward flow of dry gas for cooling.

Figure 9–14 compares regeneration with wet and dry gas. When wet gas is used (Fig. 9–14a) the line AB represents the isothermal drying step in which the desiccant loading increases from 0.2 weight percent (point A) to saturated (point B). During the heating cycle, BC, bed regeneration relies on increased temperature at constant gas

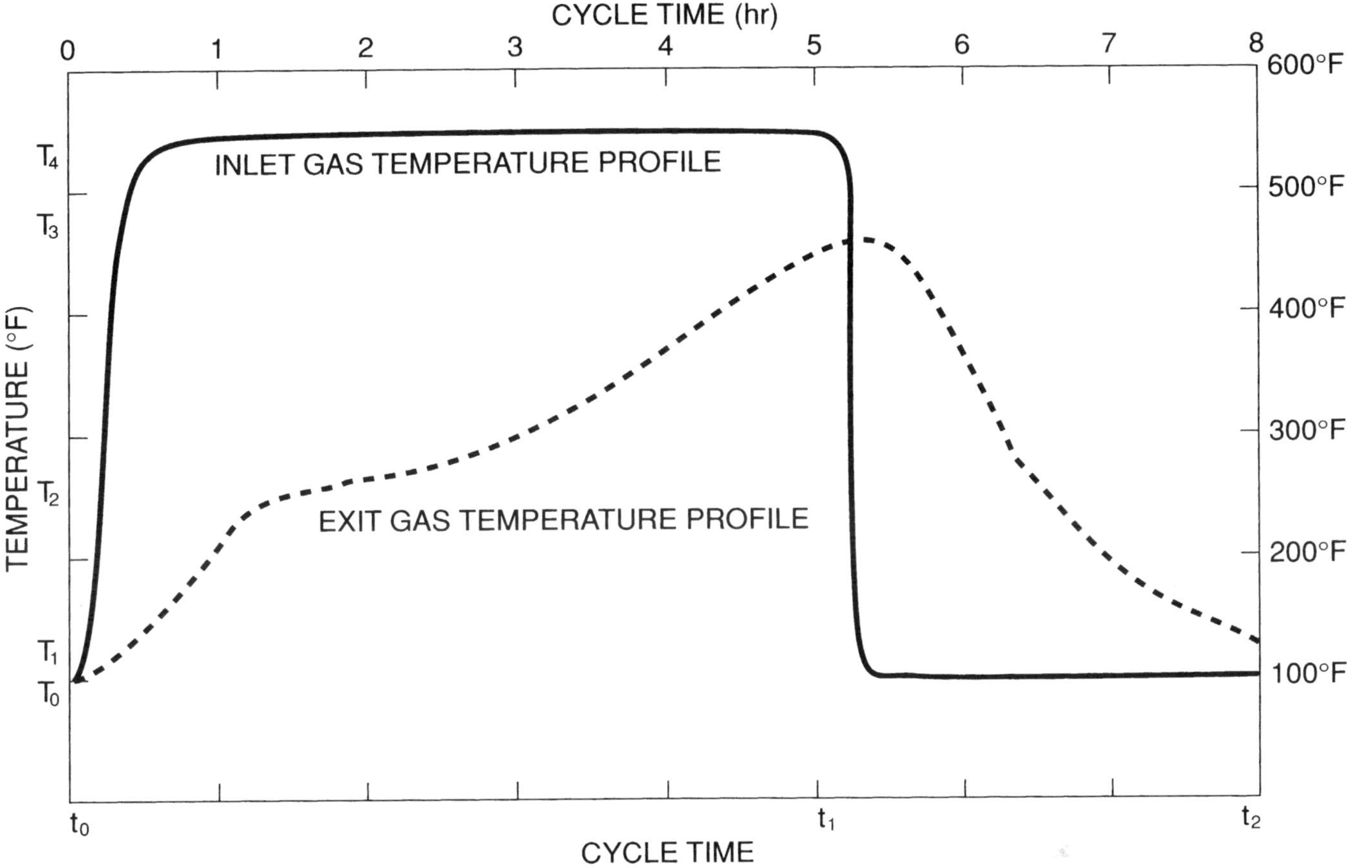

Figure 9–13.   Temperature profiles typically found in regeneration cycle.

dew point (100°F) for water removal. During the cooling step, the overall bed loading is assumed to remain unchanged (at 0.2 weight percent) as the temperature drops from 400°F (point C) to 100°F (point A). Notice that even though the bed is heated to 400°F during regeneration the minimum effluent gas dew point is −38°F (point A).

Figure 9–14b shows the advantages of using dry gas for regeneration. Again, line AB represents isothermal loading of the bed during dehydration. However, with dry gas regeneration, water removal (BC) occurs by both increased temperature (100° F to 400°F) and decreased gas dew point (100° F to −20°F). As before, regeneration cooling (CB) occurs at constant bed loading (0.003 weight percent). Note that using dry gas (dew point −20°F) results in a predicted gas dew point as low as −104°F (point A).

Operations

Desiccant installation, start up, bed switching, daily recordkeeping, and energy conservation are discussed in that order.

Desiccant Installation. Desiccant should be loaded carefully and uniformly into the vessel using a cloth or plastic sock that reaches from the top loading nozzle to the bed supports at the bottom. The desiccant normally settles 5 to 7% during the first few weeks of operation. If desired, the vessel then can be topped with additional desiccant, and then the top screen and support balls can be installed (Ballard, 1983).

Start Up. Good dry-out is essential. Hydrostatic testing makes dry-out more difficult, especially if internal insulation is used. First purge and drain all vessels, dead-ended piping, low spots, heat exchanger shells, etc. Methanol can be helpful.

Switching. If low pressure regeneration gas is used, never repressure faster than 50 psi/min or depressure faster than 50 psi/min. Notice that the repressure and depressure valves shown in Figure 9–3 are positioned so as not to lift the bed.

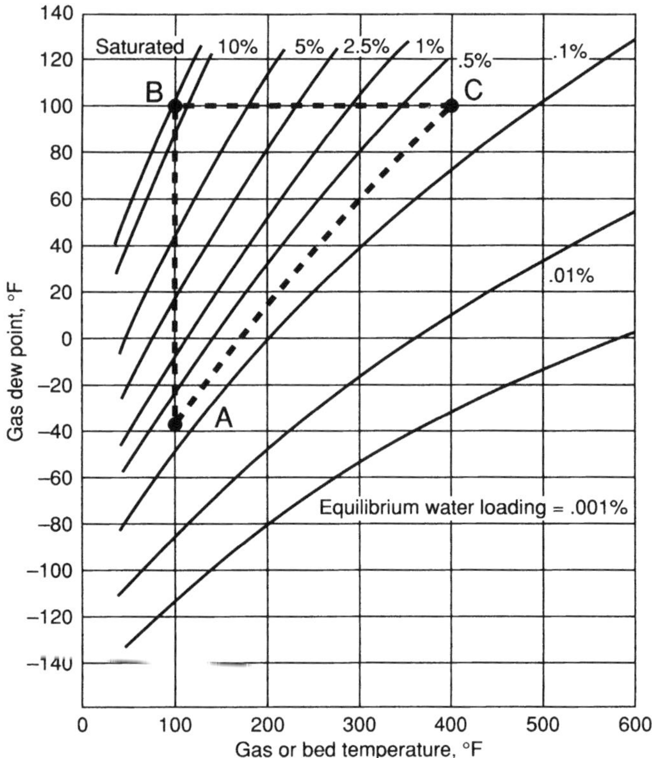

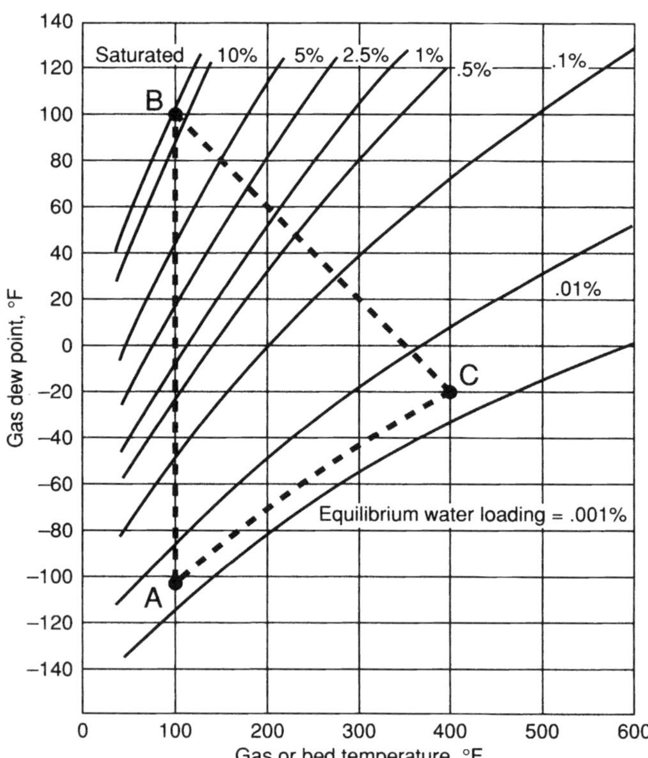

Figure 9–14.   Regeneration of F-200 activated alumina (Alcoa).

Operating Data. Adequate daily records are essential for troubleshooting. Cummings (1979) recommends taking the following data:

1. Feed gas flow rate, pressure, and temperature at least daily.
2. Complete feed-gas analysis once a month.
3. Outlet gas humidity continuously if unit feeds a cryogenic plant and at least once a shift if meeting pipeline specifications.
4. Temperature profiles of the gas entering and leaving the regeneration tower for both heating and cooling.
5. The towers should be equipped with pressure taps so that the pressure drop across the desiccant bed can be monitored as required.

Energy Conservation. Barrow and Veldman (1988) estimate that regeneration usually requires about 6500 Btu/lb water removed. They suggest the following methods for saving energy:

1. Use waste heat such as hot exhaust from the gas turbines that drive the pressure-boosting compressors. Safety considerations and operating problems can be involved.
2. The most popular cycle time is eight hours, but larger beds with 10–12 hour cycles are optimum.
3. Internal insulation of the solid-desiccant bed can save 20–30% of the required energy. A two- to three-year payback time is possible.
4. Use a compressor to recycle the regeneration gas (Fig. 9–3). This requires less energy than dropping the entire feed gas stream pressure across a control valve (valve B, Fig. 9–1) in order to drive a slip stream through the regeneration loop.
5. Desiccant capacity does decrease with use, and design is based on the relatively constant one-year to four-year-old value. Therefore, when the desiccant is new, the dehydration cycle can and should be lengthened. Installing a sample probe near the bottom of the bed permits operation at the maximum cycle time.

Oliker and Hong (1985) describe the dehydration energy savings that can be obtained by computer control. Thermocouples mounted along the tower wall are used to optimize both the drying and regeneration cycles. Maximum desiccant loading during drying and minimum regeneration heat can be realized.

Design of Solid-Desiccant Units

A rigorous mathematical simulation of adsorption in a fixed bed results in non-linear, coupled partial differential equations that are extremely difficult, if not impossible, to solve. The zone concept is, therefore, used to model the adsorption bed and design approaches are based on experimental experience obtained by desiccant vendors (e.g., Grace, 1988; Union Carbide, 1988; Zeochem, 1989) and equipment manufacturers (e.g., NATCO, 1987). Design considerations, required information, and a detailed design example are now presented.

Design Considerations

Special consideration is required if any of the following unfavorable conditions exist (Alcoa, 1983):

- Inlet feed gas temperature is above 100°F.
- Inlet feed gas pressure is below 100 psia.
- Inlet feed gas contains appreciable CO_2, H_2S, or more than 0.3 weight percent C7+.
- Feed gas contact time with desiccant is <10 sec.
- Feed gas superficial velocity exceeds recommended maximum value.
- Regeneration gas velocity is less than 10 ft/min.
- Wet gas is used for regeneration and a product dew point of less than −45°F is required.

Example 9–1.

Design Data: Gas flow rate   50 MMscfd
                                (14.696 psia, 60°F)
               Gravity   0.7
               Pressure   600 psig
               Temperature   100°F
               atm pressure   14.4 psia (elevation 600 ft)
               See Table 9–4 for a complete list of required information.

Design   Use molecular sieves, Type 4A size ⅛-in.
Criteria: extruded cylinders
               dynamic capacity at 75° = 13 weight percent
               bulk density =   44 lb/ft^3
               specific heat = .25 Btu/lb-°F

               use SA–515–70 steel plate

               spot examine welds

               CA = ¹⁄₁₆ in.

Adsorption Vessel Specifications:

1. $\dfrac{\text{lb mol gas}}{\text{min}} = \dfrac{\text{MMscf}}{\text{day}} \dfrac{\text{day}}{\text{min}} \dfrac{\text{lb mol}}{\text{MMscf}}$

$= \dfrac{50}{1} \dfrac{1}{1440} \dfrac{10^6}{379.5} = 91.5$

2. Actual gas volumetric flow, $\dot{v} = \dot{n} \, Z \, R \, T/P$

$\dfrac{\text{ft}^3}{\text{min}} = \dfrac{\text{lb mol}}{\text{min}} \dfrac{\text{psia.ft}^3}{\text{lb mol °R}}$ (°R)/psia

$= \dfrac{(91.5)(.89)(10.732)(100 + 460)}{600 + 14.4}$

acfm = 796.6

3. Establish maximum allowable superficial gas velocity, V_{max} such that pressure drop does not exceed 0.35 psi/ft.

$\mu = 0.014$ cp GPSA (1987), Figure 23–26
$\rho = P\, MW/Z\, R\, T$
$= (614.4)(.7 \times 28.96)/(.89)(10.732)(560)$
$= 2.33$ lb/ft^3

first estimate $V_{max} = 29$ ft/min (Fig. 9–11) check pressure drop using Equation 9–1 and Table 9–3.

$\Delta P/L = B\, \mu\, V + C\, \rho\, V^2$
$= (.0722)(.014)(29) + (.000124)(2.33)(29)^2$
$= 0.029 + 0.249 = 0.27$ psi/ft   low?

Table 9–4   Information Required to Design an Adsorption Dehydration Plant (Harrell 1988)

Maximum design working pressure	______psig @ ___°F
Maximum inlet gas flow rate	______MMscfd
Minimum inlet gas flow rate	______MMscfd
Maximum inlet gas temperature	______°F
Minimum inlet gas temperature	______°F
Maximum inlet gas pressure	______psig
Minimum inlet gas pressure	______psig
Gas specific gravity (see *gas analysis)	______(Air = 1.0)
**Inlet gas water content	______lb/MMscf
***Inlet gas dew point	______°F
Outlet gas water content or dew point	______lb/MMscf or °F
Outlet gas hydrocarbon dew point	______°F @______psig
Utilities available	______(Electricity, Fuel Gas, Instrumentation Air)
Site elevation	______Feed above sea level
Ambient temperatures	______°F (maximum summer)
	______°F (minimum winter)
Design wind velocity with wind load rating of	______MPH ______lb/ft^2

* Preferably gas analysis should provide vol percent N_2, CO_2, H_2S, C1, C2, C3, iC4, nC4, iC5, nC5 and C6+ (characterized by MW and SG). For sweet gases the specific gravity may suffice, however, for sour gases the CO_2 and/or H_2S content is required.
** If not at saturated inlet gas temperature and pressure conditions.
*** A calibrated hygrometer will be used to establish dew-point temperatures. Water content shall be determined by IGT Research Bulletin #8, "Equilibrium Moisture Content of Natural Gases."

4. Estimate desiccant vessel internal diameter, ID.

$$\text{Vessel area} = (\text{acfm})/(V_{max}) = (796.6)/(29)$$
$$= 27.5 \text{ ft}^2$$
$$\text{vessel ID} = 5.91 \text{ ft}$$

Check if ID = 5.5 ft is satisfactory
$$V = (796.6)/(.7854)(5.5)^2$$
$$= 33.5 \text{ ft/min}\quad\text{high?}$$
$$V_{max} = 55/(P)^{.5} = (55)/(2.33)^{.5}$$
$$= 36.0 \text{ ft/min}$$
$$\Delta P/L = B\,\mu\,V + C\,\rho\,V^2$$
$$= (.0722)(.014)(33.5)$$
$$+ (.000124)(2.33)(33.5)^2$$
$$= 0.034 + 0.325 = 0.36 \text{ psi/ft}\quad\text{high}$$

but okay

Set vessel ID = 5.5 ft.

5. Estimate water loading for 8 hr cycle.
Inlet gas contains 88 lb H_2O/MMscf   (Fig. 4–6)

$$\text{lb } H_2O/\text{cycle} = \frac{\text{lb } H_2O}{\text{MMscf}}\,\frac{\text{MMscf}}{\text{day}}\,\frac{\text{day}}{\text{hr}}\,\frac{\text{hr}}{\text{cycle}}$$
$$= (88)(50)(1/24)(8) = 1467$$

6. Estimate weight of desiccant required in vessel, m_d.

$$\frac{\text{dynamic}}{\text{capacity } 100°F} = \frac{\text{dynamic}}{\text{capacity } 75°F}\,(C_T)\,C_{ss})$$
$$= (13.0)(.93)(1) = 12.1 \text{ lb } H_2O/100 \text{ lb}$$
sieve

Note: C_T obtained from Figure 9–15
C_{ss} obtained from Figure 9–16

$$\frac{\text{lb sieve req'd}}{\text{saturation zone}} = \frac{\text{lb } H_2O \text{ loaded}}{\text{8 hr cycle}}\,\frac{\text{lb mol sieve}}{\text{lb } H_2O}$$
$$= (1467)(100/12.1) = 12,120$$
$$L_{MTZ} = (V/35)^{0.3}\,(Z)$$
$$= (33.5/35)^{0.3}\,(1.70) = 1.7 \text{ ft}$$
$$\frac{\text{lb sieve req'd}}{\text{for MTZ}} = (\pi/4)(\text{ID})^2\,(L_{MTZ})(\rho_b)$$
$$= (.7854)(5.5)^2\,(1.7)(44) = 1777$$

Total desiccant req'd, m_d = 12120 + 1777 = 13900 lb

7. Estimate vessel height, h.

$$(\pi/4)(\text{ID})^2\,(L)(\rho_b) = m_d$$
$$(.7854)(5.5)^2\,(L)(44) = 13,900$$
$$\text{bed length, } L, = 13.3 \text{ ft}$$
$$\text{vessel height} = L + \text{ID} = 19.0 \text{ ft}$$

8. Check total flow pressure drop across tower.

$$\Delta P = (\Delta P/L)(L)$$
$$= (.36)(13.3) = 4.8 \text{ psi}$$

Regeneration Heating and Cooling Requirements:

1. Adsorption vessel wall thickness is estimated using
ASME Pressure Vessel Code, Section 8, Division 1.

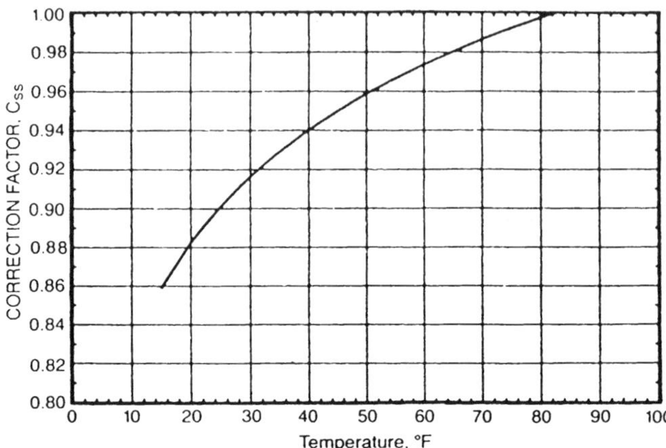

Figure 9–15.   Mole sieve capacity correction for unsaturated inlet gas (GPSA, 1987, Fig. 20–44).

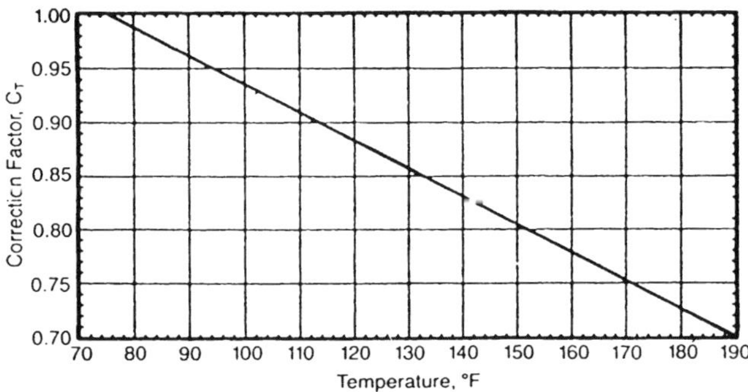

Figure 9–16.   Mole sieve capacity correction for temperature (GPSA, 1987, Fig. 20–45).

$$t_m = P_{DP}\,(\text{ID}/2)/(SE - 0.6\,P_{DP}) + CA$$
$$t_m = \text{wall thickness (in.)}$$
$$P_{DP} = \text{design pressure (700 psig)}$$
$$\text{ID} = \text{vessel inside diameter (66 in.)}$$
$$S = \text{maximum allowable stress in steel}$$
$$\quad\text{(17500 for SA–516–70 plate)}$$
$$E = \text{weld efficiency (.85 for spot-examined)}$$
$$CA = \text{corrosion allowance (1/16 in.)}$$
$$t_m = (700)(66/2)/[(17500)(.85) - 0.6(700)] + .063$$
$$= 1.598 + .063 = 1.661 \text{ in.}$$
set t_m = 1.75 in.

2. Adsorption Vessel Weight, m_v

$$\text{vessel weight, } m_v = [\pi(\text{OD})h + 2(1.09)(\text{OD})^2]\,t_m\,\rho_s$$
where $\pi(\text{OD})h$ = outside surface area of cylindrical shell
$1.09(\text{OD})^2$ = surface area of 2:1 ellipsoidal head
ρ_s = density of steel (490 lb/ft^3)
$\text{OD} = \text{ID} + 2t = 69.5$ in. = 5.79 ft
$h = L + \text{ID} = 13.3 + 5.5 = 18.8$ ft
$$m_v = [\pi(5.79)(18.8) + 2(1.09)(5.79)^2]$$
$$\times (1.75/12)(490)]$$
$$= 29700 \text{ lb.}$$

3. Estimate heat required for regeneration.
   a. Heating water to 250°F

   $$Q_{hw} = m_w \, C_{pw} \, (250 - 100)$$
   $$= (1467)(1.0)(250 - 100) = 220000 \text{ Btu}$$

   b. Vaporizing water

   $$Q_{vw} = (m_w)(1800)$$
   $$= (1467)(1800) = 2640000 \text{ Btu}$$

   c. Heating desiccant bed to 500°F

   $$Q_{hd} = (m_d)(C_{pd})(500 - 100)$$
   $$= (13,900)(.25)(500 - 100) = 1390000 \text{ Btu}$$

   d. Heating vessel

   $$Q_{hv} = (m_v)(C_{ps})(500 - 100)$$
   $$= (29700)(.12)(500 - 100) = 1426000 \text{ Btu}$$

   e. Total regeneration heat

   $$Q_{rh} = (220 + 2640 + 1390 + 1426) \text{ M Btu}$$
   $$= 5676 \text{ M Btu}$$

4. Estimate regeneration gas heat required, Q_{rg}.
   During the heating cycle the hot regeneration gas is
   cooled from the hot inlet gas temperature T_4 to the
   bed exit temperature, T. Therefore, the heat supplied
   in a differential time, dt, is:

   $$m_{rg} \, C_{pg} \, (T_4 - T) \, dt = dQ_{rh} = K \, dT$$

   where K = an assumed constant.

   Integrate between the start (t_o, T_o) and the end (t_1, T_3)
   of the heating period:

   $$m_{rg} \, C_{pg} \, (t_1 - t_o) = K \ln [(T_4 - T_o)/(T_4 - T_3)]$$
   But $Q_{rh} = K \, (T_3 - T_o)$ therefore:
   $$m_{rg} C_{pg} \, (t_1 - t_o) = [Q_{rh}/(T_3 - T_o)] \ln [(T_4 - T_o)/(T_4 - T_3)]$$

   However, the heat supplied to the regeneration gas during
   the heating cycle is:

   $$Q_{rg} = m_{rg} \, C_{pg} \, (T_4 - T_o)(t_1 - t_o)$$

   Therefore:

   $$Q_{rg} = Q_{rh} \frac{(T_4 - T_o)}{(T_3 - T_o)} \ln \left[\frac{(T_4 - T_o)}{(T_4 - T_3)} \right]$$
   $$Q_{rg} = 5685 \frac{(550 - 100)}{(500 - 100)} \ln \left[\frac{(500 - 100)}{(550 - 500)} \right]$$
   $$= 14050 \text{ M Btu}$$

5. Regeneration Gas Flowrate.

   $$Q_{rg} = \dot{m}_{rg} \, C_{pg} \, (T_{rg} - T_{fg}) \, \Delta t_h$$

where $\dot{m}_{rg}$ = regeneration gas flowrate
   (lb/min)
   C_{pg} = specific heat of natural gas
   (Btu/lb °F)
   Δt_h = duration of heating cycle (min)
   $$14050000 = \dot{m}_{rg} \, (.58)(550 - 100)(300)$$
   $$\dot{m}_{rg} = 180 \text{ lb/min}$$

6. Flow Pressure Drop During Regeneration.
   a. Density of hot regeneration gas

   $$\rho = P \, MW/Z \, R \, T$$
   $$= (6144)(20.3)/(.99)(10.732)(550 + 460)$$
   $$= 1.16 \text{ lb/ft}^3$$

   b. Superficial velocity, V, of hot regeneration gas
   $$V = \dot{m}_{rg}/\rho_{rg} \, A_{bed}$$
   $$= (160)/(1.16)(\pi 14)(5.5)^2 = 5.81 \text{ ft/s}$$

   c. Pressure drop due to regeneration-gas flow

   $$\Delta P/L = B \, \mu \, V + C \, \rho \, V^2$$
   $$= (.0722)(.016)(5.81)$$
   $$+ (.000124)(1.16)(5.81)^2$$
   $$= 0.0064 + 0.0049$$
   $$= 0.011 \text{ psi/ft}$$

Regeneration Gas Cooling Required.
1. Cooling desiccant from 500 to 125°F

   $$Q_{cd} = m_d \, C_{pd} \, (125 - 500)$$
   $$= (13900)(.25)(125 - 500) = -1303 \text{ M Btu}$$

2. Cooling adsorption vessel from 500 to 125°F
   $$Q_c = m_v \, C_{ps} \, (125 - 500)$$
   $$= (29700)(.12)(125 - 500) = -1337 \text{ M Btu}$$

3. Total regeneration cooling load, Q_{rc}

   $$Q_{rc} = Q_{cd} + Q_{cv} = -2640 \text{ M Btu}$$

4. Estimate time of cooling cycle, Δt_c

   $$Q_{rc} = \dot{m}_{rg} \, C_{pg} \, (T_{ave} - T_o)$$

   where $T_{ave} = (T_1 + T_3)/2$
   $$2640000 = (160)(.58)(312.5 - 100) \, \Delta t_c$$
   $$\Delta t_c = 134 \text{ min} = 2.23 \text{ hr}$$

Troubleshooting Solid-Desiccant Units

Most solid desiccant dehydration units are reliable and
require relatively little operating attention. However, poor
design, operation, and maintenance can cause unnecessary
expenses such as frequent desiccant changeouts, excessive
plant shutdowns, and equipment replacements.

How proper design can result in trouble-free operation
is described first. Then bed contamination—solid desiccants
can be most unforgiving—is reviewed. Finally, two common

problems—high dew point and premature breakthrough—are discussed.

Design Considerations. Neumann (1977) suggests the following *design checklist* for avoiding operating problems:

1. Can the unit handle the worst (largest) inlet water loading?
2. If dehydration and regeneration occur at different pressures, how will the desiccant beds be protected from excessive pressure swings?
3. Will the selected switching valves prevent leakage of wet regeneration gas into dried gas?
4. Are the dehydration and regeneration gas flow rates designed for proper flow distribution?
5. Is the regeneration gas heater adequately sized for the maximum load?
6. How will any hydrocarbon liquids condensed in the regenerator cooler be handled?
7. Can the water in the regeneration gas cooler freeze?
8. Does the design include suitable temperature and water-content monitors, recorders, and shut down devices?
9. Will chemical reactions between methane, hydrogen sulfide, carbon dioxide, mercaptans, and oxygen occur at the high regeneration gas temperature?
10. Has all equipment been inspected to verify compliance with design specifications?
11. Have all internal refractories been cured properly?
12. Has the desiccant been loaded according to vendor's recommendations?

Bed Contamination. The most frequent cause is incomplete removal of contaminants in the inlet gas separator. Another possibility is lubricating oil from compressors used to boost the regeneration gas pressure. Also, if the regeneration gas leaving the separator is commingled with the feed gas to the dehydrators (Fig. 9–1) then a separator malfunction can dump liquid hydrocarbons and water onto the desiccant.

High Dew Point. High dew point is one of the two common problems that can cause operating trouble. Possible causes are:

1. "Wet" inlet gas bypasses the dehydrator through cracks in the internal insulation. Cracks in a liner or in sprayed-on insulation can be detected by "hot spots" and peeling paint on the outer shell. Other symptoms are fast water breakthrough and an unusually rapid rise in the effluent gas temperature during regeneration.
2. Leaking valves also permit wet gas to bypass the dehydrators. Even a slight leak of hot gas usually produces detectable temperature rise in what should be

the cold side of the valve. Ultrasonic translators are useful also.
3. Incomplete desiccant regeneration.
4. Excessive water content in the wet feed gas due to increased flow rate (MMscfd), higher temperatures, and lower pressure.

Premature Breakthrough. Satisfactory dew points are observed at the beginning but not for the entire duration of the drying cycle. Desiccant capacity should decrease with use but should stabilize at 55–70% of the initial capacity (Ballard, 1983). However, premature symptoms of "old age" are caused by unrecognized increase in inlet water loading, increase in heavy hydrocarbons (C4+) in feed gas, methanol vapor in feed, desiccant contamination or incomplete regeneration.

INTRODUCTION TO LOW-TEMPERATURE PROCESSES

When the wellhead pressure is high enough, low-temperature separation is usually the most efficient method of:

1. Separating liquid hydrocarbons and water
2. Recovering additional liquids (condensate) from the gas stream
3. Dehydrating the gas to pipeline specifications.

If the wellhead pressure exceeds that of the pipeline, then the gas can be passed through a choke or throttled in a constant-enthalpy Joule-Thomson expansion to provide cooling. Figure 9–17 or 9–18 can be used to estimate the magnitude of this Joule-Thomson cooling.

Low temperature processes may be classified as follows:

1. Those that intentionally form and melt hydrates, e.g., NATCO's LTX.
2. Those that use hydrate inhibitors (EG), e.g., BS&B's LTS.
3. Those where additional cooling (mechanical refrigeration) is required to obtain the required gas dehydration and/or condensate recovery.

Low Temperature Separation With Hydrate Formation

Figure 9–19 shows the basic *NATCO LTX process* that consists of:

1. Indirect heater (optional)
2. Low temperature separator
3. High pressure knockout
4. Inlet gas/sales gas heat exchanger

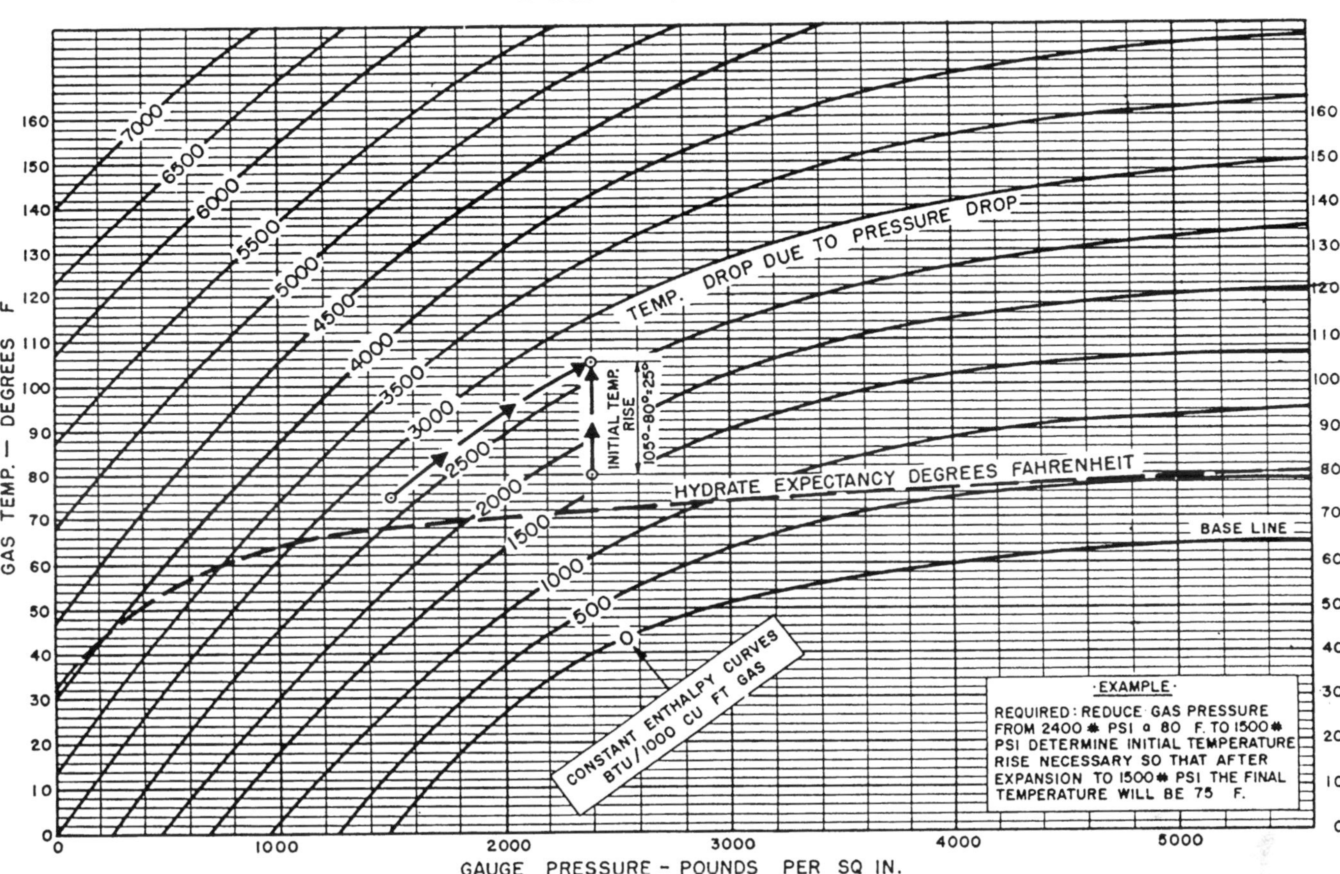

Figure 9–17. Natural gas expansion-temperature reduction curves (courtesy PETEX, 1987. Copyright © 1987 by the University of Texas at Austin).

5. Choke
6. Flash separator or stabilizer
7. Piping and controls.

The wellhead stream is first heated, if necessary, and then passes through the *hydrate melting coils* in the base of the LTX separator. This stream also provides enough heat near the choke to prevent hydrate buildup and plugging the choke entrance. The wellstream then flows into the high pressure knockout (HPKO) which removes liquids (hydrocarbon and water) condensed by the temperature drop from the wellbore. These *free liquids* may be handled in three ways (PETEX, 1987, p. 46):

1. Only water is removed in the HPKO and both the hydrocarbon condensate and gas pass through the choke.
2. All liquids are removed in the HPKO and then dumped into the lower liquid section of the LTX (Fig. 9–19).
3. All liquids are removed in HPKO, water is dumped to disposal while condensate is fed to lower liquid section

of LTX. The HPKO temperature must be warm enough (about 100°F) to prevent hydrates from forming in the liquid dump valves.

The feed gas stream leaving the HPKO then flows through the tube side of the inlet-gas/sales-gas heat exchanger where it is cooled to the lowest safe temperature above the hydrate formation temperature at the pressure upstream of the choke. This insures the lowest separation temperature in the LTX and maximum liquids recovery. A three-way bypass valve adjusts the cold sales gas flow to the shell side of the heat exchanger thus maintaining the lowest safe temperature. This also warms the sales gas before it enters the sales pipeline thus preventing other "wet" gas in the sales lines from being chilled below the hydrate point.

Low temperature separation units have been built in four shapes—horizontal (Fig. 9–20a), vertical (Fig. 9–20b), spherical, and T-bone. Current wellhead pressures are not high enough to warrant the spherical shape, and the

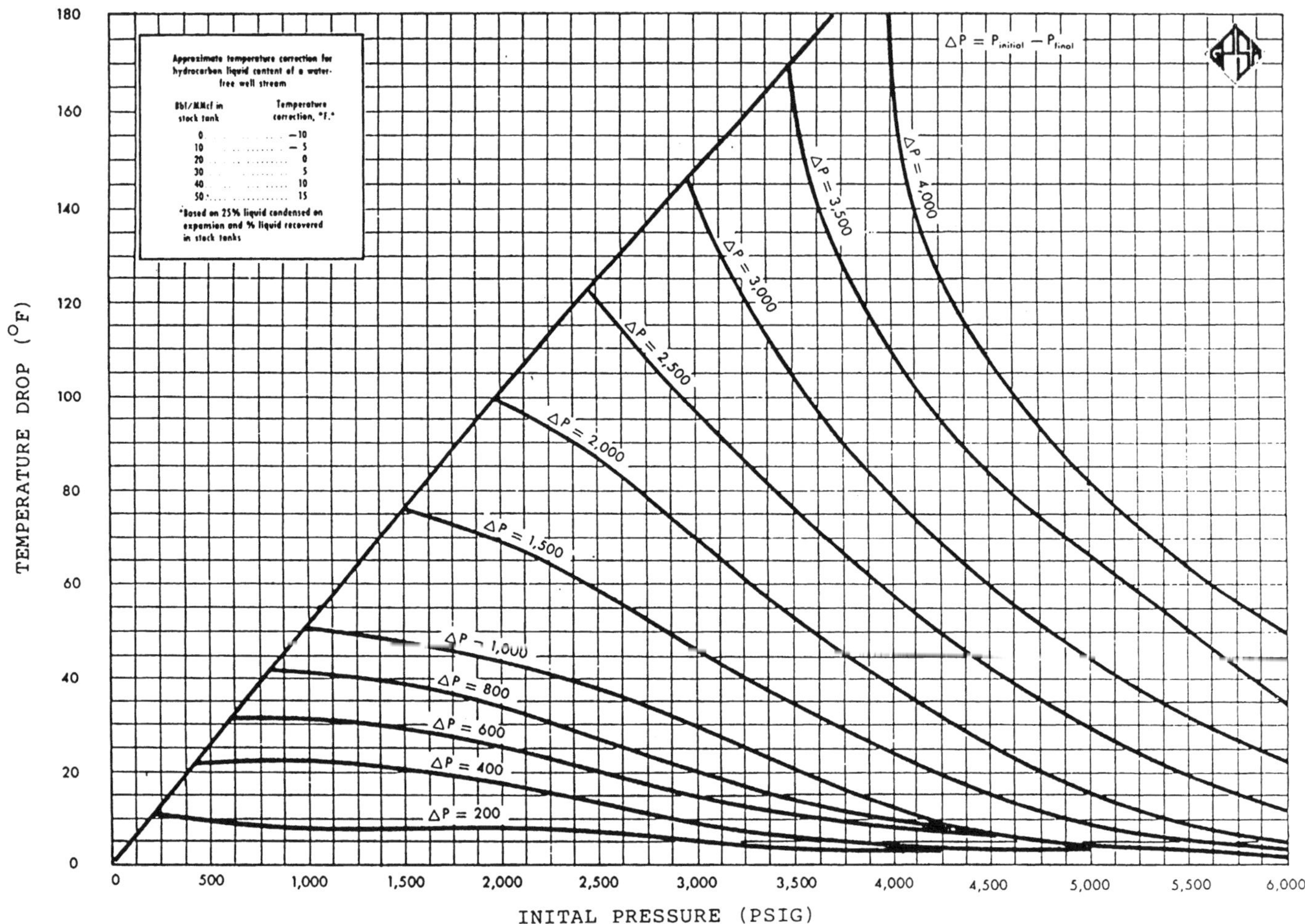

Figure 9–18. Temperature drop accompanying a given pressure drop (GPSA, 1981).

horizontal unit (Fig. 9–21) is preferred because of its advantages over the vertical.

The cooled feed gas flows through the *choke* where it expands from the high inlet wellstream pressure to the sales line pressure. This choke can be used to regulate either the pressure in the LTS or the flow rate. As the feed stream flows through the choke, the resulting J-T expansion and cooling causes some gas and most of the water vapor to liquefy. Hydrates are formed as the condensate and water fall to the lower liquid section. The *heating coil* melts the hydrates and partially stabilizes the hydrocarbon liquids.

Hydrates actually form inside the choke but are broken up and carried into the LTX by the sonic flow. This *hydrate formation* and release sounds like pellets entering the vessel. As shown in Figure 9–21, the high-velocity, vapor/liquid/solid stream leaving the choke enters tangentially into a cylindrical spinner box. This tangential entry absorbs the

inlet momentum and directs the stream onto the LTX coils. The heated spinner coil and the spinner box disengage the vapor from the liquid.

LTX operation exploits two properties of hydrates:

1. Hydrates are heavier than condensate and lighter than water.
2. When hydrates form they "extract" water from the gas.

Coils at the bottom of the LTX heat the condensate to 65–75°F and the water to 15–20°F hotter. Therefore, the hydrates sink through the condensate layer, float on top of the water, and "melt" or decompose as they flow toward the liquid outlets. This horizontal flow is the major advantage of horizontal LTX units (GPSA, 1987).

Gas-phase *insulating baffles* keep the gas leaving the LTX at the cold "choke-exit" temperature. On the other

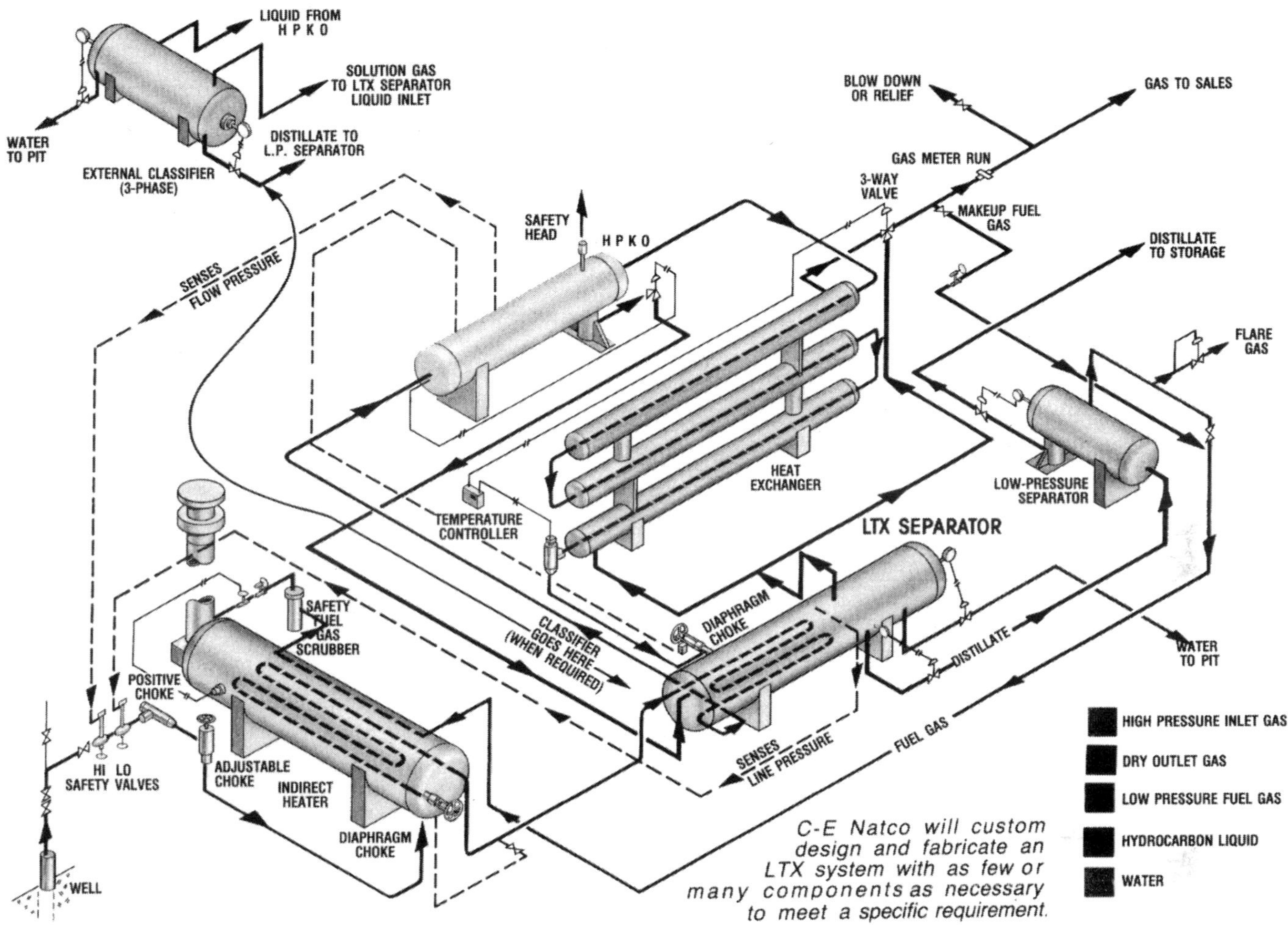

Figure 9–19. LTX low-temperature separation unit (NATCO, 1984).

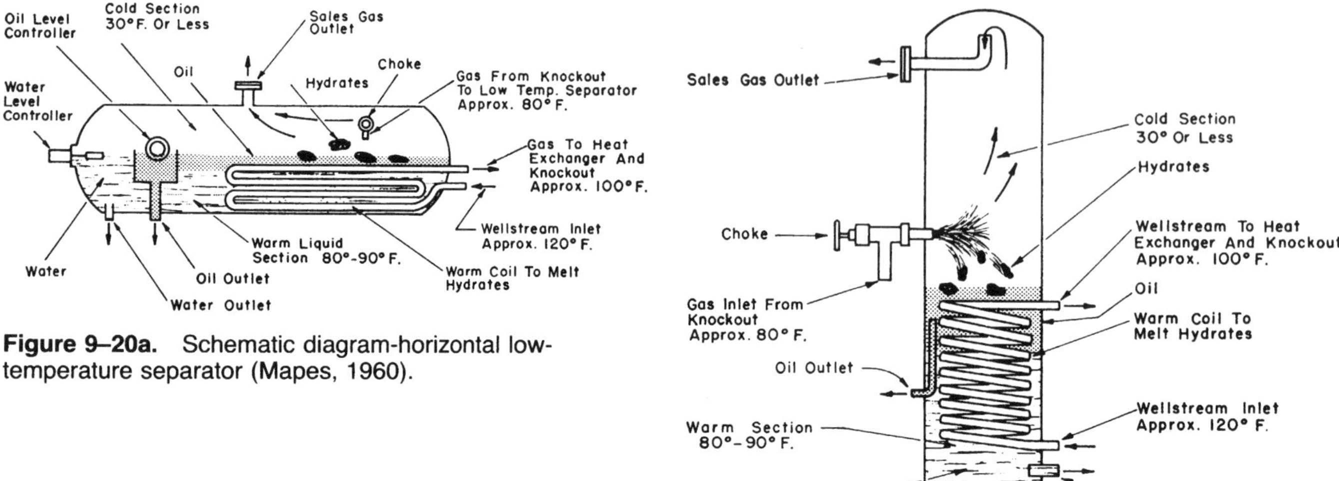

Figure 9–20a. Schematic diagram-horizontal low-temperature separator (Mapes, 1960).

Figure 9–20b. Schematic diagram-vertical low-temperature separator (Mapes, 1960).

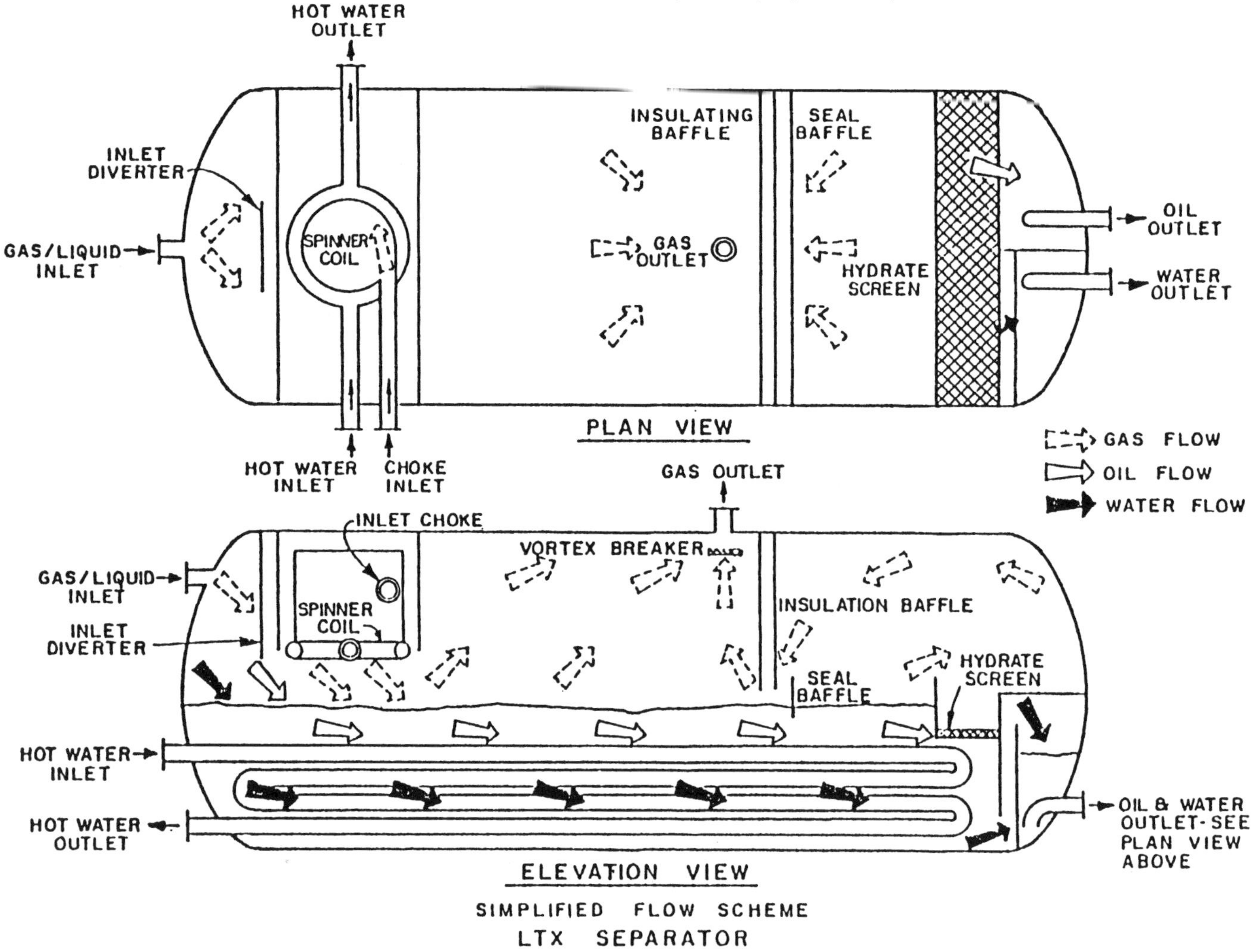

Figure 9–21. LTX separator, simplified flow scheme (NATCO, 1987).

side of the insulating baffles, vapors flashed from the condensate are quite warm.

Downstream of the insulating baffles, the surge baffle stops ''waves'' or liquid surges from swamping the condensate and water buckets.

The HPKO is generally placed ahead of the inlet gas/ sales gas heat exchanger when heat exchange is critical (the choke pressure drop and J-T cooling are barely adequate) and/or when a paraffin base condensate must be removed. Placing the HPKO downstream of the heat exchanger allows removal of additional liquids, feeds liquid-free gas to the choke, and generates the lowest expansion temperatures.

Especially in the vertical LTS, the extreme temperature difference between the cold top and hot bottom sections is very visible. The outside of the top section is often covered with ice frozen out of the atmosphere; hence the alternate name for LTS—*Hot Bottom Unit* or *HBU*.

Low Temperature Separation Without Hydrate Formation

In this LTX separator, the inlet gas stream cannot be cooled below the hydrate formation temperature in the gas-to-gas exchanger. The resulting temperature in the LTS is therefore fixed by the available pressure drop from wellstream to sales gas. When this J-T cooling cannot achieve the desired gas dehydration, then *additional cooling* of the inlet gas is required prior to the choke. Hydrate inhibitors are required, as discussed in Chapter 6.

Refrigeration-Aided Low Temperature Separation

Figure 9–22 shows the basic *mechanical refrigeration* flow sheet. Again, this process is very similar to the LTS previously described and similar to the hydrate prevention

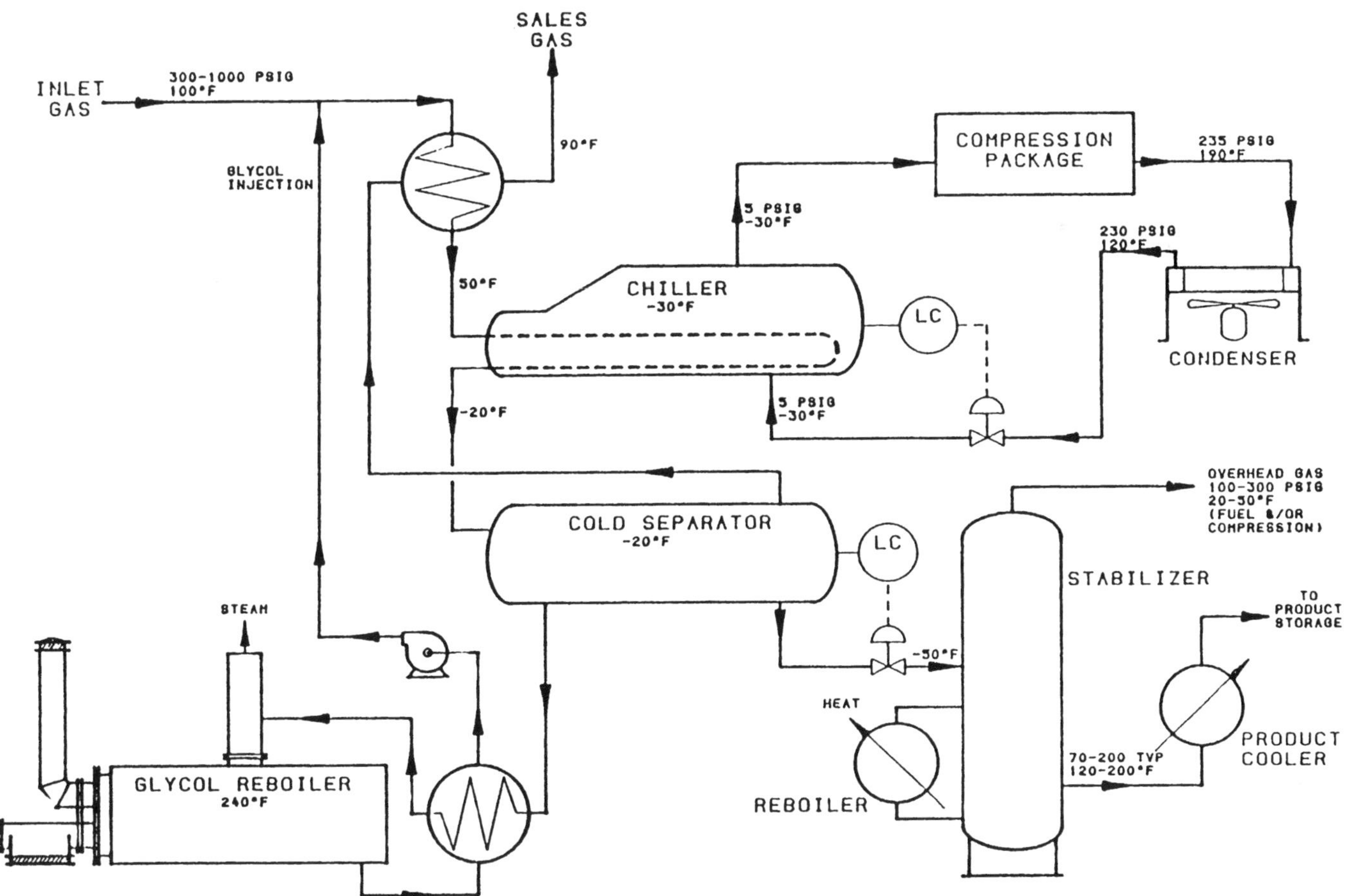

Figure 9–22. LTS using mechanical refrigeration (NATCO, 1987).

by glycol addition described in Chapter 6. External refrigeration is required when the combined cooling of the gas-to-gas exchanger and any J-T expansion is insufficient to provide the desired water and hydrocarbon dew points.

DEHYDRATION USING CALCIUM CHLORIDE

Literally hundreds of calcium chloride dehydration units are currently operating in the U.S. Calcium chloride is most attractive for small (50 Mscfd–2.5 MMscfd) streams in remote locations (Rocky Mountains) where low water dew points are not required (e.g., lift gas) and for offshore retrofits with severe space and weight limitations.

Calcium Chloride-Water Phase Behavior

Figure 9–23 shows the phase behavior for $CaCl_2$ and H_2O. Solid calcium chloride can exist in the anhydrous form, $CaCl_2$, and four levels of hydration—$CaCl_2 \cdot H_2O$, $CaCl_2 \cdot 2H_2O$, $CaCl_2 \cdot 4H_2O$ and $CaCl_2 \cdot 6H_2O$. In Figure 9–23, the line between area 8 (liquid brine) and areas 5, 6, and 7 shows the concentration (lb H_2O/lb $CaCl_2$) of saturated

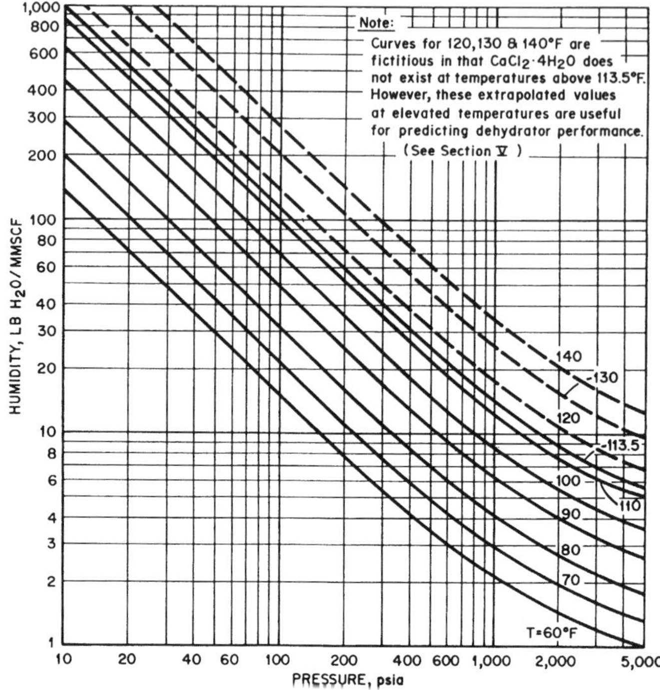

Figure 9–24. Water content of natural gas in equilibrium with $CaCl_2 \cdot 4H_2O$ (courtesy Dow Chemical Company, 1973).

brine in equilibrium with solid $CaCl_2 \cdot 2H_2O$, $CaCl_2 \cdot ^6.4H_2O$, and $CaCl_2 \cdot 6H_2O$, respectively.

Figure 9–24 (Dow, 1973) shows how the water content (lb H_2O/MMscf) of natural gas in equilibrium with solid $CaCl_2 \cdot 4H_2O$ varies with gas pressure and temperature. Similar graphs are available for the other hydrates: $CaCl_2 \cdot H_2O$, $CaCl_2 \cdot 2H_2O$, and $CaCl_2 \cdot 6H_2O$ (Dow, 1973).

Calcium Chloride Process Description

As shown in Figure 9–25, a calcium chloride dehydration unit consists of three sections: separation at the bottom, trays in the middle, and bed of $CaCl_2$ pellets at the top.

Separation Section. The wet gas and any free liquids enter at the bottom and immediately contact a horizontal baffle or other separation-aiding device. The wet gas, now free from entrained liquids, flows upward to the tray section.

Tray Section. Here the wet gas is contacted counter currently with concentrated calcium brine in three to five specially designed trays. As the brine flows downward, it is diluted continuously, and as the gas flows upward it is progressively dehydrated. The trays are specially designed so that the gas velocity (kinetic energy) is used to recirculate the brine on each tray. Because very small amounts of brine are formed from large gas volumes, liquid recirculation

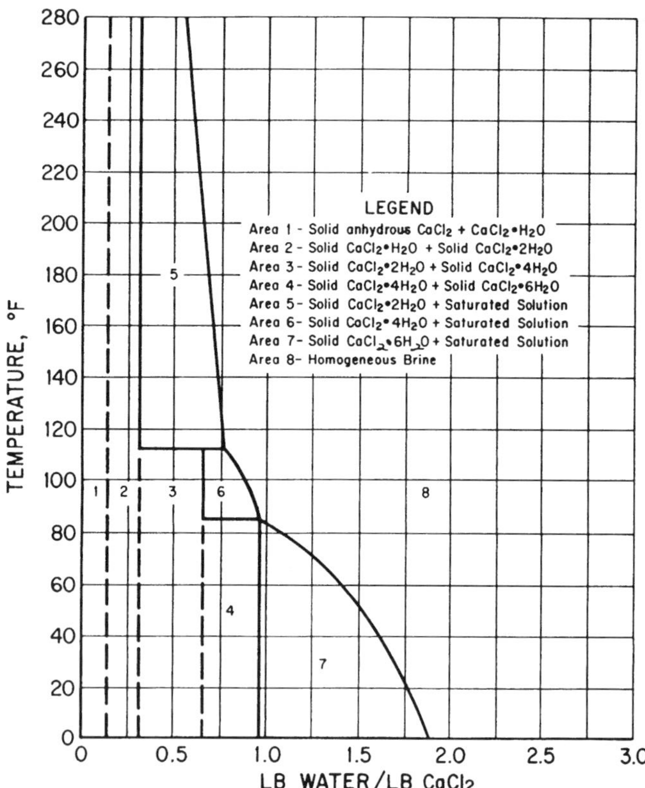

Figure 9–23. Pure calcium chloride phase diagram (courtesy Dow Chemical Company, 1973).

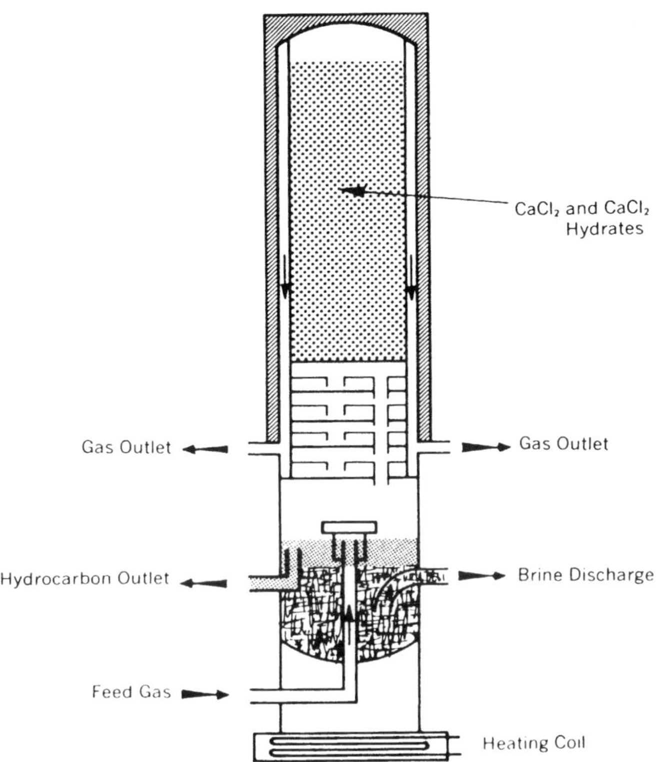

Figure 9–25. CaCl$_2$ gas dehydration unit (Dow, 1962).

is required to get good liquid-vapor contact or high tray efficiency (Hutchison, 1957; Hodgson and Martinez, 1984). The Hodgson-Martinez tray also has a 20 to 1 gas turndown ratio which is a big advantage for a small gas well.

Bed Section. Additional gas dehydration occurs as the gas flows upward through a bed of ⅜– to ¾–in. calcium chloride pellets. Here the water removed from the gas is absorbed on the surface of the pellets, forming saturated liquid brine that drips continuously down onto the trays below. Initially, the pellets consist primarily of anhydrous CaCl$_2$. Therefore, as the gas flows upward through the bed it contacts progressively drier pellets—that is, solid calcium chloride with progressively lower hydrate states.

In the lower part of the bed, the solid pellets are consumed (converted to the saturated liquid brine). The weight of the pellets above causes the bed to settle gradually. Satisfactory gas dehydration is realized as long as two feet of pellets remain in the bed. Then breakthrough (high exit gas water content) can occur, and it is time to refill the bed section with fresh anhydrous CaCl$_2$ pellets.

Figure 9–25 shows how the wellstream energy is conserved in cold climates. The dehydration unit is insulated, and the dried gas leaving the top of the bed section flows downward through a cylindrical shell between the bed-section wall and the outer insulation. Figure 9–23 shows that saturated brine solutions can freeze above 32°F.

Design of Calcium Chloride Units

Dow Chemical Company's (1973) design procedure is summarized using the following example:

Feed gas	flow rate	1.5 MMscfd
	temperature	80°F
	pressure	988 psig
Design specs	elevation	5000 ft
	atm press	12.2 psia
	Recharge CaCl$_2$ pellets every 31 days.	
CaCl$_2$ Pellets	Dow PELADOW	
	bulk density	62–65 lb/ft^3
	CaCl$_2$ content	min 91 wt %

CaCl$_2$ Dehy Capacities (Dow, 1983):

Pressure	Maximum Recommended Gas Flow MMscfd/ft^2
100 psia	0.86
500	1.91
1000	2.70
1500	3.34
2000	3.85

Inlet gas contains 33.6 lb H$_2$O/MMscf (Fig. 4–6).
Exit gas contains 4.2 lb H$_2$O/MMscf (Fig. 9–24).

Exit gas humidity is lower than the conservative estimate (Fig. 9–24) due to presence of lower hydrates (CaCl$_2$·2H2O, CaCl$_2$·H$_2$O) and anhydrous CaCl$_2$ in the freshly charged bed.

For example, a *CaCl$_2$ dehydration unit can meet the conventional 7 lb H$_2$O/MMscf spec at 1000 psia and 100°F* (Wood, 1989) even though Figure 9–24 predicts 8.5 lb H$_2$O/MMscf. Therefore, an exit gas humidity of 3.6 lb H$_2$O/MMscf is used to provide a conservative design.

$$\text{Water removed} = 33.6 - 3.6 = 30.0 \text{ lb H}_2\text{O/MMscf}$$
$$= (30)(1.5) = 45 \text{ lb H}_2\text{O/day}$$

Assume brine leaving bottom tray contains 3.5 lb H$_2$O/lb CaCl$_2$

$$\frac{\text{lb CaCl}_2 \text{ req'd}}{\text{day}} = \frac{\text{lb H}_2\text{O removed}}{\text{day}} \frac{\text{lb CaCl}_2 \text{ req'd}}{\text{lb H}_2\text{O removed}}$$
$$= (45/1)(1/3.5) = 12.9$$

$$\frac{\text{PELADOW}}{\text{req'd}} = \frac{\text{lb CaCl}_2 \text{ req'd}}{\text{day}} \frac{\text{day}}{\text{charge}} \frac{\text{lb PELADOW}}{\text{lb CaCl}_2}$$
$$= (12.9)(31)(1/0.91) = 439 \text{ lb}$$

$$\text{Volume PELADOW req'd} = 439/63 = 7.0 \text{ ft}^3$$

A 1 ft ID column has 0.7854 ft^2 flow area.

Max capacity is $(2.70)(.7854) = 2.12$ MMscfd. okay

Bed height $= (7.0/.7854) + 2 = 13$ ft

Bed is recharged when only 2 ft of pellets remain.

The concentration of brine leaving the bottom tray may be estimated by plotting a *McCabe-Thiele type diagram*. See Dow (1973) for details.

Calcium Chloride Process Operation

At typical contact temperatures (80 to 100°F), the nearly saturated brine dripping off the bottom pellets contains approximately *1 lb water per lb CaCl$_2$* (Fig. 9–23). In the tray section, this brine absorbs more water from the gas and is diluted to 20–25 weight percent CaCl$_2$. Therefore, a CaCl$_2$ dehydration unit with five trays removes about *3.5 lb water per lb CaCl$_2$ consumed.*

Normally, the pressure drop across the entire unit (trays and bed sections) is less than 8 psi. The heat of absorption can be neglected under normal temperatures (50 to 140°F) and pressures (300 to 3000 psia) (Dow, 1973).

Bridging and channeling are the main problems. *Bridging*, which is the joining or fusion of adjacent CaCl$_2$ pellets, can cause the pellets to stick to the column wall. Cyclic operation is the most common cause of bridging. For example:

1. Slight temperature drops can freeze the brine that drips off the pellets thus fusing the pellets.
2. When a dehydrator is removed from service, left idle, and then put back in service bridging can occur if the temperature has dropped while the unit was down.
3. If wet gas or free water contacts the bed, more brine is formed and upsets can cause worse cases of bridging.

Bridging alone is not serious, however, bridging usually results in channeling. In turn, *channeling* causes poor gas distribution and early breakthrough (premature poor dewpoint depression). When channeling occurs the dehydration unit must be opened and the pellets redistributed. Fortunately, bridging and channeling occur rarely.

Calcium chloride solutions are highly corrosive when air is present. In the absence of air (oxygen) corrosion is not a problem.

PROCESS SELECTION GUIDELINES

Every application certainly merits individual consideration. Nevertheless each dehydration method has its own advantages and disadvantages that define where it is especially attractive. Accordingly, the key advantages and disadvantages of the various dehy methods are now listed. Then the usual areas of application of each method are outlined.

Dehydration Method Advantages and Disadvantages

Triethylene Glycol.

Advantages
1. Lower installed costs—Kohl and Riesenfeld (1979) report that solid desiccant plants cost 50% more at 10 MMscfd and 33% more at 50 MMscfd.
2. Lower pressure drop—5 to 10 psi vs 10–50 psi for dry desiccant units.
3. Glycol dehydration is continuous rather than batch.
4. Glycol makeup is easily accomplished. Recharging solid desiccant towers is time-consuming and sometimes requires interruption of gas sales.
5. Glycol units require less regeneration heat per pound of water removed thus lowering utility costs.
6. Glycol systems will operate in the presence of materials that would foul solid desiccants.
7. Glycol units can dehydrate natural gas to 0.5 lb H$_2$O/ MMscf. Indeed, 0.25 lb H$_2$O/MMscf has been achieved on the North Slope with stripping gas and a Stahl column.
8. TEG units are simple to operate and maintain.
9. TEG units are easily automated for unattended operation in remote locations.

Disadvantages
1. Water dewpoints below -25°F require stripping gas and a Stahl column.
2. Glycol is susceptible to contamination.
3. Glycol is corrosive when contaminated or decomposed.

Solid Desiccants.

Advantages
1. Dew points as low as -150°F (1 ppmv of H$_2$O) are obtainable.
2. They are less affected by small changes in gas pressure, temperature, or flow rate.
3. They are less susceptible to corrosion or foaming.

Disadvantages
1. Higher capital costs and higher pressure drops.
2. Desiccant poisoning by heavy hydrocarbons, H$_2$S, CO$_2$, etc.
3. Mechanical breaking of desiccant particles.
4. High space and weight requirements.
5. High regeneration heat requirements and high utility costs.

Low Temperature Separation.

Advantages
1. When adequate wellhead pressure is available, LTS separation can meet pipeline specifications for water and HC dew points.
2. Power consumption is minimal, and so operating costs are low.
3. Corrosion is minimal, especially when hydrate inhibitors are not used.

Disadvantages
1. LTS processes are unattractive when adequate wellhead pressure is not available.
2. Adding hydrate inhibitors and using external refrigeration increase both capital cost and operating expenses.

Calcium Chloride.

Advantages
1. Can operate unattended until fresh desiccant is required.
2. No fire hazard, very compact, low capital cost.

Disadvantages
1. Dew-point depression is limited by $CaCl_2$-H_2O equilibrium.
2. Even with five trays, 1 lb $CaCl_2$ removes only 3.5 lb H_2O.

Dehydration Method Selection Recommendations

TEG is by far the most popular choice and is probably the best choice unless one of the following recommendations is pertinent.

Solid Desiccants. Solid desiccants are usually preferred for:

1. Natural-gas dehydration to pipeline specs (4–7 lb H_2O/ MMscf or water dew points of 10–30°F) where TEG is not suitable—for example, aboard floating production platforms where wave action disturbs glycol flow on the contactor trays (Rietveld *et al.*, 1985); or when the gas is sour (Alexander, 1988).
2. Dehydration of high-pressure (supercritical) carbon dioxide (Skopak and Phillips, 1985) because of excessive carbon dioxide solubility in TEG.
3. Dehydration of natural gas for cryogenic (below −30°F) processing.
4. Simultaneous removal of water and hydrocarbons to meet both water and hydrocarbon dew-point specifications.
5. Recovery of liquid hydrocarbons from lean (0.5 GPM C3+ or less) natural gases (often with refrigeration).

Low Temperature Separation Processes. These are attractive for:

1. Sweet gas with a wellhead pressure considerably greater than pipeline pressure.

Calcium Chloride. These units are suitable for:

1. Small gas wells in remote locations.

Review Questions

1. Describe a typical solid desiccant dehy unit. Describe how the process varies to accommodate the following sources for bed regeneration gas: inlet gas, product gas, demethanizer overhead gas. Compare the relative advantages and disadvantages of these choices of regeneration gas.
2. Name the more common types of solid desiccants. Compare their relative advantages and disadvantages.
3. List the common components of a typical solid desiccant dehy unit. Discuss the function of these components.
4. Identify the factors that govern desiccant choice and direction of gas flow in the solid desiccant bed during both adsorption and regeneration.
5. Describe procedures for desiccant installation, start up, bed switching, and energy conservation.
6. What factors should be considered in designing a solid desiccant dehy unit? What information is required? What factors determine the cycle time?
7. When should a desiccant bed be replaced? What factors determine the life of solid desiccants? How does solid desiccant performance vary with use?
8. How can solid desiccant bed contamination be avoided?
9. What are the likely causes of high dew point in the exit gas from a solid dehy unit?
10. What causes premature breakthrough in a solid desiccant bed?
11. What is "dusting" of solid desiccants?
    What causes "dusting?"
    How is "dusting" controlled?
    What problems are caused by "dusting?"
12. List the more common low-temperature dehydration processes. Compare their relative advantages and disadvantages.
13. Describe a typical calcium chloride dehydrator.
14. Compare the relative advantages and disadvantages of the two types of calcium chloride dehydrator.
15. Compare the following methods of drying natural gas: TEG, solid desiccant, refrigeration (or low temperature), and calcium chloride. Include the following aspects: areas of application, achievable

water dew points, relative advantages, relative disadvantages, and economics.

Nomenclature

A = maximum superficial velocity when $\rho = 1$ lb/ft^3 (see Equation 9–2)

acfm = gas flowrate at vessel P and T (ft^3/min)

B,C = pressure drop constants (see Table 9–3)

C_o = feed gas inlet water content (Fig. 9–7)

C_s = gas water content in equilibrium with regenerated desiccant (Fig. 9–7)

C_{ss} = capacity correction factor for unsaturated gas (Fig. 9–15)

C_T = mole sieve capacity correction for temperature (Fig. 9–16)

CA = corrosion allowance (in.)

C_{pd} = specific heat of desiccant (Btu/lb °F)

C_{pg} = specific heat of gas (Btu/lb °F)

C_{ps} = specific heat of steel (Btu/lb °F)

C_{pw} = specific heat of liquid water (Btu/lb °F)

E = double-welded butt joint efficiency
= 1.00 fully radiographed; = .85 spot examined

h = adsorption vessel height (ft)

ID = internal diameter of adsorption vessel (in. or ft)

L = desiccant bed depth or height (ft)

L_{MTZ} = depth (or height) of mass-transfer zone (Fig. 9–7)

MTZ = mass transfer zone

m_d = mass of desiccant in bed (lb)

m_v = mass of adsorption tower (lb)

m_w = mass of water adsorbed on desiccant (lb).

$\dot{m}_{rg}$ = flowrate of regeneration gas (lb/min).

$\dot{n}$ = feed gas flowrate (lb mol/min)

OD = outside diameter of adsorption vessel (in. or ft)

P = pressure (psig or psia)

P_{DP} = design pressure (psig)

ΔP = pressure drop across bed (psi)

Q_{cd} = heat required to cool desiccant (Btu)

Q_{cv} = heat required to cool vessel (Btu)

Q_{hd} = heat required to heat desiccant (Btu)

Q_{hv} = heat required to heat vessel (Btu)

Q_{hw} = heat required to heat water in bed (Btu)

Q_{rc} = total regeneration cooling load (Btu)

Q_{rg} = regeneration gas heat requirement (Btu)

Q_{rh} = total regeneration heat required (Btu)

Q_{vw} = heat required to vaporize water in bed (Btu)

T = gas temperature (°R)

T_o = temperature of feed gas (°F or °R)

T_1 = temperature of exit gas at end of cooling cycle (°F)

T_2 = temperature at which water is driven off desiccant (°F)

T_3 = temperature of exit gas at end of heating cycle (°F)

T_4 = temperature of hot regeneration gas (°F)

T_{ave} = = $(T_1 + T_3)/2$

t = time (hr or min)

t_m = adsorption vessel wall thickness (in.)

t_o = time at start of regeneration cycle

t_1 = time at end of heating and start of cooling cycle

t_2 = time at end of cooling cycle

Δt_c = cooling cycle time

Δt_h = heating cycle time

V = superficial gas velocity (ft/min)

V_{max} = maximum allowable superficial gas velocity (ft/min)

$\dot{v}$ = gas flow rate (ft^3/min)

Z = depth of mass transfer zone (ft)
= 1.70 ft for ⅛-in. pellets or 4–8 mesh
= .85 ft for ¹⁄₁₆-in. pellets or 8–12 mesh

Greek

μ = gas viscosity (cp)

ρ = gas density (lb/ft^3)

ρ_b = bulk density of desiccant bed (lb/ft^3)

ρ_s = density of steel (lb/ft^3)

References

Alcoa (1983), "Choosing an Alcoa Activated Alumina Desiccant—Basics of Dehydration Design," Aluminum Company of America, Pittsburgh, PA.

Alcoa (1986), "Product Data-Active Bed Support," Alcoa Chemicals Division, Vidalia, LA (April).

Alexander, R. A., "An Offshore Dehydration System for the Production of the Norphlet Sour Gas in Mobile Bay," Proceedings, 20th Annual Offshore Tech. Conf. Presentation 5837, pp. 273–280, Houston, TX (May 2–5).

Ausikaitis, J. P. (1983), "Trisiv Adsorbent," International Conference on Fundamentals of Adsorption, Klas, Upper Bavaria, West Germany (May 6–11).

Ballard, D. (1983), "How to Improve Cryogenic Dehydration," Petroenergy 83, Houston, TX (Sept. 12–16).

Barrow, J. A. (1982), "Simplified Molecular Sieve Designs," Petroenergy 82.

Barrow, J. A., and Ray Veldman (1988), "Proper Design Saves Energy for Molecular Sieve Dehydration Systems," Petroenergy 88, Houston, TX.

Campbell, J. M. (1979), *Gas Conditioning and Processing*, Vol. 2, Third Printing, Chapter 18, Campbell Petroleum Series, Norman, OK.

Cummings, W. P. (1979), "Natural Gas Dehydration with Solid Desiccant," 13th Annual Kansas University Heart of America Pipeline Oper. & Maint. Inst., Liberal, KS (Nov. 15–16).

Dow (1962), "Gas Conditioning Fact Book," Dow Chemical Co., Midland, MI.

Dow (1973), "PELADOW DG for Gas Dehydration," Dow Chemical Co., Midland, MI.

Fowler, O. W. (1957), "Wellhead Dehydrators in S. J. Basin," *Oil & Gas J.*, Vol. 55, No. 17, pp. 188–190 (April 29).

GPSA (1981), Engineering Data Book, 9th Ed., 5th Revision, Gas Processors Suppliers Association, 6526 East 60th St., Tulsa, OK 74145.

GPSA (1987), Engineering Data Book, 10th Ed., Vol. 2, Section 20, Gas Processors Suppliers Association, 6526 East 60th St., Tulsa, OK 74145.

GPSA (1987), Engineering Data Book, 10th Ed., Vol. 2, Chapter 23, Gas Processors Suppliers Association, 6526 E. 60th St., Tulsa, OK 74145.

Grace (1988), "Molecular Sieves," Davison Chemical Division, W. R. Grace & Co., Baltimore, MD 21203.

Hales, G. E. (1971), "Tips on Gas Dehydration," *Hydrocarbon Processing*, Vol. 50, No. 6, pp. 151–154 (June).

Harrell, B. E. (1988), Personal Communication, NATCO, P.O. Box 1710, Tulsa, OK 74101.

Hodgson, Robert A., and Sam J. Martinez (1984), "Calcium Chloride Dehydration Nozzle," U.S. Patent 4,433,983. Assigned to Maloney-Crawford Corp., Tulsa, OK (Feb. 28).

Hutchison, A. J. L. (1957), "Vapor-Liquid Contacting Apparatus," U.S. Patent 2,804,935 (Sept. 3).

Kali-Chemie (1987), "Sorbead Desiccants/Absorbents," Kali-Chemie Corp., Greenwich, CT.

Mapes, George J. (1960), "The Low Temperature Separation Unit," *World Oil*, Vol. 150, No. 1, p. 93 (Jan.).

NATCO (1984), "LTX Low-Temperature Extraction Systems," Brochure G–401–A3, NATCO, P.O. Box 1710, Tulsa, OK 74101.

NATCO (1987), "Adsorption Systems—Gas Processing," Technical Development Program, P.O. Box 1710, Tulsa, OK 74101.

Neumann, J. Michael (1977), "GPA Cryogenics Panel–1," *Oil & Gas J.*, Vol. 75, No. 29, pp. 60–64 (July 18).

Oliker, M. D., and G. T. Hong (1985), "4FS Control Cuts Drying Costs," *Hydrocarbon Processing*, Vol. 64, No. 4, pp. 72–74 (April).

PETEX (1987), *Field Handling of Natural Gas*, 4th Ed., Petroleum Extension Service, University of Texas at Austin, Austin, TX.

Petty, L. E. (1976), "Practical Aspects of Molecular Sieve Unit Design and Operation," Proceedings, 55th Annual GPA Convention, pp. 103–108, San Antonio, TX (March 22–24).

Rietveld, R. J. C., N. C. Biss, and D. Gelderblom (1985), "Tazertia Multiwell FPSU: Design of Production Facilities Integrated with the Tanker," *J. Petroleum Technology*, Vol. 37, No. 7, pp. 1091–1096 (June).

Skopak, J. E., and L. A. Phillips (1985), "Design and Operation of a High Pressure CO2 Dehydration Facility," Proc., 32nd Annual Southwestern Petroleum Short Course, pp. 496–509, Lubbock, TX (April 23–25).

Union Carbide (1979), Technical Data, Danbury, CT 06817.

Wood, Hal (1989), Personal Communication, Maloney-Crawford Corp., Tulsa, OK.

Zeochem (1982), "Molecular Sieve Type 4A", Technical Data Z12–1M–82, Zeochem, Louisville, KY 40232.

Chapter 10

Natural Gas Compression

INTRODUCTION

Compression is used in all aspects of gas processing such as gas lift, gas gathering, helium recovery, condensate recovery, transmission and distribution, reinjection for pressure maintenance, gas storage, and liquefaction for transport.

Figure 10–1 classifies the various types of compressors by their method of operation. However, in the "oil patch" the most common types are reciprocating, centrifugal, sliding vane, and rotary screw.

Reciprocating compressors are widely used in the process industry because they are flexible in throughput and discharge pressure range. Driver horsepower requirements vary from less than 1 to 10000 hp or higher. Speeds may range from 125 to 1000 rpm. Piston speeds range from 500 to 950 ft/min, the majority being 700 to 850 ft/min. Nominal gas velocities usually range from 4500 to 8000 ft/min, and discharge pressures vary from vacuum to 50000 psig (James, 1982b).

Sliding vane compressors are limited to low pressure (up to 130 psig), small flows (50 to 3500 ACFM), and clean gases. Their advantages include low cost, high compression and overall efficiencies, high volumetric efficiencies due to low slip, and relatively low maintenance. They are often used for vapor recovery from stock tanks.

Rotary screw compressors are used for air compression and vapor recovery units and have even been used in refrigeration units. Where the flood-oil/gas separation can be accomplished satisfactorily rotary-screw compressors are very attractive. Rotary screw compressors are relatively recent and promising "newcomers" to the oilfield.

Centrifugal compressors are used in a wide variety of applications in chemical plants, refineries, and in field gathering of natural gas. Centrifugal compressors can be used for outlet pressures as high as 10000 psia, thus overlapping with reciprocating compressors over a portion of the flow-rate/pressure domain. Centrifugal compressors have a definite lower limit of flow rate, however, due to the lower limit in physical size of the impeller wheels. Flow rates of 100 to 200000 cfm at inlet conditions can be handled in these devices. The lower limit of 100 cfm is for any impeller, which is usually the final stage. Mass flow rate depends on the inlet temperature and pressure, of course.

Every compressor requires a source of energy or *driver* or prime mover. The more common types are steam turbines, electric motors, gas turbines, and gas engines. Sometimes diesel engines are used, but mainly as emergency drivers.

Historically the *internal-combustion gas engine* has been the leading prime mover in the oil patch. It has maintained its popularity in the face of strong challenge by the gas turbine because of its overall efficiency and the ready availability of natural gas. In addition, gas engines and reciprocating compressors make compact integral units in sizes from 200 to over 7000 HP. Large gas-engine/ reciprocating-compressors can achieve 40% overall thermal efficiency with extensive maintenance. Gas plants with smaller units can expect about 25%. Very large steam turbines with, say, five stages of regenerative feed heating can achieve 33% thermal efficiency. Thermal efficiencies of 20 to 25% are typical of the simpler cycles used in the field. Smaller size simple-cycle gas-turbine/centrifugal-compressor units usually achieve 25% thermal efficiency. This comparison allows no credit for the heat content of the low-pressure steam or the gas turbine exhaust gases.

For about 40 years, *gas turbines* have proved to be successful prime movers in oil fields, refineries and process industries. In fact, gas turbines provide high power outputs at high speeds, need only limited space, are light weight, have good reliability and low maintenance, and offer 1000 to 100000+ HP sizes.

Steam turbines are most desirable where there is demand for low-pressure steam for process and/or heating, e.g., refineries and petrochemical plants. *Electric motors* provide convenience, efficient operation, low maintenance, and

198

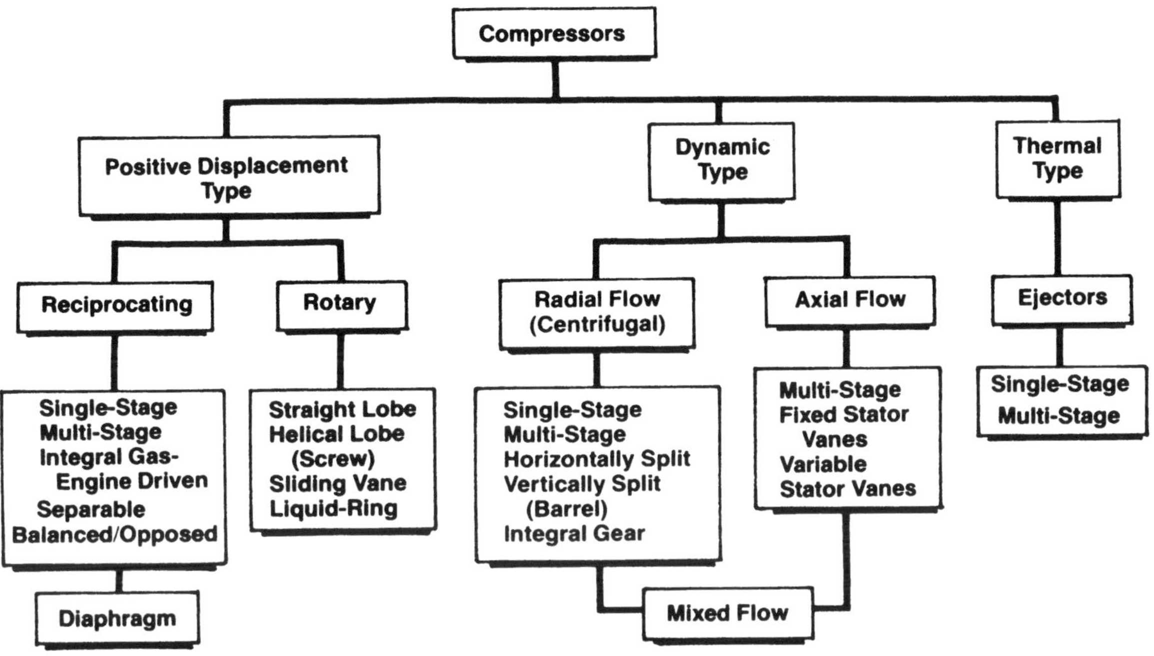

Figure 10–1. Types of compressors (GPSA, 1987, p. 13–2).

design flexibility to refineries, petrochemicals, and gas processing plants. Because these two types are not the most common drivers for oilfield compressors they are not described here; excellent reviews are available, e.g., GPSA (1987, Section 15); Bloch *et al.* (1982), or Neerken (1980). Oilfield pumps and their most common drivers—electric motors—are discussed in Volume 2.

A very important feature of any compressor installation is the *suction scrubber. It is absolutely necessary to prevent contaminants—solids or liquids—from entering a compressor.* Solids are often present in gas streams, either picked up in previous piping and equipment or rust and mill scale from the pipe and equipment walls. Salt may accumulate as brine evaporates. Liquid droplets result from condensation of water and/or process condensate or entrained corrosion inhibitors, glycol, or amine. Reciprocating compressors are particularly susceptible to dirt and liquid entry because the latter destroy cylinder oil film. Centrifugal compressors can tolerate some liquid entry in mist form but not in slug form. In any event, good practice demands an adequately sized suction scrubber or inlet vapor-liquid separator. Vapor-liquid separators are discussed in Volume 2.

Gas compression is reviewed using the following topics: reciprocating compressors, rotary compressors, centrifugal compressors, compressor design, gas engines, gas turbines, preventive maintenance, and selection of compressors and drivers.

RECIPROCATING COMPRESSORS

Compression Cycle

The main elements of a reciprocating compressor are the *piston* and *cylinder*. The driver moves the piston inside the cylinder, alternately pulling in and pushing out the process fluid (Fig. 10–2). The circular motion of the crank end of the connecting rod is translated into linear back-and-forth motion by the wrist-pin connection between the connecting and piston rods, guided by the crosshead sliding in its bearing.

As illustrated in Figure 10–3, the reciprocating cycle consists of four steps: *compression, discharge, expansion,* and *suction.* In Figure 10–3, compression starts at point 1 with the piston all the way out (bottom dead center) and both the suction and discharge valves closed. During compression (points 1 to 2), the piston moves in and compresses the gas trapped in the cylinder. At point 2 the discharge valve (D) opens, and the compressed gas is discharged between points 2 and 3. When the piston is all the way in at point 3, the discharge valve closes and the trapped gas expands from discharge to suction pressure as the piston moves back to point 4. At point 4 the suction valve (S) opens, and gas is drawn into the cylinder as the piston continues moving back to point 1, thus completing the cycle. Figure 10–3 describes the theoretical cycle based on the standard assumptions of isentropic compression and

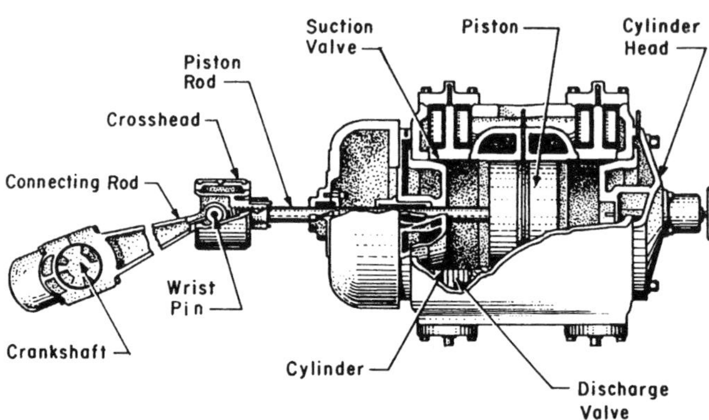

Figure 10–2.   Components of a reciprocating compressor (courtesy PETEX, 1987. Copyright © 1987 by the University of Texas at Austin).

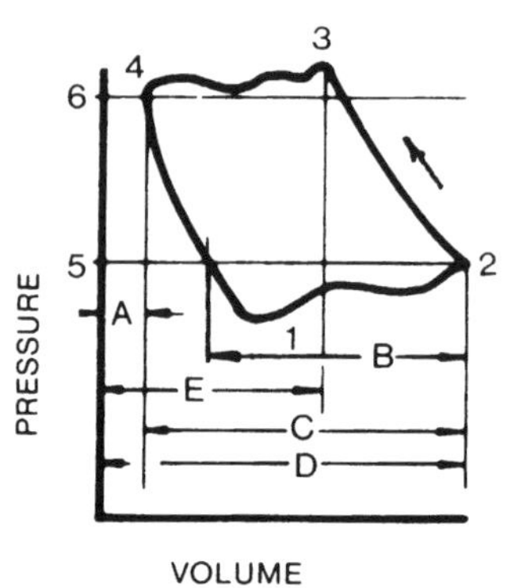

Figure 10–4.   Actual P-V diagram for reciprocating compressor (PMC/Beta Corp.).

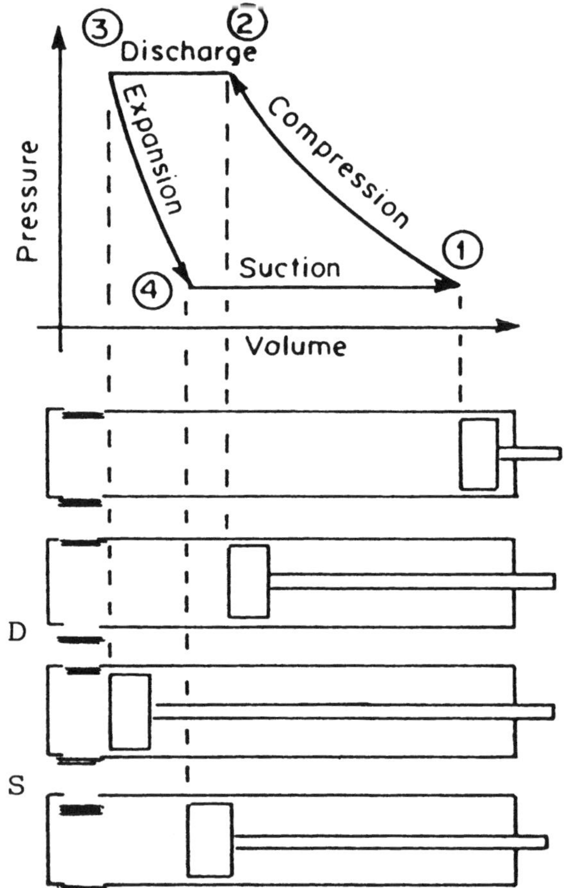

Figure 10–3.   Ideal P-V diagram for reciprocating compressor (reprinted from James, 1982b, courtesy of Marcel Dekker Inc.).

expansion and zero pressure differential in the valves. An actual cycle, shown in Figure 10–4, is obtained using a Beta analyzer.

Reciprocating compressors are prone to wear and tear during operation. The great number of moving parts, the sliding friction, the reversal of direction at rapid intervals, and the mechanical shock involved all contribute to severe wear.

A *double-acting cylinder* balances some of the load on the piston and rod components. Now both sides of the cylinder are used to compress gas, one side being in suction when the other is in compression. Double-acting pistons are often used for the first stages of compression while single-acting are common in the higher stages.

Valves

A reciprocating compressor resembles an internal combustion piston engine in construction and operation. One very important difference is the kind of valve used. Piston engines use tulip valves that are lifted by rods actuated by a camshaft. Piston compressors use automatic valves that operate from the pressure differential between line and cylinder pressure. On the suction stroke, the pressure in the cylinder must fall slightly below that of the gas entering the cylinder. Likewise, upon discharge the pressure in the cylinder must be slightly higher than that in the discharge line in order to actuate the valve. The flow efficiency of these valves is, therefore, paramount for overall compression efficiency.

Depending on the rotating speed, gas compressed and operating pressures, different valve concepts with various geometries are used (Fig. 10–5). Plate and ring valves prevail in medium and high speed gas compressors. Channel and strip valves are used in air compressors. Poppet valves

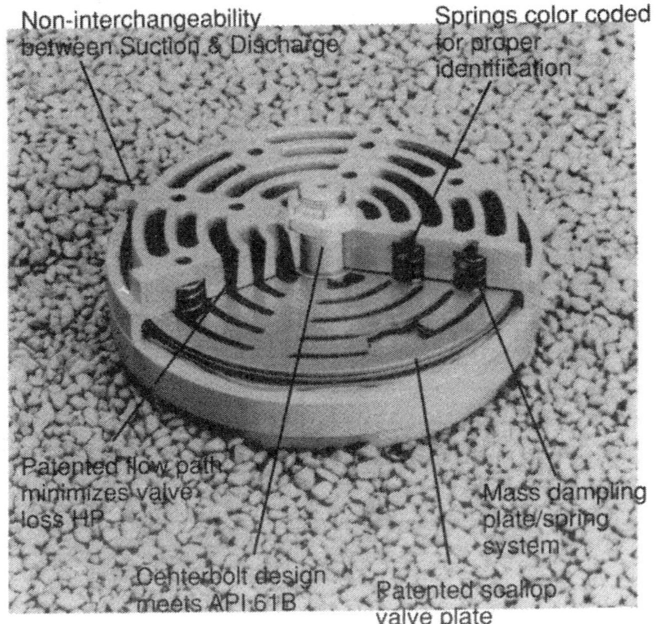

Figure 10–5a. Plate valve, Type PF (Dresser-Rand, 1990).

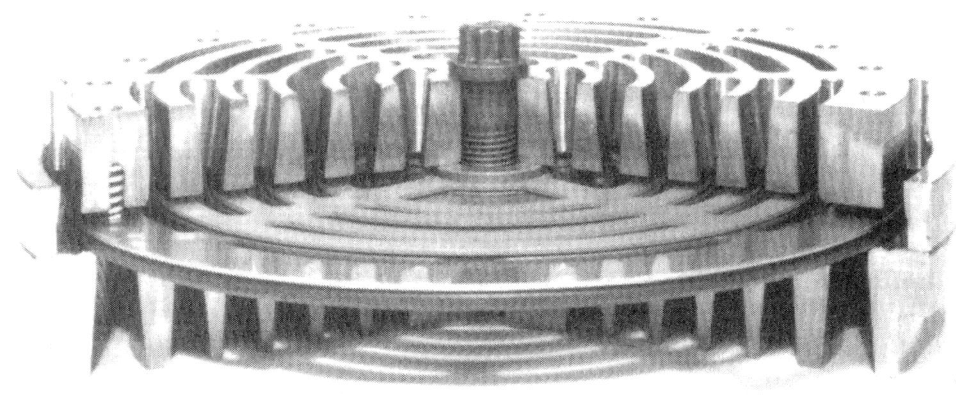

Figure 10–5b. Plate valve, Type CT (Hoerbiger, 1990).

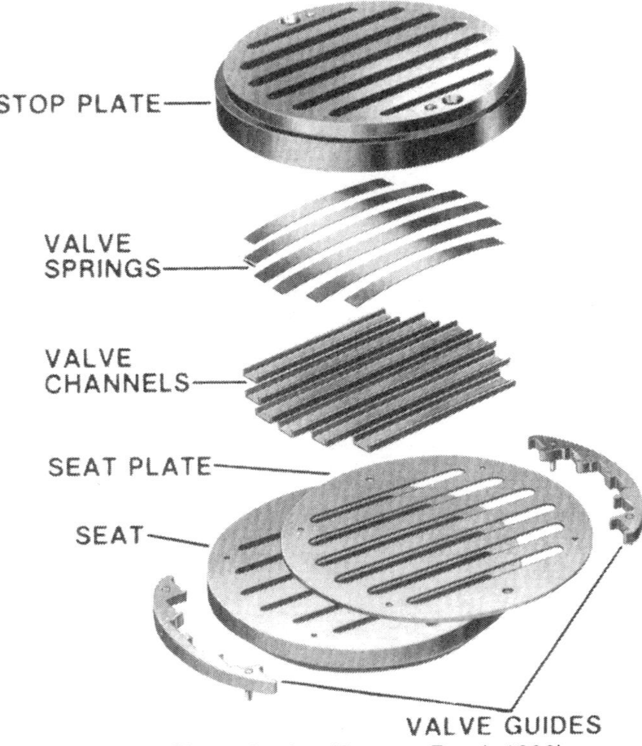

Figure 10–5c. Channel valve (Dresser-Rand, 1990).

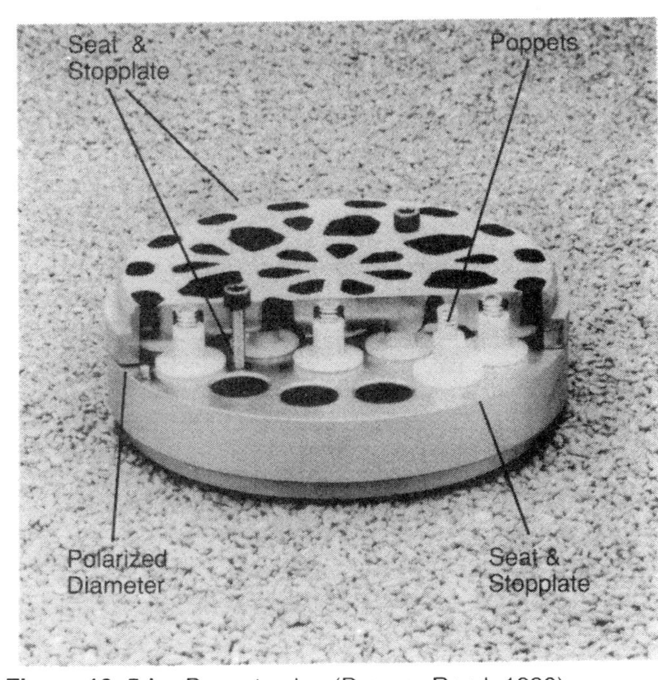

Figure 10–5d. Poppet valve (Dresser-Rand, 1990).

are found in gas transmission applications and low pressure process compression equipment.

Compressor valves need to be tailored to the operating conditions for optimum performance and valve life. The proper selection of springing is important when designing a valve for the given parameters of operation.

Although valves are considered maintenance items, a well-designed compressor valve should achieve a life cycle of uninterrupted service, depending on operating speed, usually from six months and extending to several years in slow speed and clean environment.

Clearance

Compressor throughput can be controlled by regulating the speed of the prime mover. The number of strokes per minute times the displacement volume gives the inlet (or suction) flow rate of gas delivered. When feasible, varying prime-mover speed is usually the best and simplest way to *control throughput*. When prime-mover speed cannot be varied sufficiently, the clearance in the head end of the cylinder is changed. Two types of *clearance control* are used (Fig. 10–6). In the first, a *fixed change in clearance* is obtained, either by lifting a plug using a hand-operated crank (Fig. 10–6a) or by actuating a relay-operated valve—or alternatively by changing out a plug in the head end during shutdown. In the second method, *variable clearance* is provided by a movable piston mounted in the head and adjusted using a crank-driven, threaded shaft (Fig. 10–6b). The larger the clearance volume at top dead center, the more slowly the pressure rises during the compression stroke and falls during the suction stroke—and the smaller the volume of intake gas. Figure 10–7 shows the reduction in throughput achieved by increasing clearance.

Unloaders

Sometimes the flow through a compressor must be shut off faster than can be done by shutting down the prime mover. This is accomplished using an *unloader*, which causes the intake valve to stay open throughout the stroke; no gas can be delivered because the pressure cannot increase above the suction pressure. Unloaders are often required for compressor start up to reduce starting torque. Figure 10–6b depicts a hand-operated valve-lifter or unloader.

Adjusting clearance and unloading clearly waste energy compared to slowing down or shutting off the driver, because the driver is not as efficient as at normal design conditions. This inefficiency is important when operating installations where several different compressor combinations can supply the required gas flow. Some combinations will be more energy efficient than others.

Pulsation Dampeners

The pulsating gas delivery by a reciprocating compressor creates pressure variations in the suction and discharge lines. Suction and discharge "bottles" are used to smooth out these pressure surges. The bottles furnish a quantity of gas that expands to retard pressure decreases and compresses to absorb pressure surges.

Lubrication

Proper lubrication is very important to good operation. Frame and cylinder/packing are lubricated by separate means. The extensive bearing surfaces must be provided with adequate lubrication to allow operation without excessive need for shutdown and maintenance. Lubrication is provided using an arrangement similar to that of an internal combustion engine. A sump below the compressor provides surge capacity for the oil pump that recirculates the oil through lines to small-diameter lubrication holes. Contamination of the compressed gas by lubricating oil can be a problem.

Nonlubricated units are available for unusual applications. Special packing and piston ring materials are required.

Sour Gases

When compressing sour gas:

(i) Follow NACE Standard MR0175–88 when selecting materials. In particular this means selecting a suitable high-grade stainless steel for all wetted surfaces. Also, avoid using any copper-based alloys.

(ii) Do not allow H2S to contaminate the crankcase oil. Gas leakage through the rod packings into the crankcase can be prevented either by using multiple packings or by venting any leakage to the flare.

(iii) Send a gas analysis to the cylinder lubricating oil supplier and/or the compressor vendor. A special lube oil formulation may be required to provide adequate lubrication and to prevent washing of lube oil off the cylinder wall and scoring.

(iv) Use NACE certified instrumentation, install a corrosion coupon in each cooler to detect any corrosion.

Control Devices

Typical control applications include unloading for starting, regulating the compressor output to match the demand, and protecting compressor (and engine) from overload.

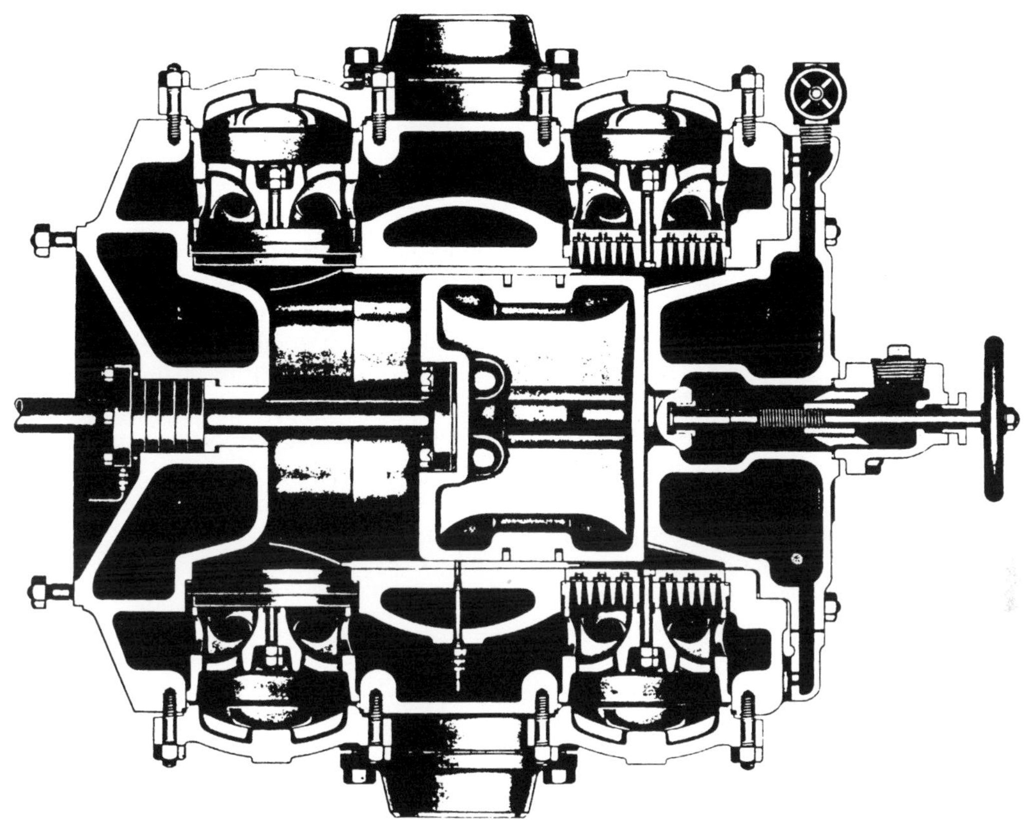

Figure 10–6a. Head-end fixed-volume clearance pocket (courtesy PETEX, 1987. Copyright © 1987 by the University of Texas at Austin).

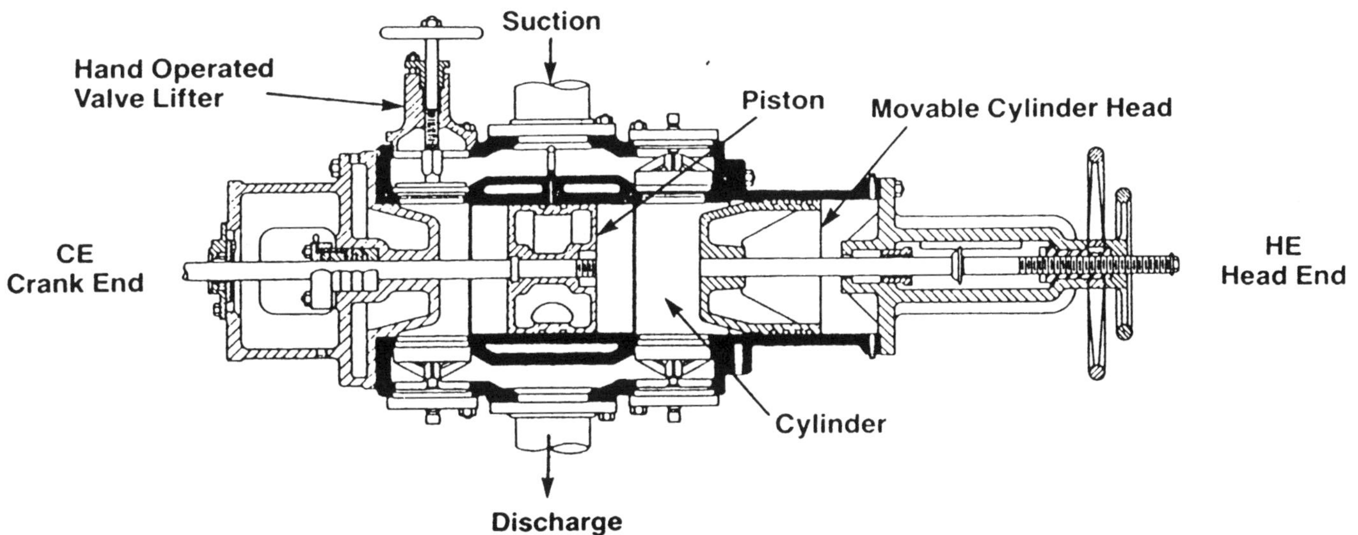

Figure 10–6b. Cylinder equipped with a hand-operated valve lifter and variable-volume clearance (GPSA, 1987, p. 13–19).

Start-up unloading is mandatory for both manual and automatic starting to avoid exceeding available driver torque. Unloading is accomplished by discharge to suction bypass, discharge venting, or using valve lifters to hold inlet valves open.

Capacity control: At constant speed and compression ratio a reciprocating compressor admits the same volume of gas. Therefore, when the inlet pressure rises, more gas (weight, moles, scf) is compressed, and more brake horsepower (Bhp) is required. When appropriate, speed control for the prime mover is used. With steam or gas-engine drivers, the capacity is readily reduced by supplying less steam or fuel. When constant-speed electric motors are used as drivers, the capacity may be reduced by increasing the clearance as shown previously in Figure 10–7. Remote operation of clearance volume can be provided when controlled by valving, the operator must go to the compressor unit to adjust crank-driven clearance.

On-off control must be provided for prime movers. Starting a reciprocating gas engine requires a source of compressed air or gas to initiate rotation of the drive shaft.

As mentioned above, in compressor systems with several compressors in series and/or parallel, driver energy requirements should be minimized by proper selection of the compressor combination.

Compression Ratio

The *compression ratio*, r, of a single compressor stage (ratio of outlet to inlet pressure in absolute units) is limited to about 5:1 to 6:1 for two reasons: rod loadings and discharge gas temperature.

First, excessive rod loadings must be avoided to prevent mechanical failure. As shown in Figure 10–8, the rod load varies cyclically from compression to tension. The *maximum rod loadings* are (Fig. 10–8)

$$\text{compression load} = P_d A_p - P_s (A_p - A_r) \quad (10\text{--}1)$$
$$= (P_d - P_s) A_p + P_s A_r$$

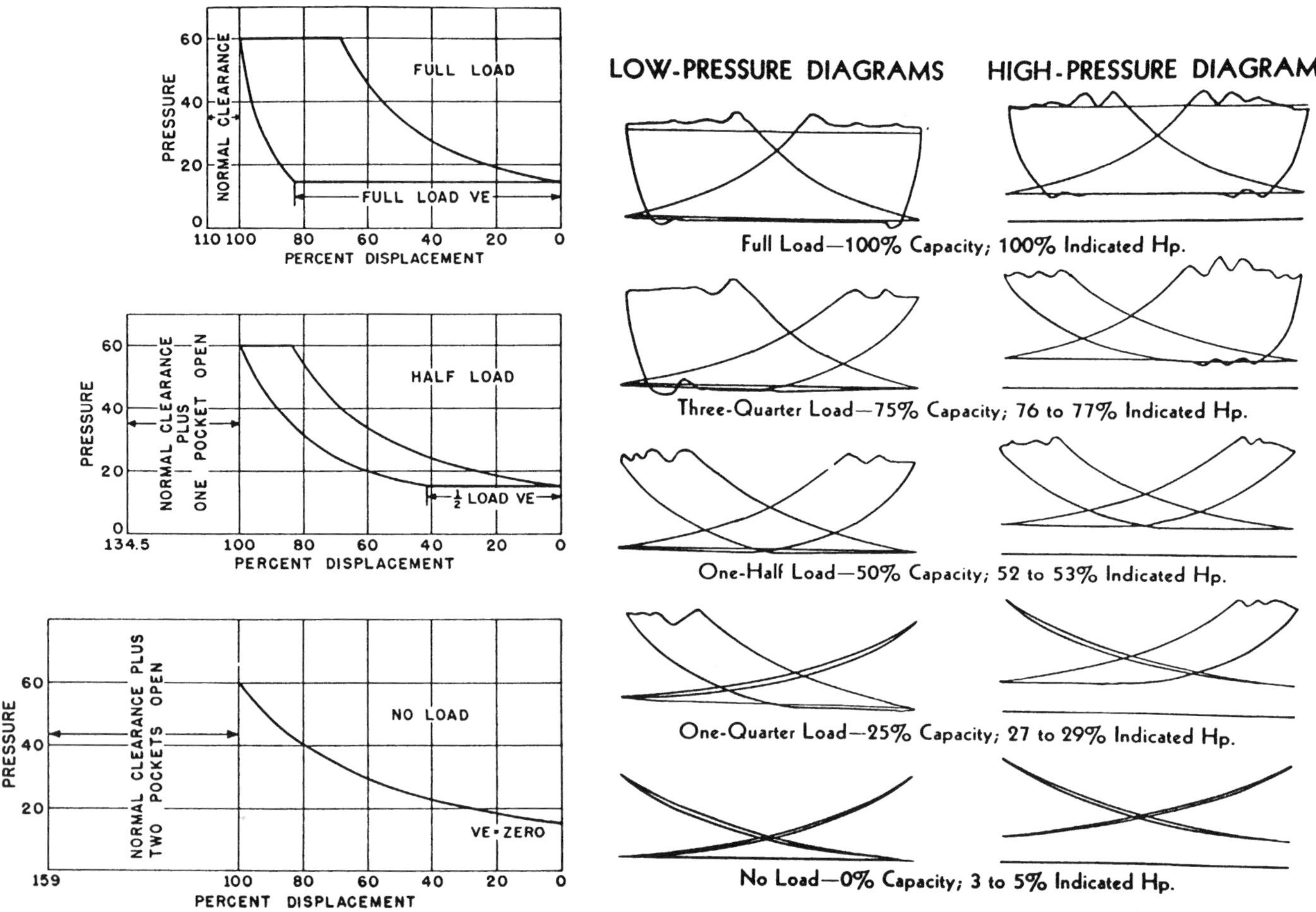

Figure 10–7. Reducing load by adding clearance (Ingersoll-Rand, 1980, pp. 4–6 and 4–7).

Rod in Compression

Direction of motion

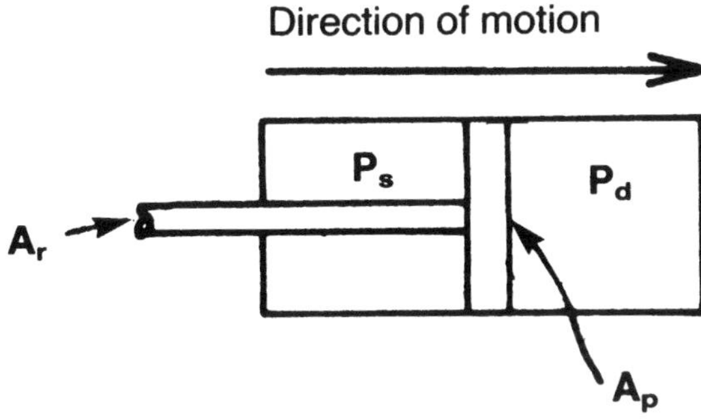

Rod in Tension

Direction of motion

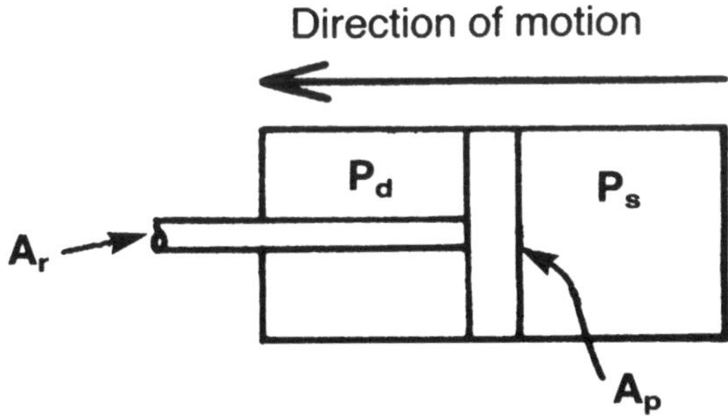

Figure 10–8. Rod loadings (GPSA, 1987, p. 13–9).

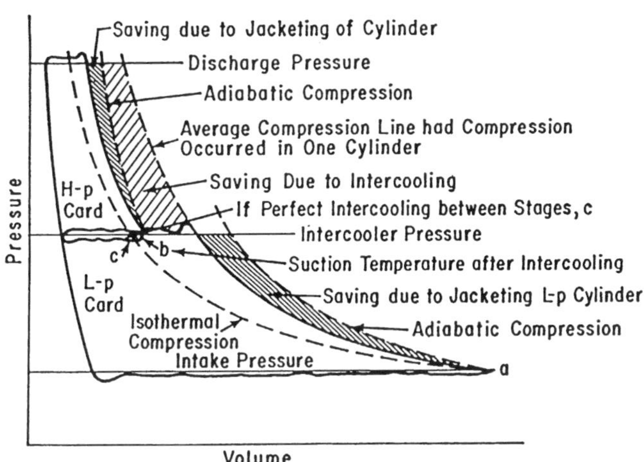

Figure 10–9. Combined indicator cards from a two-stage compressor show how cylinder water jackets and intercooler help bring compression line nearer to isothermal (Miller, 1944).

$$\text{tension load} = P_d (A_p - A_r) - P_s A_p \qquad (10\text{–}2)$$
$$= (P_d - P_s) A_p - P_d A_r$$

where P_d = discharge pressure
P_s = suction pressure
A_p = piston area
A_r = rod area.

Second, the gas is heated as it is compressed. The outlet temperature is normally limited to 300°F (150°C). Water or air cooling of the cylinders removes heat, but problems with lubricant breakdown and damage to compressor parts may still occur. If there is any doubt, consult the manufacturer.

High overall compression ratios are attained by using more than one compressor stage, or cylinder, with intermediate cooling of the process gas between stages and after the last stage. Figure 10–9 shows the P-V or indicator cards (or diagrams) for a two-stage compression. Figure 10–9 also shows how cooling (either by cylinder water jackets or by intercoolers) saves compression horsepower. Generally, intercooling savings are maximized by using equal compression ratios for all stages of compression. Figure 10–9 and the "equal r" rule-of-thumb both assume that no gas condensation occurs. When compressing natural gas, hydrocarbon liquids and possibly water can condense in the intercooler and must be completely removed in a scrubber before the next stage. Similarly, all liquids must be removed from the gas stream before it is fed to the first stage. It is very foolish to severely damage a million-dollar compressor as a result of attempting to save $1000 by undersizing the inlet separator or interstage scrubber.

Mounting and Housing

Reciprocating gas-engine/compressor units are large, noisy, and vibrating. Normal installation, especially for larger units, should be on a concrete foundation or pad. Some units are skid-mounted for portability and ease of interchange. Such an arrangement may be justifiable but will lead to increased maintenance, lubricant leakage, etc.

Compressor units are normally housed inside buildings for cleanliness and protection against the environment. Also, the noisy exhaust piping of a gas-engine driver can be directed to the outside. Gas engines can be very noisy and ear protection is often required. Small field compressors are sometimes mounted under simple roofs with no walls.

Materials of Construction

Martino (1983) reviews the materials used in reciprocating compressors. Frames and running gear are not exposed to process gas and the materials used are fairly standard. The frame is gray cast iron. Crosshead extensions also are cast iron, either integral or bolted on. Crankshafts and connecting rods are forged steel. Bearings are sleeve type and generally aluminum.

Distance pieces separate the driver end and the compressor-cylinder (gas end) and provide a seal against entry of lube oil. These pieces are generally cast iron.

Gas-end materials are more of a problem because of the temperature and pressure ranges encountered, as well as possible corrosion problems with the process gas. In addition, there are a large number of parts including cylinders, cylinder liners and heads, pistons, piston rods, valves, rings, and packings.

Previously, cylinder parts were cast but, due to environmental problems with making castings, are now forged or otherwise fabricated. Cast iron can be used for most gases up to about 1500 psig, both lubricated and nonlubricated. Nodular iron is used at pressures up to 2200 psig and steel above 2200 psig. Cylinder liners are generally gray cast iron, except at low temperatures where Ni-resist is used.

Factors other than pressure may dictate cylinder materials. Stainless or 3.5% nickel steel are used below $-60°F$. Alloyed cast iron is used from -60 to $-20°F$. Generally, sour gas is handled with the aforementioned materials if the H_2S content is 5% or less. At higher concentrations, stainless steel is used.

Large pistons are made of aluminum to decrease inertia. Smaller pistons are made of cast iron or steel. Stainless steel is used for low temperature and sour gas service. Piston rods are alloyed carbon steel, with some exceptions.

Most packing cases are steel or cast iron, with stainless steel used for special applications. Piston rings are TFE (tetrafluoroethylene) for pressures to 5000 psig and bronze at higher pressures.

Valve material selection is rather complex due to severity of service and the need for springs. Different materials are used for the various parts. See Martino (1983) for further details.

ROTARY COMPRESSORS

Rotary compressors are also positive displacement machines. Primarily a special purpose compressor, they are not used as much as reciprocating. Nevertheless, rotary compressors are important and are described briefly for completeness. The basic types of rotary compressors include lobed blowers, screw-type, sliding vane, and liquid-ring compressors.

Lobed Blowers

A common type of lobed blower is the two-impeller type, having rotors or impellers with a figure-eight shape (Fig.

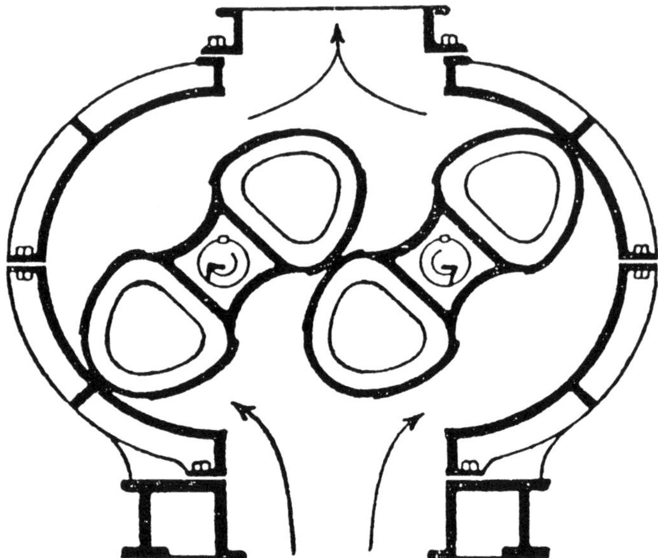

Figure 10–10. Two-impeller type of positive rotary blower (Perry and Green, 1984, p. 6–26).

10–10). The impellers rotate in opposite directions and entrap gas as shown in the figure, delivering it from suction to discharge. Contrary to the appearance of Figure 10–10, the rotors do not ride on each other or the compressor casing. One rotor is driven directly by the prime mover and the other is driven via a gear train to guarantee proper positioning. Small clearances are maintained between moving parts and the casing. Lobed blowers are inherently slow-speed, high-inertia devices. Typical applications are for rather large amounts of gases at low pressures and low compression ratios.

Rotary Screw Compressors

Screw compressors are also called *helical-* or *spiral-lobe* compressors. Figure 10–11 depicts a typical unit. Their action is much like that of the lobed blower except that the helical design demands that the two rotors not be identical in shape. One rotor has rounded tips to fit into the rounded grooves of the other rotor. The rotors do not need to have the same number of lobes since they may operate at different rotation rates.

When the two grooves encounter their respective inlet ports, the opposite ends are closed off to their discharge ports. Each groove fills with inlet gas. As the two grooves rotate toward each other, the rounded lobe encounters the rounded groove and traps the inlet gas in the groove. Further rotation reduces the volume containing the gas, compressing it. When the discharge end of the groove reaches the outlet port, the compressed gas is delivered to the discharge pipe.

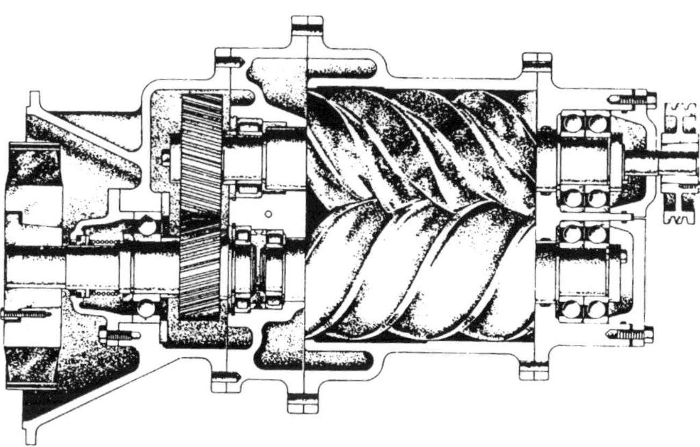

Figure 10–11. Cross section of typical single-stage oil-flooded screw compressor (Ingersoll-Rand, 1980, p. 8–2).

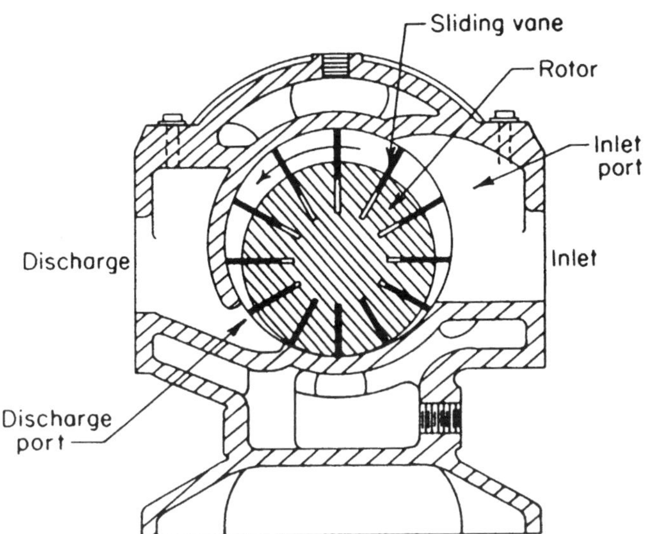

Figure 10–12. Sliding-vane type of rotary compressor (Perry and Green, 1984, p. 6–26).

There is a parasitic back leakage of gas to suction that can be deliberately increased to reduce throughput. As with the lobed blower, one of the rotors is driven directly by the prime mover and the other is driven via a gear train.

Screw compressors may be oil-flooded or dry. Oil in the former type lubricates the machine, furnishes a seal for the gas, and absorbs much of the heat generated during compression. This results in roughly ''isothermal'' compression and a lower Bhp. Oil-flooded rotary screws have 10 to 100% continuously-variable capacity control (by increasing the back leakage), easy start-up, and no pulsations. In air compression, where 2–3 ppm oil in the compressed air is acceptable for general shop requirements, rotary screws now dominate the market. Dry units have the advantage of delivering oil-free gas but have particularly high noise levels unless inlet and outlet silencers are provided. Both types are generally of low capacity, say a few hundred or thousand acfm. The units are higher in speed than the lobed blower and can attain higher compression ratios.

Sliding-Vane Compressor

The sliding-vane compressor consists of a rotor mounted eccentrically inside a casing. The rotor has longitudinal slots into which are fitted sliding vanes. Figure 10–12 is an end view of the compressor showing the action of the vanes in trapping the inlet gas and compressing it to the outlet port. In this respect, the sliding-vane compressor is like the reciprocating compressor, except that it needs no valves. Because of its design, however, the sliding-vane compressor has an internally fixed compression ratio. The exit gas, just before it goes to the exhaust port, has the same pressure regardless of the pressure in the discharge line, thus leading to inefficiency and limiting the pressure range.

Sliding-vane compressors are generally small units used for air compression or vacuum service. Speeds are moderate, roughly 600 to 1800 rpm. Drivers are usually electric motors directly coupled.

Liquid-Ring Compressors

Figure 10–13 shows a typical liquid-ring compressor. The rotor has forward-curved blades mounted on it that throw liquid against the outer casing, forming a liquid seal ring. The eccentric shape of the casing forces the liquid inward in the compression sector, compressing the gas and delivering it to the discharge line. Like the sliding-vane, the liquid-ring compressor has a fixed internal compression ratio. The liquid acts as lubricant, sealant, and coolant. Liquid-ring compressors are used for corrosive gases, especially where oil cannot be tolerated. Other services include vacuum and vapor recovery.

CENTRIFUGAL COMPRESSORS

Dynamic compressors consist essentially of one or more impellers mounted on a shaft that rotates at 2000 to 20000 rpm within a casing (Fig. 10–14). Stationary diffusing passages and guide vanes connect the flow between impellers. Each compressor ''stage'' includes one group of rotating impeller and stationary diffuser and guide vanes.

The two common types of dynamic compressors— *centrifugal* and *axial*—accelerate the fluid in the radial and the axial directions, respectively. Some blade designs place a compressor in an intermediate category called ''mixed-flow.''

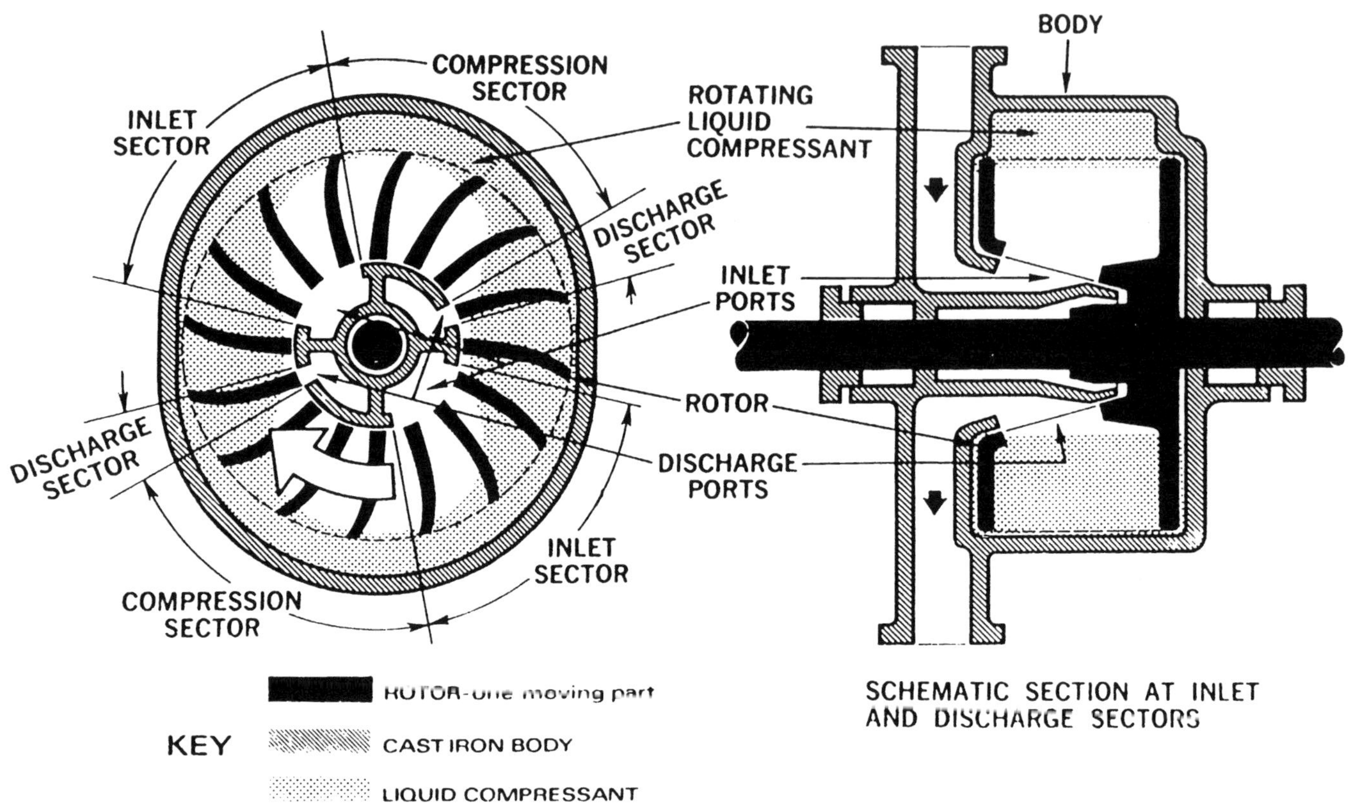

Figure 10–13. A typical liquid-piston rotary compressor (Copyright © 1970, The Nash Engineering Company).

When properly balanced and lubricated, centrifugal compressors exhibit little wear and tear. Typical operating time between shutdowns for routine maintenance is 1–2 years. The units are relatively lightweight and essentially vibration-free. Cheaper mounting is possible than for reciprocating compressors.

Mechanical Details

Figure 10–14 shows a typical multistage centrifugal compressor. Several impeller wheels are mounted on a rotating shaft supported horizontally on lubricated bearings. Casing-shaft seals are needed to prevent leakage of the process gas outwards. Since the gas flow is continuous, there is no need for inlet and outlet valves. In some applications, adjustable inlet guide vanes are used to control throughput.

Multiple wheels are often used since the compression ratio of a single wheel is limited. Up to 10 impellers may be used in a single frame. Some of these may be in series (to increase the compression ratio) and some in parallel (to increase capacity). Two or three frames in series may

be required for high-pressure applications. Intercooling of a hot gas stream can be done for a side stream from a single frame (i.e., between wheels) or between frames.

The impeller of the centrifugal compressor operates very much like that of a centrifugal pump. Gas enters in the eye of the impeller and is accelerated outward by the motion of the impeller blades. A portion of the imparted kinetic energy is transformed into pressure as the gas expands in the impeller. An additional smaller pressure rise occurs in the diffuser following the impeller.

A limitation in the operation and design of centrifugal compressors is imposed by the phenomenon known as "surge." Surge is an aerodynamic condition occurring at low flow and constant speed in which vibration may cause mechanical damage to the compressor. Obviously, it is absolutely imperative to prevent surge as is discussed under controls.

Impellers

The rotating impellers impart kinetic energy to the gas. Thus, impeller design is of great importance (Sayyed, 1976).

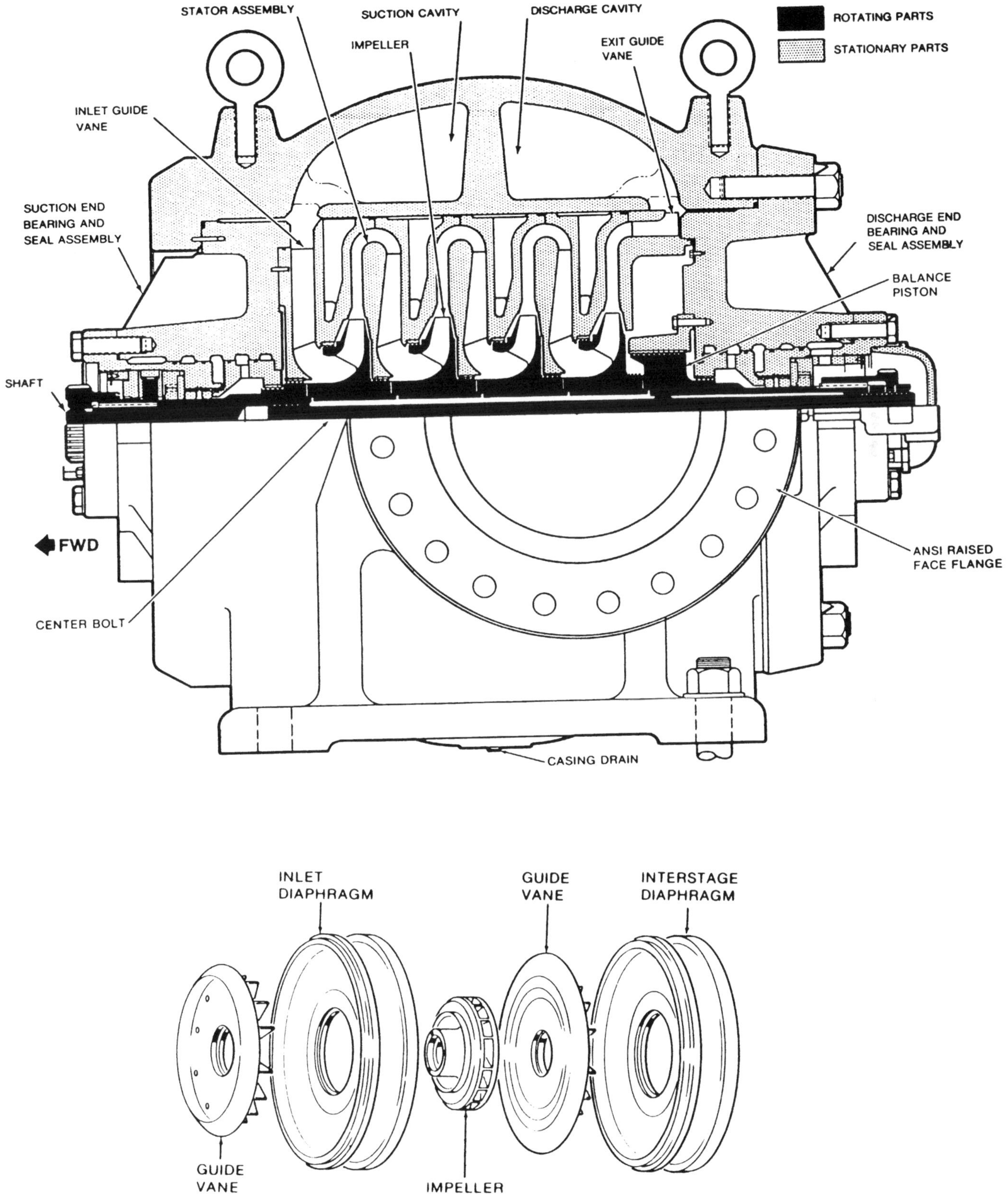

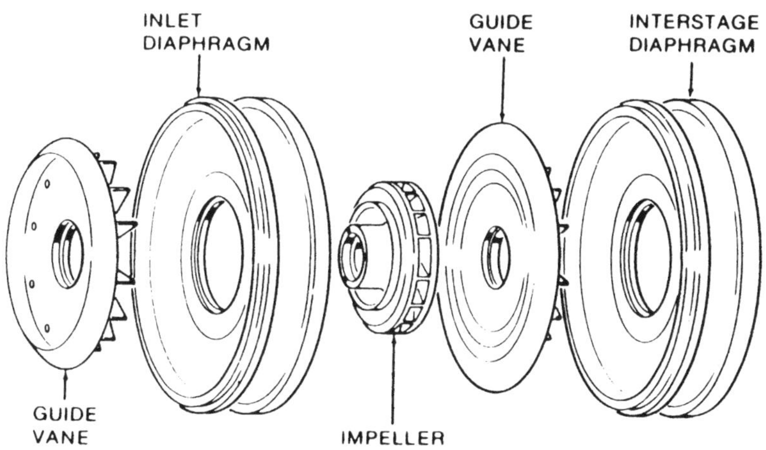

Figure 10–14. Solar centrifugal compressor (Solar Turbines, Inc., 1989).

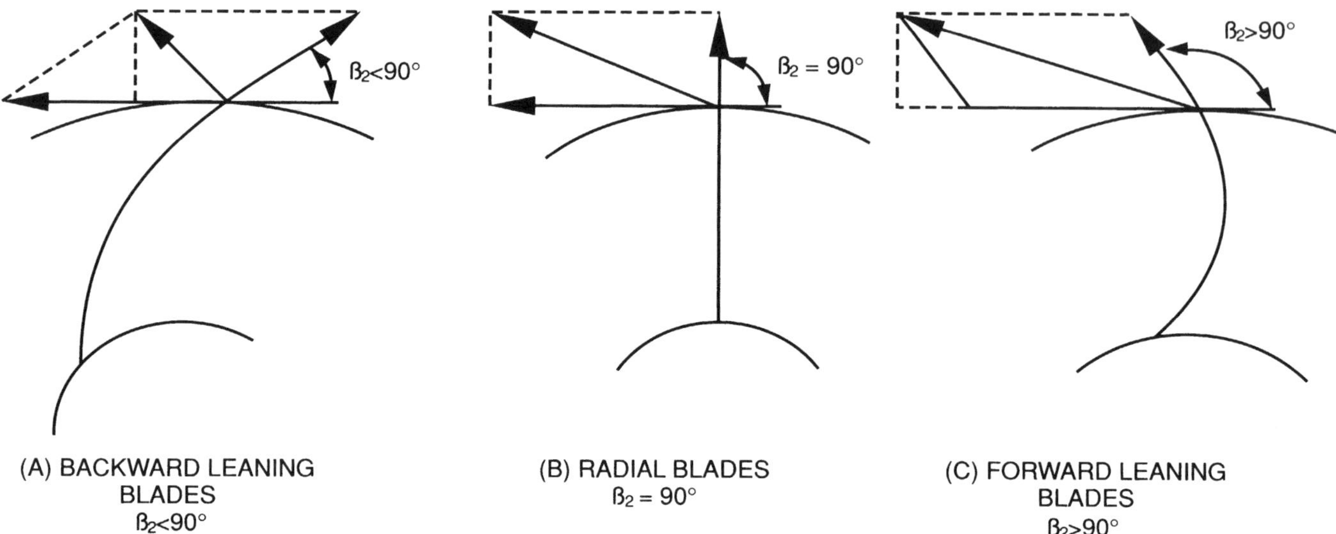

Figure 10–15. Centrifugal compressor impeller blade configurations (Sayyed, 1976).

Impeller blades usually lean backward (Fig. 10–15). The blades may be open or shrouded (Fig. 10–16). *Open impellers* have excellent operating characteristics but require closer tolerance with the stationary parts than shrouded impellers. *Shrouded or closed impellers* are made in three basic styles: three-dimensional, two-dimensional, and wedge (Fig. 10–17). Three-dimensional impellers have wider flow ranges than two-dimensional but are more expensive. Two-dimensional impellers are used mainly for low flow rates while wedge blades are used for very low flows.

Stationary Parts

Nozzles, diaphragms, diffusers, and guide vanes are included in the stationary parts. These parts guide the gas into the impellers, channel it between impellers, and deliver it to the discharge chamber.

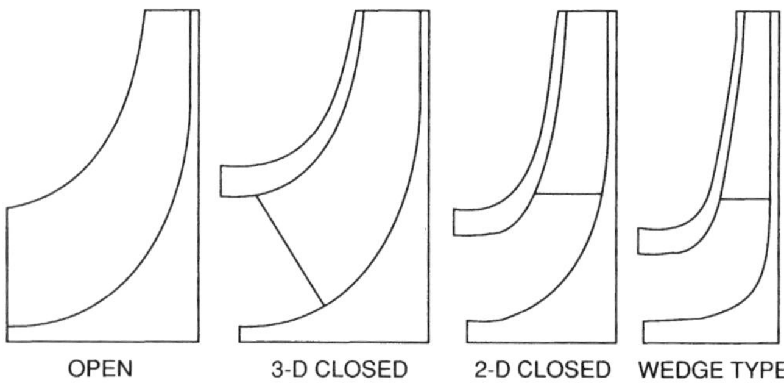

Figure 10–16. Centrifugal compressor impeller types (Sayyed, 1976).

Inlet nozzles accelerate the fluid into the impeller with minimum friction loss and provide a uniform velocity profile for the gas entering the impeller. Inlet nozzles may be axial or volute. Axial nozzles provide better velocity profiles, especially at high specific speeds, but require more space than volute nozzles (Sayyed, 1976).

Diffusers are expanding flow channels that guide the gas from the exit of one impeller to the inlet of the next and convert some of the kinetic energy imparted by the impeller into pressure energy. Diffusers may have vanes for improvement of aerodynamic behavior, but vanes limit the flow range greatly and are often omitted. Outlet nozzles are special diffusers that convey the gas to the discharge cavity and are axial or volute.

Diaphragms are the solid pieces that form the walls of the channels constituting the nozzles and diffusers.

Guide vanes are sometimes used in the inlet nozzle to provide a desired flow direction to the impeller inlet. Guide vanes may be fixed or adjustable. The latter are sometimes used to provide a measure of flow control through the compressor.

Casings

The centrifugal compressor casing houses the stationary and rotating parts and forms the outside wall of the inlet and outlet chambers. Two basic types of casing are used, the horizontally split casing and the vertically split or barrel casing.

Figure 10–18 depicts a typical *horizontally split* casing with the top half removed. The two pieces of the casing

—Wedge type impeller without cover plate.

—Two dimensional shrouded impeller.

—Three dimensional shrouded impeller.

Figure 10–17. Centrifugal compressor impeller designs (Sayyed, 1976).

are bolted together. This is the preferred type of casing for single or multistage compressors. The chief limitation on the horizontally split casing is its low pressure rating. For higher pressures, the *barrel-type* casing is superior (Fig. 10–19a). The barrel casing consists essentially of a cylindrical shell with end covers bolted at low pressures (Fig. 10–19b) and held in place by shear rings at high pressures (Fig. 10–19c).

Bearings

The *main bearings* on a centrifugal compressor rotor provide a lubricated suspension for the rotor. These bearings are generally located outside the casing to prevent oil contamination of the process gas and vice versa. The bearings are of the sleeve, ball, or roller bearing types, depending on the service. A thrust bearing is provided on

Figure 10–18. Horizontally-split centrifugal compressor (Neerken, 1975, Vol. 95, No. 10, 22–26, 1988).

the low-pressure end of the rotor to absorb unbalanced pressure force in the axial direction.

The pressures on either side of each impeller wheel are not much different. The axial thrust on the impeller is caused mostly by the larger area on the back side of the impeller than on the front side. A so-called balancing drum is placed on the discharge side of the rotor, with the high pressure on the inside and suction pressure on the outside (Fig. 10–20). This approximately balances the pressure forces exerted on the impellers. An axial thrust bearing is provided on the low-pressure end of the rotor to absorb the remaining force.

Seals

The pressure differences between the inlet and outlet sides of each impeller and between the inside and outside of the casing provide strong driving forces for leakage, thus requiring seals such as those shown in Figure 10–21. The *labyrinth seal* (Fig. 10–20) is invariably used for interstage applications. The arrangement in Figure 10–21 has the advantage that the casing-to-atmosphere seal must withstand only the suction pressure at both ends of the compressor. Because labyrinth seals always leak process or seal gas, they are not used as casing-atmosphere seals except for air whose leakage to atmosphere poses no problems.

Mechanical (contact) seals operate with zero clearance and oil pressures 35–50 psi above the internal gas pressure. Liquid film seals (pumping and cylindrical bushings) are suitable for severe service but require higher oil circulation than the mechanical contact type. The sleeves must be lined with babbit or other non-galling material that is compatible with both the gas being compressed and the circulating oil. The circulating oil acts as the sealing fluid, cools the seal area, and lubricates the sleeves.

Balancing

Centrifugal compressors are *high-speed devices*, rotating in the range of 5000–20000 rpm. Rotating shafts have critical

(a) High-pressure barrel with shear-ring closure.

(b) Low-pressure barrel compressor closure employing a ring of bolts to secure the end wall.

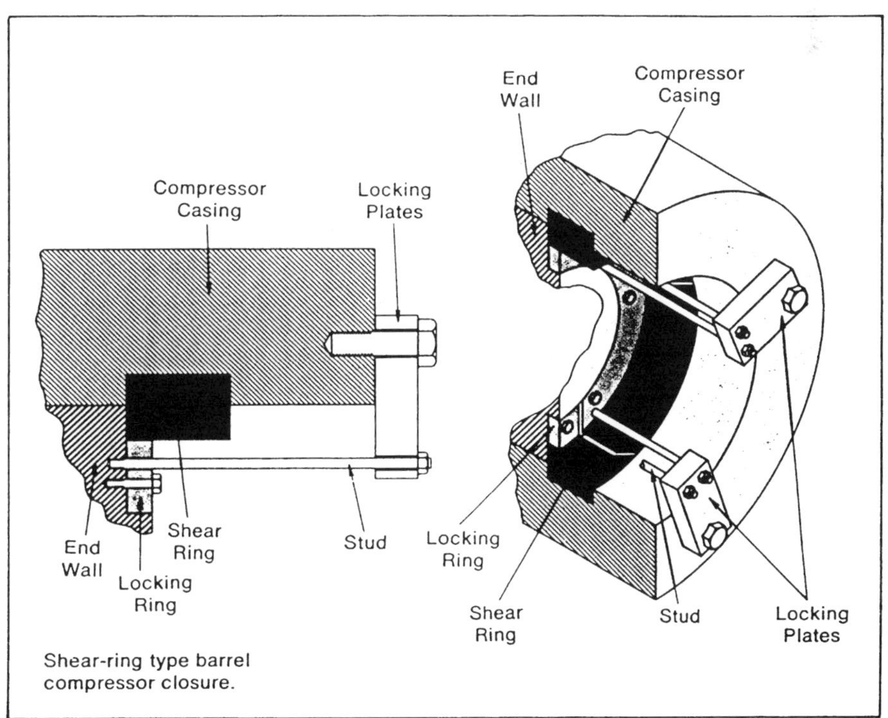

(c)

Figure 10–19.   Barrel-type casing (Elliott Co., 1981).

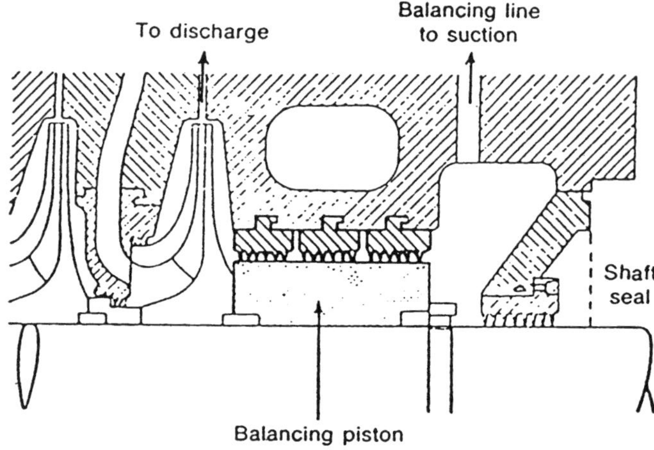

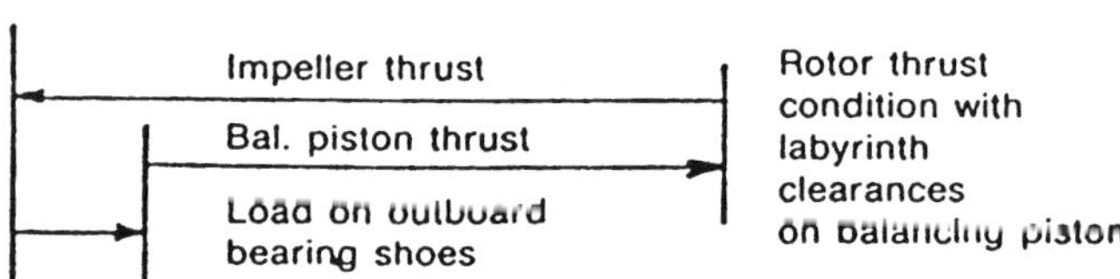

Figure 10–20. Centrifugal compressor balancing piston (Nelson, 1977).

speeds corresponding to their natural vibration frequencies. Lateral vibration at these speeds will be multiplied by resonance with the shaft, and structural failure can result. Normal rotor speeds exceed the first or higher critical speeds, thus demanding that a rotor be properly balanced when installed to minimize vibration.

Compression Ratio

The compression ratio of a single impeller is inherently limited to 1.2–1.5. However, the overall compression ratio of a compressor can be increased by adding impellers in series.

The *temperature limit* is normally about *300°*F, as dictated by compressor materials. Higher temperatures can be tolerated with proper selection of materials.

Materials

Selecting materials to use in turbomachinery (dynamic compressors and turbines) is very complex. Many factors must be considered such as operating stresses of rotating and stationary components, corrosion, and operating temperatures (Cameron *et al.* 1982). Appropriate

construction materials are listed in API Standard 617, Appendix B.

Centrifugal compressor impellers are usually made from alloy steel such as AISI 4140 or 4340. Blades can be fillet welded to the hub and cover (fully closed impeller). Impellers are heat treated to provide the desired hardness and strength. Consult Cameron *et al.* (1982) for further details.

Controls

Figure 10–22 presents typical characteristic curves for a centrifugal compressor and shows how the gas flow rate, head generated (discharge pressure–suction pressure) and speed are related. At constant speed, a compressor will not operate stably below a certain minimum flow called the *surge point*. Surge occurs when the compressor cannot generate the head required to overcome the imposed discharge/suction pressure ratio. As the flow decreases, partial stall (or minor recirculation near the impeller tip) occurs first. If the gas throughput is decreased further, violent surge or oscillation of the entire flow in the compressor occurs. **Surge is very harmful** to the compressor (Sapiro) because:

1. Rotor vibration can damage interstage labyrinth seals.
2. Flow reversal continuously increases temperatures in the compressor by bringing higher temperature gas into the compressor. Such overheating can damage seals and bearings.
3. Rapid changes in axial thrust may damage thrust bearings.

Two types of controls are needed: process and anti-surge. *Process controls* regulate the throughput, discharge pressure, and suction pressure in combinations consistent with the given compressor performance characteristics as shown in Figure 10–22. At a given compressor speed, setting any two of the three process variables fixes the third. *Anti-surge control* ensures the flow always exceeds the surge limit.

There are four basic control methods:

1. Suction throttling
2. Discharge throttling
3. Moveable inlet guide vanes in compressor
4. Driver speed regulation.

For minimum energy consumption, driver speed regulation is best. Variable-speed drivers are not always practical. Overall economics may favor constant-speed drivers, in which case the other three methods are used.

Anti-surge minimum flow limiter control continuously measures compressor flow and head to detect proximity

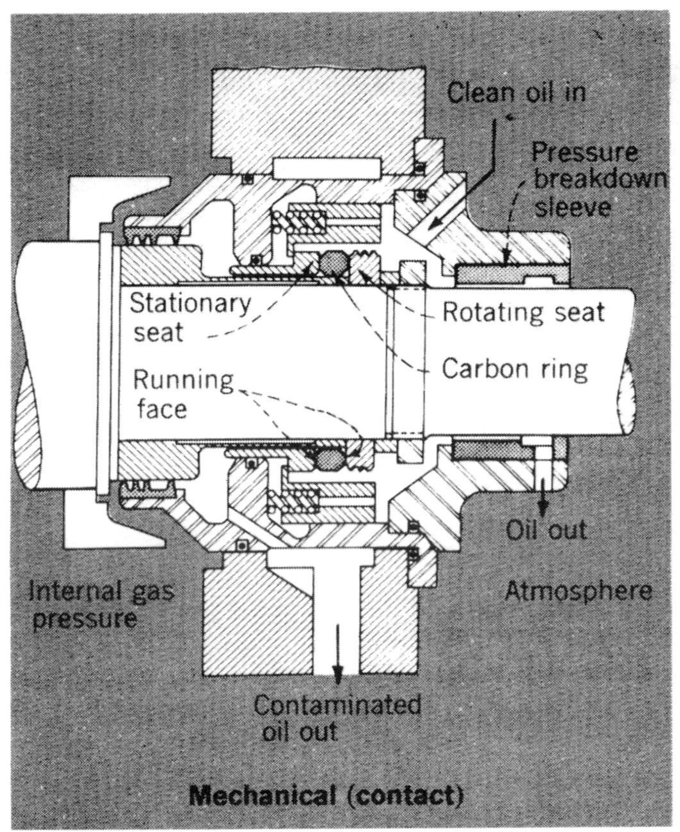

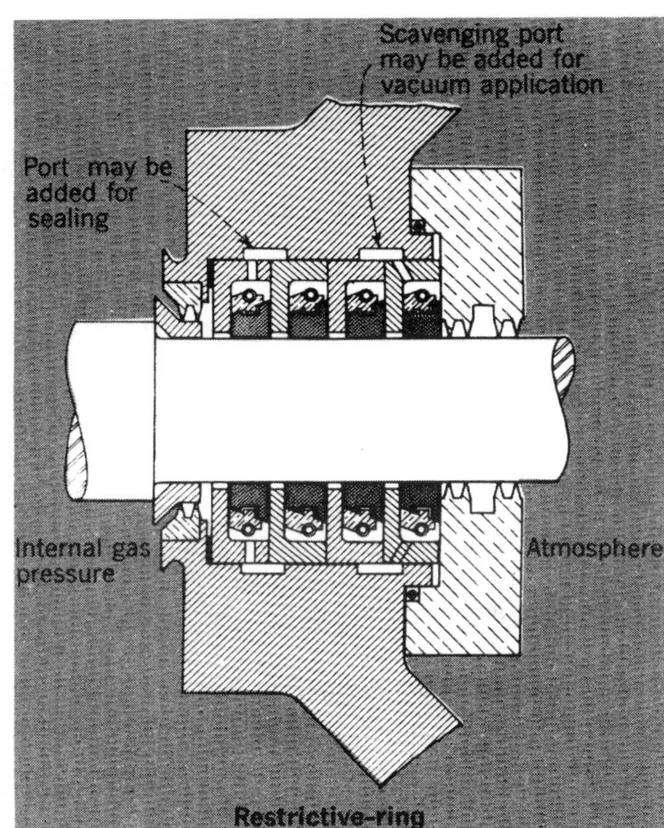

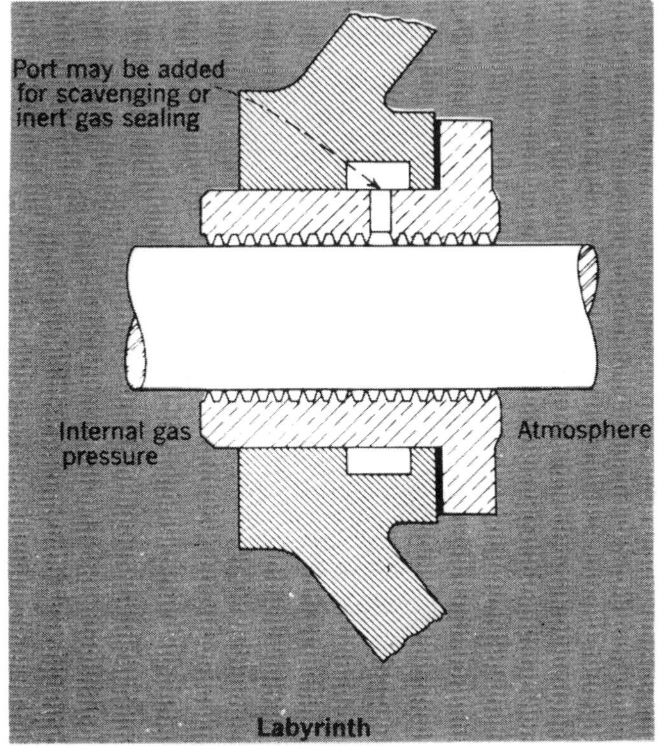

Figure 10–21.   Shaft end-seals for centrifugal compressors (API std 617).

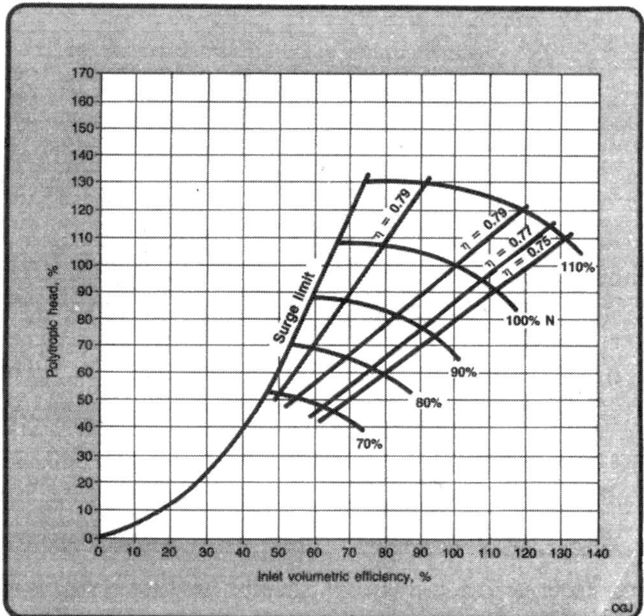

Figure 10–22. Typical centrifugal compressor head-capacity curves (Campos & Rodrigues, 1988).

to the surge line, i.e., minimum flow line. At a preset flow—surge minimum plus a safety allowance—the control opens a bypass valve downstream of the discharge cooler to recycle gas. By cooling the recycle gas, it is possible to run with 100% recycle without overheating the compressor.

Testing

Larger centrifugal compressors should pass a *performance test* in the manufacturer's shop before shipment. Performance shortfall experience suggests that the added expense is justified. The performance test should follow ASME Power Test Code 10 and be witnessed (Hunt, 1989).

COMPRESSOR DESIGN

The basic design equation is presented first, and the isotropic and polytropic compression models are described next. Then these models are applied to reciprocating and centrifugal compressors and some design examples given. Finally, volumetric efficiencies and water requirements of reciprocating compressors are discussed.

Basic Relations

Compressor design is based on the first and second laws of thermodynamics. The "first law" or *energy balance* is

written for the steady-state gas flow through the compressor. If kinetic and potential energy changes are neglected, as well as heat loss to the surroundings, then:

$$-W = h_2 - h_1 \qquad (10\text{–}3)$$

where W = compressor work, Btu/lbm or kJ/kg (negative
  because the work is done on the gas)
h_2 = discharge gas enthalpy, Btu/lbm or kJ/kg
h_1 = inlet gas enthalpy, Btu/lbm or kJ/kg

Use of Equation 10–3 requires that the inlet pressure and temperature and the outlet pressure and temperature be known to evaluate the change in enthalpy. However, in design work only the first three of these items are known. The outlet temperature cannot be arbitrarily specified. A *compression model* is needed.

Isentropic Model

This classic model assumes "second-law perfection"—the gas is compressed reversibly and adiabatically; that is, at constant entropy (isentropically).

$$-W_{is} = h_{2s} - h_1 \qquad (10\text{–}4)$$

where $-W_{is}$ = *isentropic work*, Btu/lbm or kJ/kg
h_{2s} = discharge gas enthalpy for a constant-
  entropy path, Btu/lbm or kJ/kg

An experimentally-determined compression efficiency, η_{is}, is now used to relate the theoretical isentropic work, W_{is}, to the actual work, W_{act}, required by the compressor

$$-W_{act} = -W_{is}/\eta_{is} \qquad (10\text{–}5)$$

where $-W_{act}$ = *actual compression work*, Btu/lbm or
  kJ/kg
η_{is} = *isentropic efficiency*, dimensionless

The compressor "gas" power is obtained by multiplying the work per unit mass by the mass flow rate.

$$p_g = \dot{m}\,(-W_{act}) \qquad (10\text{–}6)$$

where p_g = gas power, Btu/hr or kJ/s (kW)
$\dot{m}$ = mass flow rate, lbm/hr or kg/s

$$\textit{Gas horsepower, } Ghp = p_g\,(Btu/hr)/2544 \qquad (10\text{–}7)$$
$$= p_g\,(kW)/(.746)$$

The isentropic efficiency accounts for irreversibilities that occur as the gas is being compressed and are probably dissipated as turbulent friction within the gas. In addition, mechanical inefficiency occurs in the compressor due to rolling friction of the shaft in its bearings and the like. These inefficiencies are handled later. The total shaft power required by the compressor from the prime mover is called the *brake power* or Bhp.

The above equations may be applied directly to substances for which thermodynamic data are available in charts, tables, or computer programs.

For substances for which complete P-T-h-s data are not easily available, but for which ideal gas behavior is approached, simple equations can be derived. These equations are suitable for hand calculations. Many permanent gases approach ideal gas behavior at ambient and higher temperatures if the pressure is not too high. For an ideal gas:

$$-W_{is} = h_{2s} - h_1 = C_p(T_{2s} - T_1) \qquad (10\text{--}8)$$

where C_p = mean heat capacity at constant pressure between T_1 and T_{2s}, Btu/lbm-R or kJ/kg-K

T_1 = inlet gas temperature, °R or K

T_{2s} = discharge gas temperature, isentropic path, °R or K

The *isentropic coefficient*, k, is defined by

$$k = C_p / C_v \qquad (10\text{--}9)$$

where C_v = constant volume specific heat, Btu/lbm-R or kJ/kg-K

For an ideal gas,

$$C_v = C_p - R/MW \qquad (10\text{--}10)$$

where R = gas-law constant, energy/mol degree

MW = molecular weight of the gas

Combining Equations 10–9 and 10–10 gives:

$$C_p = R \, k / [MW \, (k\text{--}1)] \qquad (10\text{--}11)$$

Finally the *isentropic discharge temperature* for an ideal gas is:

$$T_{2s} = T_1 \, (P_2/P_1)^{(k-1)/k} \qquad (10\text{--}12)$$

where P_2 = discharge pressure, psia or kPa

P_1 = inlet or suction pressure, psia or kPa

Substituting Equations 10–11 and 10–12 into Equation 10–8 yields:

$$-W_{is} = \frac{R \, k \, T_1}{MW \, (k\text{--}1)} [(P_2/P_1)^{(k-1)/k} - 1] \qquad (10\text{--}13)$$

To calculate $-W_{is}$ in Btu/lbm, use R = 1.986 Btu/lbmol °R, T in °R, and P in psia. For $-W_{is}$ in kJ/kg, use R = 8.3145 kJ/kgmol K, T in K, and P in kPa.

Non-ideal gas behavior is allowed for empirically by:

$$-W_{is} = Z_{av} \frac{R \, k \, T_1}{MW \, (k\text{--}1)} [(P_2/P_1)^{(k-1)/k} - 1] \qquad (10\text{--}14)$$

where $Z_{av} = (Z_1 + Z_2)/2 \qquad (10\text{--}15)$

= average compressibility for the gas

Equation 10–14 is used henceforth exclusively. If the gas is approximately ideal, simply set $Z_{av} = 1$.

A commonly encountered form of Equation 10–14 involves substituting 1545 ft lbf/lbmol °R for the gas-law constant R. Thus, the work is in ft lbf/lbm, the so-called *adiabatic head*. The term "head" or "ft" is used because 1 lb mass of gas weighs 1 lb wt and so ft lbf/lbm can be considered ft lbf/lb wt or "ft."

$$H_{is} = Z_{av} \frac{1545 \, k \, T_1}{M \, (k\text{--}1)} [(P_2/P_1)^{(k-1)/k} - 1] \qquad (10\text{--}16)$$

where H_{is} = adiabatic head, ft lbf/lbm or "ft"

The gas horsepower, *Ghp*, is found by using the appropriate conversion factor.

$$Ghp = \dot{m} \, H_{is}/(1{,}980{,}000 \, \eta_{is}) \qquad (10\text{--}17)$$

Equation 10–12 does not estimate the actual discharge temperature of a compressor. A safe estimate assumes that all the irreversibilities in the compressor increase the gas enthalpy and thus the temperature. The gas *discharge temperature* is:

$$T_2 = T_1 + T_1 [(P_2/P_1)^{(k-1)/k} - 1]/\eta_{is} \qquad (10\text{--}18)$$

In practice, Equation 10–18 is often used with Equation 10–14, whether or not $Z_{av} = 1$.

Note that both isentropic and actual compressions are assumed to be adiabatic. Therefore, if the mean heat capacity, C_p, is the same over the two temperature ranges— T_1 to T_{2is} and T_1 to T_2—then:

$$\eta_{is} = (T_{2s} - T_1)/(T_2 - T_1) \qquad (10\text{--}19)$$

Some authors (Cohen *et al.*, 1975) use Equation 10–19 rather than Equation 10–5 to define η_{is}.

Polytropic Model

A second compression model uses a *polytropic coefficient*, n, in place of the isentropic coefficient, k. By definition, n provides the actual outlet or discharge gas temperature, T_2.

$$T_2 = T_1 \, (P_2/P_1)^{(n-1)/n} \qquad (10\text{--}20)$$

However, compressor manufacturers do not supply values for n; instead they provide a *polytropic* (or small-stage) *efficiency*, η_p. By definition, η_p is the isentropic efficiency of an elemental stage in the compression such that η_p is constant throughout entire compression (e.g., a multistage centrifugal compressor). Mathematically

$$\eta_p = dT_s/dT \qquad (10\text{--}21)$$

dT_s = elemental temp rise for isentropic compression, dP
dT = elemental temp rise for actual compression, dP

Integration of Equation 10–21 and use of Equation 10–12 yields:

$$T_2/T_1 = (P_2/P_1)^{(k-1)/k\eta_p} \qquad (10\text{–}22)$$

Comparison of Equations 10–20 and 10–22 shows:

$$(n-1)/n = (k-1)/(k\,\eta_p) \qquad (10\text{–}23)$$

The *adiabatic polytropic work* or head is:

$$-W_p = Z_{av}\frac{R\,n\,T_1}{MW\,(n-1)}\,[(P_2/P_1)^{(n-1)/n} - 1] \qquad (10\text{–}24)$$

Then the actual work of compression, W_{act}, is:

$$-W_{act} = -W_p/\eta_p \qquad (10\text{–}25)$$

Finally, the isentropic and polytropic efficiencies are related by:

$$\eta_{is} = \eta_p\,[(P_2/P_1)^{(k-1)/k} - 1]/[(P_2/P_1)^{(n-1)/n} - 1] \qquad (10\text{–}26)$$

Sizing (Estimating BHP)

Reciprocating compressors are usually sized using the adiabatic model and adiabatic efficiency. The mechanical efficiency is about 0.93. In other words, the brake power is given by the gas power divided by 0.93. Isentropic efficiencies for reciprocating compressors run fairly high, say 0.88–0.90. Contact a reliable manufacturer for more accurate sizing.

Reciprocating compressor horsepower can be estimated by three methods in addition to those previously described (GPSA, 1987).

1. *Quick Estimate* (GPSA, 1987, p. 13–4, Eq. 13–4)

$$BHP = (MF)(r)(N)(MMcfd)(F) \qquad (10\text{–}27)$$

where BHP = brake horsepower
MF = multiplication factor
  = 22 for large, slow speed (300–450 rpm) units with SG = .65 and r > 2.5
  = 20 when 0.8 < SG < 1.0
  = 16 to 18 when 1.5 < r < 2
r = compression ratio per stage
N = number of stages
$MMcfd$ = capacity at 14.4 psia and T_s
F = 1.0 for single-stage
  = 1.08 for two-stage
  = 1.10 for three-stage
T_s = suction temperature (°R).

2. *Quick Graphical* (see GPSA, 1987; p. 13–7, Fig. 13–9).

3. *Detailed Estimate* (GPSA, 1987; pp. 13–11 through 13–14)

   a. Calculate r for each stage.
   b. Estimate average $k = C_p/C_v$ for each stage using GPSA, 1987, Fig. 13–8, p. 13–6.
   c. Estimate discharge temperature, T_d, for each stage using Equation 10–12 or GPSA, 1987 Figure 13–11 or Equation 13–18.

$$T_d = T_s + T_s\,[r^{(k-1)/k} - 1] \qquad (10\text{–}28)$$

   d. Estimate compressibility factors at suction, Z_1, and discharge, Z_2, conditions and then calculate average $Z_{av} = (Z_1 + Z_2)/2$.
   e. Calculate Bhp using GPSA, 1987, Equation 13–21, p. 13–11.

$$Bhp = (Bhp/MMcfd)(P_L/14.4)(T_s/T_L)$$
$$\times (Z_{av})\,(MMcfd) \qquad (10\text{–}29)$$

   where Bhp/MMcfd is read from Figures 13–12, 13–13
   P_L = standard pressure, psia, for MMcfd
   T_l = standard temp, °R, for MMcfd

   f. Corrected estimate

$$BHP\,(corr) = Bhp\,(Eq.\,10\text{–}29)\,(CHS)$$
$$\times (CLIP)(CG) \qquad (10\text{–}30)$$
   where CHS = correction for high speed, p. 13–11
     CLIP = correction for low intake pressure, GPSA, 1987, Figure 13–14, p. 13–14
     CG = correction for gas gravity, GPSA, 1987, Figure 13–15, p. 13–14.

Centrifugal Compressors

Both the adiabatic and polytropic models can be used; however, the polytropic efficiency varies less with compression ratio. Centrifugal compressor efficiencies are lower than those for recips, of the order of 0.75. The Elliott table and chart (Fig. 10–23) can be used for estimation purposes. First find the polytropic efficiency in the table, then use the chart to find the adiabatic efficiency. Mechanical efficiencies are not generally used to determine brake power for centrifugal compressors. A rough rule is to simply add 50 HP for bearing and seal losses.

Pressure-Enthalpy Diagram. This method is the easiest to apply, involving straightforward use of the log P-h diagram. However, log P-h diagrams are generally available only for pure components, which limits the utility of the method. The method is now illustrated.

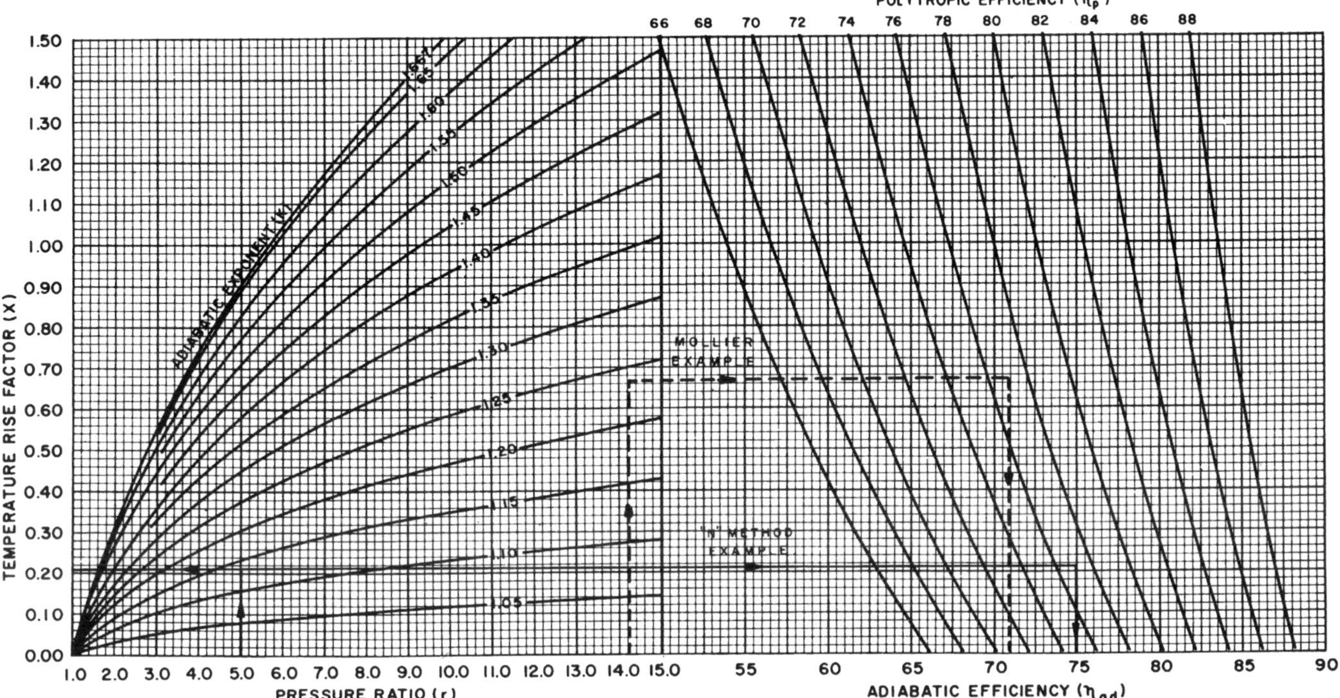

Figure 10–23. Elliott compressor data (Elliott Co., 1981).

Example 10–1. Compress 50,000 lbm/hr propane from 20 psia, 80°F, to 100 psia in a centrifugal compressor. The adiabatic efficiency is estimated to be 0.75.

Solution: See Figure 10–24 for the log P-h diagram. Locate the point 20 psia, 80 F, marked 1, for which $h_1 = -651$ Btu/lbm

Now follow a line of constant s upward and to the right to 100 psia, as indicated by point 2s in Figure 10–24.

$$h_{2s} = -611 \text{ Btu/lbm}$$
$$-W_s = h_{2s} - h_1$$
$$= -611 - (-651) = 40 \text{ Btu/lbm}$$
$$-W_{act} = 40/0.75$$
$$= 53.3 \text{ Btu/lbm}$$
$$h_2 = h_1 + (-W_{act})$$
$$= -651 + 53.3 = -598 \text{ Btu/lb}$$

This point is located on Figure 10–24 as point 2.

From the graph $T_2 = 211°F$.

The power is now computed.

$$p_c = 50,000 \times 53.5/2544 + 50$$
$$= 1100 \text{ hp} \quad \longleftarrow$$

Log P-h charts can be used for gas mixtures that are essentially pure—say 95 mol

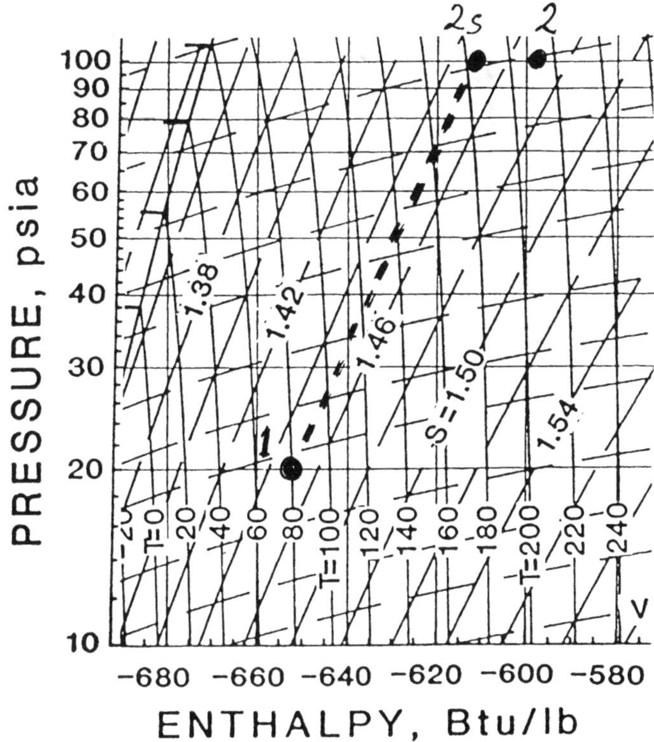

Figure 10–24. Compressor design using log P-h diagram (GPSA, 1987, p. 24–30, Fig. 24–6, drawn by B. C. Bain and J. F. Ely, NBS, Boulder, CO).

percent or better. When feasible, this method is the easiest, and most accurate.

Thermodynamic Tables

Starling (1973) presents tables of thermodynamic properties for a number of hydrocarbons. A portion of the propane table is shown below.

Psia		80°F	Psia		180°F	200°F	220°F
15	v	8.60678	100	v	1.45359	1.51021	1.56590
	h	−652.20		h			
					−614.16	−604.29	−594.23
	s	1.4608		s	1.4439	1.4592	1.4744
	f/P	0.9833		f/P	0.9372	0.9434	0.9488
25	v	5.10329					
	h	−653.16					
	s	1.4365					
	f/P	0.9720					

$$h_{2s} = -614.16$$
$$+(0.2157)(-604.29 + 614.16)$$
$$= -612.03 \text{ Btu/lb}$$
$$-W_{is} = h_{2s} - h_1 = -612.03 + 652.68$$
$$= 40.65 \text{ Btu/lbm}$$
$$-W_{act} = -W_{is}/\eta_{is} = 40.64/0.75$$
$$= 54.20 \text{ Btu/lbm}$$

Example 10–2. Solve the previous problem using the tabular data of Starling.

Solution: Find h_1 and s_1 by interpolation of the first column of data at 80°F.

Using linear interpolation for h_1:

$$h_1 = (-652.20 - 653.16)/2$$
$$= -652.68 \text{ Btu/lbm}$$

For an ideal gas $s + R \ln P$ is constant at constant temperature—therefore linear interpolation of s versus P at constant temperature is inaccurate.

s (15 psia)	= 1.4608 Btu/lbm °R	
(R/MW) ln 15	= (1.987/44) ln 15	
	= 0.1220 Btu/lbm °R	
s + (R/MW)ln 15	= 1.5828 Btu/lbm °R	

similarly

$$s + (R/MW)\ln 25 = 1.5816 \text{ Btu/lbm °R}$$

linear interpolation yields:

$$s + (R/MW) \ln 20 = 1.5822 \text{ Btu/lbm °R}$$

and finally

$$s (20 \text{ psia}, 80°F) = 1.4472 \text{ Btu/lbm °R}.$$

Find the temperature, T_{2s}, at 100 psia for $s_2 = s_1$.

$$T_{2s} = 180 + 20 \frac{(1.4472 - 1.4439)}{(1.4592 - 1.4439)}$$
$$= 180 + 20(0.2157) = 184.3 \text{ °F}$$

Find the actual outlet temperature from the outlet enthalpy by interpolation.

$$h_2 = h_1 - W_{act} = -652.68 + 54.20$$
$$= -598.48 \text{ Btu/lbm}$$
$$T_2 = 200 + (20) \frac{(-598.48 + 604.29)}{(-594.23 + 604.29)}$$
$$= 211.5 \text{ °F}$$

The power, p_c, is now computed

$$p_c = 50{,}000 \times 54.2 / 2544 + 50$$
$$= 1115 \text{ hp} \quad \leftarrow$$

We can see from the similarity of the numbers in the computations that the enthalpy chart is based on Starling's tabulated data. The results are the same within the ability to read the graph.

Computer Simulation

Flow sheet simulator packages provide quick and simple means of solving the compressor design equations. The thermodynamics properties are used in exactly the same way as in the hand calculations above. The only difference is the source of the properties, which in the simulator is provided by ideal-gas properties (v, h, s) combined with equation-of-state-derived corrections for real-fluids properties.

Example 10–3. Solve the previous example using the OPSIM flow sheet simulator.

Solution: Shown below is a print-out of the computer results.

Compressor Calculation:

Adiabatic Efficiency	=	.7500
Mechanical Efficiency	=	1.0000
Delta P	=	80.0
Gas Horsepower	=	1078.2
Compression Ratio	=	5.000

Strm No.	1	2
C3	1133.90	1133.90
Mol Hr	1133.90	1133.90
Lb Hr	50001.59	50001.59
T,Deg F	80.00	214.07
P,Psia	20.00	100.00
Mol Wt	44.10	44.10
Lb/CuFt	.1557	.6469
Btu/Hr	.71921E+07	.99362E+07
Btu/R-Hr	.66529E+05	.67569E+05
Frac. Vap	1.0000	1.0000

If 50 HP is added for bearing and seal losses, the horsepower is 1128, and the outlet temperature is 214.07 °F, which is excellent agreement with the thermodynamic tables.

Design Equations. The previously described isentropic and polytropic models generally provide reasonably good estimates.

Example 10–4. Use the Isentropic Model—Equation 10–14—to solve the previous Example 10–1.

Solution: The heat capacity of propane at an assumed average temperature of 150°F is:

$$C_p = 19.52 \text{ Btu/lbmole-°F (GPSA, 1987, p. 13–5)}$$

Use the ideal-gas relation, $C_v = C_p - R$

$$C_v = 19.52 - 1.99$$
$$= 17.53 \text{ Btu/lbmole-F}$$
$$k = C_p/C_v = 19.52 / 17.53 = 1.1135$$

Check the temperature assumption using Equation 10–18.

$$T_2 = 80 + (540)[(100/20)^{(.1135/1.1135)} - 1]/0.75$$
$$= 80 + 128 = 208 \text{ °F}$$
$$T_{av} = (80 + 208)/2 + 144°F, \text{ which is close enough}$$
to the assumed value of 150.

but $Z_{av} = 0.96$ (Starling's data)
$MW = 44.097$ (GPSA, 1987, p. 23–2, Fig. 23–2).

$$-W_s = \frac{(.96)(1.986)(1.1135)(540)}{(44.097)(0.1135)}$$
$$\times [(100/20)^{.1135/1.1135} - 1]$$
$$= 40.82 \text{ Btu/lbm}$$
$$- W_{act} = 40.82/0.75 = 54.4 \text{ Btu/lbm}$$

This is close to the previous solutions, but not as accurate.

Example 10–5. Use the polytropic model to solve Example 10–1. Assume $\eta_p = 0.77$.

As in Example 10–3, estimate $k = 1.1135$ at assumed average temperature of 150°F.

$$(n–1)/n = (k–1)/(k)(\eta_p)$$
$$= (1.1135 - 1.0)/(1.1135)(.77) = 0.1324$$
$$T_2 = T_1 (P_2/P_1)^{(n–1)/n}$$
$$= (80 + 460)(100/20)^{0.1324}$$
$$= 668°R = 208°F$$

$$-W_p = Z_{av}\frac{R n T_1}{MW (n-1)} [(P_2/P_1)^{(n–1)/n} - 1]$$

$$= (0.96)\frac{(1.986)}{(44.097)}\frac{1}{.1324}(540)[(100/20)^{.1324} - 1]$$

$$= 41.9 \text{ Btu/lbm}$$
$$-W_{act} = 41.9/0.77 = 54.4 \text{ Btu/lbm}.$$

Reciprocating Compressor Cylinder Capacity

Obviously the piston cannot travel completely to the end of the cylinder at the finish of every discharge stroke. Some *clearance volume*, CV, is required and, therefore, some gas is left in the cylinder after the discharge stroke. As the piston travels away from the head end, this residual gas expands until the pressure is slightly less than that of the suction line. Then inlet gas can enter through the intake valves. The point is that the total piston displacement, PD (stroke length times cross-sectional area), is not available for throughput. Mathematically, the percent clearance, C (%) = 100 CV/PD. In turn *volumetric efficiency*, VE(%), is defined as the actual pumping capacity (at intake conditions) divided by the piston displacement. In theory, the volumetric efficiency can be calculated from the ideal cycle

$$VE = 100 - C (r^{1/k} - 1)$$

To repeat, even in theory VE is less than 100% because the high pressure gas left in the cylinder at the end of the discharge stroke expands and so delays the opening of the suction valve.

In practice, the VE is further reduced empirically to

account for real gas behavior ($Z \neq 1$), suction and discharge valve losses, etc.

$$VE = 96 - r - C\,[(Z_s/Z_d)(r^{1/k}) - 1]$$

Deduct an additional 5% for non-lubricated compressors and still another 4% for propane or other "heavy gases."

Example 10–6. A compressor cylinder has 100 cfm displacement (displacement volume times rpm) and a clearance of 10 percent. What is the volumetric efficiency for a compression ratio of 4.0 and $k = 1.3$? Assume ideal gas behavior.

Solution:

$$VE = 96 - 4 - 10[(4)^{1/1.3} - 1]$$
$$= 96 - 4 - 10(1.905) = 73\,\% \quad \leftarrow$$

The throughput is 73 cfm (approximately).

INTERNAL-COMBUSTION GAS ENGINES

Gas engines may be described and/or classified as follows:

Combustion cycle: two-stroke
             four-stroke

Cylinder arrangement: vertical/upright
             horizontal
             V-type
             opposed
Power impulse: single acting
Speed: low (below 700 rpm)
             medium (700–1500 rpm)
             high (above 1500 rpm)
Connection: belt or separable
             direct drive
             integral

Gas engines are now discussed using the following topics: combustion cycles, engine description, and a brief comparison of two- and four-stroke engines.

Combustion Cycles

Figures 10–25 and 10–26 show schematic sketches of two-stroke and four-stroke cycle engines, respectively.

The conventional *two-stroke engine* is a compact unit consisting of gas engine and scavenging piston. Each cycle is one full revolution or two strokes. As shown in Figure 10–25, when the piston moves down (towards the crank

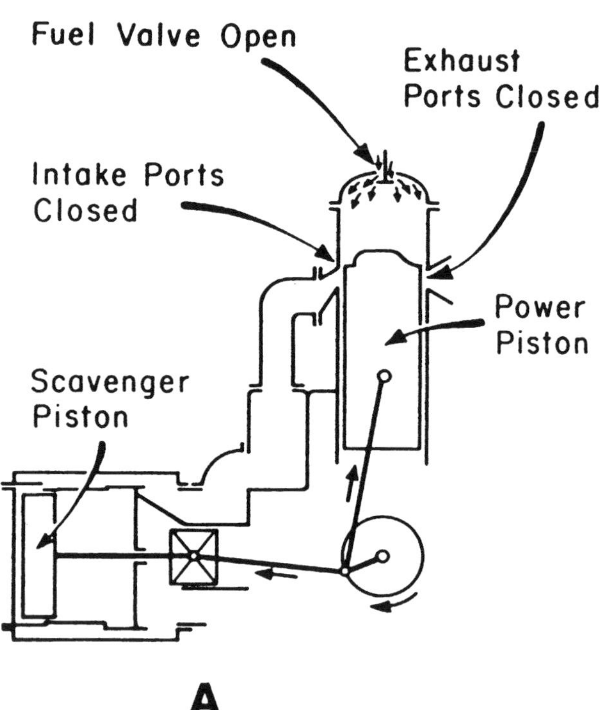

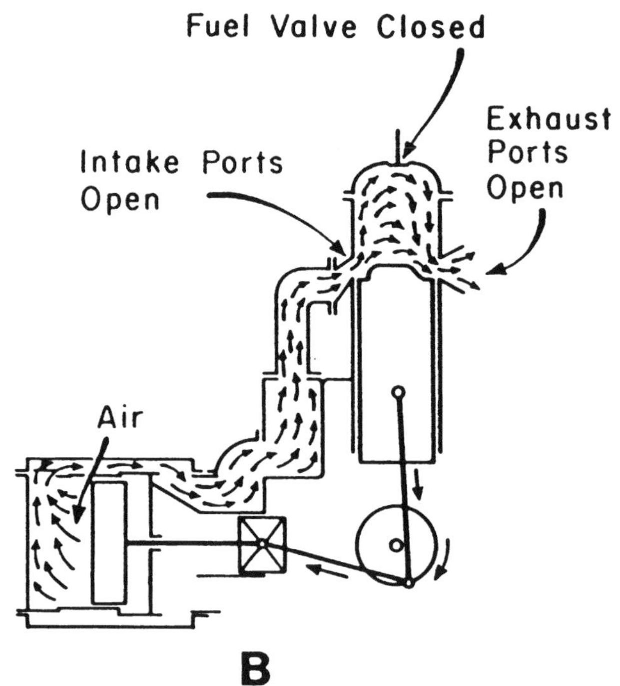

Figure 10–25. Schematic sketch of a two-stroke cycle engine operation (courtesy PETEX, 1987. Copyright © 1987 by the University of Texas at Austin).

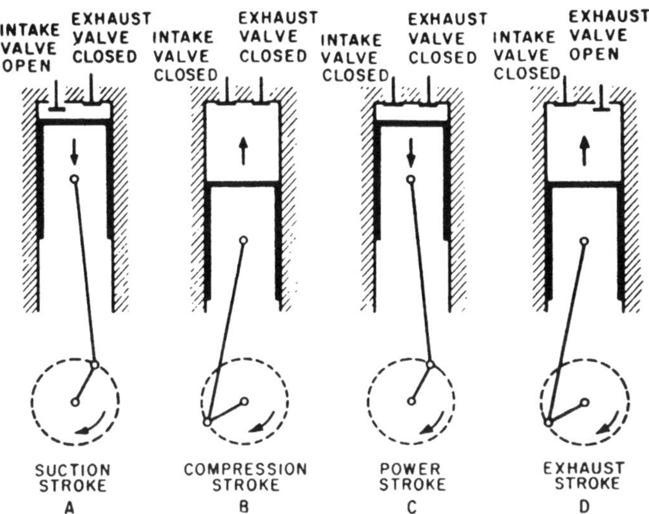

Figure 10–26. Schematic sketch of a four-stroke cycle engine operation (courtesy PETEX, 1987. Copyright © 1987 by the University of Texas at Austin).

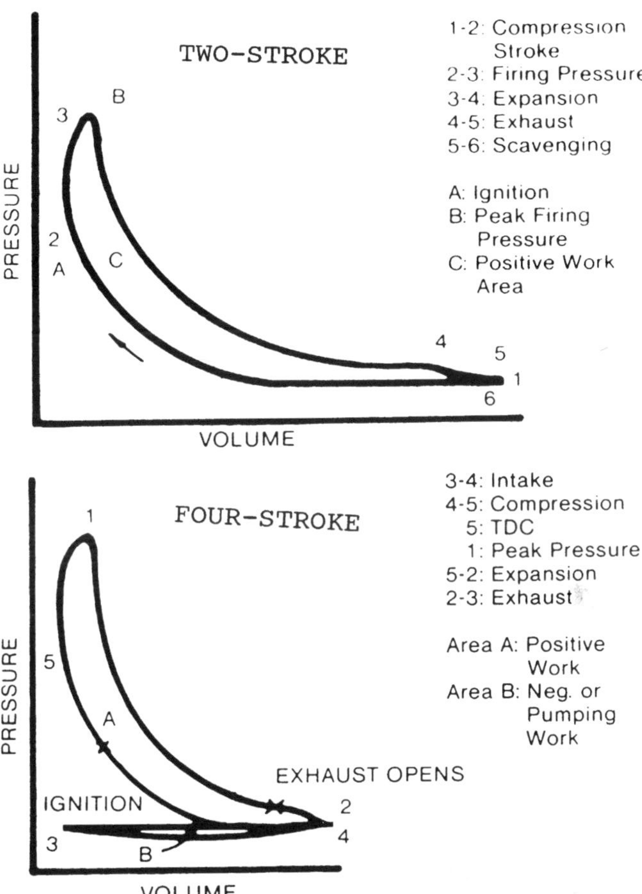

Figure 10–27. P-V diagrams (PMC/Beta Corp.).

end) in the power stroke it first uncovers (or opens) the exhaust port. This allows the exhaust gases to escape and lowers the pressure in the cylinder to atmospheric. When the intake port is opened (Fig.10–26) air, previously compressed in the scavinging cylinder to about 5 psig, flows from the scavenging cylinder through the intake port into the cylinder. This air scavenges or expels the exhaust gases. When the piston moves upward (towards the head end) it closes the intake port first, then the exhaust port, and finally compresses the air. Then fuel gas is injected into the cylinder, and the compressed air/fuel-gas mixture is ignited near the end of the compression stroke. This combustion generates a high pressure (and temperature), thus forcing the piston down and completing the cycle.

In the *four-stroke engine* each cycle is four strokes or two full revolutions. As shown in Figure 10–26A, during suction stroke the piston moves downward, sucking air into the cylinder. As the piston starts to move upward (towards the head end) the intake valve closes and the air is compressed (Fig. 10–26B). During this compression-stroke the fuel gas is injected. Near the end of compression the fuel-gas/air mixture is ignited, and the resulting explosion forces the piston downward (toward the crank end), thus generating the power stroke (Fig. 10–26C). At the end of the power stroke, the exhaust valve opens and the piston expels the exhaust gases (Fig. 10–26D) thus completing the cycle.

Figure 10–27 shows PV or indicator diagrams of the actual change in gas pressure and volume in the power-cylinder during the two- and four-stroke cycles.

Engine Description

Cylinder Arrangement. Power cylinders are usually arranged upright or vertically—in a V or opposed (Fig. 10–28).

Couplings. Internal-combustion gas engines are connected to the reciprocating compressors in one of three ways

1. Integral: A single crankshaft with multiple throws is used for both the power and compressor cylinders. Power cylinders are often vertical or V-type while the compressor cylinders are usually horizontal.
2. Direct Coupled: The driver and compressor crankshafts are coupled, therefore, speeds must be the same.
3. Belt Driven: Suitable for smaller units (less than 200 HP). Couplings also may be described as rigid or flexible.

Engine Starting. Standard L or V two-cycle engines are often started with compressed air at 250 psig by admitting

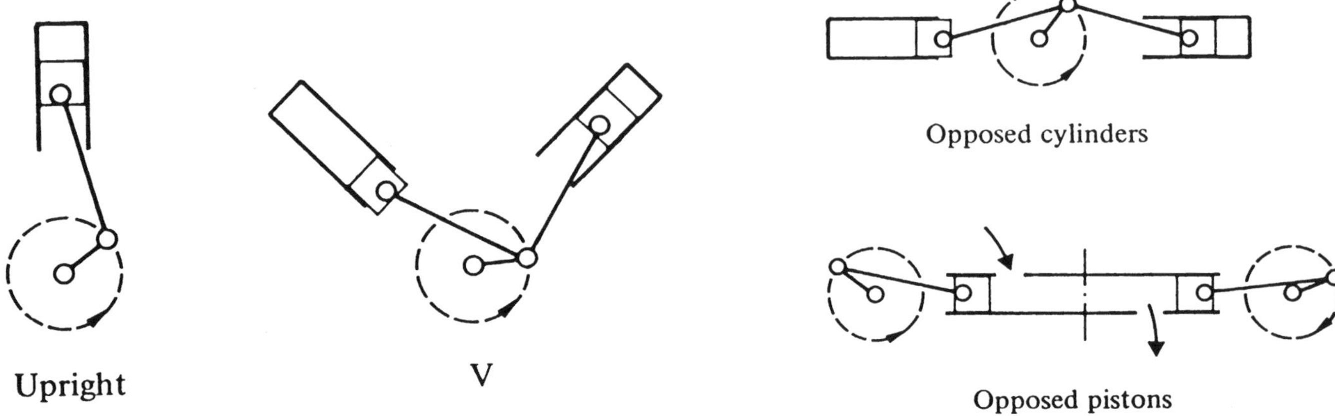

Figure 10–28.   Common cylinder arrangements.

air to the power cylinders and "rolling" the engine. Turbochargers are accelerated at start up by either starting air or hydro-mechanically.

Fuel-Gas System. Fuel gas should be fed through a pressure regulator into a header having at least 1 ft³ capacity per 100 Bhp of engine rating.

Timing Control. Automatic ignition timing control is standard, thus allowing the timing to be retarded during start up and gradually increased as the load is applied.

Air Filters. Clean air is essential for smooth operation; consult manufacturer for recommendations.

Ignition. Some of the more common ignition types are magneto-modified, dual-ignition, magneto-dual ignition, and pulse-generator dual ignition. In the pulse-generator type, low tension energy impulses are stepped up by a transformer mounted on the power cylinder head covers.

Lubricating Systems. Full-pressure lubrication is used. A crankshaft-driven, main oil, gear-type pump lubricates

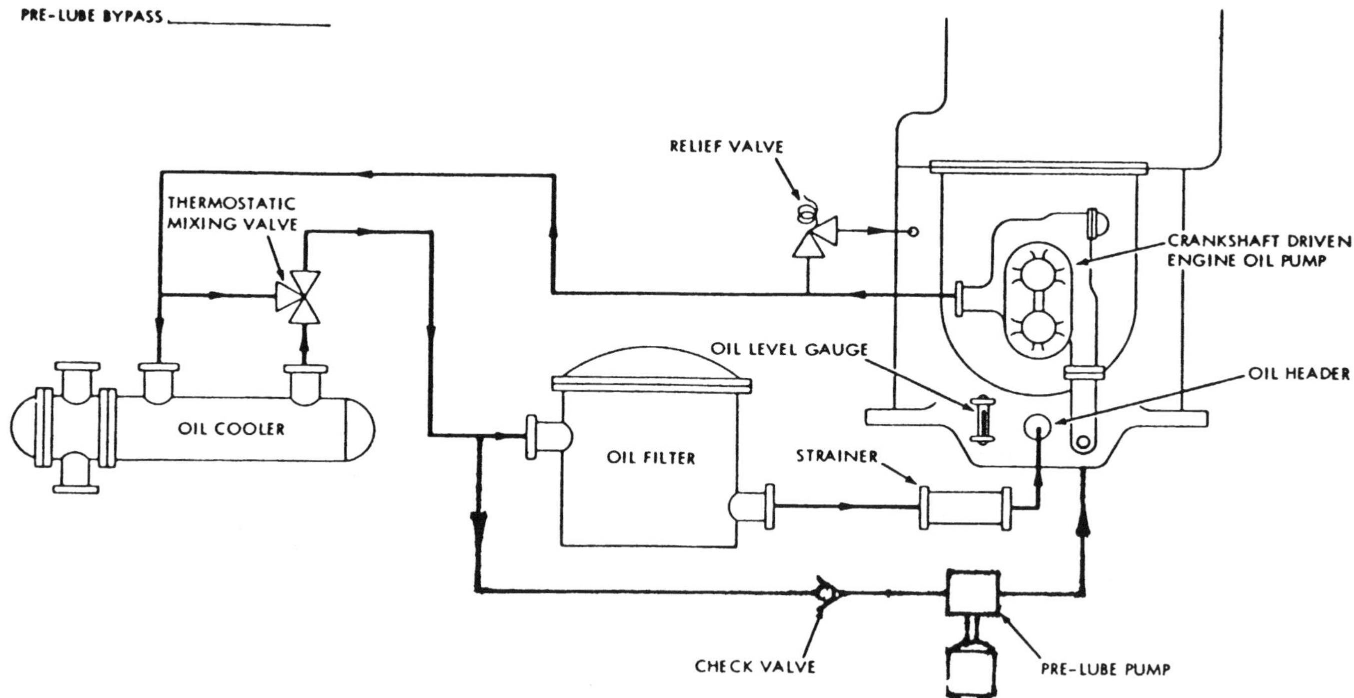

Figure 10–29.   Typical engine lubricating system (Dresser-Clark; Berger and Anderson, 1980).

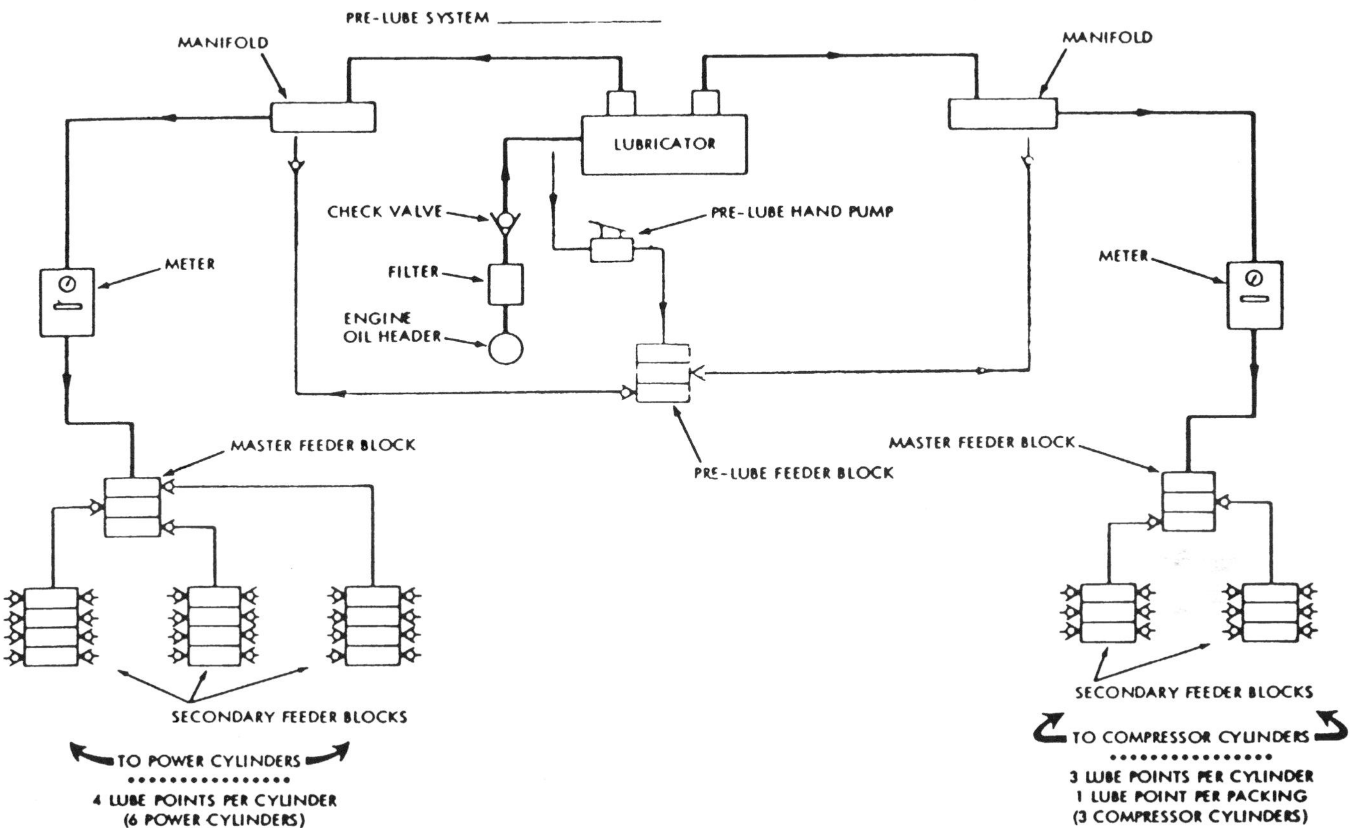

Figure 10–30. Typical cylinder and packing lubricating system (Dresser-Clark; Berger and Anderson, 1980).

the engine and any mounted turbochargers (Fig. 10–29) while a mechanical lubricator supplies oil to the engine power cylinder walls, compressor cylinder walls, and compressor piston rod packings (Fig. 10–30). Each lubricating system can flood the running surfaces prior to start up.

Summary

Four-cycle gas engines are easier to start, are smoother running, consume less lubricating oil, and have higher thermal efficiencies than two-cycle engines. Two-cycle engines have no inlet valves, often no exhaust valves, require less displacement per Bhp; therefore, they are smaller, lighter, and cheaper. However, the differences are minor, especially in the turbocharged types. GPSA (1987, Section 15) provides further details.

GAS TURBINES

Gas turbines are self-contained power plants with many advantages over steam turbines (Molich, 1980):

1. No need for boilers, condensers, boiler-feed water, etc.
2. Substantially higher thermal efficiencies.
3. Can offer environmental benefits such as reduced blowdowns of cooling and boiler-feed waters.

General Description

Figure 10–31 shows the mechanical arrangement of a *simple-cycle, single-shaft* turbine. The centrifugal or axial compressor supplies air to the combuster where the fuel is burned at constant pressure. Then the resulting power gas is expanded through the turbine. About two-thirds of the turbine output is used to drive the inlet air compressor, and the remaining one-third is converted to useful output shaft power.

Only about 20% of the injected and compressed air is used for combustion. The remaining 80% cools the combustion chamber to 1400°F for heavy-duty types and to 2200°F for some high-performance aircraft derivatives.

Figures 10–31b,c,d illustrate how the gas engine configuration can be modified to improve efficiency. *Two-* or *three-shaft* arrangements allow each shaft to rotate at the correct speed for maximum efficiency.

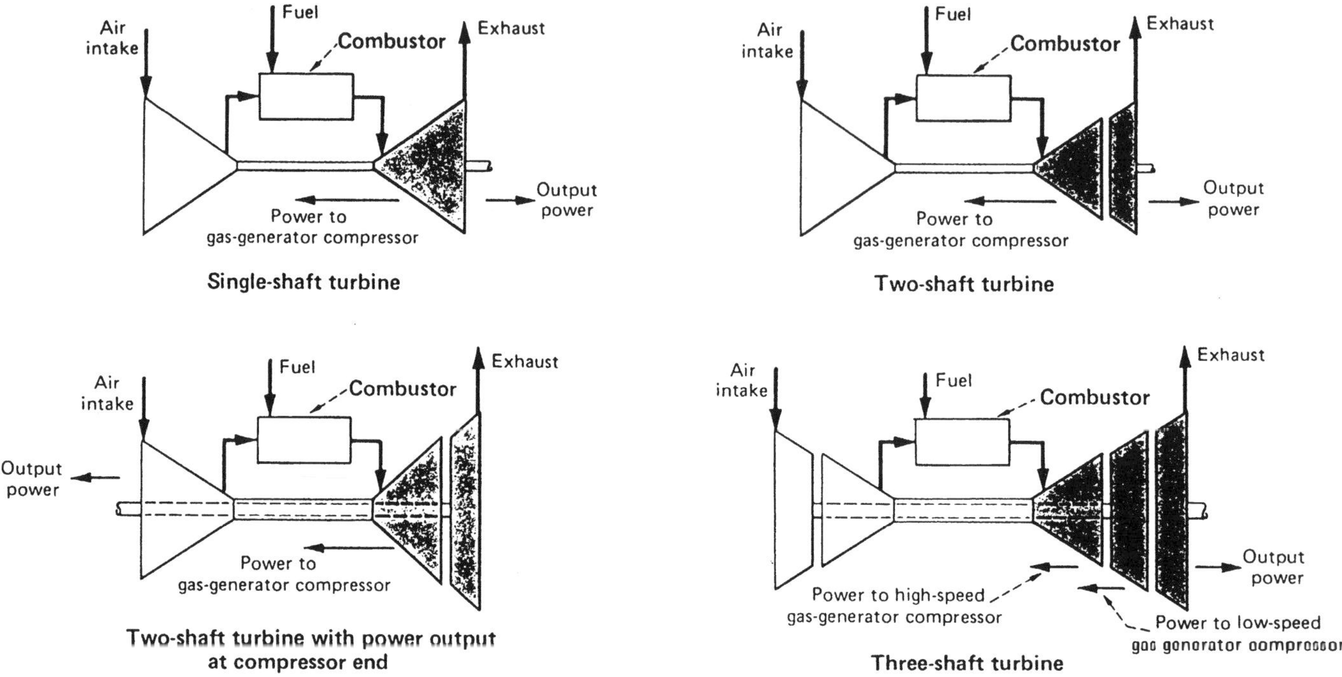

Figure 10–31. Gas turbine shaft configurations (Molich, 1980, Vol. 95, No. 10, 22–26, 1988).

The efficiency of the simple cycle (Fig. 10–31) can be improved by three modifications. First is *regenerative heating* which uses the hot turbine exhaust gases to preheat the compressor discharge air before it enters the combustor. Second, the compressor work can be reduced by *intercooling*. Third, the turbine power output can be increased by *interstage reheating*. While Figure 10–32 shows all three modifications, regenerative heating is often

used alone. The *closed-cycle gas turbine* is finding increased used—see Molich (1980) for a good description.

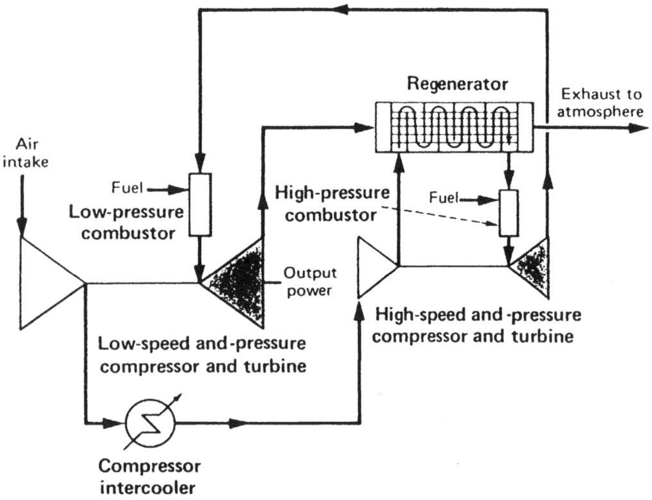

Figure 10–33 depicts a Mars turbine engine.

Gas Turbine Cycles

Figure 10–34 shows the basic gas turbine cycle or Brayton cycle. The *ideal Brayton cycle* assumes:

1. Air is compressed isentropically from 1 to 2.
2. Combustion occurs at constant pressure: 2 to 3.
3. Isentropic expansion occurs in turbine: 3 to 4.

Line 4 to 1 does not exist in practice as the exhaust gases leave the engine. Nevertheless the *thermal efficiency*, $\eta_{thermal}$, of this cycle is (Anson, p. 3–6)

$$\eta_{thermal} = \frac{\text{useful work}}{\text{heat supplied}} = 1 - (1/r)^{(k-1)/k}$$

In practice the efficiencies of the compressor, η_c, and turbine, η_T, must be considered. Now (Anson, p. 3–7)

$$\eta_{thermal} = (\eta_c \, \eta_T \, K - 1)/ \, \eta_c \, (K - 1)$$

where K = ratio of turbine to compressor ideal works.

Sorensen (1983) calculated how the thermal efficiency of gas turbine varies with the pressure ratio, r, and the maximum temperature, T_3. Figure 10–35 shows the results for representative values of inlet air temperature, T_1, and

Figure 10–32. Modifications to improve cycle efficiency (Molich, 1980, Vol. 95, No. 10, 22–26, 1988).

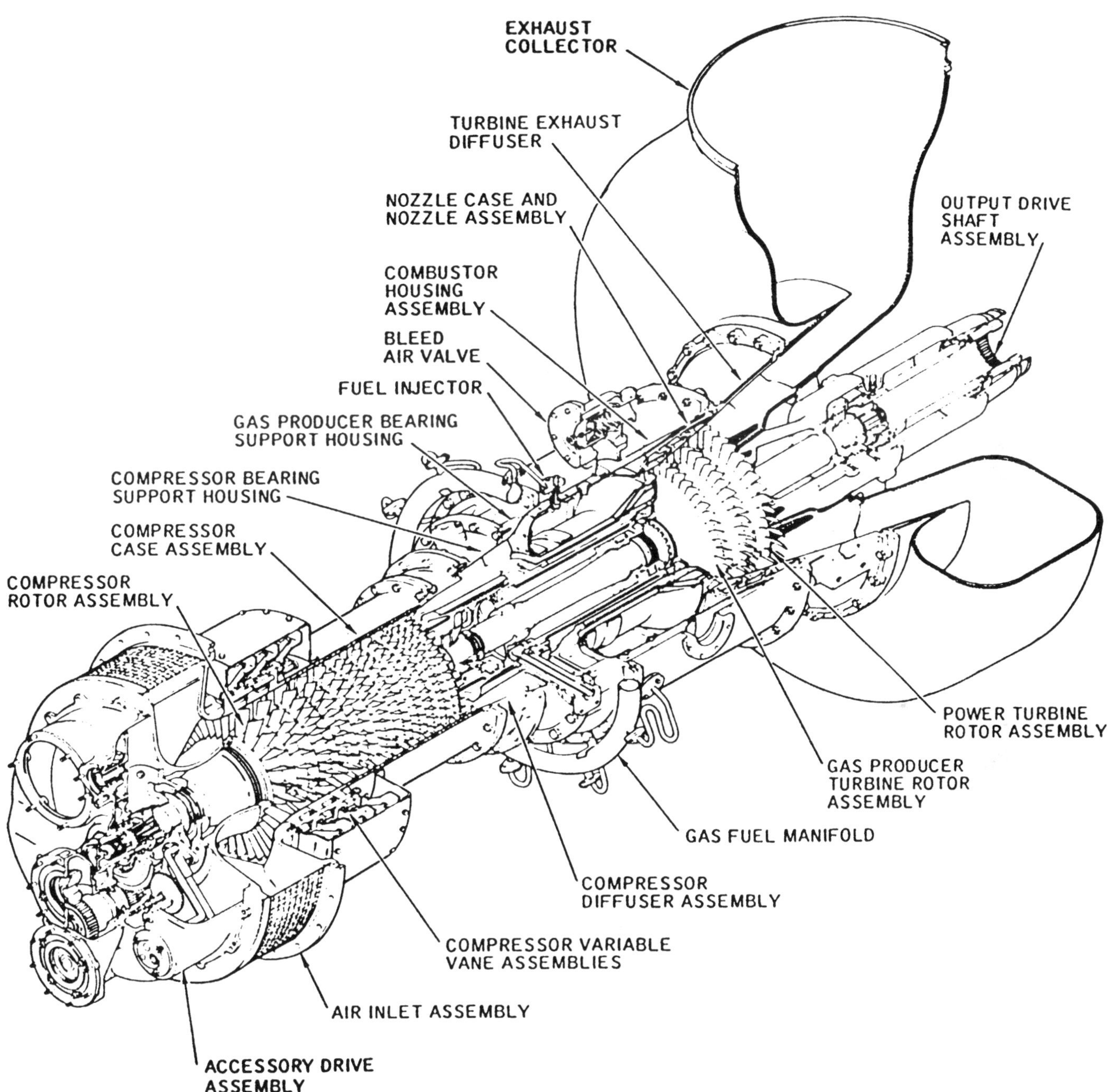

Figure 10–33. Mars turbine engine (Solar Turbines, Inc., 1989).

compressor and turbine efficiencies. Combustion, heat, pressure, mechanical, and turbine losses are also included (Sorensen, 1983).

Materials of Construction

Compressor and turbine efficiencies have reached about 90%, and major efficiency improvements are unlikely. However, maximum safe firing temperature for turbines are increasing continuously. There are three approaches (Molich, 1980):

1. Use materials with high *creep-rupture* strength for highly stressed rotating components such as first-stage turbine buckets and disks, first-stage turbine nozzles, combustor liners, and transition pieces. Nickel and cobalt based alloys are used.

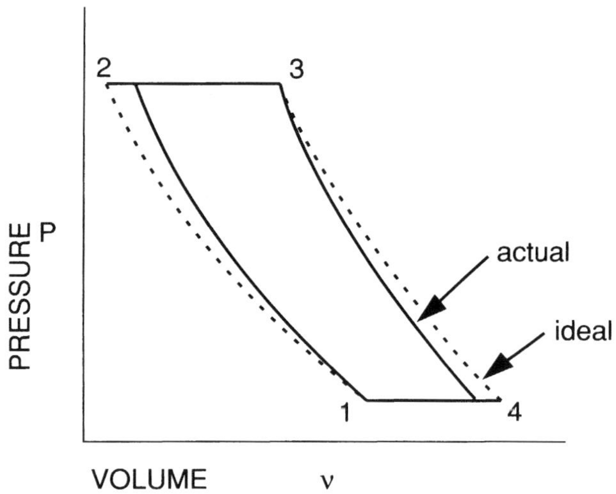
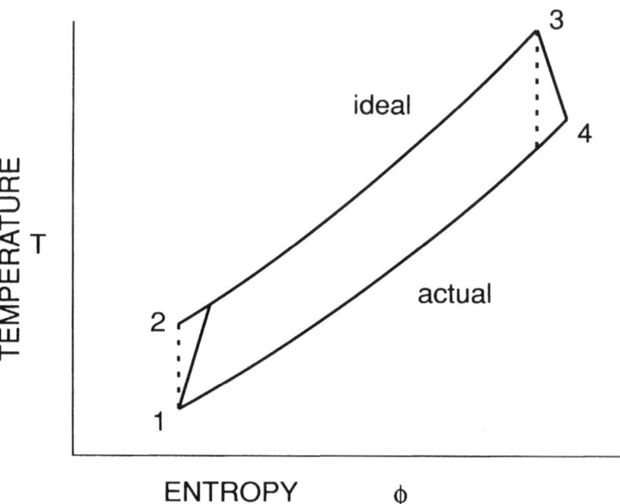

Figure 10–34. Brayton cycle (Anson, p. 3–7).

2. Protective thermal-barrier coatings for high-temperature service; for example, aluminum oxide, tungsten and chromium carbides, alloyed with Ni, Cr, Pt.
3. Air cooling for turbine nozzles, buckets and disks can keep components 250–500°F below hot-gas temperature.

Gas Turbine Types

The two types of gas turbines are the heavy duty and the aircraft derivative (Molich, 1980). The *heavy duty gas turbine* is suitable for normal industrial uses without space or weight limitations. The turbine is either single- or two-

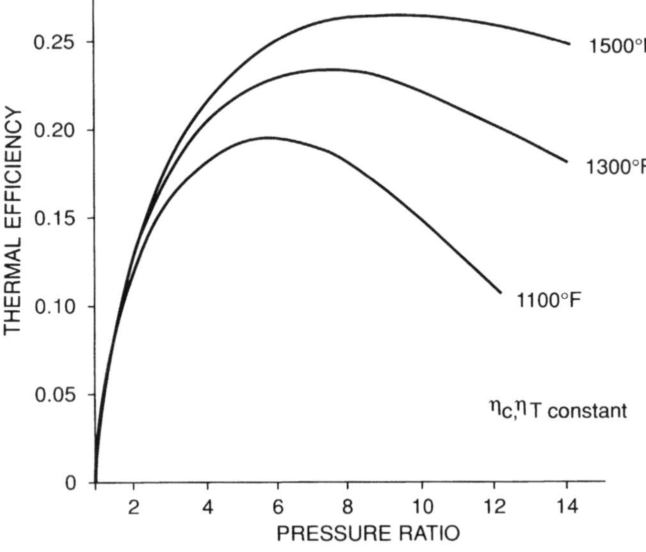

Figure 10–35. Thermal efficiency of gas turbine (Sorensen, 1983).

shaft type and the compressor, turbine blades and buckets, and turbine nozzles are of heavy design. Moderate compression ratios (6–9) and power gas temperatures (1700–1900°F) are selected for long-time (100000 hour) continuous duty between inspection and maintenance. Shaft bearings are of conventional design—sleeve or tilting pad for radial bearings and tapered land or multiple segment for thrust bearings—arranged for operation with a pressure lubrication system common to the gas turbine and driven equipment. The turbine, lubrication system, auxiliary systems, and instrumentation are stipulated by API Standards 614 and 616.

The *aircraft-derivative gas turbine* is essentially an aircraft jet engine with the jet nozzle replaced by a power turbine. Four advantages are:

1. Sophisticated aircraft technology and associated research and laboratory facilities.
2. Mass-production techniques and quality-control methods used by the aircraft industry.
3. The service centers for aircraft jets, with their rigorous certification requirements, spare-parts inventories, and test facilities are available.
4. Fast changeout of the entire gas generator element, typically 12 hours. Overhaul need not be done ''in situ.''

The aircraft-derivative gas turbine is inherently of either the two- or three-shaft configuration. The regenerative cycle cannot be applied for this turbine type. The power turbine and the gas generator (the jet engine) are separate entities, with no mechanical linkage. This separation also applies to the auxiliary systems. The power turbine is of heavy design and shares its conventional accessories, instrumentation, and lubrication system with the driven equipment. Gas generator exhibits lighter weight and compactness imposed by aircraft design, extensive use of antifriction bearings; special lubrication systems using synthetic, nonflammable lube oil; and hydraulic-type accessory and electronic-hydraulic instrumentation. When aircraft units are used in industry, the firing temperature and power output are reduced considerably to increase life span and time between overhauls.

Environmental Impact

Two distinct problems are *noise* and *stack emissions*. To meet the normal noise criterion of 85 to 90 dBa at a 3-foot distance, inlet and exhaust silencers together with acoustic enclosures are required. Noise-abatement components are now of heavy structural design and often of the free-standing type. Outer walls may be fabricated of $\frac{3}{16}$- to $\frac{1}{4}$-in.-thick steel plates and the enclosures sound insulated. High temperature corrosion-resistant materials are used for the exhaust stack. The sound-absorbing materials may be mineral or glass fiber.

Normal *stack pollutants*, such as sulfur dioxide, carbon monoxide, and particulates are controlled by fuel specifications and turbine combustor design. *Nitrogen oxides, NOx,* are much more troublesome; however, NOx emission can be controlled by injecting water of boiler-feed quality into the combustors.

In the U.S., EPA permits may be required before use of new fuel-burning engines. Obtaining these permits can take months.

Fuels

Gas turbines can use many fuels: liquid fuels include light distillates, diesel oil, gas oil, bunker C, and crude oil. Gaseous fuels include natural gas, propane, butane, refinery gas, hydrogen, carbon monoxide, and coke-oven gas.

Fuel contaminants such as Na, K, V, or S can and do cause problems. Any sulfur in the fuel forms SO_2 or SO_3 when burned; while metals such as Na, K, and V, can form low melting eutectics with some brick refractories. These problems are discussed in Volume 2 in the chapters on fired heaters and energy conservation.

Operation and Controls

The *standard control system* automatically starts, runs, monitors, and stops the unit. In addition, the controller monitors the unit and protects it from abnormal operation.

The *automatic start-up* sequence occurs in one minute. First the starter motor—an electric, hydraulic, pneumatic, or gas expansion motor—spins the machine for about half a minute. This purges the turbine and exhaust systems. Then fuel is introduced and fired by the ignitor. Finally the machine is loaded at unit speed.

Gas turbine operation requires *careful and extensive monitoring* as illustrated in Table 10–1.

Turbine performance deteriorates as the turbine and compressor blades become fouled. Clean fuel will reduce turbine fouling; however, regardless of the level of air filtration, the air compressor blades will foul. *Abrasive cleaning* by adding nut shells or rice to the air intake is often performed regularly—maybe once a week. Special nozzles are required for abrasive cleaning of the turbine (Molich, 1980).

Washing with hot water, detergent, or solvent is required if the compressor has digested oil or if ash is deposited on the hot-gas parts (fuel nozzles, combustors, transit pieces, turbine nozzles, and buckets) (Molich, 1980).

Table 10–1 also summarizes a typical *alarm and shutdown system* and provides some insight into the sophistication of the control system.

PREVENTIVE MAINTENANCE

Compressors are generally the most expensive gas-processing equipment to buy and operate. Good maintenance is essential.

The objectives of any maintenance program are to:

1. Ensure absolute safety for all personnel.
2. Keep equipment running reliably.
3. Minimize costs.

Meeting the above goals is most challenging. *Cost control* in particular is getting harder every year. Accordingly, alternate maintenance philosophies are discussed first and then the essential ingredients for a workable preventive maintenance program are reviewed.

Maintenance Philosophies

A maintenance program can follow three philosophies:

1. The ''poor boy'' marginal approach that believes fervently in ''run it 'til it quits'' or ''if it ain't broke, don't fix it.''
2. The fancy ''whole-life insurance policy,'' i.e., *regular complete overhauls* on a previously defined schedule.
3. *Planned preventive maintenance* designed to provide adequate repair and monitoring services at minimum cost.

The ''poor boy'' marginal approach is very costly in that, without adequate inspections, breakdowns occur

Table 10–1   Alarm and Shutdown Indication (Solar Turbines Inc., 1989)

Indication	Alarm (Indication Only)	Shutdown* (Indication and Shutdown)	Supplied With
Engine System			
High Engine Temperature (T_7)		L	Basic set
Fail to Crank		NL	Basic set
Fail to Start		NL	Basic set
Ignition Failure		NL	Basic set
Overspeed Power Turbine		NL	Basic set
Lube System			
High Oil Temperature		NL	Basic set
Low Oil Pressure		L	Basic set
Low Prelube Oil Pressure		L	Basic set
Low Oil Level		L	Basic set
Fuel System			
Low Gas Fuel Pressure		NL	Basic set
Compressor Systems			
Low Seal Oil ΔP		L	Basic set
Fail to Load		NL	Basic set
High Discharge Gas Pressure		NL	Optional
High Discharge Gas Temperature		NL	Basic set
Compressor Surge		NL	Optional
Yard Valve Sequence Failure		L	Basic set
Low Suction Gas Pressure		NL	Optional
Ancillary Systems			
Low Battery Voltage		NL	Basic set
Inlet Air Filter—High ΔP	X	NL	Inlet air filter
Inlet Air Filter—Blow-In Door Open	X		Inlet air filter (with blow-in door)
Inlet Air Filter—Blower Motor Failure	X		Inlet air filters that include a scavenge fan
High Enclosure Temperature	X		Enclosure (optional)
Low Enclosure Pressure	X		Pressurized enclosures (desert application)
High Gas Level (enclosure)	X	L	Gas detection system (optional)
Gas Monitor Failure (enclosure)		L	Gas detection system (optional)
Fire System Locked out	X		Fire protection system (optional)
Fire System Discharged		L	Fire protection system (optional)
Instrument Air System Failure	X		Surge control system, etc.
High Fuel Gas Scrubber Level		L	Requires input signal from external equipment
High Fuel Gas Filter—ΔP	X		Requires input signal from external equipment
High Gas Cooler Discharge Temperature		NL	Requires input signal from external equipment
High Scrubber (1) Level		NL	Requires input signal from external equipment
High Scrubber (2) Level		NL	Requires input signal from external equipment
Low Scrubber (1) Level		NL	Requires input signal from external equipment
Low Scrubber (2) Level		NL	Requires input signal from external equipment
Gas Cooler (1) Vibration		NL	Requires input signal from external equipment
Gas cooler (2) Vibration		NL	Requires input signal from external equipment

* L = Lockout, NL = Nonlockout

unexpectedly. This "*reaction maintenance*" results in unscheduled repairs, usually performed under adverse conditions. As predicted by Murphy's law, breakdowns always seem to occur at particularly inconvenient times that result in large losses of production. And catastrophic failures of compressors often cost millions of dollars to repair (Dodd, 1984).

The "whole-life" method is practiced in regulated industries, e.g., airline companies. In maximizing passenger safety, engines are overhauled and parts replaced whether needed or not. This overhauling at specified intervals regardless of condition not only wastes time by disassembling good machinery but reassembly often generates unnecessary start-up programs and can increase wear.

Planned preventive maintenance is the sensible approach. The goals are to monitor equipment so that repairs are performed only when needed. This "condition maintenance only'' plans to run the machine its last mile without wrecking it (Dodd, 1984).

James (1982) defines good maintenance by two rules:

1. If a machine is running well by analytical check, leave it alone.
2. If a machine is not running well, fix only what needs to be fixed.

Any satisfactory preventive maintenance program must involve the following:

1. All applicable governmental regulations must be followed, and industry codes (e.g., API's RPs) should be considered carefully.
2. A good lubrication system.
3. Reliable monitoring and shutdown systems.
4. Thorough on-line inspections.
5. Careful downtime inspections and servicing.
6. Adequate diagnostic testing.
7. Adequate record keeping.

These seven items are now discused. Lane *et al.*, (1988) describe in detail the successful implementation of such a preventive maintenance program in a gas plant.

Regulations and Codes

Government regulations and industry codes are designed to protect employees. Ignoring governmental regulations is not only illegal but can result in very expensive litigation and employee compensation. On the other hand API Recommended Practices are just that—recommendations— and should aid, and not control engineering judgment.

Lubrication System

A sound *lubrication program* is absolutely essential for a successful preventive maintenance program. Manufacturers' recommendations and/or company specifications should be followed closely. High-performance specialty lubricants may not be suitable for other applications and, even if suitable, might not be worth the cost.

When monitoring lubrication remember that *too much oil is almost as bad as too little.* Excess oil can cause hydroplaning, damage cylinder walls and pistons, and form excessive deposits. Dirt, abrasives, water, and diluents must be kept out of the oil. Filters should be checked regularly and changed when required. Remember that dirty gas is a frequent source of oil contamination. Lube oil must be tested regularly and replaced promptly when needed.

Monitoring and Shutdown Systems

To be satisfactory, a monitoring system must give adequate advance warning of impending trouble and shutdown the machine thereby avoiding (or at least minimizing) damages from catastrophic failures.

As a minimum the following should be monitored: engine, compressor and fan vibration, cooling water temperature and pressure, high and low gas pressures, scrubber high level, high gas discharge temperature, high engine exhaust temperature, engine overspeed, turbocharger low discharge air pressure, and lubricating oil pressure and flow. On the other side of the spectrum is the *API 670* machinery protection system that continuously monitors vibration, axial position, and thrust bearing temperatures. Dodd (1984) outlines the success of API 670 in avoiding catastrophic failure of major turbomachines.

Safety shutdowns should protect against random failure and/or operator error. To be successful a safety shutdown system needs:

1. Good design and reliable components
2. Regular component testing program (follow codes such as OCS, MMS, and API RPs)
3. On-line testing capability
4. Strict policy against bypassing safeguards.

Daily On-Line Inspections

On-line inspection programs need: reliable instrumentation; comprehensive check list; good, concise operating procedures; thorough follow-up on suspected problems; and thorough operator training. Pressure and temperature gauges should be checked daily and replaced immediately when defective. Daily inspections should include: knocks and unusual noises; excessive vibration; compressor suction and discharge pressures and temperatures for all stages; engine speed; lubrication system pressures, flows, levels, temperatures, and filter pressure drop; crankcase oil appearance; compressor rod packing vent for excessive gas leakage; compressor rod packing oil drain for discolored oil; engine manifold pressure; turbocharger speed and vibration; jacket water temperatures, flows, pressures, and levels; oil and water leaks; engine exhaust temperatures; governor oil level; scavenging air pressure; fan belts; ignition wiring; fuel gas pressure; fuel gas scrubber liquids; suction and discharge scrubber liquids; starting air receiver water level; loose anchor bolts, crosshead bolts, head bolts, and flange bolts; housekeeping; and others as recommended by manufacturers or learned from practical experience.

Downtime Inspections and Servicing

Early diagnosis of problems is the key when an engine is shut down for routine servicing such as oil change, oil filter change, spark plug or magneto change, etc., this is a ''free'' opportunity to check for potential problems. Some of the following checks should be made: compressor rods for signs of excessive wear; compressor cylinder walls for unusual wear by pulling top valve from outboard end of compressor cylinder and spotting the piston at crank end (a light and inspection mirror will help and all safety precautions should be taken); and fan belts for wear. If oil is being changed, remove cover plate from engine crankcase and check the internals—look for metal cuttings and sludge deposits in the crankcase, and look at all bearing caps for signs of overheating, loose bolts, and other evidence of possible trouble. If filters are being changed, check old filter elements for metal cuttings. Check combustion and scavenging air filters and clean when needed. With crankcase open, check power cylinder liners for scuffing and wear by spotting each piston at TDC (take all necessary safety precautions). If spark plugs are being changed, run a compression check on each power cylinder. Check valve tappet clearance; cylinder jackets or engine and compressor for scaling, corrosion, etc.; oil and jacket water coolers for scale, corrosion, leaks, etc.; and any other internal inspection points that may be accessible without excessive dismantling.

Diagnostic Testing

This field has been revolutionized in the last 10 years. Some of the instruments in current use include the following items.

Portable Computer with *CRT* (cathode ray tube with video display screen) and keyboard. This unit can be "plugged in" and will measure and report engine pressures and temperatures, movement of pistons, and time of revolutions of crankshaft. The goal is improving and maintaining fuel efficiency.

Engine/Compressor Analyzers including an oscilloscope, for reciprocating engines and four channels, inspect (PMC Beta) ignition voltage; vibrations; firing pressures; and timing. Obtainable information includes: power cylinder horsepower readout; compressor cylinder horsepower readout; engine speed readout; compressor volumetric efficiency; compressor cylinder pressure-volume oscilloscope display; power cylinder pressure-time oscilloscope display; vibration-time oscilloscope display; and ignition-time oscilloscope display. With this information, the following analyses of the engine/ compressor performance can be made: power losses, bad engine and compressor valves, defective rings and bearings, scored parts, carbon deposits in parts, scored cylinder walls, piston slap, and faulty ignition components.

A *Fiber Optic Probe*—flexible, periscope-type instrument—actually bends light to see around corners and over obstacles deep within the machines. It can inspect turbine blades and buckets as clearly as if the engine were disassembled. And, it can record views by color photography and/or television.

An *Ultrasonic Detector* complete with display screen can detect cracked bolts by "shooting" an ultrasonic wave down bolt as the engine runs. Another type of ultrasonic detector—a cube-shaped listening device or stethescope— checks for valve leakage when the engine is running.

Spectroscopic Analysis of Lubricating Oils is a quick, simple, and low-cost way of checking engine wear. Wear occurs as ultra-fine metal particles which are too small to be filtered out of lube oil. For example, Pb, Cu, Sn, Al, Ag, indicate bearing wear; Fe, Cr, Ni indicate liner or piston ring wear; Si is a contaminant from the air intake system; B, Na come from the cooling water system; P, Zn, Ca, Ba are related to lube oil additives. Oil analysis complements electronic detectors, requires no special tools, and can be subcontracted to reliable laboratories. Monthly sampling is recommended.

Ferrography—A direct reading ferrograph (Fig. 10–36a) measures the relative number of small (1–2 μm) and large (> 5 μm) metal particles in the lube oil. Figures 10–36b and 10–36c show pictures of wear particles while Figure 10–36d shows how the size and number of metallic wear particles vary with time. Laskowski (1984) documents the benefits of using ferrography.

Optical Alignment Devices—Instrument looks like a telescope, is positioned on a tripod, and is an ultra-accurate engineering level that "shoots" levels when the engine is reassembled after an overhaul. Use with an optical micrometer. Can set bearings within 0.001 in. of desired elevation; this prevents wear, especially crankshaft deflection.

Troubleshooting

The GPSA (1987) Engineering Data Book (Section 13) provides excellent one-page troubleshooting charts for reciprocating (p. 13–21) and centrifugal (p. 13–39) compressors, respectively.

Record Keeping

Without adequate record keeping it is impossible to assess performance properly or to establish the benefits of preventive maintenance. However, excessive record keeping wastes money and hurts employee morale. Broyles (1978) discusses this dilemma and recommends the following parameters for monitoring monthly compression performance:

1. Percent of available HP that is used
2. Specific fuel consumption
3. Downtime
4. Volume of gas flared

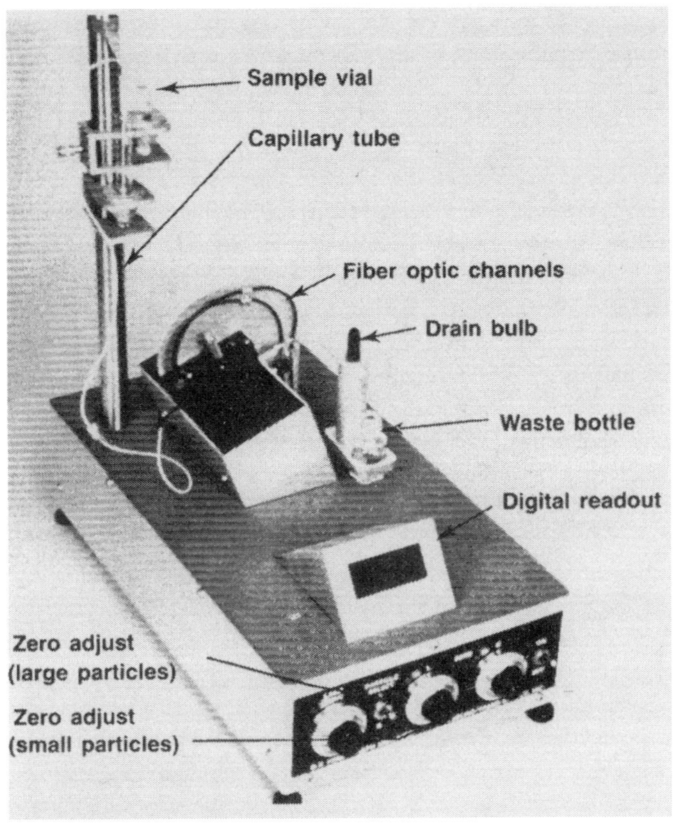

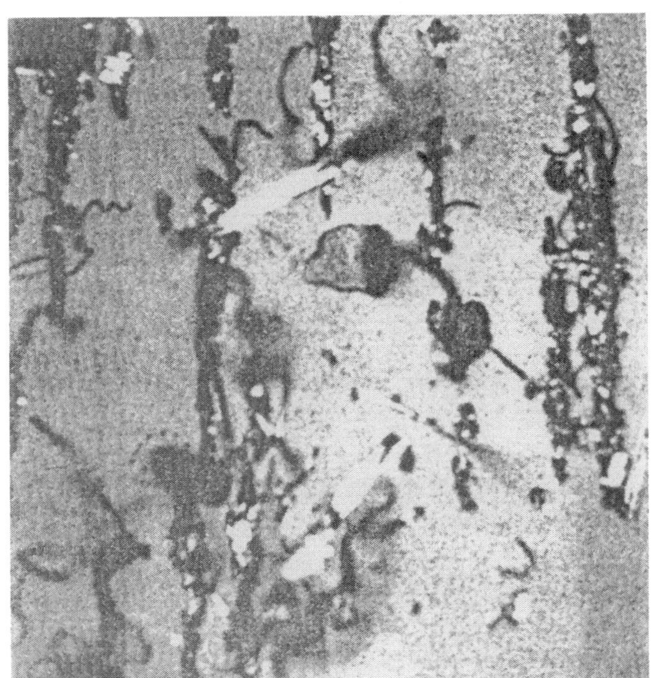

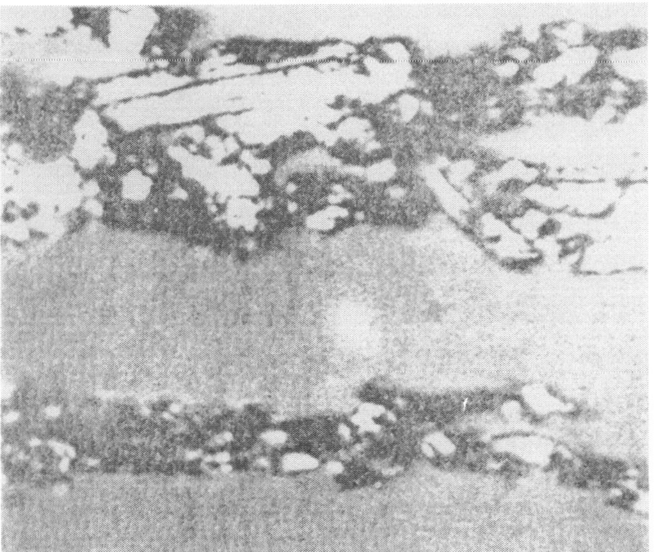

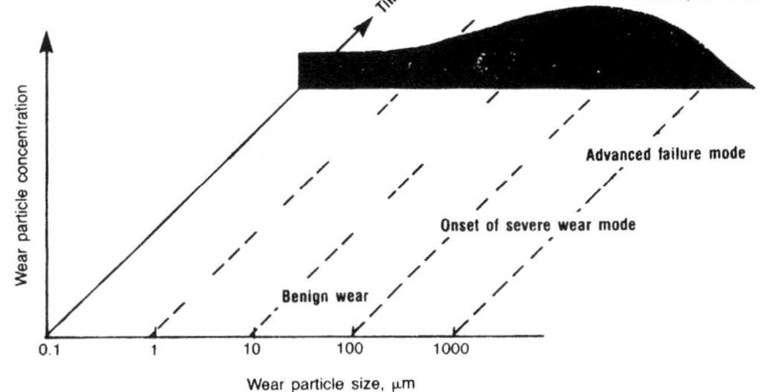

Figure 10–36.   DR ferrography (Laskowski, 1984).

SELECTION OF COMPRESSORS AND DRIVERS

Reciprocating compressors operate at relatively low speeds, say 200–800 rpm. The most common driver for small reciprocating compressors is the electrical motor. The speed of the motor is reduced by means of pulley and belt drive. The larger pulley, or sheave, is mounted on the compressor shaft and is driven by the smaller pulley on the motor shaft. In this arrangement there is no speed control possible, except perhaps by a factor of two, since AC induction motors are inherently constant in speed. Direct coupling of the motor to the compressor or connection through a gear train are less-used alternatives.

Medium- to large-sized reciprocating compressors are driven by gas engine, steam turbine, or electric motor depending on convenience and economics. In the field compression of natural gas the gas engine is almost always used. In a gas plant, steam may be used due to its availability. Connection of driver to compressor may be by coupling or gear train. In large units, the driver and compressor may be integral (both operating off the same crankshaft).

The high speed of *centrifugal compressors* limits choice of drivers somewhat. In field applications, the most common driver is the aircraft-derivative gas turbine. Gas turbines are either directly-connected or geared to the compressor. In the plant, steam turbines and electrical motors are common. Power recovery is sometimes obtained by using expansion turbines (expanders) with high-pressure process gas streams. Turbine drives have the advantage of easy speed control. Electrical motors are generally constant-speed devices. Variable-speed, variable-frequency AC motors can be used, but they are relatively expensive.

Figure 10–37 (GPSA, 1987, p. 13–3) depicts the usual

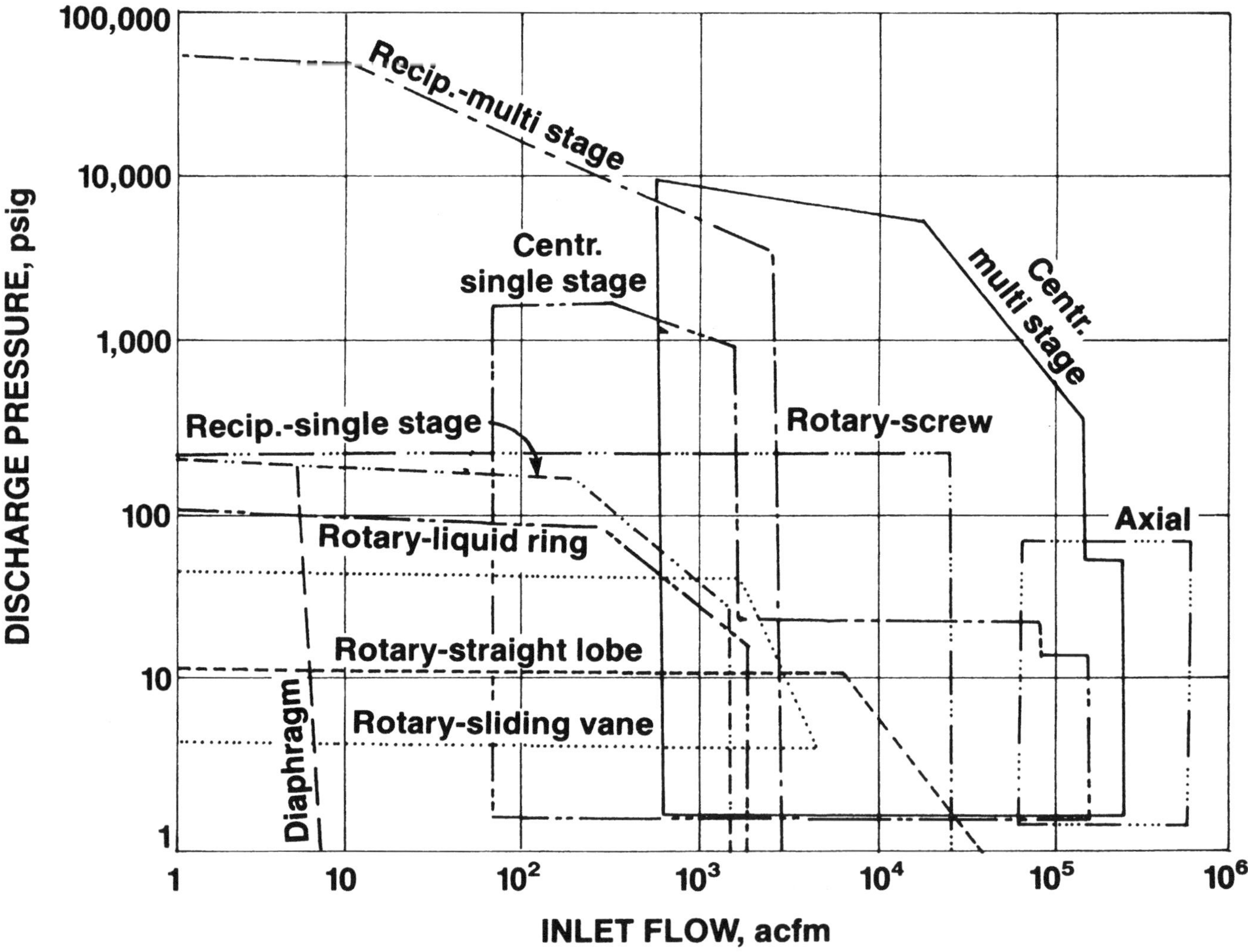

Figure 10–37. Compressor selection guidelines (GPSA, 1987, p. 13–3).

choice of compressor type as a function of discharge pressure and inlet flow (ACFM at inlet flowing pressure and temperature). Rotary compressors are generally used in special-purpose applications, such as wet or corrosive gases, vacuum applications, and the like. Jet ejectors are used almost exclusively in vacuum applications. The main choice lies between recips and centrifugals.

Reciprocating compressors can handle much lower flow rates per frame than centrifugal compressors—two orders of magnitude lower. Both types are competitive to 10000 psig discharge pressure, which covers the vast majority of applications. Note there is a definite lower limit to the flow rate of centrifugals. Axial compressors are seen to be confined to very high-flow, low-discharge-pressure applications.

Since reciprocating compressors are high-maintenance devices, it is common practice to specify two or more in parallel for a given service. Centrifugal compressors, on the other hand, are generally sized for the maximum expected load and a single machine is specified. Given these facts, it is clear that the overlap in service is broader than Figure 10–37 might imply.

Matching the driver to the compressor is, obviously, most important. Gas engines and reciprocating compressors are readily compatible because of their mutual reciprocating action and relatively slow (1000 rpm or less) speed. Offshore, high-speed gas turbines with double reduction gears are used to drive recip compressors. Gas turbines and axial compressors also pair due to their mutual high-speed (greater than 5000 rpm) rotary motion.

Centrifugal compressors operate best with high throughputs and relatively stable volume flows, suction, and discharge pressures. Multistage centrifugal compressors can generally achieve about the same compression ratios as single-stage recips. They have lower fuel efficiencies than recips. They also have fewer moving parts and, therefore, lower maintenance and lubrication costs. Centrifugal compressors have balanced, steady forces; low initial installation costs; and simple auxiliary equipment.

Reciprocating compressors are suitable for low or high compression ratios at varying rates under varying suction and discharge pressures with a wide range of through put volumes. Recips have higher overall thermal efficiencies, many moving parts, and higher maintenance.

Economic Comparison

Shaw (1980) claimed that both reciprocating and centrifugal units are capable of meeting virtually any application. Shaw further argued that the *selection choice should be made on economics*, i.e., total cost of service. Accordingly, Shaw estimated capital costs, annual operation, maintenance costs,

and fuel efficiencies and projected cost escalations. He computed the total cost (present worth) for 20 years of gas-transmission service. This analysis involved the following assumptions:

1. 95% load and 95% use factors.
2. Working capital and land will amount to 37.5% first-year maintenance and operation costs.
3. Investment financed 75% by bonds (12% interest) and 25% equity.
4. 20–year life and 15% DCF rate of return.

Shaw's computations show that the recips' higher first-year cost and higher operation and maintenance are very nearly balanced by their superior fuel efficiency. This apparent economic equality probably explains why so often the choice is made primarily on prior experience and/or personnel preference.

Conclusions

The optimum choice requires careful and complete study of all relevant plant, site, project, process, and machinery parameters. Nevertheless the following recommendations have much merit.

1. Low-speed integral reciprocating compressors are best when:
   a. High fuel cost
   b. No waste heat required
   c. Large horsepower increments are required
   d. Maximum operating flexibility is needed
   e. Block mounted installation is OK
   f. Long delivery/construction schedule OK
   g. Long project life
   h. Very high pressures involved
   i. Heavy molecular weight gases are involved.
2. High-speed separable recips are best when:
   a. Medium fuel gas cost
   b. Little or no waste heat is required
   c. Minimum initial cost is required
   d. 4000 BHP or less horsepower increments required
   e. Short project life is involved
   f. Relocation and/or conversion may be required
   g. Minimum shipping/construction schedules.
3. Gas turbine/centrifugal compressors are best when:
   a. Good process fit
   b. Waste heat is required
   c. Minimal foundation requirements are necessary
   d. Light weight desirable
   e. Maximum reliability required
   f. Short—medium delivery/construction schedule required

g. Minimized field work is required

h. Low fuel gas cost.

Review Questions

1. Compare reciprocating, screw, and centrifugal compressors. Include the following considerations:
    - equipment size, weight, and cost
    - maintenance
    - fuel efficiency
    - choice of driver
    - overall economics.
2. List five methods for sizing a compressor (see Examples 10–1 through 10–5). Compare these procedures. Include:
    - ease of calculation
    - data required
    - accuracy.
3. List and compare three maintenance philosophies.
4. What are the key factors for any good maintenance program?

Problems

1. Estimate the horsepower required to compress 2 MMscfd of gas measured at 14.65 psia and 60°F.
    Intake conditions are 36 psig and 100°F.
    Discharge pressure is 436 psig.
    Gas gravity is 0.55.
    Atmospheric pressure is 14.2 psia at 1000 ft elevation.
    Assume isentropic efficiency = 85%.
2. A three-stage reciprocating compressor compresses 4.88 MMscfd of gas from 135 psia and 110°F to 1150 psia.
    a. If the gas specific gravity is 0.863, estimate the horsepower required.
    b. Estimate how the compression horsepower varies if:
        The compressor suction pressure varies by 15%.
        The suction temperature varies by 30°F.
        The required discharge pressure decreases by 40%.
        The required discharge pressure increases by 20%.
        The pressure drops across the two intercoolers increase from 5 psi to 20 psi.
        The gas specific gravity increases to 1.
        The k value increases from 1.2 to 1.3.
        The compressor speed increases 11% to 1000 rpm.
    Mak (1987) discusses the effects of these changes in detail.

Nomenclature

Bhp = brake horsepower

C = clearance, %

C_p = specific heat at constant pressure, Btu/lb °F

C_v = specific heat at constant volume, Btu/lb °F

CG = correction factor for gas gravity, Equation 10–30

CHS = correction factor for high speed, Equation 10–30

$CLIP$ = correction factor for low intake P, Equation 10–30

CV = clearance volume

F = stage factor, Equation 10–27

Ghp = gas horsepower

H = head, ft lbf/ lbm or "ft"

h = enthalpy, Btu/lb, kJ/kg

hp = horsepower

k = ratio of C_p / C_v

$\dot{m}$ = mass flowrate, lb/hr, kg/s

MF = multiplication factor, Equation 10–27

$MMcfd$ = million cubic feet at 14.4 psia and T_s

MW = molecular weight, lb/lb mol

N = number of compression stages

n = polytropic coefficient, Equation 10–20

P = absolute pressure, psia, kPa

p = power, hp, ft lbf/sec, kW

R = ideal gas constant

r = compression ratio, P_d / P_s (absolute pressures)

T = absolute temperature

VE = volumetric efficiency, %

W = compression work, Btu/lb, kJ/kg

Z = compressibility factor

<u>Greek</u>

η = efficiency (theoretical/actual)

<u>Subscripts</u>

act = refers to actual work

av = refers to average property

c = refers to compression

d = discharge condition

g = refers to gas

is = denotes isentropic change

L = denotes standard sales conditions

p = denotes a polytropic change

s = suction condition

T = refers to turbine

$thermal$ = refers to thermal efficiency

1 = suction or inlet condition

2 = discharge or exit condition
2s = discharge condition, isentropic path

References

Anson, O. C., "Industrial Turbines and Gas Compressors," Solar Turbines Inc., San Diego, CA.

Berger, B. D., and K. E. Anderson (1980), "Gas Handling and Field Processing," PennWell Books Inc., Tulsa, OK.

Bloch, H. P. (1982), "General Type Selection Factors," Chapter 4 Compressors and Expanders, H. P. Bloch, J. A. Cameron, F. M. Danowski, Jr., R. James, Jr., J. S. Swearingen, and M. E. Weightman, Marcel Dekker, NY.

Broyles, G. W. (1978), "Mechanical Equipment Surveillance," SPE 7410, presented at 53rd Annual Fall Tech. Conference of SPE at Houston, TX.

Cameron, Joseph A., Frank M. Danouski, Jr., and Marilyn E. Weightman (1982), "Materials of Construction," Chapter in Compressors and Expanders, Marcel Dekker, Inc., New York, NY.

Campbell, John M. (1979), *Gas Conditioning and Processing*, Vol. 2, Campbell Petroleum Series, Inc., Norman, OK.

Campos, Mario Cesar M. M., and Paulo Sergio B. Rodrigues (1988), "Equations developed to accurately model centrifugal compressor performance," *Oil & Gas J.*, Vol. 86, No. 48, pp. 75–77 (Nov. 28).

Cohen, H., G. F. C. Rogers, and H. I. H. Saravanamutto (1975), "Gas Turbine Theory," Longman.

Dimoplon, William (1978), "What Process Engineers Need to Know About Compressors," *Hydrocarbon Processing*, Vol. 57, No. 5, pp. 221–227, (May).

Dodd, V. Ray (1984), "Conditioning Monitoring of Major Turbomachinery Cuts Costs Over 4–Year Period," *Oil & Gas J.*, Vol. 82, No. 12, pp. 96–102 (March 12).

Dresser-Rand (1990), Engine Process Compressor Division, Painted Post, NY, 14870.

Elliott Co. (1981), "Elliott multistage centrifugal compressors," Bulletin P–25B, Jeanette, PA.

GPSA (1987), Engineering Data Book, 10th Ed, Section 13, Gas Processors Suppliers Association, Tulsa, OK.

Hoerbiger Corp. (1990), Ft. Lauderdale, FL 33310.

Hunt, J. W. (1989), Personal Communication, Conoco Inc., Ponca City, OK.

Ingersoll-Rand (1980), Compressed Air and Gas Data, 3rd Ed., Ingersoll-Rand Co., Washington, NJ.

Ingersoll-Rand (1981), Gas Properties and Compressor Data, Ingersoll-Rand Co., Woodcliff Lake, NJ.

Ingersoll-Rand (1982), "KVGR Gas Compressors Brochure," Ingersoll-Rand Co., Woodcliff Lake, NJ.

James, Ralph Jr. (1982a), "Driver Selection," Chapter in Compressors and Expanders, H. P. Block, J. A. Cameron, F. M. Danowski, Jr., R. James, Jr., J. S. Swearingen, and M. E. Weightman, Marcel Dekker, Inc., NY.

James, Ralph Jr., (1982b), "Positive Displacement Compressors," Chapter in Compressors and Expanders, H. P. Block, J. A. Cameron, F. M. Danowski, Jr., R. James, Jr., J. S. Swearingen, and M. E. Weightman, Marcel Dekker, Inc., NY.

Lane, J. L., D. T. Reeh, and A. Niblett Jr. (1988), "The Total Gas Plant Maintenance Program," Proc. 67th Annual GPA Convention, pp. 208–209, Dallas TX (March 14–16).

Laskowski, John (1984), "Ferrography Powerful Maintenance Technique," *Oil & Gas J.*, Vol. 82, No. 5, pp. 85–92 (Nov. 5).

Ludwig, Ernest E. (1965), *Applied Process Design for Chemical and Petroleum Plants*, Vol. III, Gulf Publishing Co., Houston, TX.

Mak, H. (1987), "Process considerations in reciprocating compression," *Hydrocarbon Processing*, Vol. 66, No. 10, pp. 41–42, (Oct.)

Martino, E. (1983), "Materials for Reciprocating Process Gas Compressors," *Chem. Eng. Prog.*, Vol. 79, No. 4, pp. 77–83 (April).

Miller, H. H. (1944), "Keep Up Compressor Output by Holding Down Preventable Losses," *Power*, Vol. 88, No. 8, pp. 514–516 (Aug.).

Molich, Kai (1980), "Consider Gas Turbines for Heavy Loads," *Chemical Engineering*, Vol. 87, No. 17, pp. 79–89 (Aug. 25).

Neerken, Richard F. (1975), "Compressor Selection for the Chemical Process Industries," *Chemical Engineering*, Vol. 82, No. 2, pp. 78–94, (Jan. 20)

Neerken, Richard F. (1980), "Use Steam Turbines as Process Drivers," *Chemical Engineering*, Vol. 87, No. 17, pp. 63–78 (Aug. 25).

Nelson, W. E. (1977), "Compressor Seal Fundamentals," *Hydrocarbon Processing*, Vol. 56, No. 12, pp. 91–95 (Dec.).

Perry, Robert H., and Don Green (1984), "Perry's Chemical Engineers' Handbook," 6th Ed., McGraw-Hill Book Co., NYC, NY.

PETEX (1987), "Field Handling of Natural Gas," 4th Ed., Petroleum Extension Service, University of Texas at Austin, TX.

PMC/Beta Corp., "The New Beta 350," PMC/Beta Corp., Houston, TX.

Sapiro, Leon, "Centrifugal Gas Compressors," Solar Turbines, Inc., San Diego, CA.

Sayyed, Sulaiman (1976), "How Compressor

Components Affect Performance,'' *Hydrocarbon Processing*, Vol. 55, No. 9, pp. 273–277 (Sept.).

Shaw, Jr., H. C. (1980), ''Reciprocating, Centrifugal Compressors Compared,'' *Oil & Gas J.*, Vol. 78, No. 17, pp. 84–88 (April 28).

Solar Turbines (1989), Solar Centaur Turbine Natural Gas Compressor Set, Solar Turbines Inc., San Diego, CA.

Sorenson, H. A. (1983), ''Energy Conversion Systems,'' John Wiley and Sons, Inc., N.Y.

Starling, Kenneth E. (1973), ''Fluid Thermodynamic Properties for Light Petroleum Systems,'' Gulf Pubishing Company, Houston, TX.

Sullair Refrigeration Compressors, Michigan City, IN.

Wood, Bernard D. (1968), ''Applications of Thermodynamics,'' Addison-Wesley Pub. Co., Reading, MA.

Chapter 11

Natural Gas Measurement

In general, flow measurement is characterized by the continuous development of new and improved techniques. Therefore, the more common types of flowmeters, their current uses, and typical accuracies are reviewed in Volume 2.

Natural gas measurement is presently characterized by very extensive research programs worldwide that were, at least in part, undertaken because of differences between national and international standards. In the U.S., the American Petroleum Institute's Manual of Petroleum Measurement Standards (API, 1965–date) is the primary reference for all oilfield applications. Chapter 14, "Natural Gas Fluids Measurement," is of particular interest.

Orifice metering is described first in detail because *orifice meters* are almost always chosen for natural gas custody transfer both in the U.S. and internationally. Gas turbine meters have gained rapid and widespread use since being introduced in the U.S. in 1963. Accordingly, *turbine meters* are discussed next in detail. *Vortex* and *ultrasonic meters* find limited but increasing use, so they are reviewed also. Critical flow nozzles or sonic nozzles are often used to calibrate or prove meters and are described in Volume 2. Some selection criteria—relative advantages, disadvantages, and uncertainties of measurement—are then presented for orifice, turbine, vortex, and ultrasonic meters. Because natural gas is often odorized when metered and transported, Chapter 11 concludes with a brief summary of odorization.

ORIFICE METERS

An orifice plate is essentially a restriction in the pipe that forces the fluid to accelerate and then decelerate as it flows through the meter. Figure 11–1 shows the resulting changes in pressure as confirmed by the Bernoulli equation (ANSI/API, 1985). Flow rate is inferred from the pressure difference measured by *pressure taps* upstream and downstream of the plate. Figure 11–1 also shows that the observed pressure drop depends on the tap locations. While Table 11–1 shows the locations of the more commonly used taps, flange taps dominate in the U.S., and flange and radius (or D and D/2) are common in Europe.

Orifice metering is now described in detail using the following topics: meter piping and differential pressure (DP) devices, meter tube and fittings, sealing units, ANSI/API specifications, calculation of gas flowrates, orifice meter operation, sources of error, and current research programs.

Meter Piping and DP Devices

Figure 11–2 shows a typical arrangement for *gas meter piping*. Valves A and B are bypass valves, C the bleed valve, and D and E supply valves. While Figure 11–2 may suggest that the meter is "clamped" to the pipe, a separate, firm, reasonably level, and vibration-free support is highly recommended.

Figure 11–3 shows schematics of the mercury-manometer and bellows-type differential-pressure (DP) measuring devices. The *bellows-type* is often preferred because:

1. It is particularly adaptable to measuring wet gas due to its self-draining feature.
2. It finds wide applicability with integrators and controllers due to its rapid response and high torque output.

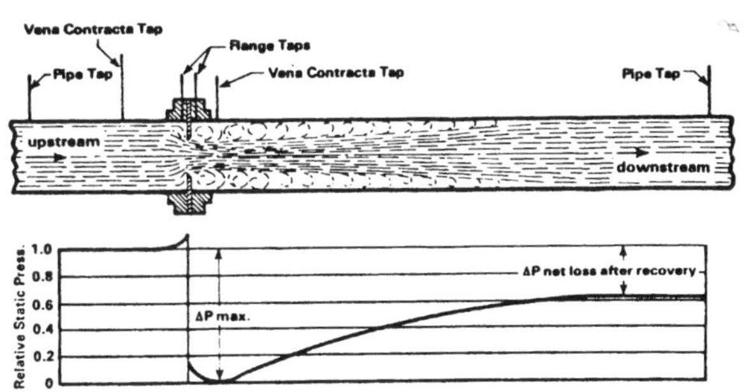

Figure 11–1. Flow and pressure patterns (North American Mfg. Co., 1978).

Table 11–1 Pressure Taps (North American Mfg Co., 1978)

Type	Upstream tap location	Downstream tap location		Advantages and disadvantages
Flange taps	1″ from upstream face	1″ from outlet face		Convenient tap locations fixed by purchasable flanges. Difficult if such flanges are not available.
Vena contracta taps	1 × D from upstream face	d/D□	from downstream face	Most sensitive (highest pressure differential). Fussy tap locations.
		0.3	0.80 × D	
		0.4	0.74 × D	
		0.5	0.66 × D	
		0.6	0.57 × D	
		0.7	0.45 × D	
Radius taps	1 × D from upstream face	½ pipe diameter from upstream face		Easier to locate than vena contracta. Meager data available on flow coefficients.
Pipe taps	2½ × D from upstream face	8 pipe diameters from upstream face		Not dependent on d/D ratio. Pressure drop = net pressure loss. Take up a lot of space.
Corner	upstream edge	downstream edge		

The bellows-type meter should always be mounted above the orifice to exploit its self-draining ability, unless a liquid-filled system with seal pots, etc., is used.

Meter Tube and Fittings

Fig. 11–4 shows typical meter tubes for orifices.

A *senior orifice fitting* (Figure 11–5a) consists essentially of two chambers separated by a lubricated slide valve. This permits removal of the orifice plate without flow interruption or bypass around the meter. Senior fittings are commonly used for custody transfer or sales meters where removal, inspection, and changing of orifice plates are required and where flow bypass is illegal.

The *junior orifice fitting* (Fig. 11–5b) uses the same type of plate carrier as the senior fitting, and the orifice plate is removed similarly by a rack and pinion. Because there is no upper chamber, the fitting must be depressurized using block and blowdown valves before removing the orifice plate.

The *simplex orifice fitting* is a single chamber device (Fig. 11–5c). Removing the orifice plate requires depressurization of the fitting by loosening the top set screws, sliding the clamping bar out, and lifting the sealing bar carrier and orifice plate.

Standard orifice flanges are still widely used to hold orifice plates; however, plate removal and replacement are time-consuming.

Sealing Units (Fig. 11–6)

If a plate seal is used, then the *dual sealing* (DS) unit is the most common. This seal consists of a molded Hycar synthetic rubber ring. A slot in the inner center holds the round universal plate firmly. In larger sizes (12 in. +) the seal is vulcanized to each side of the plate and is called a dual vulcanized seal (DVS).

At higher temperatures (275°F) a *metal* (MS) or a *teflon*

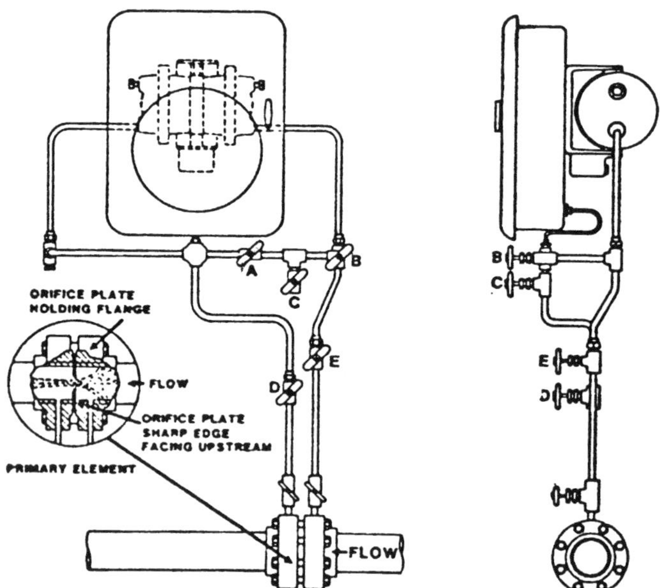

Figure 11–2. Meter piping for gas (Crabtree, 1981).

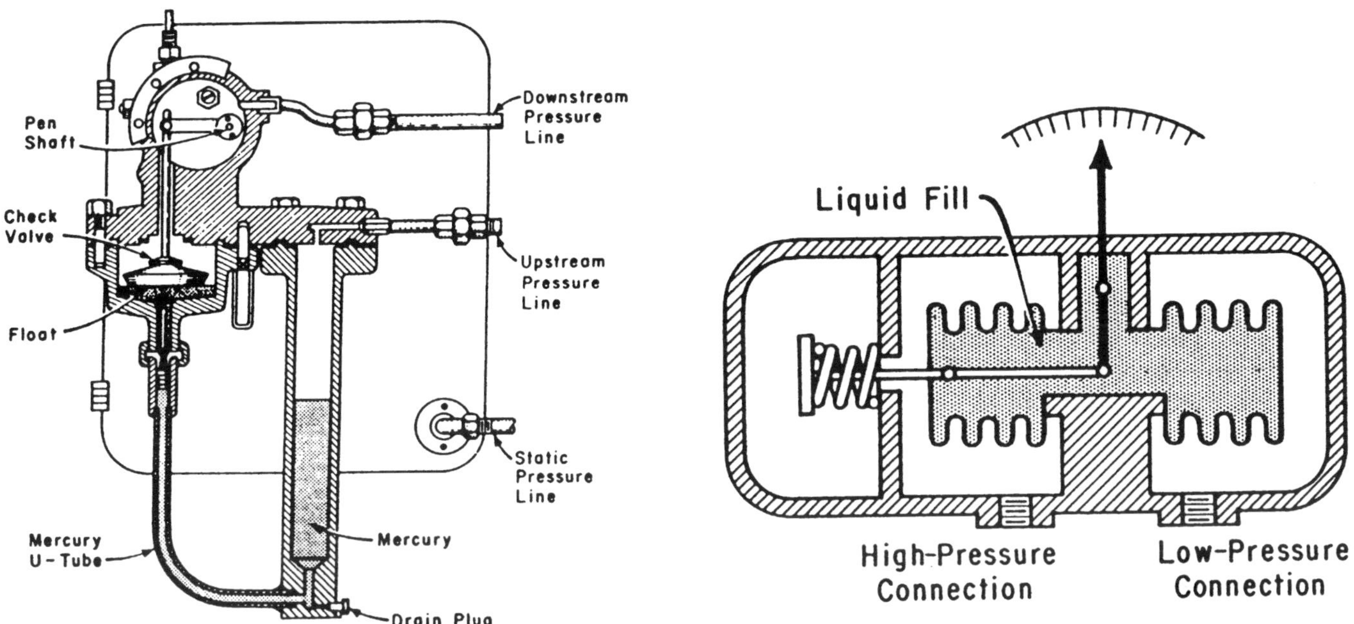

Mercury-Manometer-Type Differential-Pressure Measuring Device

Schematic View of Bellows-Type Differential-Pressure Measuring Device

Figure 11–3.   Differential pressure devices (PETEX, 1987. Copyright © 1987 by the University of Texas at Austin).

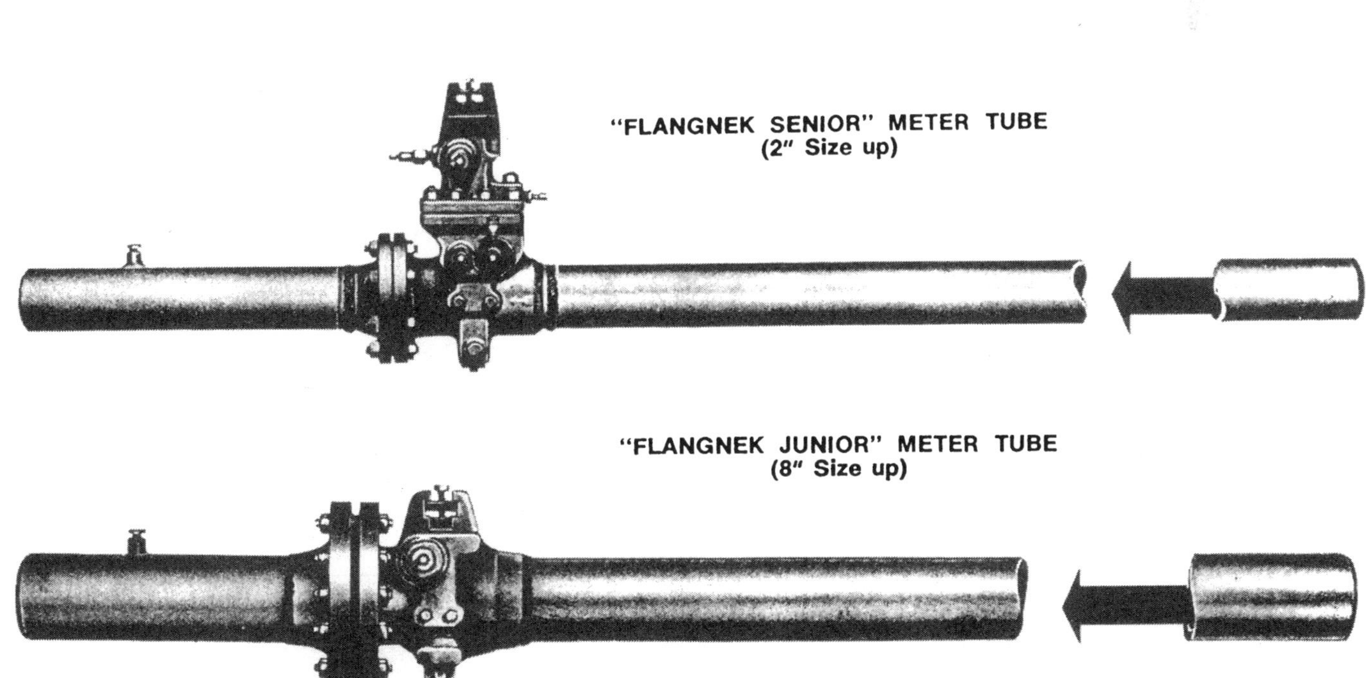

Figure 11–4.   Typical meter tubes (Daniel, 1983).

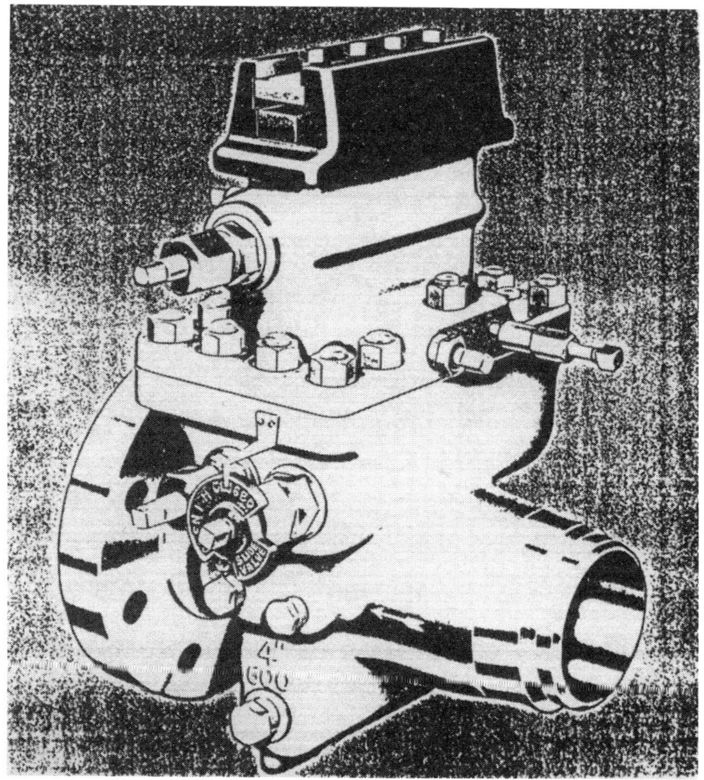

Figure 11–5a.   Senior orifice fitting (Daniel, 1983).

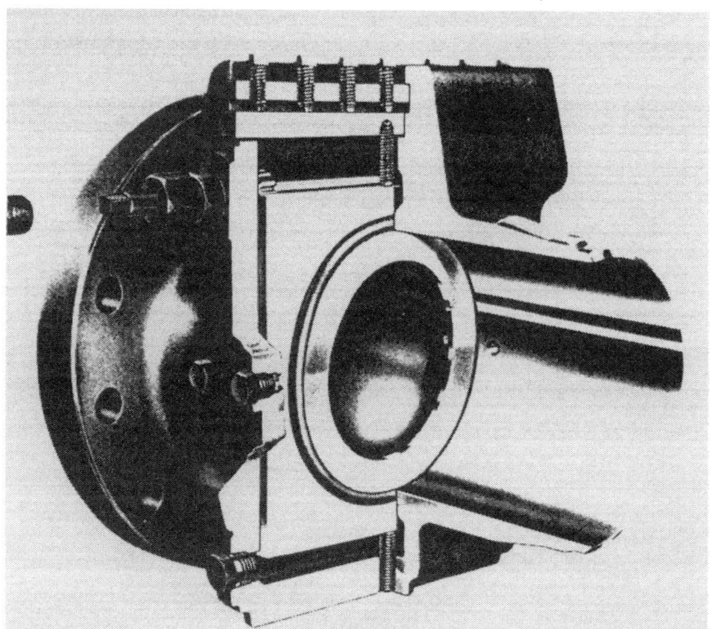

Figure 11–5b.   Junior fitting (Daniel, 1983).

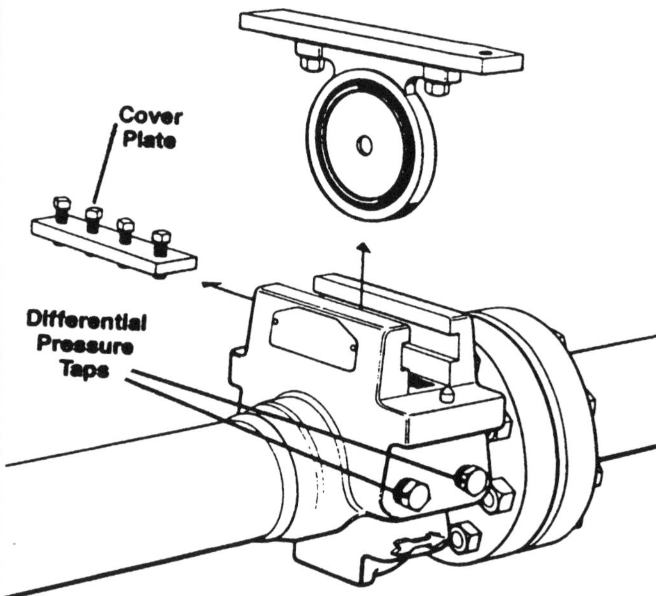

Figure 11–5c.   Simplex fitting (GPSA, 1987).

(TS) seal should be used. These seals are two-piece seals that are attached to the orifice plate by spring clips.

The *snap seal ring* consists of two metal rings having an O-ring on both sides. These two metal rings are held onto the orifice plate by a specially designed external snap ring.

ANSI/API Orifice Specifications

Because orifice plates are NOT CALIBRATED they must be manufactured and installed to meet very definite specifications. In the U.S., the standard is ANSI/API Standard 2530 (1985). It also is referred to as AGA Report No. 3, GPA 8185–85, or API Manual of Petroleum Measurement (Chapter 14, Section 3).

Orifice Plate. The plate is usually made of 304 or 316 stainless steel or monel according to the specifications listed in Figure 11–7. The upstream edge shall be flat to within 0.01 in. per inch of the dam height—(D − d)/2. The upstream edge shall be square and sharp so that it does not show any light when checked with an *orifice edge gage* (Bean, 1985), or it does not reflect a beam of light. Unofficially, a fingernail may be drawn across the orifice edge—a sharp edge will catch or "bite."

The orifice must be concentric within 3% of the metering tube inside diameter.

The *mean orifice diameter*, d, is defined as the arithmetic average of four or more evenly spaced inside diameter measurements. Practical tolerances are:

Orifice Diameter	Tolerance, plus or minus
0.250 in.	0.0003 in.
0.375	0.0004
0.500	0.0005
0.625	0.0005
0.750	0.0005
0.875	0.0005
1.000	0.0005
over 1.000	0.0005 per in. of d

The *Beta Ratio*, or orifice (d) to meter (D) diameter ratio shall be

flange taps	$0.15 < \beta < 0.70$
pipe taps	$0.20 < \beta < 0.67$

Meter Tube. The inside pipe walls shall be smooth (less than 300 microinches roughness). Grooves, scoring, pits, ridges, and offsets are not allowed.

The allowable tolerances of the actual meter tube ID to the published ID varies from 0.7% at $\beta = 0.7$ to 4.4% at $\beta = 0.15$ for flange taps (e.g., GPSA, 1987, p. 3–8).

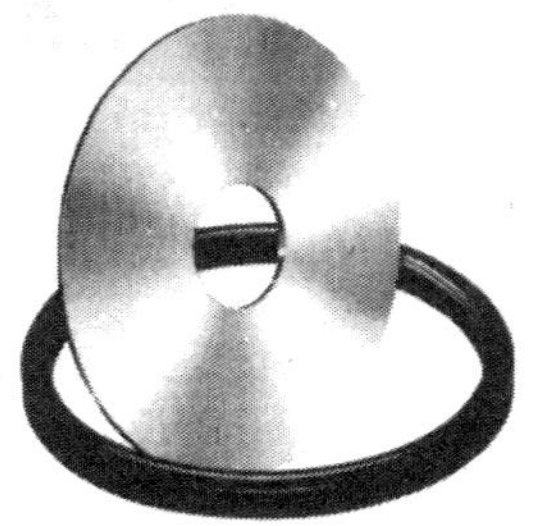

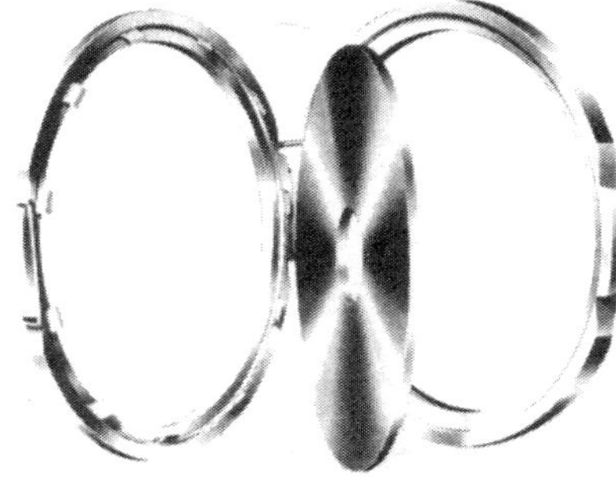

Figure 11–6. Orifice seals (Daniel, 1983).

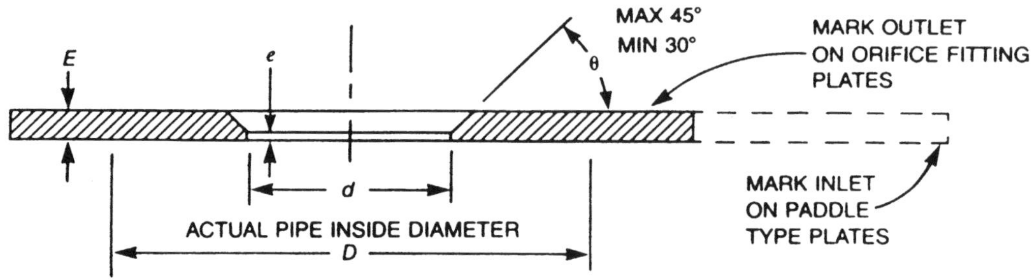

Nominal Inside Diameter, In Inches

	2	3	4	6	8	10	12	16	20	24	30
Published	1.687	2.624	3.152			9.562			18.812		
Inside	1.939	2.900	3.438	4.897 5.761	7.981	10.020	11.938	15.000	19.000	23.000	29.000
Diameter	2.067	2.300 3.068	3.826 4.026	5.187 6.065	7.625 8.071	10.136	11.374 12.090	14.688 15.250	19.250	22.624 23.250	28.750 29.250

Orifice Plate Thickness, E, In Inches

	2.067	2.300	3.068	3.826	4.026	5.187	6.065	7.625	8.071	10.136	11.374	12.090	14.688	15.250	19.250	22.624	23.250	28.750	29.250
Minimum	0.115	0.115	0.115	0.115	0.115	0.115	0.115	0.115	0.115	0.115	0.175	0.175	0.175	0.175	0.240	0.240	0.240	0.370	0.370
Maximum	0.130	0.130	0.130	0.130	0.130	0.163	0.192	0.254	0.269	0.319	0.379	0.398	0.490	0.500	0.505	0.505	0.505	0.562	0.578
Recommended	0.125	0.125	0.125	0.125	0.125	0.125	0.125	0.125	0.125	0.250	0.250	0.250	0.375	0.375	0.375	0.375	0.375	0.500	0.500

Maximum Orifice Edge Thickness, e, In Inches

Orifice Diameter d	$e \leq d/8$	2.067	2.300	3.068	3.826	4.026	5.187	6.065	7.625	8.071	10.136	11.374	12.090	14.688	15.250	19.250	22.624	23.250	28.750	29.250
0.250	1/32	×1/32	×1/32	1/32	1/32	1/32														
0.375	3/64			×3/64	3/64	3/64														
0.500	1/16				×1/16	1/16	1/16	1/16												
0.625	5/64					×5/64	5/64	5/64												
0.750	3/32						×3/32	3/32												
0.875	7/64							×7/64	7/64	7/64										
1.000	1/8								1/8	1/8	1/8									
1.125	9/64								×9/64	9/64	9/64									
1.250	5/32									×5/32	5/32	5/32	5/32							
1.375	11/64										11/64	11/64	11/64							
1.500	3/16										×3/16	3/16	3/16	3/16	3/16					
1.625	13/64											13/64	13/64	13/64	13/64					
1.750	7/32											×7/32	7/32	7/32	7/32					
1.875	15/64												×15/64	15/64	15/64					
2.000	1/4													1/4	1/4	1/4				
2.250	9/32													×9/32	9/32	9/32				
2.375	19/64														×19/64	19/64	19/64	19/64		
2.500	5/16															5/16	5/16	5/16		
2.750	11/32															11/32	11/32	11/32		
2.875	23/64															23/64	23/64	23/64	23/64	23/64
3.000	3/8															×3/8	3/8	3/8	3/8	3/8
3.250	13/32																13/32	13/32	13/32	13/32
3.500	7/16																×7/16	7/16	7/16	7/16
3.625	29/64																	×29/64	29/64	29/64
3.750	15/32																		15/32	15/32
4.000	1/2																		1/2	1/2
4.250	17/32																		17/32	17/32
4.500	9/16																		×9/16	9/16
4.625	37/64																			×37/64
4.750	19/32																			
5.000	5/8																			

Notes:

1. The maximum edge thickness is defined by $e \leq D/50$ or $e \leq d/8$, whichever is smaller.

2. Orifice edge thickness marked with × in this table is the maximum for that particular meter tube diameter and is applicable to all larger orifice diameters for that meter tube diameter.

3. Orifice diameters smaller than those marked × are defined by $e \leq d/8$.

4. Orifice plates of which the edge thickness meets the value $e \leq D/30$ need not be rebeveled unless reconditioning is required for other reasons.

5. All dimensions are in inches. For ease in machining, the next smaller value of e in even multiples of 1/16 or 1/32 inch may be used where e is given in 1/64ths.

6. Orifices used to measure dual directional flows must not be beveled. Where e exceeds the above limits, the flow constant F_b may be subject to higher uncertainty.

Figure 11–7.   Orifice plate dimensions (ANSI/API, 1985).

Length of pipe preceding and following orifice varies with the type of fittings (partially closed valve, ell, reducer, or expander) in the line—see ANSI/API, or GPSA (1987, p. 3–8, 3–9) for numerical values. For safety assume that $\beta = 0.7$ when designing new installations. Straightening vanes (e.g., GPSA, 1987, p. 3–11) reduce the required lengths of upstream pipe, and this is especially welcome for large diameter pipe when space is limited, e.g., offshore. Occasionally, vanes can cause plugging and slipping problems.

Other Specifications. Refer to ANSI/API 2530 for additional details on straightening vanes, orifice flanges, orifice fittings, pressure tap holes, and thermometer wells.

Calculating Gas Flow Rate

As defined in ANSI/API 2530 (1985), the fundamental equation for gas measurement is:

$$Q_h = C' \sqrt{h_w \, P_f} \qquad (11\text{–}1)$$

where Q_h = quantity rate of flow at base conditions in cu ft/hr

h_w = differential pressure in inches of water

P_f = absolute static pressure, psia

C' = orifice flow constant

$$C' = F_b \cdot F_r \cdot Y \cdot F_{pb} \cdot F_{tb} \cdot F_{tf} \cdot F_{gr} \cdot \qquad (11\text{–}2)$$
$$F_{pv} \cdot F_a \cdot F_{am} \cdot F_{wl} \cdot F_{wt} \cdot F_{pwl} \cdot F_{hgm} \cdot F_{hgt}$$

and F_b = basic orifice factor

F_r = Reynolds number factor

Y = expansion factor

F_{pb} = pressure base factor

F_{tb} = temperature base factor

F_{tf} = flowing temperature factor

F_{gr} = specific gravity factor

F_{pv} = supercompressibility factor

F_a = orifice thermal expansion factor

F_{am} = correction for air over the water in the water manometer during the differential instrument calibration

F_{wl} = local gravitational correction for water column calibration standard

F_{wt} = water weight correction (temperature) for water column calibration standard

F_{pwl} = local gravitational correction for dead weight tester static pressure standard

F_{hgm} = manometer factor, correction for gas column in mercury manometers

F_{hgt} = mercury manometer temperature factor

Brief descriptions of these factors follow, for more complete equations, tables, and descriptions see ANSI/API 2530.

Basic Orifice Flow Factor, F_b, is the scf of air that will flow per hour through the orifice at 60°F with a static pressure of 14.73 psia and a differential pressure of 1 in. water column. Values of F_b, which depend only on d, D, and the choice of flange taps, are given in ANSI/API 2530 and GPSA (1987, p. 3–14, 3–15).

Reynolds Number Factor, F_r, corrects for changes in C' with flowrate or Reynolds number. Values for F_r ($F_r = 1 + b/\sqrt{h_w P_f}$) are given in GPSA (1987, p. 3–16, 3–17).

Expansion Factor, Y, adjusts C' depending on whether P_f is measured from the upstream or downstream pressure tap. See GPSA (1987, p. 3–18, 3–19) for values of Y.

Pressure Base Factor, F_{pb}, adjusts the flow (scfh) from the AGA base of 14.73 psia to any other base pressure (P_b). Numerically,

$$F_{pb} = 14.73 \,/\, P_b \text{ (psia)} \qquad (11\text{–}3)$$

Temperature Base Factor, F_{tb}, adjusts the flow (scfh) from the AGA base of 60°F to any other temperature base, T_b (°F). Numerically,

$$F_{tb} = (460 + T_b)/(520) \qquad (11\text{–}4)$$

Flowing Temperature Factor, F_{tf}, adjusts Q if the flowing temperature, T_f (°F), is not 60° F. Numerically,

$$F_{tf} = \sqrt{(520)/(460 + T_f)} \qquad (11\text{–}5)$$

Specific Gravity Factor, F_{gr}, corrects the flow rate when the specific gravity of the flowing gas, SG, is not that of air. Numerically,

$$F_{gr} = \sqrt{1/SG} \qquad (11\text{–}6)$$

Supercompressibility Factor, F_{pv}. ANSI/API describes a detailed, empirical procedure for calculating F_{pv} which corrects for real gas behavior (compressibility factor, Z, $\neq$ 1). Numerically,

$$F_{pv} = \sqrt{1/Z} \qquad (11\text{–}7)$$

Orifice Thermal Expansion Factor, F_a, corrects for thermal expansion of the orifice plate. For 304 and 316 stainless steel:

$$F_a = 1 + 0.0000185 \,(T_f - T_{meas}) \qquad (11\text{–}8)$$

For monel:

$$F_a = 1 + 0.0000159 \,(T_f - T_{meas}) \qquad (11\text{–}9)$$

where T_f = flowing temperature (°F)

T_{meas} = temperature (°F) at which orifice diameter is measured.

Water Manometer Correction Factor, F_{am}, corrects for the gas leg over water when the differential gage is calibrated using a water manometer

$$F_{am} = \{[62.3663 - (P_{atm} + h_w/27.707)/ \quad (11\text{–}10)$$
$$192.4]/62.3663\}^{0.5}$$

$F_{am} = 0.99924$ when $P_{atm} = 14.73$ psia and $h_w = 100$ in.

Local Gravitational Correction for U-Tube Manometers, F_{w1}. The local weight of a manometer fluid is directly proportional to the local gravity force.

$$F_{w1} = (g/32.17405)^{0.5} \quad (11\text{–}11)$$

In turn, g may be estimated by (ANSI/API 2530):

$$g = 0.032808*(978.018\ 55 - 0.0028\ 247\ L \quad (11\text{–}12)$$
$$+ 0.002\ 029\ 9\ L_2 - 0.000\ 015\ 058\ L^3$$
$$- 0.000\ 094\ H)$$

$$\text{where } L = \text{latitude in degrees}$$
$$H = \text{elevation in feet above sea level.}$$

Water Manometer Temperature Correction Factor, F_{wt}, corrects for variations in the density of the water in the manometer used to calibrate the differential gage

$$F_{wt} = (\rho_w / 62.3663)^{0.5}$$

where ρ_w = density of water (lb/ft^3) in manometer.

Local Gravitational Correction for Weight Standards, F_{pw1}, corrects for the local gravity effect on the weights of the dead weight tester

$$F_{pw1} = (g/g_o)^{0.5} \quad (11\text{–}13)$$

where g = acceleration due to local gravity force
    g_o = acceleration of gravity used to calibrate dead weight tester.

The *Mercury Manometer Factors*, F_{hgm} and F_{hgt}, correct for the gas or vapor leg and the temperature respectively. See ANSI/API 2530 for definitions. Omit F_{hgm} and F_{hgt} for bellows meters.

Example 11–1. Calculate meter coefficient and gas flowrate. An orifice meter run has the following characteristics:

Meter Pipe	4 in. schedule 40 (4.026 in. ID)
	flange taps
	static pressure measured from upstream tap
Orifice Plate	stainless steel
	1.500 inch at 20°C (68°F)
Recorder	100 in. WC differential
	1000 psia static spring
Readings	atmospheric pressure = 14.4 psia (500 ft elevation)
	flowing temperature = 100°F
	gas gravity = 0.60
	static pressure = 651 psig
	differential = in. (bellows)

Base Conditions (Oklahoma or Texas) = 14.65 psia, 60 °F
Basic flow equation is, $Q_h = C' \sqrt{h_w\ P_f}$ (11–1)

where, $C' = F_b \cdot F_r \cdot Y \cdot F_{pb} \cdot F_{tb} \cdot F_{tf} \cdot F_{gr} \cdot F_{pv} \cdot F_a \cdot$
$F_{am} \cdot F_{w1} \cdot F_{wt} \cdot F_{pw1} \cdot F_{hgm} \cdot F_{hgt}$ (11–2)

1. Base orifice factor, F_b

$$F_b = 460.80 \quad \text{(GPSA, p. 3–14, F 3–18)}$$

Need flange taps

$$\text{pipe ID} = D = 4.026 \text{ in.}$$
$$\text{orifice diameter} = d = 1.500 \text{ in.}$$

2. Reynolds number factor, F_r

$$F_r = 1 + b/ \sqrt{h_w\ P_f}$$
$$= 1 + 0.0336/ \sqrt{(65)\ (665.4)} = 1.0002$$

Need flange taps

$$\text{pipe ID} = D = 4.026 \text{ in.}$$
$$\text{orifice diameter} = d = 1.500 \text{ in.}$$
$$b = 0.0336 \text{ (GPSA, 1987, p. 3–6, Fig 3–19)}$$

$$P_f = 665.4 \text{ psia}$$
$$h_w = 65 \text{ in W.C.}$$

3. Expansion factor, Y_1

$$\text{As defined, } \beta = d/D = 1.500/4.026 = 0.373$$
$$h_w/P_{f1} = 65/(651 + 14.4) = 0.098$$

From GPSA, p. 3–18, Fig. 3–20

	β =	0.3	(0.373)	0.4
h_w/P_{f1}	0.0	1.000		1.000
	0.1	0.9989		0.9988
Interpolating	0.098	0.9989	0.9988	0.9988

Need flange taps
static pressure from upstream tap and h_w/P_{f1}

4. Pressure base factor, F_{pb}

$$F_{pb} = 14.73/P_b$$
$$= 14.73/14.65 = 1.0055$$

Need P_b base pressure for desired scf.

5. Temperature base factor, F_{tb}

$$F_{tb} = (T_b + 460)/(60 + 460)$$
$$= (60 + 460)/(60 + 460) = 1.0000$$

Need T_b = base temperature for desired scf.

6. Flowing temperature factor, F_{tf}

$$F_{tf} = \sqrt{(520)/(T_f + 460)}$$
$$= \sqrt{(520)/(100 + 460)} = 0.9636$$

Need T_f = flowing temperature.

7. Specific gravity factor, F_{gr}

$$F_{gr} = \sqrt{1/SG} = \sqrt{(1)/(.60)} = 1.2910$$

Need SG, gas specific gravity relative to air or gas composition and calculate SG.

8. Supercompressibility factor, F_{pv}

$$F_{pv} = 1.0401, \text{ ANSI/API 2530 (1985), p. 103,}$$
$$\text{Table D–5}$$

Alternatively,

$$F_{pv} = \sqrt{1/Z} = \sqrt{1/(.917)} = 1.0443$$

Note $Z = .917$ Fig. 3–17
 or GPSA (1987, Fig. 23–8)
Need information to calculate Z; usually T_f, P_f, and gas composition.

9. Orifice thermal expansion factor, F_a

$$F_a = 1 + 0.000018 (T_f - T_{meas})$$
$$= 1 + 0.000018 (100 - 68) = 1.0006$$

Need stainless steel, T_{meas}.

10. $F_{am} = \{[62.3663 - (14.4 + 65/27.707)/192.4]/$
 $62.3663\}^{0.5}$
 $= 0.9993.$

11. $\quad g = 32.1418$ when $H = 500$ ft and $L = 35°$
 $F_{wl} = (32.1418/32.174)^{.5} = 0.9995$

12. Assume water in manometer was at 60°F during calibration so

$$F_{wt} = 1.0000.$$

13. $F_{pwl} = 0.995$ same as F_{wl}

14. Bellows meter so omit F_{hgm} and F_{hgt}.

Therefore (at last!)

$C' = F_b \cdot F_r \cdot Y_1 \cdot F_{pb} \cdot F_{tb} \cdot F_{tf} \cdot F_{gr} \cdot F_{pv} \cdot F_a \cdot F_{am} \cdot$
$\quad F_{wl} \cdot F_{wt} \cdot F_{pwl} =$
$= (460.80) (1.0002) (.9988) (1.0055) (1.000) (.9636)$
$\quad (1.2910) (1.0443) (1.0006) (.9993) (.9995) (1.0000)$
$$(.9995)$$
$= 600.66.$

$Q_h = C' \sqrt{h_w P_f}$
$\quad = (600.66) \sqrt{(65)(665.4)}$
$\quad = 125,100$ scfh (14.65 psia, 60°F).

Note $F_b \cdot F_{tf} \cdot F_g = (460.80)(.9636)(1.2910) = 573.23$

The product of $F_b \cdot F_{tf} \cdot F_g$ and C' agree within 5%; and so $F_b \cdot F_{tf} \cdot F_g$ is satisfactory for sizing a meter run.

Orifice Meter Operation

Schepers (1981) describes in detail the inspection and calibration of an orifice meter including the plate, differential

Table 11–2 Troubleshooting Orifice Meters (after Pulley, 1988)

Problem	Possible Causes
Low or no Reading	Orifice plate installed backward or oversize
	Flow blocked upstream
	Density changes in process media or reference leg
	Pressure-tap holes and/or piping plugged
	Bypass valve open or leaking
	Liquid or gases trapped in piping
	Block or shut-off valves closed
	Piping leaks, high-pressure side
	Housing filled with solids
	Gas trapped in housing in liquid service
	Liquid trapped in housing in gas service
	High-pressure (HP) housing gasket leaks
	Differential-Pressure unit (DPU) tampered with
	Loose links or movements in mechanism
	Pointer loose
	Mechanism out of calibration
	Corrosion or dirt in mechanism
Indicates High	Orifice partially restricted or too small
	Leak in low-pressure side piping
	Loss of liquid in reference leg (liquid level)
	Gas trapped in low pressure housing, liquid service
	Liquid trapped in high pressure housing, gas service
	Low-pressure (LP) housing gasket leaks
	Bellows range spring broken, or DPU tampered with
	Loose links or movements in mechanism
	Mechanism out of calibration
Erratic Indication	Flow pulsating
	Liquid trapped in gas piping
	Gas bubble in liquid piping
	Vapor generator incorrectly installed
	Reference leg gassy or liquid vaporizing
	Obstructed bellows travel
	Gas trapped in DPU, HP or LP pressure housing
	Mechanism linkage dragging or dirty
	Pointer dragging on scale plate

bellows, orifice charts, differential and static pens, etc. Table 11–2 summarizes some troubleshooting advice.

Two common types of charts used are the linear and the square root or L10 (Fig. 11–8). As the name implies, the linear chart is divided into equal sized divisions, usually 100. As is emphasized in Figure 11–8, numbers on the square root chart are merely the square root of linear chart readings.

During calibration the pens should be zeroed correctly. The differential pen is set originally at zero because the meter is calibrated with no flow. However, atmospheric pressure acts on the meter during calibration, and the static

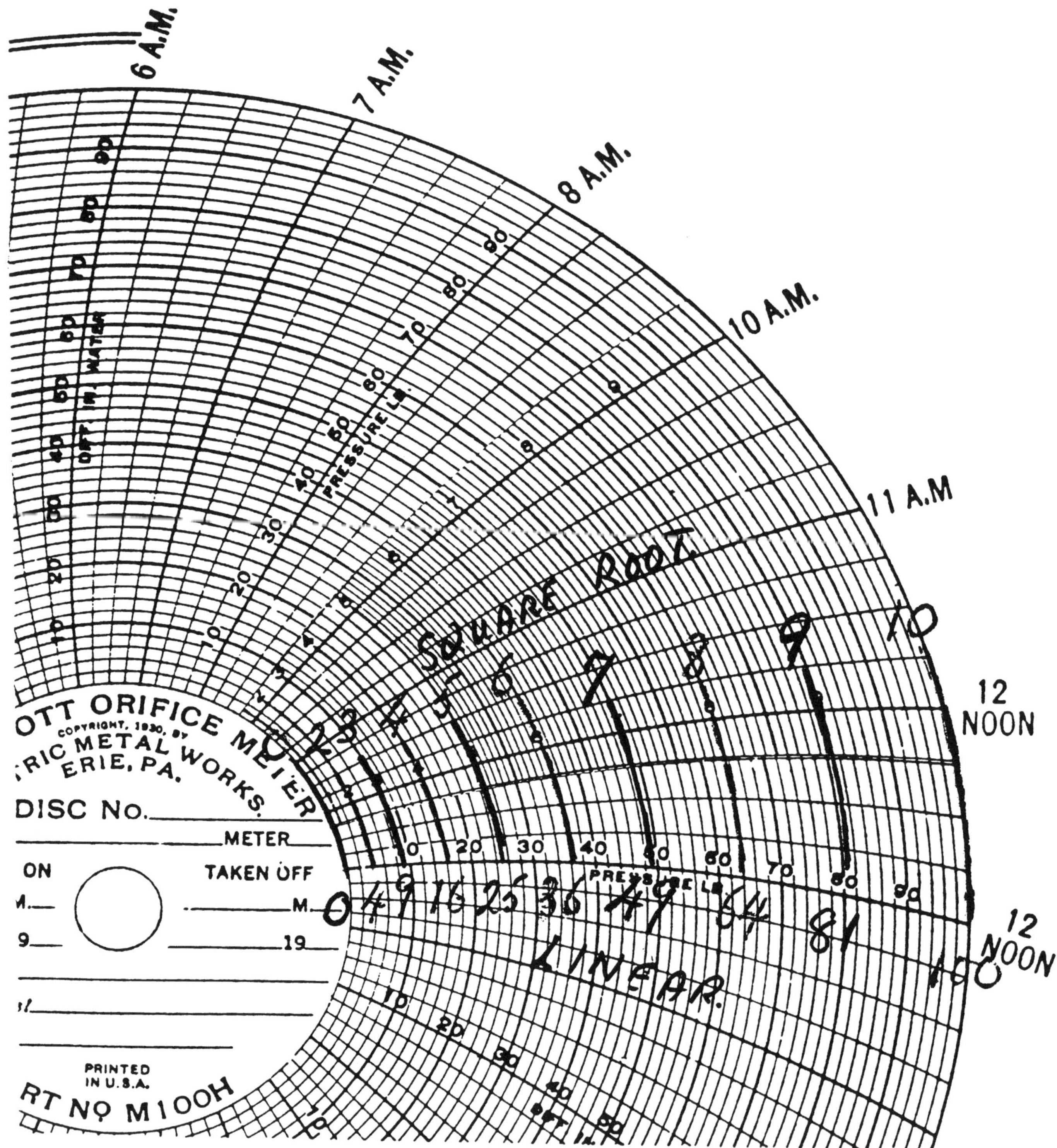

Figure 11–8.   Comparison of linear and square root charts.

pen is set at the current ambient atmospheric pressure. Numerically, the correct static pen "zero" is:

$$\text{Linear: } R_c = (100)(P_{atm})/P_{max} \qquad (11\text{–}14)$$

$$\text{Square Root } R_L = \sqrt{(R_c)} = 10\sqrt{(P_{atm}/P_{max})} \quad (11\text{–}15)$$

$$\text{where } R_c = \text{reading on linear chart}$$
$$R_L = \text{reading on L–10 chart}$$
$$P_{atm} = \text{ambient atmospheric pressure (psia)}$$
$$\text{(remember this varies with elevation)}$$
$$P_{max} = \text{pressure (psia) to cause full scale}$$
$$\text{reading on charts.}$$

Example 11–2. Error caused by incorrect zeroing of static pen. Estimate the error caused by incorrectly zeroing the static pen, i.e., incorrectly setting static pen at zero of chart instead of correctly zeroing pen at 1 atm.

Does the orifice meter read high or low? In the standard orifice equation, $Q_h = C'\sqrt{h_w}\sqrt{P_f}$, P_f should be in psia. Incorrect zeroing causes the static element to measure gage pressures: $P_f(psia) - P_{atm}$. Therefore

$$Q_h \text{ (correct)} = C'\sqrt{h_w}\sqrt{P_f \text{ (psia)}}$$
$$Q_h \text{ (incorrect)} = C'\sqrt{h_w}\sqrt{P_f - P_{atm}}$$

$$Q_h \text{ (incorrect)}/Q_h \text{ (correct)}$$
$$= \sqrt{h_f \text{ (psig)}}/\sqrt{P_f \text{ (Psia)}}$$

The magnitude of the error is now computed for several static pressures. Assume, $P_{atm} = 14.2$ psia (1000 ft elevation)

Static (psig)	Pressure (psia)	$\dfrac{Q_h \text{ (incorrect)}}{Q_h \text{ (correct)}}$	Error % Low
35.8	50	0.846	15.4
85.8	100	0.926	6.4
185.8	200	0.964	3.6
485.8	500	0.986	1.4
985.8	1000	0.993	0.7

Note the orifice meter reads low.

Sources of Error

Measurement problems and sources of error are described separately.

Measurement Problems. These include freezing, pulsating flow, slugging, and sour gas. Commonly implemented solutions are shown below (PETEX, 1987):

Freezing	• Install line heaters, heated meter house
	• Dry gas, use hydrate inhibitors
	• Enlarge meter piping to 0.5 in
Pulsations	• Locate meter away from reciprocating compressor
	• Insert capacity, restriction, or filter in line
	• Operate at as high as possible h_w
Slugging	• Remove liquids in slug catchers
	• Dry gas
Sour gas	• Use 316 stainless static spring
	• Use Teflon bearings for differential pen shaft
	• Avoid copper piping, mercury manometers

Sources of Error. Figure 11–9 illustrates some common sources of error. Installing the orifice plate *backwards* causes the meter to read low by as much as 17 to 30%. A *dull plate* can easily cause an orifice to read 1 to 2% low. Batchelder (1985) describes these and other sources of error and also documents the importance of witnessing orifice meter calibration and field testing. Burgin (1971) investigated the effects of many abnormal situations on orifice meter readings. Hoch (1983) and Studzinski (1988) also studied orifice plates under abnormal conditions. Table 11–3 summarizes the observations of Batchelder, Burgin, Hoch, and Studzinski. The importance of following all ANSI/API specifications is obvious. These findings also show that orifice meters tend to read low under abnormal conditions.

Current Research Programs

The orifice meter has been the preferred method of gas measurement and custody transfer by pipeline and production companies for over 70 years. The current (1985) ANSI/API Standard 2530 was originally issued by the American Gas Association as AGA Report No. 2 in 1935 and supplemented in 1955 as AGA Report No. 3. It is based primarily on Beitler's (1935) data. The current ANSI/API 2530 equations were developed by Buckingham and Bean in 1934 (ANSI/API 2530, p. 26). The International Organization for Standardization document, ISO–5167 (1980), adopted the Stolz (1978) equation. Fling and Spencer (1984) estimate that the Stolz equation is based on perhaps 600 flange- and corner-tap data points obtained by Beitler and by Witte in the early 1930s.

Figure 11–10 compares the ANSI/API 2530 and ISO–5167 equations. Even a very small difference in C' becomes very significant when 1 trillion scf of gas are sold.

The American Petroleum Institute (API) and the Gas Processors Association (GPA) have completed development of a new data base of coefficients of discharge for concentric, flanged-tapped, square-edged orifice meters. The program uses three, single-phase, Newtonian fluids (5–centistoke

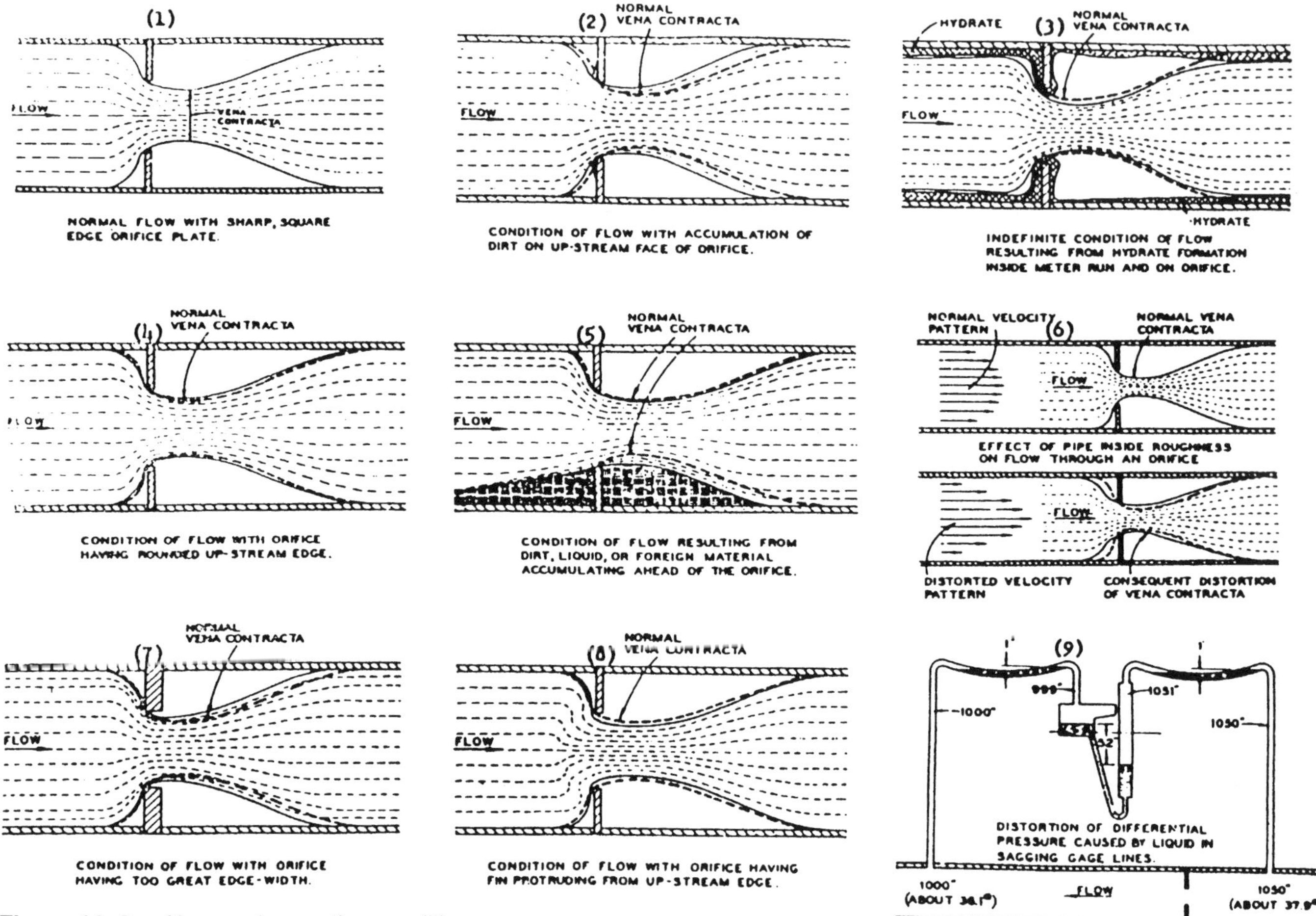

Figure 11–9. Abnormal operating conditions.

Table 11–3 Effects of Abnormal Conditions on Orifice Measurement

Condition	Error	Reference
Plate installed backwards	17% low	Batchelder (1985)
	14% low	Hoch (1983)
	up to 30% low	Schepers (1981)
Dirt upstream of plate	6% low	Hoch (1983)
	11% low	Burgin (1971)
Dirt downstream of plate	2% low	Hoch (1983)
Dirt both sides of plate	3% high	Hoch (1983)
Grease and dirt in tube	11% low	Burgin (1971)
Liquid in meter tube	11.3% low	Burgin (1971)
Dirt between pipe and plate on face of plate	23% low	Hoch (1983)
Dirt on upstream face of plate	27% low	Hoch (1983)
	0–4.7% low	Burgin (1971)
Valve lubricant on upstream face	0–4.7% low	Burgin (1971)
Valve lubricant on downstream face	1% low to 3% high	Burgin (1971)
Rounded and scratched place	11% low	Hoch (1983)
Dull plate	.5% low	Burgin (1971)
Upstream edge bevelled	2–13.5% low	Burgin (1971)
Upstream edge rounded	0–9.3% low	Burgin (1971)
Warped plate	2% low to 10% high	Burgin (1971)
Methanol injection upstream	6% low	Batchelder (1985)
Static pressure 10 psi low at 221 psig and flow temperature 7°F high at 75°F	3% low	Batchelder (1985)

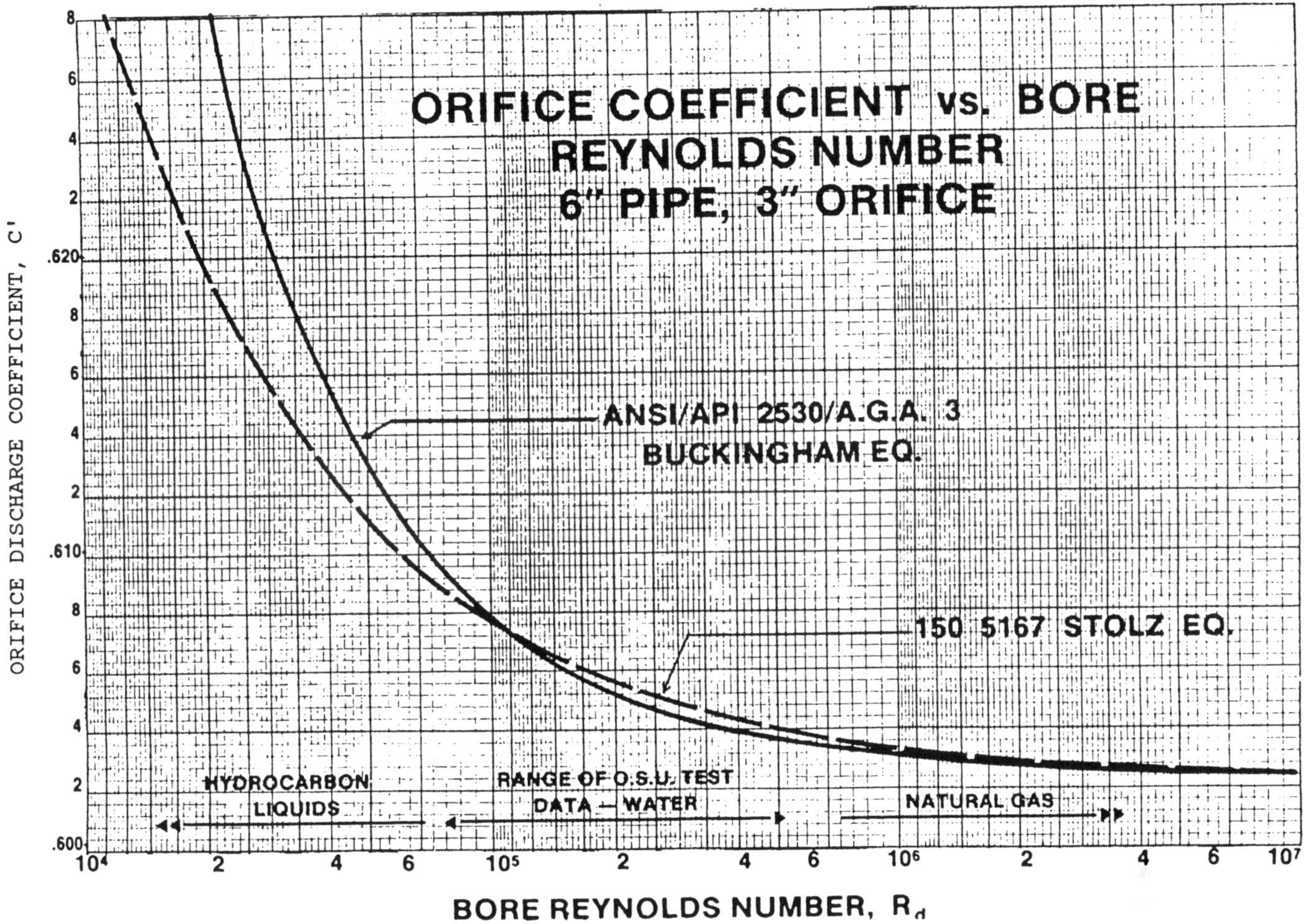

Figure 11–10. Comparison of ANSI/API 2530 and ISO 5167 (Hoglund, 1984).

white oil, water, and natural gas) flowing in 2-, 3-, 4-, 6- and 10-in. ID meter tubes, over a pipe Reynolds number range of 150 to 20000000, and yields 21000 new discharge coefficient data points. The data from this project provide a new and many times larger data base than has ever been presented before (Fling, 1986).

In the European work, orifice plates of six different diameters (51 mm–191 mm) and two thicknesses were tested at six laboratories in 100- and 250-mm pipes using water, natural gas, and air over a Reynolds number range of 5×10^3 to 3.5×10^7. Over 4000 data points were obtained for flange and D and D/2 taps (CEC, 1985; CEC, 1987).

These 25000 data points have been analyzed, and several new equations for the orifice discharge coefficient have been suggested. Hopefully, agreement will soon be reached on a new and improved equation that will replace the Buckingham and/or Stoltz equations.

A full-scale gas metering facility—the Karsto Metering & Technology Laboratory, or K-Lab—has been built in Stavager, Norway (Bosio, Wilcox, and Sembsmoen, 1988).

The project is a joint venture between Statoil and Total. The facility uses sonic nozzles as reference flowmeters and a gravimetric system for the primary calibration of the sonic nozzles. The facility operates as a closed loop, the maximum volume flowrate will be 460 million scfd; and the operating pressures in the working sections range from 145 to 2262 psia. The objective is to calibrate gas-flow meters in pipes with diameters of 4 to 24 in. with an uncertainty in mass flow of less than 0.25%.

The Gas Research Institute recently initiated the Metering Research Facility program with Southwest Research Institute. In 1987 an extensive industry survey was conducted to identify research needs, the priority of needs, and the expected benefits. This survey resulted in the following priority rating for testing meter types:

1. Turbine
2. Orifice
3. Vortex
4. Positive displacement.

The priority rating for research topics is:

1. Upstream lengths
2. Pulsation effects
3. Swirl.

See McKee (1989) for more details.

TURBINE METERS

Today *turbine meters* are available for gas measurement in sizes and working pressures ranging from 0.24 to 3.36 MMscfd and 175 to 1440 psig. The construction of turbine meters is presented first and then meter performance is reviewed by discussing the validity of two basic assumptions: first, that the rotor rotation varies linearly with the average fluid velocity, and second, that the volumetric flowrate is proportional to the average fluid velocity. Finally, meter sizing and accuracy are summarized and a troubleshooting checklist is presented.

Construction of Turbine Meters

As shown in Figure 11–11, a turbine meter consists of (Baker and Kalivoda, 1977):

1. *Housing*—a flanged pipe spool .25 to 24 in. diameter with 275–6000 psig working pressure and –20 to +500°F standard design temperature. Construction material is usually carbon steel or stainless steel for corrosive environments or low temperature applications.
2. *Upstream and Downstream Hangers* (or stators, supports)—which center and support the rotor and axially clamp the rotor thrust bearings. These hangers contain the thrust washers, provide passages for hydraulic thrust balancing of the rotor, and include blades for straightening the flow.
3. A *Rotor*—with stainless steel blades supported by tungsten carbide journal bearings and thrust washers. These tungsten carbide bearings are highly polished and have a small bearing surface to minimize drag. In modern designs the rotor "floats" between the upstream and downstream cones on a thin film of liquid that flows between the bearing and the shaft. The rotor is thrust upstream by the pressure difference across the rotor blades and downstream by the flow impinging on the outer rim of the rotor hub which is, by design, not shielded by the upstream cone.
4. A *Variable Reluctance Pick-up Coil*—which detects the rotational speed of the rotor by monitoring changes in magnetic flux passing through the coil bobbin. The rotor must have regularly spaced paramagnetic material at

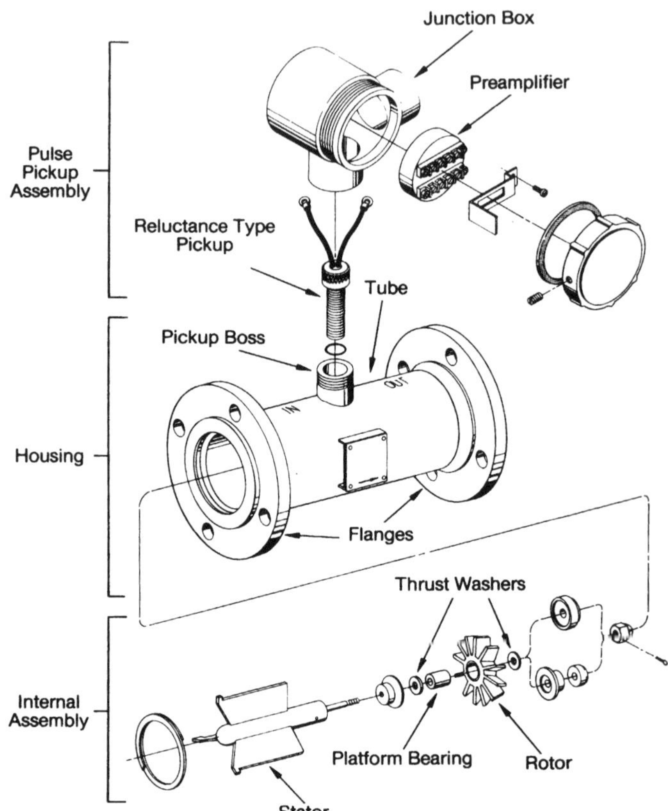

Figure 11–11.   Turbine meter construction (Laird, 1988).

the periphery for the variable reluctance pickup to work (Daniel, 1983).

Gas Turbine Meters

In principle, gas turbine meters are the same as for liquids, with a few important differences (Fig. 11–12). Since the driving torque is proportional to the density of the flowing gas, this torque is much lower than for liquids. The rotor speed is therefore maintained high by operating at high pipeline velocities and by having a high ratio of center body diameter to pipe diameter. A *nose cone* or flow deflector forces the gas to flow through an annulus having an open area approximately one-third of the open pipe area, thus providing more driving torque. The nose cone also absorbs most of the flow stream thrust that otherwise might damage the rotor bearings.

The *rotor* spins at similar speeds to those for liquids, and hence smaller blade angles are used (10 degrees) compared to liquids (35 degrees). The rotor blades are often helical rather than flat and are machined or molded as an integral part of the hub to improve strength. Because light weight improves rotor performance and bearing life, high-

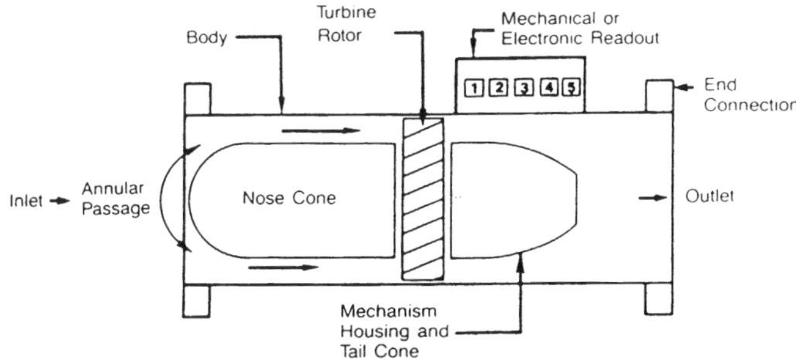

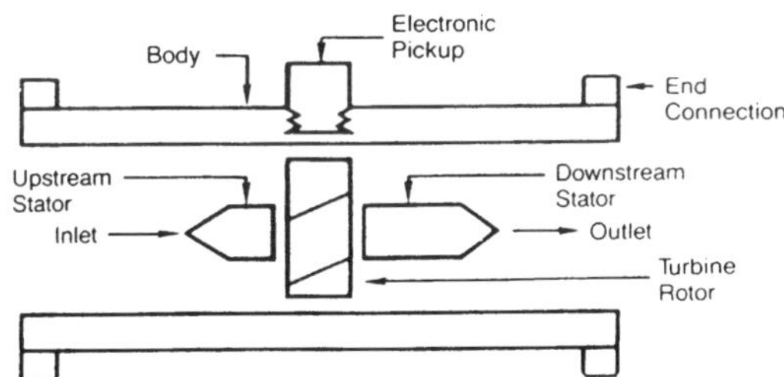

Figure 11–12.   Gas turbine meter (AGA, 1985).

strength, impact-resistant plastic or alumina is normally used.

Bearings are usually of the ball race type and small relative to the meter partly to reduce frictional drag and partly due to the high rotational speed. The nose cone usually shields the bearings from liquids, dirt, and grit entrained in the following gas. Bearings must be lubricated either permanently or periodically during operation. Most meters use a wick or felt to add lubricating oil, even when the meter is pressurized. Excess oil in the bearings flushes out any dirt and eventually evaporates into the gas stream (Allen, 1989).

There are two distinct methods of measuring rotor rotation—mechanical and electro-magnetic. Mechanical designs use gear trains connected to a counter clock. There are four types of electro-magnetic sensors—induction pickup coil, reluctance pickup coil, modulated carrier pickup coil, and light emitting diode sensor (Allen, 1989).

Many flanged meters are designed to have the internal working parts contained in a *module assembly* that is easily accessible for maintenance or replacement. All rotating parts which are likely to wear or erode are part of this module.

Due to the large contraction in area, gas meters tend to be less influenced by inlet conditions than liquid meters. Nevertheless, some *20 pipe diameters* are recommended as the *minimum inlet length* even though the flow standards that are appearing suggest 10 diameters as a minimum length.

Calculation of Gas Flow Rate

The recommended method for calibrating turbine meters is the transfer test in which a volume of gas (or air) is passed through the meter in series with a "master meter," or a critical-flow or a bell prover, or sonic nozzle. Standard factory curves are established. Tests at 100 psig are adequate to establish meter performance at elevated pressures (Kemperman, 1986).

The basic *volumetric flow rate equation* is

$$Q = (MF) \; (\text{Dial Diff}) \; (C_t) \; (C_p) \; (C_Z) \qquad (11\text{–}16)$$

where Q = scf at 14.73 psia and 60°F

MF = meter factor (normally set at 1)

Dial Diff = difference between ending and beginning meter dial readings

C_t = temperature correction factor
= 520/(460 + flowing temp °F)

C_p = pressure correction factor
= flow pressure (psia)/(14.73)

C_Z = compressibility factor correction
= (1)/(Z at flowing T and P).

In most cases the manufacturer lists the capacity and pressure drop for a 0.6 SG gas. These curves can be converted for other conditions as follows:

$$\Delta P_f = \Delta P_r \left(\frac{\rho_f}{\rho_r} \right) \left(\frac{Q_f}{Q_r} \right)^2 \qquad (11\text{–}17)$$

$$Q_{x,max} = Q_{.6,max} \left(\frac{P_f}{P_r} \right) \left(\frac{.6}{SG} \right)^{0.5} \qquad (11\text{–}18)$$

where ΔP = pressure drop across meter

ρ = gas density

Q = flow scf

P = static pressure psia

SG = gas gravity

Subscripts

f = current case

r = reference or calibration with
$G = .6$

x = refers to x gravity gas.

Operation

To avoid problems and prolong meter life (Kemperman, 1986), do the following:

1. Flush bearings regularly with high-grade, low viscosity instrument oil.

2. If necessary, install an upstream strainer and/or a filter separator to remove solid particles and liquids from the gas.

Table 11–4 "Trouble-Shooting Guide" (Furness, 1984)

Condition	Cause	Correction
No output pulses from amplifier module.	1. Input voltage to amplifier below minimum required for operation. 2. Damaged amplifier module. 3. Receiver unit not operating. 4. Turbo-meter rotor not turning.	1. Replace pick-off. 2. Replace amplifier module. 3. Refer to instruction manual on defective unit. 4. Trouble shoot rotor.
Turbo-meter rotor not turning.	1. Defective rotor shaft bearings. 2. Rotor damaged by foreign material passing through meter.	1. Replace bearings. 2. Send rotor to factory for replacement or repair.
No mechanical output (mechanical type meters such as Brooks, Rockwell, etc.)	1. Sheared coupling. 2. Rotor not turning. 3. Bevel gears worn out. 4. Gears in register cup worn out. 5. Defective accessory equipment. 6. Shaft not aligned with coupling. 7. Driven magnet slipping on shaft. 8. Broken shaft.	1. Replace coupling. 2. Trouble shoot rotor. 3. Replace bevel gears. 4. Replace defective gears in register cup. 5. Refer to instruction manual on defective unit. 6. Align shaft with coupling. 7. Tighten set screw in magnet. 8. Replace the broken shaft.
Inaccurate readout.	1. Foreign material on rotor blades. 2. Rotor blades bent. 3. Defective accessory equipment. 4. Defective magnet.	1. Check and clean blades. 2. Send to factory for replacement or repair. 3. Troubleshoot equipment. 4. Replace magnet.

3. Protect meter from weather by a suitable cabinet.
4. Operate meter within its range.

Meter testing, onsite proving, operating problems, and troubleshooting are now summarized.

Meter Testing. The "spin test" is a simple, fast, and reliable way to detect any increased friction. A spin test consists of spinning the meter rotor either by hand or with an air jet and then measuring how long it takes for the rotor to coast to a complete stop (spin time).

Onsite Proving. Beeson (1988) describes in detail how to prove a turbine meter using a sonic nozzle. An accuracy of 0.5% is claimed.

Operating Problems. Overloaded meters and/or poor gas quality are the major problems (Lofton, 1988). Do not operate the turbine meter continuously at over 70% of rated capacity. Entrained liquids—water and/or hydrocarbons—tend to wash the lubricants out of the ball bearings. Frequent lubrication and protection of the ball bearings is recommended (Lofton, 1988).

Troubleshooting. See Table 11–4.
Consult AGA (1985) for detailed specifications and recommendations for gas turbine meter construction,
installation, operation, performance characteristics, volumetric flow measurement, calibration, and field checks.

VORTEX METERS

When a fluid impinges on a bluff or nonstreamlined body, it splits into two paths. The resulting instability of the fluid flow field causes *alternating vortices* to shed from each side of the bluff body at a frequency proportional to the incident fluid velocity (Fig. 11–13). This phenomenon is readily visible when a flag waves in the breeze. The flag pole serves as the bluff obstruction and generates vortices that cause the flag to wave. Such vortex shedding can

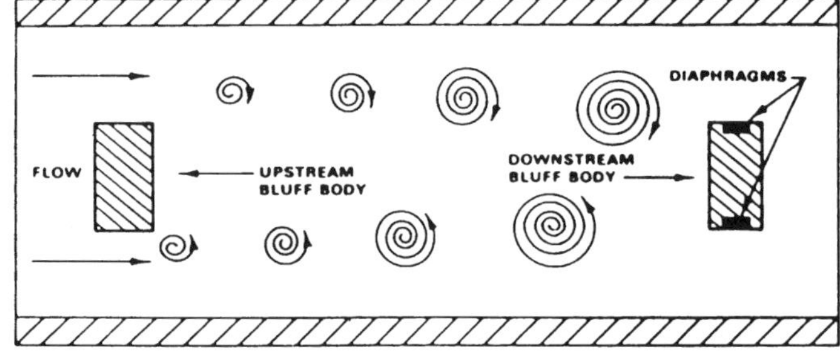

Figure 11–13. Vortex formation (Todd, Fisher Controls, 1988).

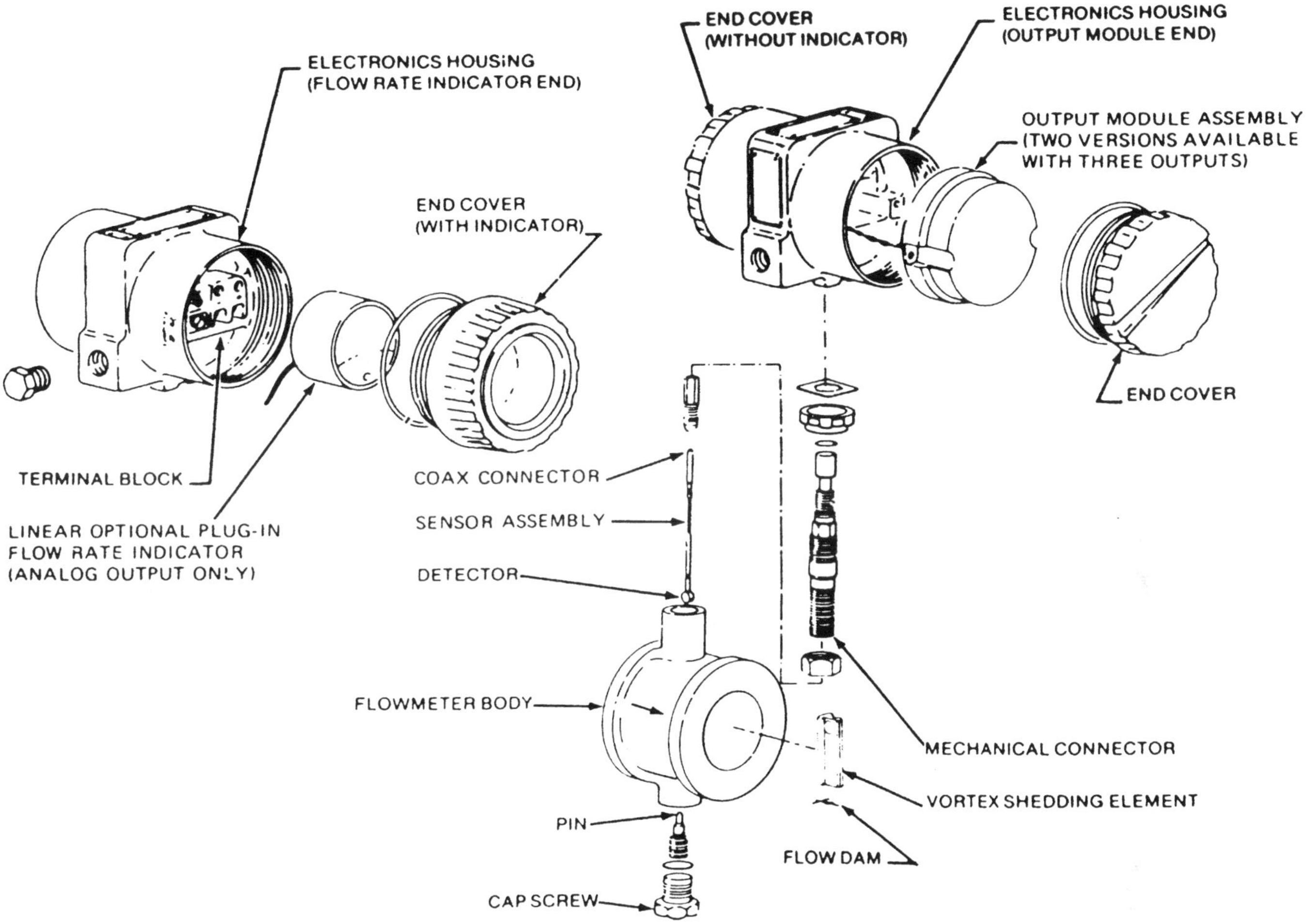

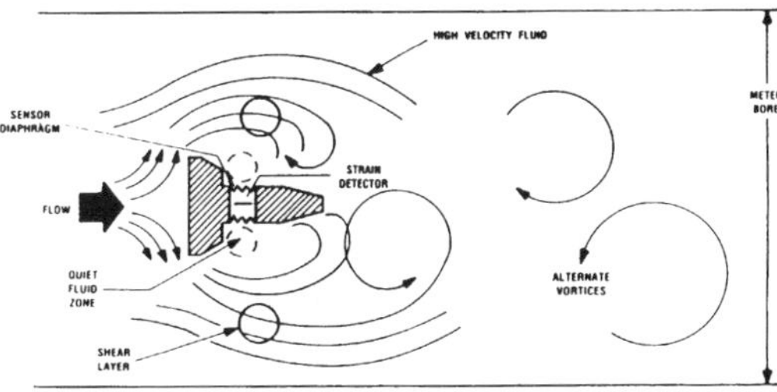

Figure 11–14. Vortex flowmeter, wafer style (Echeverria, 1985).

produce sound as when a wire vibrates and "sings" in the wind. Note that the vortices are formed alternatively, first off one side of the body and then off the other side and are 180 degrees out of phase. This so-called von Karman vortex street results in areas of alternating high and low pressure. Vortex meters usually use piezoelectric crystals that act as force-to-charge transducers to detect these pressure fluctuations. Figures 11–14 and 11–15 show Foxboro Co. vortex flowmeters in which the bluff, or vortex-shedding, body also houses the piezoelectric sensor. Fisher Controls uses a dual bluff body design (Fig. 11–13) to amplify the vortex signal, while Foxboro locates the sensor in a quiet fluid zone (Fig. 11–15). Both approaches reduce background noise.

Gas Sizing

The *Strouhal number*, St, relates the frequency of vortex formation, f, to the fluid velocity, v.

$$St = f\, d_b\, /\, v \qquad (11\text{–}19)$$

where d_b = width of bluff body.

Figure 11–16 shows that the Strouhal number remains constant when the pipe Reynolds number, $D\, v\, \rho/\mu$, ranges from 10^4 to 10^6. Vortex shedding does not occur at Re less than 3000.

While the vortex-shedding frequency varies linearly with

Figure 11–15. Vortex meter, Foxboro Co. (Echeverria, 1985).

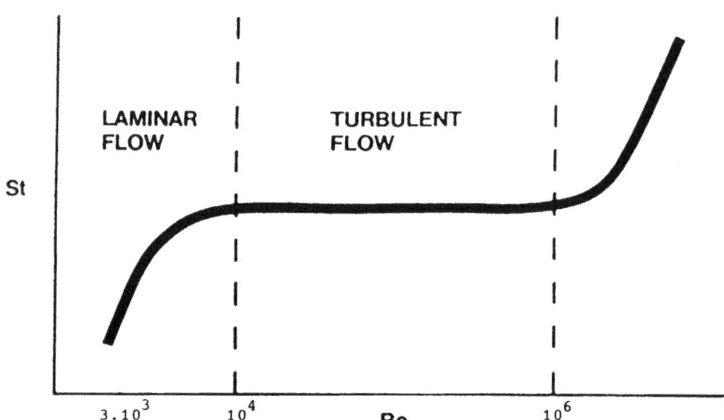

Figure 11–16. St No. vs. Re No. (Todd, 1988).

the fluid velocity over a very wide range, in practice the measuring range is limited by the ability of the piezoelectric sensor and electronics to detect and amplify the pressure pulses. Mathematically, the differential pressure is given by

$$\Delta P = C_{vm}\, \rho_f\, \bar{v}^2 \qquad (11\text{–}20)$$

where ΔP = differential pressure, psi
ρ_f = fluid density (lb/ft^3)
v = average fluid velocity, ft/s
C_{vm} = proportionality constant.

Rearranging Equation 11–20:

$$\bar{v} = \sqrt{\Delta P / C_{vm}\, \rho_f} \qquad (11\text{–}21)$$

$$\bar{v}_{min} = a/\sqrt{\rho_f} \ \text{and} \ \bar{v}_{max} = b/\sqrt{\rho_f} \qquad (11\text{–}22)$$

where $\bar{v}_{min}$ = minimum detectable velocity
$\bar{v}_{max}$ = maximum velocity that will not damage sensor
a,b = constants for a specific meter.

The limit velocities and C_{vm} are determined empirically for each specific meter design. For the Foxboro E83 vortex flowmeter (Echeverria, 1985):

a = 5 for gases,   a = 10 for liquids
b = 158 for both gases and liquids

To *size a vortex meter*, compute $\bar{v}_{max}$ using either the frequency limitation (max Re for which St is constant) or the sensor and electronics limitation, whichever is lower. Then:

$$Q_{vm} = A_b\, \bar{v} \qquad (11\text{–}23)$$

where Q_{vm} = actual flowrate (ft^3/s)
A_b = meter bore area (ft^2).

The smallest bore area provides the cheapest flowmeter with the greatest rangeability.

ULTRASONIC FLOWMETERS

Ultrasonic flowmeters can be divided into four basic types: *time of flight* (TOF), Doppler, cross-correlation, and swept-beam (Cascetta and Vigo, 1988). The best known types are the convenient but not very accurate Doppler and the more accurate and more expensive TOF. Only the TOF type is discussed further because this is the most popular oilfield type.

The TOF ultrasonic flowmeter consists of two piezoelectric sensors located 180 degrees apart and separated by an axial distance, L, as shown in Figure 11–17. When a voltage pulse is applied to sensor A, it changes its mechanical dimension—alternately expanding and contracting—and so generates an ultrasonic energy pulse. This pulse travels at sonic velocity through the fluid and is received by sensor B. The fluid velocity, v_f, is proportional to the difference between the transit times for the pulse to travel upstream, t_{BA} (from sensor B to sensor A), and downstream, t_{AB} (A to B). Velocity is given by (Kyser, 1988)

$$v_f = D(t_{BA} - t_{AB})/t^2 \sin 2\theta_f \qquad (11\text{–}24)$$

$$\text{where } t = D/\cos \theta_f\, v_{son} \qquad (11\text{–}25)$$

v_{son} = sonic velocity in fluid.

Scelzo and Munk (1987) report a three-year field test of a TOF ultrasonic flowmeter for natural gas pipelines 6 to 30 inches in diameter. Their data confirms a 50:1 turn-down ratio and a $\pm 2\%$ accuracy. Ultrasonic flowmeters also are being used to measure flare gas in the North Sea (Mylvaganam, 1988).

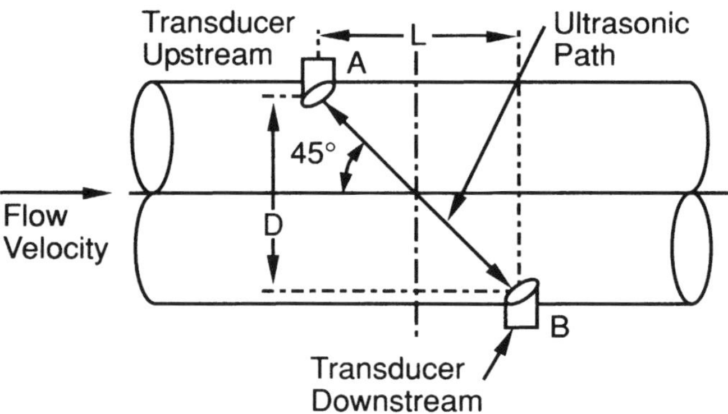

Figure 11–17. Across-the-pipe ultrasonic flow meter geometry (Scelzo and Munk, 1987).

SELECTION CRITERIA

The relative advantages and disadvantages of orifice, turbine, vortex, and ultrasonic meters are reviewed first. Then uncertainty and repeatability are reviewed.

Relative Advantages and Disadvantages

Orifice Plates.

Advantages:	Accepted as international standards
	Simple to use
	No moving parts
	Do not require frequent maintenance/ calibration
Disadvantages:	Square root head/flow relationship
	High permanent pressure drop
	Limited rangeability − 4:1
	Tends to read low under abnormal conditions

Turbine Meters.

Advantages:	Very good repeatability—± 0.1%
	High rangeability—10:1
	Low pressure drop
Disadvantages:	Moving parts subject to wear
	Bearings damaged by overspeeding, corrosion
	Rather expensive
	Dirty fluids must be filtered

Vortex Meters.

Advantages:	Good rangeability—30:1
	No moving parts
	Good linearity over operating range
Disadvantages:	Not suitable for dirty/abrasive fluids
	Appreciable pressure drop
	At lowest velocity $Re > 10^4$

TOF Ultrasonic.

Advantages:	No intrusion into pipe
	Wide rangeability − 50:1
	Easy to install ("clamp on")
	Cost almost independent of size
Disadvantages:	Periodic calibration required
	Accuracy not better than ± 2% FS
	Gases must be clean.

Uncertainty and Repeatability

Much confusion exists in some of the terminology such as accuracy, error, uncertainty, and repeatability. Figure 11–18 illustrates the difference between accuracy and repeatability.

Figure 11–18. Accuracy vs. repeatability (Hydril).

Error is the difference between the measured and true values, and is nearly always unknown unless the true value is provided, e.g., a standard sample of known, made-up composition.

Uncertainty is half the range within which the true value is expected to lie with a stated probability. Unlike error, uncertainty can be calculated using proper statistics, but the uncertainty cannot be separated from the probability or confidence level with which it is associated.

Repeatability or *precision* can detect spurious and random errors but fail to measure constant systematic errors or bias.

Variable systematic errors such as progressive wear of a turbine meter's bearings or progressive dulling of an orifice plate are also hard to detect.

Norman and Jepson (1986) estimated the measurement of orifice and turbine flowmeters as summarized in Tables 11–5 and 11–6. ANSI/API 2530 assumes that all the involved variables are independent of each other—by far the simplest case to compute. ISO 5168 (1978) accounts for the dependence of the calculated compressibility factor on the measured pressure, P_f; temperature, T_f; and gas gravity, SG.

Table 11–5 Comparison of ISO and AGA3 Uncertainties for Orifice Plate Meters (Norman and Jepson, 1986)

Variable	Percent Uncertainty of Variable ($\pm$)	Effect factor	
		Exponent	Square
F_h Basic Orifice Factor	0.5	1 (1)	0.25 (0.25)
F_r Reynolds Number Factor	0.1	1 (1)	0.01 (0.01)
Y Expansion Factor	0.1	1 (1)	0.01 (0.01)
T_f Temperature	0.4	1.84/2(½)	0.1354 (0.04)
G_r Relative Density	0.3	0.68/2(½)	0.0104 (0.0225)
Z_{fl} Compressibility	0.25	½ (½)	0.0156 (0.0156)
h_w Differential Pressure	0.5	½ (½)	0.0625 (0.0625)
P_f Absolute Static Pressure	0.5	1.21/2(½)	0.0915 (0.0625)
Sum of Squares			0.5854(0.4731)
Overall Uncertainty (Square root of Sum)			$\pm 0.765\%(\pm 0.688\%)$

(AGA3 values in brackets)

Table 11–6 Example of Uncertainty Calculations for a Turbine Meter (Norman and Jepson, 1986)

Variable	Percent Uncertainty of Variable $\pm$	Effect factor	
		Exponent	Square
Q_{act} Basic Turbine Factor	0.5	1 (1)	0.25 (0.25)
T_f Temperature	0.4	1.84 (1)	0.5417 (0.16)
G_r Relative Density	0.3	0.32 (0)	0.0092 (0)
Z_{fl} Compressibility	0.25	1 (1)	0.0625 (0.0625)
P_f Absolute Static Pressure	0.5	1.21 (1)	0.3660 (0.25)
Sum of Squares			1.2294(0.7225)
Overall Uncertainty (Square Root of Sum)			$\pm 1.11\%(\pm 0.85\%$

(Assuming variables independent in brackets)

Tables 11–5 and 11–6 readily support two conclusions:

1. The orifice plate appears to be more "accurate" than the turbine meter.
2. ISO 5168 estimates higher uncertainties than ANSI/API 2530.

While ANSI/API 2530 provides definitive specifications and recommendations for the primary element (*orifice plate* and *meter* tube) it does not address secondary elements used to measure static pressure, temperature, gravity, gas composition. Such elements as *pressure* and *temperature recorders* or *integrators* are just as important as the primary element. Poor equipment design, improper installation, and bad operating practices can, of course, cause *major error*— for example, installing the orifice plate backwards causes the meter to read low by 14 to 30%! Manual integration of the orifice charts will introduce additional uncertainty that can vary from 0.1 to 1.4% (Jones, 1986).

Batchelder (1985), Scheppers (1981), and Woods (1989) summarize correct procedures for *field testing* and *witnessing* orifice meter *calibration*. Ting *et al.* (1989) discuss and compare chart recorders and flow computers (dedicated microprocessor-based measurement and computation devices). Ting *et al.* also examined the thermal stability, hysteresis, line-pressure effect, and long-term performance of a chart recorder and a flow computer. They drew the following conclusions:

1. Both the chart recorder and flow computer met their specifications at room temperature and low pressure.
2. Long-term flow deviations due to time and P and T variations were within 0.5% for the flow computer and up to 1.9% for the chart recorder.
3. All differential pressure transducers exhibited significant temperature effects at low temperature.

The authors conclude that *"flow computer bests chart recorder."*

Birkhead (1986) reports a comparison of a single 4–in. turbine meter in series with an existing dual 4–in. orifice meter station offshore. After measuring 1.2 billion scf the turbine results were 0.05% higher than the orifice measurements.

Nearly always, gas flow meters are designed, analyzed, and tested assuming steady-state flow. However, very often these meters are exposed to *flow pulsations* caused by reciprocating compressors, pressure regulators, and other disturbances. McKee (1987) compared pulsation effects on various gas meters. Pulsations causing between 10 and 35% velocity modulation are not uncommon at field meter sites, and this level starts to cause noticeable results. Very severe pulsations—75 to 100% velocity modulation—cause large meter errors—as much as 20% high for orifice, 50% for turbine, 30% low for tee-shaped vortex, and 50% high for wedge-shaped vortex meters. Rotating vane positive displacement meter shows 1–5% error which just exceeds the uncertainty limit in the overall experiment (McKee, 1987). McKee concludes that orifice meters are less affected by pulsation than turbine and vortex meters.

GAS ODORIZATION

Methane and sweetened natural gas are odorless. Therefore, a leak could easily go unnoticed, thus causing an explosion hazard at concentrations between the lower (5 mol percent in air) and higher (15 mol percent) *flammability limits*. Suffocation occurs when the natural gas concentration in air exceeds 50 volume percent.

In the U.S., *Federal law* (CFR 49, Paragraph 192.625) *requires odorization* of natural gas and its products before transportation in pipelines. Basically, the law states that:

"A combustible gas in a distribution line must contain a natural odorant or be odorized so that at a concentration in air of one-fifth of the lower explosive limit, the gas is readily detectable by a person with a normal sense of smell."

Some State regulations are more severe and require detection at one-tenth of the lower explosion limit.

There are *some exceptions*. Gas used in the field—heater treaters, gas engines, gas lift, cycling, repressuring—need not be odorized. Gas pipelined directly to gas plants and refineries also can be exempted. See the Federal regulations for the specific exemptions.

LP-Gas Odorants

In 1931 the Bureau of Mines studied many potential odorants for natural gas (Fieldner *et al.*, 1931). This very comprehensive study formed the basis for establishing the current practice of adding 1 lb of ethyl mercaptan per 10,000 gallons of LP-gas. Today, LP-gas is usually odorized when it is distributed to wholesalers or retailers who, in turn, sell to many customers. Using multiple odorants would risk chemical interactions. Ethyl mercaptan was probably chosen because its high vapor pressure helps it vaporize as the LP-gas changes from liquid to vapor in the LP-gas tank.

Ethyl mercaptan is not suitable as a natural gas odorant because of its lower stability. Ethyl mercaptan can "FADE" or be oxidized:-

$$3Fe_2O_3 + 2C_2H_5SH = C_2H_5\text{-}S\text{-}S\text{-}C_2H_5 + 2Fe_3O_4 + H_2O.$$

The produced disulfide does not have as strong an odor. Fading notwithstanding, ethyl mercaptan is almost universally used for LP-gas which is a "clean" liquid hydrocarbon mixture.

Natural Gas Odorants

Gas pipelines can and do contain water, rust, pipline chemicals and other impurities. This more reactive environment dictates a less reactive odorant. Natural gas is usually odorized just prior to its distribution to ultimate customers. Multiple odorants can therefore be used.

Natural gas odorants are usually blends of two or more sulfur-containing hydrocarbons from three basic groups: mercaptans, alkyl sulfides, and cyclic sulfides. Johnson (1985) describes in detail the more-commonly-used component chemicals and the most popular "blends."

Odorants such as tertiary butyl mercaptan (TBM), isopropyl mercaptan (IPM), dimethyl sulfide (DMS), and tetrahydrothiophene (THT) (also called thiophane) (THT) can be readily detected at 1 ppb by a person with average sense of smell. Addition of *0.5 to 0.75 lb odorant per MMscf* is adequate to give natural gas its "gassy" smell. When a new pipeline is first used, 2 lb odorant per MMscf are added to allow for adsorption of the odorant on the pipeline wall. This adsorption provides a safety factor in the event that odorization equipment fails temporarily.

LP-gas is odorized by adding *1 lb ethyl mercaptan per 10000 gal* liquid LP-gas.

WARNING . Be extremely careful when handling liquid odorants—any spill will create a very obnoxious and long-lasting stink.

Odorizers

Odorants are added by two basic methods:

1. Evaporating the liquid odorant and mixing the resulting vapor with the natural gas,
2. Injecting liquid odorant directly into the gas pipeline.

In the *vaporization approach*, a bypass stream of natural gas absorbs odorant by flowing next to a saturated wick or over the surface of the liquid odorant. Then the resulting gas-odorant mixture is returned to the pipeline. This type is low cost, ideally suited for small flows, and requires minimum maintenance—periodic filling of the odorant storage chamber (Katuran,1989).

There are three types of *liquid injection odorizers*: drip, meter-driven pump, and positive displacement (PD) pump. The drip type is often used to provide temporary supplemental odorization in new lines. The meter-driven type uses the pressure drop across an orifice at the full gas stream pressure to drive a meter located in the odorizer compartment. In turn, this meter operates a scoop that dips

liquid odorant from a vessel and pours it into a tube leading to the gas stream. Odorant is added proportionally because meter rpm varies directly with the main gas flow. Either diaphram or piston-type PD pumps are used for high gas flows. Control instrumentation is available to add odorant in direct proportion to the gas flow (Katuran, 1989).

Contact the U.S. Department of Transportation for detailed instructions on operation, maintenance, sampling, and record keeping.

Review Questions

1. What type of flow meter is used predominantly for custody transfer of natural gas?
2. Name two other types of meters that are finding increasing use.
3. What type of pressure tap is used most frequently on orifice meters?
4. What is the chief advantage of a senior orifice fitting? What is its main disadvantage?
5. Are orifice meters calibrated in the field?
6. Are orifice plates calibrated in the field?
7. List the precautions required for accurate orifice metering.
8. The orifice flow constant, C', is, in general the product of 15 factors. List these factors.
   Which of these factors differ significantly from 1.0?
9. Describe the correct zeroing of an orifice meter.
10. Estimate the size of the error caused by incorrectly zeroing the static pen at 0 psia instead of at 1 atm.
11. List the major problems encountered in orifice measurement of natural gas.
    How are these problems handled?
12. List some of the common situations that produce abnormal conditions in orifice metering.
13. Do abnormal conditions tend to cause the orifice meter to read high or low?
    Do abnormal conditions tend to benefit the seller or the buyer?
14. How much error results if an orifice plate is installed backwards? Who is hurt—the buyer or the seller?
15. Does a dull orifice plate read high or low?
    Estimate the magnitude of the error.
16. How can orifice plate sharpness be measured?
17. What error would you expect from a dirty orifice plate?
18. Describe a gas turbine meter.
19. What precautions are recommended to avoid operating problems with a gas turbine meter in order to prolong meter life?
20. What phenomenon causes a flag to wave in the breeze?
21. Describe a gas vortex meter.

22. Describe an ultrasonic gas meter.
23. Compare orifice, turbine, vortex, and ultrasonic gas meters. List their advantages and disadvantages.
24. Define accuracy, uncertainty, and repeatability.
25. Which are more accurate—orifice or turbine meters?
26. Why is natural gas odorized?

    What types of compounds are used to odorize natural gas?

    How are odorants added?

    How much odorant is usually added?

Problems

1. Natural gas is flowing in a pipeline at a bore Reynolds number of 10^6. One billion (10^9) scf per day are sold at say \$1.50/Mscf.

   Estimate the difference in daily revenue (\$/day) if ISO 5167 is used instead of ANSI/API 2530.

   Use Figure 11–10.

2. A small lease gathers low pressure gas, compresses it, and sells it to a high-pressure pipeline. Typically, 1 MMscfd are gathered and purchased at pressures ranging from 50 to 100 psig and sold at about 800 psig.

   The proprietor loudly proclaims ''A zero is a zero is a zero.'' He insists on setting both the differential and static pens at the bottom line of the chart when he calibrates his purchase orifice meters.

   Can you offer a financial reason for the operator's reluctance to follow correct calibration procedures?

   The low-pressure, purchase orifice meters have static pressure springs with ranges of 50 or 100 psia. The pipeline custody transfer meter uses a 1000 psia static pressure spring.

3. An orifice meter run has the following characteristics:

Meter Pipe 6 in. schedule 40 (6.065 in. ID)		
	flange taps	
	static pressure measured from upstream tap	
Orifice	stainless steel	
Plate	2.00 inch at 20°C (68°F)	
Recorder	100 in. WC differential	
	1000 psia static spring	
Readings	atmospheric pressure	= 14.2 psia (1000 ft elevation)
	flowing temperature	= 90°F
	gas gravity	= 0.60
	static pressure	= 703 psig
	differential	= 59 in. (bellows)
Base Conditions	Oklahoma or Texas	14.65 psia, 60°F

Calculate the Meter Coefficient and Gas Flow rate. Estimate the error caused by incorrectly zeroing static pen, i.e., incorrectly zeroing static pen at zero of chart instead of correctly zeroing pen at 1 atm.

Nomenclature

A,B = locations of sensor in ultrasonic meter (Fig. 11–17)

A_b = vortex meter bore area (ft^2)

a = constant for ^{v}min (Eq. 11–22)

b = constant for ^{v}max (Eq. 11–22)

C' = orifice flow constant (Eqs. 11–1, 11–2)

C_p = turbine meter pressure correction (Eq. 11–16)

C_t = turbine meter temperature correction (Eq. 11–16)

C_{vm} = proportionality constant (Eq. 11–20)

C_Z = turbine meter compressibility factor (Eq. 11–16)

D = pipe inside diameter (in. or ft)

DP = differential pressure

d = orifice diameter (in. or ft)

d_b = diameter or width of bluff body (Eq. 11–19)

F_a = orifice thermal expansion factor (Eqs. 11–8, 11–9)

F_{am} = correction for air over the water in the water manometer during the differential instrument calibration

F_b = basic orifice factor

F_{gr} = orifice meter specific gravity factor (Eq. 11–6)

F_{hgm} = mercury manometer factor, correction for gas column

F_{hgt} = mercury manometer temperature factor

F_{pb} = pressure base factor (Eq. 11–3)

F_{pv} = orifice supercompressibility factor

F_{pvl} = local gravitational correction for dead weight tester static pressure standard

F_r = orifice Reynolds number factor

F_{tb} = temperature base factor (Eq. 11–4)

F_{tf} = orifice flowing temperature factor (Eq. 11–5)

F_{wl} = local gravitational correction for water column calibration standard

F_{wt} = water density correction (temperature or composition) for water column calibration standard

f = frequency of vortex formation

g = local acceleration due to gravity

H = elevation above sea level (ft)

h_{we} = orifice meter differential pressure (in water)

L = latitude, degrees (0 deg = equator, 90 deg = pole)

MF = turbine meter factor (Eq. 11–16)

P_{atm} = atmospheric pressure (psia)

P = absolute pressure (psia)

ΔP = pressure difference (psi)
Q_h = fluid flow rate (scfh)
Q_{vm} = flow rate through vortex meter (ft^3/s)
R_c = orifice reading on linear chart (Eqs. 11–14, 11–15)
R_L = orifice reading on L–10 chart (Eqs. 11–14, 11–15)
SG = gas specific gravity relative to air
T = gas temperature (°F)
T_{meas} = temperature (°F) at which orifice plate is measured
t = $D/\cos \theta_f\ v_{son}$ (Eq. 11–25)
t_{AB} = ultrasonic meter downstream transit time
t_{BA} = ultrasonic meter upstream transit time
St = Strouhal number, $f\ d_b/v$ (dimensionless)
v = velocity
v_{son} = velocity of sound in fluid (ft/s)
Y = orifice expansion factor
Y_z = expansion compressibility factor
Z = gas compressibility factor

Greek

β = d/D ratio of orifice to pipe diameter
θ_f = tan θ = L/D (Fig. 11–17)
μ = fluid viscosity (lb/ft s)
ρ = fluid density (lb/ft^3)

Superscript

— = denotes average value

Subscripts

b = denotes base or standard value
f = denotes current case
r = denotes reference or calibration with SG = .6
x = refers to gas with SG = x

References

AGA (1985), "Measurement of Gas by Turbine Meters," TMC Report No. 7, American Gas Association, Arlington, VA 22209.

Allen, C. R. (1989), "Fundamentals of Gas Turbine Meters," Proceedings of 64th International School of Hydrocarbon Measurement, pp. 8–13, University of Oklahoma, Norman, OK (May 16–18).

ANSI/API STD 2530 (1985), "Orifice Metering of Natural Gas," also published as AGA Report No. 3; GPA Std. 8185–85; API Manual of Petroleum Measurement Standards, Chapter 14, Section 3.

API (1965–date), Manual of Petroleum Measurement Standards, American Petroleum Institute, 2101 L Street, Washington, DC.

Baker, P. D. (1983), "Improving Petroleum Custody Transfer Measurement Accuracy," Geosource Inc., Smith Meter Division Tech. Paper 111, also presented at Fifth Petroleum Measurement Seminar, API, Tulsa, OK (March 10, 1983).

Baker, Philip D., and Raymond J. Kalivoda (1977), "Turbine Meters for Liquid Measurement," Geosource Inc., Smith Meter Division Tech. Paper 103A; also presented at the International School of Hydrocarbon Measurement, Norman, OK (April 1977).

Batchelder, Ned (1985), "Witnessing Orifice Meter Calibration and Field Testing," Proceedings of 60th International School of Hydrocarbon Measurement, University of Oklahoma, Norman, OK.

Bean (1985), "The Bean Orifice Edge Gage," Cumberland, Inc., P.O. Box 1296, Spring, TX 77383.

Beeson, Jim (1988), "Onsite Proving of Gas Turbine Meters," Proceedings 63rd International School of Hydrocarbon Measurement, pp. 81–87, University of Oklahoma, Norman, OK.

Beitler, Samuel (1935), "Flow of Water Through Orifices," Ohio State University Bulletin 89, Columbus, OH.

Birkhead, William (1985), "Field Experience With Turbine Meters," Proceedings 60th International School of Hydrocarbon Measurement, University of Oklahoma, Norman, OK (April 16–18).

Bosio, Jan, P. Wilson, and Ove Sembsmoen (1988), "Gas-metering Test and Research Facility to Meet North Sea Needs," *Oil & Gas J.* , Vol. 86, No. 50, pp. 33–42 (Dec. 12).

Burgin, E. J. (1971), "Effects of Abnormal Conditions on Accuracy of Orifice Measurement," Proceedings of 47th International School of Hydrocarbon Measurement, University of Oklahoma, Norman, OK.

Cascetta, F. and P. Vigo (1988), "Flowmeters—A Comprehensive Survey and Guide to Selection," Instrument Society of America, Research Triangle Park, NC.

CEC (1985), "Experimental Data for the Determination of Basic 100 mm Orifice Discharge Coefficients," Commission of the European Communities, BCR Information Report No. EUR 10027 EN, Brussels, Belgium.

CEC (1987), "Experimental Data for the Determination of Basic 350 mm Orifice Discharge Coefficients," Commission of the European Communities, BCR Information Report No. EUR 10979 EN, Brussels, Belgium.

Cheremisinoff, Nicholas P. (1979), "Applied Fluid Flow Measurement," Marcel Dekker, Inc., NY.

Crabtree, Giles M. (1981), "Bellows-Type Orifice Meters," Proceedings of 61st International School of Hydrocabon Measurement, University of Oklahoma, Norman, OK.

Daniel Industries, Inc. (1983), Flow Products Division, P.O. Box 19097, Houston, TX 77224.

Echeverria, Alfredo (1985), "Flow Measurement by Vortex Shedding Meters," Proceedings 60th International School of Hydrocarbon Measurement, pp. 190–193, University of Oklahoma, Norman, OK.

Fieldner, A. C., R. R. Sayers, W. P. Yant, S. H. Katz, J. B. Shohan, and R. D. Leitch (1931), "Warning Agents for Fuel Gases," Monograph No. 4, U.S. Bureau of Mines, Washington, DC (May).

Fling, Jr., W. A. (1986), "The API/GPA Orifice Plate Data Base," SPE 15393, presented in New Orleans, LA (Oct. 5–8).

Fling, Jr., W. A., and E. A. Spencer (1984), "Progress Made on Standardization of Plates," *Oil & Gas J.*, Vol. 82, No. 1 (Jan. 2).

Furness, R. A. (1984), "Operating Problems With Turbine Meters," presentation, University of South Hampton, England (June).

Harrison, Paul (1980), "Flow Measurement—A State of the Art Review," *Chem. Eng.*, Vol. 87, No. 1, pp. 97–104 (Jan. 24).

Hoch, Kenneth A. (1983), "Adverse Effects on Orifice Measurement," 43rd Applachian Gas Measurement Short Course, pp. 51–59, R. Morris College, Coraopolis, PA (Aug. 16–18).

Hoglund, Paul (1984), "What Does the New A.G.A. Report No. 3 Say?," presented at P.C.G.A. Transmission Conference, San Jose, CA (March).

Hydril Control Systems Division, "Selection and Application of Flow Measurement Instrumentation," Bull. 5126, 9303 Roark Rd., Alief, TX.

ISO 5168 (1978), "Estimation of the Uncertainty of a Flowrate Measurement," International Organization for Standardization, Geneva, Switzerland.

ISO 5167 (1980), "Measurement of Fluid Flow by Means of Orifice Plates, Flow Nozzles, and Venturis," International Organization for Standardization, Geneva, Switzerland.

Johnson, J. T. (1985), "Natural Gas Odorants and Their Components," Proc. 60th International School of Hydrocarbon Measurement, p. 412, University of Oklahoma, Norman OK, (April 16–18).

Jones, Jr., Emrys H. (1986), "Theoretical Uncertainty of Orifice Flow Measurement," Proceedings 61st International School of Hydrocarbon Measurement, University of Oklahoma, Norman, OK (May 20–22).

Katuran, Ira (1989), "Odorization—Think or Stink," Proc. 64th International School of Hydrocarbon Measurement, p. 325, University of Oklahoma, Norman OK, (May 16–18).

Kemperman, Bernard J. (1986), "Fundamentals of Gas Turbine Meters," Proceedings 61st International School of Hydrocarbon Measurement, pp. 41–47, University of Oklahoma, Norman, OK.

Kyser, Michael D. (1988), "Other Flow Measuring Devices," Proceedings 63rd International School of Hydrocarbon Measurement, pp. 67–69, University of Oklahoma, Norman, OK.

Laird, C. B., (1988) "Turbine Meters for Liquid Measurement," Proceedings 63rd International School of Hydrocarbon Measurement, p. 141, University of Oklahoma, Norman OK (May 24–26).

Lofton, Mike (1988), "Field Experience with Gas Turbine Meters," Proceedings 63rd International School of Hydrocarbon Measurement, pp. 78–80, University of Oklahoma, Norman, OK.

McKee, R. J. (1987), "Pulsation Effects on Various Gas Meters," Proceedings 10th Annual ASME Energy-Sources Technol. Conf. Pipeline Energy Symp., pp. 215–220, Dallas, TX (Feb. 15–18).

McKee, R. J. (1989), "A Progress Report on the GRI Metering Research Facility," Proceedings of 64th International School of Hydrocarbon Measurement, pp. 537–543, University of Oklahoma, Norman, OK (May 16–18).

Munk. W. D. (1982), "Ultrasonic Flowmeter Offers New Approach to Large-Volume Gas Measurement," *Oil & Gas J.*, Vol. 80, No. 36, pp. 111–117 (Sept. 6).

Mylvaganam, K. S. (1988), "Ultrasonic flowmeters measure flare gas in North Sea," *Oil & Gas J.*, Vol. 86, No. 42, pp. 54–56 (Oct. 17).

Norman, Roger and Peter Jepson (1986), "Unaccounted-For Gas in Natural Gas Transmission Lines," Proceedings of International Symposium on Fluid Flow Measurement, American Gas Association, Washington, DC (Nov. 16–19).

North American Manufacturing Co. (1978), "Combustion Handbook," 2nd Ed., pp. 189–190, Cleveland OH.

PETEX (1987), "Field Handling of Natural Gas," 4th Ed., Petroleum Extension Service, University of Texas at Austin, Austin, TX.

Pulley, David E. (1988), "Orifice Meters—Operation and Maintenance," Proc. 63rd International School of Hydrocarbon Measurement, pp. 428–431, University of Oklahoma, Norman, OK (May 24–26).

Scelzo, M. J. and W. D. Munk (1987), "Field Test of an Ultrasonic Flowmeter for Natural Gas Pipelines," Proceedings 10th Annual ASME *Et Al.* Energy-Sources Technology Conference, Pipeline Eng. Symp., pp. 111–114 (Feb. 15–18).

Schepers, H. H. (1981), "Operation Meter Operation," Proceedings of 56th International School of Hydrocarbon

Measurement, p. 268, University of Oklahoma, Norman, OK.

Spink, L. K. (1978), "Flow Meter Engineering," 9th Ed., Foxboro Company, Foxboro, MA.

Stolz, J. (1978), "A Universal Equation for the Calculation of Orifice Plates," Proceedings Flomeko Conference, Amsterdam, "Flow Measurement of Fluids," pp. 519–534, Dijstelbergen and Spencer (Eds.), North Holland Pub. Co.

Studzinski, Wojciech (1988), "Effects of Abnormal Conditions on Accuracy of Orifice Measurement," Proceedings 63rd International School of Hydrocarbon Measurement, pp. 70–74, University of Oklahoma, Norman, OK (May 24–26).

Ting, C. V., J. J. S. Shen, and E. H. Jones, Jr. (1989), "Flow Computer Best Chart Recorder in Metering Tests," Oil & Gas J., Vol. 87, No. 23, pp. 55–59 (June 5).

Todd, Eric (1988), "Flow Measurement by Vortex Shedding Meters," Proceedings 63rd International School of Hydrocarbon Measurement, pp. 45–49, University of Oklahoma, Norman, OK.

Woods, David (1989), "Witnessing Orifice Meter Calibration and Field Testing," Proceedings of 64th International School of Hydrocarbon Measurement, pp. 389–392, University of Oklahoma, Norman, OK (May 16–18).

Chapter 12

Heating and Cooling

Heat transfer is very important in gas processing because heat exchangers are used extensively. Generally, these exchangers fulfill one of two functions. First is heat addition or removal to accomplish a specific purpose (such as reboiling to drive out an absorbed substance, heating to prevent hydrate formation, or chilling to condense NGL). Second is energy savings, i.e., saving either heat or refrigeration in order to cut total fuel and thus save money.

Heat exchange is discussed under three headings: energy balances, exchanger design, and exchanger types. Energy balances are reviewed first since this important calculation depends on the process conditions and requirements but is independent of the type of exchanger used. Design is discussed next to provide the foundation for the next topic—comments on the various types of heat exchangers commonly used in field processing. Finally air and water cooling are compared.

ENERGY BALANCES

Heat exchangers are designed for steady state operation, so that the exchanger "duty," i.e., the energy lost by one fluid and picked up by the other fluid, is determined by energy balance. The general energy balance equation is discussed in Appendix 3. As indicated there, simplifying assumptions can be made for heat exchangers, leading to the simple form:

$$q = H_{out} - H_{in} \qquad (12\text{--}1)$$

where q = exchanger duty, Btu/hr
 H_{out} = exit enthalpy of one fluid, Btu/hr
 H_{in} = inlet enthalpy of the same fluid, Btu/hr

Heat losses to the surroundings are neglected, so that the duty is the same on both sides of the exchanger, i.e., Equation 12–1 applies for both the fluid being heated and that being cooled.

Two kinds of energy change occur in exchangers. The first is *sensible heat*, which is just the energy required to change the temperature of a fluid with no change in phase. The second is *latent heat*, the energy required to effect a change in phase of a fluid, as in boiling or condensing. In some cases the duty includes both sensible and latent heat, as in cooling and partly condensing a rich gas stream.

The GPSA (1987) Engineering Data Book—Section 24—gives enthalpy data suitable for energy-balance calculations. Computer flowsheet simulators, such as OPSIM, are ideal for determining heat duties for natural-gas process streams.

Appendix 6, Figure A6–4 gives heat capacities for sweet natural gases. These heat capacities can be used to calculate sensible heat changes, as shown by Example A3–2 in Appendix 3.

EXCHANGER DESIGN

The GPSA (1987) Engineering Data Book has three sections devoted to heat exchangers, including design considerations. Section 8 deals with fired heaters; Section 9 with shell-and-tube, double-pipe, plate-fin, and plate-and-frame exchangers; and Section 10 with air-cooled exchangers. Each section has an extensive bibliography. A comprehensive reference on heat-exchanger design is the Heat Exchanger Design Handbook (Schlunder, 1983) while Saunders (1988) is more concise. Shell-and-tube design by digital computer is outlined by Palen (1986).

The following discussion presents the basic design equation and reviews the evaluation of each term.

Design Equation

The design equation used for all heat exchangers except fired-heater radiant sections is the basic convection equation

$$q = U_o\, A_o\, \Delta T_m \qquad (12\text{--}2)$$

265

where q = duty, Btu/hr (discussed previously)
  U_o = overall heat-transfer coefficient based on the outside tube area, Btu/hr-ft^2-°F
  A_o = outside area for heat transfer, ft^2
    = πD_o L N / 4 for tubular exchangers
  D_o = outside diameter of tube(s), ft
  L = tube length, ft
  N = number of tubes
  ΔT_m = mean temperature difference, °F

From the standpoint of preliminary design or of evaluating bids, which is the approach taken here, the required area, A_o, is desired. As indicated above, the duty is obtained by energy balance. The remaining quantities are the overall heat-transfer coefficient, U_o, and the mean temperature driving force, ΔT_m. The estimation of these two is discussed later.

In the detailed design of exchangers the usual procedure is the so-called "rating." An exchanger of known geometry (and thus area) is proposed for the duty on the basis of preliminary calculations of the type described here, or by experience. If rating, i.e., detailed calculation, shows the proposed exchanger can provide the required service, it is accepted. Otherwise the geometry is modified and the process repeated. Palen (1986) summarizes the use of expensive proprietary programs for design of exchangers (as opposed to rating). Most designers use rating programs except for very difficult designs.

Actual rating of a shell-and-tube exchanger, for example, involves first specifying which fluid will be on the tube side and which on the shell side. Then a shell diameter and tube diameter and pitch (the arrangement of the tubes for cross flow: square, equilateral-triangular, rotated-square) are selected. This selection allows finding the number of tubes from standard tube-count tables (see Perry and Green, 1984; Schlunder *et al.*, 1983; or the particular manufacturer). A tube length is chosen also. Then, parameters for the shell side must be specified: baffle type, spacing, and percentage cut for longitudinal flow of the shell-side fluid. With the proposed geometry, the heat transfer can be evaluated. See Buthod (1960) for details.

Overall Heat-Transfer Coefficients

Strictly speaking the overall heat-transfer coefficient is evaluated from calculating and summing the various resistances in the heat flow path: inside-film coefficient (h_i), inside fouling (r_{fi}), metal-wall resistance (r_w), outside fouling (r_{fo}), and outside-film coefficient (h_o). Mathematically,

$$\frac{1}{U_o} = \frac{A_o}{h_i A_i} + r_{fi} \frac{A_o}{A_i} + r_w + r_{fo} + \frac{1}{h_o} \quad (12\text{–}3)$$

For present purposes, type-of-service estimates of the overall coefficient are used.

Estimating overall coefficients is well established for shell-and-tube exchangers. Typical type-of-service estimates are given in Table 12–1 (GPSA 1987) and more extensively by Bell (1978).

For air-cooled exchangers, see Table 12–2 (McGlynn) or GPSA (1987–Section 10).

Design procedures for estimating overall coefficients for the proprietary plate-fin and plate-and-frame exchangers are not readily available. Much more reliance must be placed

Table 12–1   Typical Heat-Transfer Coefficients (U_o) and Fouling Resistances (r_f) for Shell & Tube Exchangers (GPSA, 1987, p. 9–6, Fig. 9–9)

Service	r_f	U_o
Water with	(0.002)	
100 psi gas	(0.001)	35–40
300 psi gas	(0.001)	40–50
700 psi gas	(0.001)	60–70
1000 psi gas	(0.001)	80–100
Kerosene	(0.001)	80–90
MEA	(0.002)	130–150
Air	(0.002)	20–25
Water	(0.001)	180–200
Condensing with Water	(0.002)	
C3 or C4	(0.001)	125–135
Naphtha	(0.001)	70–80
Still Overhead	(0.001)	70–80
Amine	(0.002)	100–110
Rich Oil to	(0.001)	
Lean Oil	(0.002)	80–100
C3 liq/C3 liq	(0.001)	110–130
MEA/MEA	(0.002)	120–130
100 psi gas/500 psi gas		50–70
1000 psi gas/1000 psi gas		60–80
1000 psi gas/Cond. C3	(0.001)	60–80
Steam Reboiler	(0.0005)	140–160
Hot Oil Reboiler	(0.002)	90–120
Heat-Transfer Fluid		
Reboiler	(0.001)	80–110

Note: U_o is in Btu/hr ft^2 °F; r_f is in hr ft^2 °F/Btu

TABLE 12–2   Typical Heat-Transfer Coefficients for Air-Cooled Heat Exchangers (McGlynn, 1989)

Service	U_{bare}
Natural Gas Coolers	70
Regeneration Gas Coolers	60
Light Hydrocarbon Liquid Coolers	75
Light Hydrocarbon Condensers	80
Amine Still Condensers	80
Amine Coolers	70
Lube Oil Coolers (using turbulators)	35

on vendors. See Saunders (1988) for summaries of a few published procedures.

Fired heaters require experience and know-how (Radco, 1988).

Mean Temperature Driving Force (MTD)

Figure 12–1 shows a typical temperature-duty plot for a simple case of sensible heating and cooling of two streams. Note that T indicates temperatures of the warm fluid and t the cold fluid; subscript 1 refers to inlet conditions and 2 to outlet conditions.

Evaluation of the MTD is based on the countercurrent flow model, which is the actual flow geometry in a double-pipe exchanger. In countercurrent flow, if only sensible heat change is involved or latent heat with no temperature change, the correct MTD is the log-mean temperature difference (LMTD).

$$\text{LMTD} = \Delta T_{lm} = \frac{(T_1 - t_2) - (T_2 - t_1)}{\ln\,[(T_1 - t_2)/(T_2 - t_1)]} \quad (12\text{–}4)$$

In other words the correct MTD equals the LMTD when both temperature-duty lines are linear.

The heat balance is:

$$W\,C_p(T_1 - T_2) = w\,c_p\,(t_2 - t_1) \quad (12\text{–}5)$$

where W = flow rate of warm fluid, lb/hr
 w = flow rate of cold fluid, lb/hr
 C_p = mean heat capacity of warm fluid, Btu/lb-°F
 c_p = mean heat capacity of cold fluid, Btu/lb-°F

In general, the quantities WC_p and wc_p are not equal, so the slope of one of the temperature lines will be greater than the other. Therefore the "terminal" temperature difference at one end of the exchanger will be smaller than at the other end. In Figure 12–1 the smaller difference, or

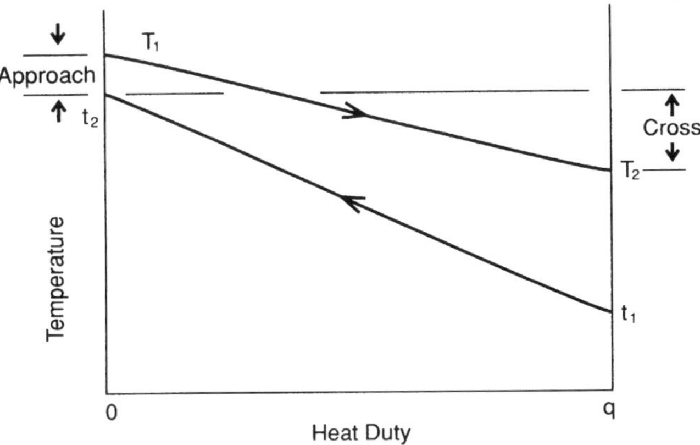

Figure 12–2. Temperature profile with "cross."

approach as it is called, is at the cold end of the exchanger. The approach can be at either end depending on the relative size of WC_p and wc_p.

Figure 12–2 shows a case where the cold-fluid outlet temperature, t_2, is higher than that of the warm fluid, T_2. In such a case a *temperature cross* is said to exist in the exchanger, as shown. If t_2 and T_2 are exactly equal the temperatures are said to *touch*.

Figure 12–3 shows the temperature profile in pure-fluid condensers, which approximates the case for a propane refrigeration system. There are three heat-transfer zones: desuperheating (where the gas is cooled to its boiling point), condensing (at constant temperature), followed by a small amount of subcooling of the liquid). The temperature profiles show that this (Fig. 12–3) type of exchanger cannot be designed in one piece: the areas of the three aforementioned sections must be estimated individually and then added (see GPSA, 1987, p. 9–11 for a detailed example).

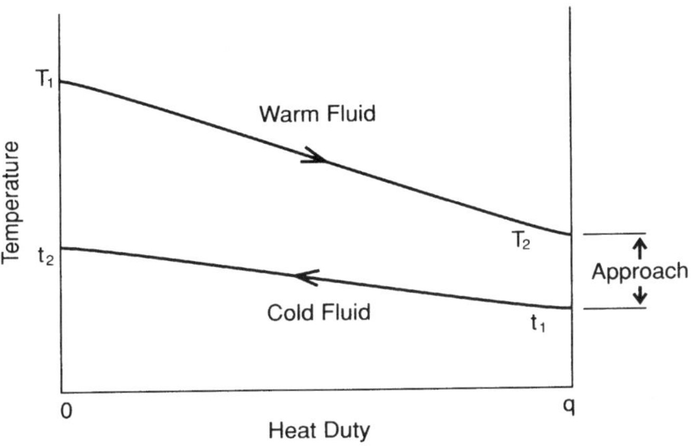

Figure 12–1. Temperature profile, ordinary case.

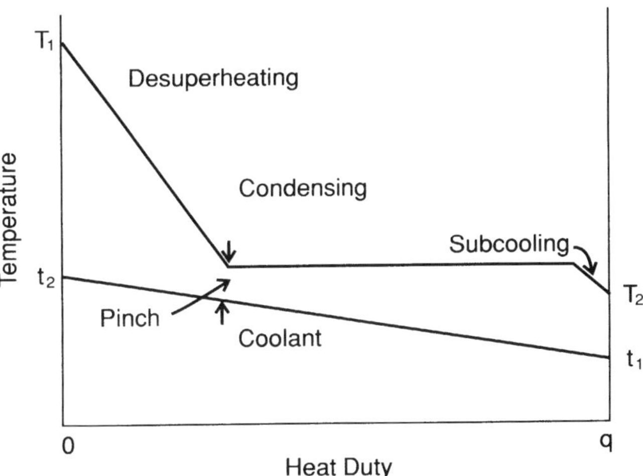

Figure 12–3. Temperature profile for condenser.

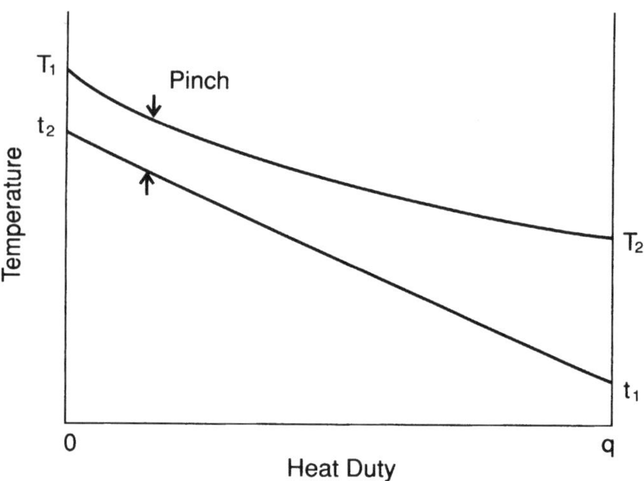

Figure 12–4. Temperature profile, gas-to-gas exchanger.

Plotting the temperature-duty profiles as shown in Figure 12–3 is very important to ensure that the coolant line does not intersect or cross over the warm fluid profile. Notice also that the closest approach in the exchanger, or *pinch*, is at an intermediate point, not at either end.

Figure 12–4 shows a case where cooling of the warm fluid causes partial condensation. The result is a curved temperature profile rather than the essentially straight profile for sensible heat changes. This is a second case in which the closest approach in the exchanger, or pinch, can be at an intermediate point in the exchanger rather than at one end or the other.

To get the best possible heat exchanger design, determine and specify the relevant thermo-physical properties as accurately and precisely as possible. As a minimum, this generally means properties for all streams at two temperatures. For condensing streams, enthalpy, fraction vapor, and vapor density should be specified as a function of temperature (McGlynn, 1989).

The geometry of most heat exchangers causes flow patterns that deviate more or less from pure countercurrent flow. A *multiplier*, F, dependent on the actual geometry and fluid temperatures, is used to correct the LMTD.

$$\Delta T_m = F \, \Delta T_{lm} \text{ or } MTD = F \times LMTD \quad (12-6)$$

Multipliers for several geometries are shown later (Fig. 12–9), and complete sets are available (TEMA, 1988).

TYPES OF HEAT EXCHANGERS

The kinds of exchangers commonly found in gas processing are now described and references given for further study. Type of service is discussed first, followed by a description of the various equipment types and their application.

Type of Service

The functions of heat exchangers are manifold. The following are typical of gas processing.

Heating. At the wellhead natural gas is often passed through a choke to regulate the flow. In some instances advantage can be taken of this to deliberately lower the gas temperature, as in the LTS process. In other cases, the gas from the choke is delivered directly into gathering lines. In this latter case, the gas must be kept above the hydrate temperature at all times. The temperature-lowering effect of the choke or Joule-Thomson expansion valve may have to be offset by heating the gas upstream of the choke.

Another instance is long gathering lines. Again the gas may be passed through heaters at various points along the line to maintain the temperature above the hydrate point.

Gas-to-Gas Exchange. This important service is often found in NGL recovery, as discussed in Chapters 5 and 14. Here the goal is to allow the cold residue gas to approach as closely as possible the temperature of the inlet gas, thus either maximizing the savings in refrigeration or allowing a lower processing temperature.

Chilling. Recovery of NGL from natural gas can be increased by cooling the gas in an exchanger with a refrigerant stream, such as liquid propane. The cold propane removes heat from the gas, vaporizing in the process.

Reboiling. This service is very similar to chilling, except that the vaporizing fluid is now the process stream and the energy source is the heating medium, which can be hot gas, hot water, steam, hot oil, or hot combustion gases.

Reboiling is required in such services as condensate stabilizers or fractionators and amine and glycol solution regenerators. Boiling also occurs in the refrigerant side of a chiller.

Inter- and After-Cooling. When gas is compressed it is heated and must be cooled prior to further compression to avoid excessive temperatures or to reduce horsepower. Hot gas must be cooled also before injection into a pipeline to avoid higher pressure drop and thus higher compression cost.

Cooling is accomplished by three media: cooling air, cooling-tower water, or tempered cooling water (Brown and Benkly, 1974). Air cooling is accomplished by fin-tubed heat exchangers, discussed below. Cooling towers are beyond the present scope; cooling towers cool recirculated water by evaporating a portion of it. Tempered cooling water is usually an ethylene glycol solution in water, which, in turn, is cooled using seawater.

Desuperheating and Condensing. After a refrigerant is vaporized in the chiller, it is compressed to a pressure and temperature at which it can be condensed by rejecting heat to the surroundings. This condensation occurs in a cooling-water or air-cooled exchanger.

Condensing. Conventional distillation columns require overhead product condensation to provide the necessary reflux and to supply the distillate product in convenient liquid form. A coolant (air, water) or refrigerant is used for this purpose.

Exchanger Materials. Like most process equipment, heat exchangers are fabricated from carbon steel where possible. Exceptions are made for low temperatures and corrosive materials.

Carbon steel becomes brittle at approximately $-20°F$. Charpy-impact-tested carbon steel can be used to $-40°F$. Between -40 to $-150°F$, 3.5% Ni steel is used, and below $-150°F$ stainless steel is used. *Stainless steel* (300 series) may be more readily available or more economical than low Ni steel. In plate-fin heat exchangers one exception is *aluminum*, which can be used at any cryogenic processing temperature.

Gas containing hydrogen sulfide and carbon dioxide can cause severe corrosion. Stainless steel is often required for this service. Carbon steel and a type of brass called Admiralty metal are sometimes used for cooling water.

Type of Equipment

The exchangers used in gas processing are of several different basic geometrical configurations or types. The more important types and their appropriate services are now reviewed.

Shell-and-Tube Exchangers. The name describes the geometry—a bundle of tubes mounted within a cylindrical shell. The "tube-side" fluid flows inside the tubes and the "shell-side" fluid passes inside the shell but outside the tubes. The two fluids exchange heat through the tube walls. Shell-and-tube exchangers are by far the most common type in gas processing.

Shell-and-tube exchangers are named by the Tubular Exchanger Manufacturers Association TEMA (1988) as shown in Figure 12–5. Letters are given to indicate the style of *head* (or front) end, *shell*, and *rear* end. Selection guidelines may be summarized as follows (McGlynn, 1989).

Head. Type B, the *bonnet* (integral cover) is often used for hazardous (H_2 or HF) in refineries, for high pressure service in gas plants, and for clean fluids. *Removable covers*,

type A or N, are used when cleaning is required. Bonnets are cheaper than removable covers and reduce leaks by eliminating one gasket.

Shell. The *one-pass* shell, type E, is most common. A close temperature approach or pinch or a temperature cross can require two or more shells in series to achieve an acceptably high LMTD (or F factor). A *two-pass* shell, type F, has a much higher LMTD F factor but has a much higher pressure drop. Fluid leakage past a longitudinal baffle (unless welded to the shell) can reduce heat transfer dramatically. Less than 0.01 in. clearance between the baffle and the shell can reduce heat transfer by 30% or more. Divided flow shells (type J) reduce shell-side pressure drop to about one-eighth of a comparable E shell. *Kettle* (type K) shells are used for reboiling or vaporizing, as in a chiller.

Rear. There are three types: first is the *fixed-tube-sheet* exchanger, shown in Figure 12–6. This figure also shows the standard TEMA nomenclature. Fixed-tube sheets are relatively hard to remove or replace; therefore, they are used for clean streams and low temperature differences.

The second rear-end type is the *floating-head* exchanger, depicted in Figure 12–7. Floating-head exchangers are used in a variety of services. Manufacture is more expensive than for the fixed-tube sheet, but the channel head permits easier access for maintenance. Also, the floating head allows large temperature difference between ambient and operating conditions without excessive thermal stress on the equipment.

The third type is the *U-tube*. Like the floating-head, the U-tube is a removable bundle and has similar advantages. However, cleaning inside the tubes is extremely difficult.

In gas processing three kinds of shell-and-tube exchangers predominate—fixed tubesheet (Fig. 12–6), floating head (Fig. 12–7), and kettle (Fig 12–8). Figure 12–6 shows exchanger type *BEM*, for bonnet front end, one-pass shell, with fixed tube sheet. Similarly, the Figure 12–7 exchanger is an *AES* for the channel head, one-pass shell, and floating head. The Figure 12–8 exchanger is an *AKT* to designate the channel head, kettle shell, and "pull-through" floating head (needed here for maintenance purposes). Kettles are also built with U-tube rear heads (e.g., *AKU*).

Exchanger size is indicated by two numbers, the inside diameter (ID) of the shell and the tube length, both in inches. For example, a 29-in. ID shell with 16-ft long tubes is referred to as size 29-192. A kettle with 23-in. ID front-end flange, a 37-in. ID kettle shell, and 16-ft long tubes is 23/37-192.

Common tube diameters are 0.75 and 1.0 in. outside diameter (OD) with varying thickness (usually 12 to 16 BWG). Standard lengths are 8, 10, 12, 16, and 20 ft,

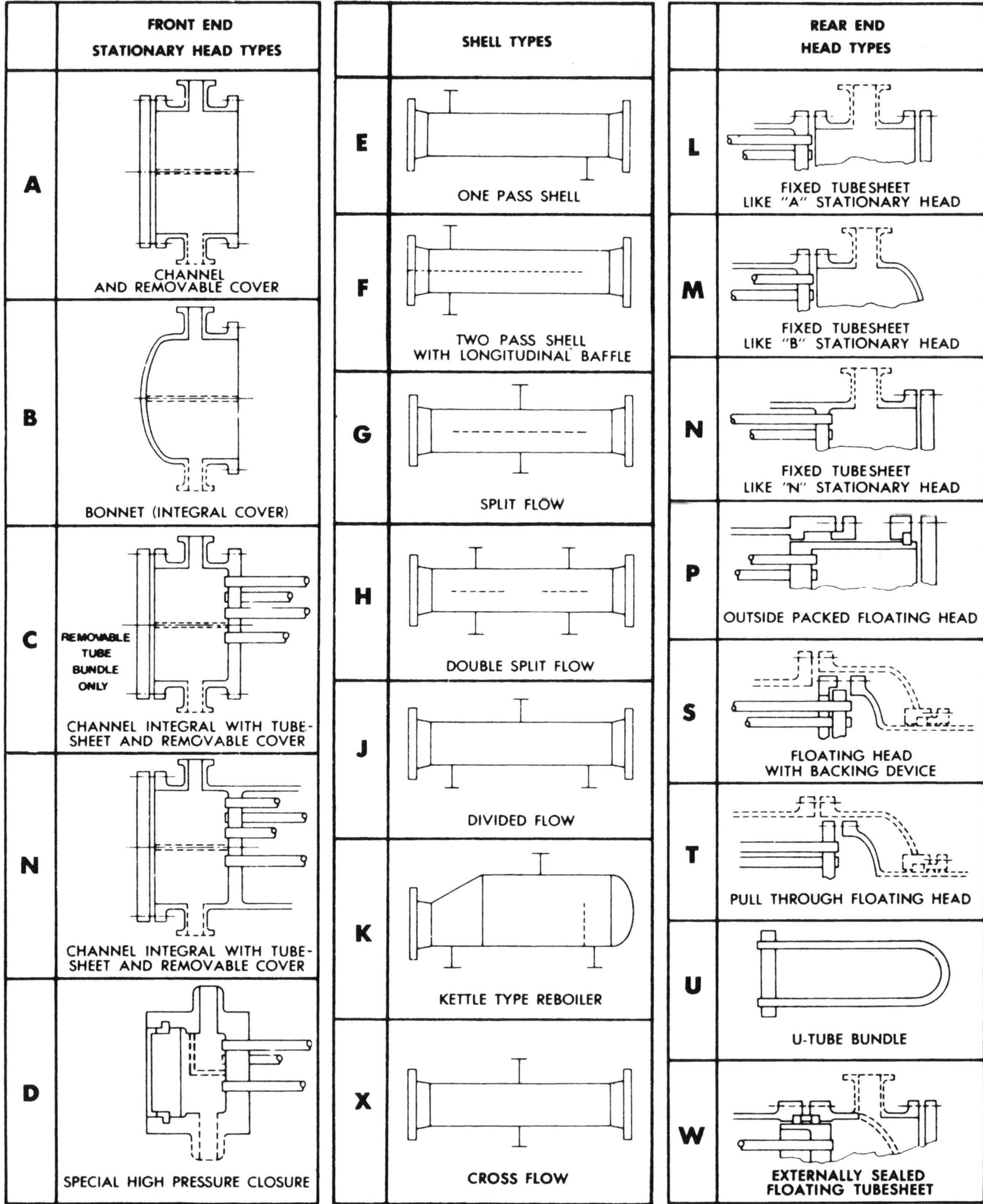

Figure 12–5. TEMA nomenclature (GPSA, 1987; TEMA, 1988).

Heat Exchanger Nomenclature

N-2 NOMENCLATURE OF HEAT EXCHANGER COMPONENTS

For the purpose of establishing standard terminology, Figure N-2 illustrates various types of heat exchangers. Typical parts and connections, for illustrative purposes only, are numbered for identification in Table N-2.

TABLE N-2

1. Stationary Head—Channel
2. Stationary Head—Bonnet
3. Stationary Head Flange—Channel or Bonnet
4. Channel Cover
5. Stationary Head Nozzle
6. Stationary Tubesheet
7. Tubes
8. Shell
9. Shell Cover
10. Shell Flange—Stationary Head End
11. Shell Flange—Rear Head End
12. Shell Nozzle
13. Shell Cover Flange
14. Expansion Joint
15. Floating Tubesheet
16. Floating Head Cover
17. Floating Head Flange
18. Floating Head Backing Device
19. Split Shear Ring

20. Slip-on Backing Flange
21. Floating Head Cover—External
22. Floating Tubesheet Skirt
23. Packing Box
24. Packing
25. Packing Gland
26. Lantern Ring
27. Tierods and Spacers
28. Transverse Baffles or Support Plates
29. Impingement Plate
30. Longitudinal Baffle
31. Pass Partition
32. Vent Connection
33. Drain Connection
34. Instrument Connection
35. Support Saddle
36. Lifting Lug
37. Support Bracket
38. Weir
39. Liquid Level Connection

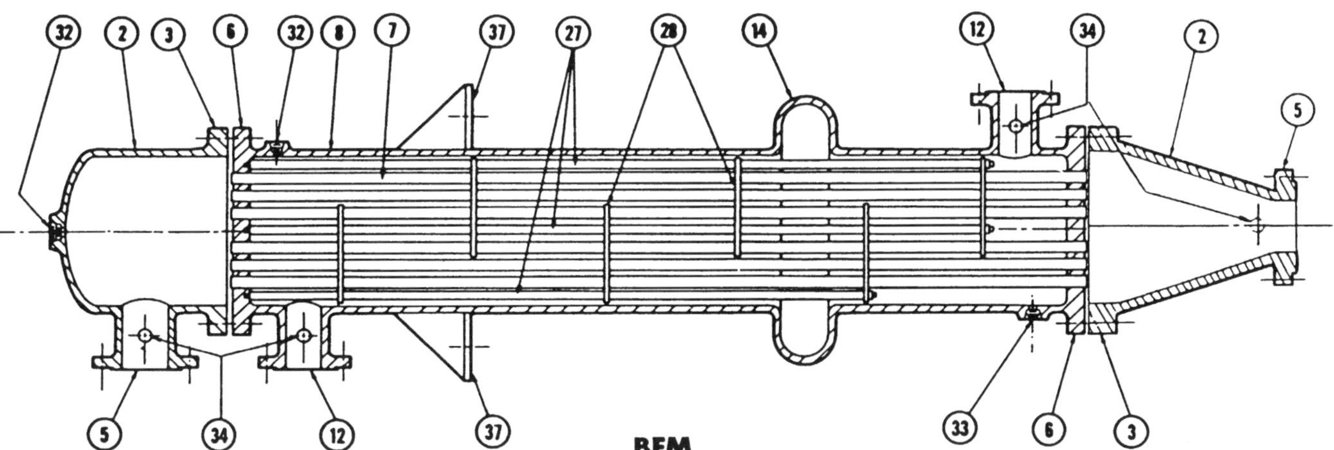

Figure 12–6.   Fixed-tube-sheet exchanger (Tubular Exchangers Manufacturers Association, TEMA: 1988).

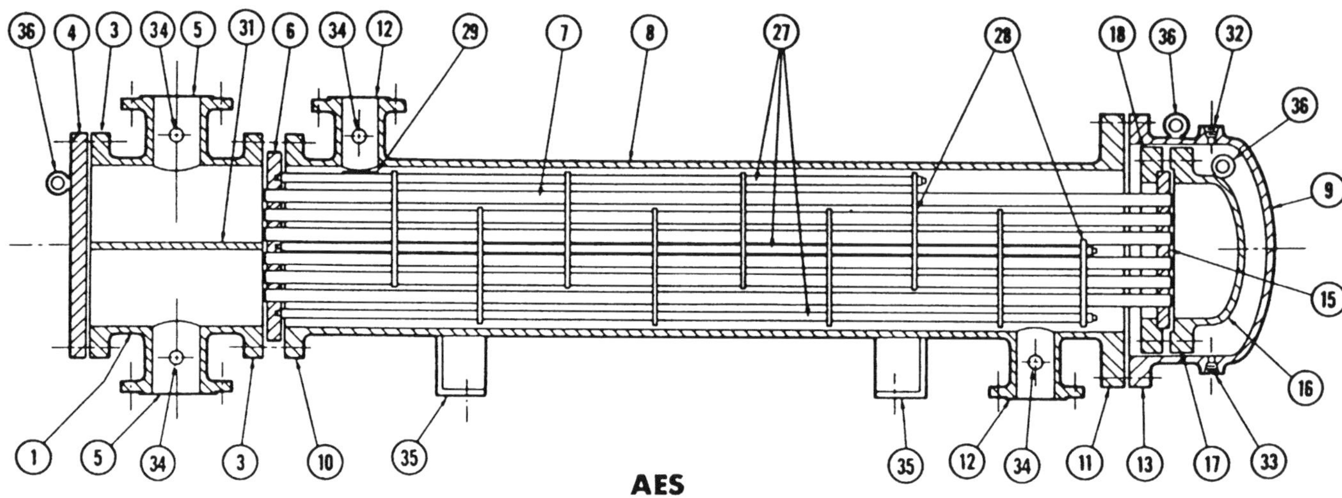

Figure 12–7. Floating-head exchanger (TEMA, 1988).

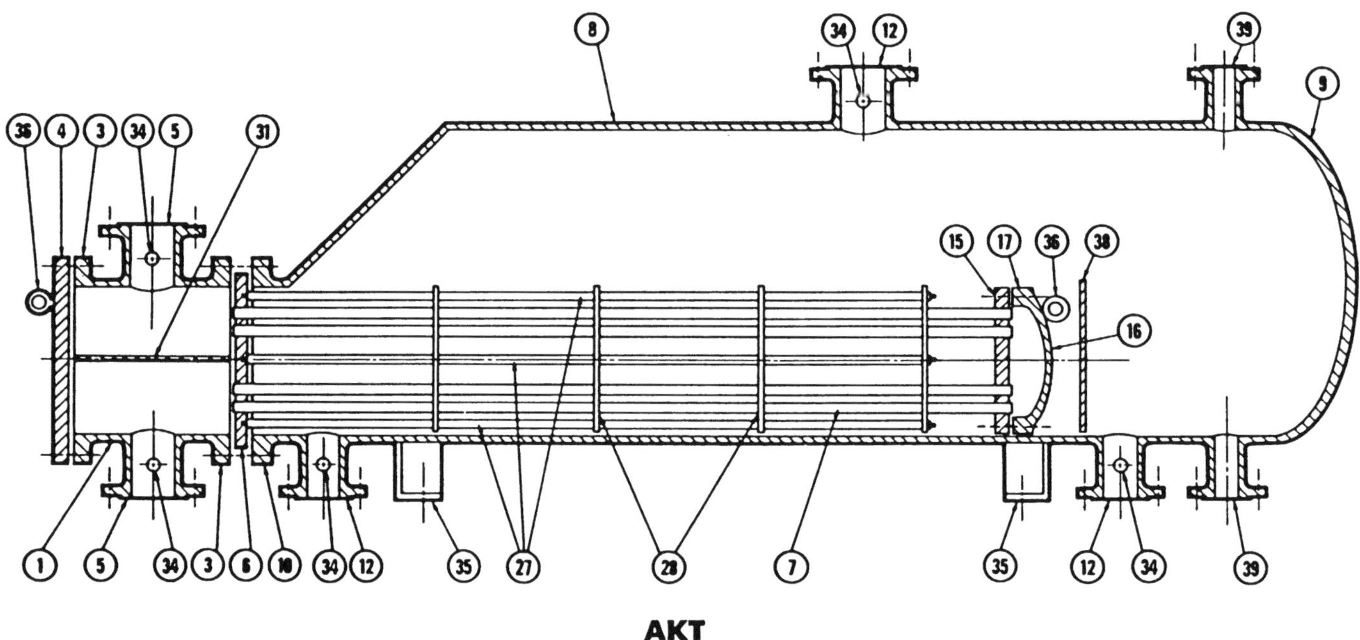

Figure 12–8. Kettle-type exchanger (TEMA, 1988).

except in some gas-to-gas exchangers where longer tubes are used. Shell diameters vary somewhat between manufacturers. Buthod (1960) lists the following, in inches: 8, 10, 12, 13, 15-¼, 17-¼, 19-¼, 21, 23, 25, 27, 29, 31, 33, 35, 37, 39, and 42.

Design, fabrication, and material standards are approved by TEMA for three classes of service, as follows.

Class R: Severe service, petroleum refinery and related applications.

Class C: Moderate service, commercial and general applications

Class B: Chemical process service

Generally, exchangers should be TEMA "C" or "R," and built to ASME Code. For large gas plants (expected life 20 years with annual shutdowns) specify TEMA "R" and National Board Registered. TEMA "C" is often used for small plants. Table 12–3 shows the standard specification sheet used for shell-and-tube exchangers (TEMA, 1988; GPSA, 1987).

Table 12–3 Heat Exchanger Specification Sheet (GPSA, 1987)

#				
1			Job No.	
2	Customer		Reference No.	
3	Address		Proposal No.	
4	Plant Location		Date	Rev.
5	Service of Unit Lean Oil to Rich Oil Exchanger		Item No. 1	
6	Size 30-360 Type NEN (Hor/Vert X)		Connected In | Parallel | Series	
7	Surf/Unit (Gross/Eff.) 4525 Sq Ft; Shells/Unit One		Surf/Shell (Gross/Eff.) 4525 Sq Ft	
8	PERFORMANCE OF ONE UNIT			
9	Fluid Allocation	Shell Side		Tube Side	
10	Fluid Name	Lean Oil		Rich Oil	
11	Fluid Quantity, Total Lb/Hr	475,723		650,860	
12	Vapor (In/Out)				
13	Liquid	475,723	475,723	650,860	650,860
14	Steam				
15	Water				
16	Noncondensable				
17	Temperature (In/Out) °F	197	100	60	124
18	Specific Gravity $(\#/ft^3)$ AVG.	41.2		38.3	
19	Viscosity, Liquid (AVG.) Cp	0.34		0.21	
20	Molecular Weight, Vapor				
21	Molecular Weight, Noncondensable				
22	Specific Heat (AVG.) Btu/Lb °F	0.542		0.600	
23	Thermal Conductivity *	0.077 (AVG)		0.078 (AVG.)	
24	Latent Heat Btu/Lb @ °F				
25	Inlet Pressure Psig	110		435	
26	Velocity Ft/S				
27	Pressure Drop, Allow./Calc. Psi	12	/ 9.58	10	/ 0.51
28	Fouling Resistance (Min.)	0.0020		0.0010	
29	Heat Exchanged 25,000,000 Btu/Hr; MTD (Corrected) 54.9				°F
30	Transfer Rate, Service 100.6 Clean 148.4 Btu/Hr Sq Ft °F				
31	CONSTRUCTION OF ONE SHELL			Sketch (Bundle/Nozzle Orientation)	
32		Shell Side		Tube Side	
33	Design/Test Pressure Psig	185 / Code		500 / Code	
34	Design Temperature °F	650		650	
35	No. Passes per Shell	One		One	
36	Corrosion Allowance In.	0.0625		0.0625	
37	Connections In	10"		14"	
38	Size & Out	10"		14"	
39	Rating Intermediate				

40 | Tube No. 784 OD 0.75 In.;Thk (Min/Avg) 0.065 In.; Length 30 Ft; Pitch 0.9375 in. ✗ 30 △ 60 ⊡ 90 ◇ 45
41 | Tube Type SA-214 (welded steel)   Material
42 | Shell SA-516-70   ID 30   OD   In.   Shell Cover   (Integ.) (Remov.)
43 | Channel or Bonnet (2) SA-516-70 (Int.)   Channel Cover (2) SA-516-70 (Rem.)
44 | Tubesheet-Stationary (2) SA-516-70   Tubesheet-Floating
45 | Floating Head Cover   Impingement Protection
46 | Baffles-Cross SA-283-C   Type Segmented   % Cut (Diam/Area) ✗ 34   Spacing: c/c 18-5/8   In.
47 | Baffles-Long   Seal Type
48 | Supports-Tube SA-283-C   U-Bend   Type
49 | Bypass Seal Arrangement Spacers   Tube-Tubesheet Joint Rolled
50 | Expansion Joint NONE   Type
51 | ρv²-Inlet Nozzle 1717   Bundle Entrance 2220   Bundle Exit 1210
52 | Gaskets-Shell Side   Tube Side
53 | -Floating Head
54 | Code Requirements ASME Sec. VIII Div. 1 (with Stamp)   TEMA Class "C"
55 | Weight/Shell 20,600   Filled with Water 29,900   Bundle   Lb
56 | Remarks
57 | * Btu/[(hr · sq ft · °F)/ft]
58 |
59 |
60 |
61 |

Generally speaking, corrosive or fouling fluids are placed on the tube side for ease of cleaning. If the fluid requires a metal other than carbon steel, it is generally placed on the tube side, as are higher-pressure, lower-flow, or more-viscous (and more likely dirty) fluids. See Blackburn (1967) for more information on heat-exchanger specifications including choice of tube-side and shell-side fluid, among other things.

Condensing vapors and boiling liquids are usually located on the shell side, as are higher-flow or lower pressure-drop fluids. High pressure drop increases the heat-transfer coefficient and self-cleans better but may use valuable pumping or compressing energy.

Shell-and-tube exchangers should be operated at reasonably high velocities (pressure drops) to improve heat transfer and reduce fouling. When specifying velocities and pressure drops be sure to consider both normal and alternate operating conditions. Tube-side velocities are affected most by the number of tube passes while baffle type and spacing determine shell-side velocity. When several exchangers are arranged in parallel and all use cooling water, a low-pressure drop exchanger may be a "water hog" that "steals" water from the other higher-pressure-drop exchangers (McGlynn, 1989).

Segmental baffles are the most common type. A baffle cut of about 25% of diameter is normal. Larger cuts are used for condensing or boiling, and cuts smaller than 15% should be questioned. Very small baffle cuts result in leakage, reduced heat transfer, and erosion at the baffle. If leakage clearances plug up, unexpectedly high pressure drops can occur, and this results in excessive pumping costs and/or reduced throughput. Unusually large baffle cuts can create "dead zones" where solids accumulation, fouling, and possibly corrosion can result. Limit baffle spacing to 80% of the TEMA specified maximum to avoid poor tube support, possible mechanical damage during handling, and vibration (McGlynn, 1989).

Triangular tube pitches usually result in smaller and therefore less expensive bundles with better heat transfer and a correspondingly higher pressure drop. Specify triangular pitches only when shell-side mechanical cleaning is not contemplated. If bundle removal and shell-side mechanical cleaning is anticipated, specify a square pitch. A *rotated square pitch* may give better heat transfer. In gas service do not use rotated square pitch with double segmental baffles because acoustic vibration may result. TEMA specifies minimum tube pitch ratios and minimum distances for cleaning lanes (McGlynn, 1989).

Type AES exchangers are not usually used for countercurrent flow but for an even number of tube passes typified by two passes. The exchanger is referred to as a 1–2, parallel-counterflow exchanger (one shell pass, two tube passes). The shell-side fluid flows counter to the flow of the fluid in one of the tube passes and parallel to the second.

The correction factor for a shell-and-tube exchanger with one tube pass is essentially 1.0. The flow pattern is actually not pure counterflow but cross flow with many crossings; this geometry gives F factors of approximately 1.0 for Equation 12–2.

The MTD correction factor, F, for an E-shell exchanger with an even number of tube passes can be much less than one depending on the temperature conditions in the exchanger, as shown in Figure 12–9. Indeed, if the fluid temperatures show a "cross" (Fig. 12–2), factor F can be low enough to render the 1–2 exchanger impractical. However, the duty can be split into two or more equal-sized exchangers of this same type connected in series, and a satisfactory F may possibly be obtained. If the temperature cross is very large, a straight countercurrent exchanger is often selected.

Example 12–1. A gas-to-gas exchanger is to be sized. The inlet temperature and pressure of the feed gas are 100°F and 1000 psia, and it is to be cooled to 30°F by cold residue gas, which is heated from −20 to 90°F at 500 psia. The heat duty is 15 MMBtu/hr. A fixed-tubesheet, countercurrent shell-and-tube exchanger will be used. Estimate the area required. Could a 1–2 exchanger be used for this service? Estimate the overall coefficient to be 60 from Table 12–1. The MTD is the LMTD.

$$\Delta T_m = \frac{(100 - 90) - (30 + 20)}{\ln[(100 - 90)/(30 + 20)]} = 24.85°F$$

$$A_o = (15000000)/(60 \times 24.85)$$
$$= 10060 \text{ ft}^2 \quad \longleftarrow$$

If the surface is provided by 0.75-in. OD tubes, 40 ft long (not unusual for this service), the tubes required are:

$$N = 10{,}060 \, / \, [\pi(0.75/12)\,(40)] = 1281$$

This appears to be a large number but could be provided by one 37-in. ID shell or two 27-in. shells in parallel.

For Figure 12–9:
$$P = (t_2 - t_1)/(T_1 - t_1)$$
$$= (90 + 20)/(100 + 20)$$
$$= 0.92$$
$$R = (T_1 - T_2)/(t_2 - t_1)$$
$$= (100 - 30)/(90 + 20)$$
$$= 0.64$$

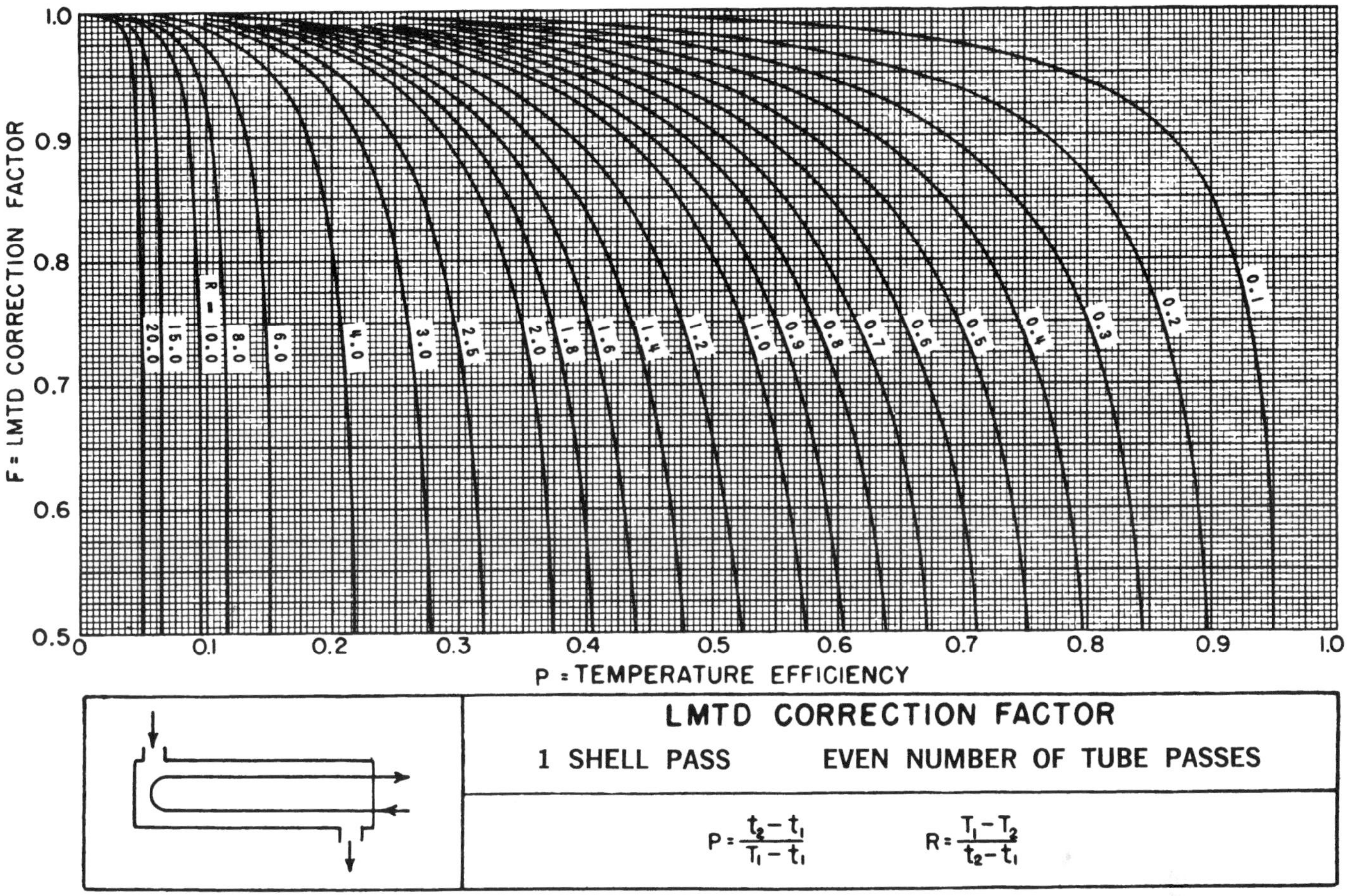

Figure 12–9. LMTD correction factor (GPSA, 1987).

Inspection of Figure 12–9 reveals that this is an impossible combination. A 1–2 exchanger is inappropriate. A single-pass (possibly an NEN type) exchanger could be used.

Shell-and-tube exchangers are used for almost any service in gas processing, including gas-to-gas exchange, heating and cooling, reboiling, chilling, and condensing.

Double-Pipe Exchangers. Figure 12–10 shows a *double-pipe*, or *hairpin*, exchanger. One fluid passes through the inside of the inner pipe and the second fluid flows through the annulus between the outside of the inner pipe and the inside of the outer pipe. The flow is countercurrent, so that the F factor is 1.0 for this type of exchanger.

Double-pipe exchangers have limited area and are used for services with small heat duties (UA < 100,000 Btu/hr °F). If one of the fluids shows a very low heat transfer coefficient, that fluid can be placed in the annulus and longitudinal fins used on the outside of the inside pipe. The extended surface of the fins provides better heat transfer for the fluid with higher resistance. These exchangers are used primarily for heating and cooling of gas streams. McDonough (1987) reviews design procedures and areas of application. Manufacturers, e.g.,

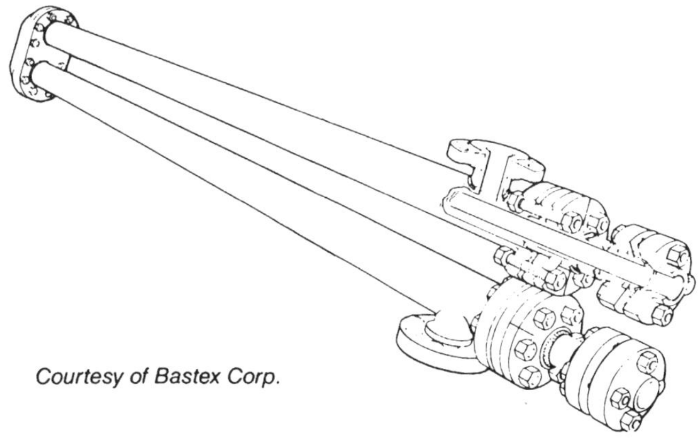

Figure 12–10. Double pipe heat exchanger (GPSA, 1987).

Brown Fintube (1989), present detailed information on fin efficiency, heat resistances, and pressure drops.

Indirect Heaters. The indirect heater is similar to a shell and tube exchanger, being composed of a large cylindrical shell with a tube bank mounted in the upper portion for heating the process fluid (Fig. 12–11). A U-shaped fire tube is located in the bottom, and the shell is filled with a heat transfer medium that transmits heat from the fire tube to the tube bank.

Heat is provided by burning gas or oil in the burner. The flaming combustion gas passes through the first section of the fire tube, giving up mainly radiant heat with a lesser amount of convective heat. The hot gases then turn and pass through the return section, in which the heat transfer is mainly by convection. The gases then flow up the stack to the atmosphere. Indirect heaters are akin to fire-tube boilers.

Water has an unequaled ability to transfer heat and so is almost always used as the heat transfer fluid for applications between 35 and 190°F. Adding ethylene glycol extends the range from −50 to 250°F. Special heating oils are used for higher temperature service, say up to 650°F; these oils have a low vapor pressure and high specific heat. Molten salt is used for high-temperature applications from 500 to 900°F. Molten salt will not decompose, as will

oils, and has good heat transfer properties. Ballard and Manning (1989) discuss the design and operation of heat-transfer-fluid systems and also discuss in detail the evaluation of heat-transfer fluids.

The conventional glycol regenerator is essentially an indirect heater converted to direct-heat service by placing the glycol outside the fire tubes. The regenerator has a small packed column mounted atop it to furnish a bit of reflux for the outgoing vapors, and thus prevent glycol vapor loss with the discharged water vapor as detailed in Chapter 8.

In gas processing, *indirect heaters* are used primarily for heating gas streams, including regeneration gas in small solid-desiccant dehydration units. Volume 2 discusses indirect heaters in detail.

Air-Cooled Exchangers. Air-cooled exchangers have the process fluid inside the tubes and ambient air on the outside, either moving by natural convection or blown by a fan. Because of the low heat-transfer coefficient for atmospheric air, fins are used on the outside of the pipes (Fig. 12–12).

Both *induced* and *forced draft* fans are used (Fig. 12–13). The latter are specified in most applications. When recirculation of cooling air is a problem, induced draft fans are used to provide positive outflow of the air.

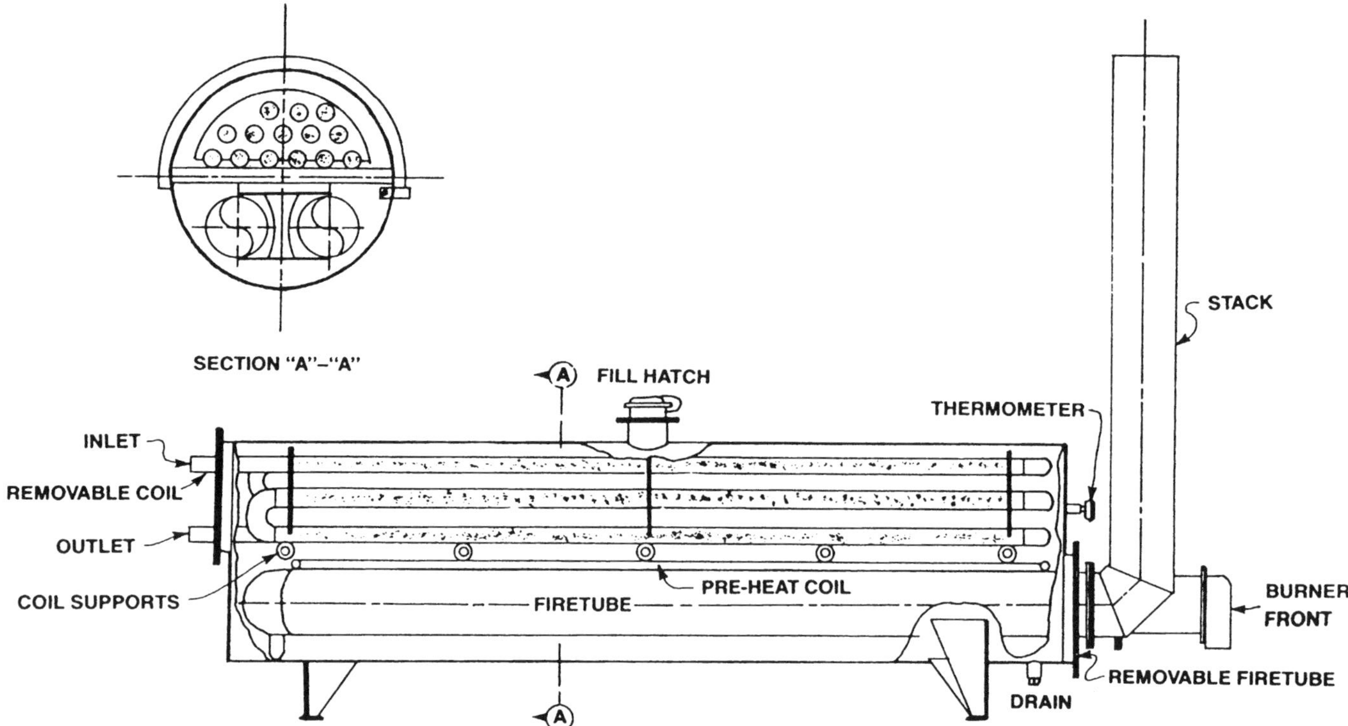

Figure 12–11. Water bath indirect heater (GPSA, 1987).

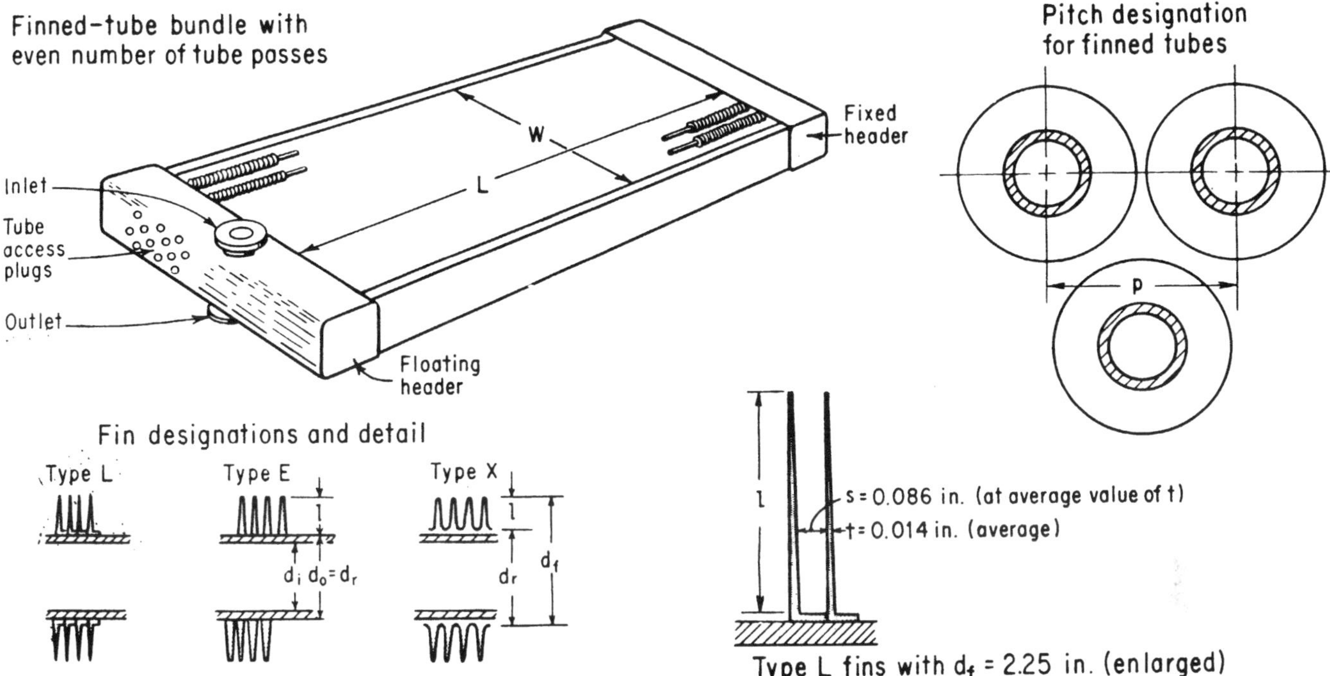

Figure 12–12. Details of finned tubes and exchanger bundle (Cook, 1964, Vol. 95, No. 10, 22–26, 1988).

Special precautions are needed for cold climates (API, 1988; Shipes, 1974; Brown and Benkly, 1974; Franklin and Munn, 1974). Wind skirts or housing may be necessary, as well as air recirculation.

The flow pattern in air-cooled exchangers is cross flow, with either an odd or even number of tube passes. Mean-temperature-difference correction factors for cross flow are given in Figure 12–14. The upper curves are for an even or odd number of side-by-side passes of the process fluid. The lower curves in Figure 12–14 apply to two-pass, over-and-under flow.

In air-cooled exchanger design, a difficulty arises with the exit air temperature, which is needed to estimate the LMTD. Air enters the bottom of an air-cooled exchanger at essentially constant temperature but is heated differently in each location across the exchanger. Rigorous estimation of the average outlet air temperature would be very complex. GPSA (1987—Section 10) details a method of estimating the outlet air temperature and designing air-cooled exchangers. Brown (1978), Ganapathy (1978), and Glass (1978) provide detailed design information.

Air-cooled exchangers are used for inter- and after-cooling of compressed gases, desuperheating and condensing refrigerant streams, and fractionator condensers.

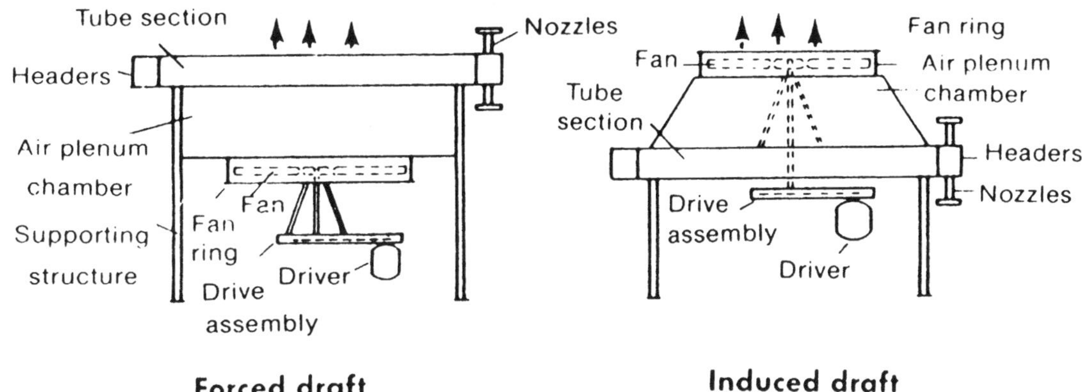

Figure 12–13. Typical side elevations of air coolers (GPSA, 1987).

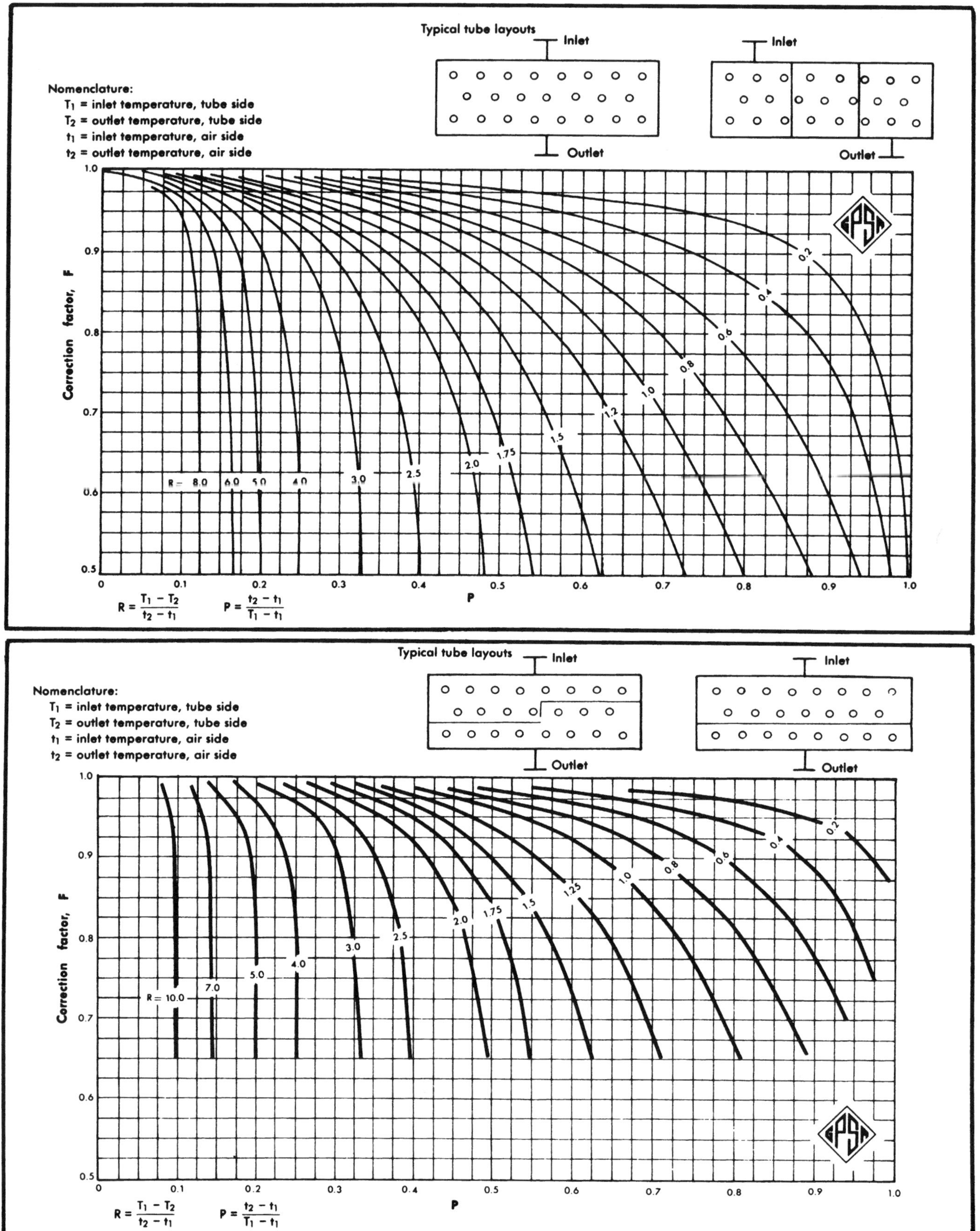

$$R = \frac{T_1 - T_2}{t_2 - t_1} \qquad P = \frac{t_2 - t_1}{T_1 - t_1}$$

Figure 12–14. Air-fin MTD correction factors (GPSA, 1987).

Plate-Fin Exchangers. Brazed-aluminum, plate-fin exchangers of the type shown in Figure 12–15 are used extensively at temperatures below −20 to −40°F, i.e., in cryogenic processing, both for gas-to-gas and chilling service. They also are used for low-temperature reboiling and condensing.

The exchanger is composed of plates separated by spacer bars, between which heat-transfer fins of various geometrical shapes are placed. Manifolding is a bit complex, as seen in the figure. Flow may be crossflow or counterflow. A large surface is contained in a relatively small volume; heat transfer is efficient. Design is discussed by Kays and London (1964).

The heat-exchanger assembly is mounted vertically in an insulating box containing vermiculite or other efficient insulating material. A single unit can handle several exchangers or services (more than two streams).

Plate-fin exchangers require very clean process fluids; it may be necessary to filter inlet gas streams from solid-desiccant dehydration units that contain fine solids from the bed. Plate-fins have definite pressure and temperature restrictions. The current pressure rating of 1400 psig is expected to increase in a few years. Maximum design temperature is normally set at 150°F but can be as high as 400°F.

Plate-and-Frame Exchangers. Plate-and-frame exchangers, also called *gasketed-plate* or *plate* exchangers, are similar to the plate-fin except that the plates contain the extended surface themselves, are separated by gaskets, and are held together by compression bolts in a frame (Fig. 12–16). These exchangers provide a large surface area in a small plan area. Pressure is limited to less than 1400 psig. Maximum temperature of about 300°F is limited by the gasketing material. Some design procedures, areas of application, and operating guidelines are available (Marriott, 1971; APV, 1981)

These exchangers are used to *save weight and space—* often offshore for interchanging heat between tempered cooling water and seawater. Untreated seawater is extremely corrosive, but titanium plates can be used. In turn, the cold tempered water is circulated through exchangers to cool process streams.

Fired Heaters. This type of exchanger is used for either the direct heating and boiling of process fluids or for heating a circulating oil, which in turn heats or boils process fluids. Fired heaters are also used to heat regeneration gas in solid-desiccant dehydration units.

Fired heaters come in two basic styles: *horizontal-tube box-type* heaters and *vertical-tube cylindrical* heaters (Fig. 12–17). Combustion of the gas or oil fuel occurs in burners at the bottom of the heater. Heat transfer in the main chamber is primarily by radiation. The hot flue gases travel upward through a second section in which the heat transfer is mainly

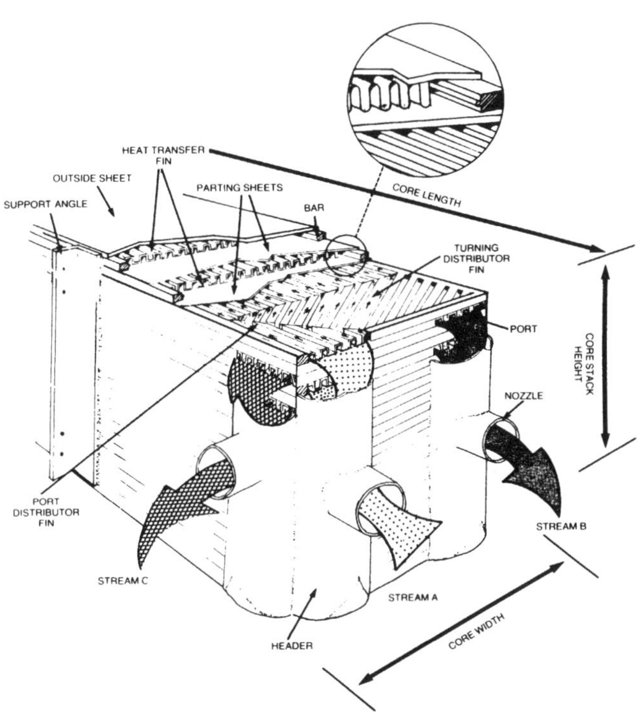

Figure 12–15. Brazed-aluminum plate-fin exchanger (GPSA, 1987).

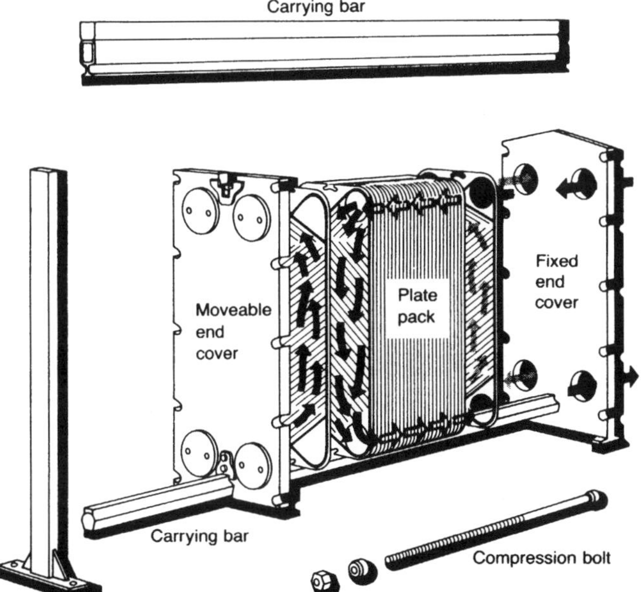

Figure 12–16. Plate heat exchanger (courtesy Thermal Division, Alfa-Laval Thermal Co.).

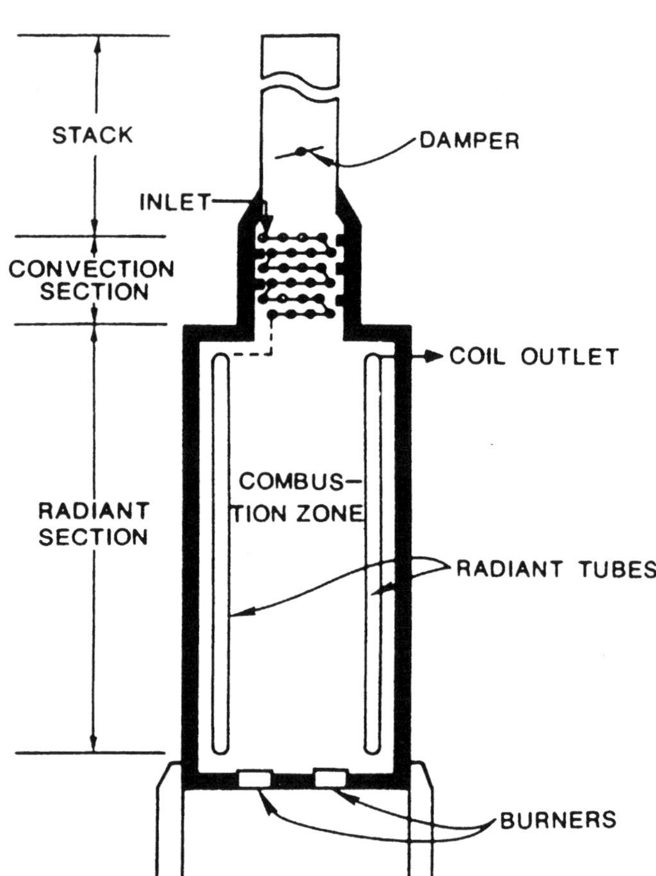

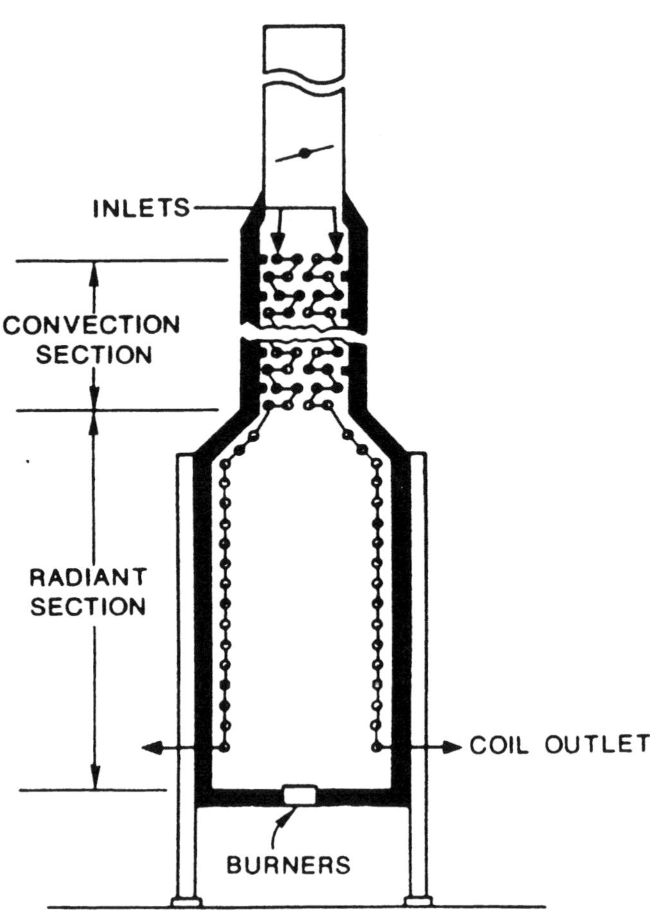

Figure 12–17. Fired heaters (Bartok et al., 1988).

by convection. The tubes in the convection section are usually finned tubes after the first two bare *shock* or *shield* rows.

Fired heaters are usually used in gas-processing plants, as opposed to field processing. Berman (1978), Melton (1978), Radco (1988), and Wimpress (1977) give many details on design and mechanical considerations. Fired heaters are discussed further in Volume 2.

AIR COOLING VS. WATER COOLING

Air and water cooling can be compared by considering the following topics.

Coolant Availability

Air is available anywhere, at no cost, requires no preparation, and therefore imposes no restrictions on plant location.

On the other hand adequate supplies of suitable cooling water are available only at restricted locations.

Design Considerations

Water can cool process fluids 5 to 10°F lower than air.

Water has a higher overall heat transfer coefficient than air, and so requires less heat transfer area.

Air coolers occupy considerably more space except when a water cooling tower is required.

Design procedures for water-cooled, shell-and-tube heat exchangers are well established. Air coolers require the specialized technology of fins and their design is less certain.

Nearby buildings and trees can cause air recirculation and so reduce air-cooler performance.

Water-cooled exchangers can be located anywhere in the plant.

Operation

Air is seldom corrosive and so air-side fouling is usually negligible.

On the other hand, water is often corrosive, scales, and deposits dirt. Water-side fouling is a frequent problem.

Cooling water often requires extensive chemical and bacteriological treatment.

Seasonal variations in air temperatures can cause inadequate cooling in summer and excessive cooling in winter (even freezing of process fluids). Sun and/or rain cause appreciable variations in air-cooler performance, as do sudden atmospheric temperature changes. Hail can damage exposed fins.

Cleaning inside the tubes is relatively easy and air-side cleaning is seldom required. Do not spray water on air-cooler finned-tubes to improve cooling during hot weather. When the water evaporates any dissolved solids will deposit on, corrode, and plug the fins. This error has ruined many air coolers.

A large temperature reduction for the tubeside fluid of an air cooler can cause the tube bank to warp due to differential thermal expansion.

Mechanical

The high-pressure fluid is always inside the tubes thus simplifying the mechanical design of air coolers.

Contamination of process fluid is greater in water coolers.

Costs

For the same duty a water-cooled shell-and-tube heat exchanger is much smaller and cheaper—the comparable air cooler can cost two to four times as much. This capital cost advantage disappears when auxiliary equipment such as a cooling tower, water-treatment plant, pumps, pump-house, etc., are required. Maintenance costs for air coolers are typically 20 to 30% of those for water coolers.

Total running costs require the classic comparison of depreciation, maintenance, power, and labor costs.

Summary

In many cases cooling water is either not available or supplies are limited or further use would require a larger cooling tower. These reasons explain why air cooling is used far more frequently than water cooling.

Review Questions

 1. What two functions do heat exchangers fulfill in gas processing?

 2. Define: sensible heat and latent heat.
 3. Summarize the differences between preliminary and detailed design of a heat exchanger.
 4. What specifications are involved in rating a shell-and-tube heat exchanger?
 5. Summarize how overall heat transfer coefficients may be obtained for shell-and-tube, air-cooled, plate-fin, and plate-and-frame heat exchangers.
 6. Define MTD, LMTD, approach, cross, touch, and pinch.
 7. List some of the types of service for heat exchangers in gas processing.
 8. How does TEMA name shell-and-tube heat exchangers?
 9. Name the three types of rear ends for shell-and-tube exchangers.
10. What types of shell-and-tube heat exchangers predominate in gas processing?
11. What TEMA class of service would you specify for a gas processing shell-and-tube heat exchanger?
12. While specifying a shell-and-tube heat exchanger would you place the following streams "shell side" or "tube side?" Why?
     a. corrosive fluid
     b. fouling fluid
     c. condensing vapor
     d. boiling liquid
     e. high pressure stream
13. When is fouling of most concern?
14. List typical gas processing applications for shell-and-tube, double-pipe or hairpin, air-cooled, plate-fin, and plate-and-frame heat exchangers.
15. List typical applications for direct and indirect fired heaters.
16. Compare air cooling and water cooling.

Problems

1. The following data apply to the gas-glycol heat exchanger in a TEG dehy unit:

$$U_o = 50 \quad \text{Btu/hr-ft}^2\text{-}°\text{F}$$
$$r_{fi} = 0.001 \ \text{hr-ft}^2\text{-}°\text{F/Btu}$$
$$A_o/A_i = 1.30$$

How does U_o change if the fouling resistance increases to 0.003?

2. The following data apply to the water-cooled condenser in a propane refrigeration unit:

$$U_o = 135 \quad \text{Btu/hr-ft}^2\text{-}°\text{F}$$
$$r_{fi} = 0.001$$
$$r_{fo} = 0.001$$
$$A_o/A_i = 1.3$$

How does U_o change if both fouling resistances increase to 0.003?

3. Consider Example 12–1.

   Could two 1–2 heat exchangers in series be used?

   Would one 2–4 heat exchanger be better?

   See GPSA (1987) or TEMA (1988) for LMTD correction factors for 2–4 exchangers.

4. In a glycol-glycol exchanger, 10 gpm of lean (98.5 weight percent) glycol enter at 380°F and leave at 200°F. The rich glycol (94.9 weight percent) enters at 120°F and leaves at 310°F.

   a. Compute the approach, temperature cross, LMTD, and CMTD.

   b. Compute UA (in Btu/hr °F) and recommend either a shell-in-tube or double-pipe exchanger.

   c. If U_o = 10 Btu/hr ft^2 °F, determine the required heat transfer area.

   Densities and specific heats for TEG solutions are given in Figures 8–21 and 8–22, respectively.

Nomenclature

A = heat transfer area, ft^2

BWG = Birmingham wire gauge (specifies tube wall thickness)

C_p = mean heat capacity of warm fluid, Btu/lb °F

c_p = mean heat capacity of cold fluid, Btu/lb °F

D = diameter of tube, ft

F = correction factor for LMTD, Equation 12–6

H = fluid enthalpy, Btu/hr

h = film heat transfer coefficient, Btu/hr-ft^2-°F

L = tube length, ft

LMTD = log mean temperature difference, °F

MTD = mean temperature difference, °F

N = number of tubes

q = exchanger duty, Btu/hr

r_f = fouling resistance, hr-ft^2-°F/Btu

T = temperature of warm fluid, °F

TEMA = Tubular Exchanger Manufacturers Association

t = temperature of cold fluid, °F

U = overall heat-transfer coefficient, Btu/hr-ft^2-°F

W = flow rate of warm fluid, lb/hr

w = flow rate of cold fluid, lb/hr

Greek

ΔT = temperature difference for heat transfer

Subscripts

bare = value based on tube area with no fins

f = refers to fouling

i = refers to inside of tube

in = refers to inlet stream

lm = log mean value

m = refers to mean or average value

o = refers to outside of tube

out = refers to outlet stream

w = refers to tube wall

1 = refers to inlet

2 = refers to outlet

References

Alfa-Laval Thermal Co., "Plate Heat Exchangers," 2115 Linwood Ave, Fort Lee, NJ 07024.

API RP 632 (1988), "Winterization of Air-Cooled Heat Exchangers," 1st Ed., American Petroleum Institute, Washington, DC.

APV (1981), "Design & Application of Paraflow Plate Heat Exchangers," 3rd Ed., A.P.V. Company, Inc., Tonawanda, NY 14150.

Ballard, Don, and William Manning (1989), "Heat Transfer Fluid Systems," and "Heat Transfer Fluids," Coastal Chemical Co. Inc., Pasadena, TX.

Bartok, W., R. K. Lyon, A. D. McIntyre, L. A. Ruth, and R. E. Sommerlad (1988), "Combustors: Applications and Design Considerations," *Chem. Eng. Prog.*, Vol. 84, No. 3, pp. 54–72 (March).

Bell, K. J.(1978), "Estimate S&T Exchanger Design Fast," *Oil & Gas J.*, Vol. 76, No. 49, pp. 59–68 (Dec. 4).

Berman, H. L. (1978), "Fired Heaters-I. Finding the Basic Design for Your Application," *Chem. Eng.*, Vol. 85, No. 14, pp. 99–104, (June 19); "II. Construction Materials, Mechanical Features, Performance Monitoring," ibid, No. 17, pp. 87–96 (July 31);"III. How Combustion Conditions Influence Design and Operation," ibid, No. 18, pp. 129–138 (Aug. 14); "IV. How to Reduce Your Fuel Bill," ibid, No. 20, pp. 165–169 (Sept. 11).

Blackburn, W. R. (1967), "How to Write Effective Heat Exchanger Specifications," *Petro/Chem Eng.*, Vol. 39, No. 9, pp. 19–24 (Aug.).

Brown, J. W., and G. J. Benkly (1974), "Heat Exchangers in Cold Service—A Contractor's View," *Chem. Eng. Prog.*, Vol. 70, No. 7, pp. 59–62 (July).

Brown, Robert (1978), "Design of Air-Cooled Exchangers—a Procedure for Preliminary Estimates," *Chemical Engineering*, Vol. 85, No. 7, pp. 108–111 (March 27).

Brown Fintube (1989), "How to Design Double Pipe Finned Tube Heat Exchangers," A Bas-Tex Corp. Houston TX.

Buthod, A. P. (1960), "How to Estimate Heat Exchangers," *Oil & Gas J.*, Vol. 58, No. 3, pp. 67–82 (Jan. 18).

Cook, E. M. (1964), "Rating Methods for Air-Cooled Heat Exchangers-Part 3," *Chemical Engineering*, Vol. 71, No. 16, p. 99, (Aug. 3)

Franklin, G. M. and W. B. Munn (1974), "Problems with Heat Exchangers in Low Temperature Environments," *Chem. Eng. Prog.*, Vol. 70, No. 7, pp. 63–67 (July).

Ganapathy, V. (1978), "Design of Air-Cooled Exchangers—Process-Design Criteria," *Chemical Engineering*, Vol. 85, No. 7, pp. 112–119 (March 27).

Gas Processors Suppliers Association, "Engineering Data Book," 10th Ed., 1987, Tulsa, OK.

Glass, John (1978), "Design of Air-Cooled Exchangers—Specifying and Rating Fans," *Chemical Engineering*, Vol. 85, No. 7, pp. 120–124 (March 27).

Kays, W. M., and A. L. London (1964), *Compact Heat Exchangers*, 2nd Ed., McGraw-Hill, NYC.

Marriott, J. (1971), "Where and How to Use Plate Heat Exchangers," *Chemical Engineering*, Vol. 78, No. 8, pp. 127–134 (April 5).

McDonough, Michael J. (1987), "Hairpin Exchangers: Double-Pipe and Multitube," *Chemical Engineering*, Vol. 94, No. 10, pp. 87–90 (July 20).

McGlynn, Robin (1989), Private Communication, Conoco, Inc., Ponca City, OK.

Melton, M. S. (1978), "Advances Tighten Fired Heater Design," *Oil & Gas J.*, Vol. 76, No. 30, pp. 62–67 (July 24).

Palen, J.W. (1986), "Designing Heat Exchangers by Computer," *Chem. Eng. Prog.*, Vol. 82, No. 7, pp. 23–27 (July).

Perry, R. H., and Don Green (Ed.) (1984), "Perry's Chemical Engineers' Handbook," 6th Ed., McGraw Hill, NY.

Radco (1988), "Fired Heater Design Manual," 6520 South Lewis, Tulsa OK, 74136.

Saunders, E. A. D. (1988), "Heat Exchangers—Selection, Design and Construction," Longman Scientific & Technical, Essex, England.

Shipes, K. V. (1974), "Air-Cooled Exchangers in Cold Climates," *Chem. Eng. Prog.*, Vol. 70, No. 7, pp. 53–58 (July).

Schlunder, E. *et al.*, Editors (1983), "Heat Exchanger Design Handbook," 5 vols., Hemisphere Publishing Corp., Washington, DC.

Tubular Exchanger Manufacturers Association, "Standards of the Tubular Exchanger Manufacturers Association," 7th Ed., NY, 1988.

Wimpress, N. (1977), "Handy Rating Method Predicts Fired-Heater Operation," *Oil & Gas J.*, Vol. 75, No. 48, pp. 144–156 (Nov. 21).

Chapter 13

Transportation of Natural Gas

INTRODUCTION

This chapter emphasizes flow of natural gas in gathering and transmission lines. Generally, flow in transmission lines is gas phase only, although some offshore pipelines are two-phase flow. As discussed in Chapter 5, hydrocarbon condensate is transported simultaneously with the gas. Field gathering lines are generally shorter than transmission lines and often the flowing gas contains liquid hydrocarbons (or condensate) and/or liquid water and even minor amounts of solids. Thus, design and operation of gathering lines is similar to transmission lines, but two-phase or even three-phase flow is more common. Hydrate formation is more common in gathering lines, because the gas may not have been dried.

General considerations regarding gas gathering and transport are discussed first. Then pipeline design is described, with primary emphasis on single-phase flow. Two-phase flow is treated also, although somewhat briefly. A detailed treatment of two- or three-phase flow is beyond the present scope and is available elsewhere (e.g., Brill and Beggs, 1986). Pipeline installation and operation is considered next. Because pipeline investment constitutes a major expenditure in the overall production and processing of natural gas, this chapter closes with a brief discussion of pipeline cost considerations.

TYPES OF PIPELINES

As stated above, there are two basic types of gas pipelines: *gathering systems* and *transmission lines*. Piping on platforms and in gas-processing facilities constitutes a third type of gas flow line from a design standpoint. Table 13–1 summarizes the types of flow and field applications (after National Tank Co., 1987). The terms ''pigging'' and ''slug catching'' in Table 13–1 are now briefly described.

Pigging

Flow line ''pigs'' or scrapers are devices used to remove condensate and solids from transmission lines, as well as to inspect such lines. These devices are described in detail later. The first use of pigs appears to have been in 1954 in Arkansas (Seymour, 1981a).

Slug Catching

Slug catchers are special piping arrangements used to catch large slugs of liquid in multiphase flow, to hold these slugs temporarily, and then to allow them to flow into downstream equipment and facilities at a rate at which the liquid can be properly handled.

Two flow situations cause *liquid slugging*. The first is high gas flow with an intermediate liquid rate. Instead of simply running along the bottom of the pipe, the liquid is caught up as slugs, propelled by the fast-moving gas. Slugging is aggravated by hilly terrain and vertical runs (such as offshore platform risers). The second slugging situation occurs when pigs are used to increase the efficiency of wet gas lines. Liquid gathers in front of the pig which helps to reduce pressure drop in the line. However, when the pigs reach the end of the line, large slugs of liquid must be handled.

Seymour (1981a) claims the first slug catcher was installed by Colorado Interstate Gas Co. in Oklahoma in 1959. Slug catchers are used commonly at the terminus of wet offshore transmission lines.

Gathering Systems

Gathering refers to the transport of the gas stream from its source to the processing facility. The gas source will be either a wellhead in the case of a gas-well gas or the field separator outlet in the case of an associated gas. In

284

Table 13–1   Gas Flow Applications
(After Natco, 1987)

Flow	Phase Condition	Application	Usual Processing Required	Remarks
Gas	HC and H_2O dew pts below minimum pipeline temperature (MPT)	High-pressure transmission lines, Compressor and heat exchanger lines	Dehydration, NGL recovery	Pigging (for cleaning & inspection
Gas & condensate	HC dew pt above, H_2O dew pt below MPT	Wet transmission lines, e.g. platform to shore	Compression, dehydration	Pigging & slug catcher
Dense phase	H_2O dw pt below MPT	Very-high-pressure transmission lines	Dehydration	Pigging & slug catcher
Gas, condensate, & free water	HC and H_2O dew pts above MPT	Wet transmission lines	Methanol injection	Pigging & slug catcher
Gas, condensate, & free water	HC and H_2O dew pts above MPT	Cold and/or high-pressure process lines	Glycol injection	Separator

both cases there is usually more than one source well or separator. Gathering systems consist of several lines, perhaps interconnected, often of relatively small diameter (say 4- to 8-in.) and low pressure (0–500 psia). Of course there are exceptions, such as high–pressure gas wells.

The configuration of the gathering system depends on the field involved. The gathering system may be new, constructed specifically to serve a newly-discovered field, or it may be an addition to an existing system (Powell, 1981). In either case, the design base must be as accurate as possible for realistic design. The production rate, temperature, and pressure must be estimated, often from inadequate data. *Deliverability* over the field life must be predicted, thus further complicating design. Koster *et al.* (1982) developed a computer program to estimate long-term gas production and deliverability.

In offshore production, several gas wells produce to one platform, so that gathering is simplified; all wells produce to a common manifold. If there are two or more platforms, processing may be done on only one central platform. If so, outlying platforms may send the total wellstream—after free-water knockout—to the main processing platform in a pipeline along the sea floor. *Two-phase flow* can occur in this line, with consequent *slugging* problems in the riser to the central platform. Nevertheless, this option is used to reduce overall processing costs. Subsea slug catchers can be used to reduce riser slugging (Huntley, 1986).

Offshore platform space is very expensive. Therefore, the cost of installing offshore separators to reduce the pressure drop in transporting fluids to shore should be compared with the costs of higher-pressure drop two-phase transmission lines (e.g., Hein, 1983).

Onshore processing often involves a number of scattered gas sources, thus requiring rather *complex gathering systems*. If the system is operated by a single producer, computer programs can be used to minimize the cost of the gathering system. Maggert and Anwar (1976) used such a program to optimize a system, including consideration of deliverability (rate of production) vs. time.

Maggert and Anwar also observe that *satellite gathering sites* are a viable alternative to the *central facility*. Each satellite site gathers from a set of wells in its vicinity. Total production from each satellite goes through a single line to the central processing facility.

In the United States, many lease holders often deliver to a common gathering system which is likely to be random in pattern. Of course, if feasible, the overall flows and resulting pressure drops—including the reservoir, wells, and surface gathering lines—should be optimized (e.g., Hein, 1984a).

Available gas pressure is important. Low-pressure fields or separators require field compression of the gas for delivery to the central plant. In *low-pressure gathering* systems suction pressure at the compressor may fall below atmospheric pressure. Great care is necessary to *prevent air leaks* into such manifolds. Small amounts of air may contribute to corrosion and low heating value. Large amounts can even result in explosive mixtures in the gathering system. Careful operation and monitoring of such a system is mandatory (McKnight, 1988)

Any of the flow types described in Table 13–1 may be found in gathering systems, although dense-phase flow is unusual. Hydrocarbon condensate and liquid water are present more often in gathering lines than in transmission lines and cause a variety of problems, from hydrate formation to excessive corrosion.

The type of pipe depends on source pressure and the distance the gas must flow. If the *pressure* is *low* and the plant close by, *plastic pipe* can be used (Anon., 1975). Plastic pipe is cheap, light, and easily coupled. *Carbonsteel* pipe is used for *high pressures* and/or *long distances*. For *sour streams*, *alloy steel* may be used to protect against corrosion that would otherwise accompany water condensation. Alternatively, line heaters with heat tracing or insulation are used to prevent water condensation and/or hydrate formation.

Choice of pipe diameter in gathering systems may be based on flow rates and pressures dictated by design conditions. For low-pressure fields, optimization of pipeline plus compression cost dictates choice of diameter.

Transmission Lines

Transmission lines are used to transport natural gas, usually dehydrated, to a final sales or processing destination. Large diameter pipe, say 12- to 48-in., is used. Pressures are typically 700–1200 psia. Dense-phase lines may have pressures as high as 2500 psia. Carbon steel is the usual material of choice. Selection of line diameter is based on optimization of pipeline and compression costs. The smaller the pipeline diameter, the lower its cost. But a smaller line will require higher compressor costs.

Process Lines

Gas flow lines in processing plants, wellhead manifolds, and offshore platforms are special cases because their function dictates the number and kind of phases present and the pressure and temperature conditions. Carbon steel is the usual material although special conditions may dictate choice of alloy or plastic materials. Because process lines are short, sizing is usually based on simple pressure drop considerations.

PIPELINE DESIGN

Design and analysis of natural-gas pipelines involve the application of mass, energy, and force balances. Compression and/or formation pressure drive the flow. Flow resistance is due primarily to *pipe friction*. Pipeline elevation gradients do not influence pressure loss very much for single-phase flow due to the relatively low density of the gas.

Elevation gradients become more important for gas-liquid flow. Frictional pressure drop also becomes proportionately greater because two or three phases are present. Friction

is dissipated not only at the pipe wall but also by *slippage* between phases. A further complication is that the frictional pressure loss depends on the angle of the flowline.

Two important facts should be remembered in pipeline design. First, *line size selection is generally not a problem.* Second, *estimating compressor bhp requirements may be a problem.*

Other factors being equal, pressure drop is inversely proportional to the pipe inside diameter to approximately the fifth power. Select a 24-in. pipe as a reference and tabulate the relative pressure drop for various sizes.

Relative Pressure Drop as a Function of Diameter (Standard Pipe)

Nominal Diam., In.	Relative Pressure Drop
16	8.24
20	2.57
24	1.000
26	0.662
28	0.452
30	0.317
36	0.125

Because pipe comes in finite size increments the correct pipe diameter is most likely chosen even if the pressure drop calculation is in error by 15 or 20 percent.

On the other hand, *compressor size selection* and energy requirement are directly dependent on the pressure-drop calculation. Errors can be more important. Even for single-phase pressure drop prediction, significant error can be introduced because of incorrect estimation of pipe roughness. Pipe roughness is not really known until the pipeline is built and commissioned. Also, two-phase pressure drop predictions can easily be in error by up to 30 percent; such errors can have significant impact on compression requirements. Each case is different, depending on the total flow rate, the pressure levels involved, and the line length.

In pipe design, the internal pressure must not exceed the maximum allowable working pressure (MAWP) of the pipe. The latter factor will be discussed first, followed by treatment of pressure drop in single-phase flow, then two-phase flow, and finally for process lines.

Pressure Level

Figure 13–1 shows the *pressure profile* in a pipeline or gathering system. The outlet pressure of a piping segment is generally known or set by the pressure required

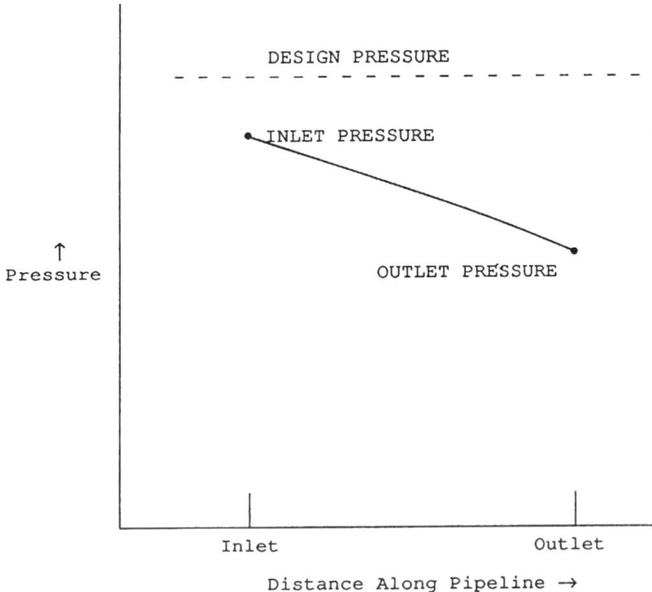

Figure 13–1. Pipeline pressure profile.

downstream, such as the pressure necessary to deliver the gas to a sales line or processing plant. Generally, a fairly high pipeline pressure is desirable because the gas density will be greater and the required pipe diameter smaller.

The pressure drop in the line determines the inlet pressure for a fixed outlet pressure. Again, pipe diameter is important since the pressure drop for a given flow varies inversely with the fifth power of the inside diameter (approximately). *Pipe selection* is influenced by the allowable pressure, which depends on the pipe fabrication method, material, and dimensions.

Pipe Fabrication Methods

Carbon steel pipe is manufactured either in seamless or longitudinally welded form. Welded pipe has lower allowable stress than seamless pipe because of the potential weakness of the welded seam.

Generally, *seamless pipe* is made by heating bar stock, then piercing it with a mandrel, and rolling and/or drawing the hot steel over the mandrel to the desired thickness. The inside pipe diameter is governed by the outside diameter of the mandrel.

Longitudinal seams are welded by three main commercial methods: electric-resistance welding (ERW), submerged electric arc welding (SAW), or continuous welding (CW). In electric resistance welding, sheet stock is rolled to the proper diameter, then welded by squeezing the edges together with simultaneous induction heating. SAW pipe

is made by first bending the pipe to a circular form in presses, followed by conventional submerged arc welding of the longitudinal seam. In double submerged arc welded (DSAW) pipe the seam is welded twice, once on the inside and once on the outside. Continuous-welded pipe is formed by rolling the hot stock into cylindrical shape, followed by pressing the very hot edges together to accomplish the weld.

API RP 14E (1984) recommends *ASTM A106, Grade B* (available only in seamless) and *API 5L, Grade B* (seamless, ERW, or SAW) for non-corrosive hydrocarbon service. Also for hydrocarbon service, ANSI B31–3 specifically excludes ASTM A120, furnace lap-weld, furnace butt weld, fusion weld per ASTM A134 and A139, and spiral weld except API 5L. API RP 14E (1984) prefers seamless pipe "due to its consistent quality" and also recommends "100% radiography for welding in accordance with API Std 1104."

Size ranges for the various fabrication techniques are as follows:

Type of Pipe	Size Range, in.	Max. Length, ft.
Seamless	2–26	44
ERW	4–20	80
SAW	20–48	40
CW	1/8–4	50

Fabrication techniques similar to those for carbon steel are used for other metal pipe (Mendel, 1981).

Materials of Construction

Carbon steel is by far the *most commonly used* pipe material. Alloy steel, which contains relatively high chromium and/or nickel, may be required for gas streams with high H_2S content (NACE Std MR–01–75).

Carbon steel is referred to as if it were a single material. In fact, there are different grades of carbon steel, depending on chemical composition. Table 13–2 (API, 1990) shows chemical analyses for varying grades of steel line pipe. Table 13–3 shows the widely varying minimum yield and tensile strengths for these pipe. Appropriate specifications are used to determine allowable stresses for alloy steels.

Plastic and glass-fiber reinforced plastic pipe (FRP) are light, easily handled, and do not corrode. Allowable pressures and temperatures are low. The latter limitations dictate fairly small diameters.

Plastic piping may have threaded or flanged joints but can also be joined by adhesive bonding, butt-fusing, heat fusing, or solvent cementing (Mendel, 1981). Available materials include polyethylene (PE), polypropylene (PP), and polyvinyl chloride (PVC).

Table 13–2   Chemical Requirements for Heat Analyses, percent

1	2	3	4		5		6	7	8	9
Type of Pipe	Grade and Class	Carbon Max.[2]	Manganese		Phosphorus		Sulfur Max.	Cb Min.	V Min.	Ti Min.
			Min.	Max.[2]	Min.	Max.				
Seamless										
Non-expanded or cold expanded	A25, CI I	.21	.30	.60	—	.045	.06	—	—	—
"	A25, CI II[1]	.21	.30	.60	.045	.080	.06	—	—	—
"	A	.22	—	.90	—	.04	.05	—	—	—
"	B[7]	.27	—	1.15	—	.04	.05	—	—	—
Non-expanded	X42[7]	.29	—	1.25	—	.04	.05	—	—	—
"	X46[7], X52[7]	.31	—	1.35	—	.04	.05	—	—	—
Cold expanded	X42[7], X46[7], X52[7]	.29[3]	—	1.25	—	.04	.05	—	—	—
Non-expanded or cold expanded	X56[4], X60[4]	.26	—	1.35	—	.04	.05	.005[5]	.005[5]	.005[5]
"	X65, X70, X80	(By Agreement)								
Welded										
Electric-welded or Continuous welded only	A25, CI II	.21	.30	.60	—	.045	.06	—	—	—
"	A25, CI II	.21	.30	.60	.045	.080	.06	—	—	—
Non-expanded or cold expanded	A	.21	—	.90	—	.04	.05	—	—	—
"	B[7]	.26	—	1.15	—	.04	.05	—	—	—
"	X42[7]	.28	—	1.25	—	.04	.05	—	—	—
Non-expanded	X46[7], X52[7]	.30	—	1.35	—	.04	05	—	—	—
Cold-expanded	X46[7], X52[7]	.28	—	1.25	—	.04	.05	—	—	—
Non-expanded or cold expanded	X56[4], X60[4]	.26	—	1.35	—	.04	.05	.005[5]	.005[5]	.005[5]
"	X65[6]	.26	—	1.40	—	.04	.05	.005[5]	.005[5]	.005[5]
"	X70[4]	.23[8]	—	1.60[8]	—	.04	.05	—	—	—
Non-expanded or cold expanded	X80[4]	0.18[8,9]	—	1.80[8,9]		0.030[9]	0.018[9]			

[1] Class II Steel is rephosphorized (see Par. 1.1 for note on bending and threading properties).

[2] In Grades X42 through X65, for each reduction of 0.01 percent below the maximum carbon content, an increase of 0.05 percent manganese above the specified maximum is permissible, up to a maximum of 1.45 percent for Grades X52 and lower and up to a maximum of 1.60 percent for Grades higher than X52.

[3] For cold expanded seamless pipe in Sizes 20 in. O.D. and larger, the maximum carbon content shall be 0.28 percent.

[4] Other chemical analyses may be furnished by agreement between purchaser and manufacturer.

[5] Either columbium, vanadium or titanium, or a combination thereof, shall be used at the discretion of the manufacturer.

[6] For Grades X65 in Sizes 16 in. and larger with wall thicknesses .500 in. *(12.7 mm)* and less, the chemical composition shall be as shown or as agreed upon between the purchaser and manufacturer. For other sizes and wall thicknesses, the chemical composition shall be as agreed upon between purchaser and manufactuer.

[7] Columbium, titanium, vanadium, or a combination thereof, may be used be agreement between purchaser and manufacturer.

[8] Manganese level may be increased by 0.05 percent for each 0.01 percent decrease in carbon up to a max manganese level of 2.00 percent.

[9] Levels are product analysis limits eliminating the need for check tolerances in Par. 3.3.

Dimensional Standards

Carbon and alloy steel pipe dimensions are specified in American National Standards Institute (*ANSI*) *Standard B36.10*. Ten different pipe "schedules" are defined (10, 20, 30, 40, 60, 80, 100, 120, 140, and 160), the wall thickness increasing as the schedule number increases. Schedule 40 is referred to as "standard" pipe, abbreviated "S," for 1/8- through 10-in. pipe. Schedule 80 is referred to as "extra strong," indicated often as "XS," for 1/8- through 8-in. pipe. Also used is a double extra strong designation "XXS" for 1/2- through 8-in. pipe. These latter three are old NPS designations.

American Petroleum Institute (API) *5L* is a second standard for dimensional requirements. Standard 5L agrees with ANSI 36.10 for the most part, with a few more wall thicknesses for several diameters.

Stainless steel pipe dimensions are designated by ANSI B36.19 and are much the same as for carbon steel pipe.

Allowable Stress

Two ANSI codes govern allowable stress for various types and grades of steel. The more stringent requirements of ANSI B31.3 apply to piping in processing plants and on platforms—the so-called "within fences" standard. ANSI

Table 13–3   Tensile Requirements

1	2		3		4		5
Grade	Yield Strength, Min.		Ultimate Tensile Strength, Min.		Ultimate Tensile Strength, Max.		Elongation, Min. Percent in 2 in. (50.80 mm)[1]
	PSI	MPa	PSI	MPa	PSI	MPa	
A25	25,000	(172)	45,000	(310)			
A	30,000	(207)	48,000	(331)			
B	35,000	(241)	60,000	(413)			
X42	42,000	(289)	60,000	(413)			
X46	46,000	(317)	63,000	(434)			
X52	52,000	(358)	66,000	(455)			
X56	56,000	(386)	71,000	(489)			
X60	60,000	(413)	75,000	(517)			
X65	65,000	(448)	77,000	(530)			
X70	70,000	(482)	82,000	(565)			
X80	80,000	(551)	90,000	(620)	120,000	(827)	

[1] The minimum elongation in 2 in. (50.80 mm) shall be that determined by the following formula:

English Formula	Metric Formula
$e = 625,000 \dfrac{A^{0.2}}{U^{0.9}}$	$e = 1942.57 \dfrac{A^{0.2}}{U^{0.9}}$

Where: e = minimum elongation in 2 in. (50.80 mm) in percent to nearest ½ percent.
A = cross-sectional area of the tensile test specimen in sq. in (mm) based on specified outside diameter or nominal specimen width and specified wall thickness rounded to the nearest 0.01 sq. in. (6.5 mm^2) or .75 sq. in. (484 mm^2), whichever is smaller.
U = specified minimum ultimate tensile strength, psi, (MPa).

See Appendix C for minimum elongation values for various size tensile specimens and grades. The minimum elongations for both round bar tensile specimens (0.350 in. (8.9 mm) diameter with 1.4 in. (35 mm) gage length, and the 0.500 in. (12.5 mm) diameter with 2.00 in. (50.8 mm) gage length shall be that shown in the Area A line of .20 sq. in. in the Elongation Table of Appendix C.

B31.8 applies to cross-country piping. Equations and tables for both codes are presented. Consult the codes themselves for complete details.

ANSI B31.3. The thickness required for "within fences" piping is:

$$t_m = CA + P_i d_o / [2(S'E' + P_i Y')] \qquad (13\text{–}1)$$

where t_m = minimum required thickness, in.
CA = mechanical, corrosion, and/or erosion allowance, in. (typically 0.064 in.)
P_i = internal design pressure, psig
d_o = outside diameter of pipe, in.
S' = allowable stress, psi
E' = longitudinal joint factor
  = 1.0 for seamless pipe
  = 0.85 for electric resistance welded pipe
Y' = temperature factor for ferritic steels
  = 0.4 up to 900°F
  = 0.5 for 950°F
  = 0.7 for 1000°F and above

Allowable stress values for carbon and low alloy steels are given in Table 13–4.

Equation 13–1 can be rearranged to solve for the pressure.

$$P_i = 2 (t_m - CA) S'E' / [d_o - 2 (t_m - CA)Y'] \qquad (13\text{–}2)$$

ANSI B31.8. The required thickness for gas transmission and distribution piping is:

$$t_m = P_i d_o / (2 S'' E'' F'' T'') \qquad (13\text{–}3)$$

where t_m = nominal wall thickness, in.
P_i = design pressure, psig
d_o = outside diameter of pipe, in.
S'' = specified minimum yield strength, psi
E'' = longitudinal joint factor
  = 1.0 for seamless and welded pipe, except
  = 0.80 for Fusion Welded A 134 and A 139
  = 0.80 for Spiral Welded A 211
  = 0.60 for Furnace Butt-Welded ASTM-A53, API-5L
F'' = construction design factor
  = 0.72 for cross-country locations, etc.
  = 0.60 for fringe areas near cities and towns, etc.

Table 13–4 Allowable Stresses in Tension for Materials (GPSA, 1987, p. 17–24)

Allowable Stresses in Tension for Materials (1)
(Excerpted from ANSI B31.3a-1985, Appendix A, Tables A-1, A-1B)

BASIC ALLOWABLE STRESSES IN TENSION (PSI)(1)

Material	Specification	Grade	Class	Factor (E)	Tensile Strength Min. PSI	Yield Strength Min. PSI	Notes	(6) Min. Temp.	To 100	200	300	400	500	600	650	700	750	800	850	900	950	1000	1050	1100
CARBON STEEL																								
Seamless Pipe	ASTM A53	A	Type S		48000	30000	57, 59	−20	16000	16000	16000	16000	16000	14800	14500	14400	10700	9300|	7900	6500	4500	2500	1600	1000
	ASTM A53	B	Type S		60000	35000	57, 59	−20	20000	20000	20000	20000	18900	17300	17000	16500	13000	10800|	8700	6500	4500	2500	1600	1000
	ASTM A106	A			48000	30000	57	−20	16000	16000	16000	16000	16000	14800	14500	14400	10700	9300|	7900	6500	4500	2500	1600	1000
	ASTM A106	B			60000	35000	57	−20	20000	20000	20000	20000	18900	17300	17000	16500	13000	10800|	8700	6500	4500	2500	1600	1000
	ASTM A106	C			70000	40000	57	−20	23300	23300	23300	22900	21600	19700	19400	19200	14800	12000|						
	ASTM A120						8	−20	||12000	11400		||												
	ASTM A333	1			55000	30000	57, 59	−50	18300	18300	17000	17200	16200	14800	14500	14400	12000	10200|	8300	6500	4500	2500	1600	1000
	ASTM A334	1			55000	30000	57, 59	−50	18300	18300	17000	17200	16200	14800	14500	14400	12000	10200|	8300	6500	4500	2500	1600	1000
	ASTM A333	6			60000	35000	57	−50	20000	20000	20000	20000	18900	17300	17000	16500	13000	10800|	8700	6500	4500	2500	1600	1000
	ASTM A334	6			60000	35000	57	−50	20000	20000	20000	20000	18900	17300	17000	16500	13000	10800|	8700	6500	4500	2500	1600	1000
	API 5L	A			48000	30000	57, 59	−20	16000	16000	16000	16000	16000	14800	14500	14400	10700	9300|	7900	6500	4500	2500	1600	1000
	API 5L	B			60000	35000	57, 59	−20	20000	20000	20000	20000||	18900	17300	17000	16500	13000	10800|	8700	6500	4500	2500	1600	1000
	API 5LX	X42			60000	42000	51, 55	−20	20000	20000	20000	20000||												
	API 5LX	X46			63000	46000	51, 55	−20	21000	21000	21000	21000||												
	API 5LX	X52			66000	52000	51, 55	−20	22000	22000	22000	22000||												
	API 5LX	X52			72000	52000	51, 55	−20	24000	24000	24000	24000||												
Electric Resistance Welded Pipe																									
	ASTM A53	A	Type E	0.85	48000	30000	57, 59	−20	13600	13600	13600	13600	13600	12600	12300	12250	9100	7900|	6700	5500	3800	2150	1350	850
	ASTM A53	B	Type E	0.85	60000	35000	57, 59	−20	17000	17000	17000	17000	16100	14700	14500	14000	11000	9200|	7350	5500	3800	2150	1350	850
	ASTM A120			0.85			8	−20	10200	9800														
LOW AND INTERMEDIATE ALLOY STEEL & STAINLESS STEEL (4,40)																									
Seamless Pipe																									
3½ Ni	ASTM A333	3			65000	35000		−150	21700	19600	19500	18700	17800	16800	16300	15500	13900	11400	9000	6500	4500	2500	1600	1000
3½ Ni	ASTM A334	3			65000	35000		−150	21700	19600	19500	18700	17800	16800	16300	15500	13900	11400	9000	6500	4500	2500	1600	1000
Ni-Cr-Cu-Al	ASTM A333	4			60000	35000		−150	20000	19100	18200	17300	16400	15500	15000									
2¼ Ni	ASTM A333	7			65000	35000		−100	21700	19600	19500	18700	17600	16800	16300	15500	13900	11400	9000	6500	4500	2500	1600	1000
2¼ Ni	ASTM A334	7			65000	35000		−100	21700	19600	19500	18700	17600	16800	16300	15500	13900	11400	9000	6500	4500	2500	1600	1000
9 Ni	ASTM A333	8			100000	75000	47	−320	31700	31700														
9 Ni	ASTM A334	8			100000	75000		−320	31700	31700														
18Cr-8Ni Pipe	ASTM A376	TP304			75000	30000	20, 26, 28, 31, 36	−425	20000	20000	20000	18700	17500	16400	16200	16000	15600	15200	14900	14600	14400	13800	12200	9700
18Cr-8Ni Pipe	ASTM A376	TP304H			75000	30000	26, 31, 36	−325	20000	20000	20000	18700	17500	16400	16200	16000	15600	15200	14900	14600	14400	13800	12200	9700

NOTE: These Notes are requirements of the Code. Those marked with an asterisk (*) restate requirements found in the text of the Code. The other Notes are limitations or special requirements applicable to particular materials. Full interpretation of these notes will require reference to ANSI/ASME B31.3a — 1984 Edition.

(1)* The stress values in Table A-1 and the design stress values in Table A-2 are basic allowable stresses in tension in accordance with 302.3.1(a). For pressure design, the stress values from Table A-1 are multiplied by the appropriate quality factor E (E_c from Table A-1A, or E_j from Table A-1B). Stress values in shear and bearing are stated in 302.3.1(b); those in compression in 302.3.1(c).

(3)* This casting quality factor can be enhanced by supplementary examination in accordance with 302.3.3(c) and Table 302.3.3C. The higher factor from Table 302.3.3C may be substituted for this factor in pressure design equations.

(4)* In shaded areas, stress values printed in *italics* exceed two-thirds of the expected yield strength at temperature. All other stress values in shaded areas are equal to 90% of expected yield strength at temperature. See 302.3.2(d)(4) and 302.3.2(d) [Note (3)].

(6)* The minimum temperature shown is that design minimum temperature for which the material is normally suitable without impact testing other than that required by the material specification. However, the use of a material at a design minimum temperature below −20°F (−29°C) is established by rules elsewhere in this Code, including any necessary impact test requirements.

(7)* A single bar (\|) in these Stress Tables indicates there are conditions other than stress which affect usage above or below the temperature, as described in other referenced Notes. A double bar (\|\|) after a tabled stress indicates that use of the material is prohibited above that temperature. A double bar (\|\|) before the stress value for "Min. Temp. to 100°F" indicates that the use of the material is prohibited below the listed minimum temperature. At temperatures where there are no stress values, the material may be used in accordance with 323.2 unless prohibited by a double bar (\|\|).

(8)* There are restrictions on the use of this material in the text of the Code.

(20) For pipe sizes NPS 8 and larger and for wall thicknesses of Schedule 140 or heavier, the minimum specified tensile strength is 70.0 ksi (483 MPa).

(26) These unstabilized grades of stainless steel have increasing tendency to intergranular carbide precipitation as the carbon content increases above 0.03%.

(28) For temperatures above 1000°F (538°C), these stress values apply only when the carbon content is 0.04% or higher.

(30) For temperatures above 1000°F (538°C), these stress values may be used only if the material has been heat treated at a temperature of 2000°F (1090°C) minimum.

(31) For temperatures above 1000°F (538°C), these stress values may be used only if the material has been heat treated by heating to a minimum temperature of 1900°F (1040°C) and quenching in water or rapidly cooling by other means.

(36) The specification permits this material to be furnished without solution heat treatment or with other than a solution heat treatment. When the material has not been solution heat treated, the minimum temperature shall be −20°F (−29°C) unless the material is impact tested per 323.3.

(40) The stress values for austenitic stainless steels in these Tables may not be applicable if the material has been given a final heat treatment other than that required by the material specification and any overriding requirements of this Code called for by Note (30) or (31).

(47) If no welding is employed in fabrication of piping from these materials, the stress values may be increased to 33.3 ksi (230 MPa).

(51) Special P-Numbers SP-1, SP-2, and SP-3 of carbon steels are not included in P-No. 1 because of possible high carbon, high manganese combination which would require special consideration in qualification. Qualification of any high carbon, high manganese grade may be extended to other grades in its group.

(55) Pipe produced to this specification is not intended for high temperature service. The stress values apply to either non-expanded or cold expanded material in the as-rolled, normalized, or normalized and tempered condition.

(57)* Conversion of carbides to graphite may occur after prolonged exposure to temperatures over 800°F (425°C) (see Appendix F).

(59)* For temperatures above 900°F (480°C), consider the advantages of killed steel (see Appendix F).

$$= 0.50 \text{ for commercial and residential}$$
areas, etc.
$$= 0.40 \text{ for areas with multistory}$$
buildings, etc.
$$T'' = \text{temperature factor}$$
$$= 1.000 \text{ for } 250°F \text{ or less}$$
$$= 0.967 \text{ for } 300°F$$
$$= 0.933 \text{ for } 350°F$$
$$= 0.900 \text{ for } 400°F$$
$$= 0.867 \text{ for } 450°F$$

Equation 13–3 can be rearranged to solve for the pressure.

$$P_i = 2 \, S'' \, E'' \, F'' \, T'' \, t_m/d_o \qquad (13\text{–}4)$$

Table 13–5 gives dimensions and allowable pressures for ASTM A 106, API 5L, and API 5LX seamless transmission pipe of various minimum yield strengths.

In design calculations, the pressure shall never exceed the allowable working pressure. Pressure sources are the reservoir, compressors, and to a lesser extent elevation head. In downhill sections of pipelines the elevation head increases in the flow direction, although frictional drag tends to decrease pressure.

Accidental pressure excursions are unusual in gas transport. Such excursions may cause the operating pipeline pressure to exceed the MAWP for a short time. Fortunately, the code provides a safety factor and hence the pipe may not burst. However, the designer cannot claim this safety factor. Line pipe contains inhomogeneities that cause weak spots. Such inhomogeneities occur unavoidably in pipe manufacture, even with good practice; hence the need for safety factors. Field welding is also a source of inhomogeneities. In any case, pipelines must be *hydrostatically tested* with water at 150% of the MAWP to ensure integrity before gas flow is allowed.

In spite of the above precautions, *pipelines occasionally burst*. Typical reasons are the unfortunate piercing of the line by heavy equipment such as bulldozers and progressive corrosion that weakens the pipe. Precautions taken to reduce corrosion include protective coatings and/or cathodic protection.

Mass Balance

The steady-state relationship derived from mass-balance considerations for a pipe is termed the "continuity equation."

$$m = \rho \, V \, A \qquad (13\text{–}5)$$

where m = mass flow rate of the fluid
ρ = density of the flowing fluid at line conditions
V = average velocity of fluid in the pipe
A = cross-sectional area of the pipe

The continuity equation states that the mass flow rate is constant at steady state. If the pipe diameter is constant, then A is also constant. The fluid density is a strong function of pressure, however, decreasing as the pressure drops along a pipeline. Therefore, the average velocity of the fluid at steady state will vary inversely with pressure, increasing as the pressure drops.

Fluid density decreases as temperature rises. Temperature is a less important factor in pipe flow calculations since the temperature is usually relatively constant.

Liquids are relatively incompressible, and so the velocity is approximately constant throughout the flow line.

The actual volumetric flow rate, Q, in a pipe is given by

$$Q = m / \rho \qquad (13\text{–}6)$$

Equations 13–5 and 13–6 can be combined to obtain

$$V = Q / A \qquad (13\text{–}7)$$

Equation 13–7 can be used to calculate the average velocity from the volumetric flow rate and vice versa.

In the gas industry it is customary to express mass flow rate in terms of so-called *standard cubic feet* or *standard cubic meters*—the volume the gas would occupy if it were an ideal gas at an arbitrary pressure and temperature base. The relation between mass and volume from the ideal gas law is

$$P_b \, Q_s = n \, R \, T_b = m \, R \, T_b / MW \qquad (13\text{–}8)$$
where P_b = pressure base for standard volume (absolute)
Q_s = standard volumetric flow rate
n = molar flow rate
R = universal gas constant in appropriate units
T_b = temperature base for standard volume (absolute)
m = mass flow rate of gas
MW = average molecular weight of gas

Thus, the *standard volumetric rate* in terms of the mass rate is

$$Q_s = (m / MW) \, R \, T_b / P_b \qquad (13\text{–}9)$$

Care of the units must be taken and the pressure and temperature bases clearly defined when Equation (13–9) is used.

Energy Balance

The *mechanical energy* or *Bernoulli equation* can be written between upstream point 1 and downstream point 2 in a pipeline as follows:

$$P_1/\rho_1 + \alpha_1 V_1^2 /2g_c + gz_1/g_c$$
$$= P_2/\rho_2 + \alpha_2 V_2^2/2g_c + gz_2/g_c + h_L \qquad (13\text{–}10)$$

Table 13–5

Gas transmission and distribution piping
Code for pressure piping ANSI B31.8-1982
Carbon steel and high yield strength pipe
Values apply to A106, API 5L and API5LX pipe having the same
Specified minimum yield strength as shown

ALLOWABLE WORKING PRESSURES UP TO 250°F, IN PSIG

CONSTRUCTION TYPE DESIGN FACTORS

GR.B applies to the 35,000 column of each construction type.

NOM PIPE SIZE	O.D.	WALL THK.	TYPE A, F = 0.72 *					TYPE B, F = 0.60					TYPE C, F = 0.50					TYPE D, F = 0.40				
			35,000	42,000	46,000	52,000	60,000	35,000	42,000	46,000	52,000	60,000	35,000	42,000	46,000	52,000	60,000	35,000	42,000	46,000	52,000	60,000
2	2.375 (STD)	.154	3268					2723					2270					1816				
		.218	4626					3855					3213					2570				
3	3.500 (STD)	.125	1800					1500					1250					1000				
		.156	2246					1872					1560					1248				
		.188	2707					2256					1880					1504				
		.216	3110					2592					2160					1728				
		.250	3600					3000					2500					2000				
		.281	4046					3372					2810					2248				
		.300	4320					3600					3000					2400				
4	4.500 (STD)	.125	1400	1680	1840			1167	1400	1533			973	1167	1278			778	933	1022		
		.156	1747	2097	2296			1456	1747	1913			1214	1456	1595			971	1165	1276		
		.188	2105	2526	2767			1754	2105	2306			1462	1755	1922			1170	1404	1537		
		.219	2453	2943	3224			2044	2453	2686			1703	2044	2239			1363	1635	1791		
		.237	2654	3185	3488			2212	2654	2907			1844	2212	2423			1475	1770	1938		
		.250	2800	3360	3680			2333	2800	3067			1945	2333	2556			1556	1869	2044		
		.281	3147	3776	4136			2623	3147	3447			2186	2622	2873			1748	2098	2298		
		.312	3494	4193	4593			2912	3494	3827			2427	2912	3190			1941	2330	2552		
		.337	3774	4530	4961			3145	3775	4134			2621	3146	3445			2097	2516	2756		
6	6.625 (STD)	.156	1187	1424	1560	1763		989	1187	1300	1469		824	989	1083	1224		659	791	866	980	
		.188	1429	1716	1880	2124		1192	1430	1567	1770		993	1192	1306	1475		794	954	1044	1180	
		.219	1666	2000	2190	2475		1388	1666	1825	2063		1157	1389	1521	1719		926	1111	1216	1375	
		.250	1902	2282	2500	2826		1585	1902	2083	2355		1321	1585	1736	1963		1057	1268	1389	1570	
		.280	2130	2556	2799	3164		1775	2130	2333	2637		1479	1775	1944	2198		1183	1420	1555	1758	
		.312	2373	2848	3120	3527		1978	2374	2600	2903		1649	1978	2167	2449		1319	1582	1733	1959	
		.375	2853	3424	3750	4237		2377	2853	3125	3531		1981	2378	2604	2943		1585	1902	2083	2354	
		.432	3287	3943	4319	4883		2739	3286	3599	4069		2283	2738	3000	3391		1826	2191	2400	2713	
8	8.625 (STD)	.156	912	1094	1198	1354		760	912	998	1128		633	760	832	940		506	608	666	752	
		.188	1098	1318	1444	1632		915	1098	1203	1360		763	915	1003	1133		610	732	802	907	
		.203	1186	1424	1559	1762		989	1186	1299	1469		824	989	1083	1224		659	791	866	979	
		.219	1280	1535	1681	1901		1067	1280	1401	1584		889	1067	1168	1320		711	853	934	1056	
		.250	1461	1753	1920	2170		1217	1461	1600	1809		1014	1217	1333	1507		812	974	1067	1206	
		.277	1618	1942	2128	2405		1349	1618	1773	2004		1124	1349	1478	1670		899	1079	1182	1336	
		.312	1823	2189	2396	2709		1520	1823	1997	2258		1266	1520	1664	1881		1013	1216	1331	1505	
		.322	1882	2258	2473	2796		1568	1882	2061	2329		1307	1568	1717	1941		1045	1254	1374	1553	
		.344	2011	2412	2642	2988		1676	2011	2202	2490		1396	1676	1835	2075		1117	1340	1468	1660	
		.375	2191	2628	2880	3256		1826	2191	2399	2713		1521	1826	1999	2261		1217	1460	1599	1808	
		.438	2560	3071	3364	3803		2133	2560	2804	3170		1778	2133	2336	2641		1422	1706	1869	2113	
		.500	2922	3506	3840	4341		2435	2922	3200	3617		2029	2435	2667	3014		1623	1948	2133	2412	
10	10.750 (STD)	.188	881	1058	1158	1310		733	881	965	1091		612	735	804	909		490	588	644	728	
		.203	952	1143	1251	1415		794	952	1043	1179		661	794	869	983		529	635	695	786	
		.219	1026	1231	1348	1525		855	1026	1124	1271		713	855	936	1059		570	684	749	847	
		.250	1172	1407	1540	1741		977	1172	1284	1451		814	977	1070	1209		651	781	856	967	
		.279	1309	1570	1719	1944		1091	1309	1433	1620		909	1091	1194	1350		727	872	955	1080	
		.307	1440	1728	1892	2138		1200	1440	1577	1782		1000	1200	1314	1486		800	960	1051	1189	
		.344	1613	1935	2120	2396		1344	1613	1767	1997		1120	1344	1473	1664		896	1075	1178	1331	
		.365	1711	2054	2249	2542		1426	1711	1874	2119		1188	1426	1562	1766		951	1141	1249	1412	
		.438	2054	2464	2700	3051		1712	2054	2250	2543		1426	1712	1875	2119		1141	1369	1500	1695	
		.500	2344	2813	3081	3483		1953	2344	2567	2902		1628	1953	2140	2419		1302	1563	1712	1935	
12	12.750 (STD)	.188	743	892	977	1104		619	743	814	920		516	619	678	767		413	495	543	613	
		.203	803	963	1055	1193		669	803	879	995		558	669	733	829		446	535	586	663	
		.219	866	1039	1138	1287		722	866	948	1073		601	722	790	894		481	577	632	715	
		.250	988	1186	1299	1468		824	988	1082	1224		686	824	902	1020		549	659	722	816	
		.281	1111	1332	1460	1651		926	1111	1217	1376		771	926	1014	1146		617	740	811	917	
		.312	1233	1480	1620	1832		1028	1233	1350	1527		856	1028	1125	1273		685	822	900	1018	
		.330	1305	1566	1715	1939		1088	1305	1430	1616		906	1088	1191	1346		725	870	953	1077	
		.344	1359	1631	1786	2020		1133	1359	1488	1683		944	1133	1240	1403		755	906	992	1122	
		.375	1482	1779	1948	2202		1235	1482	1624	1835		1029	1235	1353	1529		824	988	1082	1224	
		.406	1606	1926	2110	2385		1338	1606	1758	1988		1115	1338	1465	1656		892	1070	1172	1325	
		.438	1732	2077	2275	2572		1443	1732	1896	2144		1203	1443	1580	1786		962	1154	1264	1429	
		.500	1976	2372	2598	2936		1647	1976	2165	2447		1373	1647	1804	2039		1098	1318	1443	1631	

* Type A construction also applicable to "Liquid Petroleum Transportation Piping Code," ANSI B31.4-1979

where P = pressure
 ρ = density
 α = velocity profile correction factor
 V = average velocity of flowing fluid
 g = acceleration of gravity
 g_c = conversion factor = 32.174 ft lbm/s^2 lbf
 z = elevation
 h_L = frictional energy loss due to pipe skin friction
    and form drag of fittings

Turbulent flow usually prevails for gas flow, and the velocity profile correction factors are approximately one. Also, in many cases V_1 and V_2 are not greatly different, and evaluation of the α's is unimportant. These factors are set equal to one in the remainder of the chapter.

In gas pipeline design and analysis, the elevation terms are not generally very important but will be retained for rigor.

The *friction loss* term, h_L, is given by

$$h_L = 4 f_F (L/D) (V^2/2g_c) + \Sigma K (V^2/2g_c) \quad (13\text{--}11)$$

where f_F = Fanning friction factor, a function of the
    Reynolds number and the pipe roughness
 D = inside diameter of the pipe
 ΣK = sum of the resistance coefficients for the
    fittings in the line

Friction factors must be used with care. There are two commonly defined friction factors, the *Fanning friction factor*, f_F, shown above and the *Moody* or *Darcy-Weisbach friction factor*, f, which is four times the Fanning friction factor.

In cross-country pipelines, there are few fittings (elbows, valves, etc.) and the second term on the right hand side of Equation 13–11 will be neglected.

A working formula for flow rate can be obtained from Equations 13–5, 13–9, 13–10, and 13–11, with proper allowance for units (Finch and Ko, 1988). The assumptions inherent in the derivation are that the kinetic energy terms $(V^2/2g_c)$ can be neglected, and that an average density can be used for the gas with little error. See Johnson and Berwald (1935) for all the details.

$$Q_s = 38.77 \left(\frac{T_b}{P_b}\right) F d^{2.5}$$

$$\left[\frac{P_1^2 - P_2^2 - (0.0375 \, SG \, \Delta z \, P_m^2/Z_m T)}{SG \, T \, Z_m \, L}\right]^{0.5} \quad (13\text{--}12)$$

where Q_s = standard cubic feet per day, scfd
 T_b = temperature base, °R, usually 519.67 (60°F)
 P_b = pressure base, psia, usually 14.696
 F = transmission factor, $1/\sqrt{f_F}$
 d = pipe inside diameter, in.
 P_1 = inlet pressure, psia

 P_2 = outlet pressure, psia
 SG = gas specific gravity, MW / 28.9625
 MW = average molecular weight of gas
 Δz = change in elevation (outlet minus inlet), ft
 P_m = average pressure, psia
 Z_m = compressibility factor of gas at T and P_m
 T = flowing temperature of gas, °R
 L = pipe length, miles

The average pressure to be used in the computations is given by the following relation:

$$P_m = (2/3) \{P_1 + P_2 - [P_1 P_2/(P_1 + P_2)]\} \quad (13\text{--}13)$$

If P_1 and P_2 are not much different, the arithmetic average can be used.

Friction Factor Correlation

In the past, several *empirical flow equations* have been used to predict the friction factor, and thus transmission factor, for the flow equation (Eq. 13–12). Notable examples are the *Weymouth* equation and the *Panhandle A and B* formulas. While the friction factor does not actually appear in these equations, Finch and Ko (1988), Rhoads (1983), and others have shown that the equivalent friction factors inherently implied in these equations are not correct. **These equations are not recommended** even though they are still used.

The Moody correlation shown in Figure 13–2 is recommended. Figure 13–2 presents the Darcy-Weisbach friction factor, indicated here simply as f with no subscript. Remember

$$f_F = f/4 \quad (13\text{--}14)$$

The *transmission factor* for use in Equation 13–12 is given by

$$F = 1/\sqrt{f_F} \quad (13\text{--}15)$$

The *Moody chart* shows that the friction factor is a function of the flow regime (as reflected by the dimensionless *Reynolds number*) and the pipe roughness (as reflected by the relative roughness).

The Reynolds number, Re, is given by

$$Re = D V \rho / \mu \quad (13\text{--}16)$$

where D = pipe inside diameter
 V = average flow velocity in the pipe
 ρ = fluid density
 μ = fluid absolute viscosity

The units must be consistent to give a dimensionless ratio. The kinematic viscosity, ν, is related to the absolute viscosity:

$$\nu = \mu / \rho \quad (13\text{--}17)$$

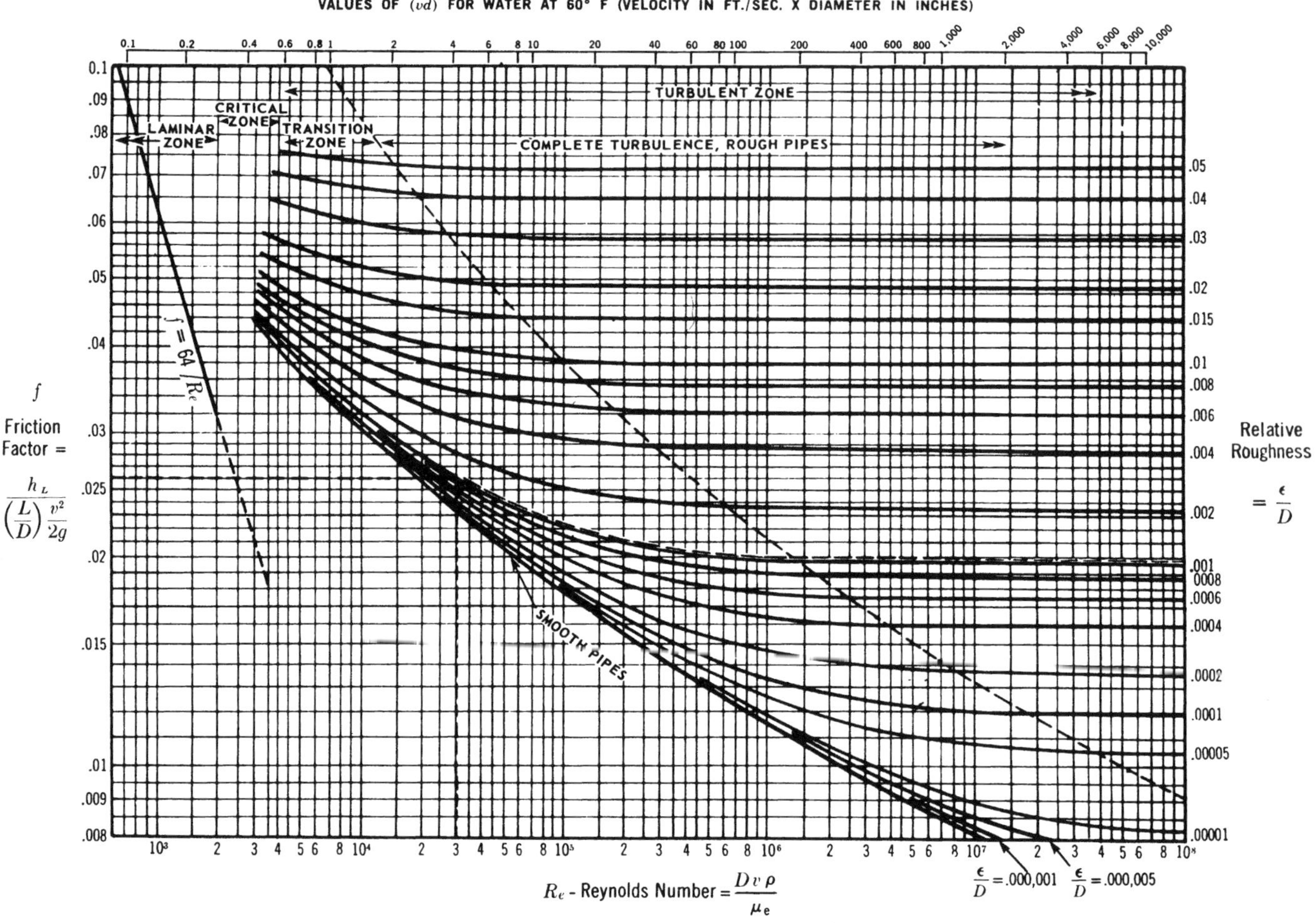

Figure 13–2. Moody friction factor (Crane Company, 1985).

so that the Reynolds number can also be written as

$$Re = D V / v \tag{13–18}$$

The continuity equation, Equation 13–5, can be used to eliminate V ρ in Equation 13–16.

$$Re = D m / (A \mu) \tag{13–19}$$

The mass velocity, G_m, is defined by

$$G_m = m / A = V \rho \tag{13–20}$$

An alternative expression for the Reynolds number is thus

$$Re = D G_m / \mu \tag{13–21}$$

Finally, for a cylindrical pipe $A = \pi D^2 / 4$, which can be introduced into Equation 13–19 to give

$$Re = 4 m / (\pi D \mu) \tag{13–22}$$

The choice of expression for computing Reynolds number depends on the available data.

Pipe roughness varies considerably from pipe to pipe and makes the accurate use of the Moody correlation difficult, since the user has the burden of choosing a value of roughness that may or may not correspond to that of the actual pipe. For an existing pipe, the value of the roughness can be "backed out" from good-quality operating data. For a new pipe the value must be estimated. Moody indicates a value of 0.00015 ft for new, commercial-steel pipe. Rau and Hein (1982) use a value of 0.0000625 ft, which accords better with the value of 0.0000833 ft given by Rhoads (1983).

The Moody chart is a plot of the Colebrook equation, an implicit relation for the Darcy-Weisbach friction factor.

$$\frac{1}{\sqrt{f}} = - 2 \log_{10}\left(\frac{e}{3.7D} + \frac{2.51}{Re \sqrt{f}}\right) \tag{13–23}$$

The American Gas Association developed a correlation of friction factors for transmission lines based on measured data for *cross-country pipelines* (Uhl, 1967). In the *AGA correlation*, two transmission factors are calculated and the lesser is used. The procedure gives a result very nearly the same as the Moody chart, as will be shown.

In the AGA correlation, the "smooth-pipe" transmission factor is calculated first.

$$F_{SP} = 4 \log_{10} (Re/F_{SP}) - 0.6 \qquad (13\text{–}24)$$

The above equation is implicit in F_{SP} and requires trial-and-error solution. Table 13–6 will aid in the computation, which converges rapidly with a hand calculator.

Next determine the *drag factor*, F_f, from Figure 13–3. The partially-turbulent transmission factor is given by

$$F_{PT} = 4 F_f \log_{10} [Re/(1.4 F_{SP})] \qquad (13\text{–}25)$$

The partially-turbulent transmission factor is compared with the fully-turbulent transmission factor, which is calculated by

$$F_{FT} = 4 \log_{10} (3.7 D/e) \qquad (13\text{–}26)$$

The smaller of the two is used.

Recall that the transmission factor F is $1/\sqrt{f_F}$, or $2/\sqrt{f}$. Thus, if Equation 13–23 is used, F is given by

$$F = -4 \log_{10} \left(\frac{e}{3.7D} + \frac{1.255 F}{Re} \right) \qquad (13\text{–}27)$$

If the Reynolds number is very high, the second term in the argument of the logarithm approaches zero, and Equation 13–27 becomes the same as Equation 13–26. In other words, the Moody diagram and the AGA correlation are the same at high Reynolds numbers. Figure 13–4 shows the equivalent friction factor of the AGA correlation for a relative roughness of 0.0001 superimposed on the Moody chart. The two correlations nearly have the same friction factor. For this reason, the *Moody chart* is used for the rest of the chapter and *is recommended* for general use, especially since it also applies for liquids.

Gas properties needed for the computations include density, compressibility factor, and viscosity. The GPSA Engineering Data Book (1987) contains natural gas viscosity data. The density and the compressibility factor are directly related, of course.

$$\rho = P\,MW\,/\,Z\,R\,T \qquad (13\text{–}28)$$

The Z-charts in Chapter 3 can be used for hydrocarbon gas calculations. Do not confuse the compressibility factor with the "supercompressibility" factor, F_{pv}, used in metering work. The relation between the two is

$$F_{pv} = 1\,/\,\sqrt{Z} \qquad (13\text{–}29)$$

Two examples illustrate pressure-drop calculations. The first example is for a high-pressure transmission line and the second for a low-pressure gathering line.

Example 13–1. A 0.55-gravity gas flowing at 100 MMscfd is to be pipelined from a platform to shore through a 65-mile long pipeline. The platform pressure is 1000 psia, and the pressure arriving onshore is to be 800 psia. The temperature of the gas is 45°F. What pipe size and thickness do you recommend? Note that the change in elevation is negligible. Assume Type A, Grade B, API 5L, seamless pipe. A design pressure of 1200 psig is desired. Assume a pipe roughness of 0.001 in.

Solution: Use Equation 13–12.

Equation 13–13:
$P_m = (\frac{2}{3}) (1000 + 800 - 800000/1800)$
     $= 904$ psia

From Figure 3–16, at 45°F and 904 psia,
$Z_m = 0.87$

$MW = 28.9625 (0.55) = 15.93$
$T = 45 + 460 = 505°R$
$\rho_m = P_m\,MW\,/\,(Z_m\,R\,T)$
$\quad = \dfrac{(904)(15.93)}{(0.87 \times 10.73 \times 505)}$
$\quad = 3.05$ lb/ft^3

$m = \dfrac{(100000000)(15.93)}{(379.5 \times 24 \times 3600)}$
$\quad = 48.6$ lb/s

From Appendix 6, Figure A6–5,
$\mu = (0.013) (0.000672)$
$\quad = 8.74 \times 10^{-6}$ lb/ft-s
Equation 13–22.
$Re = \dfrac{(4) (48.6)}{(3.14159)\,D\,(8.74 \times 10^{-6})}$
$\quad = 7.08 \times 10^6\,/\,D$

Table 13–6 Smooth-Pipe Transmission Factor, F_{SP}
(Uhl, *Pipe Line Industry*, p. 47, Jan., 1967)

$10^{-6}Re$	F_{SP}	$10^{-6}Re$	F_{SP}	$10^{-6}Re$	F_{SP}
0.1	14.727	1.0	18.346	6.0	21.207
0.2	15.809	1.5	18.990	7.0	21.454
0.3	16.444	2.0	19.449	8.0	21.669
0.4	16.897	2.5	19.805	10.0	22.028
0.5	17.249	3.0	20.096	15.0	22.682
0.6	17.537	3.5	20.343	20.0	23.146
0.7	17.781	4.0	20.556	30.0	23.802
0.8	17.992	4.5	20.745	40.0	24.268
0.9	18.179	5.0	20.914	50.0	24.630

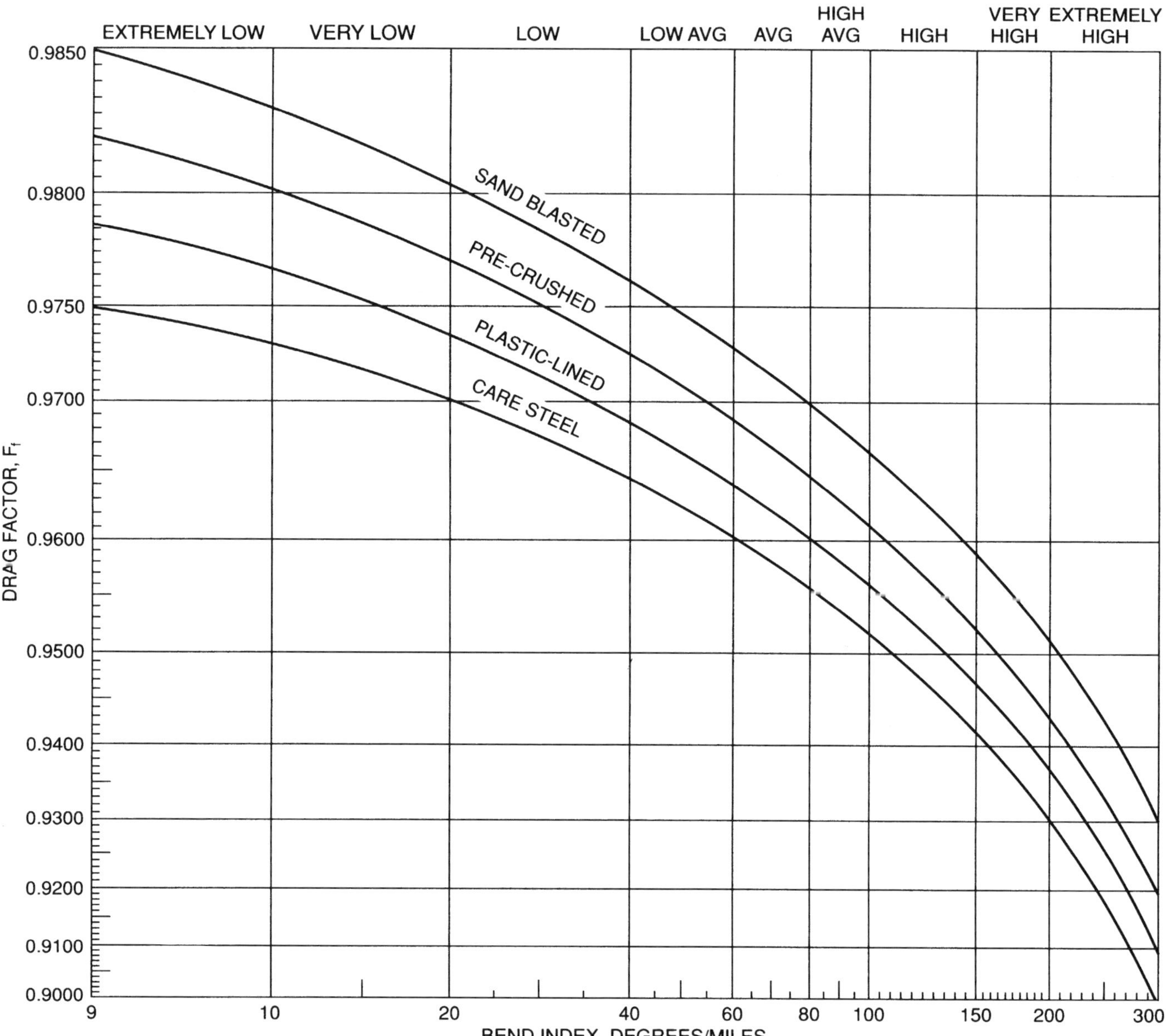

Figure 13–3. Drag factor, F_f (Uh1, *Pipe Line Industry,* p. 58, March 1967).

Bracketed term in Equation 13–12:

$$[(1000^2 - 800^2)/(0.55 \times 505 \times 0.87 \times 65)]^{0.5} = 4.79$$

Substitute known quantities into Equation 13–12.

$$100000000 = 38.77 (520 / 14.7)Fd^{2.5}(4.79), \text{ or}$$

$$Fd^{2.5} = 1.523 \times 10^4$$

Assume $f = 0.015$

$$F = 2/(0.015)^{0.5} = 16.33$$
$$d^{2.5} = 1.523 \times 10^4/16.33$$
$$d = 15.4 \text{ in.}; D = 1.284 \text{ ft}$$
$$Re = 7.08 \times 10^6/1.284$$
$$= 5.5 \times 10^6$$
$$e/D = 0.001/15.4 = 0.000065$$

Figure 13–2 gives f = 0.0113; repeat the calculation

$$F = 2/(0.0113)^{0.5} = 18.81$$
$$d^{2.5} = 1.523 \times 10^4/18.81$$
$$d = 14.56 \text{ in.}$$

From Table 13–5, select a 16-in. pipe with 0.438-in wall thickness, allowable pressure 1379 psi.   ←

Verify the maximum allowable working pressure by Equation 13–4. From the table, $S'' = 35000$ psi

Seamless pipe has $E'' = 1.0$; Type A has $F'' = 0.72$;

$T'' = 1.0$ for temperature less than 250°F

$$P_i = 2(35000) \ (1.0) \ (0.72) \ (1.0) \ (0.438)/16.0$$
$$= 1379 \text{ psi} \quad \text{(Check)} \quad ←$$

Example 13–2. A gathering line delivers 0.70 gravity gas to a processing plant at 100 psia. The temperature of the gas is 80°F and the flow rate is 2.3 MMscfd. A 4-in., Sch 40 line 8.5 miles long is available. There is a drop in elevation of 500 ft from the inlet to the plant fence. What will the pressure be at the inlet to the line? Assume a roughness of 0.001 in.

Solution: Assume $P_m = 150$ psia;
$$T = 80 + 460 \ = 540°R$$
From Figure 3–19 $Z_m = 0.98$

$$MW = 0.7 \ (28.9625) = 20.27$$
$$m = \frac{(2,300,000) \ (20.27)}{(379.5) \ (24) \ (3600)}$$
$$= 1.422 \text{ lb/s}$$
$$\rho_m = (150) \ (20.27)/$$
$$(0.98 \times 10.73 \times 540)$$
$$= 0.5355 \text{ lb/ft}^3$$

From Application 6,
$$\mu = (0.011) \ (0.000672)$$
$$= 7.39 \times 10^{-6} \text{ lb/ft-s}$$
Table 13–5 gives $d_o = 4.500$ in.,
$$t = 0.237 \text{ in.}$$

$$d = 4.500 \ -2(0.237) = 4.026 \text{ in.};$$
$$D = 0.3355 \text{ ft}$$
$$Re = \frac{(4) \ (1.422)}{(3.14159 \times 0.3355 \times 7.39 \times 10^{-6})}$$
$$= 7.30 \times 10^5$$
$$e/D = 0.001/4.026 = 0.000248$$

Figure 13–5 gives f = 0.0153 so
$$F = 2/(0.0153)^{0.5}$$
$$= 16.17$$
Elevation term in Equation 13–12 is

$$\frac{(0.0375) \ (0.70) \ (-500) \ (150)^2}{(0.98 \times 540)} = -558$$
$$SG \ T \ Z_m \ L = (0.7) \ (540) \ (0.98) \ (8.5)$$
$$= 3149$$

$$38.77 \ (T_b/P_b) \ F \ d^{2.5}$$
$$= 38.77 \ (520/14.7) \ (16.17) \ (4.026)^{2.5}$$
$$= 7.212 \times 10^5$$
Equation 13–12 gives

$$2300000 = 7.212 \times 10^5\{[P_1{}^2 - 100^2-(-558)]/3140\}^{0.5}$$
$$\text{which gives } P_1 = 204 \text{ psia} \quad ←$$

The mean pressure used checks closely.

Compared to real design applications, the above calculations are simple but do serve to illustrate the basic factors required in pipeline design.

Process Lines

The recommendations of API RP 14E are appropriate for the sizing of gas lines in process plants and on platforms (as opposed to gathering or transmission lines). Line size selection is actually an economic compromise between an expensive large pipe with negligible pressure drop and an inexpensive small pipe with costly high pressure drop. The following table lists economical pressure loss as a function of operating pressure.

Acceptable Pressure Drops For Single-Phase Gas Process Lines
(API RP 14E, 1984)

Operating Pressure, psig	Acceptable Pressure Drop, psi/100 ft
0–100	0.05–0.20
100–500	0.20–0.50
500–2000	0.50–1.2

Reciprocating and centrifugal compressor piping should be sized to minimize noise, vibration, and pulsation. Each specific application should be studied individually.

The GPSA Engineering Data Book gives a very simple method for calculating pressure drop per 100 ft.

$$\Delta P_{100} = C_1 \ C_2 \ / \ \rho \qquad (13\text{–}30)$$

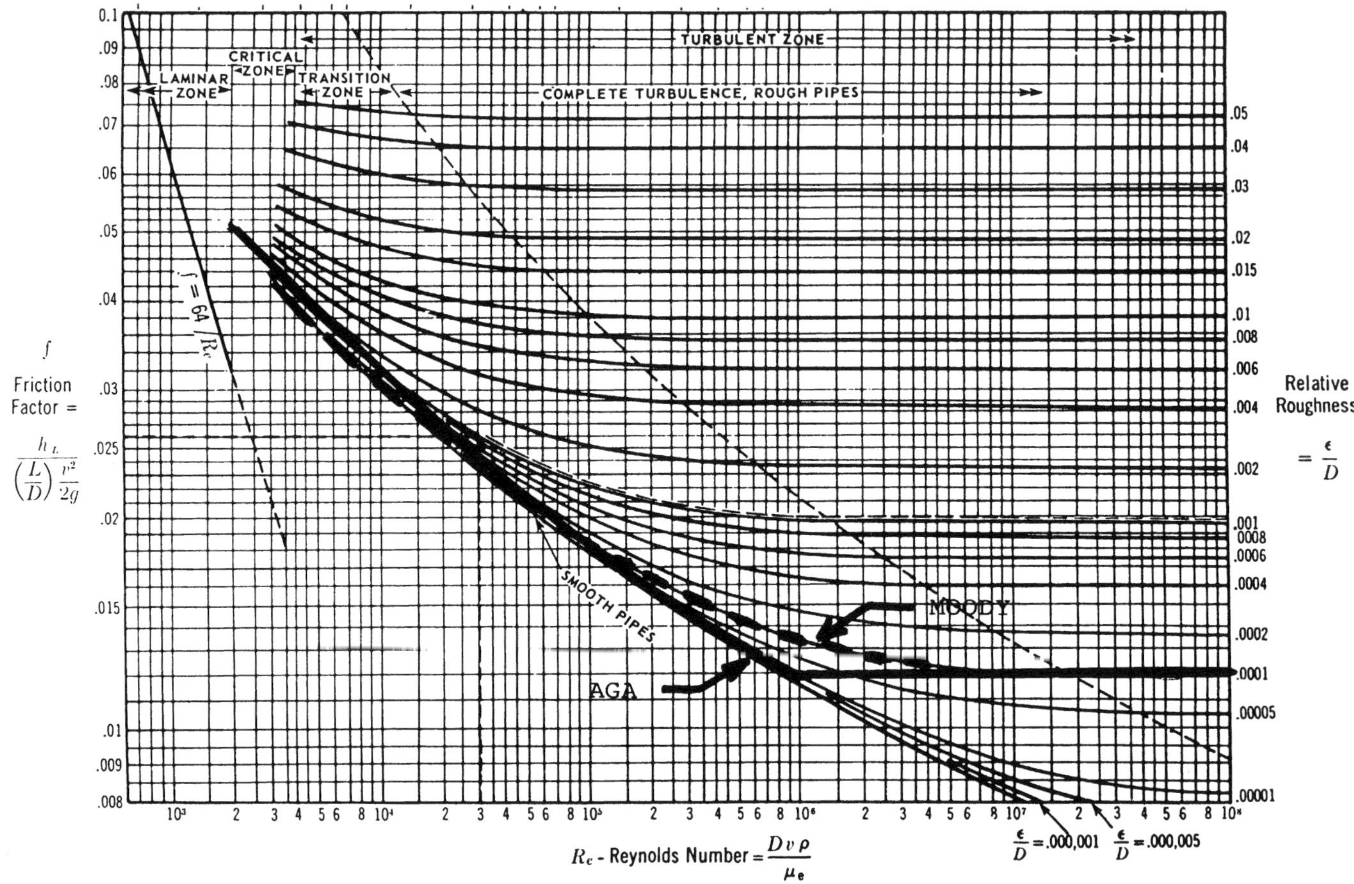

Figure 13–4. Comparison of AGA and Moody friction factors for a relative roughness of 0.0001.

where ΔP_{100} = pressure drop in psi per 100 ft pipe
$C_1 = 0.013\ m^2$
m = gas flow rate, lb/sec
$C_2 = 336000\ f / d^5$
f = Darcy-Weisbach friction factor
d = pipe inside diameter, in.
ρ = gas density, lb/ft^3

For sizing purposes, select the friction factor for well-developed turbulence—that is, at very high Reynolds number. The usual case will involve selecting ΔP_{100} from the table above, obtaining m and ρ from the given flow conditions, and solving for C_2. A little trial and error then will be required to select a pipe size which has C_2 equal to or less than the previously calculated value.

Example 13–3. Size a manifold for the 100–psia plant inlet gas of Example 13.2. Assume barometer is 14.0 psia.
Solution: Gage pressure = 100 − 14 = 86 psi
Set $\Delta P_{100} = 0.18$, from tabulated recommendations

$\rho = (100)(20.27)/(0.98 \times 10.73 \times 540)$
$= 0.357\ lb/ft^3$
$C_1 = 0.013\ m^2 = 0.013(1.422^2) = 0.0263$
$C_2 = \rho\ \Delta P_{100}/C_1 = 0.357(0.18)/0.0263$
$= 2.44$

d	e/d	f	336000 f/d⁵
4.026	0.00025	0.0142	4.51
6.065	0.00016	0.0129	0.53

Choose a 6-in. pipe.   ←

Two-Phase Flow

Simultaneous flow of gas and liquid in a pipe is very complex. This topic is addressed briefly here, looking only at horizontal flow and considering only the Beggs and Brill (1987) correlation. A qualitative discussion of the nature of two-phase flow is given first, including consideration of the various flow regimes. Then the determination of liquid "holdup" is described, followed by calculation of the friction loss.

Nature of Two-Phase Flow. In gas-liquid flow, inertial effects cause the liquid to tend to lag behind the gas phase. In turn, the gas phase dissipates energy by trying to impart momentum to the liquid, i.e., drag it along. The result is that the pressure drop is increased rather dramatically over single-phase flow and that the liquid tends to stack up in the pipe. The latter phenomenon is known as "holdup."

Actual flow behavior depends on the absolute and relative flow rates of the gas and liquid. Figure 13–5 shows one concept of the resulting types of flow. The seven patterns are divided into three general categories for correlation of liquid holdup. These are:

1. *Segregated,* in which the two phases travel essentially separately.
2. *Intermittent,* characterized by plugs or slugs of vapor moving along in the otherwise continuous liquid phase, or vice versa.
3. *Distributed,* involving the flow of a distribution of particles of one phase in the other, continuous phase.

The complex nature of these flows indicates the difficulty of modeling the flow by a mechanistic model. Empirical correlations are used almost exclusively.

Prediction of Holdup. The Beggs-Brill holdup correlation is based on determining first the flow regime, as shown in the flow regime map, Figure 13–6. The full lines indicate the regions of flow as determined experimentally. For numerical simulation purposes, the zones are approximated by the dashed lines in Figure 13–6. Notice that small region IV represents an intermediate region, which is taken to be segregated and intermittent flow.

The abscissa variable is λ_L, the *volumetric flowing fraction* of the liquid. The ordinate variable is the *Froude number*, Fr, for the liquid.

$$Fr = V_m^2 / (g\,D) \qquad (13\text{–}31)$$

where V_m = mixture superficial velocity
$\qquad$ g = acceleration of gravity
$\qquad$ D = inside diameter of pipe

The following relations can be shown to be correct for V_m and λ_L.

$$V_{sL} = m_L / (\rho_L\,A) \qquad (13\text{–}32)$$

$$V_{sG} = m_G / (\rho_G\,A) \qquad (13\text{–}33)$$

$$V_m = V_{sL} + V_{sG} \qquad (13\text{–}34)$$

$$\lambda_L = V_{sL} / (V_{sL} + V_{sG}) \qquad (13\text{–}35)$$

SEGREGATED

INTERMITTENT

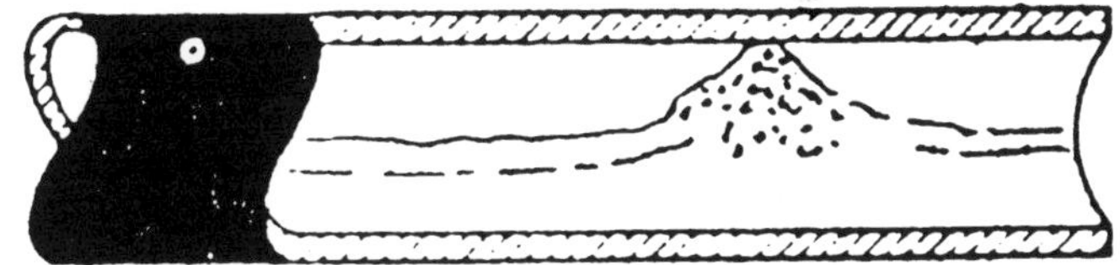

DISTRIBUTED

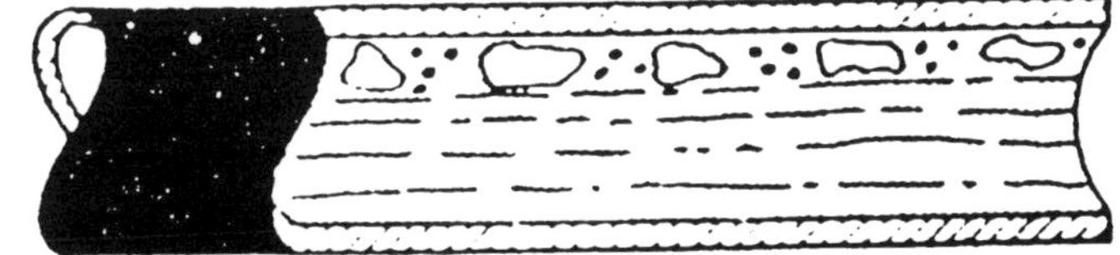

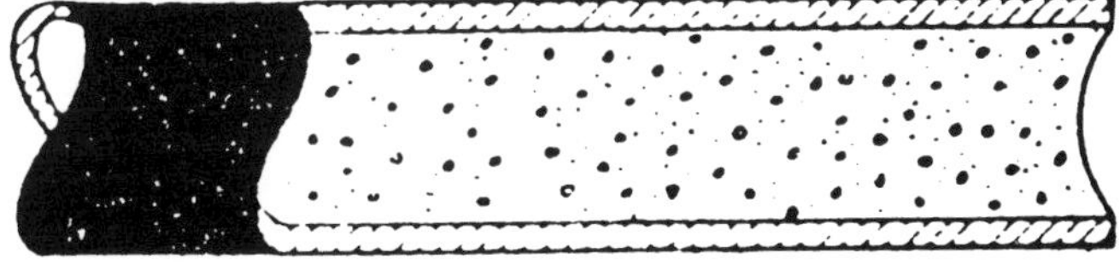

Figure 13–5. Horizontal flow patterns (Brill and Beggs, 1986).

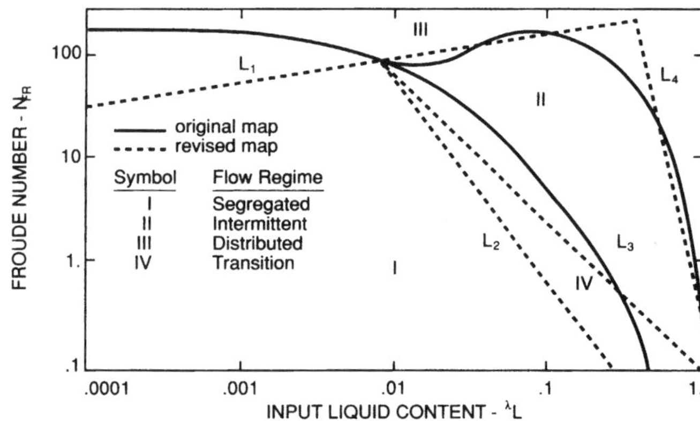

Figure 13–6. Horizontal flow pattern map (Brill and Beggs, 1986).

where λ_L = liquid volumetric flowing fraction
 V_s = superficial velocity
 m = mass flow rate
 ρ = density
 A = cross-sectional area of pipe
  = $\pi D^2 / 4$
 sub L refers to liquid
 sub G refers to gas

The analytical method for determining the *flow regime* is based on a vertical line in Figure 13–6 at the λ_L in question. The intersection of the vertical line with line L_1, L_2, L_3, or L_4 locates the corresponding Froude number, which can be compared to the actual Froude number to obtain the flow regime. The *formal rules* for the procedure and the equations for the L's are given below.

$$L_1 = 316 \, \lambda_L^{0.302} \qquad (13\text{–}36)$$

$$L_2 = 0.0009252 \, \lambda_L^{-2.4684} \qquad (13\text{–}37)$$

$$L_3 = 0.10 \, \lambda_L^{-1.4516} \qquad (13\text{–}38)$$

$$L_4 = 0.5 \, \lambda_L^{-6.738} \qquad (13\text{–}39)$$

Segregated Flow (see Fig. 13–6)

$$\lambda_L < 0.01 \text{ and } Fr < L_1$$
or $$\lambda_L \geq 0.01 \text{ and } Fr < L_2$$

Transition Flow

$$\lambda_L \geq 0.01 \text{ and } L_2 \leq Fr \leq L_3$$

Intermittent Flow

$$0.01 \leq \lambda_L < 0.4 \text{ and } L_3 < Fr \leq L_1$$
or $$\lambda_L \geq 0.4 \text{ and } L_3 < Fr \leq L_4$$

Distributed Flow

$$\lambda_L < 0.4 \text{ and } Fr \geq L_1$$
or $$\lambda_L \geq 0.4 \text{ and } Fr > L_4$$

The *holdup*, H_L, is calculated by

$$H_L = a \, \lambda_L^b / Fr^c \qquad (13\text{–}40)$$

with

Flow Pattern	a	b	c
Segregated	0.98	0.4846	0.0868
Intermittent	0.845	0.5351	0.0173
Distributed	1.065	0.5824	0.0609

provided

$$H_L \geq \lambda_L$$

In the *transition region,* calculate H_L as follows:

$$H_{L,\text{trans}} = \delta \, H_{L,\text{segr}} + \gamma \, H_{L,\text{inter}} \qquad (13\text{–}41)$$

where $\qquad \delta = (L_3 - Fr) / (L_3 - L_2)$

and $\qquad \gamma = 1 - \delta$

Frictional Pressure Drop. First calculate the *no-slip density and viscosity.*

$$\rho_n = \lambda_L \, \rho_L + (1 - \lambda_L) \, \rho_G \qquad (13\text{–}42)$$

$$\mu_n = \lambda_L \, \mu_L + (1 - \lambda_L) \, \mu_G) \qquad (13\text{–}43)$$

Next calculate the *no-slip Reynolds number.*

$$Re_n = \rho_n \, V_m \, D / \mu_n \qquad (13\text{–}44)$$

Now obtain the *no-slip friction factor*, either from the smooth pipe curve in Figure 13–2 or by the following equation.

$$f_n = 1 / \{2 \log_{10} [Re_n / (4.5223 \log_{10} Re_n - 3.8215)]\}^2 \qquad (13\text{–}45)$$

The ratio of the two-phase and no-slip friction factors is given by the following relation.

$$f_{tp} / f_n = \exp(s) \qquad (13\text{–}46)$$

where $s = \ln(y) / \{-0.0523 + 3.182 \ln(y)$

$$-0.8725 \, [\ln(y)]^2 \qquad (13\text{–}47)$$
$$+ 0.01853 \, [\ln(y)]^4\}$$

and $\quad y = \lambda_L / H_L^2 \qquad (13\text{–}48)$

For $1 < y < 1.2$, calculate s by

$$s = \ln(2.2y - 1.2) \qquad (13\text{–}49)$$

Neglecting elevation and acceleration effects, the two-phase friction factor is now calculated.

$$f_{tp} = f_n \, (f_{tp}/f_n) \qquad (13\text{–}50)$$

The rate of pressure drop with distance along the pipe is:

$$(dP/dX)_f = f_{tp}\ \rho_n\ V_m^2\ /\ (2\ g_c D) \qquad (13\text{--}51)$$

where X = distance along the pipeline

Solution Procedure. A *trial-and-error procedure* is required to use the relations just given. The first step is to divide the pipe under consideration into segments. The calculations for all segments follow the same pattern. First, guess the average conditions along the segment, evaluate the derivative of Equation 13–50, and then estimate the outlet conditions for the segment. Repeat the calculation as necessary. Then go on to the next segment.

Example 13–4. It is discovered that the gas of Example 13–2 actually contains 20 bbl/MMscf of condensate. The condensate has the properties of n-butane. Estimate the pressure drop for the two-phase flow.

Solution: 20 bbl/MMscf $\times$ 2.3MMscf/day $\times$ 42 gal/bbl $\times$ 1 ft^3/7.4805 gal $\times$ 1 day/86400 s = 0.00299 ft^3/s

From Example 13–2, m_G = 1.422 lb/s

Assume P_m = 167 psia

From Figure 3–19, z_m = 0.96

$$\rho_G = (167)\ (20.27)\ /\ (0.96 \times 10.73 \times 540)$$
$$= 0.611\ \text{lb/ft}^3$$
$$A = \pi (4.026/12)^2/4 = 0.0884\ \text{ft}^2$$

GPSA Figure 23–13 gives

$$\rho_L = 0.571\ (62.4)$$
$$= 35.6\ \text{lb/ft}^3$$
$$m_L = 0.00299\ \text{ft}^3/\text{s} \times 35.6\ \text{lb/ft}^3$$
$$= 0.1064\ \text{lb/s}$$
$$V_{sL} = 0.1064/(35.6 \times 0.0884)$$
$$= 0.0338\ \text{ft/s}$$
$$V_{sG} = 1.422/(0.611 \times 0.0884)$$
$$= 26.34\ \text{ft/s}$$
$$V_m = 26.34 + 0.03 = 26.37\ \text{ft/s}$$
$$\lambda_L = 0.0338/26.37 = 0.00128$$
$$Fr = 26.37^2/(32.174 \times 0.3355) = 64.4$$
$$L_1 = 316\ (0.00128)^{0.302} = 42.3$$

$\lambda_L < 0.04$ and $Fr > L_1$, so flow is zone III, distributed (see Fig. 13–6)

$$H_L = 1.065(0.00130)^{0.5824}\ /(62.6^{\ 0.0609})$$
$$= 0.0171$$
$$\rho_n = (0.00128)(35.6)$$
$$+ (1 - 0.00128)(0.611) = 0.656$$

GPSA Figure 23–35 gives

$$\mu_L = (0.15)(0.000672)$$
$$= 1.01 \times 10^{-4}\ \text{lb/ft-s}$$
$$\mu_n = (0.00128)(1.01 \times 10^{-4})$$
$$+ (1 - 0.00128)(7.39 \times 10^{-6})$$
$$= 7.51 \times 10^{-6}\ \text{lb/ft-s}$$

$$Re_n = (0.3355)(26.37)(0.656)/$$
$$(7.51 \times 10^{-6})$$
$$= 7.72 \times 10^5$$
$$f_n = 0.0122\ (\text{Fig. 13–2, smooth pipe})$$
$$y = 0.00128/(0.0171^2) = 4.39;\ \ln\ (y)$$
$$= 1.48$$

Equation 13–47 gives s = 0.522

$$f_{tp}/f_n = \exp(0.522) = 1.685$$
$$f_{tp} = (1.685)(0.0122) = 0.0205$$

Equation 13–51 gives

$$(dP/dX)_f = \frac{(0.0205)(0.656)(26.37^2)}{(2 \times 32.174 \times 0.3355)\ (144)}$$

$$= 0.00301\frac{\text{psi}}{\text{ft}}$$

Assume this applicable across the whole distance (not a good assumption)

$$L = 8.5\ \text{mile} \times 5280\ \text{ft/mile} = 44880\ \text{ft}$$
$$\Delta P = (44880)\ (0.00301) = 135\ \text{psi}$$
$$P_1 = 100 + 135 = 235\ \text{psia}$$
$$P_m \cong 167.5,\ \text{so initial assumption OK}$$

This result compares with the 104 psi drop for the dry gas. A better solution will result if we divide the line into several segments and calculate each one separately, repeating the calculation across each segment until the average conditions check.

The calculation indicates the importance of a small amount of condensate, which if omitted can lead to a line size much smaller than that required for a reasonable pressure drop.

Erosional Velocity

Gas/liquid two-phase lines and manifolds are often sized to avoid *erosion-corrosion*. The recommendation of API RP 14E (1984) is that the velocity be limited to:

$$V_e = k\ /\ \sqrt{\rho_m} \qquad (13\text{--}52)$$

where V_e = limiting erosional velocity, ft/sec

$\quad k$ = empirical constant

$\quad\quad$ = 125 for intermittent service

$\quad\quad$ = 100 for continuous service

$\quad \rho_m$ = gas/liquid mixture density at flowing conditions, lb/ft^3

Example 13–5. The wet gas of Example 13–4 enters a manifold at 200 psia, 80°F. Estimate the manifold diameter based on the erosional velocity.

Solution:

$$V_e = 100/\sqrt{\rho_m} = 100/\sqrt{0.656}$$
$$= 123 \text{ ft/s}$$
$$Q = V_m A = 26.37 \times 0.0884$$
$$= 2.33 \text{ ft}^3/\text{s}$$
$$A_e = Q / V_e = 2.33 / 123$$
$$= 0.0190 \text{ ft}^2$$
$$D = \sqrt{4 A_e / \pi} = 0.155 \text{ ft} = 1.86 \text{ in.}$$

Use at least a 2-in. pipe.   ←

Temperature Profiles

Temperature is assumed constant in the above discussion. The *pipeline pressure profile* may be significantly influenced by the *temperature profile* and vice versa. In extreme cases it may be desirable to solve the pressure and thermal energy equations simultaneously, although this is not usually required for gases. In dense-phase (e.g., CO_2) pipelines, the fluid densities may be large enough to cause significant compression heating when the pipeline elevation decreases (Hein, 1986).

The *temperature change equation* is now presented for gas flow in pipelines. Heat transfer by conduction to the soil or by convection to sea water, air, etc., is obtained by substituting the appropriate heat-transfer rate equation into the energy balance equation. Combination of these two relations leads to the equation presented below. Note the heat-transfer coefficient or thermal conductivity is assumed constant.

The basic formula is (Hein, 1984b)

$$\frac{T_2 - (T_s + \theta)}{T_1 - (T_s + \theta)} = \exp(-\beta L) \qquad (13\text{--}53)$$

where T_2 = temperature at distance L along line, °F
T_1 = temperature at line inlet, °F
β = distance gradient factor, mile^{-1}
L = distance along line, mile
θ = pressure-effect factor, °F
T_s = temperature of undisturbed soil at pipe centerline depth (for conduction to soil) or temperature of air, sea water, or other surrounding material (for convection), °F.

Factor β depends on heat transfer *and* flow conditions.

$$\beta = 5280 \text{ UA} / (3600 \text{ m } C_p)$$
$$= 1.467 \text{ UA} / \text{m } C_p \qquad (13\text{--}54)$$

where U = overall heat-transfer coefficient, Btu/hr-ft^2-°F
A = heat-transfer area, ft^2 *per ft* of pipe
m = fluid flow rate, lb/s
C_p = fluid heat capacity, Btu/lb-°F.

The product UA is given by

$$UA = (\pi D_p) / (R_f + R_d + R_p + R_i + R_s) \qquad (13\text{--}55)$$

where D_p = outside diameter of pipe, ft.

Table 13–7 gives the various *heat-transfer resistance terms* for pipes with and without insulation. Convective heat-

Table 13–7   Heat-Transfer Resistance Terms

Resistance	Mechanism	Formula
Fluid, R_f	Convection	$D_p / (h_f D_{pi})$
Inside dirt or scale, R_d	Conduction	$D_p r_d / D_{pi}$
Pipe wall, R_p	Conduction	$D_p \ln(D_p/D_{pi}) / (2 k_p)$
Insulation, R_i	Conduction	$D_p \ln(D_i/D_{pi}) / (2 k_i)$
Surroundings, R_s:		
Soil	Conduction	$D_p \ln\left(\dfrac{B + \sqrt{B^2 - (D_i/2)^2}}{(D_i/2)}\right) \Big/ 2 k_s$
or, Air	Convection	$D_p / (D_i h_s)^*$

* If there is no pipe insulation, replace D_i by D_p in these formulae.

Nomenclature: D_{pi} Inside diameter of pipe, ft or m
               D_p Outside diameter of pipe, ft or m
               D_i Outside diameter of insulation, ft or m
               B Depth of burial to centerline of pipe, ft or m
               h_f Inside fluid convective heat-transfer coefficient, Btu/hr-ft^2-°F or W/m^2-°C
               r_d Inside fluid fouling factor, hr-ft^2-°F/Btu or m^2-°C/W
               h_s Convective heat-transfer coefficient for sea water or air, Btu/hr-ft^2-°F or W/m^2-°C
               k_p Thermal conductivity of pipewall, Btu/hr-ft-°F or W/m-°C
               k_i Thermal conductivity of insulation, Btu/hr-ft-°F or W/m-°C
               k_s Thermal conductivity of soil, Btu/hr-ft-°F or W/m-°C

Table 13–8 Soil Thermal Conductivities

Substance	Thermal Conductivity Btu/hr-ft-°F*	Ref.
Calcareous earth, 43% water	0.41	1
Quartz sand, medium fine, dry	0.15	1
Quartz sand, 8.3% water	0.34	1
Sandy clay, 15% water	0.53	1
Soil, very dry	0.1–0.2	1
Soil, wet	0.8–2.0	1
Mud, wet	0.5	1
Misc. sand, 4% moisture	0.71–1.17	2
Chena river gravel, 4% moisture	0.75–1.08	2
Misc. sandy loams, 4% moisture	0.37–0.79	2
Misc. silty loams, 10% moisture	0.33–0.75	2
Sand (Libyan desert)	0.3	3
Rock	1.25	3

* To convert to W/m-°C, multiply by 1.7307
1. Ingersoll, Zobel, and Ingersoll, "Heat Conduction," University of Wisconsin Press, 1954.
2. McAdams, "Heat Transmission," McGraw-Hill, 3rd Ed., 1954.
3. Ford *et al*, Oil & Gas J., April 26, 1965, p. 107–109.

transfer coefficients can be estimated by standard techniques given in Kreith (1973), for example.

It is difficult to estimate soil thermal conductivities and sea water convection coefficients from first principles, partly because the heat-transfer coefficients are assumed constant, which is an oversimplification. Table 13–8 lists a few soil thermal conductivities; use with caution. Sea-water convection coefficients are low, say 0.5–3.0 Btu/hr-ft^2 -°F. Experience with similar situations is the best guide.

Natural gas heat capacities for the β factor can be read from Buthod's correlation (App. 6, Fig. A6–4).

Estimation of θ is somewhat more difficult. The equation is

$$\theta = [\eta (P_2 - P_1) / \beta L] + [(z_1 - z_2) / (778 C_p \beta L)] \qquad (13\text{–}56)$$

where η = the Joule-Thomson coefficient of the gas, °F/psi

P_2 = outlet pressure, psia

P_1 = inlet pressure, psia

z_1 = inlet elevation, ft

z_2 = outlet elevation, ft

The Joule-Thomson coefficient is defined as

$$\eta = (\partial T / \partial P)_H \qquad (13\text{–}57)$$

where H = enthalpy

Values of the Joule-Thomson coefficient can be estimated from the enthalpy-temperature charts in the GPSA Engineering Data Book (1987). Example 13–6 illustrates the procedure.

$$\eta = [(T_1 - T_2)/(P_1 - P_2)]_H \qquad (13\text{–}58)$$

The variables in the pressure-drop equation, Equation 13–12, and the temperature-drop equation, Equation 13–53, are coupled. In general, a trial-and-error calculation is required, with the flow line broken into segments for best accuracy. The problem is solved most easily on a digital computer or programmable calculator. See Hein (1984b) for more details and a calculator program.

Example 13–6. The gas of Example 13–1 actually leaves the platform at a temperature of 120°F. How far will the gas travel before the temperature falls to 45°F, the temperature of the sea water? Assume an overall heat-transfer coefficient, U, of 1.0 Btu/hr-ft^2-°F.

Solution:

$$d = 16 - 2(0.438) = 15.124 \text{ in.}$$
$$D = 15.124/12 = 1.260 \text{ ft}$$
$$A = \pi D = 3.14159 \times 1.26$$
$$= 3.96 \text{ ft}^2/\text{ft}$$

The heat capacity, from Appendix 6, Figure A6–4a, is $C_p = 0.65$ at 1000 psia, 82.5°F (average T)

$$m = 48.6 \text{ lb/s (Ex. 1)}$$
$$\beta = 1.467 \ UA/mC_p$$
$$= (1.467)(1)(3.96)/(48.6 \times 0.65)$$
$$= 0.184$$

Evaluate θ for the line length. Use GPSA (1987) Figure 24–12 and 24–13 to evaluate η. At 1000 psia and 120°F, H = 265 Btu/lb. At 800 psia, H = 265 Btu/lb, T = 115°F.

$$\eta \cong (115 - 120) / (800 - 1000)$$
$$= 0.025°\text{F/psi}$$

Neglect elevation change term in Equation 13–56.

$$\theta = 0.025 \, (800 - 1000) / (0.184 \times 65)$$
$$= -0.4°\text{F}$$
$$T_s + \theta = 45.0 - 0.4 = 44.6°\text{F}$$

Equation 13–54 is then,

$$\frac{T_2 - 44.6}{120 - 44.6} = \exp \, (-0.184 \, L)$$
$$T_2 = 44.6 + 75.4 \, \exp \, (-0.184 \, L)$$

Tabulate a few points.

L, mi	exp (−0.184 L)	T_2	
0	1.000	120.0	
5	0.3985	74.6	
10	0.1588	56.6	
20	0.0252	46.5	
25	0.0100	45.4	
28.5	0.0053	45.0	←
65	0.0000	44.6	

The gas falls to sea-water temperature at 28.5 miles. The interesting result is that the *gas actually falls below sea-water temperature due to the Joule-Thomson effect* of the pressure drop.

In practice, two-phase pipeline flows are designed using *computer simulations* which should include choice of appropriate pressure-drop correlation, PVT equation of state, mass transfer between phases (as in retrograde condensation), and thermal energy balances.

PIPELINE INSTALLATION

Such an apparently simple system as a pipeline presents many important construction considerations, including pipe laying, corrosion prevention, anchoring, crack arresting, hydrostatic testing, and line cleaning. These topics are detailed below.

Pipe Laying

Installation of steel line pipe in a ditch below ground is a complex process involving the following items (Smith, 1981a):

- Clearing and grading of right-of-way
- Pipe hauling and stringing
- Ditching or trenching
- Pipe bending
- Boring and tieing in
- Welding
- X-ray testing
- Coating and wrapping
- Lowering in
- Backfilling
- Hydrostatic testing
- Final cleanup
- Crossings
    River or stream
    Highway
    Railway
- Valves and fittings
- Cathodic protection

In addition to the very important factors of elevation and grade, other *terrain factors* affecting construction include presence of rock, swampy or marshy location, and highly-congested construction areas.

Offshore pipeline laying is done by three methods (Smith, 1981a): lay barge, reel barge, and pulling. The *lay barge method* is the most common method and constitutes in essence a floating counterpart of the usual onshore method (Figs. 13–7 and 13–8). Because of the stresses on the pipe during laying, the designer may have to select the pipe grade and thickness to withstand installation loading (Smith, 1981a) rather than flowing pressure. Also, stresses induced in the unusual, ocean-bottom environment must be considered. Lay barges have been used to lay pipe up to 42-in. diameter in depths to 1000 ft.

Reel barges are similar to lay barges except that reels 40 to 60 ft in diameter are used. The pipe is pre-welded and wound onto the reel on shore. The pipe is laid from the reel, thus allowing lower-cost, rapid installation for smaller line sizes, say 2- to 12-in. There is apparently no detrimental effect due to the bending and straightening.

Pulling consists of welding the pipe onshore, then pulling it into the water to the desired location using a pull barge.

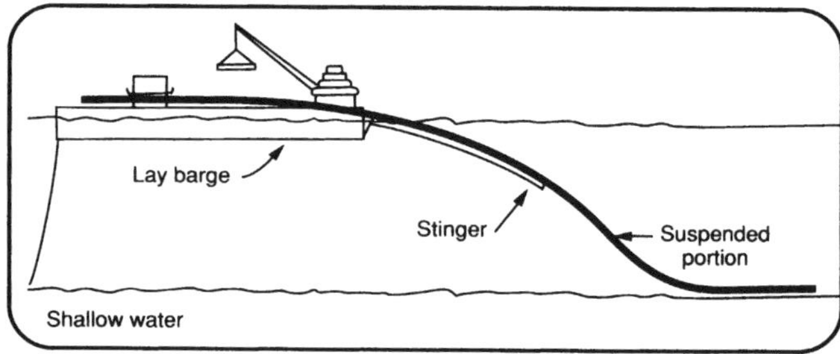

Figure 13–7. Lay barge pipeline construction method (Smith, 1981a).

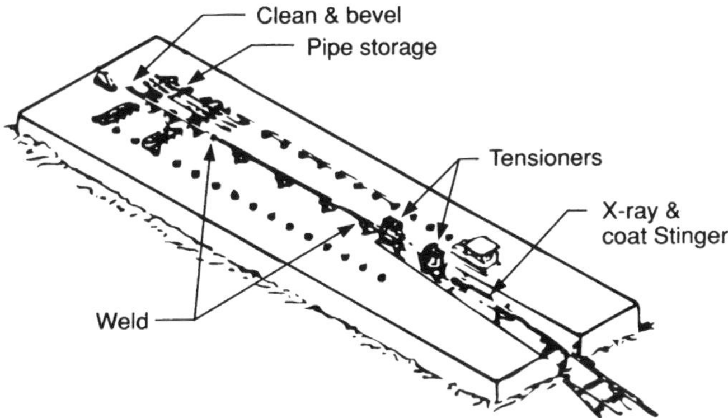

Figure 13–8. Lay barge work stations (Smith, 1981a).

This technique is practical only for short lines, say 10–15 miles or less, with pipe sizes from 16 to 24 inches.

Buoyancy must be overcome for the pipeline to be installed and remain in place. When installed, the line is filled with air; after installation, it is filled with gas which has relatively low density. The weight of the steel is not enough to overcome the buoyant force of fresh or sea water. Additional weight is provided by external coating. For example, the Ekofisk-Teeside oil line was coated with a minimum of 1.86-in.-thick concrete of 190 lb/ft³ density (Massey, 1975). The net specific gravity was 1.15 over most of the route and was increased to 1.6 (4.79 in.) near shore where the line passed through the surf zone and beach area. Water may be used as ballast before operation of an undersea line, but gas lines will likely require external coating for weighting as well as corrosion resistance.

Smith (1981b) has detailed techniques for *platform riser installation*, which is generally a separate operation from the pipeline installation. The conventional method consists of lifting the riser by pickup cables attached to winches on a barge. This method is only good to about 300 ft of depth and in relatively calm seas.

Trenching and *backfilling* may be required by law or simply to protect the line. The peculiar problems of undersea trenching and backfilling are discussed by Seymour (1981b). Brown (1981) discusses underwater plows used for this purpose.

Corrosion Prevention

Gathering and transmission line pipe must be protected against corrosion (Uhlig, 1971). Pipeline corrosion is an electrochemical process and can be inhibited by several means:

- Pipe material selection
- Injection of corrosion inhibitors
- Cathodic protection
- External and/or internal protective coatings

Material Selection. Carbon steel pipe is adequate for long, trouble-free service with sweet, dry natural gas. However, as pointed out in API RP 14E (1984),

Production process streams containing water, brine, carbon dioxide (CO_2), hydrogen sulfide (H_2S), or oxygen (O_2), or combination of these may be corrosive to metals used in system components. The type of attack (uniform metal loss, pitting, corrosion/erosion, etc.) as well as the specific corrosion rate may vary in the same system, and may vary with time. The corrosivity of a process stream is a complex function of many variables including (1) hydrocarbon, water, salt, and corrosive gas content; (2) hydrocarbon wettability; (3) flow velocity, flow regime, and piping configuration; (4) temperature, pressure, pH; and (5) solids content (sand, mud, bacterial slime and microorganisms, corrosion products, and scale).

Two distinct types of stress corrosion cracking occur, that due to *water and hydrogen sulfide* and that due to *chlorides.* Different materials are required for these two services. See the latest editions of ANSI B31.3 and B31.8, API RP 14E, and NACE MR–01–75 for specific recommendations concerning materials selection.

American National Standards Institute:

ANSI B31.3, Petroleum Refinery Piping
ANSI B31.8, Gas Transmission and Distribution Piping Systems

Americal Petroleum Institute:

API RP 14E, Design and Installation of Offshore Production Platform Piping Systems

National Association of Corrosion Engineers:

NACE Std MR–01–75, Sulfide Stress Cracking Resistant Metallic Materials for Oil Field Equipment.

Corrosion Inhibitors. Chemicals that passivate steel surfaces and make them resistant to corrosion may be injected into gas lines, especially gathering lines. These compounds are generally mixtures of organic amines. Exact formulations are not always known, as the vendors are usually secretive about analyses and often refer simply to their products by trade names.

Important parameters in corrosion inhibition are filming, injection rate, and monitoring (Gatlin and EnDean, 1975).

Gatlin and EnDean distinguish two types of injection: batch and continuous. *Periodic filming* may be quite adequate if there is no liquid water and/or hydrocarbon condensate. Inhibitor is introduced to the line in a batch between two squeegee-type pigs, which are swept through the line. This arrangement assures good contacting of the inhibitor and the pipe wall and establishes a good film. The pigs also will clean the pipe walls and clear out debris that might otherwise prevent good filming. Enough inhibitor should be introduced to produce a wall coating 1–4 mil thick.

For lines containing liquids, *steady injection* is required to maintain the film on the pipe wall. Periodic pigging will be desirable to keep the wall clean and free of corrosion products. Gatlin and EnDean recommend beginning with an injection ratio of 1 to 2 pints inhibitor per MMscf of gas, then reducing the amount stepwise to the minimum that provides satisfactory protection.

Monitoring provides evidence of the treatment efficacy. Flush-mounted, electronic, corrosion-measuring devices are used, as well as hydrogen probes. Chemical testing of water samples for iron and manganese can reveal intensity of corrosion. Chloride content indicates carryover of produced brine.

Cathodic Protection. Electrochemical protection is provided by reversing the normal current flow of corrosion, thereby safeguarding the pipe. Two means of cathodic protection are used: sacrificial anodes and applied dc current.

Sacrificial anodes (usually magnesium or zinc) are attached electrically to the line and the soil and, because of their higher activity in the galvanic series, reverse the current flow and are preferentially corroded. One anode can protect only a limited length of pipe and must be replaced when eaten away.

Applied dc current is provided by rectifiers attached to the pipeline (cathode) and to a steel or carbon anode that is in electrical contact with the soil. The applied current is low. With sufficient voltage, one rectifier can protect a long pipeline.

Protective Coatings. *External coatings* have been used to inhibit corrosion in pipelines for many years. The oldest and still used type, consists of hot-applied bituminous material (from petroleum or coal tar) wrapped with an appropriate covering (Polignano, 1982). Mineral filler may be mixed with the bitumin to impart strength. The coating is applied either in the shop or onsite. Generally, two layers of coating and wrapping are applied. Glass fiber fabric is a popular wrapping material. In addition to bituminous coats, epoxy powder and polyethylene have been used. Det norske Veritas has studied coatings extensively (Askheim and Eliassen, 1983).

Kornfeld (1967) mentions use of butyl-rubber and PVC laminated tape for a desert line. Rhodes (1982) reports the thermal instability of both polyethylene and PVC tape, and recommends using fusion-bonded epoxy coatings.

A second approach to corrosion control is *internal coating*. Rhodes (1982) reports the use of many such coatings, including

Polyurethane
Phenolic resin
Phenolic epoxy
Fusion-bonded epoxy powder

Kipin (1980) reports the use of polyamide-cured epoxies and amine adducts. These materials also have the benefit of *reducing the pipe roughness* and *flow pressure drop*.

External coatings may, in addition to providing corrosion protection, provide negative buoyancy for undersea pipe as previously mentioned.

Anchoring

Flowlines are subject to longitudinal stress due to changes in internal pressure and temperature. *Longitudinal stress causes snaking* of above-ground lines (Brock, 1967) and is also important for buried lines (ASA, 1955):

> For long lines, the friction of the earth will prevent changes in length from these stresses, except for several hundred feet adjacent to bends or ends. At these locations the movement, if unrestrained, may be of considerable magnitude. If connections are made at such a location to a relatively unyielding line, or other fixed object, it is essential that the interconnection should have ample flexibility. . . in, or, that the line shall be provided with an anchor . . .

Webb (1983) presents information on the design and use of mechanical and density *pipeline anchors*.

Massey (1980) reports ''pipe creep'' in undersea lines that actually pushed riser pipes into Ekofisk platforms. Movement was from a few inches to 4 or 5 feet. Several corrective methods have been employed. In one technique a short section of pipeline was cut out, and the riser cold sprung away from the platform and rewelded to the line. For new lines, Massey recommends use of Z-shaped expansion bends near the platform or an anchor system some distance from the platform (1600 ft).

Crack Arresting

Tise (1983) points out that high-tensile-strength line pipe may crack and, if so, the velocity of propagation of the crack can exceed the velocity of propagation of the high pressure, resulting in the loss of long segments of the pipeline. *Crack arrestors* are rings installed around the line to slow the crack propagation. Arresting rings are installed at intervals of 2500 ft.

Hydrostatic Testing and Line Cleaning

McClure *et al.* (1966) presented details on hydrostatic testing of pipelines. As they state:

> The purpose of testing after construction is to test and remove from the line any defects that escape inspection procedures in the pipe mill and any that are inadvertently produced during shipping, handling and construction. Minimum proof test pressures are established by such codes as ANSI B31.8 and ANSI B31.4 for gas, oil, and products pipelines.

Maximum proof pressures have not been established officially; each operator must establish his own.

In hydrostatic testing the line is filled with fresh or sea water. Dye, corrosion inhibitors, and biocides are often added. Care must be taken to eliminate gas trapped in high spots. All valves and other openings are carefully sealed. Pressure is applied with a hydraulic jack and gauges are used to observe the pressure-time behavior. Initial testing will involve leak detection and removal. Then the pressure may be raised to a suitable testing level and the system observed for a period of time. For example, Shell/Esso's Western Leg gas pipeline (off Scotland) was tested at 1.25 times maximum design pressure for 24 hours followed by a tightness test for 6 days at 1.1 times maximum design pressure (Pullin and Daniels, 1981).

Associated with hydrostatic testing is the task of *cleaning the pipeline* before and/or after the introduction of water. Ridding the pipeline of water is an obvious requirement before gas is flowed in the line. Various methods have been used, but basically they involve pushing the water out ahead of pigs followed by a cleanup procedure.

Hydrate formation problems generally occur during or immediately after commissioning new natural gas lines that have been hydrostatically tested. A 0.1 to 0.15 mm (4–6 mils) thick film of residual liquid water remains on the line. After testing, the line can be dried by flushing with inert gas (e.g., N_2), by establishing a vacuum, by flowing natural gas, by flowing very dry air, or by flowing methanol. The last two methods have proved superior (Kopp, 1981).

Methanol drying involves flowing a slug of methanol (held between two or more pigs) through the line. A general rule of thumb assumes that an 85 wt % methanol-water mix at the end of the line provides satisfactory drying (Kopp, 1981).

Diab (1983a) develops detailed equations to calculate the required volume of the methanol slug held between two pigs or spheres. Diab (1983b) also shows how the required volume can be reduced by using three pigs to divide the methanol slug into two equal portions.

Again, for *Shell/Esso's Western Leg line*, the procedure was as follows:

1. Water expelled by a 420-bbl slug of cleaning gel between pigs (compressed air as propellant)
2. Free water cleanup with series of air-propelled swabbing pigs
3. Vacuum drawn on line to remove residual water

In the *Shell FLAGS gas line* from the Brent field, the line contained not only water but considerable quantities of debris such as mill scale, rust, silt, welding rods, and welding slag (Scott and Zijlstra, 1981). The estimated quantity of debris in the 281-mile line was 733000 lb (367 ton). Multiple pigging was considered—as well as a high-velocity water flush—for removing the debris. The technique finally used was to pass a 22000-ft long cleaning train of Kelzan XC polymer in fresh water ahead of and between pigs. The polymer gel suspended the debris effectively. The line was not only cleaned but the treatment smoothed the pipe wall and also improved the flow.

Of course any natural gas pipeline must be *safely purged* of air before being put into service. One technique is injection of a slug of inert gas such as N_2. Perkins and Euchner (1988) present a step-by-step procedure (based on the appropriate flammability limits) for sizing the inert gas slug.

PIPELINE OPERATION

After a pipeline is installed, efficient operation is the next order of business. Automatic controls must be provided to produce safe operation in the face of unexpected upsets. Use of SCADA (supervisory control and data acquisition) systems is standard industrial practice today. Pigging and leak detection are additional important activities.

Pipeline Pigging

Pigging refers to the passage of a "pig" or scraper through the line along with the flowing fluid. The purpose is to remove foreign material from the line. Gauging or caliper pigs (so-called "intelligent" pigs) are used to detect dents, buckles, or excessive corrosion (Webb, 1978). As stated above, pigs are also useful in dewatering and drying of pipelines. Two popular pig styles are shown in Figure 13–9. Hard-rubber or plastic spheres are used also.

A pig or scraper is moved through the line by the differential pressure existing across it, caused by the process fluid moving by it at a higher velocity. For cleaning

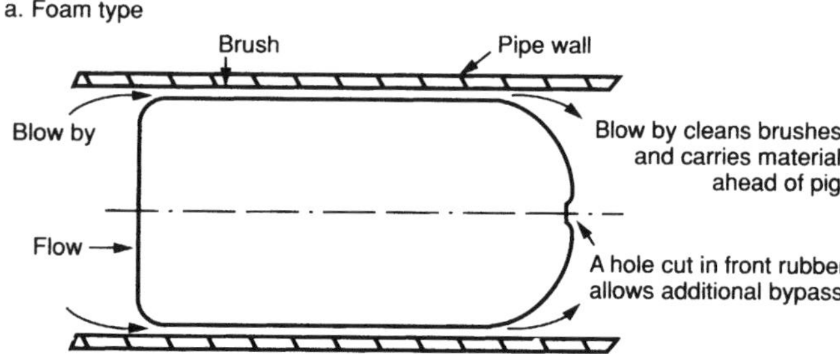

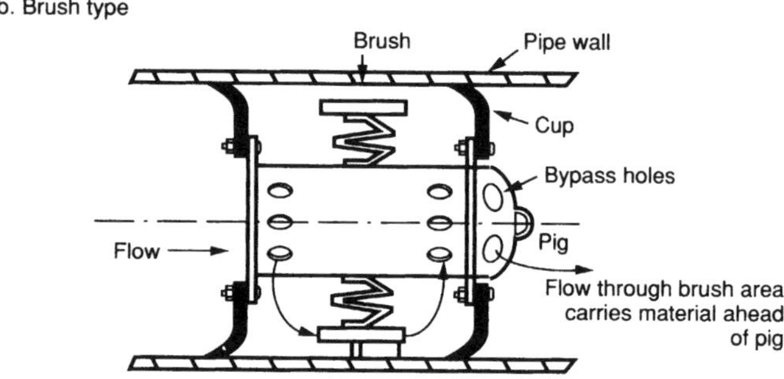

Figure 13–9.   Cleaning pigs (Webb, 1970).

purposes, a fairly good seal is required between the pig and the pipewall.

Figure 13–10 shows *typical pig launching and receiving facilities* (Webb, 1978). Operation is fairly straightforward. A pig indicator or ''pig sig'' is used to detect the passage of the pig and alert the operator by tripping a suitable warning device. Booster compressor stations must have a bypass arrangement for pigs or, alternatively, a receiver and a launcher.

A recently developed concept in pigging (Mitchell and Purinton, 1987) is the use of *gelled fluid* as a pig. Gelled fluid has the following advantages over a mechanical pig:

1. Performs effectively in lines of varying diameter
2. Passes line restrictions, intrusions, or probes
3. Maintains a good seal over long distances
4. Moves large amount of solid debris without sticking or plugging

Gelled-fluid pigs can be injected into a pipeline through a 2–in. line; they can be used for a variety of services, including dewatering and drying. To be propelled by gas, gelled-fluid pigs must have an intervening mechanical pig. See Mitchell and Purinton (1987) for additional details.

Leak Detection

Leak detection in operating pipelines is an important function. This refers to the routine detection of small leaks and not the emergency location of massive line breaks.

For gas lines, methods of leak detection are more limited than for oil. Visual inspection, for example, will not be of any help with small gas leaks. The two main methods are air sampling adjacent to the line and, for transmission lines, computer calculations built into the SCADA system.

Eynon (1980) discusses the use of sampling probes for gas pipeline leak detection by external inspection. The difficulty is that natural gas tends to rise, so the probe must be applied directly over the line. The process is slow and tedious.

Sjoen (1987) reports the use of unsteady-state gas line simulation on a SCADA system to aid in leak detection for the Statpipe system off Norway.

PIPELINE COST

Pipeline cost depends on *line size and terrain,* as well as *location.* The table below shows U.S. gas pipeline costs in 1988 as a function of size (True, 1988). Cost generally increases with size and is much higher in densely populated areas.

Onshore USA Gas Pipeline Cost, 1988 (True, 1988)					
Pipe Size In.	Average Cost 1000$/mile	Range 1000$/mile Low	High	No. of Lines	Total Miles
8	194	116	220	4	12
12	496	262	1874	10	87
16	360	203	2256	13	215
20	304	246	2776	12	363
24	787	431	4289	15	786
30	586	527	1585	35	2470
36	1099	759	2625	30	905

Three offshore gas lines averaged $755,000 per mile, essentially independent of size from 16-in. to 30-in. All three were in the Gulf of Mexico, 0.3, 13 and 388 mi.

The *breakdown on pipeline cost* is approximately as follows:

U.S. Onshore Natural Gas Pipeline Costs (True, 1988)	
Item	Percent of Total Cost
Land and Right of Way	6
Material	33
Labor	43
Misc.*	18
	100

* Engineering, supervision, interest, administration and overhead, contingencies, Afudc, and FERC filing fees.

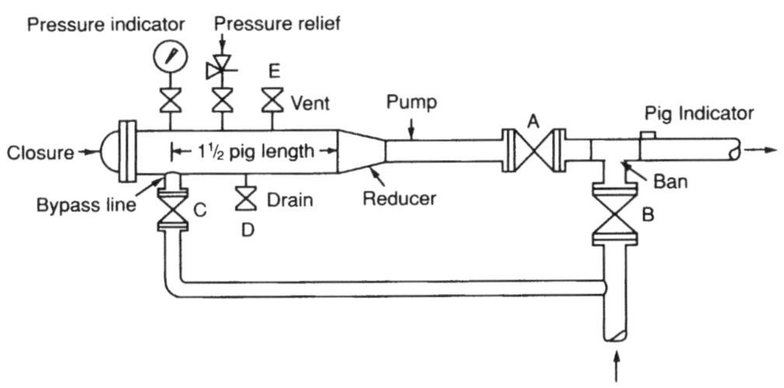

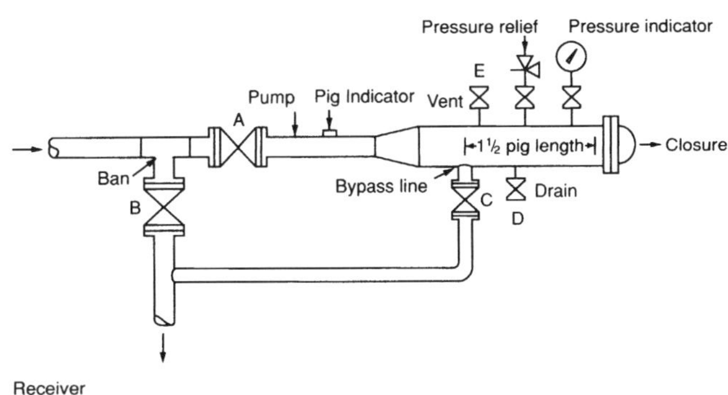

Figure 13–10. Pig launcher and receiver (Webb, 1978).

Inflation has been an important factor in pipeline cost. The OGJ-Morgan Pipeline Cost Index (Morgan, 1985) appeared to have leveled at 190 (Index = 100 in 1974) in 1982–83, but then continued to rise slowly to 200 in 1985. Construction of 58 gas lines of 0.14 mi to 254 mi in 1988 totalled 5118 mi. Pipeline costs doubled over the period 1976–85.

In addition to installation cost, True reported gas *transmission expenses* (1985–86 basis). These operating expenses are obviously important in any economic evaluation. Abidi (1975) gave shipping costs and construction costs; but these costs are rather old and therefore difficult to extrapolate to the present. Olorunniwo (1987) gives much more recent figures but very little information about them; his original Ph.D. thesis may have more details.

Review Questions

1. Name the two basic types of gas pipelines.
2. Identify the common types of gas/liquid flow and their usual applications.
3. Which is more difficult—selecting a pipe ID or specifying a compressor horsepower? Why?
4. What types of carbon steel pipe are used for non-corrosive HC service?
5. When are alloy-steel and plastic or FRP pipe used?
6. What ANSI codes govern MAWP for gas line? What factors are involved in these codes?
7. Define Moody friction factor, Fanning friction factor, pipe roughness, transmission factor.
8. Estimate an acceptable pressure drop for single-phase gas process lines.
9. Describe the seven patterns found in two-phase flow.
10. How is liquid holdup correlated?
11. How is the erosional velocity limit estimated?
12. What factors should be included in a rigorous computer simulation of two-phase pipeline flow?
13. What construction/commissioning considerations are involved in pipeline installation?
14. What methods are used to prevent corrosion of pipelines?
15. What activities are involved in pipeline operation?
16. How can air leak into pipelines? What potential problems can arise?
17. What factors influence pipeline cost?

Problems

1. Compare the flow capacities of 6-, 8-, and 10-inch pipelines. State any assumptions.
2. The pipeline described in Example 13–1 is coated internally with a very smooth 5-mil-thick fusion-bonded epoxy coating.
   a. Will the pressure drop change? If so, how much?
   b. Are there any other benefits to internally coating the pipeline?
3. To reduce the pressure drop in Example 13–2, it is suggested that 4.25 miles of the 8.5-mile line be replaced with a 6-in., Sch. 40 line.
   a. Comment on this idea.
   b. If implemented which 4.25 miles would you replace—the first, middle, or last section?
4. In Example 13–4 the condensate was underestimated —the correct flow is 40 bbl/MMscf. Estimate the pressure drop.
5. Repeat Example 13–5 using 40 bbl/MMscf of condensate.

Nomenclature

A = cross-sectional flow area of pipe, or area per length for heat transfer through pipe wall

a = constant in Equation 13–40, function of two-phase flow regime

B = depth of burial to centerline of pipe
b = constant in Equation 13–40, function of two-phase flow regime
C_p = fluid heat capacity at constant pressure, Btu/lb-°F
C_1 = 0.013 m² (Eq. 13–30)
C_2 = 336000 f / d⁵ (Eq. 13–30)
CA = corrosion allowance in Equations 13–1 and 13–2
c = constant in Equation 13–40, function of two-phase flow regime
D = inside pipe diameter, ft
d = inside pipe diameter, in.
d_o = outside pipe diameter, in.
D_i = outside diameter of insulation
D_p = outside diameter of pipe
D_{pi} = inside diameter of pipe
E' = longitudinal joint factor (ANSI B31.3)
E'' = longitudinal joint factor (ANSI B31.8)
e = pipe roughness, ft (Fig. 13–2)
F = transmission factor (Eq. 13–15)
F_f = drag factor (Fig. 13–3)
F_{FT} = fully turbulent transmission factor (Eq. 13–26)
F_{PT} = partially turbulent transmission factor (Eq. 13–25)
F_{pv} = supercompressibility factor (Eq. 13–29)
F_{SP} = smooth pipe transmission factor (Eq. 13–24)
F'' = construction design factor (ANSI B31.8)
f = Moody friction factor
f_F = Fanning friction factor
f_n = no-slip friction factor defined by Equation 13–45
f_{tp} = two-phase friction factor (Eq. 13–46)
Fr = Froude number, dimensionless (Eq. 13–31)
G_m = mass velocity = $V\rho$ (Eq. 13–20)
g = acceleration due to gravity, ft/s²
g_c = conversion factor—32.174 ft-lbm/s² -lbf
H = enthalpy, Btu/lb
H_L = liquid holdup (Eq. 13–40)
h_f = inside fluid heat transfer coefficient, Btu/hr-ft²-°F
h_L = frictional head loss, ft-lbf/lbm
h_s = convective heat transfer coefficient for sea or air Btu/hr-ft²-°F
k = empirical erosion velocity constant (Eq. 13–52)
k_i = thermal conductivity of insulation, Btu/hr-ft-°F
k_p = thermal conductivity of pipe, Btu/hr-ft-°F
k_s = thermal conductivity of soil, Btu/hr-ft-°F
ΣK = sum of resistance coefficients for fittings
L = pipe length, mile
L_1 = parameter in Figure 13–6
L_2 = parameter in Figure 13–6
L_3 = parameter in Figure 13–6
L_4 = parameter in Figure 13–6
MW = molecular weight
m = mass flow rate, lb/s
n = molar flow rate, lbmol/s

P = pressure, psia
P_b = pressure base for standard volume, psia
P_m = average pressure in line, psia
P_i = internal design pressure, psig
P_1 = inlet pressure, psia
P_2 = exit pressure, psia
Q = volumetric flow rate, ft³/s
Q_s = standard volumetric flow, e.g., MMscfd
R = universal gas constant, 10.73 ft³-psia/lbmol-°R
R_d = heat transfer resistance of inside dirt or scale
R_f = heat transfer resistance of fluid inside pipe
R_i = heat transfer resistance of insulation
R_p = heat transfer resistance of pipe
R_s = heat transfer resistance of soil or air
Re = Reynolds number, $D V \rho / \mu$, dimensionless
Re_n = no-slip Reynolds number (Eq. 13–43)
r_d = inside-fluid fouling factor (Table 13–7)
S' = allowable stress (ANSI B31.3)
S'' = minimum yield stress (ANSI B31.8)
s = constant defined by Equation 13–47 or 13–48
SG = gas specific gravity, MW/28.9625
T = temperature of flowing gas, °R
T_b = temperature base for standard volume, °R
T_s = temperature of undisturbed soil or surrounding medium (if not soil), °R
T'' = temperature factor (ANSI B31.8)
t_m = minimum pipe thickness (ANSI B 31.3 or 31.8), in.
U = overall heat transfer coefficient, Btu/hr-ft²-°F
V = average flow velocity, ft/s
V_e = limiting erosional velocity, ft/s (Eq. 13–52)
V_m = no-slip, two-phase mixture velocity, ft/s
V_s = superficial velocity, ft/s
Y' = temperature factor (ANSI B31.3)
X = distance along pipeline, mi
y = λ_L / H_L^2 (Eq. 13–48)
Z = compressibility factor
z = elevation, ft

Greek Letters

α = velocity profile correction factor
β = gradient distance factor (Eq. 13–54)
Δ = denotes difference, e.g., $\Delta z = z_2 - z_1$
δ = constant in Equation 13–41
γ = constant in Equation 13–41
η = Joule-Thomson coefficient (Eq. 13–57)
θ = pressure effect factor (Eq. 13–56)
λ_L = volumetric flowing fraction (Eq. 13–34)
μ = fluid viscosity
ν = kinematic viscosity, μ / ρ (Eq. 13–17)
π = 3.14159 . . .
ρ = fluid density

Subscripts

e   denotes erosional property
f   denotes frictional effect
G   denotes gas phase
L   denotes liquid phase
m   denotes a mean or average property
n   denotes a no-slip property
tp   denotes two-phase property
1   denotes inlet property
2   denotes exit property

References

API RP 14E (1984), "Design and Installation of Offshore Production Platform Piping Systems," 4th Ed, American Petroleum Institute, Production Department, Dallas, TX (April 15).

API Specification 5L (1990), "Specification 5 L (Spec 5L)," American Petroleum Institute, 1220 L Street NW, Washington, DC, 38th Ed., (May 1).

Abidi, F. (1975), "Pipeline-system Costs Estimated," *Oil & Gas J.*, Vol. 73, No. 41, p. 99–100 (Oct. 13).

Anonymous (1975), "Sun Installs Plastic Gathering Line," *Oil & Gas J.*, Vol. 73, No. 38, p. 109 (Sept. 22).

Askheim, N. E. and S. Eliassen (1983), "External pipeline coatings given critical review," *Oil & Gas J.*, Vol. 81, No. 32, pp. 82–84, 86 (Aug. 8).

Brill, J. P., and H. D. Beggs (1986),"Two-Phase Flow in Pipes," The University of Tulsa, Tulsa, OK, 5th Ed.

Brock, J. E. (1967), "Snaking of Pipelines," *Heating, Piping, and Ventilation*, Vol. 39, No. 1, p.144–147 (Jan.).

Brown, R. J. (1981), "Use of Underwater Plowing May Cut Offshore Pipelaying Costs," *Oil & Gas J.*, Vol. 79, No. 18, pp. 133–137 (May 4).

Diab, Saleh Y. (1983a), "Determining volume and concentration to dry gas lines," *Oil & Gas J.*, Vol. 81, No. 9, p. 80 (Feb. 28).

Diab, Saleh Y. (1983b), "Techniques for drying pipelines by the three-spheres method," *Oil & Gas J.*, Vol. 81, No. 10, p. 113 (March 7).

Eynon, Stuart B. (1980), "Line leak-detection methods updated," *Oil & Gas J.*, Vol. 78, No. 37, pp. 205–206 (Sept. 15).

Finch, J. C., and D. W. Ko (1983), "Tutorial—Fluid Flow Formulas," Proceedings, Pipeline Simulation Interest Group Annual Meeting, Toronto, ONT (Oct. 20–21).

GPSA Engineering Data Book (1987), Gas Processors Suppliers Association, Tulsa, OK, 10th Ed.

Gatlin, L. W., and H. J. EnDean (1975), "Corrosion from Wet Gas Controlled," *Oil Gas J.*, Vol. 73, No. 40, p. 63–68 (Oct. 6).

Hein, Mike (1983), "Here are methods for sizing offshore pipelines," *Oil & Gas J.*, Vol. 81, No. 18, p. 146 (May 2).

Hein, Michael A. (1984a), "Analysis of total multiphase flow systems demonstrated," *Oil & Gas J.*, Vol. 82, No. 33, p. 78. (Aug. 13).

Hein, M. A. (1984b), "Incorporating Rigorous Heat Balance Prevents Overdesign of Gas Pipelines," *Oil & Gas J.*, Vol. 82, No. 38, p. 96–100 (Sept. 17).

Hein, Mike (1986), "Pipeline design model addresses CO2's challenging behavior," *Oil & Gas J.*, Vol. 84, No. 22, p. 71 (June 2).

Huntley, Allan R. (1986), "Flexible Subsea Slug Catcher Designed for Use in North Sea's Troll Field," *Oil & Gas J.*, Vol. 84, No. 30, p. 84–86 (July 28).

Johnson, T. W., and W. B. Berwald, (1935), "Flow of Natural Gas Through High-Pressure Transmission Lines," U. S. Bureau Of Mines Monograph No. 6.

Kipin, Peter (1980), "Internal pipeline coatings pay off," *Oil & Gas J.*, Vol. 78, No. 15, pp. 158, 160 (April 14).

Kopp, Gunnar, Jr. (1981), "Why and how to dry gas pipelines," *Pipe Line Industry*, Vol. 55, No. 1, p. 35 (Oct.).

Kornfeld, Joseph A. (1967), "BP-Hunt Lays Libyan Crude Line Through Rock and Desert," *Pipe Line Ind.*, Vol. 27, No. 5, pp. 47–50 (Nov.).

Koster, W. M. P. B., K. Aziz, and G. A. Gregory (1982), "FORGAS: A Versatile Program for Gas Field Deliverability Forecasting and Development Scheduling," Proceedings of Pipeline Simulation Interest Group Annual Meeting, St. Louis, MO (Oct. 14–15).

Kreith, F. (1973), "Principles of Heat Transfer," Harper & Row, N. Y., 3rd Ed.

Maggert, Kent W., and A. S. M. Anwar (1976), "Numerical Model Simplifies Gas Gathering and Sales," *Oil & Gas J.*, Vol. 74, No. 32, p. 72–78 (Aug. 9).

Massey, Phillip S. (1975), "Eofisk-Teeside Line to Operate Continuously," *Oil & Gas J.*, Vol. 73, No. 8, pp. 85–89 (Feb. 24).

Massey, P. S. (1980), "The Ekofisk Area Pipeline; A Decade of Challenge," Proc. 59th Ann. GPA Mtg., Houston (March 17–19).

McClure, G. M., T. J. Atterbury, and A. R. Duffy (1966), "High-Pressure Hydrostatic Testing Eliminates More Line-Pipe Defects," *Oil & Gas J.*, Vol. 64, No. 28, pp. 101–105 (July 11).

McKnight, J. E. (1988), "Air Exclusion Key to Gathering-System Upkeep," *Oil Gas J.*, Vol. 86, No. 6, p. 41–42 (Feb. 8).

Mendel, Otto (1981), "Practical Piping Handbook," PennWell Books, Tulsa, OK.

Mitchell, S., and R. J. Purinton (1987), "Pipeline Pigging

Technology," Ch. 6.1, edited by J. N. H. Tiratsoo, Scientific Surveys Ltd., Beaconsfield, U.K.

Morgan, Joseph M. (1985), "Construction Costs Continue Flat While Offshore Activity Shows Promise," *Oil & Gas J.*, Vol. 83, No. 48, pp. 82–84 (Nov. 25).

National Association of Corrosion Engineers (1975), "Sulfide Stress Cracking Resistant Metallic Materials for Oil Field Equipment," NACE Std MR–01–75.

National Tank Co., 1987, Technical Development Training Program, Tulsa, OK.

Olorunniwo, F. O. (1987), "An Analysis of Pipe and Compressor Cost Functions in Natural Gas Transmission Lines," *Journal of Pipelines*, Vol. 7, p. 1–13.

Perkins, T. K., and J. A. Euchner (1988), "Safe Purging of Natural Gas Pipelines," *SPE Production Eng.*, Vol. 3, No. 4, pp. 663–668 (Nov.).

Polignano, R. (1982), "Value of glass-fiber fabrics proven for bituminous coatings," *Oil & Gas J.*, Vol. 80, No. 41, pp. 156–158, 160 (Oct. 11).

Powell, J. R. (1981), "How Western Makes Its Gathering Systems Work," *Oil & Gas J.*, Vol. 79, No. 3, p. 73–74 (Jan. 19).

Pullin, Kenneth, and Myer Daniels (1981), "The Western Leg Gas Gathering Pipeline," *J. Petroleum Technology*, Vol. 33 (April).

Rau, J. N., and M. Hein (1982), "AGA Gas Flow Equations Can Be Quickly Done with Program for Hand-Held Calculators," *Oil & Gas J.*, Vol. 80, No. 10, p. 233–241 (March 8).

Rhoads, G. A. (1983), "Which Flow Equation—Does It Matter?," Proceedings, Pipeline Simulation Interest Group Annual Meeting, Detroit, MI (Oct. 27–28).

Rhodes, K. I. (1982), "Pipeline protective coatings used in Saudi Arabia," *Oil & Gas J.*, Vol. 80, No. 31, pp. 123–127 (Aug. 2).

Scott, P. R., and Kor N. Zijlstra (1981), "FLAGGS Gas-Line Sediment Removed Using Gel-Plug Technology," *Oil & Gas J.*, Vol. 79, No. 43, pp. 97–104, 109 (Oct. 26).

Seymour, E. V. (1981a), "Design Detailed for Australia's North West Shelf Pipeline," *Oil & Gas J.*, Vol. 79, No. 35, p. 51–79 (Aug. 31).

Seymour, E. V. (1981b), "Design and Construction Plans Detailed for North West Shelf Line in Australia," *Oil & Gas J.*, Vol. 79, No. 37, pp. 75–79 (Sept. 14).

Sjoen, K. (1987), "Real-Time Simulation on the Statpipe System," *Pipe Line Ind.*, Vol. 67, No. 5, pp. 46, 48, 51, 68, 127 (Nov.).

Smith, Bill (1981a), "Offshore Line Construction Methods Examined," *Oil & Gas J.*, Vol. 79, No. 18, pp. 154, 158, 160 (May 4).

Smith, Bill (1981b), "A Look at Pipeline Riser Installation Techniques," *Oil & Gas J.*, Vol. 79, No. 19, pp. 105–108 (May 11).

Tise, T. C. (1983), "Construction of the Northern Border Pipeline," Pipeline Engineering Symposium, pp. 25–37, ASME, NY, NY.

True, W. R. (1988), "New Construction Plans Up; Revenues, Incomes Continue to Decrease," *Oil & Gas J.*, Vol. 86, No. 48, p. 33–60 (Nov. 28).

Uhl, (1967), "Steady Flow in Gas Pipelines," *Pipe Line Ind.*, Vol. 26, Part 3: No. 1, p. 47 (Jan.); Part 4: No. 3 , p. 58 (March).

Uhlig, H. H. (1971), "Corrosion and Corrosion Control," 2nd Ed., John Wiley and Sons.

Webb, B. C. (1978), "Guidelines set out for pipeline pigging," *Oil & Gas J.*, Vol. 76, No. 46, pp. 196–200 (Nov. 13); No. 48, pp. 74–78 (Nov. 27).

Webb, B. C. (1983), "Art of Pipeline Anchoring—Update 1982," Pipeline Engineering Symposium, p. 25–37, ASME, NY, NY.

CHAPTER 14

Natural Gas Liquids Recovery

INTRODUCTION

Recovery of natural-gas liquids (NGL) from natural gas is quite common in natural gas processing, and can, at times assume great economic importance. Huge quantities of solution gas produced from crude oil used to be flared routinely. Phasing out this wasteful practice has required greatly increased NGL recovery.

Prior to 1984, the value of ethane as a raw material for petrochemicals was sufficiently high that its recovery was economically desirable. Figure 14–1 (Mehra, 1987) shows recent trends in sales gas and product value in the United States. The value of sales gas declined from about $3.50/MMBtu in early 1984 to about $1.50/MMBtu in late 1986 (West-Texas intrastate prices). As a liquid, propane was worth sufficiently more than its value in sales gas to warrant investment in liquid recovery for the period 1984–85, but during 1986 there was very little economic incentive for propane recovery. During most of this period liquid was worth less than its value in sales gas.

One factor in this price reversal is that relatively-cheap naphtha, as well as n-butane, presently compete as raw materials for ethylene production. However, some ethylene plants cannot crack naphtha. Even though there is little economic incentive to recover ethane, petrochemical plant expansion is currently occurring in the United States, and this may lead to a better ethylene (and thus ethane) markets.

The value of n-butane also varies widely. Normal butane is currently used to control the vapor pressure of automotive gasoline. Butane is desirable for this purpose since it has a high vapor pressure of 52 psia and a road octane number of 91.7. The U. S. Environmental Protection Agency (EPA) has ruled that, starting in 1989, the vapor pressure of automotive gasoline must be reduced, which means less blending of n-butane. In the United States, refineries purchase essentially all current n-butane production from natural-gas processing plants for use as a blending stock. With this mandated reduction in vapor pressure, refineries will be potential sellers of n-butane instead of purchasers. At present, other uses for n-butane include isomerization to i-butane, feed stock in ethylene manufacture, and boiler fuel. In summary, the present outlook for natural-gas liquids sales in the United States is soft, although King (1988) indicates that prices have stabilized somewhat. Figure 14–2 reports recent NGL prices (Brewster, 1986–89) compared with natural gas prices (*Oil & Gas J.*, 1986–89) on an energy-equivalent basis.

Because ethane is sometimes more valuable as NGL and at other times worth more as sales gas, *NGL-recovery plants need to be versatile*—capable of either producing or rejecting ethane liquid product.

As stated in Chapter 5, NGL recovery is not totally an economic matter. Production of gas that meets stringent sales specifications (e.g., HC dew point) may require NGL extraction regardless of economics. Thus, a good sales-gas market can compensate for a poor demand for NGL. Finally, it may be desirable to produce a gas that can be pipelined without hydrocarbon condensate formation, thus requiring NGL removal.

This chapter discusses NGL recovery beginning with a brief review of the processing objectives (see Chapter 5 for more detail). The economic value of NGL is presented in quantitative terms to underscore the relative merit of selling them as liquids or as sales gas. Various liquid-recovery processes are reviewed briefly as an introduction to process selection. Then the more important liquid-recovery processes are discussed in greater detail, prefaced by remarks on modern flow-sheet simulation. Handling and separation of the liquid products complete the chapter.

PROCESSING OBJECTIVES

The *three alternatives* of liquid-recovery processing are: to produce transportable gas, to meet sales-gas specifications, or to maximize liquid recovery (Odello, 1981).

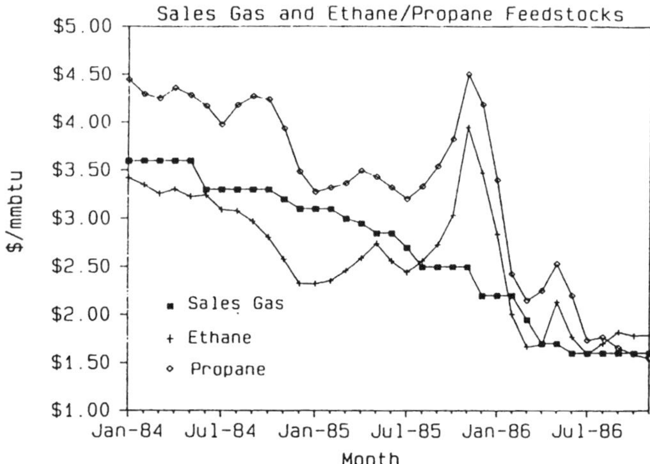

Figure 14–1. Energy equivalent prices (Mehra, 1987. Reproduced by permission of the American Institute of Chemical Engineers).

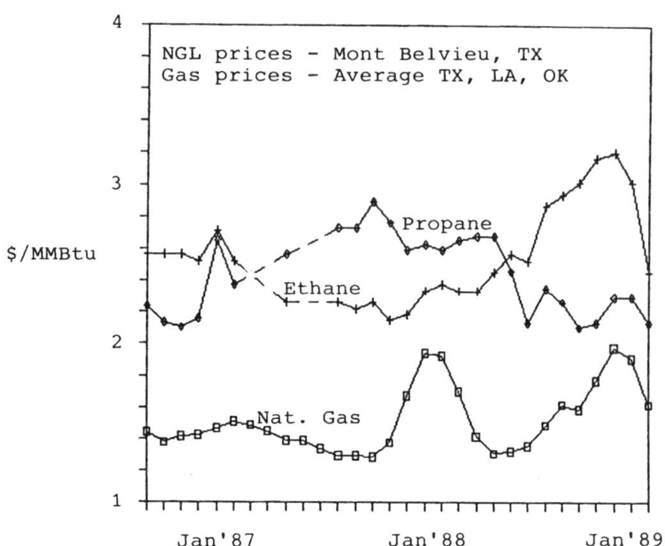

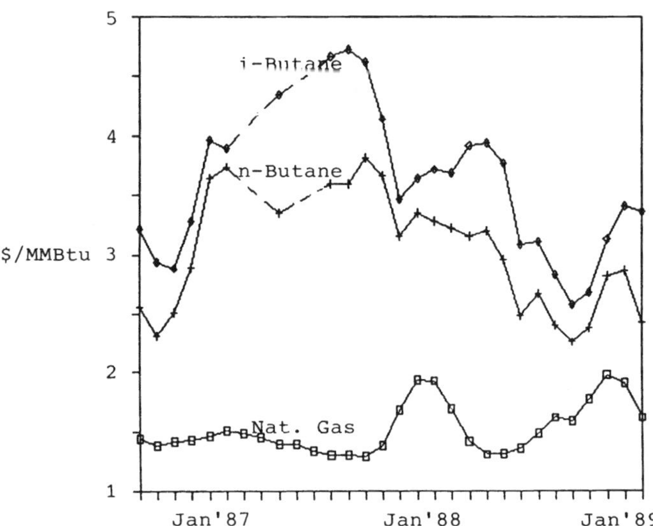

Figure 14–2. Relative price of gas and natural gas liquids.

Transportable Gas

For remote locations such as offshore or the outback, the only alternative is to produce gas that can be pipelined without condensation. Condensation has two drawbacks. First, two-phase flow requires a larger pipe diameter than single-phase flow for the same pressure drop. Second, when the two-phase stream arrives at its destination, elaborate slug-catchers may be required to protect the equipment downstream (Huntley and Silvester, 1983).

There are two expensive alternatives: NGL recovery at the remote site or dense-fluid pipelining. However, depending on the economics of the particular case, the least expensive option for the remote location can be any one of these *three options*:

1. Recover only enough NGL to obtain single-phase flow.
2. Transport the gas in two-phase flow.
3. Transport the gas in dense-phase flow.

These options are explored further in Chapter 5. Note that minimum NGL extraction is a practical objective in some cases.

Sales Gas

Perhaps producing sales gas with acceptable specifications is the most common choice. Generally there is a minimum gross heating value (GHV) specification and possibly a HC dew-point requirement. If more valuable as liquids, NGL removal should be maximized while still satisfying the minimum heating value specification. If NGL is worth more in the gas phase, it may be desirable to retain them as gas, subject to the HC dew-point requirement.

Maximum NGL Recovery

A normal heating-value specification of about 1000 Btu/scf can be met with methane alone as is shown below.

Hydrocarbon	GHV (Btu/scf)
Methane	1010
Ethane	1770
Propane	2516

However, many natural gas streams contain N_2 and/or CO_2, and these incombustible gases can require presence of ethane to provide the required heating value. If the heavier hydrocarbons are more valuable as liquids, essentially complete liquefaction of propane and heavier hydrocarbons and partial ethane recovery is desirable.

A second case for maximum liquid recovery is the *cycling* of natural gas in a condensate reservoir. Now all or a portion of the produced gas is reinjected to keep the reservoir pressure above the gas dew point and thus maximize ultimate NGL recovery. If the reservoir pressure is allowed to fall into the two-phase region, valuable liquids are condensed and will not be recovered.

VALUE OF LIQUEFIABLE COMPONENTS OF NATURAL GAS

Potentially, ethane and heavier HC components (C2+) can be liquefied. At one time, ethane was not considered a valuable liquefiable component, mainly because no economic method existed for high-yield liquid recovery. The advent of the expander plant, coupled with high demand for ethane to manufacture ethylene, motivated high ethane recovery.

Relative liquid- and gas-phase values of hydrocarbons are now illustrated for propane. In sales gas, propane is worth the contract price of the gas, assuming it can be left in the gas and sold for its GHV. Suppose gas is worth $3.00/MMBtu. The heating value of propane is 2516.1 Btu/scf (GPSA, 1987). Then

1000000 Btu x 1 scf/2516.1 Btu = 397.4 scf propane

If this propane is liquefied the amount of liquid recovered at 60°F is

397.4 scf x 1 gal/36.375 scf = 10.9 gal propane

where 36.375 scf/gal is read from Appendix 6 (Table A6–1) or Figure 23–2 (GPSA, 1987). The equivalent value of this propane as a liquid is, therefore,

$$\$3.00/10.9 \text{ gal} = \$0.275/\text{gal}$$

when gas is worth $3.00/MMBtu.

If liquid propane can be sold for more than $ 0.275/gal plus the cost of liquefaction, there is an economic incentive for propane liquifaction when gas is worth $3.00/MMbtu. Figure 14–3 shows the equivalence for NGL components as a function of gas price.

A similar argument can sometimes be made for "spiking" the condensate recovered from an associated gas back into the crude oil. Crude spiking not only increases the total barrels of oil but also raises the API gravity, which may increase the sales price per barrel. Again, crude value is important; $0.275/gal converts to $11.55/bbl. The crude

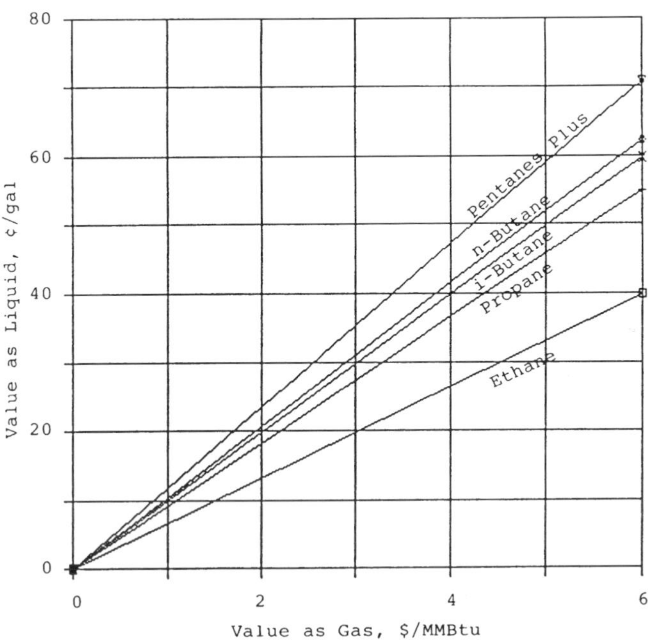

Figure 14–3.   Energy equivalent value of natural gas liquids.

would have to be worth more than $11.55/bbl to make the spiking economical.

As stated above, the extent of condensate removal may be limited by the sales-gas GHV specification, particularly if appreciable N_2 and/or CO_2 are present. Figure 14–4 (Ewan *et al.*, 1975) shows how ethane recovery is limited by inert gas content. The graph is approximate, but illustrates the principle involved.

LIQUID RECOVERY PROCESSES

Figure 14–5 shows the phase diagram for a natural gas, either a gas-well or associated gas. Obviously any cooling will induce condensation and yield NGL. The higher the pressure, the more condensation, other factors being equal. Another NGL recovery technique is the use of so-called mass-transfer agents (MTA); two MTA processes are described later.

The basic NGL liquefaction processes are now described and related to Fig. 14–5, where possible. Additional details for several processes are given later.

Refrigeration

Refrigeration, shown in Figure 14–6a, is the simplest and most direct process for NGL recovery. *External* or *mechanical refrigeration* is supplied by a vapor-compression cycle that typically uses propane as the refrigerant or working fluid. Mechanical refrigeration is described in detail below.

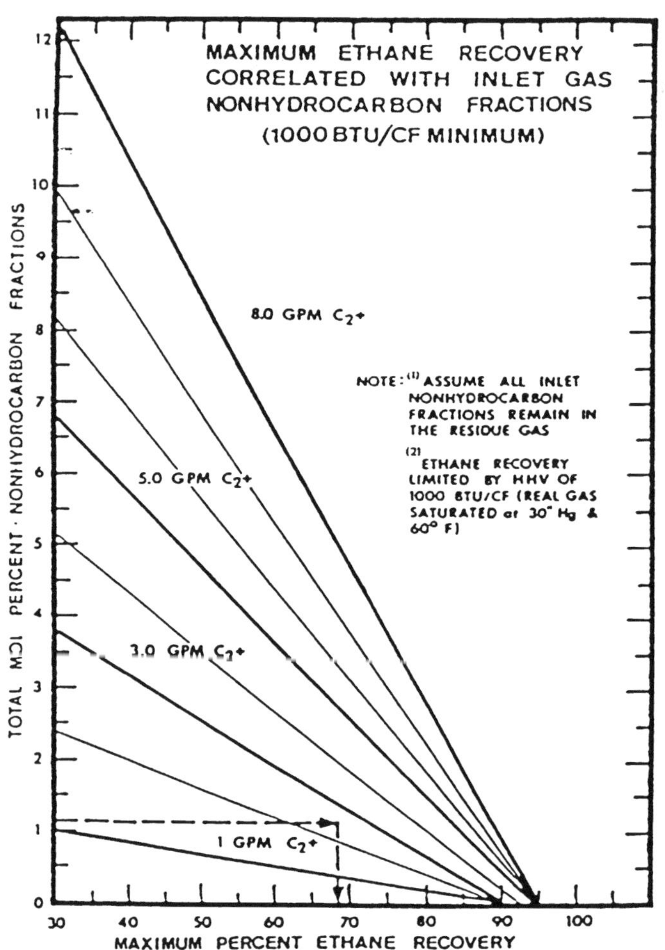

Figure 14–4. Maximum ethane recovery correlated with inlet gas nonhydrocarbon fractions (Ewan et al., 1975).

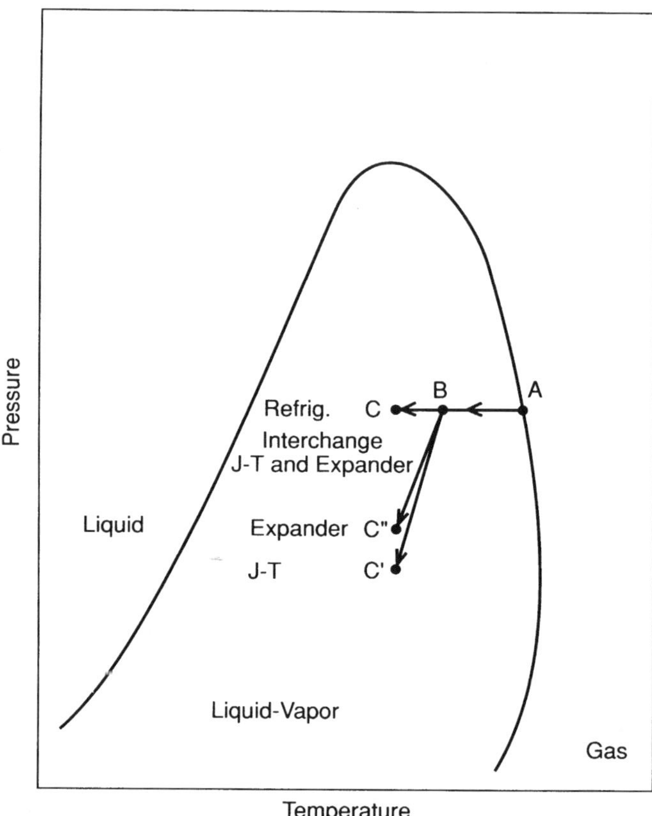

Figure 14–5. Phase-diagram paths for NGL recovery.

In Figure 14–6a, the *gas-to-gas heat exchanger* recovers additional refrigeration by passing the gas leaving the cold separator countercurrent to the warm inlet gas. The temperature of the cold gas stream leaving this exchanger "approaches" that of the warm inlet gas. Economically, this approach can be as close as 5°F or slightly less.

The *chiller* in Figure 14–6a is typically a shell-and-tube, kettle-type unit. The process gas flows inside the tubes and gives up its energy to the liquid refrigerant surrounding the tubes. The refrigerant (often C3) boils off and leaves the chiller vapor space essentially as a saturated vapor.

The refrigeration process is shown as line ABC in Figure 14–5. From A to B indicates gas-to-gas exchange; from B to C, chilling. Gas-to-gas exchange is very common in NGL recovery processes. Figure 14–6a is deceptively "simple," in that the condensed liquid must be fed to a *stabilizer* (or fractionation column) to meet typical specifications such as vapor pressure or composition (%C2 if recovering C3+).

Ethylene glycol is often injected at the inlet of the gas-to-gas exchanger and/or chiller to *prevent hydrate formation*, or freeze-up, in these exchangers. Freeze-up will partially block exchanger tubes, thus increasing pressure drop and decreasing heat exchange. The weak glycol solution, containing condensed water, is separated in the cold separator, reconcentrated, and recycled. Glycol injection is described in detail in Chapter 6.

Joule-Thomson Expansion (LTS)

Figure 14–6b shows the Joule-Thomson (JT), or low-temperature separation (LTS), process. The inlet gas passes first through the gas-to-gas exchanger, thence to an expansion or "choke" valve. The expansion is essentially a constant enthalpy process. Nonideal behavior of the inlet gas causes the temperature to fall with the pressure reduction, as shown by line ABC' in Figure 14–5. The temperature change depends primarily on the pressure drop. The JT process is a "self-refrigeration" process, as opposed to

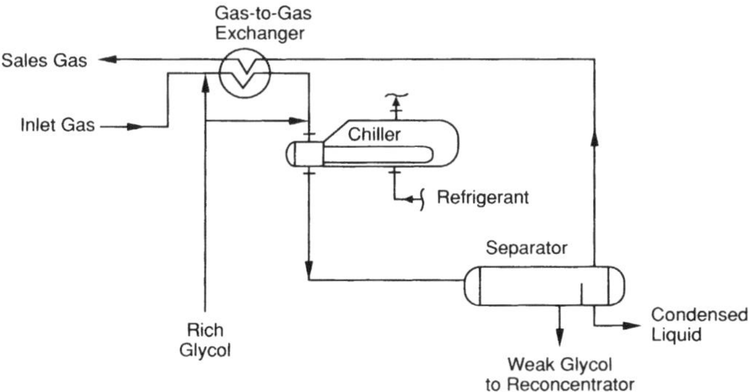

Figure 14–6a.   Refrigeration.

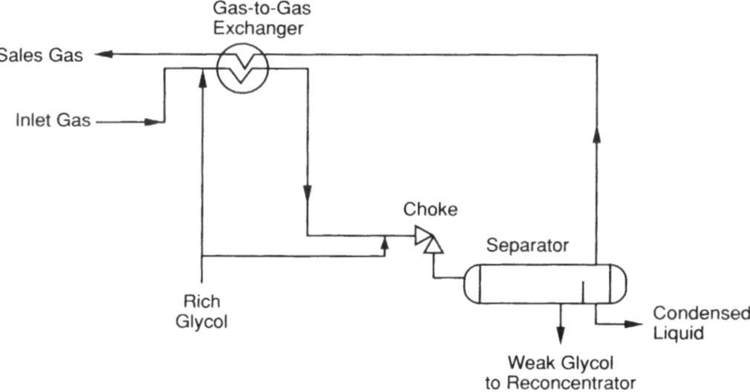

Figure 14–6b.   Low-temperature separation.

"external refrigeration." Again, the condensed liquids must be fractionated to meet vapor pressure and composition specifications.

Generally, sales gas must be *recompressed* to sales pipeline pressure if it has been expanded to a lower pressure. Thus this JT process is most favored when the wellhead gas is produced at a very high pressure and can be expanded to sales-line pressure with no recompression. If the gas must be recompressed, the JT process is penalized by the recompression horsepower requirement.

Expansion Turbine

In the so-called *cryogenic* or *expander plant*, the chiller or J-T valve is replaced by an *expansion turbine*. As the entering gas expands it supplies work to the turbine shaft, thus reducing the gas enthalpy. This decrease in enthalpy causes a much larger temperature drop than that found in the simple J-T (constant enthalpy) process. The expansion process is indicated as line ABC" in Figure 14–5. Figure 14–7 shows an expander-plant flowsheet that features two gas-to-gas exchangers and a C3 chiller.

Feed gas is cooled first in the high-temperature, gas-to-gas exchanger (A) and second in the propane chiller (B). Condensed liquids then are removed in a separator (C).

Gas from this separator is cooled further in the low-temperature gas-to-gas exchanger (D) and fed into a second cold separator (E). Gas from the cold separator expands through the expansion turbine (F) to the pressure at the top of the demethanizer (G), which varies from 100–450 psia. During the expansion, condensate is formed but the turbine is designed to withstand this condensation. Typically 6 to 8% of the feed gas is condensed in the cold separator which is usually at −30 to −60°F. The expander lowers the pressure from the inlet gas value, say 600 to 900 psia, to the demethanizer pressure of 100 to 450 psia. Typical inlet-gas temperatures to the demethanizer are −130 to −150°F, sufficiently low that a great deal of the ethane is liquefied. Expander plants typically condense 8 to 12% of the feed gas.

The demethanizer tower stabilizes the raw liquid product by reducing the methane content to a suitably low level. The molar ratio of methane to ethane in the bottom product is typically 0.02 to 0.03. The bottom product temperature is often below ambient, so that feed gas may be used as the heat-transfer medium for the reboiler. This provides additional refrigeration to the feed and yields higher ethane recovery, typically 80%. Additional refrigeration also can be obtained by using the feed gas for side-reboiling heat in the demethanizer.

If the feed gas is lean, say < 2.5 to 3 gal NGL/ Mscf (GPM), auto refrigeration (gas-to-gas heat exchange) usually suffices. For moderately rich feeds, say >3 GPM, mechanical refrigeration should be considered to obtain high ethane recovery most economically—shown in dotted lines in Figure 14–7. In fact, mechanical refrigeration alone is probably the most attractive NGL recovery process for very rich (> 10 GPM) feed gases.

In expander plants, the low temperatures make material selection very important (McKee, 1986). Carbon steel is suitable down to −20°F but becomes brittle at lower temperatures. Charpy-impact-tested carbon steel can be used as low as −50°F. Between −50 and −150°F, 3.5% Ni steel is used, and stainless steel below −150°F.

The extremely low temperatures also require a very low water content in the inlet gas to prevent hydrate formation throughout the cryogenic process. *Solid-desiccant dehydration* is required (see Chapter 9 for details).

If the feed gas contains more than about 0.5–1.0 mol percent CO_2, solid CO_2 may form in the expander outlet gas or in the demethanizer near the top. This may dictate CO_2 removal from feed when the heating-value specification would not dictate such.

Approximately 98% of the energy lost in the expander is recovered in a *brake compressor* that partially recompresses the residue gas. Additional recompression is normally required to restore the gas to sales-line pressure.

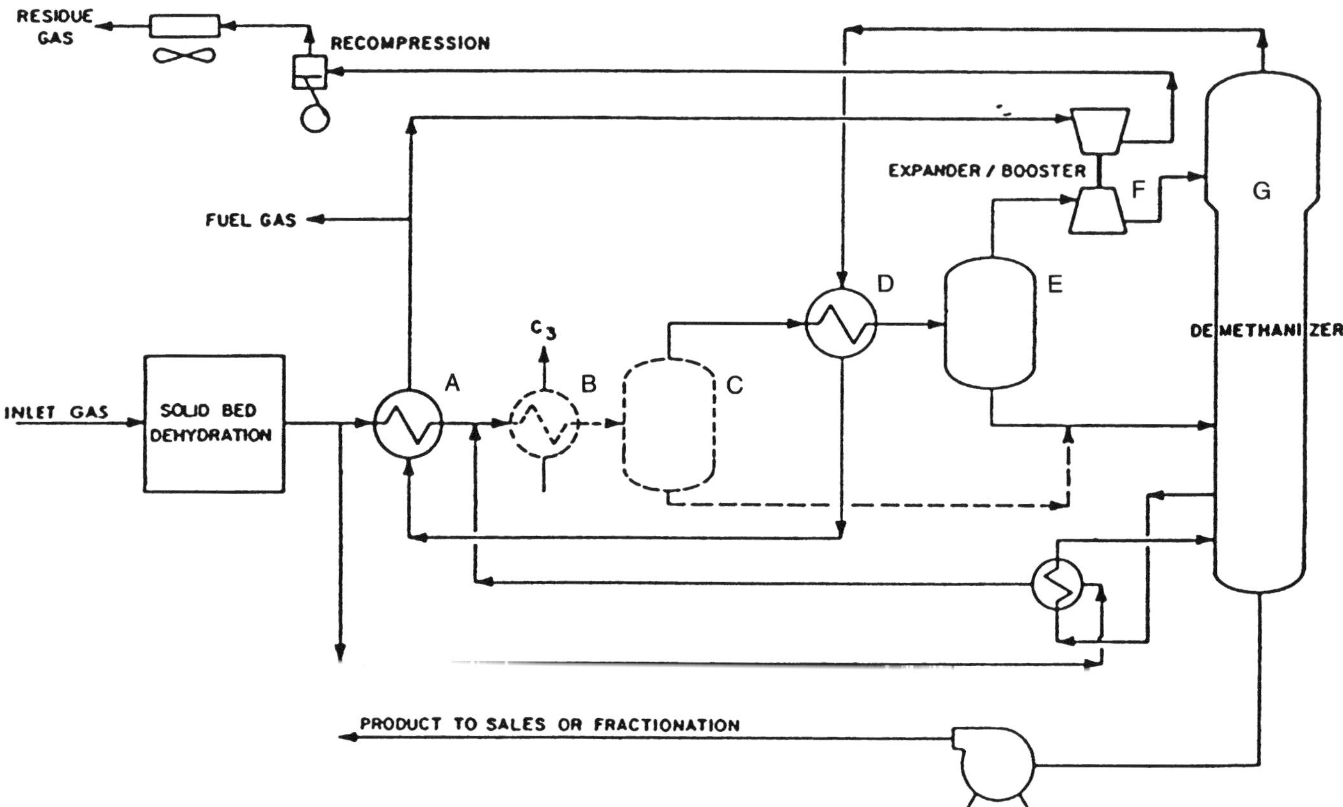

Figure 14–7. Process flow diagram turbo-expander gas liquids extraction plant (Ewan *et al.*, GPA Proceedings, 1975).

Oil Absorption

Absorber oil can be used as a mass-transfer agent (MTA) to recover liquids from natural gas with only a 5–10 psi pressure drop in the absorption tower (Fig. 14–8). If the low-pressure feed gas must be compressed to sales-gas line pressure, the tower pressure can be selected to optimize the absorption process. If, on the other hand, the feed gas is available at sales-line pressure, then absorption is carried out at pipeline pressure, thus requiring very little, if any, recompression. With no recompression, 30 to 50 psi are required to flow the feed through the plant. A condensate recovery plant that operates on a high-pressure pipeline and returns the residue gas at pipeline pressure is sometimes called a ''straddle'' plant.

The trade-off for low pressure drop is that a complex recovery process is required to separate the absorbed NGL from the absorber oil and to remove dissolved methane and/or ethane from the absorber oil (Fig. 14–8). The process has high fuel requirement, even with efficient energy recovery. Absorption oil plants still remain in operation, but many have been replaced by simpler, more-efficient expander plants.

The absorber oil and the inlet rich gas are almost always chilled to increase NGL recovery. Typically such ''cold oil'' plants operate with temperatures of +30 to −30°F for the streams entering the absorber.

As noted above, lean-oil recovery starts with the removal of dissolved methane. The demethanizer (*DEC1ER*) tower is reboiled for methane stripping, but valuable liquefied ethane and propane are vaporized also, thus requiring a cold-oil reabsorber section in the top of the tower. In cold-oil plants that do not feature ethane recovery, the demethanizer tower becomes a deethanizer tower.

Second, the absorbed condensate is separated from the lean oil in the still. The raw condensate passes overhead. The hot lean oil from the still bottom is returned through the demethanizer reboiler(s) for energy conservation, then cooled and chilled.

Absorption oil is a clear HC liquid similar to gasoline or kerosine, but with a narrower distillation range. Absorption oil has an average molecular weight (MW) of 100 to 140, depending on the temperature level in the absorber column. For −30°F inlet temperature, a 100 MW oil is satisfactory. For +30°F inlet temperature, 140 MW is appropriate. Table 14–1 gives a typical analysis for a cold-oil-plant lean oil.

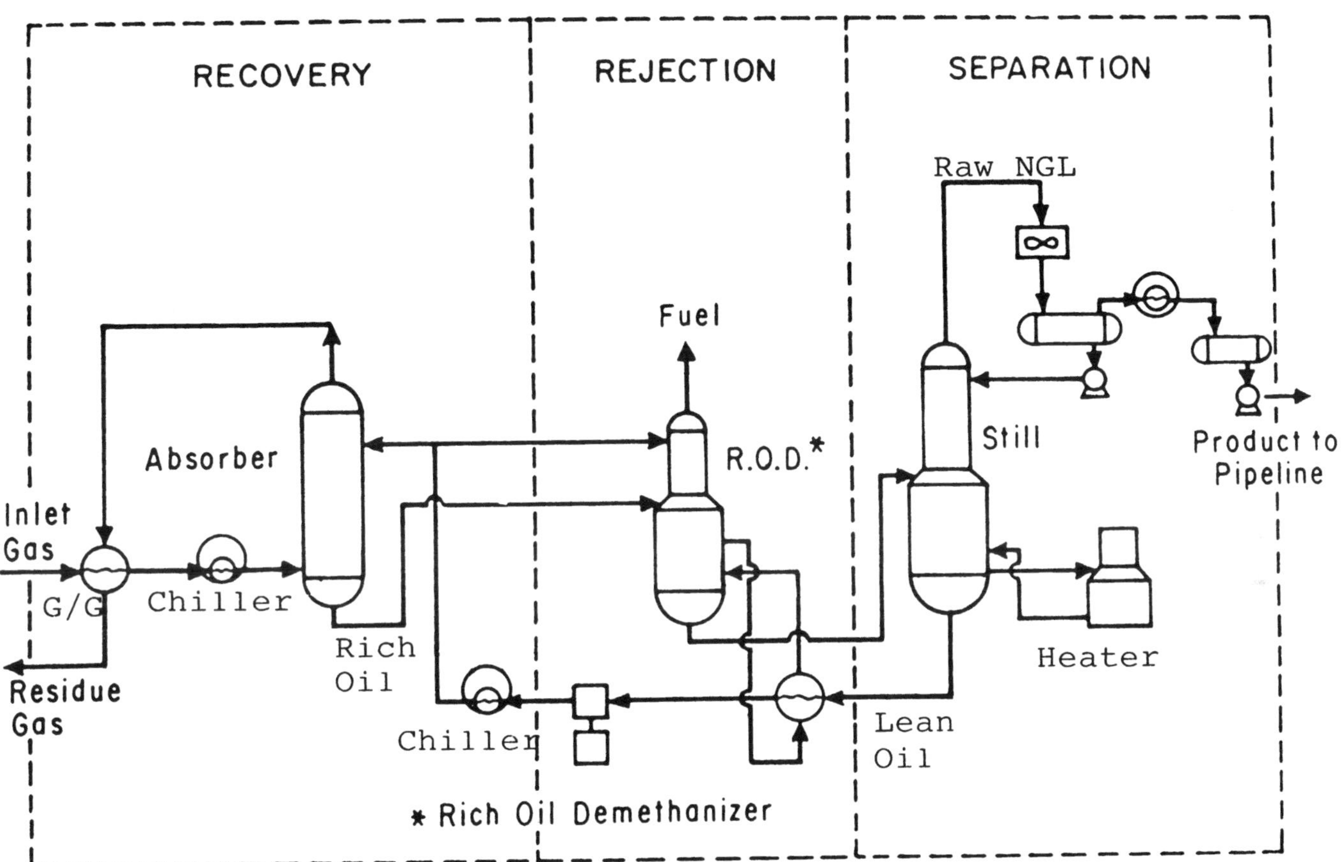

Figure 14–8.   Oil absorption plant (PETEX, 1974. Copyright © 1974 by the University of Texas at Austin).

Table 14–1   Properties of a Typical Absorber Lean Oil

Components	Analysis Mole %	Average Carbon Number of Cut	ASTM Distillation	
			Vol. %	Temp °F
Propane	0.0061	3	IBP	242
i-Butane	0.0092	4	5	270
n-Butane	0.013	4	10	283
i-Pentane	0.023	5	20	294
n-Pentane	0.017	5	30	300
Hexanes	0.45	5.92	40	306
Heptanes	2.09	6.91	50	312
Octanes	10.20	7.71	60	319
Nonanes	32.96	8.71	70	328
Decanes	34.00	9.64	80	340
Undecanes Plus	20.23	—	90	361
			95	386
			EP	435
			Recov.	98.6
			Loss	1.4

PONA Analysis:				
	Paraffins	52.2	Sp. Gr. @ 60°F/60°F	0.7703
	Olefins	0.3	Molecular Weight	130
	Naphthenes	35.6		
	Aromatics	11.9		

Data from a report by P-V-T, Inc., Houston, TX

Solid-Bed Adsorption

Solid desiccants, such as *silica* or *molecular sieves*, and *activated charcoal* can be used as an MTA for condensate recovery (Fig. 14–9). The process is continuous with respect to the gas but cyclical with respect to the adsorbent bed because the latter must be regenerated when it becomes saturated with condensate. Regeneration is accomplished by passing heated recycle or bypass gas through the bed. The condensate is recovered from the regeneration gas by cooling, condensation, and phase separation. A typical cycle time for the process is two or three hours. For this reason the process is sometimes referred to as "*quick cycle*" or "*short cycle*" adsorption. The units are also called HRU (hydrocarbon recovery units). See Chapter 9 for more details on solid-desiccant adsorption.

Short cycle units are energy intensive because of the regeneration process. The adsorption beds are heavy and expensive. This process option is not often used but may be considered in special applications, such as HC dew-point control in remote locations.

PROCESS SELECTION

Quick generalizations regarding selection of a condensate recovery process are difficult. In some cases, an economic comparison between viable alternatives will be required, however, a few guidelines can be offered.

If the NGL content of the feed gas is low, the expander process will probably be the method of choice. *For gases very rich in NGL*, simple refrigeration is probably the best choice, while oil absorption or expansion are usually not satisfactory.

If the inlet gas pressure is very high, low temperature separation (LTS) may be attractive; this process has been used in recent years, both on and off shore. It is important that the reservoir pressure remain high for the intended life of the plant. *Low inlet gas pressure* favors an expander plant or straight refrigeration (if the gas is very rich).

Very low gas rates may justify only a very simple process, such as an automatically-operated JT unit. *Large flow rates* justify a more complex plant with more complex controls and more operating personnel.

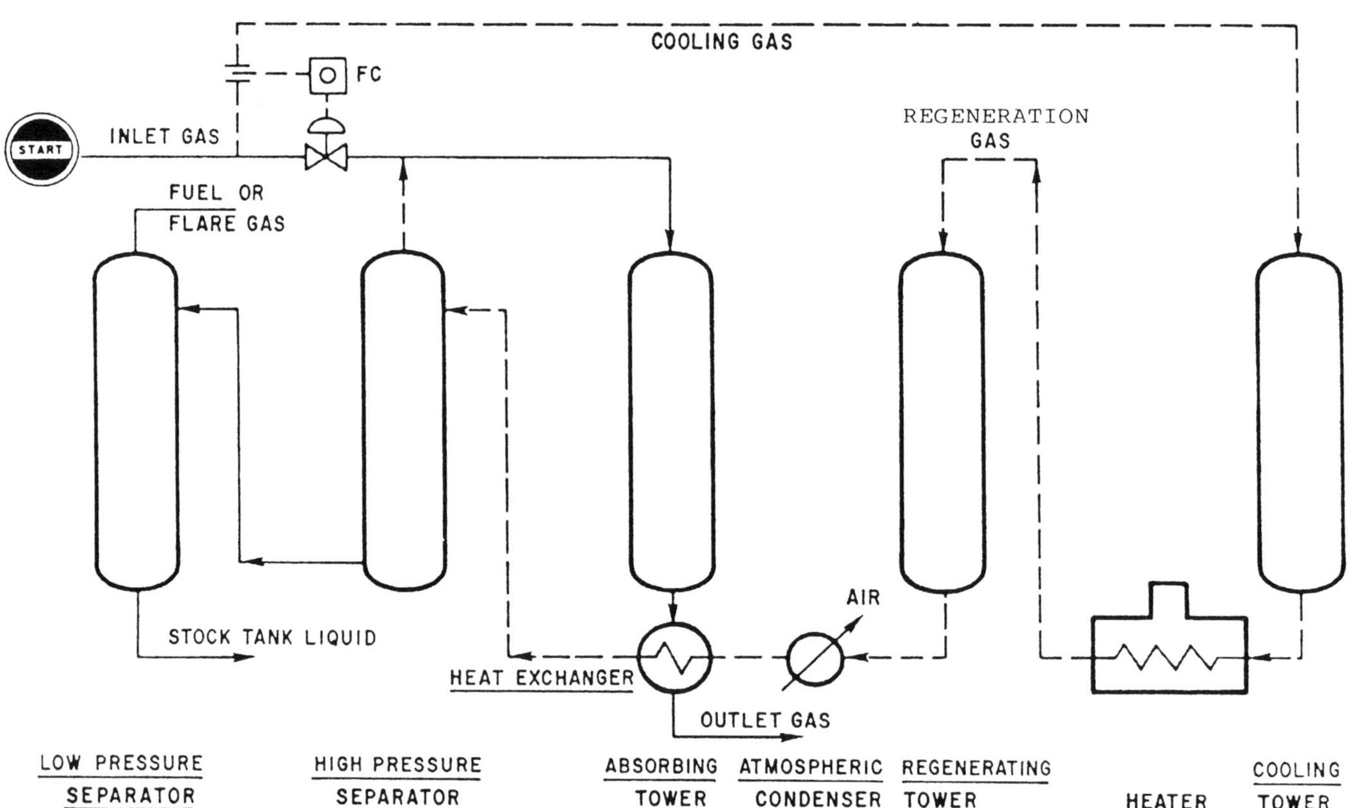

Figure 14–9. Solid-desiccant adsorption (Ballard, 1965).

For the enthalpy values shown in Figure 14–10, the value of x is

$$x = (-778 + 862) / (-690 + 862) = 0.488$$

The remaining liquid fraction is available to be boiled and thus absorb process heat and so provide the desired refrigeration. This vaporization occurs at constant pressure and results in a saturated-vapor exit stream (point 3). The enthalpy change for this step represents the available chiller duty per each pound of refrigerant circulated.

$$Q_{ch} = h_3 - h_2 = -690 + 778 = 88 \text{ Btu/lb}$$

If the total duty, q_{ch}, is 1 MMBtu/hr, then the circulation rate is:

$$m = q_{ch} / Q_{ch} = 10^6 \text{ Btu/hr} \div 88 \text{ Btu/lb}$$
$$= 11363 \text{ lb/hr}$$

At the next step, vapor (point 3) is compressed back to the inlet condenser pressure (point 4), which is taken to be the outlet pressure plus a typical allowance for pressure drop in the condenser (5 psi). Compression is calculated in two steps. First, an isentropic path is followed to the final pressure (point 4,s). This theoretical, isentropic work of compression, W_{is}, equals the change in enthalpy.

$$-W_{is} = h_{4,s} - h_3 = -645 + 690 = 45 \text{ Btu/lb}$$

This theoretical, isentropic work is now divided by an isentropic, or adiabatic, efficiency to estimate the actual work, W_{act}, done by the compressor.

$$-W_{act} = 45/0.75 = 60 \text{ Btu/lb}$$

(The 0.75 efficiency is a typical value for a centrifugal compressor. A value of 0.85 might be typical for a reciprocating compressor.) The actual final enthalpy, h_4, is calculated by assuming that the actual compression is adiabatic—all the work done on the gas ends up as enthalpy.

$$h_4 = -W_{act} + h_3 = 60 - 690 = -630 \text{ Btu/lb}$$

The actual temperature of the gas leaving the compressor is obtained by interpolation of the diagram at the final pressure and enthalpy and is approximately 175°F.

The gas (or fluid) horsepower is the work per lb multiplied by the circulation rate, divided by an appropriate conversion factor, in this case 2544 Btu/hp-hr.

$$m (-W_{act}) = (11363)(60)/2544 = 268 \text{ HP}$$

The brake horsepower (input to the compressor shaft) is greater than the gas horsepower. For centrifugal compressors a quick estimate of the bearing and seal losses is simply estimated and added to the gas horsepower to obtain the brake horsepower. For reciprocating compressors it is customary to divide the gas horsepower by a mechanical

efficiency to obtain the brake horsepower. See Chapter 10 for more information.

The final step in the cycle is the cooling and condensing of the compressed gas, shown by the approximately horizontal line (5 psi pressure drop) from point 4 to point 1. The condenser duty is

$$q_{cond} = m (h_1 - h_4) = (11363)(-778 + 630)$$
$$= -1.68 \text{ MMBtu/hr}$$

Economizer Cycle. The efficiency of the simple cycle can be improved by introducing a *flash economizer* (Fig. 14–11). As before, start with the saturated liquid from the condenser, but flash it to an intermediate pressure rather than all the way down to the chiller pressure. A portion of the recirculating refrigerant, m_1, vaporizes in the economizer drum and is separated from the remaining liquid,

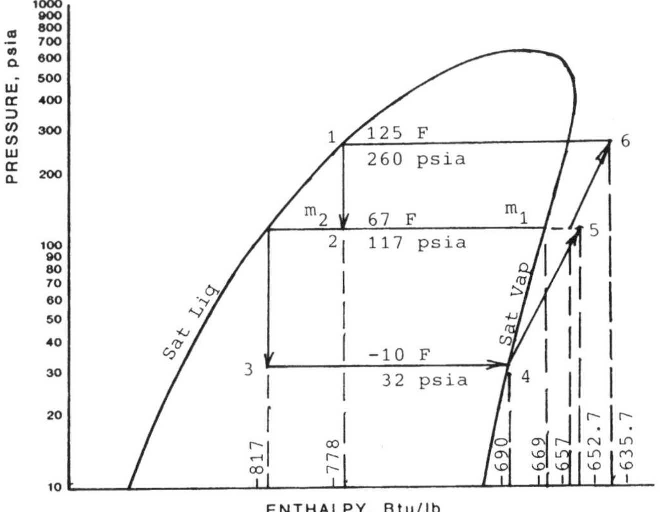

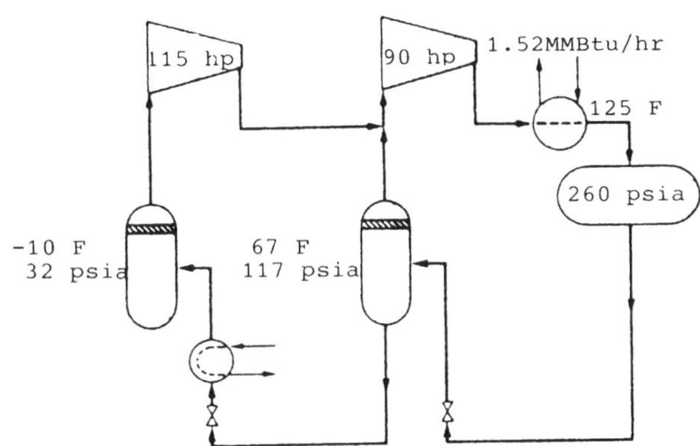

Figure 14–11. Propane refrigeration economizer cycle.

m_2. Fraser (1984) found that the optimum economizer pressure is as follows.

$$P_{econ} = P_{ch} (P_{cond,out}/P_{ch})^{0.614}$$
$$= 32 (265/32)^{0.614} = 117 \text{ psia}$$

Fraser points out that operation at pressures differing slightly from the optimum does not change the total horsepower greatly. The enthalpy change across the valve is zero and the enthalpy balance gives

$$m_1 = m (h_1 - h_{2,sat\ liq}) / (h_{2,sat\ vap} - h_{2,sat\ liq})$$
$$= m (-778 + 817) / (-669 + 817)$$
$$= 0.264\ m$$

where m_1 = vapor flow from economizer drum
m_2 = liquid flow from economizer drum
m = liquid flow from condenser accumulator

But $m = m_1 + m_2$

So $m_1 = 0.264 (m_1 + m_2)$

or $m_1 = 0.359\ m_2$

Liquid m_2, saturated at the intermediate pressure, then is flashed to the chiller, producing proportionately less vapor than in the simple cycle. Rate m_2 is

$$m_2 = q_{ch} / (h_4 - h_3)$$
$$= 1 \cdot 10^6 \text{ Btu/hr} / (-690 + 817) \text{ Btu/lb}$$
$$= 7874 \text{ lb/hr}$$

Thus $m_1 = 0.359 (7874) = 2827 \text{ lb/hr}$
$m = 2827 + 7874 = 10701 \text{ lb/hr}$

The vapor from the chiller is compressed to the economizer pressure. As before, this compression work cycle is found in two steps.

$$-W_{is} = -662 + 690 = 28 \text{ Btu/lb}$$
$$-W_{act} = 28 / 0.75 = 37.3 \text{ Btu/lb}$$
$$m_2 (-W_{act}) = (7874) (37.3)/2544 = 115 \text{ HP}$$
$$h_5 = -690 + 37.3 = -652.7 \text{ Btu/lb}$$

The chiller vapor then is combined with the economizer vapor. The enthalpy of the mixture is found by enthalpy balance for this adiabatic mixing process.

$$m_1 h_{2,sat\ vap} + m_2 h_5 = m h_{5,m}$$
$$(2827) (-669) + (7874) (-652.7) = 10701\ h_{5,m}$$
$$h_{5,m} = -657 \text{ Btu/lb}$$

The combined vapor stream is now compressed to the condenser inlet pressure. The work is:

$$-W_{is} = -657 + 641 = 16 \text{ Btu/lb}$$
$$-W_{act} = 16 / 0.75 = 21.3 \text{ Btu/lb}$$
$$m(-W_{act}) = (10701) (21.3)/2544 = 90 \text{ HP}$$
$$\text{Total HP} = 115 + 90 = 205$$
$$h_6 = -657 + 21.3 = -635.7 \text{ Btu/lb}$$

Finally, the condenser duty is calculated.

$$q_{cond} = (10701) (-778 + 635.7) = -1.52 \text{ MMBtu/hr}$$

Considerable savings in circulation rate and in horsepower is achieved (205 vs. 268 hp). The importance of the horsepower savings lies primarily in the fuel savings for the compressor driver. For a reciprocating compressor there might be no reduction in frame size and virtually the same investment cost. In spite of the potential savings for the economizer cycle, the simple cycle is used a great deal in industry.

The GPSA Engineering Data Book (1987) has charts in Section 14 for quick estimation of both the simple and economizer cycles. These charts assume an isentropic efficiency of approximately 0.75 and so imply use of a centrifugal compressor.

The simple and economizer cycles solved above using hand calculations and a pure propane log P-h diagram are now simulated using OPSIM (App. 4). The refrigerant chosen for OPSIM simulation is similar to that of a refrigerant-grade propane—2.0 mol percent C2, 97.5 mol percent C3, and 0.5 mol percent iC4.

The following OPSIM computer results compare favorably with the previous hand calculations for both cycles.

	Simple Cycle			Economizer Cycle	
Variable	Hand Calc.	Computer	Variable	Hand Calc.	Computer
m, lb/hr	11363	11171	m_1, lb/hr	2827	2950
$-w$, hp	268	260	m_2, lb/hr	7874	7551
q_{cond},			hp_1	115	106
MMBtu/hr	1.68	1.66	hp_2	90	101
P_{cond}, psia	260	267.4	Total hp	205	207
P_{ch}, psia	32	31.2	q_{cond},		
			MMBtu/hr	1.52	1.52
			P_{econ}	117	117

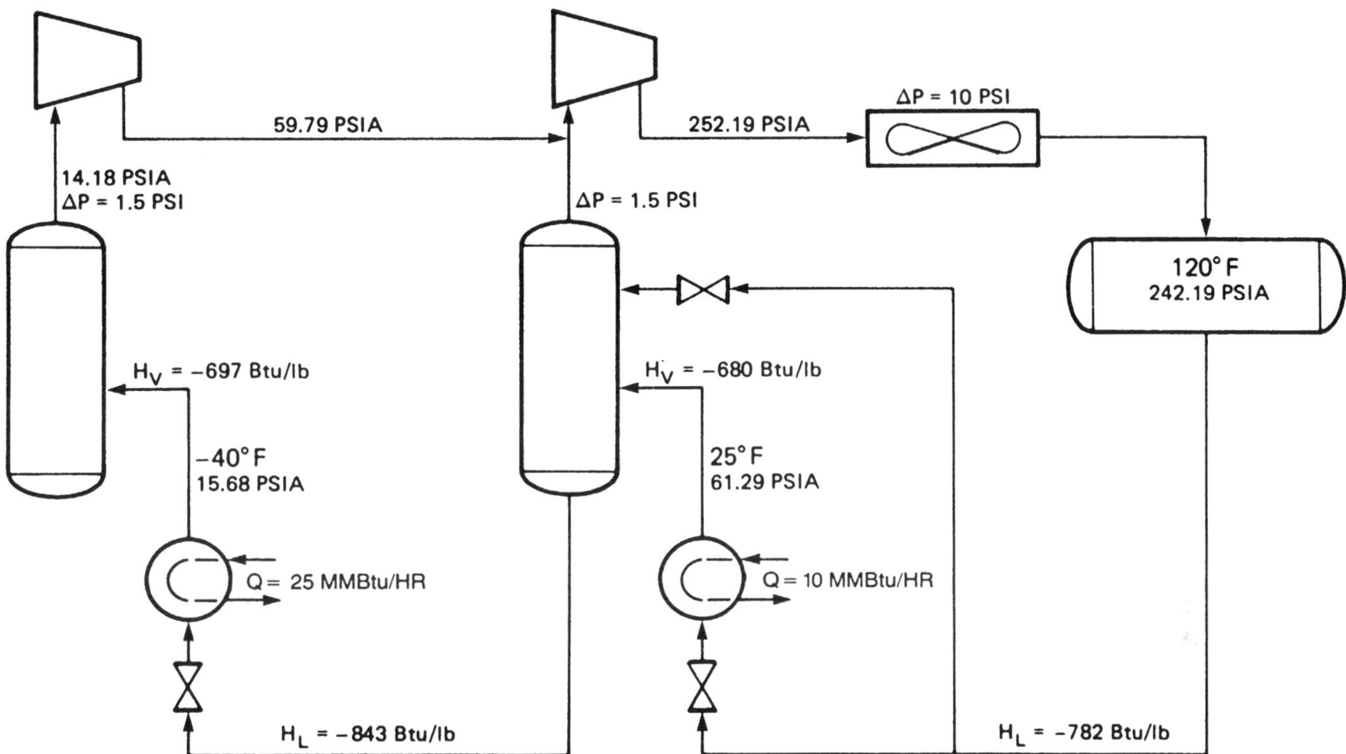

Figure 14–12. Two-stage refrigeration system (GPSA, 1987, Fig. 14–6).

Split-Level Refrigeration. A second potential improvement over the simple refrigeration cycle is encountered when two cooling duties are required at different temperature levels in a process, or when the duty can be divided into two parts, each done at a separate temperature level. Both duties can be provided by one *two-stage refrigeration cycle* as shown in Figure 14–12. The advantage of this arrangement is similar to that of the economizer cycle; only a portion of the refrigerant is circulated to the lower-pressure chiller, thus saving horsepower. On the other hand, two chillers are required rather than one, and the total area for heat exchange will be higher because the mean temperature driving force is less.

If split-level cooling is used on the front end of an expander plant, then the arrangement shown in Figure 14–13 probably provides the best heat economy. Not only is a second chiller required but also a third gas-to-gas exchanger. Moreover, a two-stage unit requires compressor load-balancing control (Fraser, 1984).

Gas-engine-driven reciprocating compressors rather than centrifugal units power many, if not most, gas-processing refrigeration units. Also, two stages of compression are usually used because the high compression ratio would create too large a compressor rod loading. The example above has an overall compression ratio of 260/32 = 8.1, far above

the usual limit of approximately 6.0. Thus, from a compressor standpoint, the addition of the economizer or use of split-level cooling involves very little additional investment—only one vessel or chiller and the previously mentioned load-balance control.

Fraser (1984) compared the split-level process and the

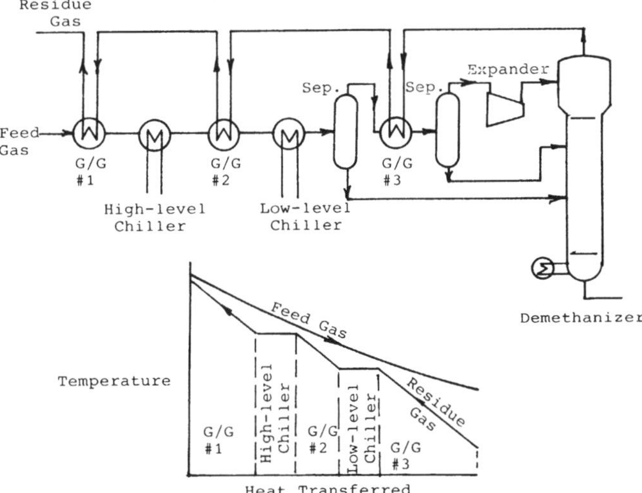

Figure 14–13. Split-level refrigeration for expander-plant feed: flow scheme and temperature pattern.

economizer cycle with the simple cycle. As previously indicated, the economizer cycle is superior to the simple cycle. Fraser concludes that split-level refrigeration is a viable alternative in many plants, especially if the two refrigeration levels differ by more than 60°F and if cooling duties exceed about 3 MMBtu/hr.

Apparently, split-level refrigeration is not commonly used. Blackburn and George (1979) show two such units. The process introduces complexity but has a large potential for saving energy costs.

Pure Refrigerant Limitations. Propane is not used for refrigeration below –40°F because its vapor pressure is less than one atmosphere. Air leakage into the system is a danger.

Cascade Refrigeration. As shown in Figure 14–14, refrigerant systems can be "stacked" to provide cooling below –40°F. The low stage uses methane as the working fluid. The methane is boiled in a chiller at near its atmospheric boiling point (–259°F), is compressed and condensed, rejecting heat to an ethylene evaporator at approximately –150°F. In turn, the ethylene is compressed and condensed, rejecting heat to a propane chiller at approximately –30°F. Finally, the propane rejects its heat to the surroundings. In this bootstrapping operation, heat can be absorbed at –250°F or so and rejected at ambient temperatures of about 100°F.

Cascade refrigeration has two drawbacks. First, a great deal of compression horsepower is required. Second, cascade systems are complex, with attendent operating and maintenance problems. A few two-stage systems are used in condensate recovery, but the process is not usually economically competitive. Cascade refrigeration *is* used in liquefied natural gas (LNG) plants.

Mixed Refrigerants. Geist (1985) describes the use of mixed refrigerants in LNG plants. *Mixed refrigerant* contains the hydrocarbons methane through pentane. This mixture evaporates over a wide range of temperatures rather than at one constant temperature. The vapor is compressed as one stream and rejects heat to cooling water or air. Compression requirements are thus lower than for a cascade system.

MacKenzie and Donnelly (1985) describe a mixed-refrigerant system used for condensate recovery. Further details are given later in the section on external refrigeration for condensate recovery.

Refrigerant Purity. As indicated in the previous OPSIM calculation, propane refrigerant used in gas plants is not pure, but typically contains small amounts of ethane and isobutane. Blackburn and George (1979) simulated three actual refrigeration cycles and concluded that *ethane content* of the refrigerant is particularly important. To control ethane content, they recommend purging, either at the cold end of the refrigerant condenser or on the vapor section of the surge tank.

The above simple refrigeration cycle is simulated using OPSIM for three different propane refrigerants having varying amounts of ethane and isobutane as impurities.

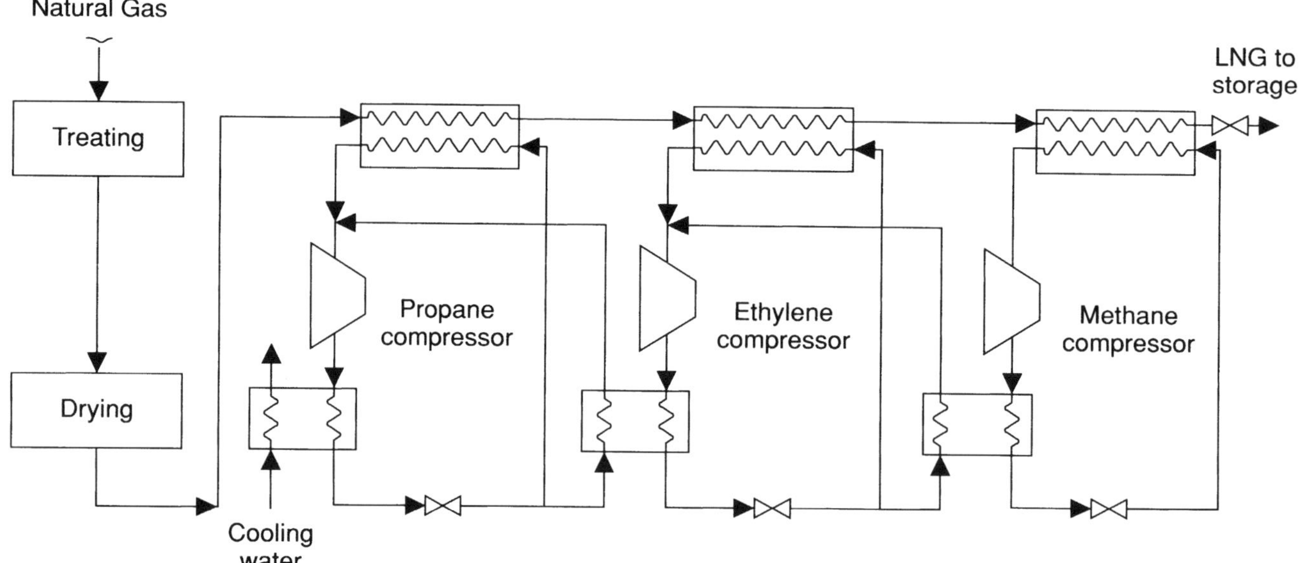

Figure 14–14. Cascade cycle (Geist, 1985).

The results tabulated below clearly confirm the undesirable effects of ethane and isobutane.

Refrigerant Composition, Mol %			Compressor Horsepower
Ethane	Propane	Isobutane	
2.0	97.5	0.5	260
4.0	95.5	0.5	266
2.0	96.5	1.5	262

Blackburn and George (1979) also note that compressor lubricating oil is a bothersome contaminant that can gather in the chiller and even coat the chiller tubes, thus reducing heat transfer. They recommend mounting the chiller on a slight angle from horizontal to facilitate drainage of the oil. The drained oil is sent to a reboiled separator to flash off propane and is then discarded.

External Refrigeration

Straight refrigeration for condensate recovery is a simple process. If propane refrigeration is used, the cold separator temperature is set straightforwardly.

If the *hydrocarbon dew point* is the controlling design parameter, then the feed gas must be refrigerated to a temperature low enough to provide a dew point lower than that encountered at any point in the downstream pipe line and/or the sales-gas specification. Practice is generally on the safe side. In other words, if a dew point of 30°F is desired, the gas will normally be cooled to a temperature safely lower than that required to obtained the desired hydrocarbon dew point. This allows for the possibility of retrograde condensation.

If *maximum NGL* recovery is desired, set the chiller temperature at the coldest practical with propane refrigeration, −20 to −40°F, depending on the steel used.

Russell (1977) noted that NGL recovery by refrigeration represents, in fact, a *flexible* process. As an example, he cites that the same plant can recover 60% of the entering ethane from a small flow of a rich gas stream (7 GPM C2+) or 30% of the ethane from three times as much of a leaner gas (3 GPM). Russell also shows how the effects of changing conditions can be estimated quickly.

Figure 14–15 shows the importance of the *condensate stabilizer* in basic refrigeration processes. The four variations shown differ in heat-exchange patterns and in disposal of the stabilizer overhead vapor. Types A and B differ only in the warming of the liquid feed to the top tray of the stabilizer. The warmer stabilizer overhead of type A requires liquid recycle to the feed, while the colder overhead of type B allows exchange with refrigerant and discharge to the residue gas.

The type C process features a refluxed stabilizer, which allows better product separation. Type D allows stabilization at a higher pressure (and thus temperature), and so removes the need for a recompressor.

In some gas plants the stabilizer overhead vapor is used as a fuel and therefore is not recycled or sent to the sales gas.

Russell states that for recovery of C3+, types B and C are better than A at higher pressure and lower temperature, or for higher GPM. Type C is more appropriate for large plants, where its higher recovery efficiency is well utilized. For C2+ recovery, type B is favored over type A.

MacKenzie and Donnelly (1985) describe a *mixed-refrigerant system* for the recovery of 79% of the propane in a feed gas by refrigeration to −100°F. For this particular case the mixed-refrigerant is less expensive to install and operate than an equivalent turboexpander plant. The plant is designed for a fairly lean feed gas (2.63 GPM), which yields less than 10% propane recovery when refrigerated to –30°F. In other words, ordinary propane refrigeration is not even a viable choice.

Figure 14–16 shows a mixed-refrigerant system. The inlet gas is refrigerated to –100°F and separated. The separator gas passes back through the gas-to-gas exchanger and thence to sales. The cold liquid is further heat-exchanged and fed to a deethanizer column. The bottom liquid goes to storage while the overhead vapor is recompressed, mixed with the cold separator gas, and sent to sales.

Mixed refrigerant is essentially 30/25/35/10 mol percent methane-ethane-propane-butane. The refrigerant flow is not described in detail, but Figure 14–16 reveals the basic pattern: partial condensation followed by expansion and vaporization.

Turboexpander Plants

McKee (1977) describes the expander plant *evolution* from the early, simple plants to today's energy-efficient units. The expander plant shown in Figure 14–7 illustrates the operation very well but is somewhat misleading. Many *variations* occur in the heat recovery schemes (refrigeration recovery, actually). These and other items important to expander plant design and operation are now discussed, beginning with two important equipment items— turboexpanders and plate-fin heat exchangers.

Turboexpanders. Turboexpanders are rotary devices through which gas flows from high to low pressure. Work is recovered from the rotating shaft. An energy balance shows that this work is obtained from the decrease in gas enthalpy (if potential and kinetic energy changes are neglected and if the expander is considered adiabatic). This

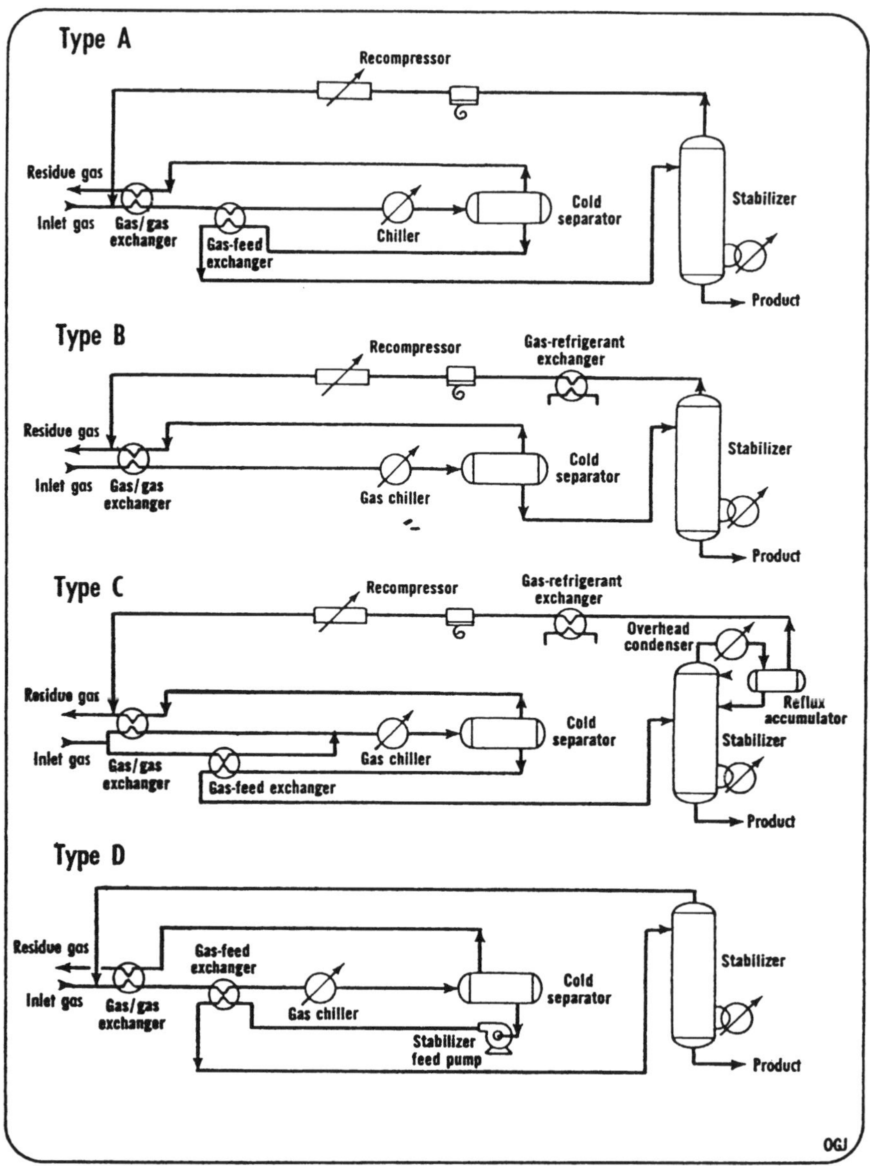

Figure 14–15. Refrigeration processes (Russell, 1977).

large decrease in enthalpy produces a correspondingly large temperature drop in the gas. This device produces low temperatures conducive to high condensate recovery.

Figure 14–17 is a cross-sectional drawing of a typical turboexpander. Guided by adjustable inlet vanes, the feed gas flows radially inward in a tangential direction parallel to the rotor vanes. The gas spirals into the central area of the rotor (B) where another set of guide vanes directs the gas axially out without a rotating or vortex motion. Expanders are constructed to handle a condensing gas stream, with up to 10 weight percent liquid in the outlet (Valdes, 1984). However, solids such as solid CO_2 erode the rotor and should be avoided.

In most expanders, the angle of the inlet guide vanes is adjustable. This allows the flow to be varied at a given inlet pressure, thus controlling the outlet pressure.

In Figure 14–18, as in most applications, the *brake* for the expander is a compressor, mounted on the same shaft. The unit is composed of a central bearing section with attached expander and compressor rotors. The expander and compressor covers can be removed easily for inspection and repair of the central section. The latter is the heart of the machine; it contains the impellers, seals, and bearings. This unit is easily replaced.

The expander *shaft* is stiff and rotates at 10 to 50 thousand or more RPM. The tip speed of the expander rotor is typically

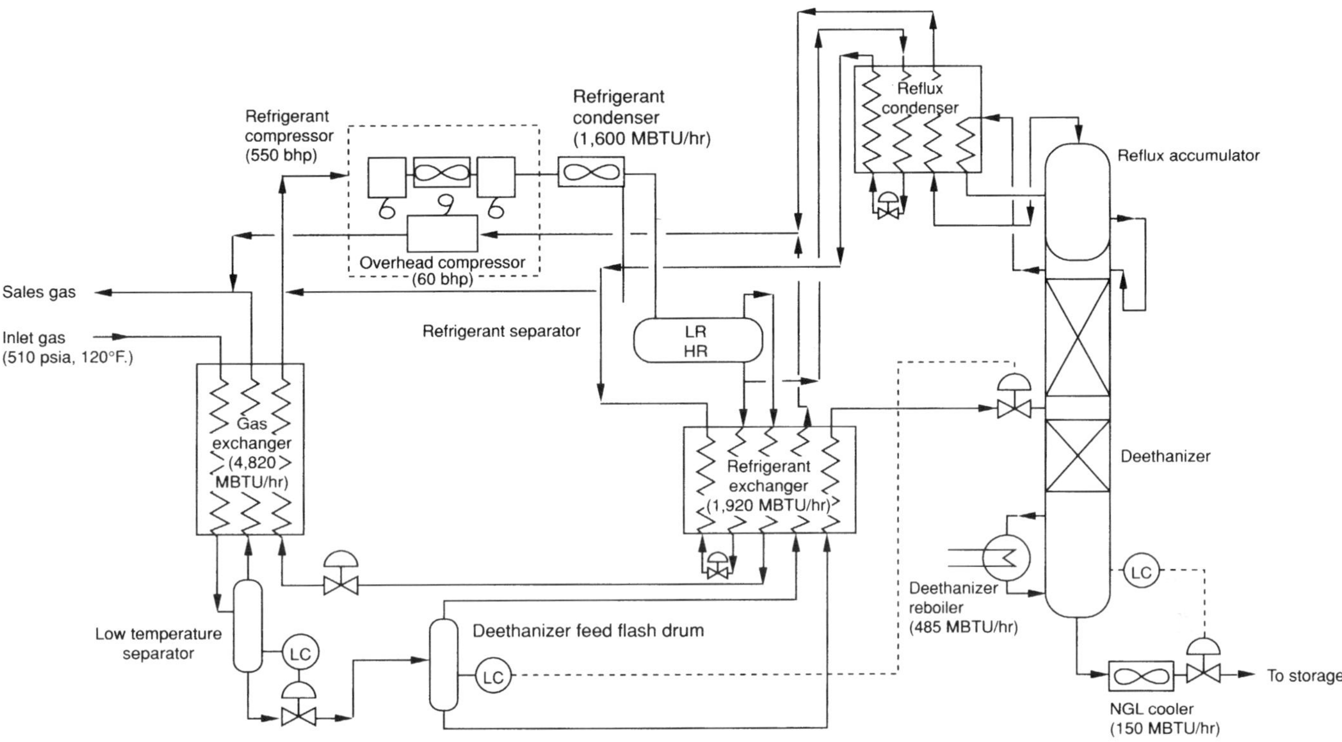

Figure 14–16. Mixed refrigerant flow (MacKenzie and Donnelly, 1985).

1500 ft/sec (this and many of the other details listed below were discussed by McWilliams, 1985). The thrust and journal bearings are bronze and are oil-lubricated. The rotor is aluminum.

Seal gas is supplied to *labyrinth seals* at a pressure about 50 psi greater than the process stream, preventing oil leakage to the gas. At the same time the lubricating oil for the bearings is protected from the cold temperature of the process stream. The seal gas must be dry. If the seal gas supply fails, process gas goes into the oil system and could ruin the expander.

The *start-up sequence* for an expander is to (1) establish seal gas pressure, (2) start the oil lubricant flow, and (3) introduce the process gas. Of course shutdown is in the opposite sequence.

A channel between the two sides of the compressor impeller is used to equalize the pressure on the two sides. Axial thrust of the impeller shaft is thus minimized. Control of this thrust has been the biggest operating problem. Improvement of automatic control systems has gone a long way towards solving the problem (McWilliams, 1985).

Because *quick shutdown is a must*, the process-gas inlet valve must be capable of closing very quickly—on the order of 0.5 sec. Air-operated, spring-loaded valves are used.

Auxiliary systems needed are: a seal-gas supply system and a lube-oil recirculation system with filtering and cooling. The oil system can be equipped with an emergency feature that guarantees oil circulation for a short time while the expander is shut down.

Operating problems include (GPSA, p. 13–42, 13–44):

- Ice or solid CO_2
- Abrasive solids
- Rotor resonance and bearing vibration
- Condensing streams
- Variable flow
- Oily feed gas
- High pressure
- Temperature shock

Special bearings handle resonance problems. This helps with variable flows. Channels in the rotor entry region help direct dust in the inlet stream to avoid problems from that source.

Expanders can handle services ranging from very low to as high as 10000 HP. One limitation is the expansion ratio per stage, normally 2.5 or less. However, expanders can be cascaded, with two or even three stages of expansion on the same or separate shafts. Another limitation is the work per stage, which is usually confined to 50 Btu/lb or less.

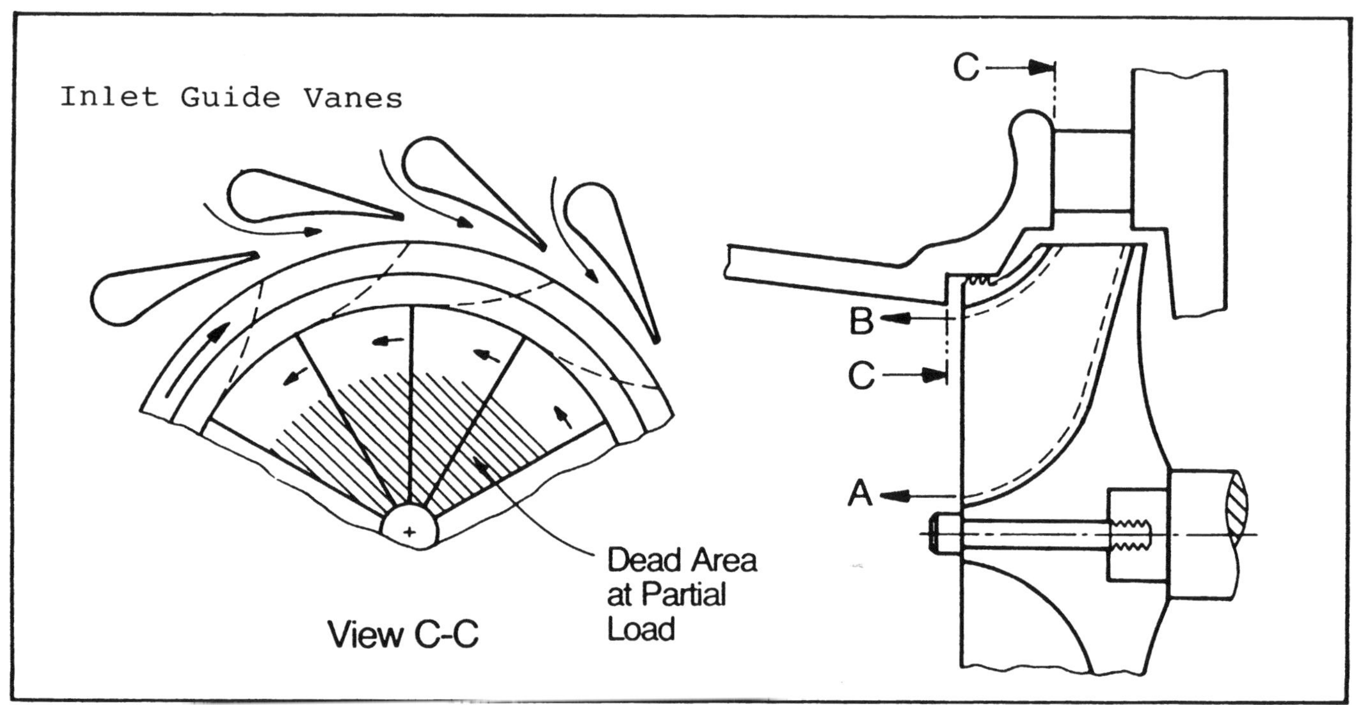

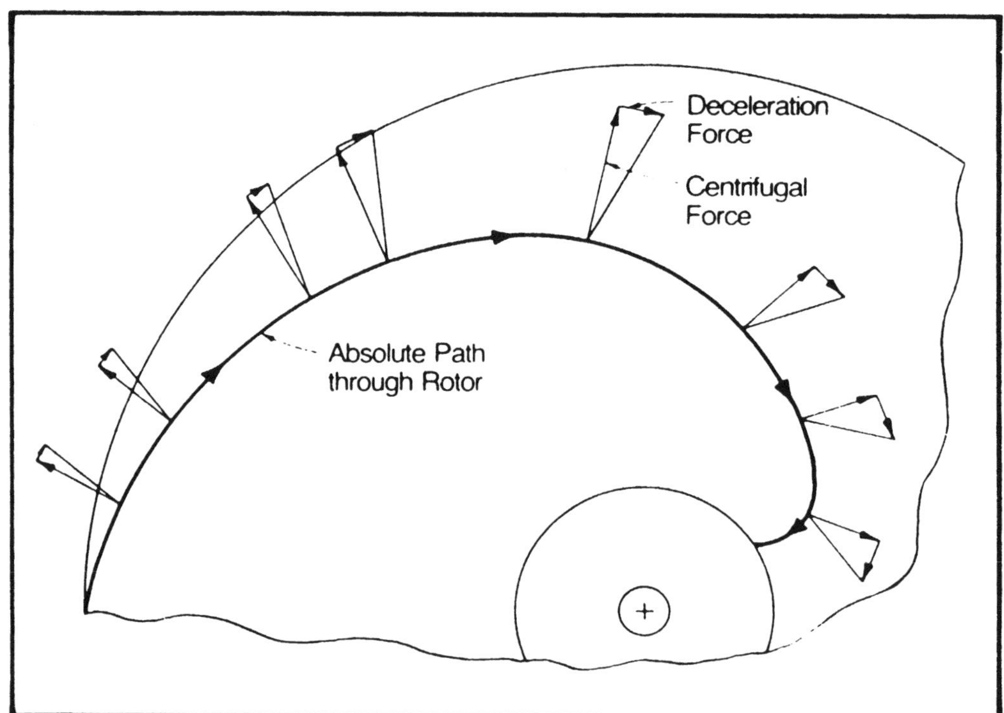

Figure 14–17. a) Cross section of turboexpander rotor and b) path of fluid in rotor (Swearingen, 1972. Reproduced by permission of the American Institute of Chemical Engineers).

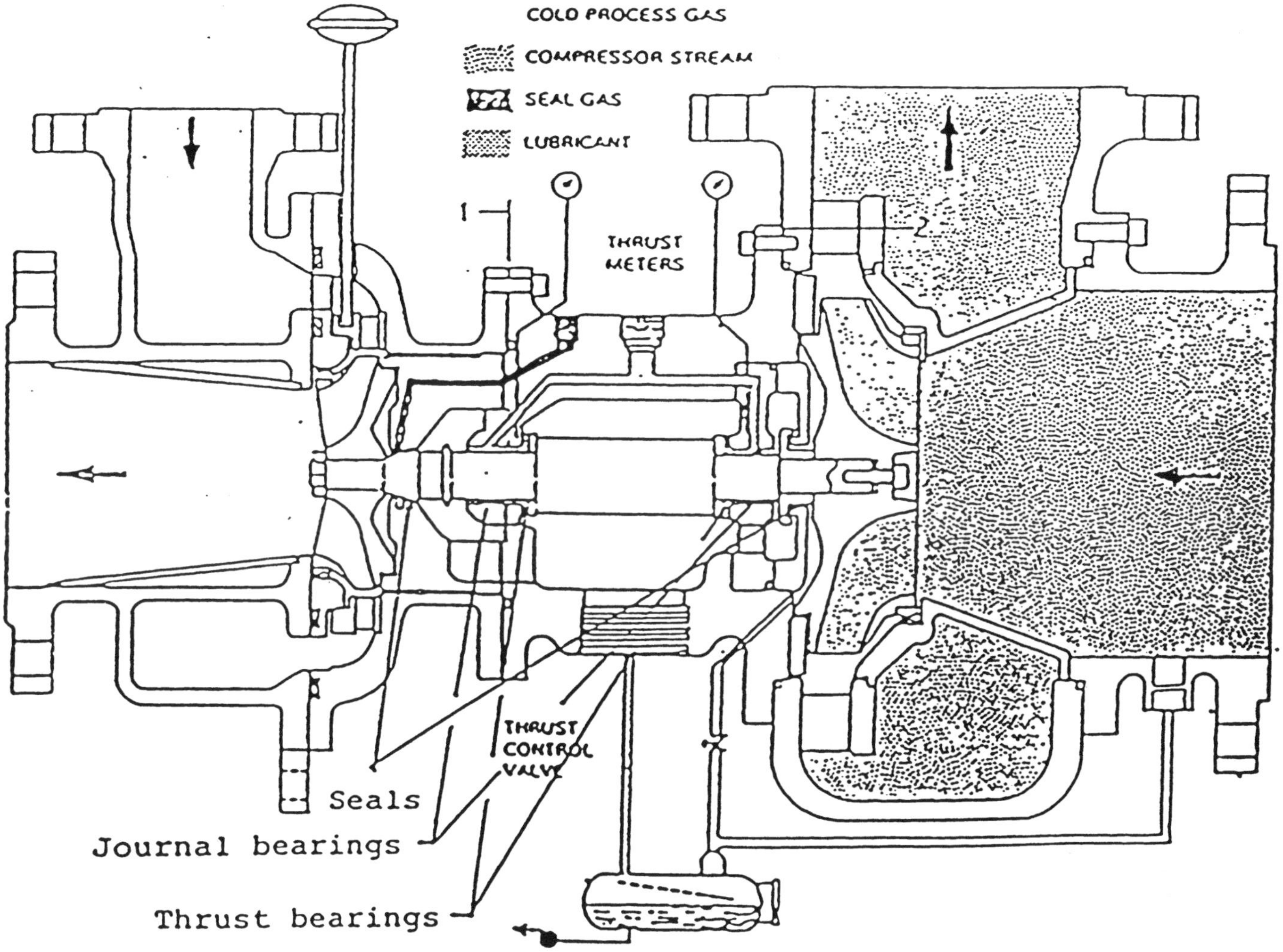

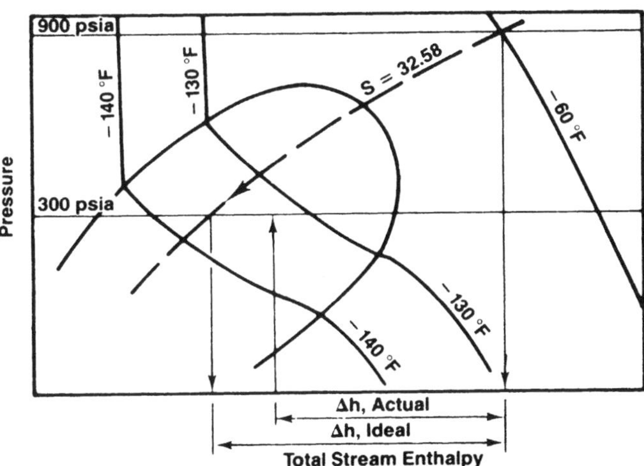

Figure 14–18. Schematic diagram of turboexpander (Holm, 1986. Reproduced by permission of the American Institute of Chemical Engineers).

As with compressors, *expander design* uses the isentropic model. This is easily illustrated by a pressure-enthalpy plot Figure 14–19. For expanders the reversible and adiabatic (isentropic) path slopes downwards and to the left on the p-H diagram. The enthalpy change for this path is the isentropic work. The actual work available from the gas is less than the ideal work because of irreversibilities in the gas as it flows through the expander.

The *actual work* recovered from the expander is thus the isentropic work *times* the isentropic efficiency. This efficiency is typically about 75%. As shown in Figure 14–19, the enthalpy change of the gas is less than the isentropic change. As a consequence, the temperature of the gas leaving the expander is somewhat higher than the isentropic value.

Not all of the work provided by the process fluid is available at the compressor end of the shaft because of *friction* in the bearings and seals. The process industry

Figure 14–19. Turbo-expander model using pressure-enthalpy coordinates (GPSA, 1987, Fig. 13–63).

typically assumes that 98% of the work is recovered in the compressor.

Plate-Fin Exchangers. An important type of heat exchanger used in expander plants is the *plate-fin* exchanger. These special-purpose exchangers (see Chapter 12, Fig. 12–14) are made of brazed aluminum, which can operate satisfactorily at very low temperatures. These exchangers are compact and efficient. Their chief drawback is the need to prevent solids from entering them. The narrow channels become clogged easily and the exchanger then performs poorly. Temperature shocks must be avoided also. Of course, shell-and-tube heat exchangers can be and are also used.

Design Procedure. The process objective is first assumed to be high percentage recovery of the entering ethane in the liquid product. Ethane rejection is discussed later.

Turboexpander or "cryogenic" plants operate with temperatures of -130 to $-160°F$ at the top of the demethanizer tower. To avoid hydrate formation and damage to equipment, the **feed gas must be dehydrated with solid desiccant**, usually molecular sieves. Also, expander plants must be thoroughly de-watered by evacuation or other procedure before commissioning and operation.

The following discussion is based on McKee's (1986) presentation. Other sources will be mentioned directly.

Start the design of a new plant by considering carefully the given information: pressure level, plant altitude, temperature level, flow rate, and composition.

The *outlet pressure* is ordinarily the delivery pressure to the sales-gas line, which ranges from 700 to 1000 psia, although occasionally it may be lower. If less than sales-gas delivery pressure, the inlet pressure is important because compression may be required to the expander inlet pressure.

Plant altitude is important because it has an appreciable effect on compressor driver horsepower, especially for gas turbine drives.

Feed-gas temperature is important since sometimes the feed can serve as the heat source to the demethanizer bottom reboiler. This saves energy and, at the same time, cools the inlet gas.

Flow rate is important for equipment sizing; the maximum expected rate should be used for this purpose. The lowest design rate also is important to the design of the recompressors. If the recompressors are centrifugal units, they have a definite surge limitation on the feed rate. If wide swings in feed rate must be provided for, it may be necessary to specify two parallel recompressors instead of one.

Composition is important for several reasons. First, CO_2 in the feed must be limited to, say, 0.5–1.0%. Otherwise solid CO_2 may form in the unit with subsequent equipment damage.

The *condensate content* is important also. High recovery of ethane (say 80%) without external refrigeration is possible only with a lean feed gas (up to 2.5 GPM) according to Wang (1985). Moderately rich gases require external refrigeration to obtain high recovery. And in the latter case, refrigeration horsepower is more effective than recompression horsepower (Wang, 1985). Rich gases will normally be processed by external refrigeration.

The next big decision is the *percentage ethane recovery in the NGL* . This may be dictated by external considerations, i.e., demand, profitability, management fiat. If the gas is very lean and contains inerts, then the sales-gas heating-value specification may control the ethane recovery. For lean gases at high cold-separator pressure, say 1000 psia, very little or no liquid may condense (in the upper retrograde region of the inlet-gas phase envelope), in which case the overall ethane recovery may be limited to 75–80%. One stage of turboexpansion will cool the feed gas enough to extract perhaps 80% of the ethane. Two expander stages will allow deeper extraction (say 90%) at the cost of increased recompression. Another technique for obtaining 90% recovery is the use of a refluxed demethanizer (see below).

A very important decision in expander-plant design is the *recompressor*. Compressors come in finite frame sizes, so that not just any pressure level can be chosen arbitrarily for the demethanizer. A compressor selection is made, and then the recompressor is calculated to determine its inlet pressure, from which the other important pressures can be selected (see McKee for details). McKee made definite recommendations for the isentropic efficiencies of the various rotating equipment:

Item	Isentropic Efficiency
Centrifugal recompressor:	
Small	0.72
Large	0.75
Reciprocating recompressor:	
High speed	0.80
Medium speed	0.82
Low speed	0.85
Refrigeration compressors	
Normal	0.70
Warmer	0.75
Expander	0.80

Refrigeration recovery is very important. If at all possible use the feed as the energy supply for reboiling. Feed temperature may preclude it as a heat source for bottom reboiling. On the other hand, side reboilers may do up to 60% of the total reboiling. These reboilers operate at very low temperatures on the process side as compared to ambient temperatures. Therefore it is possible to use feed gas as a heat source. This practice reduces the required refrigeration. Again, see McKee for details. Figure 14–20 shows a typical expander plant with side reboiling.

A simple example of the *advantage of side reboiling* is the sequencing of the inlet gas-to-gas exchangers in the absence of side reboiling. The two possible flow arrangements are shown in Figure 14–21. Note the position of the refrigeration chiller in the heat-exchanger train. Placing the chiller *between* the two gas-to-gas exchangers and the two side-reboiler exchangers allows the cold-separator temperature to be considerably lower than the chiller outlet temperature. This is a much more efficient use of the available refrigeration.

One expander plant variation presently used to a limited extent is the *refluxed demethanizer* (Fig. 14–22)—the flow sheet is proprietary and its use requires royalty payment. A portion of the chilled gas from the cold separator, say 30%, is bypassed around the expander and, instead, is cooled with demethanizer overhead vapor. The partly-condensed stream is then passed through a valve, further cooling and condensing it at a temperature lower than the outlet temperature of the expander. The bypass-stream, which is now colder than the expander outlet stream, goes into the

top of the demethanizer as cold, mostly liquid reflux. The expander outlet stream enters at a lower point in the demethanizer. The net effect is higher C2+ recovery than is possible without the reflux, up to 90%.

Complete design of an expander plant involves trial and error. See McKee (1986) for details.

Ethane Rejection

Because the market price for ethane can vary widely, it is advantageous to have a *versatile process* capable of either retaining ethane or rejecting it. Such flexibility can be built into the expander plant as shown in Figure 14–23 (Valdes, 1984). In the *ethane recovery mode*, a portion of the inlet gas goes to the gas-to-gas exchanger, and the remainder through the product exchanger and the bottom and side reboilers. The fractionator operates as a demethanizer, its bottoms product being cold enough to provide refrigeration for the feed gas.

In the *ethane rejection mode*, all the inlet gas passes through the product exchanger and the bottom and side reboilers. External heating is provided for the feed gas in order that it can be warm enough to effect ethane rejection in the stabilizer. The gas then passes through the gas-to-gas exchanger, thence to the cold separator, etc. The source of heat for the feed gas can be the hot residue gas from the recompressor. Alternatively, hot oil can be used.

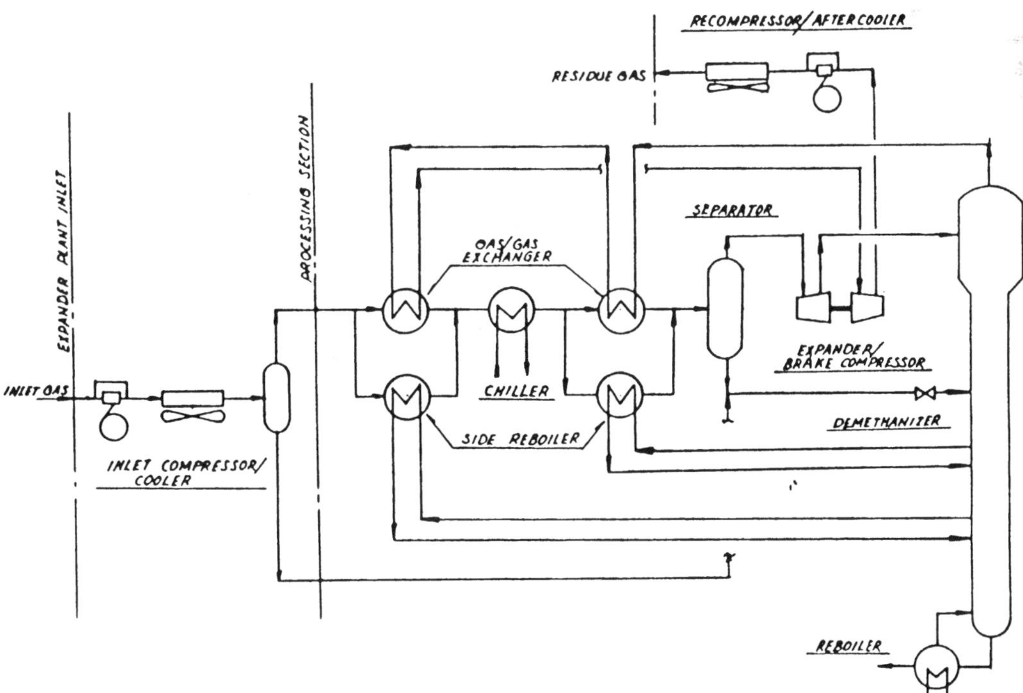

Figure 14–20. Expander plant with side reboiling (Wang, 1985).

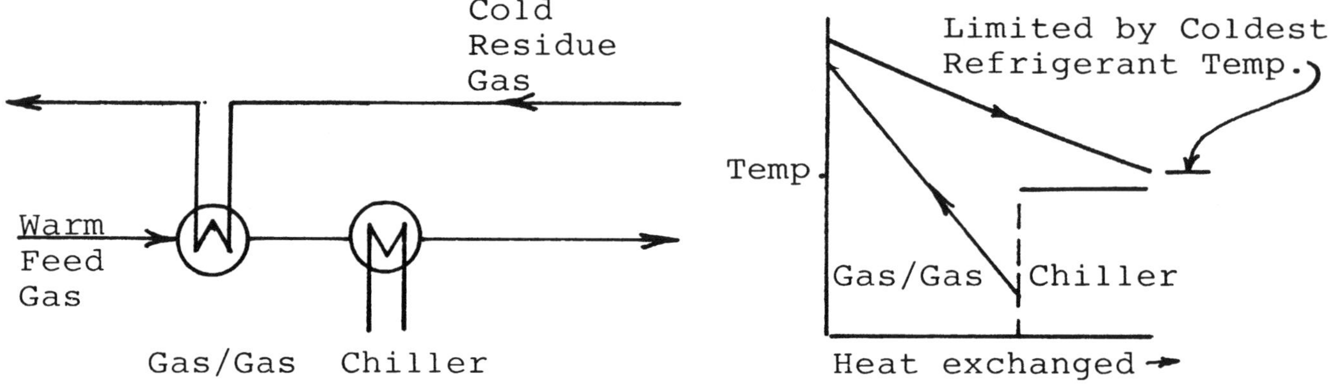

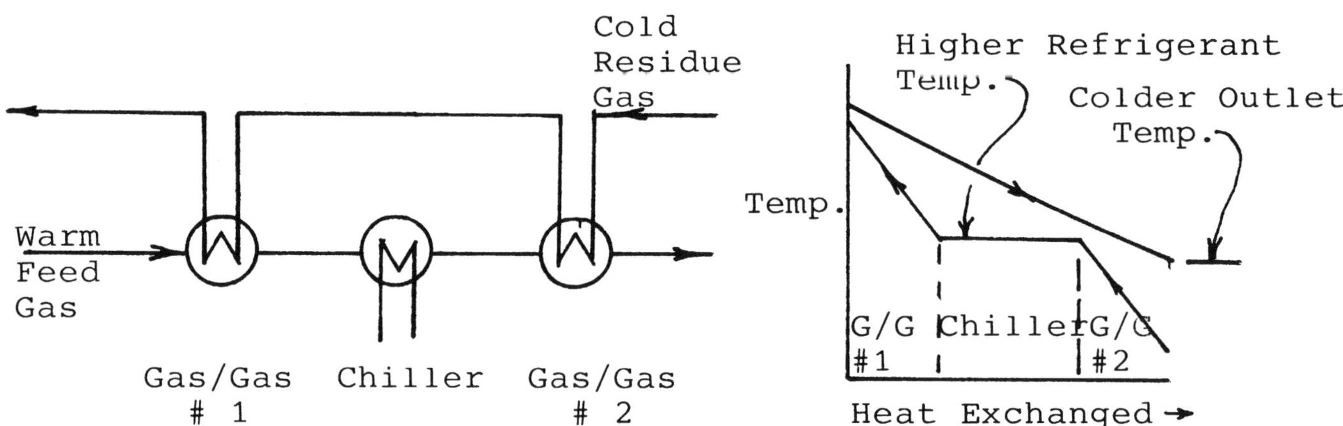

Figure 14–21. Gas-to-gas/chiller arrangements.

Evans (1980) outlines alternate ethane recovery/rejection schemes.

Oil Absorption

Oil absorption plants are more complex than the simple depiction of Figure 14–8. An old plant, no longer in operation, but typical in its complexity is the Empress Plant in Alberta (Barber *et al.*, 1965) shown in Figure 14–24. Some features of this plant are discussed to clarify the flow sheet.

First, note the *rich-oil flash tank*. This is fairly common in cold-oil plants. The idea is to heat the rich oil a bit and flash off some of the methane and ethane that would otherwise enter the rich-oil deethanizer (ROD), taking some of the load off that column. The flashed gas is recycled back to the plant feed gas.

Note next the series of *three reboilers* on the ROD (two are side reboilers). This arrangement is designed to maximize heat recovery from the hot lean oil, consistent with the temperature levels in the hot oil and the rich oil demethanizer.

The hot lean oil leaving the ROD reboilers passes through a series of two coolers (first air, then water). Then it passes through the rich-oil preheaters, being cooled to a temperature lower than ambient. These *three coolers* (air cooler, water cooler, and rich-oil preheater) minimize the load on the C3 chiller.

Note particularly that the lean oil next flows to the top of the ROD, where it contacts the methane-ethane-rich gas

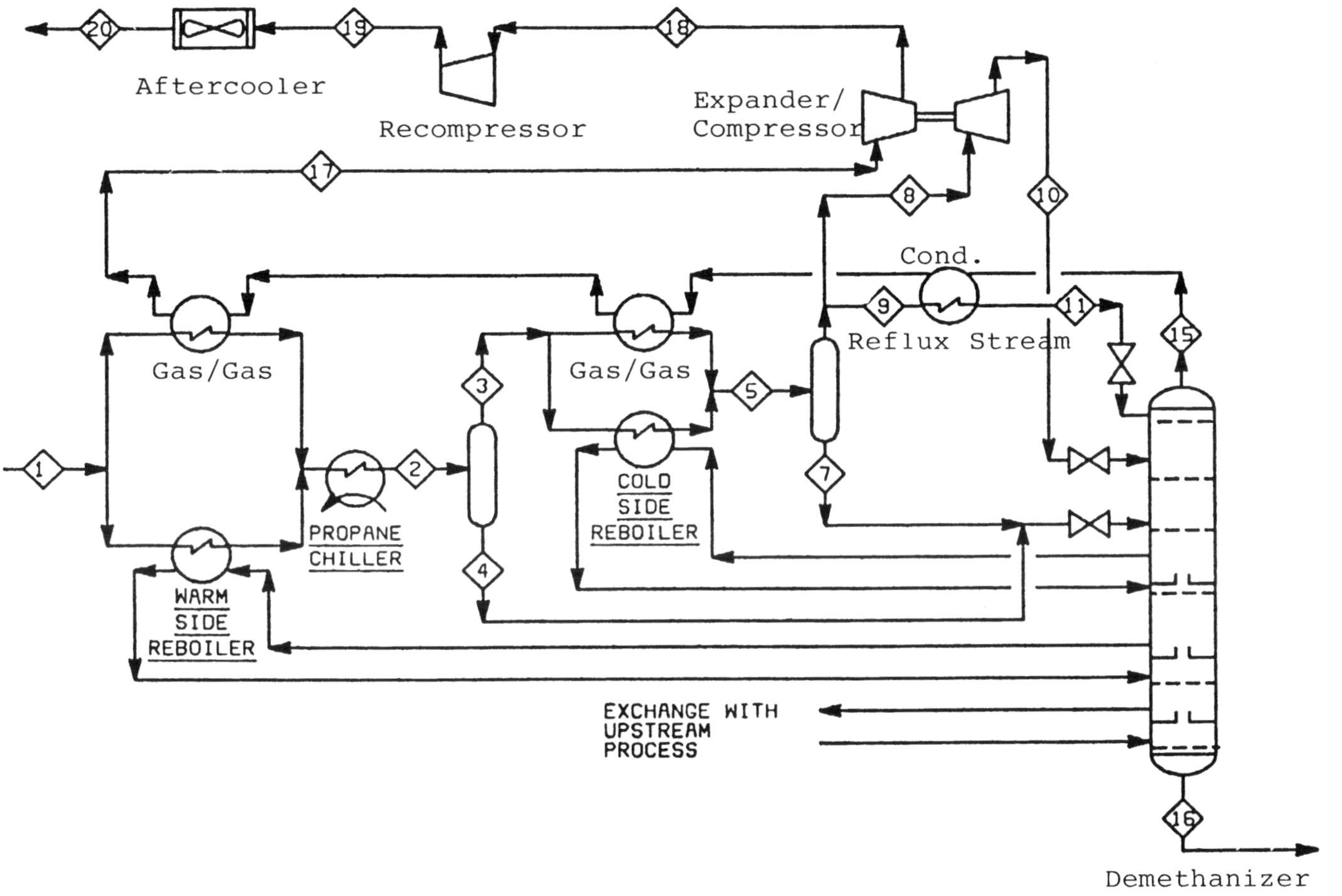

Figure 14–22. Expander plant with refluxed demethanizer.

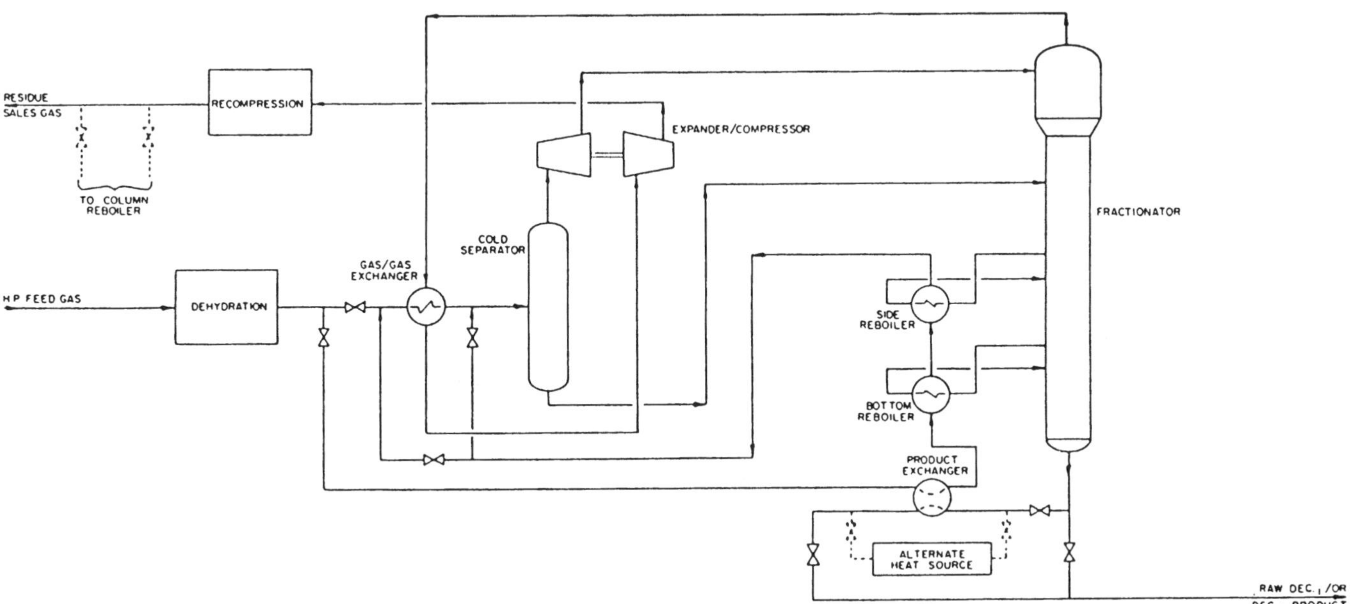

Figure 14–23. Turboexpander cycle designed for either ethane extraction or rejection (Valdes, 1984. Reprinted courtesy of Marcel Dekker, Inc.).

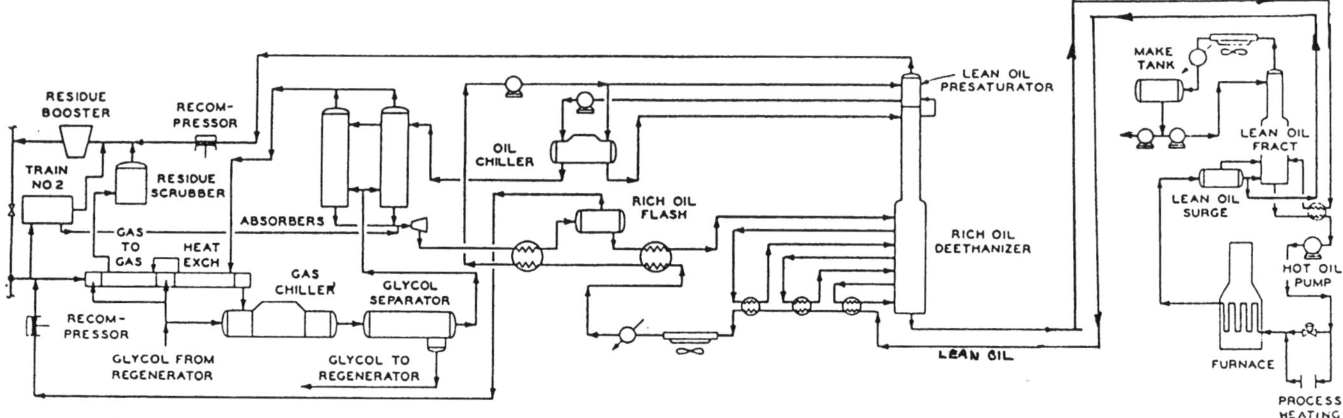

Figure 14–24. Empress cold-oil absorption plant (Barber et al., 1965).

stream leaving the top of the ROD. This so-called *lean-oil presaturator* saturates the lean oil with methane and ethane before it is chilled and fed to the top of the absorber tower. This procedure avoids much of the temperature rise otherwise encountered on the top absorber tray due to absorption of methane and ethane. Presaturation also reduces the temperature in the whole tower, thereby allowing absorption of the desired propane with less lean oil flow than would otherwise be required. The lower the lean oil flow, the lower the heat duty of the still, and the lower the energy costs. Note: the *still* is the lean-oil fractionator.

The *optimum pressure* for the absorption towers depends on two factors: the inlet pressure of the plant feed gas and the phase behavior of the key component, propane. Observation of the effect of pressure on the vapor-liquid equilibrium K-value of propane (as on p. 25–93, GPSA, 1987) indicates that the optimum pressure will be in the range of 500–900 psia. The propane K-value at typical absorber temperatures goes through a minimum in this pressure range. If the pressure of the feed gas is low, it should be compressed to this range for absorption. If the feed pressure is at pipeline pressure, then the absorption should be done at that level to avoid recompressing the whole product gas stream.

The pressure of the ROD is determined by the separation of ethane and propane. If a ratio of 0.02 C2/C3 is used, then the composition of the ROD bottoms can be estimated closely. The temperature is set by consideration of the heat transfer to the lean oil from the still. A bubble-point calculation at the assigned temperature will set the pressure.

The lean-oil fractionator bottoms is taken to be at or near 550°F, a temperature which should not cause cracking of the lean oil in the furnace. A bubble-point calculation on the lean-oil composition at the assigned temperature yields the bottoms pressure.

In practice the true lean-oil composition is determined by the nature of the C6+ materials in the plant feed gas and how the still is operated. Regardless of the nature of the lean oil originally charged to the absorber plant after several days or weeks of steady state operation, the *absorber oil will change* to reflect the gas flowing into the plant and the still operation. The molecular weight of the lean oil should be kept as low as possible without producing excessive loss in the residue gas. A lean oil "reclaimer" is used (usually on a slip stream) to remove heavier molecular weight fractions from the lean oil. The lower the molecular weight, the lower the mass circulation rate for the same number of moles (which is what controls the absorption in the cold-oil absorber). Again, lower mass flow rate means better energy economy in the furnace.

Solid-Bed Adsorption

Barrere (1968) presents the following design equation for NGL recovery using *solid-bed adsorption* units.

$$w = 100\, n\, t\, y\, MW\, /\, L$$

where w = the weight of adsorbent in the bed.

    n = feed gas molar flow rate

    t = cycle time, min (generally 2–4 hr)

    y = mol fraction of key component in feed gas, usually propane

  MW = molecular weight of key component

    L = experimental loading of key component at breakthrough time, mass/mass

The bed volume is found by dividing the adsorbent weight by its apparent density. Judicious choice of bed length to diameter ratio provides uniform flow across the bed without excessive pressure drop. Barrere describes a method for determining the equilibrium capacity of the adsorbent material. See Chapter 9 for more details on solid desiccant adsorption.

Gas Processing in Enhanced Oil Recovery (EOR)

The recovery of crude oil from reservoirs can be enhanced in certain instances by injecting either nitrogen or carbon dioxide gas. Both of these processes have influenced gas processing. In both cases, the injected gas eventually shows up in the well stream. The produced gas therefore contains large amounts of nitrogen or carbon dioxide that lower the heating value below sales-gas specifications. Complex separation operations are required to reject N_2 and/or CO_2 from the hydrocarbon portion of the effluent gas.

Ryan-Holmes Process. A process suitable for CO_2 recovery is the Ryan-Holmes process (McCann *et al.*, 1987). The difficulty with removing CO_2 from the hydrocarbon stream is the fact that the volatility of CO_2 lies intermediate to that of methane and ethane.

Figure 14–25 shows the process. Briefly, the low-pressure inlet gas containing hydrocarbon gas (including recoverable condensate) and CO_2 is dehydrated with glycol, compressed, and dehydrated thoroughly with mol sieves. The dry gas

enters a chilled absorber where recycled C5+ oil absorbs propane and heavier hydrocarbons. The gas from the column is compressed and chilled to separate methane from CO_2 and ethane. The methane-rich stream has remaining CO_2 absorbed in the demethanizer, the overhead gas going to sales.

The rich absorber oil from the propane recovery column flows to the additive recovery unit, where it is separated into a light overhead stream and a bottoms stream suitable for recycling to the propane recovery column. This novel process separates the feed gas into residue gas, C3+ condensate, and CO_2–C_2 streams. The latter stream is reinjected into the oil reservoir.

Nitrogen Rejection. Nitrogen rejection differs from the Ryan-Holmes process in that the condensate is first removed from the residue gas in conventional recovery processes. The residue gas now contains a large amount of nitrogen that would render the gas unsalable. Nitrogen rejection schemes allow separation of the hydrocarbon gases and nitrogen. The nitrogen may or may not be recycled to the oil reservoir.

Figure 14–26 shows a *nitrogen rejection process* (Alvarez *et al.*, 1984). This cryogenic process involves extensive refrigeration economy. The feed gas is thoroughly dried

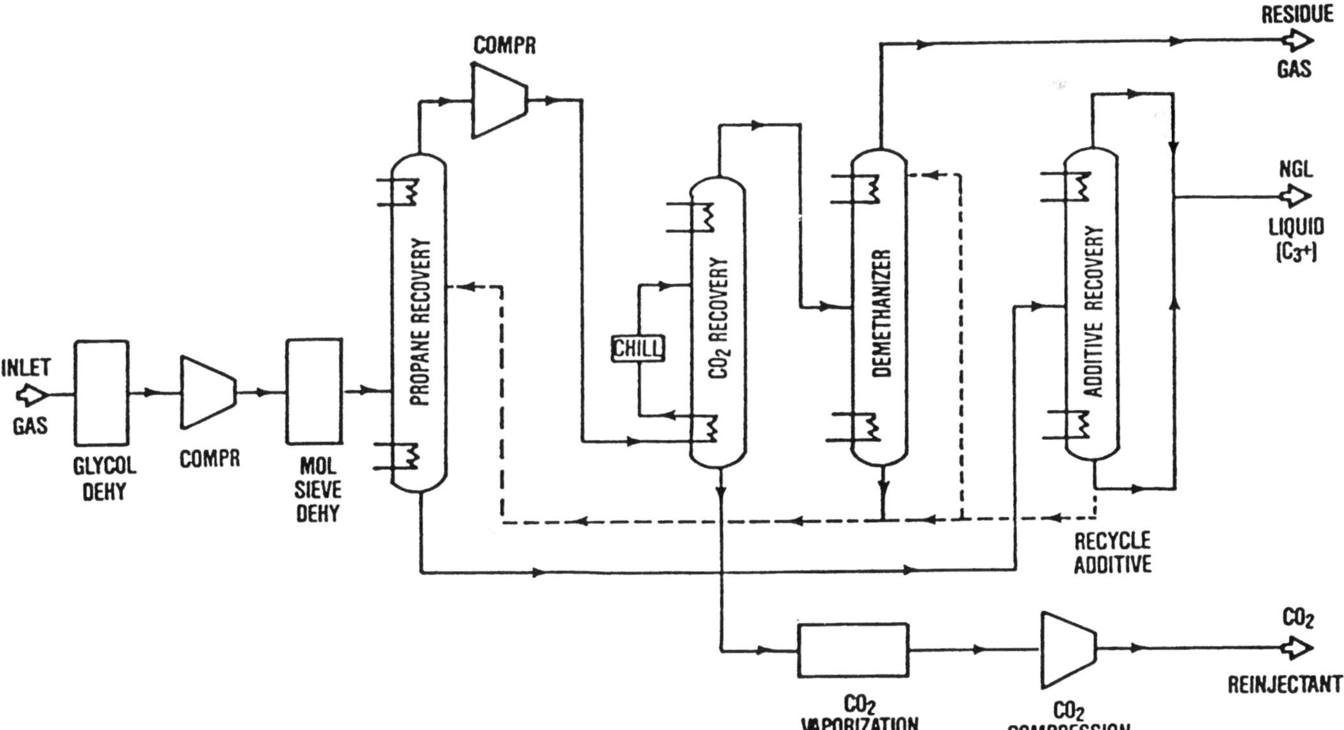

Figure 14–25. Ryan-Holmes process (McCann *et al.*, 1987. Reproduced by permission of the American Institute of Chemical Engineers).

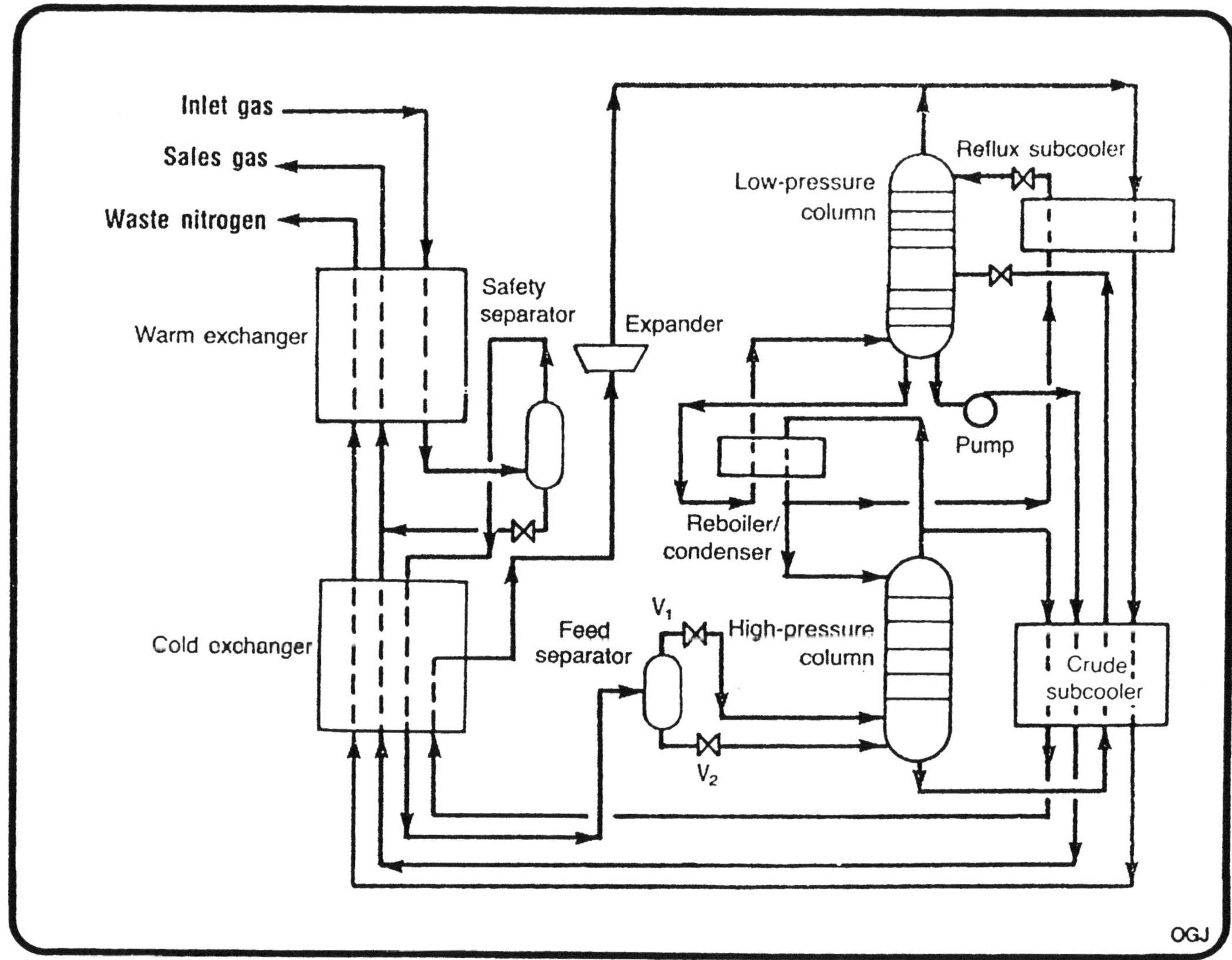

Figure 14–26.   Two-column nitrogen rejection process (Alvarez *et al.*, 1984).

with molecular sieves, then heat-exchanged with outgoing cold streams. Liquid is separated next and the gas is cooled further and introduced to the high-pressure column (approximately 400 psia). The high-pressure column is refluxed by condensing the overhead almost-pure nitrogen stream against vaporizing methane in the low-pressure column reboiler (approximately 20 psia). The crude methane stream from the bottom of the high-pressure tower is fed to the middle of the low-pressure column. The overhead nitrogen from the high-pressure column is heat-exchanged, then passes through an expander to produce a stream cold enough to provide reflux condensation for the low-pressure column. Almost pure hydrocarbon liquid is pumped from the bottom of the low-pressure column to sales-gas pressure, then exchanged with incoming streams to vaporize it. The total nitrogen from the low- and high-pressure columns is

rejected to the atmosphere after heat exchange. Davis *et al.* (1989) describe in detail a very large combined NGL recovery/nitrogen rejection facility.

Helium Recovery. Because helium is very difficult to find and recover, it is one of the most expensive industrial gases. Processes used to recover helium from natural gases include: stand-alone cryogenic, recently developed stand-alone membrane units, and combinations of cryogenic, pressure swing adsorption (PSA), and membrane units (Choe *et al.*, 1988)

LIQUID HANDLING AND TREATING

To this point, natural gas liquids have been referred to simply as a *mixed liquid product*. In older gas-processing plants, the usual practice was to have a fractionator train for separating the raw condensate into the usual products: propane, butanes (possibly separate i- and n-butane), and natural gasoline (C5+). Present practice is generally to defer the fractionation of the raw NGL product to large central units. Raw product is pipelined or otherwise transported to the fractionation plant, where the individual products are produced. Storage at the production unit will be for a high-vapor-pressure product of broad boiling range. Treating for the removal of sulfur compounds is deferred also. The liquid is stored temporarily and pumped into the pipeline, truck, or tank car.

The separation of raw product into its constituents, the treating of these products, and their subsequent disposal is now addressed.

Fractionation

The raw liquid product is separated into individual products in a series of columns or towers in which the top product is the most volatile component in the feed. The first product to be removed, if present, is ethane. Columns for the removal of any component are referred to as "de-component-ers." The first column is the deethanizer or DEC2ER followed by the depropanizer or DEC3ER, and finally the debutanizer, DEC4ER. The bottoms from the debutanizer is the pentanes-plus (C5+) or natural gasoline product. The butane tower overhead product commonly flows to a butane *splitter*, where the isobutane goes overhead and the normal butane to the bottoms. Each of the fractionated products is relatively pure. The actual purity depends on the sales specifications.

Sulfur compound removal and drying for water removal are required when sulfur-compound and water contents exceed the product specifications. Water is usually present, and the presence of sulfur compounds is determined by the feed gas analysis. If the feed gas is sour (contains H2S), mercaptans are likely to be present also.

Figure 14–27 shows the *fractionation train* used in a large, integrated plant that illustrates not only the essential elements but also a few added embellishments, such as the isobutane depropanizer and the propane deoiler. Note that the propane is treated and dried, as are the combined butanes. The natural gasoline is treated also. The fractionation sequence of Figure 14–27 will be the optimum one for most such plants. This particular plant has a deethanized feed; plants with ethane in the raw feed must have a deethanizer tower as well.

The distillation process takes advantage of the difference in volatility of the various hydrocarbons. Heat is introduced to the bottom of each column to drive the more volatile components present overhead into the distillate product. The towers are provided with condensed liquid reflux at the top to "knock back" the less volatile components into the bottoms product. Of course the distillate products are not pure but have some acceptable amount of the next-heavier component in them (heavier meaning less volatile).

Distillation columns contain trays with overflow weirs and downcomers. The reflux at the top of the tower flows across the top tray, into the downcomer, then across the next tray, etc. The vapor passes up through the tower through holes in the trays. In this manner, the vapor and liquid phases are thoroughly contacted, allowing the components

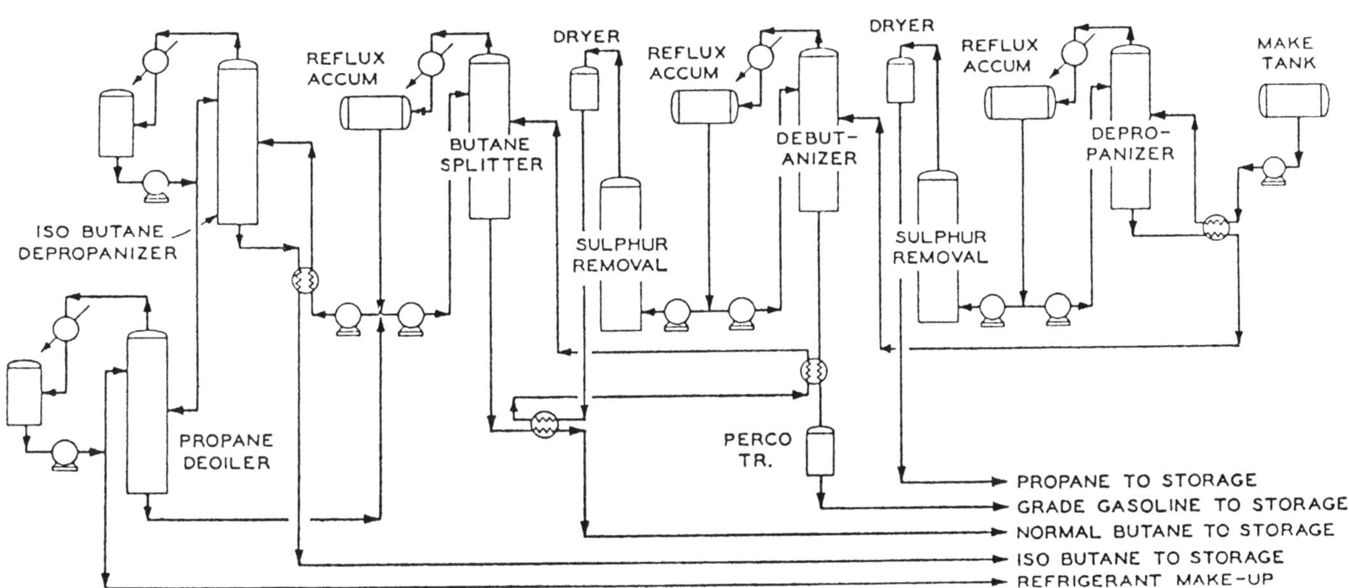

Figure 14–27. Product fractionator train (Barber *et al.*, 1965).

to vaporize and condense easily and accomplish the desired separation.

Two types of trays are used commonly in fractionator plants: sieve trays and valve trays. *Sieve trays* are simply flat plates with holes punched in them. *Valve trays* have holes or slots into which movable plates are fastened. These plates, or valves, rise as vapor flows through them. Valve trays have wider operating ranges of vapor and liquid flow rate than do sieve trays.

Each column pressure is set by the temperature of the available cooling medium (cooling water or air). The tower pressure must be high enough to condense the reflux at a temperature a bit higher than that of the cooling medium (to allow a reasonable temperature approach in the condenser). A depropanizer operates at approximately 200 psia. Debutanizers and butane splitters operate at about 70 psia. If a deethanizer is used, the distillate product and/or reflux will have to be condensed by use of propane refrigerant in order that the tower can operate at a pressure of about 400 psia. Separation in a distillation tower becomes more difficult as the operating pressure goes up due to reduced relative volatility and density difference between the liquid and vapor.

Design of the distillation columns involves setting the number of trays and reflux rate, as well as the column diameter. The desired product purities determine the number of stages and reflux. The total feed rate controls the column diameter. The complex process calculations needed for this purpose are done routinely on the digital computer. Experience results in efficient design.

Treating

Tuttle and Allen (1976) approached treating liquid products by summarizing the products and their likely contaminants (Fig. 14–28). The volatilities or vapor-liquid equilibrium K-values of the various contaminants determine where they concentrate. Hydrogen sulfide, for example, has a volatility between ethane and propane and thus tends to appear in both the ethane and propane products. Predicting the likely contaminants in each product is a logical approach since the treating processes for the various sulfur compounds differ. Removal of hydrogen sulfide is done by a different method than removal of heavy mercaptans, etc.

Table 14–4 (Tuttle and Allen, 1976) summmaries the potential treating processes for the various contaminants. Masson and Davidson (1977), in their summary of treating methods used in Canada, state the aims of treatment in

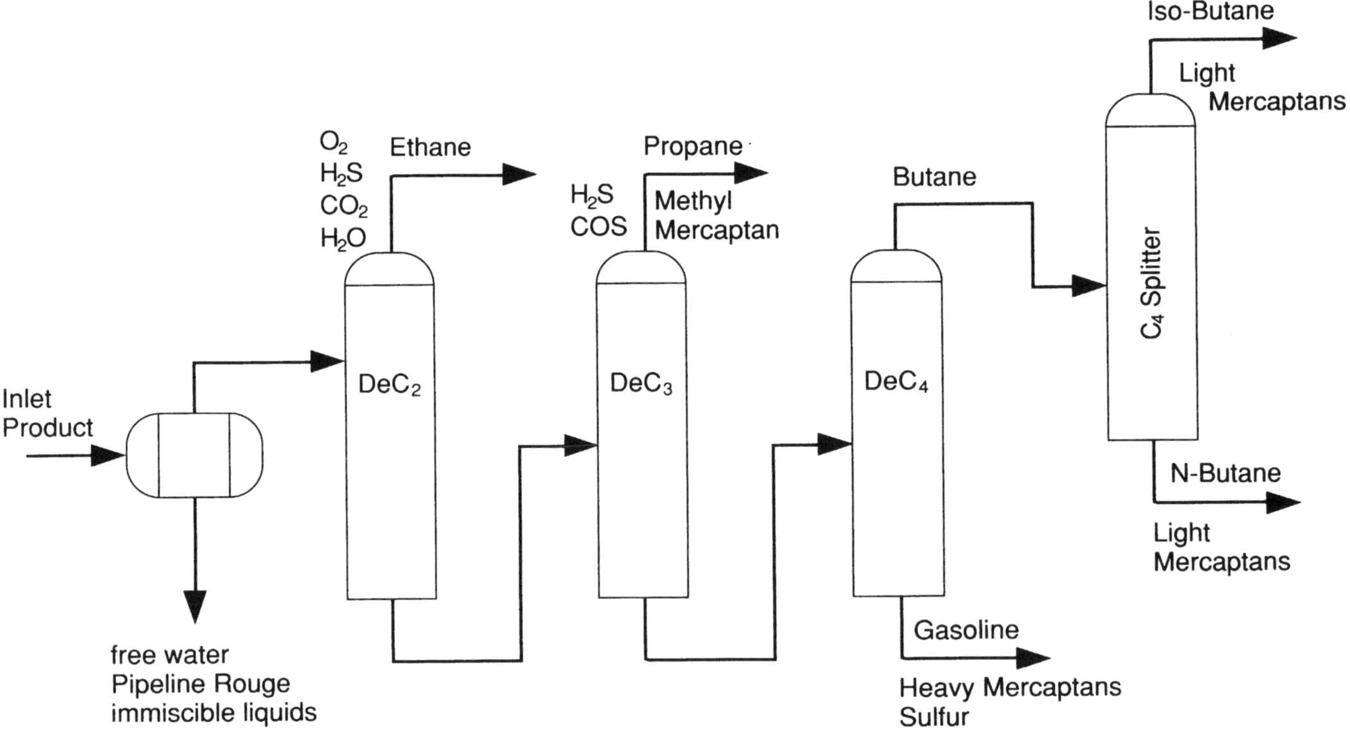

Figure 14–28. Fractionation plant contaminants (Tuttle and Allen, 1976).

Table 14–4 NGL Treating Processes (Tuttle and Allen, 1976)

	Containment						
				Mercaptans		Free	
	H_2O	CO_2	H_2S	Low MW	High MW	Sulfur	COS
Glycol: Counterflow	X						
Cocurrent	X						
Mole Sieve	X	X	X	X			X
Amine: MEA		X	X				X
DEA		X	X				X
Caustic: Wash		X	X	X		X*	
Regen'd.				X			
Merox: Extraction				X			
Sweetening				X	X		
Perco				X	X		

* If H_2S not present, free sulfur will not be removed and an Na_2S wash can be used.

terms of the GPA specifications for propane and butane products:

- HD–5 propane specification:
  Dryness—must meet the cobalt bromide test for water content
  Volatile sulfur—10 grains per 100 cu ft, max
  Corrosion—No. 1 copper strip, max
- Butane specification:
  Dryness—must contain no free water
  Volatile sulfur—15 grains per 100 cu ft, max
  Corrosion—No. 1 copper strip, max

Masson and Davidson (1977) summarize the processes used to remove the various contaminants and confirm the recommendations of Tuttle and Allen.

The processes referred to Table 14–4 are, for the most part, described in the GPSA Engineering Data Book (1987). Tuttle and Allen (1976) also furnish references for further study.

PUMPING AND STORAGE

Centrifugal pumps are generally used for liquid products. The usual care must be taken that *product pumps* be installed with adequate suction head to avoid cavitation problems. Products are at their bubble-points in the surge or storage tanks; therefore any significant pressure drop in the flow lines below the bubble-point pressure will cause vaporization and pump cavitation.

Liquid-hydrocarbon storage is discussed in Section 8 of the GPSA Engineering Data Book (1987). Horizontal cylindrical vessels, either insulated or bare, are generally used. Insulated tanks are preferred for the more volatile

products, such as a demethanized or deethanized raw product liquid.

Review Questions

1. Why should a NGL recovery plant be capable of either recovering or rejecting C2?
2. List the three objectives of NGL recovery processing.
3. What is the usual processing objective for a remote gas well? What are the expensive processing alternatives?
4. List the common NGL recovery processes.
5. Which NGL recovery processes are likely to be most attractive for:
   - lean gas
   - very rich gas
   - high pressure gas
   - low pressure gas
   - low production rates
   - remote locations?
6. How can a basic or simple refrigeration cycle be made more efficient?
7. In expander plants, cooling the feed gas with the residue gas is very important. Describe how this "refrigeration" can be made most efficient.
8. List some of the more common problems encountered while operating expanders.
9. What modifications are required for an expander plant to reject ethane?

Problems

1. Assume crude oil sells for $18 per bbl and natural gas for $1.50/Mscf.

Estimate and compare the cost of energy in $/MMBtu for both crude oil and natural gas.

List the major intangibles (no need to discuss).

2. Chapter 14 presents process calculations for the following refrigeration cycles: simple cycle, economizer cycle, and split-level refrigeration. These calculations assume that the propane refrigerant is condensed at 125°F (air cooling). Assume that suitable cooling water is available to condense the propane at 100°F.

Estimate the efficiency of the simple, economizer, and split-level refrigeration cycles.

Nomenclature

$C2+$ = ethane and heavier hydrocarbons
$C3+$ = propane and heavier hydrocarbons
EPA = U.S. Environmental Protection Agency
GHV = gross heating value
   h = enthalpy (Btu/lb)
 HC = hydrocarbon
 J-T = Joule-Thompson
   L = loading of key component on solid bed, lb/lb bed
LTS = low temperature separator
  m = refrigerant circulation rate, lb/hr
m_1 = vapor flow from economizer, lb/hr
m_2 = liquid flow from economizer, lb/hr
MTA = mass transfer agent, e.g., absorber oil or solid desiccant
MW = molecular weight, lb/lb mol
  n = feed gas flow rate, lb mol/hr
  P = pressure, psia
PSA = pressure swing adsorption
  Q = specific heat duty, Btu/lb
  q = heat duty, Btu/hr
  W = compression work, Btu/lb
  w = weight of adsorbent in bed, lb
  x = quality or mass fraction vapor
  y = mol fraction of key component in feed gas

Subscripts

  act = denotes actual work
  ch = refers to chiller
cond = refers to condenser
econ = denotes most economical value
  is = denotes isentropic (constant entropy) compression
  m = denotes adiabatically mixed value
out = denotes outlet stream value
1, 2, 3,
4, 5, 6 = see Figures 14–10 and 14–11

References

Alvarez, M. R., M. F. Hilton, and H. L. Vines (1984), "Dome's NRU Is Successfully Treating Gas from an EOR Project," *Oil & Gas J.*, Vol. 82, No. 34, p. 95–99 (Aug. 20).

Ballard, Don (1965), "How to Operate Quick-Cycle Plants," *Hydrocarbon Processing and Petroleum Refiner*, Vol. 44, No. 4, p. 131 (April 26).

Barber, H. W., J. E. Schneider, and F. K. Kennedy (1965), "How Empress Plant Was Designed," *Hydrocarbon Processing and Petroleum Refiner*, Vol. 44, No. 7, p. 127–130 (July).

Barrere, C. A., Jr. (1968), "Natural Gas Adsorption Plant Design," *Hydro. Proc.*, Vol. 47, No. 10, p. 141–144 (Oct.).

Blackburn, G. A., and B. A. George (1979), "Studies Show Effects of Composition on Propane Refrigeration," *Oil & Gas J.*, Vol. 77, No. 9, p. 127–128 (Feb. 26).

Brewster, Jack, Editor, "Gas Pressors Report," Shreveport, LA, various issues (1986–1989).

Choe, J. S., R. Agrawal, S. R. Auvil, and T. R. White, (1988), "Membrane/Cryogenic Hybrid Systems for Helium Purification," Proc. 67th Annual GPA Convention, pp. 251–255, Dallas TX, (March 14–16).

Davis, R. A., G. N. Gottier, G. A. Blackburn, and C. R. Root (1989), "Integrated Nitrogen Rejection Facility Design and Operation," Proc. 68th Annual GPA Convention, San Antonio, TX, (March 13–14).

Evans, Larry, (1980), "Flexibility can boost profits in cryogenic gas processing," *Oil & Gas J.*, Vol. 78, No. 28, p. 76 (July 14).

Ewan, D. N., J. B. Laurence, C. L. Rambo, and R. Tonne (1975), "Why Cryogenic Processing?," Proc. 54th Ann. Conv. GPA, Houston, TX, (March 10–12).

Fraser, R. W. (1985), "Simulation and Evaluation of Split-Level Propane Refrigeration," M. S. Thesis, The University of Tulsa, Tulsa, OK.

Geist, J. M. (1985), "Refrigeration Cycles for the Future," *Oil & Gas J.*, Vol. 83, No. 5, p. 56–60 (Feb. 4).

GPSA Engineering Data Book (1987), Gas Processors Suppliers Association, Tulsa, OK (10th Ed.).

Holm, John (1986), "Turboexpanders in Energy Saving Processes," *Energy Progress*, Vol. 6, No. 3, pp. 187–190 (Sept.).

Huntley, A. R., and R. S. Silvester (1983), "Hydrodynamic Analysis Aids Slug Catcher Design," *Oil & Gas J.*, Vol. 81, No. 38, p. 95–100 (Sept. 19).

King, W. H. (1988), "Olefins and Gasoline Shaping Future of NGL's," *Oil & Gas J.*, Vol. 86, No. 28, pp. 40–42 (July 11).

MacKenzie, D. H., and S. T. Donnelly (1985), "Mixed Refrigerants Proven Efficient in Natural-Gas-Liquids Recovery Process," *Oil & Gas J.,* Vol. 83, No. 9, pp. 116–120 (March 4).

Masson, W. B. and K. B. Davidson (1979), "Current Natural Gas Liquid Treating Methods in Canada," Proc. 58th Ann. Conv. GPA, Denver, CO (March 19–21).

McCann, P., J. M. Ryan, and J. V. O'Brien (1987), "Propane Recovery: The Ryan/Holmes Process," *Energy Progress,* Vol. 7, No. 4, p. 230–240 (Dec.).

McKee, R. L., (1977), "Evolution in Design," Proc. 56th Annual GPA Convention, p. 123, Dallas TX (March 21–23).

McKee, R. L. (1986), "Expander Plant Design," Paper presented at the AIChE Spring National Meeting, New Orleans, LA (April 8).

McWilliams, Bill (1985), Rotoflow Corporation, Presentation to The University of Tulsa AIChE Student Chapter (Oct. 16).

Mehra, Y. R. (1987), "New Process Flexibility Improves Gas Processing Margins," *Energy Progress,* Vol. 7, No. 3, pp. 150–156 (Sept.).

Odello, R. (1981), "Systematic Method Aids Choice of Field Gas Treatment," *Oil & Gas J.,* Vol. 79, No. 6, pp. 103–110 (Feb. 9).

Oil & Gas J., "Spot Gas prices," (source: Natural Gas Clearinghouse Inc.) various issues (1986–1989).

Peng, D-Y, and D. B. Robinson (1976), "A New Two-Constant Equation of State," *Ind. Eng. Chem. Fundam.,* Vol. 15, No. 1, pp. 59–64.

PETEX (1974), "Plant Processing of Natural Gas," Petroleum Extension Service, University of Texas, Austin, TX.

Pitzer, K. S., and L. Brewer (1961), "Thermodynamics," McGraw-Hill Book Co., NY (2nd Ed.).

Redlich, O., and J. N. S. Kwong (1949), "On the Thermodynamics of Solutions. V—An Equation of State. Fugacities of Gaseous Solutions," *Chem. Rev.,* Vol. 44, pp. 233–244.

Russell, T. H. (1977), "Straight Refrigeration Still Offers Processing Flexibility," *Oil & Gas J.,* Vol. 75, No. 4, p. 66–72 (Jan. 24).

Soave, G. (1972), "Equilibrium Constants from a Modified Redlich-Kwong Equation of State," *Chem. Eng. Sci.,* Vol. 27, pp. 1197–1203.

Swearingen, J. S. (1972), "Turboexpanders and Processes That Use Them," *Chem. Eng. Progress,* Vol. 68, No. 7, pp. 95–102 (July).

Tuttle, R., and K. Allen (1976), "Treating System Selection for Fractionation Plants," Proc. 55th Ann. Conv. GPA., San Antonio, TX (March 22–24).

Valdes, A. R. (1984), "Encyclopedia of Chemical Processing and Design," John J. McKetta and William A. Cunningham, Ed., Vol. 21, pp. 1–18, Marcel Dekker, Inc., New York.

Wang, W-B (1985), "Optimization of Expander Plants," Ph.D. Dissertation, The University of Tulsa, Tulsa, OK.

Chapter 15

Design of Oilfield Gas Processing Units

INTRODUCTION

Two examples—Example 5–1, Platform Associated Gas and Example 5–3, Low-pressure Rich Gas—are now used to illustrate previously-described design procedures. Each illustration starts with an OPSIM simulation that provides detailed material and energy balances and sizes the compressors, pumps, and heat exchangers. Then some of the major process modules or equipment are sized using hand-held methods. The methods are appropriate for preliminary cost estimates, evaluating bid tenders, and the like. Chapter 15 is outlined as follows:

Example 15–1. *Platform Associated Gas*
    OPSIM Simulation
        Feed and Product Summary
        Stream Summary (material and energy
          balances)
        Compressor Summary
        Exchanger Summary
        Comments on OPSIM Simulation
    Hand-held Methods
        Sales-Line Orifice Meter
        Platform Gas Lines
        Sales Gas Line
        Middle-Stage Compressor (centrifugal)
        TEG Unit
        Middle-Stage Exchanger (air cooler)

Example 15–2. *Low-Pressure Rich Gas*
    OPSIM Simulation
        Feed and Product Summary
        Stream Summary (material and energy
          balances)
        Compressor and Pump Summary
        Exchanger Summary
        Comments on OPSIM Simulation

    Hand-held Methods
        First-Stage Compressor (reciprocating)
        First-Stage Intercooler (water-cooled)
        Amine Unit
        Hydrate Inhibition

EXAMPLE 15–1
OPSIM SIMULATION OF PLATFORM PROCESSING OF ASSOCIATED GAS

Figure 15–1 shows the *detailed flowsheet* and the stream numbers used in the OPSIM simulation (see App. 4 for an in-depth description of OPSIM simulation). Table 15–1 presents the OPSIM output for all 30 *process streams,* as well as summaries for the *compressors* and *coolers.*

Comments on OPSIM Simulation

Several comments are now offered on the simulation presented in Figure 15–1 and Table 15–1.

Feed Flow Rate and Characterization of Pseudocomponents. The feed quantity was adjusted to obtain the desired outlet oil rate of *100000 bpd* as an *arbitrary basis.* This procedure is typical of industry practice. A reasonable production rate is chosen and the overall design based on this rate.

A detailed analysis of the reservoir fluid is required. The breakup shown is based on a chromatographic analysis of the lower molecular weight components, plus a true-boiling-point curve (TBP) for the heavier materials. The TBP was broken up into five pseudocomponents; more than five could have been used, but five are sufficient for present purposes.

Stage Separation. The pressure levels for the separators are arbitrary. In an actual design study it would be necessary

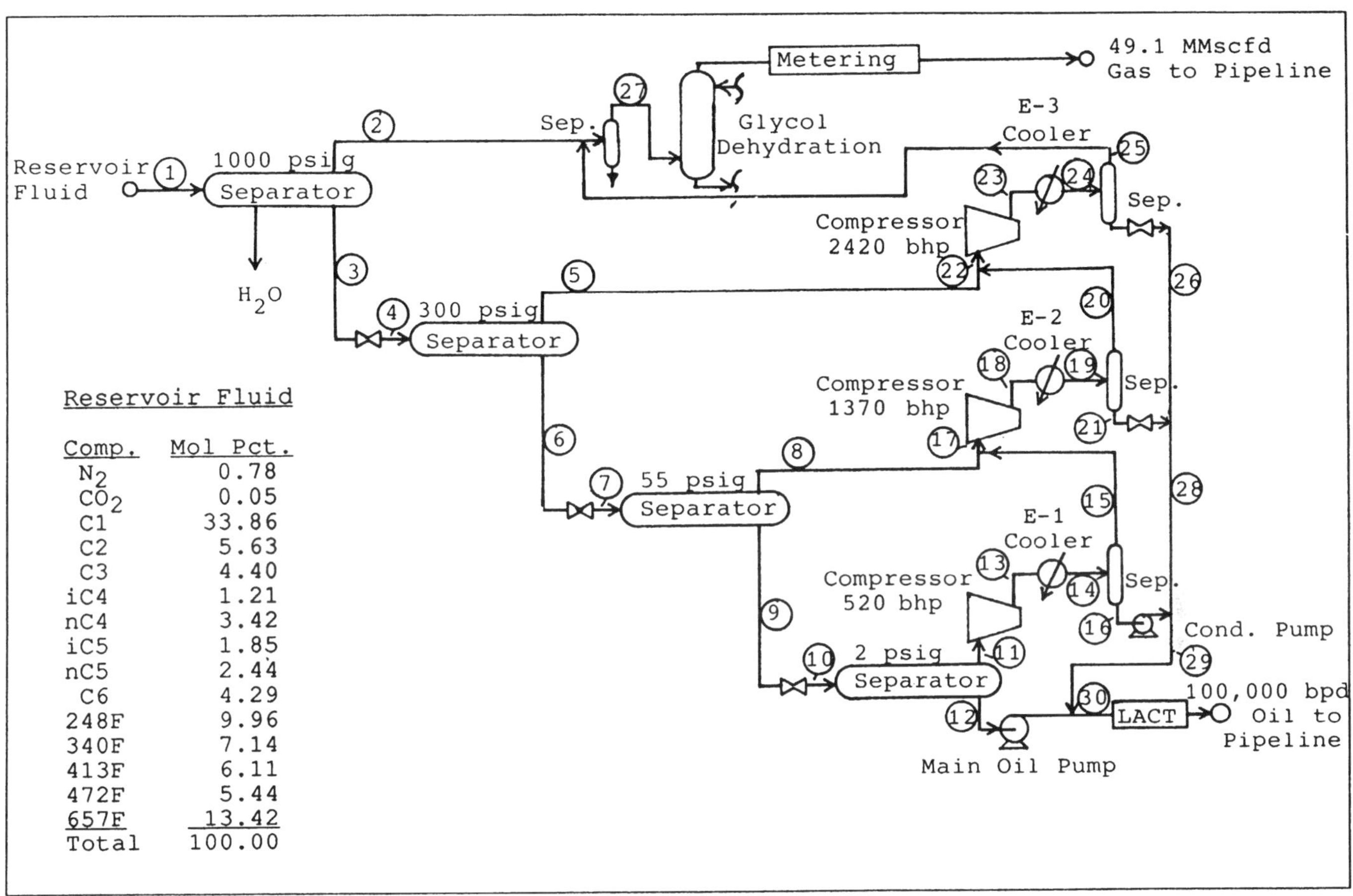

Figure 15–1. Platform processing of associated gas.

to vary these pressure levels, probably to *minimize the total compressor horsepower*. The oil recovery will vary somewhat, but with the number of stages used here, the oil recovery will not be changed a great deal by varying these pressures by small amounts.

Notice that the quantity of gas decreases in succeeding stages, with an increase in molecular weight, or gas gravity.

Stage	Separator Gas		
	Lbmole/hr	Mol Wt	Sp Gr
1	2239	18.79	0.649
2	2038	20.39	0.704
3	870	27.44	0.947
4	475	43.13	1.489

This behavior is typical of stage separation trains as is discussed in detail in Volume 2.

Compression. Compressor data are summarized in Table 15–2. Calculations are based on efficiencies typical of centrifugal compressors. Brake horsepower is estimated by adding 50 HP to allow for bearing and seal losses. The resulting Bhp's are rounded to the nearest 10 HP.

Stage Press.	ACFM	BHP	Compr. Ratio	Disch. Temp., °F
Low	2834	520	4.47	237
Intermed.	1851	1370	4.59	281
High	950	2420	3.26	285
Total		4310		

Both centrifugal and reciprocating compressors are used offshore, but the mass and space requirements are much lower for centrifugal than for reciprocating units. Vibration and maintenance also are lower. The present flow rates are a bit on the low side. Reciprocating compressors might be a better choice. An economic comparison is in order, but the choice is sometimes based on technical constraints only. *Cost is important*, since compressors cost on the order of $800–1500 per installed horsepower. Compressor

Table 15–1 Example 15–1 OPSIM Computer Simulation Results

SUMMARY OF FEED DATA:

STRM NO.	1
N2	95.92
CO2	6.15
C1	4164.02
C2	692.36
C3	541.10
IC4	148.80
NC4	420.58
IC5	227.51
NC5	300.07
C6	527.57
248F	1224.86
340F	878.06
413F	751.39
472F	669.00
657F	1650.36
MOL/HR	12297.76
LB/HR	1302320.00
T,DEG F	120.00
P,PSIA	1019.70
MOL WT	105.90
LB/CUFT	33.2885
BTU/HR	.41638E+08
BTU/R-HR	−.67638E+06
FRAC. VAP	.1821
BPD @60F	.00

PSEUDOCOMPONENT PROPERTIES:

CUT	248F	340F	413F	472F	657F
BP,F	248.00	340.00	413.00	472.00	657.00
SP GR	.7620	.7930	.8170	.8350	.8870
TC,F	573.00	673.00	750.00	807.00	975.00
PC, PSIA	442.00	370.00	350.00	295.00	220.00
MOL WT	110.00	141.00	170.00	195.00	336.00

PRODUCT SUMMARY:

COMBINED LIQUIDS = 100,000 BBL/DAY
BUBBLE POINT PRESSURE AT 100 F = 28.09 PSIA

PIPELINE GAS = 5395.21 LBMOL/HR × 379.5 SCF/LBMOL × 24 HR/DAY
= 49.14 MMSCFD

STREAM SUMMARY:

	1	2	3	4
N2	95.92	64.30	31.62	31.62
CO2	6.15	1.95	4.20	4.20
C1	4164.02	1948.96	2215.06	2215.06
C2	692.36	135.95	556.42	556.42
C3	541.10	47.62	493.48	493.48
IC4	148.80	7.50	141.31	141.31
NC4	420.58	16.26	404.33	404.33
IC5	227.51	4.85	222.66	222.66
NC5	300.07	5.11	294.96	294.96
C6	527.57	4.02	523.55	523.55
248F	1224.86	2.06	1222.79	1222.79
340F	878.06	.32	877.74	877.74
413F	751.39	.06	751.33	751.33
472F	669.00	.02	668.98	668.98
657F	1650.36	.00	1650.36	1650.36
MOL/HR	12297.76	2238.98	10058.78	10058.78
LB/HR	1302320.00	42074.95	1260245.00	1260245.00
T,DEG F	120.00	120.00	120.00	115.86
P,PSIA	1019.70	1019.70	1019.70	314.70
MOL WT	105.90	18.79	125.29	125.29
LB/CUFT	33.2885	3.4488	46.8108	19.9549
BTU/HR,	.41638E+08	.96661E+07	.31972E+08	.31972E+08
BTU/R-HR	−.67638E+06	.75263E+05	−.75164E+06	−.74256E+06
FRAC. VAP	.1821	1.0000	.0000	.2026
BPD @60F	.00	.00	111807.90	.00

	5	6	7	8
N2	27.94	3.68	3.68	3.47
CO2	2.42	1.78	1.78	1.30
C1	1645.58	569.49	569.49	487.44
C2	215.97	340.45	340.45	184.17
C3	84.69	408.79	408.79	107.15
IC4	12.70	128.61	128.61	17.69
NC4	27.10	377.23	377.23	38.59
IC5	7.25	215.41	215.41	10.30
NC5	7.38	287.58	287.58	10.41
C6	4.91	518.64	518.64	6.55
248F	1.95	1220.85	1220.85	2.34
340F	.21	877.53	877.53	.22
413F	.03	751.30	751.30	.03
472F	.00	668.97	668.97	.00
657F	.00	1650.36	1650.36	.00
MOL/HR	2038.13	8020.65	8020.65	869.66
LB/HR	41560.16	1218685.00	1218684.00	23861.69
T,DEG F	115.86	115.86	111.45	111.45
P,PSIA	314.70	314.70	69.70	69.70
MOL WT	20.39	151.94	151.94	27.44
LB/CUFT	1.0927	48.5122	12.3135	.3194
BTU/HR,	.97710E+07	.22201E+08	.22201E+08	.48245E+07
BTU/R-HR	.77439E+05	−.82000E+06	−.81561E+06	.41190E+05
FRAC. VAP	1.0000	.0000	.1084	1.0000
BPD @60F	.00	104667.30	.00	.00

	9	10	11	12
N2	.21	.21	.21	.00
CO2	.49	.49	.42	.07
C1	82.05	82.05	76.65	5.39
C2	156.27	156.27	113.88	42.39
C3	301.64	301.64	132.40	169.23
IC4	110.92	110.92	28.32	82.61
NC4	338.63	338.63	66.11	272.52
IC5	205.11	205.11	19.33	185.78
NC5	277.17	277.17	19.84	257.33
C6	512.09	512.09	12.68	499.40
248F	1218.50	1218.50	4.39	1214.12
340F	877.31	877.31	.38	876.93
413F	751.27	751.27	.05	751.23
472F	668.96	668.96	.00	668.95
657F	1650.36	1650.36	.00	1650.36
MOL/HR	7150.99	7150.99	474.67	6676.32
LB/HR	1194823.00	1194823.00	20471.21	1174352.00
T,DEG F	111.45	106.22	106.22	106.22
P,PSIA	69.70	16.70	16.70	16.70
MOL WT	167.09	167.09	43.13	175.90
LB/CUFT	49.2493	6.1676	.1204	49.6938
BTU/HR,	.17376E+08	.17376E+08	.33672E+07	.14009E+08
BTU/R-HR	−.85680E+06	−.85436E+06	.29611E+05	−.88397E+06
FRAC. VAP	.0000	.0664	1.0000	.0000
BPD @60F	101141.70	.00	.00	98533.16

	13	14	15	16
N2	.21	.21	.21	.00
CO2	.42	.42	.42	.00
C1	76.65	76.65	76.54	.11
C2	113.88	113.88	113.07	.81
C3	132.40	132.40	129.37	3.03
IC4	28.32	28.32	26.88	1.43
NC4	66.11	66.11	61.57	4.54
IC5	19.33	19.33	16.51	2.82
NC5	19.84	19.84	16.18	3.66
C6	12.68	12.68	7.61	5.08
248F	4.39	4.39	.83	3.56
340F	.38	.38	.01	.37
413F	.05	.05	.00	.05
472F	.00	.00	.00	.00
657F	.00	.00	.00	.00
MOL/HR	474.67	474.67	449.21	25.47
LB/HR	20471.21	20471.21	18605.55	1865.66
T,DEG F	236.54	100.00	100.00	100.00
P,PSIA	74.70	69.70	69.70	69.70
MOL WT	43.13	43.13	41.42	73.26
LB/CUFT	.4468	.5618	.5113	38.5636
BTU/HR,	.45555E+07	.29572E+07	.29697E+07	−.12497E+05
BTU/R-HR	.30119E+05	.27589E+05	.25976E+05	.16133E+04
FRAC. VAP	1.0000	.9464	1.0000	.0000
BPD @60F	.00	.00	.00	199.99

	17	18	19	20
N2	3.67	3.67	3.67	3.64
CO2	1.72	1.72	1.72	1.65
C1	563.98	563.98	563.98	552.99
C2	297.25	297.25	297.25	275.08
C3	236.53	236.53	236.53	192.48
IC4	44.57	44.57	44.57	30.17
NC4	100.17	100.17	100.17	61.43
IC5	26.81	26.81	26.81	11.50
NC5	26.59	26.59	26.59	9.86
C6	14.16	14.16	14.16	2.61
248F	3.17	3.17	3.17	.14
340F	.23	.23	.23	.00
413F	.03	.03	.03	.00
472F	.00	.00	.00	.00
657F	.00	.00	.00	.00
MOL/HR	1318.87	1318.87	1318.87	1141.54
LB/HR	42467.25	42467.00	42467.00	32910.73
T,DEG F	106.27	280.91	100.00	100.00
P,PSIA	69.70	319.70	314.70	314.70
MOL WT	32.20	32.20	32.20	28.83
LB/CUFT	.3824	1.3792	2.2098	1.7386
BTU/HR,	.77943E+07	.11152E+08	.58249E+07	.58149E+07
BTU/R-HR	.67393E+05	.68747E+05	.60337E+05	.51088E+05
FRAC. VAP	1.0000	1.0000	.8655	1.0000
BPD @60F	.00	.00	.00	.00

	21	22	23	24
N2	.03	31.58	31.58	31.58
CO2	.07	4.07	4.07	4.07
C1	10.99	2198.57	2198.57	2198.57
C2	22.17	491.05	491.05	491.05
C3	44.04	277.17	277.17	277.17
IC4	14.40	42.86	42.86	42.86
NC4	38.74	88.53	88.53	88.53
IC5	15.31	18.75	18.75	18.75
NC5	16.73	17.24	17.23	17.23
C6	11.55	7.52	7.52	7.52
248F	3.04	2.08	2.08	2.08
340F	.23	.21	.21	.21
413F	.03	.03	.03	.03
472F	.00	.00	.00	.00
657F	.00	.00	.00	.00
MOL/HR	177.32	3179.67	3179.66	3179.66
LB/HR	9556.27	74470.88	74470.02	74470.02
T,DEG F	100.00	107.94	285.05	100.00
P,PSIA	314.70	314.70	1024.70	1019.70
MOL WT	53.89	23.42	23.42	23.42
LB/CUFT	33.1098	1.3066	3.2209	5.2923
BTU/HR,	.10027E+05	.15586E+08	.21624E+08	.12699E+08
BTU/R-HR	.92491E+04	.12892E+06	.13124E+06	.11744E+06
FRAC. VAP	.0000	1.0000	1.0000	.9926
BPD @60F	1172.60	.00	.00	.00

	25	26	27	28
N2	31.53	.05	95.84	.08
CO2	4.05	.02	5.99	.09
C1	2192.03	6.54	4140.98	17.53
C2	487.06	3.99	623.01	26.16
C3	272.49	4.68	320.11	48.73
IC4	41.66	1.20	49.16	15.60
NC4	85.51	3.02	101.77	41.76
IC5	17.69	1.06	22.54	16.37
NC5	16.09	1.14	21.20	17.88
C6	6.60	.92	10.63	12.47
248F	1.44	.64	3.50	3.68
340F	.08	.13	.40	.36
413F	.00	.03	.07	.05
472F	.00	.00	.02	.01
657F	.00	.00	.00	.00
MOL/HR	3156.23	23.43	5395.21	200.75
LB/HR	73458.36	1011.66	115533.30	10567.94
T,DEG F	100.00	100.00	104.90	95.43
P,PSIA	1019.70	1019.70	1019.70	314.70
MOL WT	23.27	43.18	21.41	52.64
LB/CUFT	5.2341	27.3248	4.4035	22.1321
BTU/HR,	.12676E+08	.23032E+05	.22342E+08	.33058E+05
BTU/R-HR	.11636E+06	.10770E+04	.19193E+06	.10346E+05
FRAC. VAP	1.0000	.0000	1.0000	.0504
BPD @60F	.00	144.60	.00	.00

	29	30
N2	.08	.09
CO2	.09	.16
C1	17.64	23.04
C2	26.96	69.36
C3	51.75	220.99
IC4	17.04	99.64
NC4	46.30	318.82
IC5	19.19	204.97
NC5	21.54	278.87
C6	17.54	516.95
248F	7.24	1221.35
340F	.73	877.66
413F	.10	751.33
472F	.02	668.98
657F	.00	1650.36
MOL/HR	226.22	6902.53
LB/HR	12433.60	1186785.00
T,DEG F	97.93	104.75
P,PSIA	314.70	314.70
MOL WT	54.96	171.93
LB/CUFT	26.6243	49.4813
BTU/HR,	.20562E+05	.14029E+08
BTU/R-HR	.11973E+05	$-$.87290E+06
FRAC. VAP	.0275	.0000
BPD @60F	.00	100000.10

```
COMPRESSOR SUMMARY:

FOURTH-STAGE-GAS COMPRESSOR
ADIABATIC EFFICIENCY    =        .7100
MECHANICAL EFFICIENCY =       1.0000
DELTA P                 =      58.0  PSI:
BRAKE HORSEPOWER        =     466.9
COMPRESSION RATIO       =       4.473

THIRD-STAGE-GAS COMPRESSOR
ADIABATIC EFFICIENCY    =        .7100
MECHANICAL EFFICIENCY =       1.0000
DELTA P                 =     250.0  PSI:
BRAKE HORSEPOWER        =    1319.4
COMPRESSION RATIO       =       4.587

SECOND-STAGE-GAS COMPRESSOR
ADIABATIC EFFICIENCY    =        .7200
MECHANICAL EFFICIENCY =       1.0000
DELTA P                 =     710.0  PSI:
BRAKE HORSEPOWER        =    2372.6
COMPRESSION RATIO       =       3.256

HEAT EXCHANGER SUMMARY:

AFTERCOOLER E-1
   DELTA P        =            -5.0 PSI
   INLET TEMP.    =           236.5 DEG F
   OUTLET TEMP. =             100.0 DEG F
   HEAT ADDED     =    -.15982850E+07 BTU/HR

AFTERCOOLER E-2
   DELTA P        =            -5.0 PSI
   INLET TEMP.    =           280.9 DEG F
   OUTLET TEMP. =             100.0 DEG F
   HEAT ADDED     =    -.53272820E+07 BTU/HR

AFTERCOOLER E-3
   DELTA P        =            -5.0 PSI
   INLET TEMP.    =           285.1 DEG F
   OUTLET TEMP. =             100.0 DEG F
   HEAT ADDED     =    -.89250600E+07 BTU/HR
```

investment is very high in almost any gas-processing unit requiring compression.

Compression ratios are not equal, but then the amounts of gas are not the same either, so that the usual rule of approximately equal compression ratio for optimum horsepower does not apply. Again, an optimization should be performed. Discharge temperatures from each stage are reasonable—less than 300°F.

Heat Exchangers. The exchangers were assumed to be air-cooled exchangers capable of lowering the temperature of the gas to 100°F in the hottest part of the year. This is reasonable for a North Sea platform. A pressure drop of 5 psi is specified. This drop should be kept low since it contributes to higher compression cost.

Sales Orifice Meter for Platform Associated Gas

Assume sales gas (stream 27) leaves TEG unit at 1009.7 psia and 106°F. This allows a 5-psi drop across the TEG contactor.

Use AGA scf which are measured at 14.73 psia and 60°F (see page A2–6) because ANSI/API and GPSA measurement tables use AGA scf.

$$\text{scf/lb mole} = R\, T/\, P = (10.7315)(519.67)/(14.73)$$
$$= 378.60 \text{ AGA scf (ideal gas)/lb mol}$$
$$Q \text{ (stream 27)} = 5395.21 \text{ lb mol/hr}$$
$$= 2.042627 \text{ MMscfh (AGA scf)}$$
$$= 49.023 \text{ MMscfd (AGA scf)}$$

The basic orifice equation (Eq. 11–1) is:

$$Q = C' \sqrt{P_f} \sqrt{h_w}$$
$$P_f = 1009.7 \text{ psia}; \quad \sqrt{P_f} = 31.776 \qquad (15\text{–}1)$$

Use a differential bellows meter with 100-in. water column (WC) as full scale. Assume at 49 MMscfd, $h_w = 80$ in. to get a good, high reading without excursions off scale.

$$h_w = 80 \text{ in. } H_2O; \quad \sqrt{h_w} = 8.944$$

Substitute into Equation 15–1.

$$2042627 = C' \ (31.776)(8.944)$$
$$C' = 7186 \text{ for correctly-sized orifice.}$$

For meter sizing purposes one may assume:

$$C' = F_b \ F_{tf} \ F_{gr} \ F_{pv} \qquad (15\text{–}2)$$

(Other factors in Eq. 11–1 are all quite close to 1)

$$F_{tf} = (519.67/565.67)^{0.5} = 0.9585$$
$$SG = MW/28.9625$$
$$= (21.41)/(28.9625) = 0.7392$$
$$F_{gr} = \sqrt{1/SG} = 1.1631$$

Estimate Z for stream 27.

$$Z = (P)(MW)/(\rho)(R)(T)$$
$$= (1009.7)(21.41)/$$
$$(4.4035)(10.7315)(564.58)$$
$$= 0.8103$$
$$F_{pv} = 1/\sqrt{Z} = 1.1109$$

Substitute into Equation 15–2

$$7186 = (F_b)(0.9585)(1.1631)(1.1109)$$
$$F_b = 5802$$

Use ANSI/API 2530 (1985), pp. 67–68, or GPSA (1987), p. 3–14, for orifice data and Crane Co. Technical Paper No. 410 for pipe information:

Nominal Pipe Size (in.)	8	8	10
Schedule	40	80	80
D (in.) (Crane TP 410)	7.981	7.625	9.562
d (in.)	5.000	5.000	5.250
β (d/D)	0.626	0.656	0.549
F_b (GPSA, 1987, p. 3–14)	5551.2	5650.2	5897.8
C' (Eq. 15–2)	6875.	6998.	7304.
h_w (in. WC) (Eq. 15–1)	87.4	84.4	77.5

Recommend   8–in., Sch. 80 pipe
  5.000-in. orifice if design flow is a maximum estimate
  MAWP for 8-in. Sch. 40 pipe is a bit low.
Recommend   10–in., Sch. 80 pipe
  5.25-in. orifice if design flow is an average estimate

Process Piping for Platform Associated Gas
Assume $Z = 0.81$ as above.

$$Q = \dot{n} Z R T/P$$
$$= (5395)(0.81)(10.7315)(565.7)/(1009.7)$$
$$= 26300 \text{ acfh} = 7.30 \text{ acfs at pipeline P and T}$$

Estimate pressure drop (D P/L) for 8-in. Sch. 80.

$$D = 7.625 \text{ in. } = 0.6354 \text{ ft}$$
$$A = (.7854)(7.625)^2 = 45.66 \text{ in}^2 = 0.3171 \text{ ft}^2$$
$$V = Q/A = 23.0 \text{ ft/sec}$$
$$\rho = (P)(MW)/(Z)(R)(T)$$
$$= (1009.7)(21.41)/(.810)(10.7315)(565.7)$$
$$= 4.39 \text{ lb/ft}^3$$
$$\mu = 0.0132 \text{ cp} = 8.87 \times 10^{-6} \text{ lb/ft sec}$$
$$Re = D V \rho/\mu$$
$$= (0.6354)(23.0)(4.39)/(8.87 \times 10^{-6})$$
$$= 7.23 \times 10^6$$

Use pipe roughness $= (0.00008333 + 0.0000625)/2 = 0.000073$ ft per Chapter 13 recommendations. Then

$$\text{Relative roughness} = 0.000073/0.6354 = 0.000115$$

$$f = 0.0124 \text{ (Fig. 13–2)}$$
$$\Delta P = f \ (L/D) \ \rho \ V^2 \ /(2 \ g_c)$$
$$= (0.0124) \ (100/0.6354) \ (4.39)$$
$$(23.0)^2 \ /[(2)(32.174)]$$
$$= 70.4 \text{ lbf/ft}^2 \ x \ 1 \ ft^2 \ /144 \ in^2$$
$$\Delta P/L = 0.49 \text{ psi/100 ft   OKAY for platform piping.}$$

Estimate pressure drop (Δ P/L) for 10–in. Sch. 80.

$$D = 9.562 \text{ in. } = 0.7968 \text{ ft}$$
$$A = (.7854)(9.562)^2 = 71.81 \text{ in}^2 = 0.4987 \text{ ft}^2$$
$$V = Q/A = 14.6 \text{ ft/sec}$$
$$\rho = 4.39 \text{ lb/ft}^3$$
$$\mu = 0.0132 \text{ cp} = 8.87 \ x \ 10^{-6} \text{ lb/ft sec}$$
$$Re = D V \rho/\mu$$
$$= (.7968)(14.6)(4.39)/(8.87 \ x \ 10^{-6})$$
$$= 5.76 \ x \ 10^6$$

$$\text{Relative roughness} = 0.000092$$

$$f = 0.0119 \text{   (Fig. 13–12)}$$
$$\Delta P = f \ (L/D) \ \rho \ V^2 \ /(2 \ g_c)$$
$$= (0.0119) \ (100/0.7968) \ (4.39) \ (14.6)^2 \ /$$
$$[(2)(32.174)]$$
$$= 21.7 \text{ lbf/ft}^2$$
$$\Delta P/L = 0.15 \text{ psi/100 ft}$$

This is on the low side for process piping. *Platform pipe can be 8–in. Sch. 80.*

Sales Gas Line for Platform Associated Gas

The size of the sales gas line depends on the length of the line, the allowable outlet pressure, and whether or not there is to be recompression along the line. Often an economic study is required to determine the size. The following conditions are assumed for the purpose of an example calculation:

Pipe length = L = 10 mi
Minimum outlet pressure = 800 psia
Sea water temperature = Ts = 50°F
No recompression
Negligible elevation change

Estimate pressure is 900 psia at 5 miles and 800 psia at 10 miles. Make two-phase flow calculations to verify selection of pipe size. Perform OPSIM flash calculations to obtain necessary properties for two-phase flow calculations.

	Mixture	Gas	Liquid	Mixture	Gas	Liquid
P, psia	900	900	900	800	800	800
T, °F	60.3	60.3	60.3	48.8	48.8	48.8
ρ, lb/ft^3	4.480	4.298	30.55	4.055	3.791	30.98
m, lb/hr	115534	110053	5481	115534	106970	8564
SG	—	0.721	—	—	0.711	—
MW	—	—	42.98	—	—	42.70

Assume a 10–in., Sch. 80 pipe (to be confirmed by the calculations).

First, estimate the temperature profile.

$$\frac{T_2 - (T_s + \theta)}{T_1 - (T_s + \theta)} = \exp(-\beta L) \tag{13–53}$$

$$\beta = 5820\, U\, A_o / (3600\, m\, c_p) \tag{13–54}$$

$$A_o = \pi\, D_o = \pi\,(10.75/12) = 2.814\ \text{ft}^2\,/\text{ft}$$

Estimate U = 1.5 Btu/(hr-ft^2-°F)

$$\dot{m} = 115533/3600 = 32.09\ \text{lb/sec}$$

Estimate C_p assuming no condensation and average T and P of 50°F and 900 psia from Figure A6–4. SG = 21.41/28.9625 or 0.739.

$$C_p = 0.71$$
$$\beta = (5280)(1.5)(2.814)/[(3600)(32.09)(0.71)]$$
$$= 0.272$$
$$\theta = \eta\,(P_2 - P_1)/(\beta\, L) \tag{13–56}$$

Use OPSIM computer simulation to estimate η.

$$\eta = (T_1 - T_2)_H/(P_1 - P_2)$$
$$= (104.9 - 90.9)/(1000 - 800)$$
$$= 0.070$$
$$\theta = (0.070)(800 - 1000) / [(0.272)(10)]$$
$$= -5.15$$

Substitute into Equation 13–54.

$$T_2 = 44.85 + 60.05 \exp(-0.272\, L)$$

L, Mi	T$_2$,°F
0	
	104.9
5	
	60.3
10	
	48.8

For the 900–psia case:

$$V_{sL} = 5481/[(30.55)(0.4989)(3600)] = 0.0999\ \text{ft/s} \tag{13–32}$$

$$V_{sG} = 110053/[(4.298)(0.4989)(3600)] = 14.26\ \text{ft/sec} \tag{13–33}$$

$$V_m = 0.0999 + 14.26 = 14.36\ \text{ft/sec} \tag{13–34}$$

$$\lambda_L = 0.0999 / 14.36 = 0.00696 \tag{13–35}$$

$$Fr = (14.36)^2/[(32.174)(0.7968)] = 8.04 \tag{13–31}$$

From Figure 13–6, regime is segregated.

$$H_L = 0.98(0.00696)^{0.4846}/(8.04)^{0.0868} = 0.0737 \tag{13–40}$$

$$\rho_n = (0.00696)(30.55) + (1 - 0.00696)(4.298)) = 4.481\ \text{lb/ft}^3 \tag{13–42}$$

From Figure A6–5, $\mu\, G = 0.0130$ cp
Treat liquid as propane (MW = 44); use Figure 23–35, GPSA

$$\mu_L = 0.105 \tag{13–43}$$
$$\mu_n = (0.00696)((0.105) + (1 - 0.00696)(0.0130)) = 0.0136\ \text{cp} = 9.17 \times 10^{-6}\ \text{lb/ft-sec}$$

$$Re_n = (0.7968)(4.481)(14.36)/9.17 \times 10^{-6} \tag{13–44}$$
$$= 5.59 \times 10^6$$

$$f_n = 0.00883 \tag{13–45}$$

$$y = (0.00696)/(0.0737)^2 = 1.281 \tag{13–48}$$

$$s = 0.363 \tag{13–47}$$

$$f_{tp}/f_n = \exp(0.363) = 1.438 \tag{13–46}$$

$$f_{tp} = (0.00883)(1.438) = 0.0127 \tag{13–50}$$

$$(dP/dX)_f = \frac{(0.0127)(4.481)(14.36)^2}{(2)(32.174)(0.7968)}$$

$$= 0.229\ \text{lbf/ft}^2\,/\text{ft} \tag{13–51}$$
$$= 0.229/144 = 0.00159\ \text{psi/ft}$$
$$= 8.39\ \text{psi/mile}$$

A similar calculation for 800 psi gives 8.60 psi/mi. The drop at the pipe inlet is 0.153 psi/100 ft or 8.05 psi/mi. An average value for the 10–mile pipeline is 8.34 psi/mi or an overall drop of 83.4 psi, less than the 200 psi maximum drop required.

If an 8–in., Sch. 80 line were used, the pressure drop would be approximately

$$\Delta P = 83.4 \ (9.562/7.625)^5 = 259 \text{ psi}$$

This latter figure exceeds the 200 psi maximum drop. A **10–in., Sch. 80 pipeline** is recommended for the stated conditions.

TEG Unit for Platform Associated Gas Example 8–1 details this calculation.

Middle-Stage Compressor

A *centrifugal compressor* is selected. Use *polytropic model* with the OPSIM results for initial estimates:

$Z = P \ (MW)/ \ \rho \ R \ T$
$Z_1 = (69.7) \ (32.20)/(.3824) \ (10.7315) \ (459.7 + 106.3)$
$\quad = 0.966$
$Z_2 = (319.7) \ (32.20)/(1.379) \ (10.7315) \ (459.7 + 280.9)$
$\quad = 0.939$

$T_1 = 106.3°F$	$Z_1 = 0.966$
$T_2 = 280.9°F$	$Z_2 = 0.939$
$T_{av} = 193.6°F$	$Z_{av} = 0.953$

Average $k = 1.15$ (GPSA, 1987, Fig. 13–8)

$Q = (lb/hr)/(lb/ft^3)$
$\quad = (42467)/(.3824) = 1.111 \times 10^5 \ ft^3 \ /hr$
$\quad = 1851 \ acfm$

Select Elliott Frame 29M (Fig. 10–23)

$\eta_p = 0.76$ (Fig. 10–23)
$(n–1)/n = (k–1)/(k \ \eta_p)$ (10–23)
$\quad = (1.15 - 1.00)/(1.15)(0.76) = 0.172$

$$-W_p = Z_{ave} \frac{R}{MW} \frac{n}{n-1} T_1 \left[\left(\frac{P_2}{P_1}\right)^{(n-1)/n} - 1 \right] \quad (10–24)$$

$$= (0.953) \frac{1545}{32.2} \frac{1}{0.172} (566.0) \left[\left(\frac{319.7}{69.7}\right)^{0.172} - 1 \right]$$

$\quad = 45070 \ ft\text{-}lbf/lbm$

$$\dot{W}_{act} = W_p \left(\frac{1}{\eta_p}\right) \left(\dot{m} \ \frac{lb}{hr}\right) \left(\frac{hr}{min}\right) \left(\frac{hp \ min}{ft \ lbf}\right)$$

$$= (45070) \frac{1}{0.76} \frac{42467 \ lb/hr}{60 \ min/hr} \frac{1 \ hp\text{-}min}{33000 \ ft \ lbf}$$

$\quad = 1270 + 50 = 1320 \ HP$

Middle-Stage Air Cooler

Natural gas enters at 280.9°F
$\qquad$ leaves at 100°F
$\qquad$ pressure 316 psia

Follow procedure outlined in GPSA (1987, Section 10): Select following criteria for air-cooler.

Heat load	5.33 MMBtu/hr
Flow rate	42467 lb gas/hr
Temperature in	281°F
$\qquad$ out	100°F
Fouling Factor r_d	$= 0.001$
Allowable ΔP	$= 5$ psi
Ambient air temperature	75°F
Elevation	sea level

(Figures typical of North Sea offshore)

Select the following:

Type	Forced draft
Fintube	1-in. OD with ⅝-in. high fins
Tube pitch	2 ¼-in. triangular (D)
Bundle layout	6 rows, three tube passes for compactness

Estimate approximate *overall transfer coefficient* from GPSA (1987), Figure 10–10, say $U_x = 2.6$ Btu/hr-ft²-°F. Estimate approximate *air temperature rise:*

$$\Delta t_a = \left(\frac{U_x + 1}{10}\right) \left(\frac{T_1 + T_2}{2} - t_1\right)$$

$$= \left(\frac{2.6 + 1}{10}\right) \left(\frac{281 + 100}{2} - 75\right) = 41.6°F$$

Calculate *CMTD:*

Hydrocarbon gas	280.9 $\rightarrow$ 100°F	
Air	116.6 $\leftarrow$	75°F
Difference	164.3	25

$$LMTD = (164.3 - 25)/\ln (164.3/25) = 74.0°F$$

Correction factor approximately 1.0 for three tube passes

$$CMTD = (74)(1.00) = 74°F$$

Estimate *required surface area*, A_x (finned area)

$$Q = U_x \ A_x \ CMTD$$
$$5330000 = (2.6)(A_x)(74)$$
$$A_x = 27700 \ ft^2$$

Calculate *face area*, F_a, using APSF factor from GPSA (1987), Figure 10–11

$$F_a = A_x / \text{APSF} = 27700/178.2 = 155 \text{ ft}^2$$

Estimate *width of unit*. Assume length, L, of 20 ft

$$\text{Width} = F_a / L = 155/20 = 7.8 \text{ ft}$$

Round answer to L = 20 ft, W = 8 ft, F_a = 160 ft^2
With this shape of unit, two fans will be indicated.

The above result is a rough estimate of the unit. To complete a detailed design, the number of tubes is estimated, and the tube-side heat-transfer coefficient and pressure drop calculated. The air-side and overall coefficients then can be estimated to determine if the assumed U_x is reasonable. If the pressure drop is excessive, the number of tube passes may have to be decreased, which will affect the overall heat-transfer coefficient. When the heat-transfer and pressure-drop calculations are satisfactory, the diameter of the fans and fan-motor horsepower are estimated, as outlined in the Data Book.

EXAMPLE 15–2
OPSIM SIMULATION OF LOW-PRESSURE RICH GAS

Figure 15–2 shows the detailed flowsheet and the stream numbers used in the OPSIM simulation of the low-pressure rich gas case presented in Chapter 5. Table 15–2 presents the OPSIM output for all streams as well as summaries for the compressors, pump, and coolers.

Comments on OPSIM Simulation

Several comments are offered on the simulation results shown in Table 15–2.

Pseudocomponent Characterization. Information is not available for the C7+ fraction; it is treated as n-C8.

Compression. The compression ratios of the three stages are similar, and the horsepower is approximately balanced.

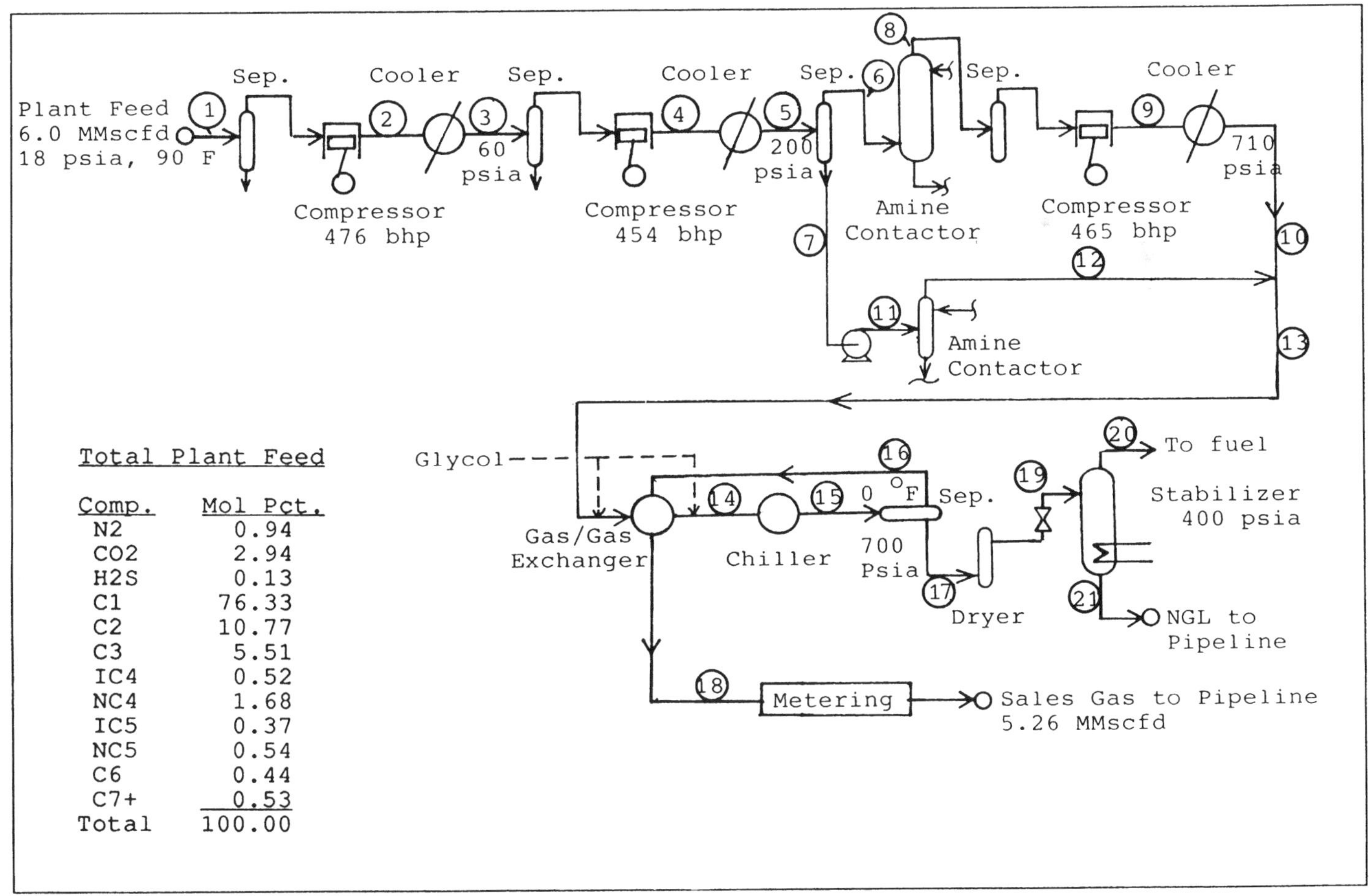

Comp.	Mol Pct.
N2	0.94
CO2	2.94
H2S	0.13
C1	76.33
C2	10.77
C3	5.51
IC4	0.52
NC4	1.68
IC5	0.37
NC5	0.54
C6	0.44
C7+	0.53
Total	100.00

Figure 15–2. Low-pressure rich gas processing.

An adiabatic efficiency of 89 is used, corresponding to a *reciprocating compressor*. The latter is appropriate for this service due to the small amount of gas being processed. A mechanical efficiency of 96% is used.

Treating. Amine treating is specified for the gas and liquid at the intermediate pressure level of 200 psia. Essentially all of both H_2S and CO_2 are completely removed. There actually may be a small amount of CO_2 remaining, but this is neglected.

Condensate Recovery. A simple refrigeration process is deemed suitable for this application. The feed gas contains 5.86 gal C2+/Mscf, making it a rich gas. The refrigeration temperature level chosen is 0°F. Glycol is injected upstream of the gas-to-gas exchanger and the chiller. A three-phase separator is used to remove dilute glycol to regeneration and condensate to stabilization.

A molecular sieve dehydrator is indicated to remove residual water from the condensate prior to stabilization. The stabilized liquid goes to a liquid products pipeline. The gas product is warmed in the gas-to-gas exchanger, then is metered into the sales-gas pipeline. The heating value of the sales gas is 1140 Btu/scf.

Exchangers. Cooling water is assumed available at 80°F. A 15°F approach is used for a process stream outlet temperature of 95°F. Five-psi pressure drops are used for each exchanger, with piping losses ignored.

TABLE 15–2.   Example 15–2 OPSIM Computer Simulation Results

SUMMARY OF FEED DATA:

STRM NO.	1
N2	6.19
CO2	14.76
H2S	.86
C1	502.83
C2	70.95
C3	36.30
IC4	3.43
NC4	11.07
IC5	2.44
NC5	3.56
C6	2.90
C7+	3.49
MOL/HR	658.76
LB/HR	14575.75
T,DEG F	90.00
P,PSIA	18.00
MOL WT	22.13
LB/CUFT	.0678
BTU/HR,	.31936E+07
BTU/R-HR	.29509E+05
FRAC. VAP	1.0000

C7+ PROPERTIES

BP,F	258.00
SP GR	.7070
TC,F	564.00
PC, PSIA	361.00
MOL WT	114.00

PRODUCT SUMMARY:

THE HEATING VALUE OF THE SALES GAS IS 1139.8 BTU/SCF.

BUBBLE POINT OF DEMETHANIZER BOTTOMS:
TEMPERATURE = 100.00   DEG F
        PRESSURE = 212.54 PSIA

STREAM SUMMARY:

	1	2	3	4
N2	6.19	6.19	6.19	6.19
CO2	14.76	14.76	14.76	14.76
H2S	.86	.86	.86	.86
C1	502.83	502.83	502.83	502.83
C2	70.95	70.95	70.95	70.95
C3	36.30	36.30	36.30	36.30
IC4	3.43	3.43	3.43	3.43
NC4	11.07	11.07	11.07	11.07
IC5	2.44	2.44	2.44	2.44
NC5	3.56	3.56	3.56	3.56
C6	2.90	2.90	2.90	2.90
C7+	3.49	3.49	3.49	3.49
MOL/HR	658.76	658.76	658.76	658.76
LB/HR	14575.75	14575.75	14575.75	14575.75
T,DEG F	90.00	249.69	95.00	250.15
P,PSIA	18.00	65.00	60.00	205.00
MOL WT	22.13	22.13	22.13	22.13
LB/CUFT	.0678	.1899	.2259	.6052
BTU/HR,	.31936E+07	.43576E+07	.32057E+07	.43133E+07
BTU/R-HR	.29509E+05	.29692E+05	.27966E+05	.28140E+05
FRAC. VAP	1.0000	1.0000	1.0000	1.0000

	5	6	7	8
N2	6.19	6.19	.00	6.19
CO2	14.76	14.74	.01	.00
H2S	.86	.85	.00	.00
C1	502.83	502.64	.19	502.64
C2	70.95	70.82	.13	70.82
C3	36.30	36.08	.22	36.08
IC4	3.43	3.38	.05	3.38
NC4	11.07	10.86	.20	10.86
IC5	2.44	2.34	.10	2.34
NC5	3.56	3.36	.19	3.36
C6	2.90	2.48	.42	2.48
C7+	3.49	1.50	2.00	1.50
MOL/HR	658.76	655.25	3.51	639.71
LB/HR	14575.75	14259.52	316.23	13582.81
T,DEG F	95.00	95.00	95.00	95.00
P,PSIA	200.00	200.00	200.00	195.00
MOL WT	22.13	21.76	90.13	21.23
LB/CUFT	.7790	.7624	40.5381	.7235
BTU/HR,	.30842E+07	.30868E+07	−.25496E+04	.30306E+07
BTU/R-HR	.26212E+05	.25961E+05	.25140E+03	.25220E+05
FRAC. VAP	.9947	1.0000	.0000	1.0000

	9	10	11	12
N2	6.19	6.19	.00	.00
CO2	.00	.00	.01	.00
H2S	.00	.00	.00	.00
C1	502.69	502.69	.19	.19
C2	70.82	70.82	.13	.13
C3	36.08	36.08	.22	.22
IC4	3.38	3.38	.05	.05
NC4	10.86	10.86	.20	.20
IC5	2.34	2.34	.10	.10
NC5	3.36	3.36	.19	.19
C6	2.48	2.48	.42	.42
C7+	1.50	1.50	1.99	1.99
MOL/HR	639.71	639.71	3.51	3.49
LB/HR	13582.81	13582.81	316.23	315.60
T,DEG F	267.99	95.00	105.08	95.00
P,PSIA	715.00	710.00	715.00	710.00
MOL WT	21.23	21.23	90.13	90.34
LB/CUFT	2.0286	2.9472	40.6420	40.9342
BTU/HR,	.41502E+07	.27168E+07	-.40891E+03	−.21133E+04
BTU/R-HR	.25391E+05	.23144E+05	.25396E+03	.25021E+03
FRAC. VAP	1.0000	.9924	.0000	.0000

	13	14	15	16
N2	6.19	6.19	6.19	6.07
CO2	.00	.00	.00	.00
H2S	.00	.00	.00	.00
C1	502.88	502.88	502.88	476.24
C2	70.96	70.96	70.96	54.53
C3	36.30	36.30	36.30	18.11
IC4	3.43	3.43	3.43	1.01
NC4	11.07	11.07	11.07	2.51
IC5	2.44	2.44	2.44	.27
NC5	3.56	3.56	3.56	.30
C6	2.90	2.90	2.90	.08
C7+	3.49	3.49	3.49	.01
MOL/HR	643.21	643.21	643.21	559.13
LB/HR	13898.41	13898.41	13898.41	10502.98
T,DEG F	96.81	40.68	.00	.00
P,PSIA	710.00	705.00	700.00	700.00
MOL WT	21.61	21.61	21.61	18.78
LB/CUFT	3.0030	3.6243	4.3461	3.3965
BTU/HR,	.27147E+07	.21131E+07	.16049E+07	.17290E+07
BTU/R-HR	.23397E+05	.22262E+05	.21208E+05	.17911E+05
FRAC. VAP	.9813	.9362	.8693	1.0000

	17	18	19	20
N2	.13	6.07	.13	.13
CO2	.00	.00	.00	.00
H2S	.00	.00	.00	.00
C1	26.64	476.24	26.64	26.39
C2	16.42	54.53	16.42	3.73
C3	18.19	18.11	18.19	1.04
IC4	2.42	1.01	2.42	.05
NC4	8.56	2.51	8.56	.12
IC5	2.17	.27	2.17	.01
NC5	3.26	.30	3.26	.01
C6	2.81	.08	2.81	.00
C7+	3.48	.01	3.48	.00
MOL/HR	84.07	559.13	84.07	31.48
LB/HR	3395.43	10502.97	3395.43	596.50
T,DEG F	.00	91.81	−15.89	−15.72
P,PSIA	700.00	695.00	400.00	400.00
MOL WT	40.39	18.78	40.39	18.95
LB/CUFT	32.1549	2.4576	13.8581	1.8534
BTU/HR	−.12406E+06	.23306E+07	−.12406E+06	.10209E+06
BTU/R-HR	.32970E+04	.19113E+05	.33167E+04	.10528E+04
FRAC. VAP	.0000	1.0000	.1820	1.0000

	21
N2	.00
CO2	.00
H2S	.00
C1	.25
C2	12.69
C3	17.15
IC4	2.37
NC4	8.44
IC5	2.16
NC5	3.25
C6	2.81
C7+	3.48
MOL/HR	52.60
LB/HR	2779.26
T,DEG F	162.14
P,PSIA	400.00
MOL WT	53.22
LB/CUFT	28.5466
BTU/HR,	.11519E+06
BTU/R-HR	.29031E+04
FRAC. VAP	.0000

COMPRESSOR AND PUMP SUMMARY

1ST-STAGE COMPRESSOR CALCULATION:
ADIABATIC EFFICIENCY   =   .8900
MECHANICAL EFFICIENCY =   .9600
DELTA P   = 47.0   PSI:
BRAKE HORSEPOWER   = 476.4
COMPRESSION RATIO   =   3.611

2ND-STAGE COMPRESSOR CALCULATION:
ADIABATIC EFFICIENCY   =   .8900
MECHANICAL EFFICIENCY =   .9600
DELTA P   = 145.0   PSI:
BRAKE HORSEPOWER   = 453.4
COMPRESSION RATIO   =   3.417

```
3RD-STAGE COMPRESSOR CALCULATION:
ADIABATIC EFFICIENCY     =    .8900
MECHANICAL EFFICIENCY =    .9600
DELTA P                  = 520.0   PSI:
BRAKE HORSEPOWER         = 458.2
COMPRESSION RATIO        =   3.667

CONDENSATE CENTRIFUGAL PUMP:
DELTA P                  = 515.0   PSI:
BRAKE HORSEPOWER         =   0.8
FLOW RATE                =   1.0   GPM
PUMP EFFICIENCY          =  35.0

EXCHANGER SUMMARY:

FIRST INTERCOOLER E-1
   DELTA P                    =      -5.0  PSI
   INLET TEMP.                =     249.7  DEG F
   OUTLET TEMP.               =      95.0  DEG F
   HEAT ADDED                 = -.11519560E+07  BTU/HR

SECOND INTERCOOLER E-2:
   DELTA P                    =      -5.0  PSI
   INLET TEMP.                =     250.2  DEG F
   OUTLET TEMP.               =      95.0  DEG F
   HEAT ADDED                 = -.12290860E+07  BTU/HR

AFTERCOOLER E-3
   DELTA P                    -      -5.0  PSI
   INLET TEMP.                =     268.0  DEG F
   OUTLET TEMP.               =      95.0  DEG F
   HEAT ADDED                 = -.14333620E+07  BTU/HR

GAS-TO-GAS EXCHANGER E-4 (COLD SIDE)
   DELTA P                    =      -5.0  PSI
   INLET TEMP.                =        .0  DEG F
   OUTLET TEMP.               =      91.8  DEG F
   HEAT ADDED                 =  .60159530E+06  BTU/HR

GAS-TO GAS EXCHANGER E-4 (WARM SIDE)
   DELTA P                    =      -5.0  PSI
   INLET TEMP.                =      96.8  DEG F
   OUTLET TEMP.               =      40.7  DEG F
   F HEAT ADDED               = -.60159530E+06  BTU/HR

CHILLER E-5
   DELTA P                    =      -5.0  PSI
   INLET TEMP.                =      40.7  DEG F
   OUTLET TEMP.               =        .0  DEG F
   HEAT ADDED                 = -.50817200E+06  BTU/HR

DEMETHANIZER REBOILER DUTY =      0.34134E+06  BTU/HR
```

First-Stage Reciprocating Compressor

Design Data:

Gas flowrate	= 658.8 lb mol hr
	= 6.0 MMscfd
Gravity	= 0.764
Suction pressure	= 18.0 psia
Suction temp.	= 90°F
Discharge pressure	= 65 psia

Method 1—Isentropic Compression

Adiabatic efficiency	89%
Mechanical efficiency	96%

Assume discharge temperature is 250°F.

$$T_{av} = (250 + 90) / 2 = 170°F$$

From GPSA (1987, Fig. 13–8, p. 13–6)

$k = 1.22$ at 170°F and MW $= 22.1$

$$T_{2s} = T_1 (P_2/P_1)^{(k-1)/k} \qquad (10\text{--}12)$$
$$= (90 + 459.7)(65.0/18.0)^{(1.22-1)/1.22}$$
$$= 693°R = 233°F$$

$$T_2 = T_1 + (T_{2s} - T_1)/\eta_{is} \qquad (10\text{--}18)$$
$$= 90 + (233 - 90)/0.89 = 251°F$$

Assumed T_2 of 250°F is okay.
Use average $k = 1.22$ for calculations.

$$-W_{is} = \frac{R\,k\,T_1}{MW\,(k-1)}\,[(P_2/P_1)^{(k-1)/k} - 1] \qquad (10\text{--}13)$$

$$= \frac{(1.987)(1.22)(90 + 459.7)}{(22.1)(1.22 - 1)}[(65/18)^{0.22/1.22} - 1]$$

$$= 71.4 \text{ Btu/lb}$$

Ghp $= $ (Btu/lb)(lb/hr)/(2545)(h_{ad})
$$= (71.4)(14576)/(2545)(.89)$$
$$= 460$$

Bhp $= $ Ghp/η_{mech}
$$= 460/0.96 = 479$$

Method 2—GPSA Quick Estimate

MMcfd $= $ (MMscfd)$(P_L/14.4)[(T_1 + 459.6)/520]$
$$= (6.0)(14.7/14.4)(549.6/520)$$
$$= 6.47$$

MF $= 21$; $r > 2.5$ but SG is nearly 0.8
Bhp $= $ (MF)(r)(N)(MMcfd)(F) (Eq. 10–25)
$$= (21)(65/18)(1)(6.47)(1.0)$$
$$= 490$$

Method 3—Detailed Estimate (GPSA, 1987, p. 13–7 through 13–14)

$$r = P_2/P_1 = 65/18 = 3.61$$
k (average) $= 1.22$ as in Method 1

$T_d = 251°F$ as in Method 1
$Z_1 = (P_1)(MW)/(\rho_1)(R)(T_1)$
$$= (18)(22.13)/(0.0678)(10.731)(549.7)$$
$$= 0.995$$
$Z_2 = (P_2)(MW)/(\rho_2)(R)(T_2)$
$$= (65)(22.13)/(0.1899)(10.731)(251 + 459.7)$$
$$= 0.995$$
$Z_{av} = (Z_1 + Z_2)/2 = 0.995$

BHP/MMcfd $= 75$ Figure 13–13, GPSA
Bhp $= $ (Bhp/MMcfd)$(P_L/14.4)\,T_s/T_L\,(Z_{av})$(MMcfd)
$$= (75)(14.7/14.4)(549.7/519.7)(.995)(6.0)$$
$$= 484 \text{ HP}$$

High-speed correction: CHS $= 1$(assume low speed)
Intake pressure correction: CLIP $= 1$(Fig. 13–14, GPSA)
Gas gravity correction: CG $= 1$(Fig. 13–15, GPSA)

Bhp (corrected) $= $ Bhp (Eq. 10–29) (CHS)
                  (CLIP) (CG)
$$= (484)(1.0)(1.0)(1.0)$$
$$= 484 \text{ HP}$$

Water-Cooled Heat Exchanger

Design Data:

$$q = \text{Duty} = 1.15 \text{ MMBtu/hr}$$

Natural gas	Pressure (psia)	Temp (°F)
inlet	65	249.7
exit	60	95.0

Assume cooling water enters at 85°F (West Texas location) and leaves at 120°F.

$$LMTD = \frac{[(250 - 120) - (95-85)]}{\ln[(250 - 120)/(95 - 85)]} \qquad (12\text{--}4)$$

$$= (130 - 10)/\ln (130/10) = 46.8°F$$

Use nomenclature given in Figure 12–9

$$P = (t_2 - t_1)/(T_1 - t_1)$$
$$= (120 - 85)/(250 - 85) = 0.212$$
$$R = (T_2 - T_1)/(t_2 - t_1)$$
$$= (250 - 95)/(120 - 85) = 4.43$$

Cannot use a one-shell-pass heat exchanger because there is a temperature cross, and the LMTD correction factor is too low.

Use a two-shell-pass, shell-and-tube exchanger

$$F_2 = 0.875 \text{ GPSA (1987, Fig. 9–5) or TEMA}$$
$$CMTD = (0.875)(46.8) = 41°F$$
$$U_o = 30 \text{ Btu/hr-ft}^2\text{-°F (Table 12–1)}$$
$$q = U_o\,A_o\,(CMTD) \qquad (12\text{--}2)$$
$$1150000 = (30)(A_o)(41)$$
$$A_o = 935 \text{ ft}^2$$

Amine Unit for Low Pressure Rich Gas

Amine Contacting Unit—Stream 6

$$MMscfd = \left(\frac{mol}{hr}\right)\left(\frac{scf}{mol}\right)\left(\frac{hr}{day}\right)$$
$$= (655.25)(379.5)(24) = 5.97\ MMscfd$$

$$lb\ H_2S/day = \left(\frac{mol\ HS}{hr}\right)\left(\frac{lb\ S}{mol\ H_2S}\right)\left(\frac{hr}{day}\right)$$
$$= (0.85)(32)(24) = 653\ lb/day$$

S content > 100 lb/day, so use amine.
Select DEA.

$$mol\ percent\ AG = 100\ \frac{(mol\ H_2S + mol\ CO_2)}{(total\ mol)}$$

$$= 100\ (0.85 + 14.74)/(655.25)$$
$$= 1.07$$

From Figure 7–9 at 6 MMscfd, 185 psig,

Contact tower diameter = 31–in. ID

Use 36–in. OD tower.
Amine circulation rate is given by:

$$gpm = K\ (MMscfd)\ (mol\ percent\ AG\ removed)$$
$$= (1.45)(5.97)(2.38 - 0)$$
$$= 9.3\ gpm$$

Use 10 gpm.
Regeneration Equipment

Item	Duty (MBtu/hr)	Area (ft^2)
Reboiler	720	113
Solution exchanger	450	113
Aerial	150	102
Reflux	300	52

Motor	Horsepower
circulation	1.2
booster	0.6
reflux	0.6
aerial cooler	3.6

Hydrate Inhibition

Design Information:

$$MMscfd = \left(\frac{mol}{hr}\right)\left(\frac{scf}{mol}\right)\left(\frac{hr}{day}\right)$$
$$= (657)(379.5)(24) = 5.984\ MMscfd$$

From Figure 4–6:
Water content at 710 psia and 103°F is 86 lb H_2O/MMscf.
Water content at 700 psia and 0°F is 2.7 lb H_2O/MMscf.

SG gas = MW/28.9625 = (22.11)/(28.9625) = 0.76

Hydrate formation temperature using Baillie and Wickert method (Fig. 4–18):

Main graph T = 58.5°F

C3 correction at 5.57 mol percent C3, 0% H_2S
and .7 x 10^{-3} psia = −5°F

Hydrate formation temperature = 53.5°F.
If methanol is used:
Minimum liquid-phase methanol concentration, W, is:

$$W = (d)(MW)(100)/\{(d)(MW) + K_H\} \quad (6-1)$$
$$= (53.5 - 0)(32)(100)/\{(53.5)(32) + 2335\}$$
$$= 42.3\ weight\ percent$$

If methanol enters pure (W_{in} = 100%) and is diluted to Hammerschmidt minimum (W_{out} = 42.3 weight percent) then:

$$w_G = I\ (K_H)/(d)\ (MW)$$
$$(86 - 2.7) = I\ (2335)/(53.5)(32) \quad (6-6)$$
$$I = 61\ lb\ pure\ methanol/MMscf.$$

Use Figure 6–3 to obtain methanol losses in gas phase. Extrapolation is required. At 700 psia:

T,°F	Lb MeOH/MMscf/W_{out}	
50	1.65	
45	1.41	
40	1.19	
35	1.01	
30	0.85	
25	0.725	
0	0.25	extrapolated

Methanol lost in vapor phase = (0.25)(42.3)
= 11 lb pure MeOH/MMscf
Total methanol required = 61 + 11 = 72 lb/MMscf
= 430 lb/day
= 65 gal/day

Note: Hammerschmidt method is conservative (see Fig. 6–4) so no extra methanol addition "to be safe" is required.
If EG (ethylene glycol) is used:
EG is selected over DEG because the stream is cooled below 20°F.

Assume that the EG enters at 75 weight percent and leaves at 65 weight percent. This prevents any crystallization (see Fig 6–8). In addition, this "dilution requirement" is more demanding than the Hammerschmidt minimum liquid-phase concentration.

$$w_G = I\ [(100/W_{out}) - (100/W_{in})]$$
$$83.3 = I\ [(100/65) - (100/75)] \quad (6-4)$$
lb pure EG required, I = 406 lb/MMscf
= 2,430 lb/day

75 weight percent EG at 96°F weighs 9.05 lb/gal

75 weight percent EG circulation = 261 gal/day.

The choice between methanol and EG deserves careful consideration. However, one rule of thumb is to use methanol if the injection rate is less than 30 gal per hour. This assumes no problems in downstream equipment due to use of methanol.

Problems

1. Size the high-stage compressor for Example 15–1.
2. Size the high-stage compressor aftercooler for Example 15–1.
3. Size the second-stage compressor for Example 15–2.
4. Size the second-stage compressor aftercooler for Example 15–2.

APPENDIX 1
GLOSSARY OF TERMS

Absolute pressure	Pressure measured from a perfect vacuum expressed in psia.
Absolute temperature	Temperature measured from absolute zero temperature (-273.15 K, $-459.67°R$) expressed in K or °R.
Absorbent	Liquid used to remove specific components from a gas stream. Common examples are: TEG to remove water; absorption oil to recover heavy hydrocarbons from natural gas; and amines to remove carbon dioxide or hydrogen sulfide.
Acid gas	H_2S and/or CO_2 contained in or extracted from a natural gas.
Accumulator	Vessel used to collect and store liquids.
Adsorbent	Solid pellets used to remove specific components from natural gas.
Adsorption	Removal of certain components from natural gas by attraction to the surface of porous solid adsorbent pellets in a packed bed. Common examples include: acid gases, water vapor, or heavier hydrocarbon vapors.
Air cooler	Heat exchanger in which air is blown over a bundle of tubes to cool the fluid in the tubes.
Amine	Basic compounds containg the amino (NH_2) or a substituted amino group (RNH, RNR'). Aqueous amine solutions are generally used to remove hydrogen sulfide and carbon dioxide from gas and liquid streams.
°API gravity	An arbitrary scale expressing the relative density of liquid petroleum products. The scale, degrees API (°API), is calculated by:

$$°API = \frac{141.5}{Sp.\ Gr.\ 60°F/60°F} - 131.5$$

Associated gas	Gas produced with crude oil and separated in a gas-oil separator. Also called casinghead gas.
Atmospheric pressure	Pressure exerted by the earth's atmosphere. Varies with elevation. A pressure of 760 mm of mercury, 29.92 inches of mercury, or 14.696 psia, which is sometimes called one standard atmosphere, refers to the absolute ambient pressure at sea level.
Blanket gas	Gas introduced into a vessel above a liquid phase usually to protect the liquid from air contamination, for reducing the detonation hazard, or for pressuring the liquid. The source of the gas is external to the vessel.
Boiling point	Temperature at which the vapor pressure of a liquid equals the external pressure. When external pressure is 1 atm (14.696 psia, 101.325 kPa), temperature is called the normal boiling point.
Bubble point	Temperature and pressure at which the first bubble of vapor forms in a liquid.
Bypass	Pipe connection around a valve (or other equipment) to permit flow to continue while the valve (or equipment) is repaired.
Calorimeter	The equipment used to measure heating value of a combustible material.
Carbonyl sulfide	A compound containing a carbonyl group and sulfur group (COS). It is a contaminant in natural gas and NGL and is usually removed to meet sulfur specifications.
Casinghead (or casinghead gasoline)	Unstabilized natural gasoline condensed in lines coming from a well casinghead. A term used synonymously with natural gasoline.
Charcoal test	Standard test (AGA & GPA) for measuring the natural gasoline content in natural gas. The gasoline is adsorbed from the gas on activated charcoal and then recovered by distillation.
Chromatography	A technique for sample analysis where individual components of a batch sample, carried by an inert gas stream, are selectively adsorbed and desorbed on an adsorbent column at different rates. Separated components are detected quantitatively at the column exit.
Claus process	A process to convert hydrogen sulfide into elemental sulfur by selective oxidation.
Cleaning	Removing liquids and solids from natural gas.
Compressibility factor (Z)	Dimensionless ratio of the actual volume occupied by a gas to that predicted using the ideal gas law.

364

Compression ratio	The ratio of the absolute discharge pressure from a compressor to the absolute suction or intake pressure.
Condensate	Hydrocarbon liquids condensed from natural gas.
Conditioning	Operations used to make produced gas and/or oil suitable for storage and transport.
Convergence	A pressure used to characterize hydrocarbon mixtures in order to estimate equilibrium K-values. K-values for all components in the mixture tend to become unity at this pressure.
Copper strip test	A test using a small strip of pure copper to determine qualitatively the corrosivity of a product. Refer to GPA LP-gas corrosion test (see copper test method)–ASTM D–1838 test procedure.
Cricondenbar	The highest pressure at which vapor and liquid phases can coexist in a multicomponent mixture.
Cricondentherm	The highest temperature at which vapor and liquid phases can coexist in a multicomponent mixture.
Critical pressure	The pressure necessary to condense a vapor at its critical temperature.
Critical temperature	The highest temperature at which a fluid can exist as a liquid. Above this temperature, the fluid is a gas and cannot be liquefied regardless of the pressure applied.
Crude oil	Unrefined liquid petroleum ranging in gravity from 9 to 55°API and in color from yellow to black (see Tables 2.1 and 2.2).
Cryogenic plant	Gas processing plant capable of recovering NGL including ethane using a turboexpander to generate very low operating temperatures.
Custody transfer	Sale (transfer of ownership) of oil or gas.
Cycling	Process in which gas produced from a gas reservoir is processed or separated, and the remaining or residue gas returned to the reservoir. Also called "recycling."
Dead oil	Oil containing virtually no dissolved gas.
Dehydration	The act or process of removing water from gases or liquids.
Desiccant	An adsorbent used to dry a fluid.
Desulfurization	Removal of sulfur and sulfur compounds from gases or liquid hydrocarbons.
Dew point	The temperature and pressure at which liquid initially condenses from a gas or vapor. Specifically applied to the temperature at which water vapor (water dew point), or hydrocarbons (hydrocarbon dew point) condense.
Doctor test	Qualitative method for detecting hydrogen sulfide and/or mercaptans in petroleum distillates. The test distinguishes between "sour" and "sweet" products.
Dry gas	(1) Gas containing little or no hydrocarbons commercially recoverable as liquid product. Preferably such gas should be called "lean." (2) Gas with water content reduced by dehydration.
EP-mix (ethane-propane)	Product mixture consisting of essentially ethane and propane.
Field processing	All operations used to make crude oil and/or natural gas suitable for storage and transport.
Flash point	Lowest temperature at which vapors from a hydrocarbon liquid will ignite (see ASTM D–56).
FWKO	Free-water knockout—separator that removes water from crude oil or natural gas.
Gage	Variation of gauge.
Gauge pressure	Pressure measured from atmospheric pressure, expressed in psig.
Gas-condensate field	Petroleum reservoir in which the hydrocarbons exist at high pressure. Lowering the pressure causes condensation of the heavier hydrocarbons, which are then not produced.
Gas constant (R)	Dimensional constant defined by ideal gas law: $PV = nRT$.
Gas hydrate	See "hydrate."
Gas processing	All the field operations used to separate the constituents of and remove the contaminants in a natural gas. Purpose is making salable products and/or meeting sales specifications.
Gas-well gas	Natural gas produced from a formation containing little or no hydrocarbon liquid.
Gathering system	Network of pipelines used to carry gas from the wells to a central processing plant.

Handling	See "conditioning."
Heating value (heat of combustion)	Heat produced by the complete combustion of a unit quantity of fuel. For natural gas: (1) Gross or higher heating value (GHV or HHV) is that measured in a calorimeter when all the water produced during combustion is condensed. The heat of condensation is included in the total measured heat. (2) Net or lower heating value (NHV or LHV) is that obtained when the water produced by combustion is not condensed. This occurs usually in equipment burning fuel gas. Heating values are usually expressed in Btu/scf at designated T & P on a wet or dry basis.
Heavy ends	Relative term used to compare those hydrocarbons that have more carbon atoms than others. For instance, in a mixture containing methane, ethane, propane, butanes, and pentanes; the pentanes (and even the butane) could be referred to as the heavy ends.
Heptane plus (C7+)	All hydrocarbon components in a fluid mixture with seven or more carbon atoms in the molecule.
Hexane plus (C6+)	All hydrocarbon components in a fluid mixture with six or more carbon atoms in the molecule.
Hydrate	A solid material resulting from the combination of a hydrocarbon with water under pressure.
Hydro-test	Applying hydraulic pressure to check pipe and/or fittings.
Injection of gas	Putting gas into the formation by force (pressure).
Inert Gas	Carbon dioxide, nitrogen, or helium.
Joint	A single length of pipe.
Joule-Thomson effect	The change in gas temperature that occurs when the gas is expanded at constant enthalpy from a high to low pressure. Usually the initial temperature and pressure are adjusted so that the gas is cooled.
Junior orifice fitting	A one-piece orifice fitting.
Lead acetate test	A method for detecting the presence of hydrogen sulfide in a fluid by discoloration of paper which has been moistened with lead acetate solution.
Lean gas	(1) Residue gas after recovery of NGL. (2) Unprocessed gas containing little or no recoverable NGL.
Lean oil	Absorption oil from which all absorbed NGL have been stripped.
Light ends	The low-boiling, easily evaporated components of a hydrocarbon mixture.
LTX unit (low temperature extraction unit)	Unit using the cooling effect of a J-T expansion for improved recovery of liquids from streams from high pressure gas-condensate reservoirs.
Manifold	A pipe with one or more inlets and two or more outlets, or vice versa.
Mercaptan	Series of compounds of the general formula RSH, analogous to the alcohols and phenols, but containing sulfur in place of oxygen.
Mole (mol)	Mass or weight numerically equal to molecular weight of substance; e.g., one mole of water (H_2O) equals 18 lb, or 18 g, or 18 kg.
Molecule	Smallest unit quantity of substance that can exist by itself; e.g., one molecule of water (H_2O) contains 2 hydrogen atoms and 1 oxygen atom.
Molecular sieve	Synthetic zeolite (essentially silica-alumina) used in adsorption processes.
Molecular weight (MW)	Sum of atomic weights (AW) in one molecule; e.g., MW for water (H_2O) is 18 because AW for each H is 1 and AW for 0 is 16.
Natural gas	Gaseous hydrocarbon mixture found in nature. Principal HC components are C1, C2, C3, C4's, and C5+. May also contain N_2, CO_2, H_2O, H_2S (see Table 2-3).
Natural gasoline	A mixture of pentanes and heavier liquid hydrocarbons extracted from natural gas.
OCS orders	Rules and regulations set by MMS (formerly USGS) that govern offshore production operations in U.S. waters.
Odorant	Highly odiferous chemical, usually a light mercaptan, added to a gas or LP-gas to impart to it a distinctive odor for safety precautions and to detect leaks.
Oil formation volume factor	Dimensionless ratio of reservoir crude-oil volume to stock tank oil volume.

Plate-fin exchangers	Heat exchangers using thin sheets of metal to separate the hot and cold fluids tubes.
Peak shaving	Use of fuels and equipment to generate or manufacture gas to supplement normal supplies of pipeline gas during periods of extremely high demand.
Pentane-plus (C5+)	All hydrocarbon components in a fluid mixture with five or more carbon atoms in the molecule.
Pigging	Procedure of forcing a solid object through a pipeline for cleaning or other purposes.
Pipeline gas	Gas which meets a transmission company's minimum specifications.
Pressure maintenance	Injection of gas into a formation to maintain reservoir pressure.
Prime mover	Power source (electric motor, internal combustion engine) for a compressor, pump, etc.
Processing	See "conditioning."
Producing string	Tubing through which gas is produced from the formation to the surface.
Pig	Device forced through a pipeline to clean out debris or inspect line.
Raw gas	Unprocessed gas or the inlet gas to a plant.
Raw mix liquids	A mixture of natural gas liquids prior to fractionation. Also called "raw make."
Recompressor	A compressor used for some particular service, such as compressing residue gas; implies restoring of pressure after pressure reduction.
Reid vapor pressure (RVP)	Vapor pressure for liquid products using ASTM test procedure D–323. RVP is reported as psia at 100°F. The RVP is always less than the true vapor pressure at 100°F.
Relief system	System for temporarily releasing excess fluid, usually gas, to avoid a pressure in excess of the design pressure for the particular equipment.
Residue gas	Gas remaining after the recovery of liquid products.
Retrograde condensation (vaporization)	Condensation or vaporization that is the reverse of usual behavior. Condensation caused by a decrease in pressure or increase in temperature. Vaporization caused by an increase in pressure or decrease in temperature. Can only occur in mixtures.
Rich gas	Gas suitable as feed to a gas-processing plant from which liquid products can be extracted.
Rich oil	The oil leaving the bottom of an absorber. Lean oil plus all the absorbed constituents.
ROD, rich-oil demethanizer	Fractionating column used to remove methane from rich oil.
Sales gas	Same as pipeline gas.
Scrubber	Vapor-liquid separator for small amounts of liquid.
Senior orifice fitting	One-piece orifice fitting that allows changing of orifice plate without stopping gas flow.
Separator	Vessel used to split a multi-phase stream into a gas stream and one or more liquid streams.
Separator gas	Same as associated gas.
Shrinkage	Reduction in volume of oil as gas is evolved from it.
Slug catcher	Piping arrangement used to catch liquid slugs.
Solution gas	Gas dissolved in oil in a reservoir or the producing equipment.
Sour gas or oil	Gas or oil containing H_2S or mercaptans above a specified concentration level.
Specific gravity	Dimensionless ratio of the mass of a given volume of a substance to that of an equal volume of another standard substance. Usually air is used as the standard for gases and water for liquids and the volumes are measured at 60°F and atmospheric pressure (or bubble-point pressure).
Spiking	Reinjecting NGL into crude oil.
Stabilized condensate	Liquid mixture of hydrocarbons containing iC4, nC4, iC5, nC5, C6, and C7+.
Stabilizer	Fractionation unit that stabilizes any liquid; i.e., reduces the vapor pressure so that the resulting liquid is less volatile.
Stock-tank oil	Oil remaining after stage-separation train or stabilization; i.e., after dissolved gas has been released.
Straddle plant	NGL recovery plant that receives feed gas and returns sales gas at pipeline pressure.
Sulfur	Yellow, non-metallic element; called "free sulfur" in its elemental state. In many gas streams, sulfur is found

	as volatile sulfur compounds, i.e., hydrogen sulfide, sulfur oxides, mercaptans, carbonyl sulfide. Desulfurization is often required for corrosion control and health and safety reasons.
Sweet	This refers to the near or absolute absence of objectionable sulfur compounds in either gas or liquid as defined by a given specification standard.
Sweet gas	Gas which has no more than the maximum sulfur content defined by (1) the specifications for the sales gas from a plant, (2) the definition by a legal body such as the Railroad Commission of Texas.
Tail gas	The exit gas from a plant.
Treating	Sweetening, i.e., removing H_2S and/or CO_2.
Turboexpander	A device which converts the energy of a gas or vapor stream into mechanical work by expanding the gas or vapor through a turbine.
Vapor	Gas that can be liquified by compression and/or cooling.
Vapor pressure	Pressure exerted by a liquid when in equilibrium with a vapor.
Volumetric efficiency	Actual fluid volume taken in by a compressor divided by volume displaced by its piston or pistons.
Wet gas	Natural gas that yields hydrocarbon condensate (does not usually refer to water content). Also called ''rich'' gas.
Working pressure	Maximum pressure at which equipment can be used.

Common Abbreviations

acf	Actual cubic feet
acfm	Actual cubic feet per minute
AG	Acid gas
AGA	American Gas Association
AIME	American Institute of Mining, Metallurgical, and Petroleum Engineers—parent group of Society of Petroleum Engineers (SPE)
AISI	American Iron & Steel Institute
ANSI	American National Standards Institute
API	American Petroleum Institute—National trade association of U.S. petroleum industry, a private standardizing organization
ASME	American Society of Mechanical Engineers
ASTM	American Society for Testing and Materials
atm	Atmosphere
bbl	Barrel (42 U.S. gallons). The oil industry standard for volumes of oil and its products. Always reduced to 60°F and vapor pressure of the liquid
Bhp	Brake horsepower
BLM	Bureau of Land Management—U.S. government agency that regulates petroleum production onshore
BLPD	Barrels of liquid per day
BOPD	Barrels of oil per day
Brf	Barrels of reservoir fluid
BS&W	Basic sediment and water; water and other contaminants present in crude oil
Bscf	Billions of standard cubic feet
Bsto	Barrels of stock-tank oil
Btu	British thermal unit
BWPD	Barrels of water per day
C1	Methane
C2	Ethane
C3	Propane
C4's	Butanes
C5's	Pentanes
C6	Hexane
C6+	Hexane and heavier
C7	Heptane
C7+	Heptane and heavier
C8	Octane
cfm	Cubic feet per minute
CW	Continuous-welded
DEC1ER	Demethanizer
DEC2ER	Deethanizer
DEC3ER	Depropanizer
DEC4ER	Debutanizer
DEA	Diethanolamine
DEG	Diethylene glycol
°API	Degrees API gravity
°F	Degrees Fahrenheit
DGA	Diglycolamine
DIPA	Diisopropanolamine
DOT	Department of Transportation
DPU	Differential pressure unit
EG	Ethylene glycol
EP	Ethane-propane mix
ERW	Electric resistance welded
ft/sec	Feet per second
FERC	Federal Energy Regulatory Commission
FRP	Fiber-reinforced plastic
FWKO	Free water knock out
gal	U.S. gallon
GHV	Gross heating value
GLC	Gas-liquid chromatography

GLR	Gas liquid ratio, expressed as scf/bbl
GOR	Gas-oil ratio, combined gas released from stage separation of oil, expressed as scf/Bsto
gpm/GPM	(1) gpm (gallons per minute): flowrate in gallons per minute (2) GPM—Preferably gal/Mcf: natural gas liquids content in gallons of liquid products per thousand cubic feet.
gph	Gallons per hour
gr	Grain (7000 gr = 1 lb)
GSC	Gas-solid chromatography
HC	Hydrocarbon
HHV	Higher heating value
HP	High pressure
hp	Horsepower
hp-h, hp-hr	Horsepower-hour
H_2O	Water
H_2S	Hydrogen sulfide
HSS	Heat stable salts
iC4	Isobutane
iC5	Isopentane
ISA	Instrument Society of America
J-T	Joule-Thompson (constant enthalpy) expansion
kW	Kilowatts
LACT	Lease automatic custody transfer
lb	Pounds
LED	Light emitting diode
LET	Lowest expected temperature
LHV	Lower heating value
LMTD	Log mean temperature difference
LNG	Liquified natural gas. Primarily methane with minor amounts of ethane and propane.
LP	Low pressure
LPG	Liquified petroleum gas. Liquid mixture of C2, C3, iC4, and nC4.
MAWP	Maximum allowable pressure
Mcf	Sloppy equivalent for Mscf
Mcfd	Thousand cubic feet per calendar day
MDEA	Methyldiethanolamine
MEA	Monoethanolamine
MMcf	Same as MMscf
MMcfd	Million cubic feet per calendar day
MMS	Minerals Management Service—U.S. Government Agency that regulates petroleum production offshore.
MMscf	Millions of standard cubic feet
MPT	Minimum pipeline temperature
MTA	Mass transfer agent, e.g., absorber oil
MTD	Mean temperature difference
MTZ	Mass transfer zone
MW	Molecular weight
N, N_2	Nitrogen
NACE	National Association of Corrosion Engineers
NBS	National Bureau of Standards, now NIST
nC4	Normal butane
nC5	Normal pentane
NGL	Natural gas liquids: includes ethane, propane, butanes, pentanes, or mixtures of these.
NHV	Net heating value
NIST	National Institute for Standards and Technology, formerly NBS
NPS	National pipe standard
NPSH	Net positive suction head
OCS	Outer continental shelf
OSHA	Occupational Safety and Health Administration
PD	Positive displacement
PE	Polyethylene
PP	Polypropylene
ppm	Parts per million
ppmv	Parts per million by volume
ppmw	Parts per million by weight
psi	Pounds per square inch
psia	Pounds per square inch absolute
psig	Pounds per square inch gauge
PVC	Polyvinyl chloride
ROD	Rich oil demethanizer
RP	Recommended practice, e.g., API RP 14E
rpm	Revolutions per minute
RVP	Reid vapor pressure
S	Sulfur
SAW	Submerged arc welded
S&W	See BS&W
SCADA	Supervisory control and data acquisition
scf	Standard cubic foot; means of expressing volume of natural and other gases. The volume at 60°F and 14.696 psia (ideal gas) for process calculations. For sales purposes, it may be defined differently by law as in some states in the USA (see App. 2).
scfm	Standard cubic feet per minute
SG	Specific gravity
SI	Abbreviation (1) shut in, (2) Système International (French for International System of Units)
SO_2	Sulfur dioxide
SPE	Society of Petroleum Engineers
STB	Stock tank barrel
TEG	Triethylene glycol
TEMA	Tubular Exchanger Manufacturers Association
USGS	United States Geological Survey
Δ	Increment or difference
VLE	Vapor liquid equilibrium
WC	Water column, e.g., h_w = 80 in. WC

APPENDIX 2
MATERIAL BALANCES

Basic Relations

The material balance and the energy balance (discussed in App. 3) are the basic tools of the process engineer. These conservation equations, plus rate relations and equilibrium relations, will solve most of his technical problems.

The most general statement of the mass or material balance is given in words below. The system referred to is a control volume selected for convenience in any particular application.

$$\begin{array}{c} \text{Rate of flow} \\ \text{of Mass into} \\ \text{system} \end{array} + \begin{array}{c} \text{Rate of production in} \\ \text{system by chemical} \\ \text{reaction} \end{array} - \begin{array}{c} \text{Rate of flow} \\ \text{of Mass out of} \\ \text{system} \end{array}$$

$$= \begin{array}{c} \text{Rate of accumulation} \\ \text{of Mass in system} \end{array}$$

If no chemical reaction occurs and if steady state is assumed, then the mass balance assumes the obvious form

$$\begin{array}{c} \text{Rate of flow} \\ \text{of Mass into} \\ \text{system} \end{array} = \begin{array}{c} \text{Rate of flow} \\ \text{of Mass out of} \\ \text{system} \end{array}$$

This relation may not be as trivial as it appears, for example in case a single stream enters an apparatus and two streams of different composition leave. This latter is the case for a separator, for example:

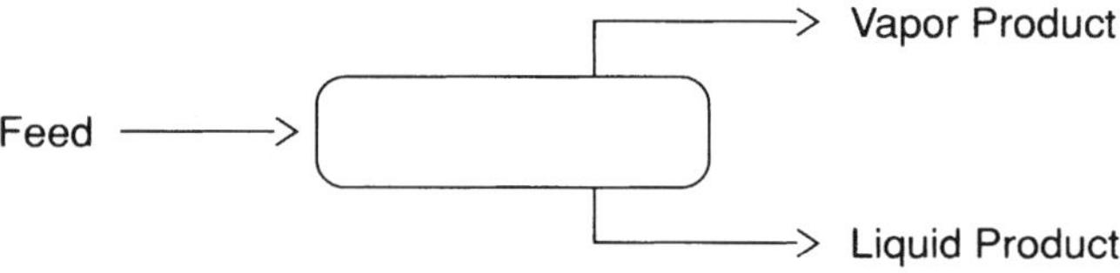

We recognize that two different sorts of material balances can be written: (1) an overall stream balance or (2) component balances.

$$F = F_V + F_L \qquad \text{Overall stream balance}$$
$$Fz = F_V y + F_L x \qquad \text{Component balance}$$

where F = flow rate of feed, moles/time
F_L = flow rate of liquid, moles/time
F_V = flow rate of vapor, moles/time
z = mole fraction of any component in the feed
x = mole fraction of any component in the liquid
y = mole fraction of any component in the vapor.

Process engineers use moles routinely in their work, whereas other engineers seldom use them. What is a mole? It is simply the mass of a substance expressed in units the size of its molecular weight. Strictly speaking, a prefix is needed to indicate whether the mole is a lbmole, gmole, or kgmole. The molecular weight of a substance tells how many lb in a lbmole (lb per lbmole), g per gmole, or kg per kgmole.

Example A2–1. There are 20 pounds of methane in a container. How many lbmole of methane are there?

Solution: The molecular weight of methane is 16.043 lb/lmole (Table A6–1). Thus,

$$\frac{20 \text{ lb}}{16.043 \text{ lb/lb mole}}$$
$$= 1.25 \text{ lbmole CH}_4 \qquad \longleftarrow$$

Example A2–2. How many kgmole of CH_4 are there in the above example?

Solution:
$$1.25 \text{ lbmole} \times \frac{1 \text{ kgmole}}{2.2046 \text{ lbmole}}$$
$$= 0.565 \text{ kgmole CH}_4 \qquad \longleftarrow$$

What is the advantage of the molar unit? Every mole of every substance has the same number of molecules in it. The molar unit expresses equivalence in terms of molecules. The mole is a useful unit for gas-law calculations because the gas law constant R in molal units is the same for every substance. The molal unit is also used in phase-equilibrium calculations.

The Continuity Equation

As a stream flows along a pipe in steady flow it may undergo changes in temperature or pressure. Although its mass rate of flow, w, is constant, the volumetric rate may change because of density changes. The average velocity of the fluid will vary also. These facts are contained in the continuity equation.

$$w = V A \rho$$

where w = mass flow rate, mass/time
V = average velocity of fluid in the conduit
A = cross-sectional area of the conduit
ρ = density of the fluid

The volumetric flow rate, Q, is given by

$$Q = V A = w / \rho$$

Example A2–3. A liquid of density 50 lb/ft^3 flows through a 4–in. Sched. 40 pipe (inside diameter 4.026 in.) at the rate of 80000 lb/hr. What are the average velocity and volumetric flow rates?

Solution:
$$A = \pi D^2/4 = (3.14159)(4.026/12)^2/4$$
$$= 0.0884 \text{ ft}^2$$

$$Q = w/\rho = \frac{80000 \text{ lb/hr}}{(50 \text{ lb/ft}^3)(3600 \text{ s/hr})}$$

$$Q = 0.4444 \text{ ft}^3/\text{s} \qquad \longleftarrow$$

$$V = Q/A = \frac{0.4444 \text{ ft}^3/\text{s}}{0.0884 \text{ ft}^2} = 5.03 \text{ ft/s} \qquad \longleftarrow$$

Density is required in continuity equation calculations. Calculation of liquid and gas density is discussed next.

Liquid Density. Liquid density is expressed not only in density units but also as API gravity and specific gravity.

Specific gravity

$$= \frac{\text{density of liquid}}{\text{density of water @ reference temperature}}$$

Different reference temperatures may be used, including the following:

Reference Temperature	Density of Water	
	lb/ft^3	kg/m^3
15.56°C (60.0°F)	62.37	999.0
16°C (60.8°F)	62.37	999.0
20°C (68.0°F)	62.32	998.3
25°C (77.0°F)	62.25	997.1

Example A2–4. The specific gravity of an oil at 60°F/60°F (i.e., the oil is at 60°F and the water reference temperature is also 60°F) is 0.850. What is the density of the oil?

Solution:
$$\text{density} = 0.850 \times 62.37$$
$$= 53.0 \text{ lb/ft}^3 \qquad \longleftarrow$$

The commonly used oil unit is degrees API, or API gravity.

$$°\text{API} = \frac{141.5}{\text{sp.gr. } 60°\text{F}/60°\text{F}} - 131.5$$

Example A2–5. What is the API gravity of the above oil?

Solution:
$$°\text{API} = (141.5/0.850) - 131.5$$
$$= 35.0 \qquad \longleftarrow$$

Example A2–6. What is the density of a 44.0°API oil at 60°F?

Solution:
$$\text{Sp Gr @ } 60°\text{F}/60°\text{F} = \frac{141.5}{°\text{API} + 131.5}$$
$$= \frac{141.5}{175.5} = 0.8063$$
$$\text{Density} = 0.8063 (62.37)$$
$$= 50.3 \text{ lb/ft}^3 \qquad \longleftarrow$$
$$= \frac{50.3 \text{ lb/ft}^3}{7.48052 \text{ gal/ft}^3}$$
$$= 6.72 \text{ lb/gal} \qquad \longleftarrow$$

In most practical problems the temperature is not at the reference condition, so a method is needed to convert densities from one temperature to another. A satisfactory correlation for crude oils is given in Figure A6–1. This figure is based on the temperature-density relations described by Hankinson, *et al.*, (*Oil & Gas J.*, Dec. 24, 1979, pp. 66–70).

Example A2–7. The 35°API oil in Example 6 is actually at 120°F. What is its density?

Solution:
From Figure A6–1 the density correction factor, DCF, is 0.9745.
$$\text{Density @ } 120°\text{F} = (\text{DCF})(\text{Density @ } 60°\text{F})$$
$$= (0.9745)(53.0)$$
$$= 51.6 \text{ lb/ft3} \qquad \longleftarrow$$

A correlation of light paraffin hydrocarbon liquid densities is given in Figure A6–2.

Example A2–8. A natural gas condensate has a molecular weight of 80. What is its density at 100°F?

Solution:
Figure A6–2. Draw a straight line starting from 100°F on the temperature scale to a molecular weight of 80. Extend this straight line to the specific gravity scale and read 0.628.

$$\text{Density} = 0.628 \times 62.43$$
$$= 39.2 \text{ lb/ft}^3 \qquad \longleftarrow$$

(we assume the reference water density is 62.43).

Oil Volumes. As seen above, the volume of a given mass of oil is dependent on the temperature. For custody transfer, oil volumes are always converted to a standard temperature . . . 60°F in the United States. Thus, in all U.S. literature, unless there is a statement to the contrary, volumes should be considered to be 60°F volumes. If the statement is made that the volume is an "actual" or a "flowing" volume, then it should be taken to be the volume at "actual" or "flowing" temperature (which must have been stated).

Figure A6–1 can be used to reduce volumes at temperatures other than 60°F to 60°F. The reciprocal relation holds here.

$$V_T/V_{60} = \rho_{60}/\rho_T = 1/DCF$$

Example A2–9. We have 1000 bbl of the above 35 °API oil at 120°F. What is the volume at 60°F?

$$V_{120}/V_{60} = 1/DCF$$
$$V_{60} = (DCF)(V_{120}) = (0.9745)(1000)$$
$$= 974.5 \text{ bbl} \quad \longleftarrow$$

Gas Volumes. Gas volumes are generally quoted in the literature in scf (standard cubic feet) or often simply cf (cubic feet). Invariably (unless specifically stated to the contrary) the cf really means scf; standard conditions are 60°F and 14.696 psia in English engineering units, 15°C and 101.325 kPa in SI units. An ideal gas at standard conditions occupies 379.50 cubic feet/lbmole.

$$V = \frac{nRT}{P} = \frac{(1)(10.732)(459.67 + 60)}{14.696}$$
$$= 379.50 \text{ ft}^3$$

The volume in scf is therefore converted to a mass unit by dividing by 379.5.

Example A2–10. A natural gas flows to a platform at 10 MMscfd. What is the flow rate in lbmole/hr?

Solution:
$$\frac{10000000 \text{ scf/day}}{(24 \text{ hr/day})(379.5 \text{ scf/lbmole})}$$
$$= 1098 \text{ lbmole/hr} \quad \longleftarrow$$

For natural gas processing design calculations the standard conditions mentioned above are invariably used. However, for sales or custody transfer, other standard conditions may be used. While 60°F is used for contracts in the United States, the standard pressure varies as follows:

14.65 psia—AR, IL, KS, OK, TX
14.73 psia—CA, MI, and by the AGA
14.85 psia—WV
15.025 psia—CO, LA, MS, NM, UT, and WY.

Furthermore, the volume may be dry or saturated with water vapor.

"Actual" or "flowing" gas volumes must be determined from the actual or flowing density of the gas, as will be illustrated below.

Gas Composition. Gas composition is normally expressed as mole percent, which is the same as volume percent for an ideal gas, or approximately so for a real gas at or near atmospheric pressure (within engineering accuracy).

The composition may be expressed in other ways, as shown in the next example.

Example A2–11. The gas in the previous example has the composition below. Find its average molecular weight and its flow rate in lb/hr.

Component	Mole %
C1 (methane, CH_4)	90
C2 (ethane, C_2H_6)	8
C3 (propane, C_3H_8)	2
	100

Solution: Basis—1 lbmole mixture.

Component	lbmole	Mol Wt.* lb/lbmole	lb
C1	0.900	16.043	14.439
C2	0.080	30.070	2.406
C3	0.020	44.097	0.882
			17.727 lb/lbmole

*Table A6–1
$$(1098 \text{ lbmole/hr})(17.727 \text{ lb/lbmole})$$
$$= 19460 \text{ lb/hr} \quad \longleftarrow$$

Note that 17.727 is the average molecular weight of the gas. The specific gravity of a gas is its molecular weight divided by the molecular weight of air (28.9625).

Example A2–12. Find the specific gravity of the gas in the previous problem.

$$Sp \ gr = 17.727/28.9625$$
$$= 0.612 \quad \longleftarrow$$

The composition in weight percent can be calculated from the data presented above.

Example A2–13. Find the composition in weight percent of the gas in the previous example.

Component	lb	Weight %
C1	14.439	81.45
C2	2.406	13.57
C3	0.882	4.98
	17.727	100.00

APPENDIX 3
ENERGY BALANCES

So-called enthalpy balances are just special cases of the general energy balance for a given system. The energy balance itself is a mathematical statement of the first law of thermodynamics. The first law states that energy is conserved in real processes and does not simply appear or disappear.

The energy balance is useful in field processing analyses involving fuel and its efficient use. The general energy balance equation for a steady-state system with single inlet and outlet streams is as follows:

$$(H_2 - H_1) + \frac{g(Z_2 - Z_1)}{g_c \, J} + \frac{(V_2{}^2 - V_1{}^2)}{2g_c \, J} \qquad \text{(A3–1)}$$
$$= Q - W_s - \Delta Hcomb$$

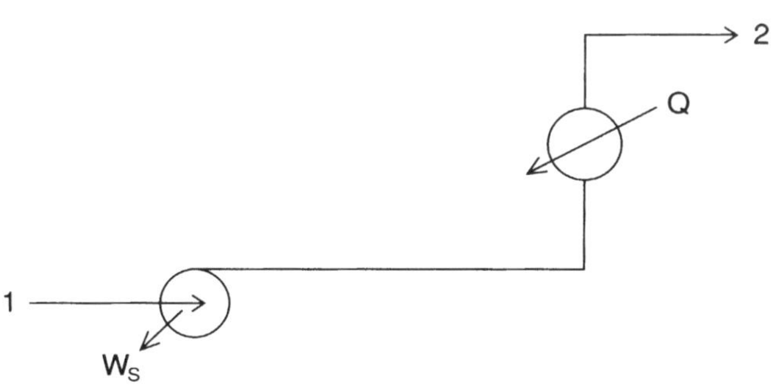

where, H = stream enthalpy, energy/mass
    Z = elevation above an arbitrary datum, length
    V = average velocity of fluid in conduit, length/time
    Q = energy added to system by virtue of a difference in temperature (heat transfer), energy/mass
    W_s = energy expended by system to perform mechanical work (shaft work), energy/mass
    ΔH_{comb} = heat of combustion of fuel burned, energy/mass
    g = acceleration of gravity, length/(time)^2
    g_c = conversion factor between mass-length-time and force units
    = 1 kg-m/(s^2-N)
    = 32.174 ft-lbm/(s^2-lbf)
    J = conversion factor between force-distance and energy units
    = 1.0 N-m/J
    = 778.17 ft-lbf/Btu.

Real systems are rarely at true steady state. Nevertheless, analysis by Equation A3–1 is sufficiently accurate for many applications. Each of the terms in Equation A3–1 will be described briefly below.

Enthalpy Change. The term $H_2 - H_1$ is the net change in enthalpy of the flowing streams. Enthalpy is composed of the sum of the internal energy of the fluid and the PV energy added to the fluid in order to force it into the system. Internal energy represents the energy resident in a substance that cannot be seen but is known to be there. Microscopically, internal energy is the sum of the kinetic and potential energies of the particles making up the fluid.

Potential Energy. The term $g(Z_2 - Z_1)/g_c \, J$, is the change in potential energy relative to the earth's gravitational field. Other types of potential energy, such as electrical or magnetic, are generally negligible in field processing.

Kinetic Energy. The change in kinetic energy of the flowing fluid is given by $(V_2{}^2 - V_1{}^2)/2g_c \, J$. This term will be important only if the change in velocity is extremely large, not usually the case in field processing applications.

Heat Transfer. Heat transfer, Q, to or from the system by conduction, convection, or radiation is often an important term, especially where fuel usage is involved. The *rate of heat transfer* is,

$$q = w \, Q$$

    where, q = heat transfer rate, energy/time
    w = flow rate of fluid, mass/time.

The heat transfer rate may be essentially negligible in some processes. A process for which $Q = 0$ is termed "adiabatic."

Where heat transfer is important it will generally be calculated by the heat transfer rate equation:

$$q = U \, A \, \Delta T_m$$

where U = overall coefficient of heat transfer, energy/[(area)(time)(temperature driving force)]
    A = heat transfer area
    ΔT_m = temperature driving force.

Methods for calculating q for different geometries and with different mechanisms are presented in standard texts on heat transfer such as Kreith, "Principles of Heat Transfer," Intext, 3rd Ed., 1973.

Shaft Work. Pumps, compressors, and turbines involve the transfer of fluid energy via a shaft to or from the surroundings. The power or rate of doing work, P_s, is the fluid flow rate times the shaft work.

$$P_s = w \, W_s$$

Heat of Combustion. In combustion devices, the flowing fluid on the combustion side is air and fuel. A portion of the fluid burns, releasing energy. This energy is represented in the heating value term, ΔH_{comb}. When the ΔH_{comb} term appears in the energy balance, all stream enthalpies should be calculated relative to 60°F, 1 atm, the datum conditions for the heat of combustion.

Applications of the Energy Balance

In field processing operations, the energy balance can often be reduced to simple forms. Some of these are summarized below. For all but combustion devices, the heat of combustion term is zero.

Heat Exchangers. Changes in potential and kinetic energy are negligible. No work is done. Equation A3–1 becomes:

$$Q = H_2 - H_1$$

The heat added or removed is equal to the change in enthalpy between the inlet and outlet streams. Quantity Q is positive if heat is added and negative if heat is removed from the fluid stream.

Valves. Changes in potential and kinetic energy are negligible. No work is done. The process is essentially adiabatic.

$$H_2 - H_1 = 0$$

The process is "isenthalpic." The expansion is referred to as a "Joule-Thomson" expansion.

Separators. Separators operate the same as a valve (isenthalpic). However, the exit enthalpy, H_2, is the sum of the exit vapor and liquid enthalpies.

$$H_2 - H_1 = 0$$

Compressors, Pumps, and Turbines. Fluid power devices usually have negligible potential and kinetic energy changes. The operation is essentially adiabatic.

$$W_s = -(H_2 - H_1)$$

The shaft work is the negative of the change in enthalpy. The shaft work is positive if work is done on the surroundings (expander or turbine) and negative if done on the fluid (pump or compressor).

Heaters and Furnaces. Kinetic and potential energy changes are negligible. No work is done. The heat of combustion contributes to the increase in combustion gas enthalpy and to heat transfer.

$$-\Delta H_{comb} = -Q + (H_2 - H_1)$$

The heat exchanged, Q, is composed of two terms: heat picked up by the fluid being heated, Q_{proc}, and heat lost to the surroundings, Q_{loss}.

$$-\Delta H_{comb} = -Q_{proc} - Q_{loss} + (H_2 - H_1)$$

Note that we are treating the heat of combustion as a negative quantity here (the thermodynamic convention).

Ideally, it is desirable to increase the first term on the right side and decrease the other two. We insulate to prevent heat loss and discharge the flue gas as cool as possible. Recovery of greater than 80% of the heating value of the fuel as process heat is common.

Combustion Engines. Gas turbines and gas engines, as well as diesel and gasoline engines, are similar to heaters except that shaft work is done.

$$-\Delta H_{comb} = W_s - Q_{loss} + (H_2 - H_1)$$

Here it is desired to increase work and decrease heat loss and exhaust gas temperature (enthalpy). From the practical standpoint W_s is limited to about 30% of the heat of combustion of the fuel. Energy savings can be obtained by introducing heat exchangers in the exhaust stack, thus recovering process heat for other purposes, such as glycol regeneration.

$$-\Delta H_{comb} = W_s - Q_{loss} - Q_{proc} + (H_2 - H_1)$$

Calculation of Enthalpy

From the above discussion it can be seen that enthalpy is an important property for process calculations. Suitable techniques for evaluating enthalpy are given in the following discussion.

Basic Relations. From thermodynamics it is known that enthalpy is evaluated for sensible heat changes (changes where temperature is raised or lowered with no phase change) by,

$$\Delta H = H_2 - H_1 = \int_{T_1}^{T_2} C_p \, dT \ (@ \text{ constant pressure}) \quad (A3\text{–}2)$$

Heat capacity, C_p, is a function of T. For many applications we can use an average value for C_p and assume it constant over the range of interest. Then,

$$\Delta H = C_p (T_2 - T_1) \quad (A3\text{–}3)$$

Enthalpy datum selection (temperature and physical state) is arbitrary. Different datums are used by different references. The main thing to remember is that if a chemical reaction takes place, the datum for enthalpy should be the same for each component. If no chemical reaction takes place, different datums can be used for different components as long as the selected datums are used consistently.

Latent heats are the enthalpy changes occurring with phase change, such as melting or vaporization.

Enthalpy charts are developed from heat capacity and latent heat data after a suitable datum is selected.

Sensible Heats. If the only heat effect in a heat exchanger is sensible heat (no phase change), then the best hand calculation method is by Equation A3–3 above. A chart is available for crude oils, Figure A6–3, and a series of graphs for natural gases, Figure A6–4.

Example A3–1. A 40°API gravity crude oil (characterization factor = 12.0) is to be heated from 100 to 130°F. What will be the enthalpy change per pound? The oil flows at the rate of 4000 bpd. What is the exchanger duty, q?

Solution: The average temperature is $(100 + 130)/2 = 115°F$. The base C_p is 0.49 Btu/lb $-$ °F and the characterization factor correction is 1.01 (Fig. A6–3).

$$C_p = 1.01 \,(0.49) = 0.495 \text{ Btu/lb–°F}$$

By Equation A3–3, the enthalpy change is:

$$Q = \Delta H = (0.495 \text{ Btu/lb–°F})(130 - 100)°F = 14.8 \text{ Btu/lb}$$
$$\text{S.G.} = (141.5)/(131.5 + 40) = 0.825$$
$$\text{(specific gravity)}$$
$$\rho = (0.825)(62.37) = 51.5 \text{ lb/ft}^3$$
$$= \frac{(51.5)(\text{lb/ft}^3)(42 \text{ gal/bbl})}{7.48 \text{ gal/ft}^3}$$
$$= 289 \text{ lb/bbl}$$
$$w = \frac{(289 \text{ lb/bbl})(4000 \text{ bbl/day})}{24 \text{ hr/day}}$$
$$= 48200 \text{ lb/hr}$$

$$q = w\,Q = (48200 \text{ lb/hr})(14.8 \text{ Btu/lb})$$
$$= 713000 \text{ Btu/hr} \qquad \longleftarrow$$

Example A3–2. A natural gas of specific gravity 0.6 is to be cooled from 280°F to 80°F at 600 psia. The gas rate is 20 MMscfd. What will be the change in enthalpy? What is the exchanger duty?

Solution: The average temperature is 180°F. At 180°F, 600 psia, the heat capacity is 0.60 Btu/lb°F from Figure A6–4. The change in enthalpy is,

$$Q = \Delta H = (0.60 \text{ Btu/lb–°F})$$
$$(280 - 80) = 120 \text{ Btu/lb}$$
$$MW = 28.9625 \text{ (S.G.)}$$
$$= (28.9625)(0.6) = 17.38 \text{ lb/lbmole}$$
$$w = \frac{(20000000 \text{ scf/day})(17.38 \text{ lb/lbmole})}{(379.5 \text{ scf/lb kmole})(24 \text{ hr/day})}$$
$$= 38200 \text{ lb/hr}$$
$$q = w\,Q = (38200 \text{ lb/hr})(120 \text{ Btu/lb})$$
$$= 4580000 \text{ Btu/hr} \qquad \longleftarrow$$

Gas and Gas Liquids Enthalpy Charts. Hand calculations for natural gas and natural gas liquids can be made with good accuracy using the charts in the GPSA Engineering Data Book, Figures 24–8 through 24–16. These charts use molecular weight to characterize mixtures.

Example A3–3. Calculate ΔH in the above example for the 0.6 gravity gas but using the GPSA enthalpy charts.

Solution: The molecular weight of the gas is 17.38, as above. The enthalpies are 235 and 353 at 80°F and 280°F, respectively, from GPSA Figure 24–11.

$$\Delta H = 353 - 235 = 118 \text{ Btu/lb}$$

This is the same as the previous result within engineering accuracy.

APPENDIX 4
OPSIM

**An Interactive Flowsheet Simulator Program
for Crude Oil and Natural Gas Processing**

by Paul Buthod
and Richard Thompson
The University of Tulsa, 1989

Introduction

OPSIM consists of a driver program and a set of subroutines used to simulate steady-state processing operations. The programs are written in FORTRAN 77. The package was originally developed by Emmanuel Udegbunam (1978) under the direction of Paul Buthod and Richard Thompson, with later additions made by Shueh Cheng (1982) and Franklin Chukwuma (1981). The interactive version was written by Robert Fraser (1984) and modified by Buthod and Thompson. The latter program was originally used on the Honeywell CP–6 main-frame at The University of Tulsa; it was modified for use on IBM AT and compatible microcomputers.

OPSIM performs phase-equilibrium and material and energy balance calculations for hydrocarbons and a few gases, plus pseudocomponents for which certain physical properties are provided. These pseudocomponents are referred to generally in natural gas processing as C7+ components and in crude oil processing as boiling-point components.

The executable program occupies approximately 130 kbytes of storage. The program can be executed on an IBM PC, XT, or AT compatible microcomputer. Fast computation is obtained only on an AT with math coprocessor.

Several items of process equipment can be simulated, including valves, compressors, expanders, heat exchangers, pumps, and vapor-liquid separators. Shortcut procedures also are provided for simple fractionators and demethanizers of the type used in gas-processing facilities. The heating value of a mixture can be computed.

The properties generated in the program are the fundamental thermodynamic properties needed for phase-equilibrium and energy-balance calculations. Enthalpies and entropies are calculated for the ideal gas state from API Refinery Data Book (1970) heat-capacity equations for the library components. A heat-capacity equation due to Stevens and Thodos (1973) is used for the C7+ or boiling-point components. The Usdin-McAuliffe modification (1976) of the Soave-Redlich-Kwong equation of state is used to calculate component fugacities (and hence vapor-liquid K-values) and mixture enthalpy and entropy deviations from ideal-gas behavior. The UMSRK equation was programmed by Perez (1978).

The program does not presently have the capability of handling pure components or water. Pure components can be approximated by a mixture containing mostly the substance of interest. In most industrial applications, this is in fact the case. For example, in gas processing "propane" is used as the working fluid in the vapor-compression refrigeration cycle; this propane invariably contains small amounts of ethane and isobutane.

It would be convenient to be able to handle water as a component, as can be done in commercially-available packages. The same can be said for the ability to predict the formation of water-hydrocarbon hydrates and CO_2 "frost." These complex phase equilibria are not treated in OPSIM. Liquid and vapor water can be handled by using the partial pressure of the hydrocarbon phase in place of total pressure, then adding the water by hand calculations using, for example, the McKetta-Wiehe chart in the GPSA Engineering Data Book (1987).

The structure of the program is important to the user who may wish to add to the package operations not presently contained in it. This task is not particularly difficult once the structure is understood. Program structure is outlined so as to allow easier reading of the FORTRAN source programs.

The average user simply wants to know how the package works, so that topic is discussed first.

Use of the Program

Two steps are involved in the use of OPSIM. First, the problem to be analyzed should be laid out in convenient form. Then the program can be called up on the computer and the actual calculations performed.

Preparation of a problem involves two steps: (1) definition of the feed or feeds, and (2) definition of the process flow diagram.

Feed Definition. Complete definition is provided by the feed composition, pressure, temperature, and flow rate.

Feed composition must be given by identifying the components of the feed and then by specifying the mole percent of each.

Library components include the following:

Nitrogen, N_2
Carbon dioxide, CO_2
Hydrogen sulfide, H_2S
Methane, C1

Ethane, C2
Propane, C3
Isobutane, iC4
Normal butane, nC4
Isopentane, iC5
Normal pentane, nC5
Hexane, C6

The hydrocarbons all contain hydrogen, of course, as in methane, CH_4. The C-H ratio is fixed for the paraffin hydrocarbons, so the carbon number is used to designate each: C1, C2, etc. Isopentane could be either 2–methyl butane or neopentane in theory; very little neopentane is found in naturally occurring mixtures, so 2–methyl butane is used. Hexane refers to all hexanes lumped as n-hexane.

As many as nine pseudocomponents can be used. For each such component the user must provide a 1-to-4 character identifier and some or all of the following properties:

Normal boiling point, °F	—Required
Liquid specific gravity at 60°F	—Required
Critical Temperature, °F	—Optional
Critical Pressure, psia	—Optional
Molecular Weight	—Optional

The optional properties are calculated by built-in correlations based on boiling point and specific gravity if not supplied by the user.

Feed temperature (°F), pressure (psia), and flow rate (lbmole/hr) are required. In actual process simulations these properties will be known. In some simulations these conditions may not be known or even germane, as in the case of building a phase envelope; in such cases the user should supply any convenient set of values, such as 100°F, 100 psia, 100 lbmole/hr.

Flow-Diagram Definition. OPSIM is a block-oriented program. A defined feed is introduced to a process module or block, along with the necessary parameters to define the module. Appropriate calculations simulate the process equipment and totally define the product stream or streams leaving the module. These streams then can be used for further operations.

The particular simulation should be sketched in a manner convenient to the use of OPSIM, i.e., as a flow diagram with streams represented by lines and operations or processes represented by convenient symbols. Each distinct stream should be numbered from 1 to 40 (the allowable stream numbers in OPSIM). In OPSIM a stream is really just a 30–item array that contains all the pertinent stream properties. The array can be reused as necessary; the data are stored and are available to be read or replaced as the user desires.

Example 1. A stream flowing at 1×10^6 scf/day contains 80 mole percent methane, 16 mole percent propane, and 4 mole percent n-heptane. The feed stream is available at 80°F and 15 psia. It is desired to compress this stream to 60 psia, cool it to 80°F with a heat exchanger having a 5 psi pressure drop, then separate out any condensate formed. The compressor is estimated to have an isentropic efficiency of 0.90 and a mechanical efficiency of 0.95. Prepare this information for use with OPSIM.

Feed Definition. Methane and propane are library components, so they can simply be selected from the library list. Normal heptane is not in the library but can be handled as a pseudocomponent. For this purpose its properties are obtained from a convenient reference, in this case the GPSA Engineering Data Book, 1987, p. 23–2 and 3).

nC7—Boiling point	209.16 F
Sp. Gr.	0.6882
Crit. Temp.	512.7 F
Crit. Press.	396.8 psia
Mol. Wt.	100.204

The feed temperature and pressure are given. The flow rate is given but is not in the proper units. At standard conditions of 60°F and 14.696 psia, one lbmole is equivalent to 379.5 scf (standard cubic feet). Thus the flow rate is

$$1 \times 10^6 \text{ scf/day} \times 1 \text{ day/24 hr} \times 1 \text{ lbmole/scf}$$
$$= 109.794 \text{ lbmole/hr}$$

Flow Diagram Definition. The flow diagram is sketched as shown below with the streams numbered as shown. The user is now prepared to simulate the process.

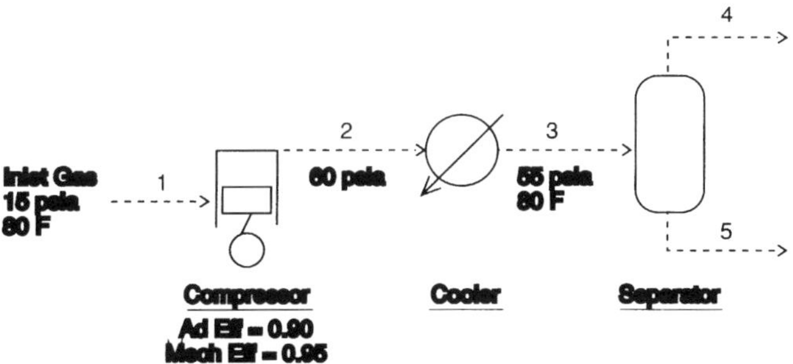

Example 2. It is desired to obtain a phase diagram for a mixture containing 80 mole percent C1, 16 mole percent C3, and 4 mole percent of a mixture considered to have the boiling point and specific gravity of nC7.

Feed Definition. Methane and propane are library components and can be selected from the library list, as in Example 1. The pseudocomponent has the same boiling point and specific gravity as in Example 1, but the remaining properties are not known. Let OPSIM calculate them by the built-in correlations. The feed temperature, pressure, and flow rate are not given but are not really needed. Use any reasonable values for the purpose of starting the simulation.

Flow Diagram Definition. There is no flow diagram required here, since the objective is simply to run a series of bubble-point, dew-point, and flash calculations.

Example 3. It is desired to simulate the stage separation of a crude oil of the composition given below. (Note that the heavy ends of the crude are lumped into one C7+ component for simplicity in this example; it would be better to break this cut into a number of components). The feed is at 2062 psia, 120°F upstream of the first-stage separator. Three stages of separation will be used; the stage pressures will be 315, 70, and 17 psia. The feed flow rate will be taken as 1000 lbmole/hr.

Component	Mole Percent
N2	0.78
CO2	0.05
C1	33.86
C2	5.63
C3	4.40
IC4	1.21
NC4	3.42
IC5	1.85
NC5	2.44
C6	4.29
C7+	42.07 (BP = 499°F, Sp Gr = 0.8426)
Total	100.00

Feed Definition. All components except C7+ are library components and can be selected from the library list, as in Example 1. The C7+ pseudocomponent has the boiling point and specific gravity given from some source, but the remaining properties are not given. In this case the remaining properties were estimated from charts in the API Refinery Data Book (1970). The critical temperature was estimated as 800°F, the critical pressure as 280 psia, and the molecular weight as 207.

Flow Diagram Definition. The process involves the flashing flow of the crude oil stream through a series of valves. Valves are treated as isenthalpic flashes, that is, adiabatic flashes (heat transfer equal to zero). Each valve is followed by a vapor-liquid separator. The flow diagram is given in Figure 1.

OPSIM Execution

Having made the proper preparations, the user is ready to execute OPSIM and obtain the desired results. Depress "CAPS LOCK" since replies to program prompts are generally in capital letters. Then select the disk or directory that contains the OPSIM.EXE file as the working directory. Type "OPSIM" and depress ENTER. The following message should be printed:

WELCOME TO INTERACTIVE OPSIM
THE UNIVERSITY OF TULSA

DO YOU WISH TO SAVE THE OUTPUT IN A FILE? (Y/N)

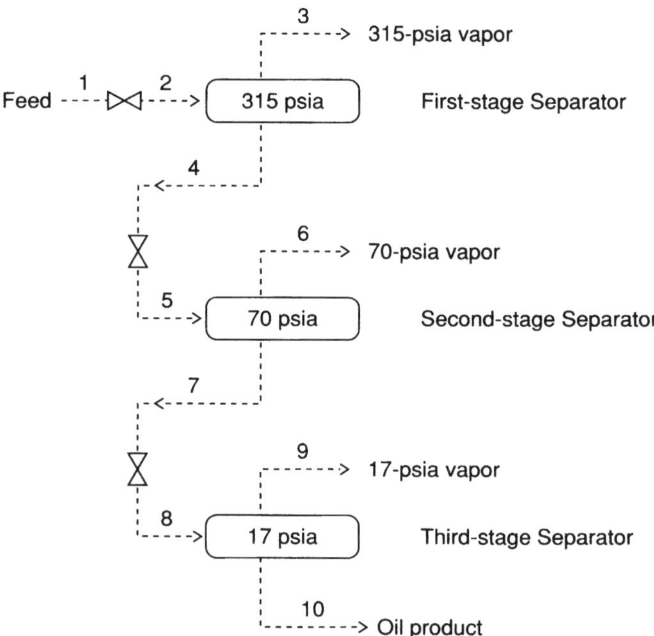

Figure A4–1.   Flow diagram for Exercise 3.

If you desire to save the output, type "Y." The file will contain a record of the important results of the simulation that can be scanned or printed later if desired. Otherwise the results will simply appear on the monitor screen and scroll off as replaced.

Note: Most of the OPSIM prompts are stated in such a way that the normal reply to "Y/N ?" is "N." It is *unnecessary* to type the "N." Simply depress "ENTER." This feature makes replying much more rapid than if a character must be entered for each such question. (To the programmer: The FORTRAN source program tests for "Y" in every case, and a carriage return is not considered to be a "Y.") If the reply to the above prompt is "Y," the following message is printed.

TYPE IN THE NAME OF THE FILE (1–14 CHAR.).

At this point type in the desired name, which can be of the maximum length allowed in DOS (letter indicating drive, colon, an eight-character name, a period, and a three-character extension). For example:

A:DISKFILE.OUT

The next prompt is:

DO YOU HAVE A PREVIOUSLY-PREPARED DATA FILE? (Y/N)

The first time a particular simulation is performed there will not be a previously-prepared input data file. However, after the feed data are typed in, the user is given the opportunity to save this information. It is a good idea to do so in order to save time if the simulation must be repeated for any reason. If the reply to the above prompt is "Y," the program will ask the user for the file name and read it, printing out the stream summary for the data (described below). If your reply is "N," then the program prompts selection of library components and/or entry of pseudocomponents. The example that follows selects the components of Example 1 above.

FEED COMPONENT ENTRY
IS THE FOLLOWING COMPONENT IN ANY OF
THE FEED STREAMS? (Y/N)
   N2   [A return is interpreted as "N"]
   CO2
   H2S
   C1
Y
   C2
   C3
Y
   IC4
   NC4
   IC5

   NC5
   C6
DO YOU WANT TO ENTER ANY "C7+" COMPONENTS? (Y/N)
MAXIMUM NUMBER IS 9
Y
ENTER NAME OF COMPONENT (4 CHARACTER MAX)
   nC7   [lower-case letters can be used here]
ENTER NORMAL BOILING POINT, DEG F: (NECESSARY)
209.16   [Numerical input is free format]
LIQUID SPECIFIC GRAVITY: (NECESSARY)
0.6882
CRITICAL TEMP., DEG F: (IF NOT KNOWN TYPE ZERO)
512.7
CRITICAL PRESS., PSIA: (IF NOT KNOWN TYPE ZERO)
396.8
ENTER MOLECULAR WEIGHT: (IF NOT KNOWN TYPE ZERO)
100.204
nC7 PROPERTIES   [Properties are echoed back]
BP, F   209.16
SP GR   .6882
TC, F   512.7
PC, PSIA   384.10
MOL WT   100.204

DO YOU WISH TO ENTER ANY MORE "C7+" COMPONENTS? (Y/N)
THE COMPONENT SET YOU HAVE CHOSEN IS:
C1
C3
nC7

DO YOU WISH TO MAKE ANY CHANGES IN THESE DATA? (Y/N)

At this point the user can go back and redefine the component set or the C7+ component name or properties. Otherwise the remainder of the feed information is requested by typing RETURN.

HOW MANY FEED STREAMS DO YOU WISH TO DEFINE?
1   [Usually just one]
TYPE IN THE STREAM NUMBER OF FEED 1
1   [Your choice, but 1 is simple]
ENTER FEED STREAM PRESSURE (PSIA) AND TEMPERATURE (F)
15 80
FEED COMPONENT ENTRY

ENTER THE MOLE % OF EACH COMPONENT IN THE FEED.

A ZERO (0) MUST BE ENTERED FOR THOSE COMPONENTS THAT ARE NOT PRESENT.

C1

80

C3

16

nC7

4

ENTER THE TOTAL FLOW RATE (LB MOL/HR) OF THE FEED:

109.794

GIVEN FEED MOLE %:

C1	80.00
C3	16.00
nC7	4.00
TOTAL	100.00
FLOW	109.79 MOL/HR
TEMP	80.00 DEG. F
PRESS	15.00 PSIA

DO YOU WISH TO MAKE ANY CHANGES IN THESE DATA? (Y/N)

The program will branch back to the feed component entry if so directed, otherwise it will continue.

IS THIS FEED A CRUDE OIL? (Y/N)

The UMSRK equation does not give very good densities for crude oils, so a different method is used for liquid density.

A flash calculation is now made for the given feed conditions to determine the complete property set of the feed. This information is presented in the *stream summary*.

ACTUAL FEED CONDITION BY FLASH

STRM NO.	1
C1	87.84
C3	17.57
nC7	4.39
MOL/HR	109.79
LB/HR	2623.87
T,DEG F	80.00
P,PSIA	15.00
MOL WT	23.90
LB/CUFT	.0622
BTU/HR,	.55225E+06
BTU/R-HR	.50403E+04
FRAC. VAP	1.0000

TO DISPLAY NEXT OUTPUT PAGE, TYPE "RETURN"

The first part of the stream summary displays the component and total stream flow rates, all in lbmol/hr. Next follow the mass flow rate (lb/hr), temperature (deg F), pressure (psia), and mol wt. The stream density appears next (lb/cu ft). If the stream is partly liquid and partly vapor, the *combined* density is given. Next come the total enthalpy (Btu/hr) and entropy (Btu/R-hr) of the stream, followed by the mol fraction of vapor present. If the vapor fraction is other than 1.0 (all vapor) or 0.0 (all liquid), then the stream is partly vapor and partly liquid. The stream can be separated into its vapor and liquid portions, giving the individual phase properties (as shown below in Ex. 1).

The stream summary totally defines the stream. The last message (TO DISPLAY . . .) allows the user to view the stream array and then proceed at his/her leisure.

DO YOU WISH TO SAVE YOUR FEED INFORMATION FOR USE AT SOME FUTURE TIME? (Y/N)

At this time the user can define the name of the file in which the feed data will be stored. The program is now ready to perform operations.

SELECT THE DESIRED OPERATION

CODE	OPERATION	CODE	OPERATION
1	SPLIT	10	BUBBLE POINT
2	SUM	11	DEW POINT
3	COMPRESSOR DESIGN	12	SEPARATOR
4	COMPRESSOR ANALYSIS	13	PUMP
5	ISOTHERMAL FLASH	14	SCALE STREAM
6	ADIABATIC FLASH	15	DISPLAY STREAM
7	HEAT EXCHANGER	16	HEATING VALUE
8	EXPANDER	17	FRACTIONATOR
9	GAL C2+ LIQ/MSCF	18	DEMETHANIZER
		19	END OF RUN

The user chooses an operation and then must identify the number of the stream(s) that feed the operation. Allowable stream numbers are 1 to 40. The user then is asked to name the output stream number(s) and pertinent parameters for the particular operation. The simulation is performed and the results presented on the monitor screen in tabular form.

Then the user is given the opportunity to choose another operation. In this manner, a complete flow sheet can be built including, if desired, recycle streams (although this choice is time-consuming). Feed streams may be used any number of times; they are not modified by the program unless so directed by the user.

The input data prompts for the modules are mostly self-explanatory. Each of the modules is described in detail at the end of this section, with the calculation technique, special input data needs, etc., listed.

Example Problems

The results of each of the three example problems outlined previously are presented below. These examples should provide the user with sufficient guidance to allow use of OPSIM. The output files only will be shown, in some cases shortened to save space.

Example 1. The feed gas is to be compressed and cooled; any liquid dropout in the condenser is to be separated. The results follow.
COMPRESSOR CALCULATION:

ADIABATIC EFFICIENCY =   .9000
MECHANICAL EFFICIENCY =   .9500
DELTA P = 45.0 PSI:
BRAKE HORSEPOWER = 83.5
COMPRESSION RATIO =   4.000

[Notice that the gas has been heated to 233.51°F in the compressor]
HEAT EXCHANGER

DELTA P = −5.0 PSI
INLET TEMP. = 233.5 DEG F
OUTLET TEMP. = 80.0 DEG F
HEAT ADDED = −.24506010E+06
BTU/HR [Negative sign indicates heat removal]

STRM NO.	2	3
C1	87.84	87.84
C3	17.57	17.57
nC7	4.39	4.39
MOL/HR	109.79	109.79
LB/HR	2623.87	2623.87
T,DEG F	233.51	80.00
P,PSIA	60.00	55.00
MOL WT	23.90	23.90
LB/CUFT	.1941	.2357
BTU/HR	.75414E+06	.50908E+06
BTU/R-HR	.50700E+04	.46815E+04
FRAC. VAP	1.0000	.9758

INLET AND OUTLET STREAMS OF SEPARATOR:

STRM NO.	3	4	5
C1	87.84	87.79	.04
C3	17.57	17.39	.18
nC7	4.39	1.96	2.44
MOL/HR	109.79	107.13	2.66
LB/HR	2623.87	2371.07	252.80
T,DEG F	80.00	80.00	80.00
P,PSIA	55.00	55.00	55.00
MOL WT	23.90	22.13	94.99
LB/CUFT	.2357	.2131	41.6890
BTU/HR,	.50908E+06	.51382E+06	−.47342E+04
BTU/R-HR	.46815E+04	.44915E+04	.19000E+03
FRAC. VAP	.9758	1.0000	.0000

STRM NO.	1	2
C1	87.84	87.84
C3	17.57	17.57
nC7	4.39	4.39
MOL/HR	109.79	109.79
LB/HR	2623.87	2623.87
T,DEG F	80.00	233.51
P,PSIA	15.00	60.00
MOL WT	23.90	23.90
LB/CUFT	.0622	.1941
BTU/HR,	.55225E+06	.75414E+06
BTU/R-HR	.50403E+04	.50700E+04
FRAC. VAP	1.0000	1.0000

Example 2. Calculation of a phase diagram involves mainly bubble and dew calculations. However, these calculations sometimes fail in the critical region. This failure is not the fault of the equation of state but of the algorithm used to solve it. One shouldn't be overly critical of failure of the algorithm. Simultaneous solution of the many complex, simultaneous, transcendental equations is not a trivial task. Any equation-of-state-based equilibrium calculation program will occasionally fail to converge. Note that all bubble-point, dew-

point, and flash calculation results are accompanied by a list of the computed vapor-liquid equilbrium K-values. If these K's appear to be reasonable, then the solution is acceptable. However, if the K's are all approximately 1.0, then the solution should be rejected; it is a false solution.

A series of bubble- and dew-point calculations were carried out for the mixture of Example 2 beginning at 100 psia and proceeding upward in pressure. The results of the successful runs are tabulated below. All these were run by specifying the pressure and finding the corresponding temperature.

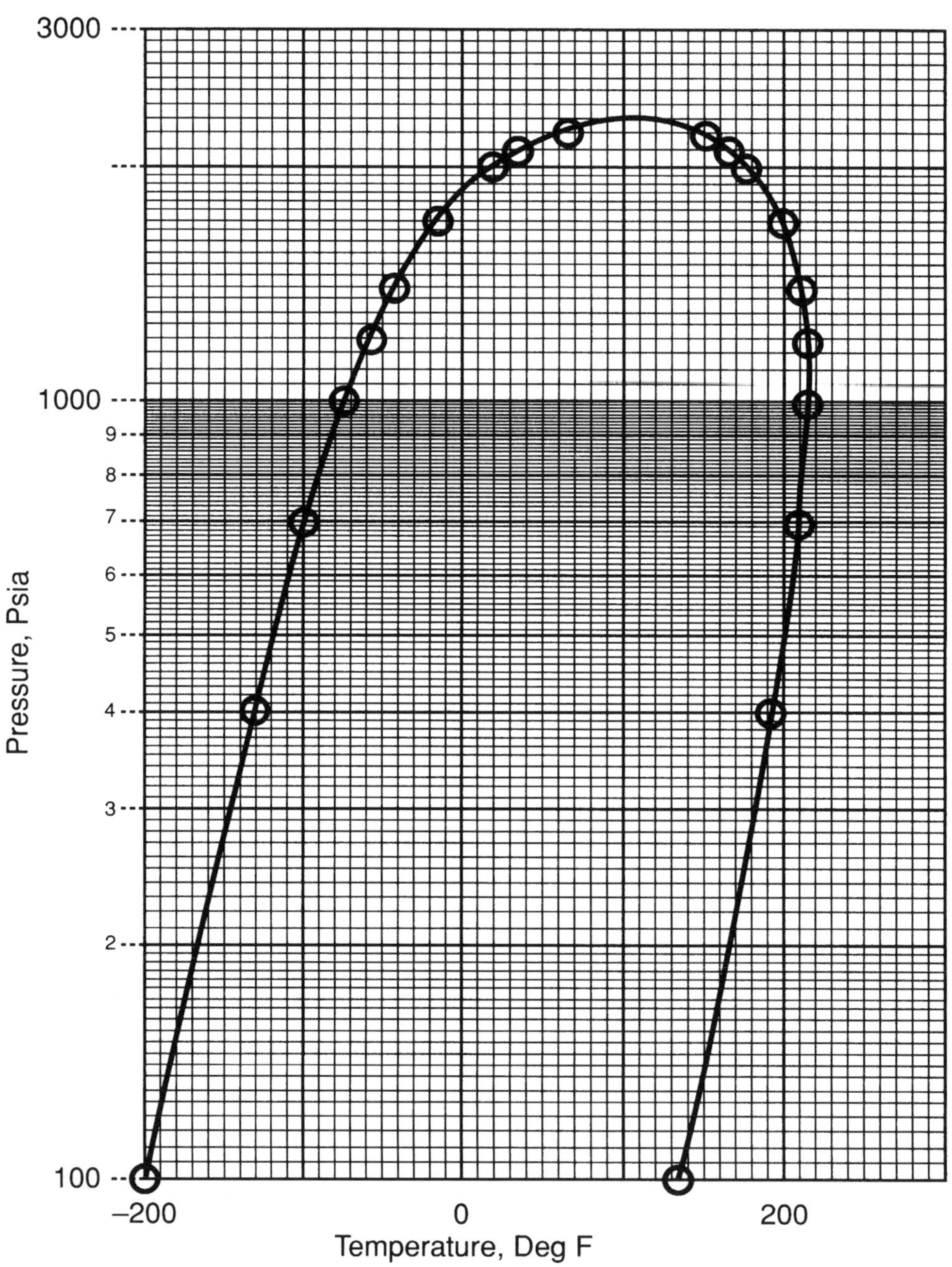

Figure A4–2. Phase envelope for Exercise 2.

This is generally the best method, since guessing the temperature range to specify is very difficult.

Pressure Psia	Temperature, Bubble	Deg F Dew
100	−199.1	133.3
400	−134.8	191.4
700	−100.6	208.9
1000	− 74.6	213.4
1200	− 59.1	211.8
1400	− 43.2*	207.3
1700	− 16.6*	194.8
2000	16.8*	173.0
2100	32.3*	162.3
2200	62.8*	148.7

*At pressures above 1250 psia, the flash loop did not converge in 50 trials, which means that there was a composition convergence problem.

The data are plotted in Figure A4–2. As stated in the note above, there was a convergence problem in the near-critical region, but the phase envelope has a reasonable appearance, and the K-values appeared to be reasonable.

Example 3. The stream summaries for this problem are presented below. The only remark to make is that the liquid-stream volumes are given as equivalent barrels per day at 60°F for convenience in crude oil problems.

SUMMARY OF INPUT DATA FILE

C7+ PROPERTIES	
BP,F	499.00
SP GR	.8426
TC,F	800.00
PC, PSIA	280.00
MOL WT	207.00

STRM NO.	1
N2	29.64
CO2	1.90
C1	1286.68
C2	213.94
C3	167.20
IC4	45.98
NC4	129.96
IC5	70.30
NC5	92.72
C6	163.02
C7+	1598.66

MOL/HR	3800.00
LB/HR	402322.00
T,DEG F	120.00
P,PSIA	2062.00
MOL WT	105.87
LB/CUFT	48.4449
BTU/HR	.52547E+07
BTU/R-HR	.27583E+06
FRAC. VAP	.0000
BPD @ 60f	36997.08

ADIABATIC FLASH:

OUTLET PRESSURE   = 315.0 PSIA
HEAT TRANSFERRED = .0 MMBTU/HR

STRM NO.	1	2
N2	29.64	29.64
CO2	1.90	1.90
C1	1286.68	1286.68
C2	213.94	213.94
C3	167.20	167.20
IC4	45.98	45.98
NC4	129.96	129.96
IC5	70.30	70.30
NC5	92.72	92.72
C6	163.02	163.02
C7+	1598.66	1598.66
MOL/HR	3800.00	3800.00
LB/HR	402322.00	402322.10
T,DEG F	120.00	114.02
P,PSIA	2062.00	315.00
MOL WT	105.87	105.87
LB/CUFT	48.4449	12.3210
BTU/HR	.52547E+07	.52546E+07
BTU/R-HR	.27583E+06	.28304E+06
FRAC. VAP	.0000	.3617
BPD @ 60F	36997.08	

P=315.00 PSIA
T=114.02 F

COMP.	K-VALUE
N2	.334045E+02
CO2	.562733E+01
C1	.122511E+02
C2	.258481E+01
C3	.828922E+00
IC4	.395504E+00
NC4	.283679E+00
IC5	.132715E+00
NC5	.100250E+00
C6	.366914E−01
C7+	.831335E−05

INLET AND OUTLET STREAMS OF
SEPARATOR:

STRM NO.	2	3	4
N2	29.64	28.15	1.49
CO2	1.90	1.45	.45
C1	1286.68	1124.68	162.00
C2	213.94	127.14	86.80
C3	167.20	53.44	113.76
IC4	45.98	8.42	37.56
NC4	129.96	18.00	111.96
IC5	70.30	4.92	65.38
NC5	92.72	4.98	87.74
C6	163.02	3.32	159.70
C7+	1598.66	.00	1598.65
MOL/HR	3800.00	1374.50	2425.50
LB/HR	402322.10	27612.67	374709.40
T,DEG F	114.02	114.02	114.02
P,PSIA	315.00	315.00	315.00
MOL WT	105.87	20.09	154.49
LB/CUFT	12.3210	1.0787	53.1112
BTU/HR,	.52546E+07	.65216E+07	−.12670E+07
BTU/R-H	.28304E+06	.51757E+05	.23128E+06
FRAC. VAP	.3617	1.0000	.0000
BPD @ 60F	.00	.00	32099.32

ADIABATIC FLASH:

OUTLET PRESSURE = 70.0 PSIA
HEAT TRANSFERRED = .0 MMBTU/HR

STRM NO.	4	5
N2	1.49	1.49
CO2	.45	.45
C1	162.00	162.00
C2	86.80	86.80
C3	113.76	113.76
IC4	37.56	37.56
NC4	111.96	111.96
IC5	65.38	65.38
NC5	87.74	87.74
C6	159.70	159.70
C7+	1598.65	1598.65
MOL/HR	2425.50	2425.50
LB/HR	374709.40	374709.40
T,DEG F	114.02	110.65
P,PSIA	315.00	70.00
MOL WT	154.49	154.49
LB/CUFT	53.1112	13.9477
BTU/HR,	−.12670E+07	−.12670E+07
BTU/R-HR	.23128E+06	.23218E+06
FRAC. VAP	.0000	.0967
BPD @ 60F	32099.32	.00

P = 70.00 PSIA
T = 110.65 F

COMP.	K-VALUE
N2.	.150752E+03
CO2	.230227E+02
C1	.525535E+02
C2	.100858E+02
C3	.299927E+01
IC4	.135244E+01
NC4	.954000E+00
IC5	.421048E+00
NC5	.312553E+00
C6	.106201E+00
C7+	.134182E−04

INLET AND OUTLET STREAMS OF
SEPARATOR:

STRM NO.	5	6	7
N2	1.49	1.40	.09
CO2	.45	.32	.13
C1	162.00	137.55	24.45
C2	86.80	45.07	41.73
C3	113.76	27.65	86.11
IC4	37.56	4.75	32.81
NC4	111.96	10.38	101.59
IC5	65.38	2.82	62.56
NC5	87.74	2.84	84.89
C6	159.70	1.80	157.90
C7+	1598.65	.00	1598.65
MOL/HR	2425.50	234.58	2190.91
LB/HR	374709.40	6277.75	368431.60
T,DEG F	110.65	110.65	110.65
P,PSIA	70.00	70.00	70.00
MOL WT	154.49	26.76	168.16
LB/CUFT	13.9477	.3129	54.1469
BTU/HR,	$-.12670E+07$	$.12842E+07$	$-.25512E+07$
BTU/R-HR	$.23218E+06$	$.10968E+05$	$.22121E+06$
FRAC. VAP	.0967	1.0000	.0000
BPD @ 60F	.00	.00	31163.73

ADIABATIC FLASH:

OUTLET PRESSURE = 17.0 PSIA
HEAT TRANSFERRED = .0 MMBTU/HR

STRM NO.	7	8
N2	.09	.09
CO2	.13	.13
C1	24.45	24.45
C2	41.73	41.73
C3	86.11	86.11
IC4	32.81	32.81
NC4	101.59	101.59
IC5	62.56	62.56
NC5	84.89	84.89
C6	157.90	157.90
C7+	1598.65	1598.65
MOL/HR	2190.91	2190.91
LB/HR	368431.60	368431.70
T,DEG F	110.65	106.36
P,PSIA	70.00	17.00
MOL WT	168.16	168.16
LB/CUFT	54.1469	7.1297
BTU/HR,	$-.25512E+07$	$-.25511E+07$
BTU/R-H	$.22121E+06$	$.22145E+06$
FRAC. VAP	.0000	.0584
BPD @ 60F	31163.73	.00

P = 17.00 PSIA
T = 106.36 F

COMP.	K-VALUE
N2	$.624394E+03$
CO2	$.911935E+02$
C1	$.212703E+03$
C2	$.390981E+02$
C3	$.112753E+02$
IC4	$.497604E+01$
NC4	$.347949E+01$
IC5	$.149991E+01$
NC5	$.110382E+01$
C6	$.363945E+00$
C7+	$.358892E-04$

INLET AND OUTLET STREAMS OF
SEPARATOR:

STRM NO.	8	9	10
N2	.09	.08	.00
CO2	.13	.11	.02
C1	24.45	22.72	1.72
C2	41.73	29.54	12.19
C3	86.11	35.43	50.68
IC4	32.81	7.74	25.08
NC4	101.59	18.02	83.56
IC5	62.56	5.32	57.24
NC5	84.89	5.44	79.46
C6	157.90	3.48	154.42
C7+	1598.65	.00	1598.65
MOL/HR	2190.91	127.90	2063.02
LB/HR	368431.70	5396.97	363034.70
T,DEG F	106.36	106.36	106.36
P,PSIA	17.00	17.00	17.00
MOL WT	168.16	42.20	175.97
LB/CUFT	7.1297	.1198	54.7861
BTU/HR,	$-.25511E+07$	$.89366E+06$	$-.34448E+07$
BTU/R-HR	$.22145E+06$	$.78817E+04$	$.21357E+06$
FRAC. VAP	.0584	1.0000	.0000
BPD @ 60F	.00	.00	30469.69

Description of the Modules

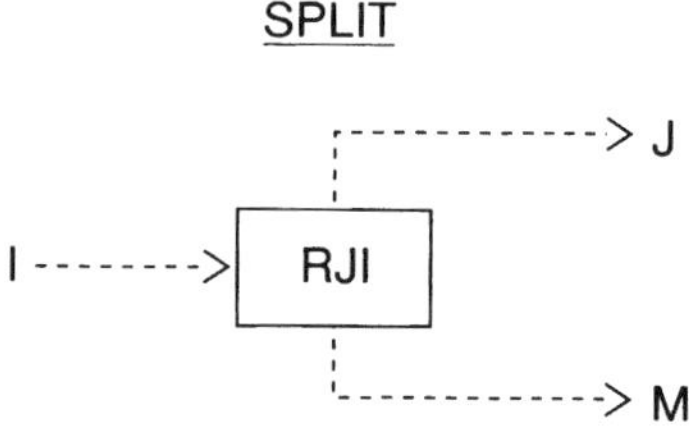

The *Split* module splits stream I into streams J and M.
Parameter RJI is the ratio of the flow rate of product stream
J to that of feed stream I. The calculation is straightforward.

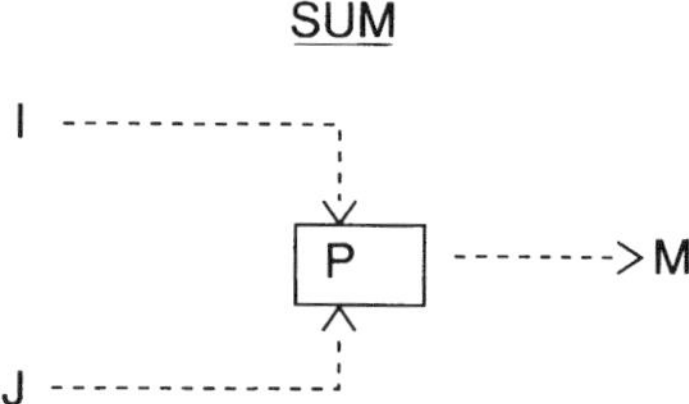

Module *Sum* adds two streams adiabatically. Parameter
P is the pressure of combined stream M. The subroutine
performs the material balance, then flashes the combined
stream at pressure with the same total enthalpy as the two
feeds (see the ADIABATIC FLASH module below).

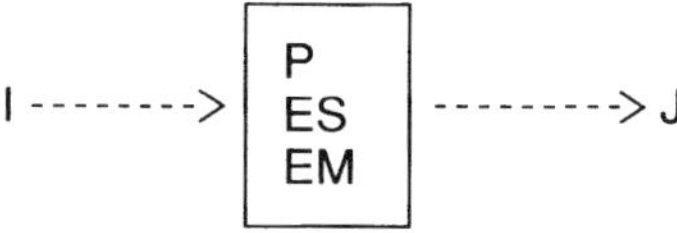

The *Compressor Design* module compresses given stream
I to stream J at the pressure given by parameter P. Parameter
ES is the adiabatic efficiency and EM is the mechanical
efficiency. The module provides the stream summary for
stream J, as well as the brake horsepower.

The computation is carried out as follows. First, a trial-
and-error calculation is made to find the temperature for
stream J that gives the same entropy as feed stream I; in
this manner the outlet enthalpy for the isentropic
compression is determined, call it $H_{J,S}$. A first estimate of
the isentropic outlet temperature is made by assuming ideal-
gas behavior and using the thermodynamic formula for the
isentropic compression of an ideal gas. The trial-and-error
calculation is carried out by means of the Muller method
(Wang and Henke, 1966). The work for the isentropic path,
W_S, is given by

$$W_S = H_{J,S} - H_I$$

The actual work received by the gas is the isentropic work divided by the adiabatic efficiency and is also equal to the actual change in enthalpy.

$$W = W_S/ES = H_J - H_I$$

Rearrange to find H_J, the actual enthalpy of the outlet stream.

$$H_J = H_I + W_S/EM$$

A second trial-and-error calculation is required to find the temperature, T_J, that yields the required value of H_J. The first guess of T_J is the outlet temperature of the isentropic compression; the Muller method is again used.

The gas horsepower is given by the mass flow rate of the gas times the work, divided by the appropriate conversion factor for English engineering units.

$$GHP = \dot{m}\, W/2545$$

The brake horsepower is given by the gas horsepower divided by the mechanical efficiency.

$$BHP = GHP/EM$$

If the user requires only the gas horsepower, a mechanical efficiency of 100% is entered.

COMPRESSOR ANALYSIS

$$I \dashrightarrow \boxed{\begin{array}{c} ES \\ EM \\ BHP \end{array}} \dashrightarrow J$$

Module *Compressor Analysis* compresses stream I to an outlet condition such that the brake horsepower is equal to that specified by parameter BHP for the adiabatic efficiency given by parameter ES and mechanical efficiency given by parameter EM.

The computation is trial-and-error to find the pressure for stream M that gives the specified brake horsepower, using the subroutine invoked by module *Compressor Design* to make the horsepower calculation. First, the outlet enthalpy for the isentropic compression is determined by multiplying the given horsepower by the adiabatic and mechanical efficiencies. A first estimate of the isentropic outlet pressure is made by assuming ideal-gas behavior and using the thermodynamic formula for the isentropic compression of an ideal gas (the isentropic and actual outlet pressure will be the same, of course). The correct value of the outlet pressure is solved for by means of the secant method.

ISOTHERMAL FLASH

$$I \dashrightarrow \boxed{P,\ T} \dashrightarrow J$$

The *Isothermal Flash* module performs a conventional flash calculation on given feed I at the pressure and temperature given by parameters P and T. See Buthod *et al.* (1978) for the method used. The name of the module is somewhat misleading, since the temperature of stream J is not the same as that of stream I (as would be true for an isothermal process) but is that given by parameter T. This nomenclature is common in the process industry, hence its use here.

The vapor-liquid K-values are printed following the stream summaries. The main reason for printing the K's is to assure that the calculation is meaningful. Occasionally the method used to solve the equation of state will converge to a set of K's that are all very nearly equal to 1.0. Such a result is not meaningful and should be rejected.

ADIABATIC FLASH

$$I \dashrightarrow \boxed{P,\ Q} \dashrightarrow J$$

The *Adiabatic Flash* module performs a flash calculation on a given feed at the pressure given by parameter P at the temperature, T, that gives the enthalpy change from I to J the same as energy input Q. The name of this module is also somewhat misleading, since the heat input Q is not necessarily zero (as it would be for an adiabatic process), although it can be. This module is used with Q = 0 to model a conventional valve, since the expansion for a valve is essentially adiabatic.

The calculation is trial and error. A temperature is assumed for stream J and the subroutine invoked by module *Isothermal Flash* is called to find the enthalpy of stream J. Then the enthalpy change is compared to that specified by Q. The computation is repeated as necessary, using the Muller method for convergence. The initial estimate of the outlet temperature is the inlet temperature (temperature of stream I).

The vapor-liquid K-values are printed following the stream summaries. The main reason for printing the K's is to assure that the calculation is meaningful. Occasionally the method used to solve the equation of state will converge to a set of K's that are all very nearly equal to 1.0. Such a result is not meaningful and should be rejected.

HEAT EXCHANGER

$$I \dashrightarrow \boxed{P,\ Q} \dashrightarrow J$$

Module *Heat Exchanger* simulates the heat exchange for a fluid on one side of an exchanger. The pressure of stream J is always an input parameter. There are four options for the remaining parameter.

Option	Parameter
1	Temperature, T, of stream J
2	Heat, Q, added to stream I to obtain stream J
3	Boiler
4	Condenser

For option 1, a flash calculation is performed at the specified P and T. The duty of the exchanger is just the difference in the enthalpies of streams I and J. For option 2, the subroutine invoked by module *Adiabatic Flash* is used for the specified heat duty, Q, to obtain stream M. For option 3, a dew-point calculation is performed for stream I to obtain the temperature of stream J (leaving the boiler as a saturated vapor); then option 1 is run. For option 4, a bubble-point calculation is performed to obtain the temperature of stream J (leaving the condenser as a saturated liquid); then option 1 is run.

EXPANDER

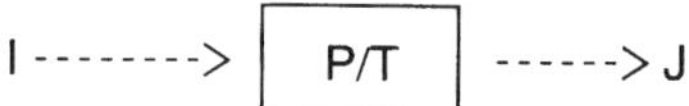

The *Expander Design* module expands given stream I with work production to stream J at the pressure given by parameter P. Parameter ES is the adiabatic efficiency. This module is used to simulate an expansion turbine.

The computation is carried out as follows. First, a trial-and-error calculation is made to find the temperature for stream J that gives the same entropy as feed stream I; in this manner the outlet enthalpy for the isentropic expansion is determined, call it $H_{J,S}$. A first estimate of the isentropic outlet temperature is made by assuming ideal-gas behavior and using the thermodynamic formula for the isentropic expansion of an ideal gas. The trial-and-error calculation is carried out by means of the Muller method. The work for the isentropic path, W_S, is given by

$$W_S = H_{J,S} - H_I$$

The actual work received by the gas is the isentropic work divided by the adiabatic efficiency and is also equal to the actual change in enthalpy.

$$W = W_S/ES = H_J - H_I$$

Rearrange to find H_J, the actual enthalpy of the outlet stream.

$$H_J = H_I + W_S/EM$$

A second trial-and-error calculation is required to find the temperature, T_J, that yields the required value of H_J. The first guess of T_J is the outlet temperature of the isentropic expansion; the Muller method is again used.

The gas horsepower is given by the mass flow rate of the gas times the work, divided by the appropriate conversion factor for English engineering units.

$$GHP = \dot{m}W/2545$$

The negative of the work is used and the horsepower is reported as a positive number, although the thermodynamic convention used herein would give a negative number (signifying the working being done by the system on the surroundings).

GAL C2 + LIQ/MSCF

This module computes the potentially recoverable liquid of a natural gas stream and reports it as gallons of liquid per thousand standard cubic feet, or as is used in the gas-processing industry, gal C2+ / Mscf. In the computation it is assumed that all hydrocarbons from ethane to the highest boiling component are totally liquefied. This quantity is used as a correlating parameter in module *Demethanizer*.

BUBBLE POINT

Module *Bubble Point* finds the bubble-point temperature of stream I at the pressure given by parameter P or the bubble-point pressure at the temperature given by parameter T. The user selects at the time of execution which parameter (P or T) is to be held constant. Stream J contains the bubble-point stream with its full stream array of properties. The equilibrium vapor stream is not generated. However, the vapor-liquid K-values are printed. The main reason for printing the K's is to assure that the calculation is meaningful. Occasionally the method used to solve the equation of state will converge to a set of K's that are all very nearly equal to 1.0. Such a result is not meaningful and should be rejected.

The method used to find the bubble point is not the conventional technique used in textbooks because of the manner in which the equation of state is used. (The vapor-liquid K values are dependent on composition as well as temperature and pressure. For this reason it was found convenient to program the equation of state to make an isothermal flash calculation, with special provisions for the finding of a single-phase product from the flash.) In addition, it was desired to have a robust algorithm for reliability of convergence. The trial-and-error method starts with a user-supplied estimate of the bubble-point temperature or pressure and searches to find a 5-°F or 5–psia span in which the phase changes from liquid to vapor, as indicated by the results of flash calculations. Then the method of bisection is applied to the interval for 14 trials to give a result within acceptable precision. After the bubble point is thus determined, a bias of -0.01°F or $+0.01$ psi is added to insure that the flash routine will give a quality of 0.000000.

DEW POINT

Module *Dew Point* finds the dew-point temperature of stream I at the pressure given by parameter P or the dew-point pressure at the temperature given by parameter T. The user selects at the time of execution which parameter (P or T) is to be held constant. Stream J contains the dew-point stream with its full stream array of properties. The equilibrium liquid stream is not generated. However, the vapor-liquid K-values are printed. The main reason for printing the K's is to assure that the calculation is meaningful. Occasionally the method used to solve the equation of state will converge to a set of K's that are all very nearly equal to 1.0. Such a result is not meaningful and should be rejected.

The method used to find the dew point is the same as for the bubble point, as described above, with the exception that the final temperature is increased by 0.01°F or the pressure reduced by 0.01 psi to assure an all-vapor product.

SEPARATOR

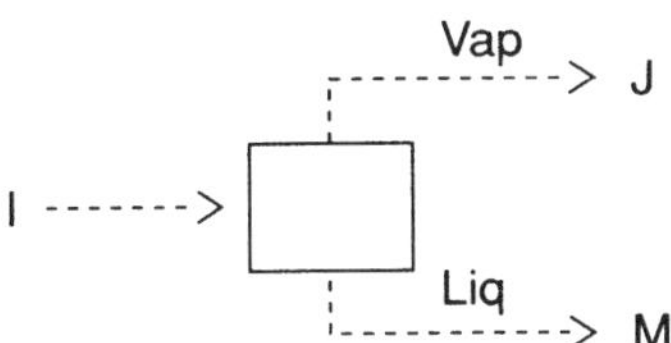

The *Separator* module performs a phase separation. Feed stream I is separated into liquid stream M and its equilibrium vapor stream J, thus simulating a perfect vapor-liquid separator. If stream I is all liquid, stream J will have nothing in it. Similarly, if stream I is all vapor, stream M will have nothing in it. The computation is straightforward.

PUMP

Module *Pump* simulates the action of a liquid pump. Liquid stream I is pumped to the pressure given by parameter P with pump efficiency given by parameter EFF. If EFF is specified to be zero, a value is estimated by a interpolation of a chart given by the GPSA Engineering Data Book (8th Ed., 1966, p. 106, Tulsa, OK).

The pump calculations are done by the standard methods of fluid mechanics. The formulae are from the GPSA Engineering Data Book (10th Ed., 1987, p. 12–3, Tulsa, OK). The pump power is given by the volumetric flow rate times the pressure rise, divided by the efficiency and an appropriate conversion factor. If the flow rate, q, is in gallons per minute and the pressure rise, DP, in psi, the formula is

$$Bhp = q \, DP \, / \, (1714 \, EFF)$$

The temperature rise in the pump casing due to the dissipation of frictional energy, DTF, is

$$DTF = H \, (1/EFF - 1) \, / \, (778 \, CP)$$

where H is the pump head (DP divided by the density, in consistent units), and CP is the heat capacity, Btu/lb$-$°F. The heat capacity is estimated by the change in liquid enthalpy for a one degree temperature change.

In addition to the temperature rise due to friction, the liquid temperature is increased by the increase in pressure. This effect is estimated from a curve fit of experimental data (Jennings and Meade, 1956).

$$DTP = DP \, \exp \, (3.717 - 3.66 \, SG)/1000$$

where SG is the specific gravity of the liquid compared to water.

The temperature of stream J is

$$T_J = T_I + DTF + DTP$$

The properties of J are evaluated at P and T_J.

SCALE STREAM

The *Scale Stream* module simply multiplies stream I by parameter FAC. The module is useful when it is desired to change the mass flow rate basis of a problem *after* defining the feed, for example to match a certain flow rate in volumetric or mass units rather than molal units. The scaled stream then can be used to complete the simulation.

DISPLAY STREAM

Module *Display Stream* allows the user to review the progress of a simulation at any time. From 1 to 4 streams can be displayed at a time. To display more than 4 requires repeated application of the module. The module parameters are the four stream numbers to be displayed; unused numbers must be designated by zeroes. For example, if the user desires to display streams 3 and 4 only, the response to the stream number prompt is:

3 4 0 0   or   3,4,0,0

HEATING VALUE

This module computes the heating value for a given stream. The heating value of a gas stream is sometimes desired in gas-processing simulations because it is a common sales-gas specification. The heating value is the molar average of the heating values of the constituents. A heating value of 5502.5 Btu/scf is used for all hypothetical components (the heating value of n-heptane). The result is given in Btu/scf, where the conditions for the standard cubic foot (scf) are 60°F, 14.696 psia.

The *Fractionator* module simulates a single-feed, two-product distillation column with overhead partial or total condenser and partial reboiler by a shortcut method. Many of the parameters of the module are shown above, including NF (feed stream number), ND (distillate product stream number), NB (bottoms product stream number), N (total number of equilibrium stages; the partial reboiler, if present, is stage 1), M (number of the equilibrium stage just below the feed introduction point), PSIA (tower pressure), QC (condenser duty, Btu/hr), and QR (reboiler duty, Btu/hr). Additional parameters that must be supplied are:

DSUM   Distillate product rate, D, mol/hr
R   Reflux ratio, L_o/D
TOVHD Estimate of distillate temperature
TBTMS Estimate of bottoms temperature
NCOND Condenser type (1 = total condenser,
                                        2 = partial condenser)

The module returns the product stream summaries and the reboiler and condenser duties.

The calculation method is the group method of Edmister (Henley and Seader, 1981) solved by a technique suggested by Erbar (1965). The first step in the calculations is to flash the feed adiabatically to column pressure. If the user wishes a particular feed condition at the column pressure, he/she must take care of this prior to calling the module. Liquid and vapor rates in the two sections are established next, using the given D, R, and feed condition information and *assuming constant molar overflow*. An average

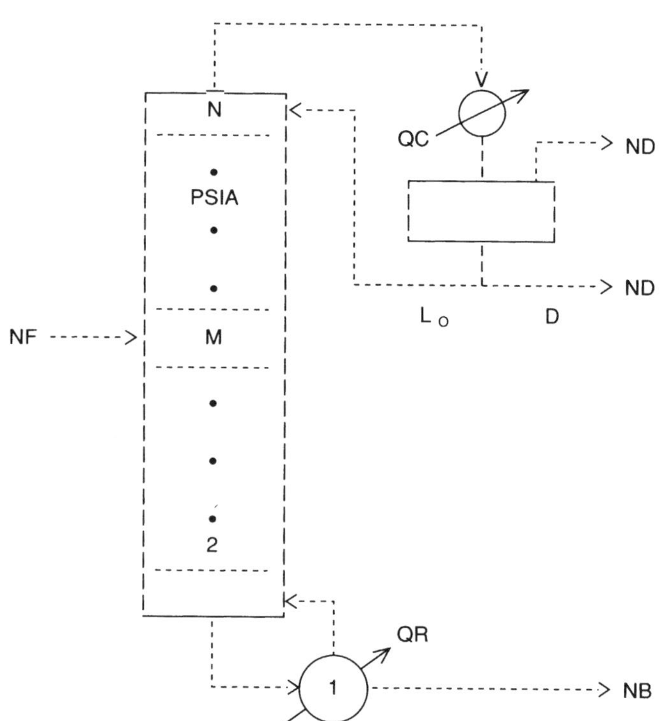

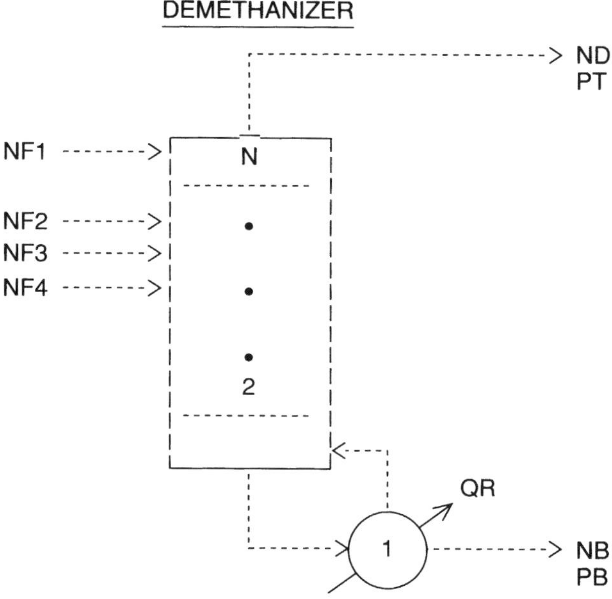

temperature halfway between the feed-tray temperature, T_M, and the product temperature is used for each section of the column. The average temperature is used to evaluate absorption and stripping factors for the particular section. Then the Edmister group equations are applied to find the product distributions. The procedure is iterated to determine the value of T_M that yields the overhead distillate rate equal to the specified value. Product temperatures are estimated by bubble- or dew-point calculations. The condenser duty is found by energy balance around the condenser. The reboiler duty is then found by energy balance around the fractionator.

The *Demethanizer* module simulates a gas-processing plant demethanizer column with partial reboiler by a shortcut method. Up to four feeds can be used. Many of the parameters of the module are shown above, including NF1–NF4 (feed stream numbers), ND (distillate product stream number), NB (bottoms product stream number), N (total number of equilibrium stages; the partial reboiler is stage 1), PT (tower top pressure, psia), PB (tower bottom pressure, psia), and QR (reboiler duty, Btu/hr). Additional parameters that must be supplied are:

C1C2SPEC Ratio of moles methane in bottoms to moles ethane in bottom (typically 0.02)
GPM Potential liquid recovery of ethane and heavier components in processing plant feed (see module GAL C2+ LIQ/MSCF)

The module provides the product stream summaries and the reboiler duty.

The calculation method was developed by Wang (1985) and is not, in fact, a fractionator calculation. The method is based on the assumption that the ratio of mol component to mol methane in the overhead product vapor is the same as in the flashed feed vapor. The overhead temperature was correlated as a function of feed temperature, tower top pressure, and the recoverable liquids in the overall plant feed. The availability of this temperature allows a dew-point calculation to be made that provides the product distributions (see Wang, 1985). Note: this method is only appropriate for the flow diagram shown above and *is not* for use with a cold-refluxed demethanizer.

COMPONENT SEPARATOR

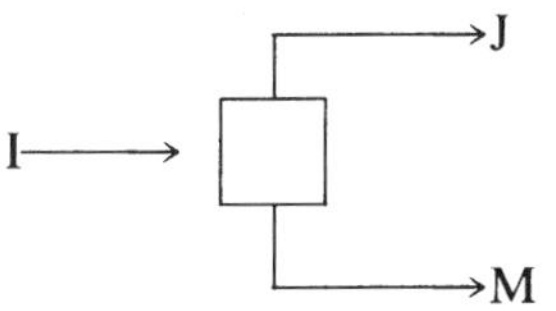

Module *Component Separator* divides each component of entering stream I into portions in each of two product streams J and M. The fraction of each component entering in stream I that is to leave in stream J is specified by the user, and the remaining amount of the component goes to stream M. The temperature and pressure of streams J and M are also supplied by the user. This module can be used to simulate a chemical reactor or other device in which one or more components are removed from a feed, such as a gas treating unit.

STORE STREAM

The *Store Stream* module is used at any time in a simulation when it is desired to save a stream for possible future use in a simulation. The user specifies the name of the file into which the particular stream array is to be stored. The file name can contain up to 14 characters.

Structure of OPSIM

The basic structure of OPSIM was suggested by the technique used by Franks (1972) in his DYFLO package. The important variables are shared between FORTRAN subroutines in a labelled COMMON called OPS.

COMMON/OPS/STRM(90,30), DATA(5,9), ID(20),NC,NH By invoking OPS any subroutine has available to it the the basic data items required.

The STRM array contains 30 defining items for each of 90 streams. For stream K, the items are as follows:

STRM(K,J),J=1,30 ARE STREAM PROPERTIES AS FOLLOWS:

J = 1	MOL	FRACTION	N2
2	"	"	CO2
3	"	"	H2S
4	"	"	C1
5	"	"	C2
6	"	"	C3
7	"	"	IC4
8	"	"	NC4
9	"	"	IC5
10	"	"	NC5
11	"	"	C6
12	"	"	C7+ NO. 1
13	"	"	C7+ NO. 2
14	'	"	C7+ NO. 3
15	"	"	C7+ NO. 4
16	"	"	C7+ NO. 5
17	"	"	C7+ NO. 6
18	"	"	C7+ NO. 7
19	"	"	C7+ NO. 8
20	"	"	C7+ NO. 9
21	DENSITY(LB/CU FT)		
22	PHASE INDICATOR (LEAVE BLANK)		
23	LB/HR		
24	MOL WT.		
25	SPARE		
26	LBMOL/HR		

J = 1	MOL	FRACTION	N2
27	DEG F		
28	BTU/HR		
29	PSIA		
30	ENTROPY(BTU/HR -DEG R)		

$$\text{PHASE INDICATOR} = 1.0 \text{ FOR GAS OR VAPOR}$$
$$= 2.0 \text{ FOR GAS PORTION OF}$$
$$\text{TWO-PHASE MIXTURE}$$
$$\text{(LIQUID OF STRM(I) IS}$$
$$\text{STORED AS STRM(I+40))}$$
$$= 3.0 \text{ FOR LIQUID}$$

Single-phase streams are stored in streams 1 to 40. If the stream contains two phases, the vapor phase is stored as stream K and the liquid phase as stream K + 40. In this manner the two phases are always available separately for manipulation. The user is not aware of this usage because the stream printer subroutine combines the two streams for the stream summary. Thus, streams 1 to 80 are required to handle 40 streams. The remaining 10 streams, 81 through 90, are utility streams used for temporary storage of streams in several of the subroutines.

The DATA array contains the five base properties for each of the hypothetical components.

FOR DATA(I,J)

I	PROPERTY
1	BP (DEGR F)
2	LIQ SP GRAV 60 F
3	TC (DEGR F)
4	PC (PSIA)
5	MOL WT

where J is the hypothetical component number (9 maximum).

Array ID(20) contains the names of the components, primarily needed for printing the standard stream summary. Item NC is the number of components present and NH the number of hypothetical components.

The evaluation of physical properties is based on the use of a subroutine called SOAVE that proceeds as if to perform a flash calculation, directing the flow of computations as required to find the vapor-liquid equilibrium K-values from the Usdin-McAuliffe modification of the Soave-Redlich-Kwong equation of state (1976). The procedure has the built-in capability of determining the existence of a single phase for the conditions given. At the same time that the K-values are being obtained, the enthalpy and entropy deviations from ideal-gas behavior are evaluated. These deviations are applied to the ideal-gas enthalpy and entropy calculated by the methods of the API Data Book (1970). The technique used to solve the many equations involved is detailed in Buthod et al. (1978).

Given the basic SOAVE subroutine and the OPS data, a given subroutine can be easily written to simulate the various modules. Any properties required are obtained by a call of SOAVE or inquiry from the STRM array.

References

Udegbunan, E. O., "A Simulator Program for Natural Gas Processing," M. S. Thesis, The University of Tulsa (1978).

Cheng, S. H., Independent Study Report, The University of Tulsa (1982).

Chukwuma, F. O., "Fractionator Program for Wide-Boiling Mixtures," M. S. Thesis, The University of Tulsa (1981).

Fraser, R., "The Split-Level Refrigeration Cycle," M. S. Thesis, The University of Tulsa (1984).

American Petroleum Institute, "Technical Data Book—Petroleum Refining," Washington, D.C., 1970 (2nd Ed.).

Stevens, W. F., and G. Thodos, "Estimation of Enthalpies: Multicomponent Hydrocarbon Mixtures at Their Saturated Vapor and Liquid States," *A.I.Ch.E. J. 9* (3), pp. 293–296 (May 1973).

Usdin, E., and J. C. McAuliffe, *Chem. Eng. Sci. 31*, pp. 1077–1084 (1976).

Perez-Blanco, N. L. (1978), "Computer Solution of the SRK Equation," M. S. Thesis, The University of Tulsa (1978).

Gas Processors Suppliers Association, "Engineering Data Book," p. 20–4, Tulsa, OK, 1987 (10th Ed.).

Wang, J. C., and G. E. Henke, "Tridiagonal Matrix for Distillation," *Hydro. Proc. 45* (8), pp. 155–163 (Aug. 1966).

Buthod, P., N. Perez, and R. Thompson, "Program, Constants Developed for SRK Equation," *Oil & Gas J.*, pp. 60–64 (Nov. 27, 1978).

Jennings, G. P. and L. P. Meade, "Determination of Pump Efficiencies from Fluid Temperature Rise," *Amer. Petrol. Inst. Proc. 36* (V), Division of Transportation, pp. 36–39 (1956).

Henley, E. J., and J. D. Seader, "Equilibrium-Stage Separation Operations in Chemical Engineering," John Wiley & Sons, New York (1981).

Erbar, J. H., Private communication (1965).

Wang, Wen-Bohr, "Optimization of Expander Plants," Ph. D. Thesis, The University of Tulsa (1985).

Franks, R. G. E., "Modeling and Simulation in Chemical Engineering," Wiley-Interscience, New York (1972).

APPENDIX 5
CONVERSION OF UNITS

Useful conversion factors are given in the attached table. Examples of their use are given below.

Example A5–1. A pressure gauge indicates 50.0 psig. Barometric or atmospheric pressure at this location is 740 mm Hg at the time the gauge was read. What is the pressure in the vessel in psia and kPa?

Solution:

$$\text{Absolute Pressure} = \text{Gauge Pressure} + \text{Atmospheric Pressure}$$
$$= 50.0 \text{ psi} + 740 \text{ mm Hg} \times \frac{14.696 \text{ psi}}{760 \text{ mm Hg}}$$
$$= 50.0 + 14.3$$
$$= 64.3 \text{ psia} \quad \longleftarrow$$

$$64.3 \text{ psia} \times \frac{101.3 \text{ kPa}}{14.696 \text{ psi}} = 443 \text{ kPa} \quad \longleftarrow$$

Example A5–2. A thermometer reads 45°C. What is the temperature in °F? °R? °K?
$$°F = 1.8°C + 32$$
$$= 1.8(45) + 32$$
$$= 113°F \quad \longleftarrow$$
$$°R = 113 + 459.67 = 573°R \quad \longleftarrow$$
$$°K = 273.15 + 45 = 318°K \quad \longleftarrow$$

Example A5–3. A compressor is rated at 500 HP. What is the rating in J/s? W? kW?

$$500 \text{ HP} \times \frac{2544 \text{ Btu}}{1 \text{ hp-hr}} \times \frac{1055 \text{ J}}{1 \text{ Btu}} \times \frac{1 \text{ hr}}{3600 \text{ sec}}$$
$$= 373000 \text{ J/s} \quad \longleftarrow$$
$$= 373000 \text{ W} \quad \longleftarrow$$
$$= 373 \text{ kW} \quad \longleftarrow$$

Example A5–4. An oil has a density of 835 kg/m³. What is its density in g/cm³? lbm/ft³?

$$835 \frac{\text{kg}}{\text{m}^3} \times \frac{\text{m}^3}{(100)^3 \text{ cm}^3} \times \frac{1000 \text{ g}}{\text{kg}}$$
$$= 0.835 \text{ g/cm}^3 \quad \longleftarrow$$

$$\frac{0.835 \text{ g}}{\text{cm}^3} \times \frac{62.43 \text{ lbm/ft}^3}{1 \text{ g/cm}^3}$$
$$= 52.1 \text{ lbm/ft}^3 \quad \longleftarrow$$

BASIC SI UNITS

Force	= newton, n
Length	= meter, m
Mass	= kilogram, kg
Mole	= kilogram mole, kmol
Temperature	= kelvin = K
Time	= second, s
Pressure	= newton/meter², N/m²
	= pascal, Pa
Energy	= newton-meter, N m
	= joule, J
Power	= newton-meter/second, N m/s
	= watt, W
Frequency	= 1/second, s^{-1}
	= hertz, Hz

BASIC ENGLISH ENGINEERING UNITS

Force	= pound force, lbf
Length	= foot, ft
Mass	= pound mass, lbm
Mole	= pound mole, lbmol
Temperature	= Rankine, °R
	= 459.67 + °F
	= Fahrenheit, °F
Pressure	= pounds force per square inch, lbf/in², psia, psig
Energy	= British thermal unit, Btu
Power	= horsepower, hp
Frequency	= hertz, Hz

PREFIXES FOR SI UNITS

Amount	Multiple	Prefix	Symbol
1,000,000	10^6	mega	M
1,000	10^3	kilo	k
100	10^2	hecto	h
10	10	deka	da
0.1	10^{-1}	deci	d
0.01	10^{-2}	centi	c
0.001	10^{-3}	milli	m
0.000001	10^{-6}	micro	μ
0.000000001	10^{-9}	nano	n

PREFIXES USED IN OIL INDUSTRY

Amount	Multiple	Prefix	Symbol
1,000,000	10^6	mega	MM
1,000	10^3	kilo	M

CONVERSION FACTORS

Length

1 m $\quad = 39.37$ in $= 10^6$ μm $= 10^{10}$ A$^\circ$
1 in. $\quad = 2.54$ cm
1 ft $\quad = 30.48$ cm $= 0.3048$ m
1 mi $\quad = 5280$ ft $= 1760$ yds $= 1609.344$ m
1 nautical mile
$\quad\quad = 6076$ ft

Mass

1 lbm $\quad = 453.6$ g $= 0.4536$ kg $= 7000$ gr (grain)
1 kg $\quad = 1000$ g $= 2.2046$ lbm
1 slug $\quad = 1$ lbf s^2/ft $= 32.174$ lbm
1 US ton $= 2000$ lbm (also called short ton)
1 long ton $= 2240$ lbm (also called British ton)
1 tonne $\quad = 1000$ kg (also called metric ton)

Force

1 lbf $= 4.448$ N $= 4.448 \times 10^5$ dynes
$\quad\quad = 32.174$ poundals $= 32.174$ lbm ft/s^2
$\quad\quad = 1$ lbw

Pressure

1 atm $= 14.696$ psia $= 2116$ lbf/ft^2
$\quad\quad = 29.92$ in Hg $= 760$ mm Hg $= 760$ Torr.
$\quad\quad = 1.013$ bar
$\quad\quad = 33.9$ ft H$_2$O $= 1.013 \times 10^5$ Pa
$\quad\quad = 101.3$ kPa
1 Pa $\quad = 1$ N/m^2 $= 10^{-5}$ bars

Volume

1 ft^3 $\quad = 7.4805$ U.S. gal $= 6.23$ Imperial gal
$\quad\quad = 28.317$ L
1 m^3 $\quad = 1000$ L $= 264.2$ U.S. gal $= 35.31$ ft^3
1 bbl $= 42$ U.S. gal (oil) $= 5.615$ ft^3

Density

water $\quad = 62.43$ lbm/ft^3 $= 1000$ kg/m^3 $= 1$ g/cm^3
$\quad\quad = 8.346$ lbm/US gal
mercury $= 13.6$ g/cm^3

Velocity

1 knot $= 1$ nautical mile/hr

Temperature

$^\circ$F $= 1.8$ ($^\circ$C) $+ 32$
$^\circ$R $= ^\circ$F $+ 459.67 = 1.8$ (K)

Energy

1 J $\quad\quad = 1$ W s $= 1$ kg m^2/s $= 1$ N m3
$\quad\quad\quad = 10^7$ dyne cm $= 10^7$ erg
$\quad\quad\quad = 1$ V C $= 1$ V A s
1 Btu $\quad = 778$ ft lbf $= 252$ cal $= 0.252$ kcal $= 0.252$ Cal
$\quad\quad\quad = 1055$ J $= 10.41$ L atm
1 HP hr $= 2545$ Btu
1 kW hr $= 3412$ Btu $= 1.341$ HP hr

Power

1 HP $= 550$ ft lbf/s $= 33000$ ft lbf/min
$\quad\quad = 746$ W $= 0.746$ kW

Gas Constant

R $= 1.9859$ Btu/lb mol $^\circ$R $= 1.9859$ cal/g mol K
$\quad = 0.73024$ atm ft^3/lbmol $^\circ$R
$\quad = 1545.3$ ft lbf/lb mol $^\circ$R
$\quad = 10.732$ psia ft^3/lb mol $^\circ$R
$\quad = 0.082057$ L atm/g mol K
$\quad = 82.057$ atm cm^3/g mol K
$\quad = 8314.5$ Pa m^3/kg mol K or J/kg mol K
$\quad = 8.3145$ kJ/kg mol K

Viscosity (dynamic) (absolute)

1 cP $\quad = 0.01$ Poise $= 0.01$ g/cm s $= 0.01$ dyne s/cm^2
$\quad\quad = 0.001$ kg/m s $= 0.001$ Pa s $= 0.001$ N s/m^2
$\quad\quad = 2.42$ lbm/ft hr $= 0.0752$ slug/ft hr
$\quad\quad = 6.72 \times 10^{-4}$ lbm/ft s $= 2.09 \times 10^{-5}$ lbf s/ft^2
1 Pa s $= 0.0209$ lbf s/ft^2
$\quad\quad = 0.672$ lbm/ft s

Kinematic Viscosity

1 St $\quad = 1$ cm^2/s $= 0.0001$ m^2/s
1 ft^2/s $= 929$ St $= 0.0929$ m^2/s

Force-mass conversion factor

$g_c = 1$ kg m/s^2 N $= 1$ g cm/s^2 dyne
$\quad = 32.174$ lbm ft/s^2 lbf $= 1$ slug ft/s^2 lbf

Acceleration due to gravity

$g = 32.2$ ft/s^2 $= 9.81$ m/s^2 $= 981$ cm/s^2 (varies very slightly
$\quad\quad$ with longitude and elevation)

APPENDIX 6
PHYSICAL PROPERTIES OF FLUIDS

Table A6–1 Properties of Hydrocarbons and Common Gases. Source: GPSA Engineering Data Book, 10th Ed., 1987

Compound	Formula	Mol Wt	Boiling Point °F (1 atm)	Vapor Press @ 100 F Psia	Critical Press. Psia	Critical Temp. Degr F	Liquid Spec. Grav. 60/60 F	Volume Ratio scf gas/ gal liq.
Methane	CH_4	16.043	−258.73	(5000)[a]	666.4	−116.67	(0.3)[a]	(59.135)[a]
Ethane	C_2H_6	30.070	−127.49	(800)[a]	706.5	89.92	0.35619[c]	37.476[c]
Propane	C_3H_8	44.097	−43.75	188.64	616.0	206.06	0.50699[c]	36.375[c]
Iso-Butane	C_4H_{10}	58.123	10.78	72.581	527.9	274.46	0.56287[c]	30.639[c]
Normal Butane	C_4H_{10}	58.123	31.08	51.706	550.6	305.62	0.58401[c]	31.790[c]
Iso-Pentane	C_5H_{12}	72.150	82.12	20.445	490.4	369.10	0.62470	27.393
Normal Pentane	C_5H_{12}	72.150	96.92	15.574	488.6	385.8	0.63112	27.674
n-Hexane	C_6H_{14}	86.177	155.72	4.960	436.9	453.6	0.66383	24.371
n-Heptane	C_7H_{16}	100.204	209.16	1.620	396.8	512.7	0.68820	21.729
n-Octane	C_8H_{18}	114.231	258.21	0.537	360.7	564.2	0.70696	19.580
n-Decane	$C_{10}H_{22}$	142.285	345.48	0.061	305.2	652.0	0.73421	16.326
Nitrogen	N_2	28.013	−320.45	—	493.1	−232.51	0.80940[d]	91.413[e]
Oxygen	O_2	31.999	−297.33	—	731.4	−181.43	1.1421[d]	112.93[e]
Carbon Dioxide	CO_2	44.010	−109.26[b]	—	1071	87.91	0.81802[c]	58.807[c]
Hydrogen Sulfide	H_2S	34.08	−76.50	394.59	1300	212.45	0.80144[c]	74.401[c]
Water	H_2O	18.0153	212.00	0.950	3198.8	705.16	1.00000[d]	175.62[e]
Air	Mixture	28.9625	−317.8	—	546.9	−221.31	0.87476[d]	95.557[e]

Compound	Acentric Factor, ω	Flammability Limits, vol% in Air Mixt. Lower	Upper	Heating Value 60°F, 1 atm Btu/scf Net	Gross	Freezing Point @ 1 atm Degr F	Heat of Vap'n @ 1 atm Btu/lb
Methane	0.0104	5.0	15.0	909.4	1010.0	−296.44[f]	219.45
Ethane	0.0979	2.9	13.0	1618.7	1769.6	−297.04[f]	211.14
Propane	0.1522	2.0	9.5	2314.9	2516.1	−305.73[f]	183.01
Iso-Butane	0.1852	1.8	8.5	3000.4	3251.9	−255.28	157.23
Normal Butane	0.1995	1.5	9.0	3010.8	3262.3	−217.05	165.93
Iso-Pentane	0.2280	1.3	8.0	3699.0	4000.9	−255.82	147.12
Normal Pentane	0.2514	1.4	8.3	3706.9	4008.9	−217.05	153.57
n-Hexane	0.2994	1.1	7.7	4403.8	4755.9	−139.58	143.94
n-Heptane	0.3494	1.0	7.0	5100.8	5502.5	−131.05	136.00
n-Octane	0.3977	0.8	6.5	5796.1	6248.9	−70.18	129.52
n-Decane	0.4898	0.7	5.4	7189.6	7742.9	−21.36	119.65
Nitrogen	0.0372	—	—	—	—	−346.00[f]	85.59
Oxygen	0.0216	—	—	—	—	−361.82[f]	91.59
Carbon Dioxide	0.2667	—	—	—	—	−69.83	246.47
Hydrogen Sulfide	0.0948	4.3	45.5	586.8	637.1	−121.88[f]	235.63
Water	0.3443	—	—	—	—	32.00	970.18
Air	—	—	—	—	—	—	88.20

a. Above critical point, extrapolated or estimated.
b. Sublimation point.
c. At saturation pressure, 60°F.
d. At normal boiling point.
e. Gas at 60°F, liquid at normal boiling point.
f. At the triple point pressure.

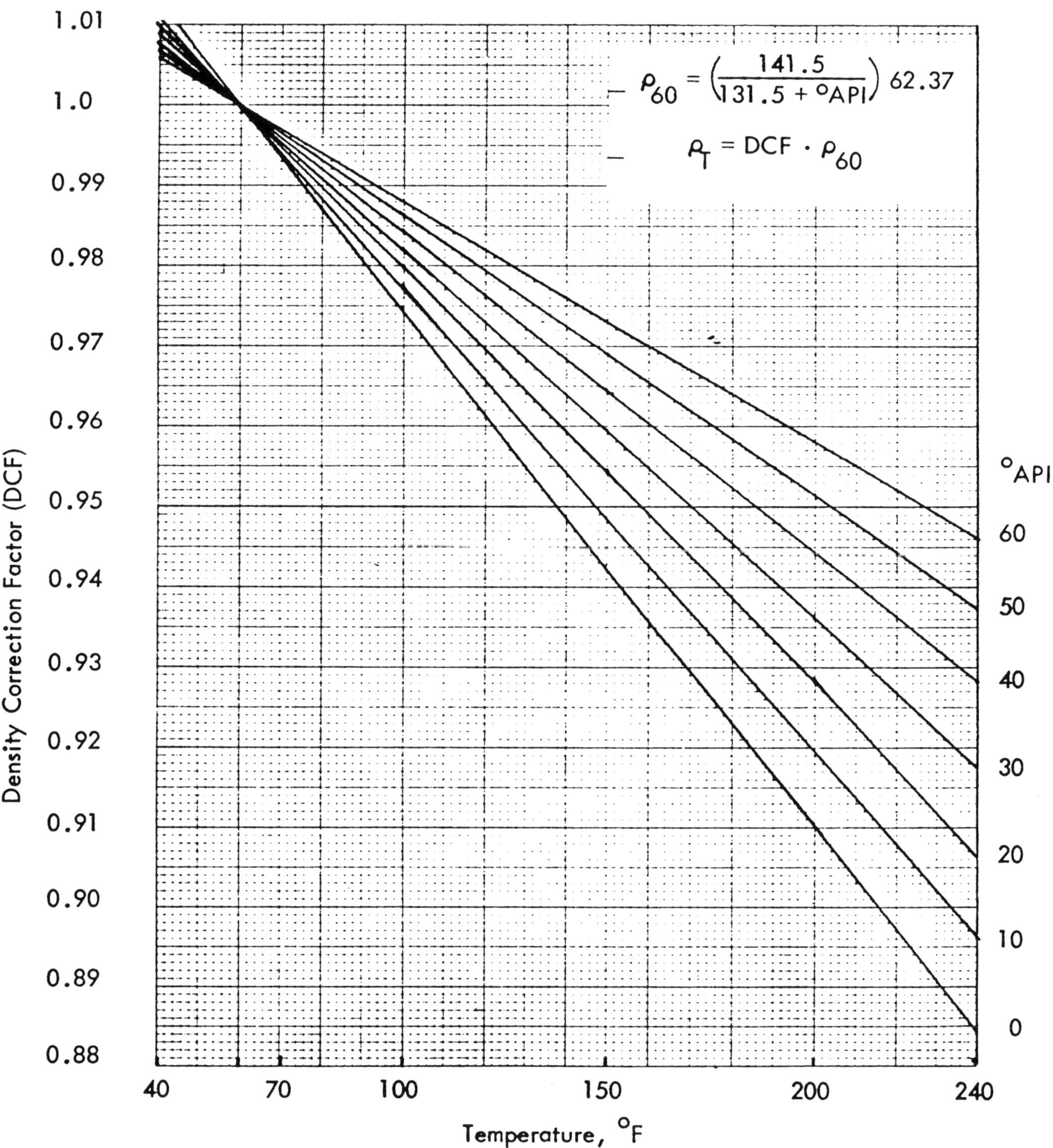

Figure A6–1. Effect of temperature on density for crude oils (Hankinson *et al.*, *Oil & Gas J.*, Dec. 24, 1979, p. 66–70).

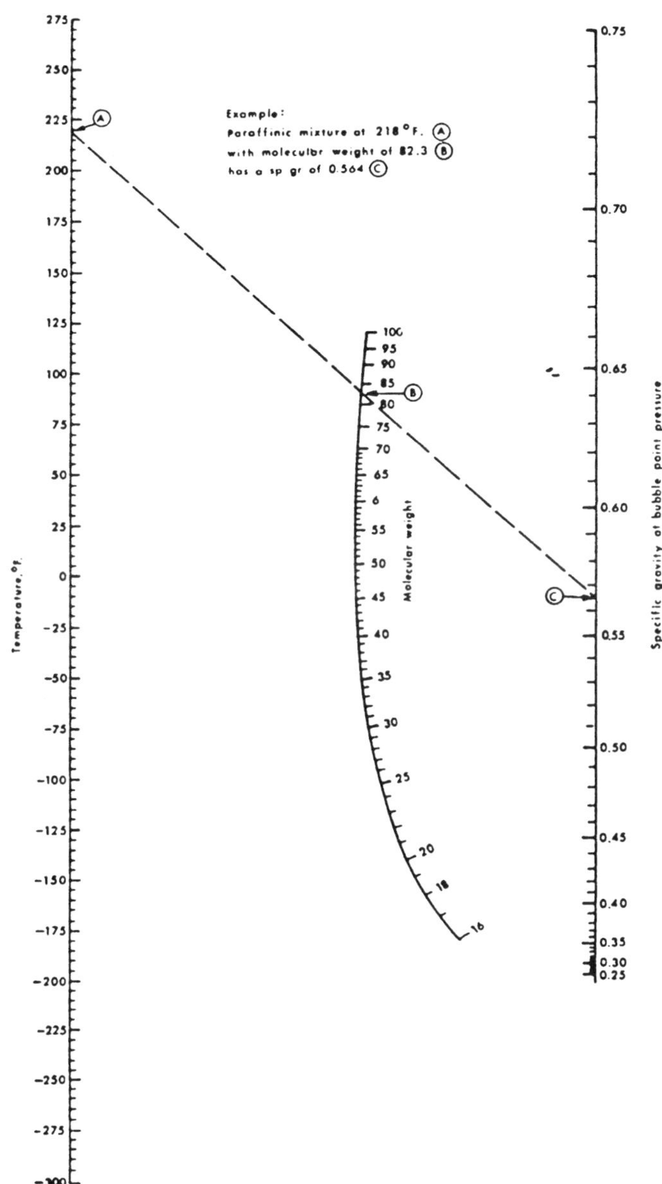

Figure A6–2.   Specific gravity of paraffinic hydrocarbon mixtures (*Gas Processors Association Engineering Data Book,* Sec. 12 1981.)

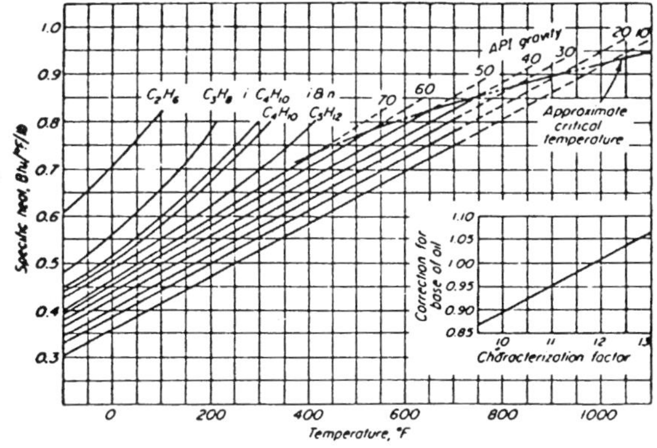

Fig. A6-3.   . Specific Heats of Liquid Oils

Typical Characterization Factors	
Natural Gas Liquids*	12.5 - 15.0
Oklahoma Crude Oils	11.7 - 12.0
East Texas Crude Oils	11.5 - 11.9
West Texas Crude Oils	11.6 - 12.0
Wyoming Crude Oils	11.6 - 11.8

*Interpolate specific heat of light hydrocarbons shown by molecular weight. Then there is no characterization factor correction.

Figure A6–3.   Specific heats of liquid oils (Ref: Nelson, *Petroleum Refinery Engineering,* McGraw-Hill).

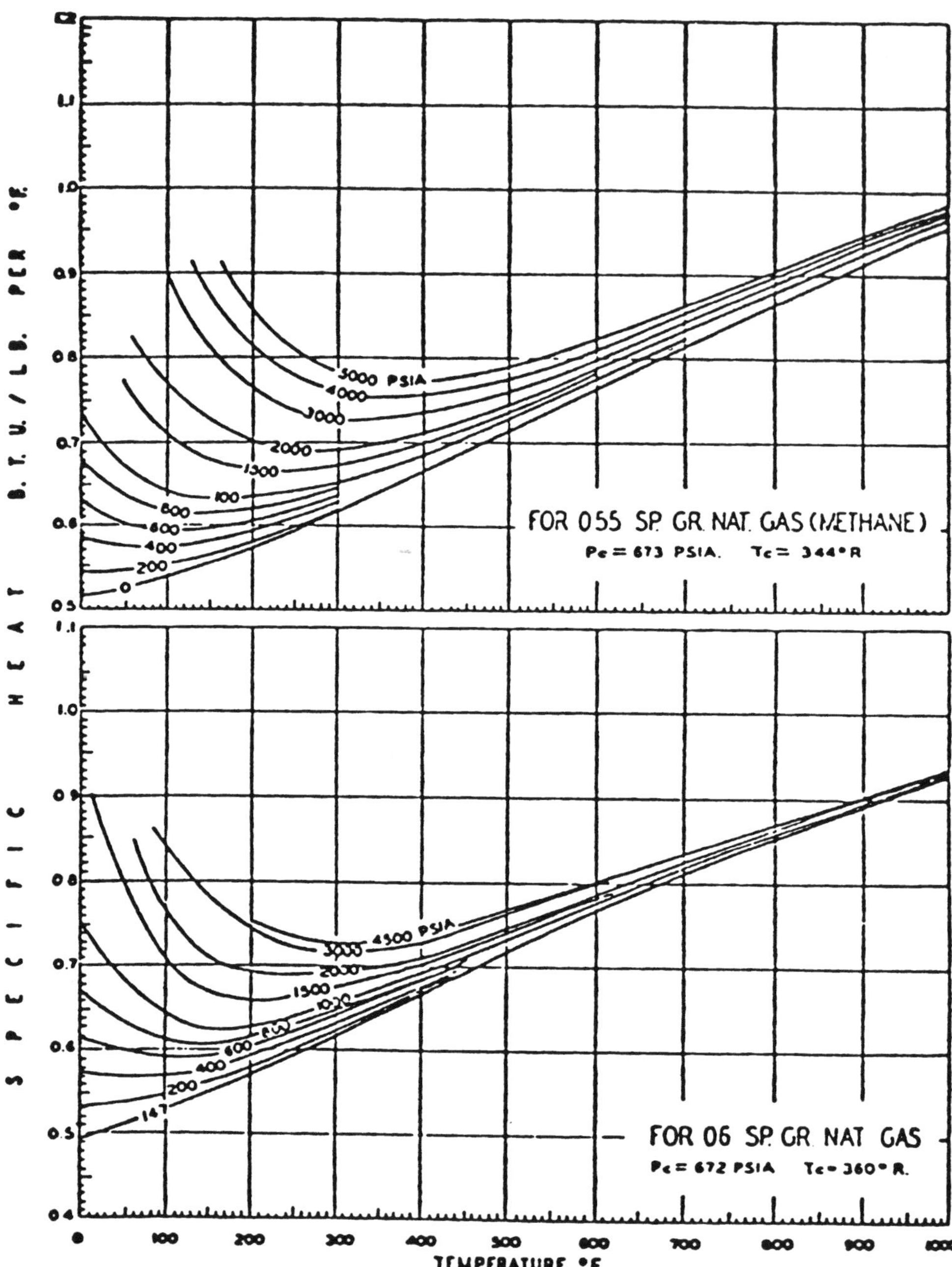

Figure A6–4a. Natural Gas Specific Heat Charts. (Buthod, A. P., *Oil & Gas J.,* Sept. 29, 1949)

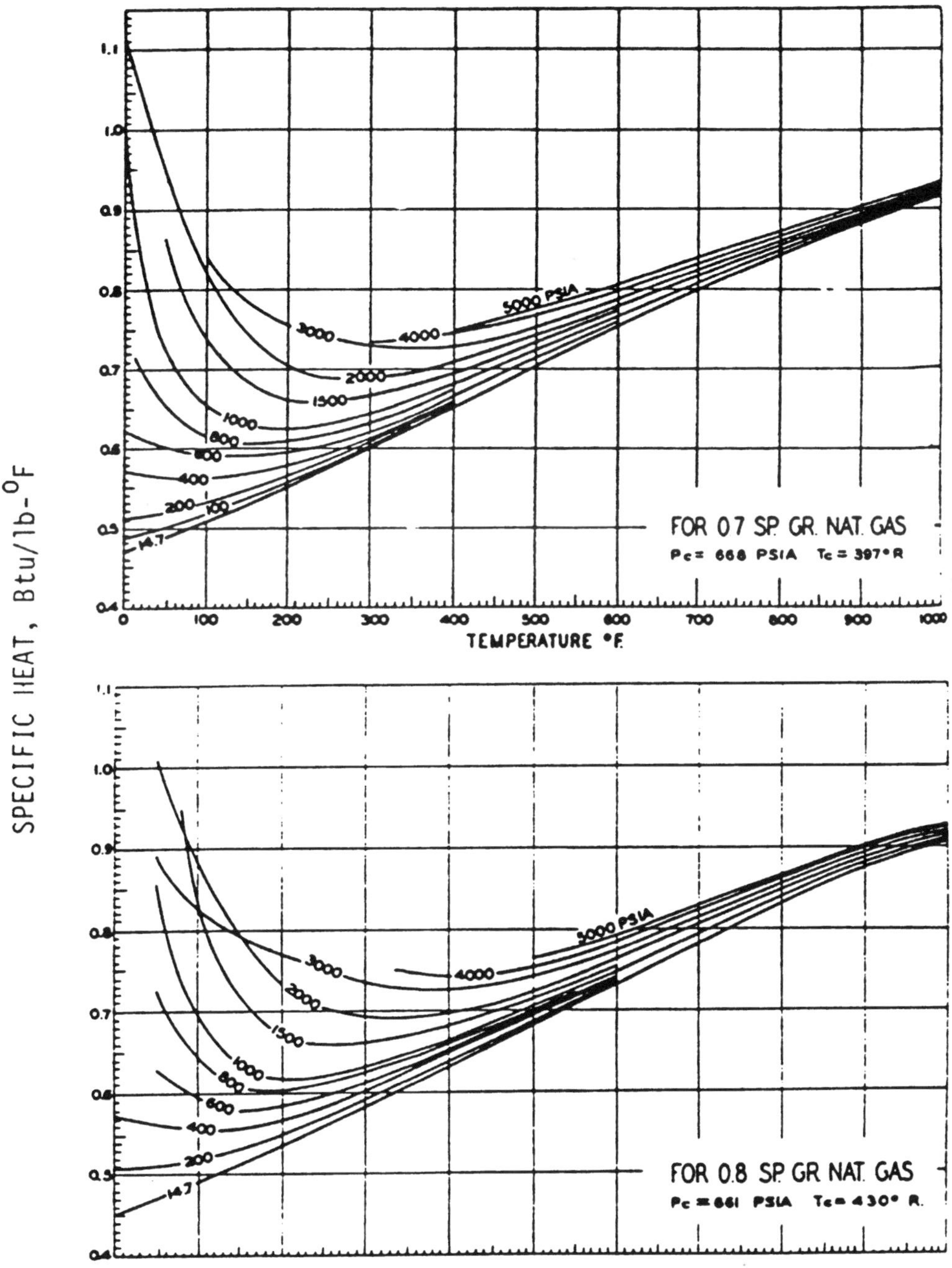

Figure A6–4b. Natural Gas Specific Heat Charts. (Buthod, A. P., *Oil & Gas J.,* Sept. 29, 1949)

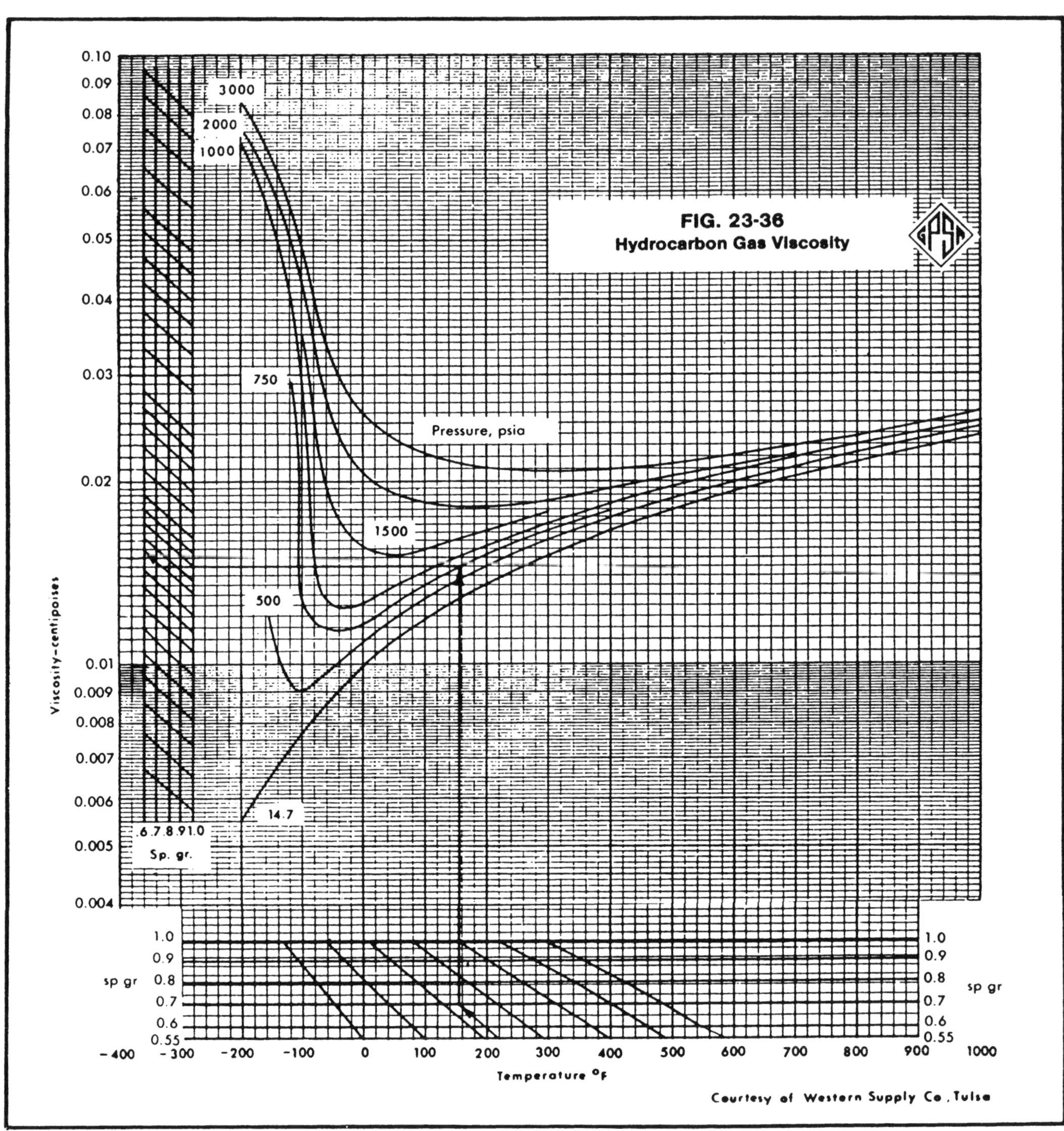

Figure A6–5.   "Hydrocarbon Gas Viscosity." (GPSA, 1987)

Table A6–2   U.S. Standard Atmosphere

Altitude (ft)	Temperature (°F)	Pressure (psia)	Density (lb/ft^3)
0	59.0	14.696	0.0765
1000	55.4	14.173	0.0743
2000	51.9	13.664	0.0721
3000	48.3	13.172	0.0700
4000	44.8	12.693	0.0679
5000	41.2	12.228	0.0659
6000	37.6	11.777	0.0639
7000	34.1	11.341	0.0620
8000	30.5	10.916	0.0601
9000	27.0	10.506	0.0583
10000	23.4	10.108	0.0565

Source: U.S. Standard Atmosphere, 1962 U.S. Government Printing Office, Washington, DC.

Interpolation Formulae:

$$T(z) = 59.0 - B\,z$$
$$P(z) = 14.696\,[1 - (B\,z/518.69)]^{5.26}$$

where: z = elevation (ft)
T = temperature (°F)
P = pressure (psia)
B = 0.003566 (°F/ft)

Index